PROCEEDINGS OF SPIE

Ground-based and Airborne Telescopes IV

Larry M. Stepp
Roberto Gilmozzi
Helen J. Hall
Editors

1–6 July 2012
Amsterdam, Netherlands

Sponsored by
SPIE

Cooperating Organizations
American Astronomical Society (United States) • Netherlands Institute for Radio Astronomy (ASTRON) (Netherlands) • Ball Aerospace & Technologies Corporation (United States) Canadian Astronomical Society (CASCA) (Canada) • European Astronomical Society (Switzerland) • ESO—European Southern Observatory (Germany) • International Astronomical Union • Korea Astronomy and Space Science Institute (KASI) (Republic of Korea) • National Radio Astronomy Observatory • POPSud (France) • TNO (Netherlands)

Published by
SPIE

Part Two of Three Parts

Volume 8444

Proceedings of SPIE 0277-786-786X, V.8444

SPIE is an international society advancing an interdisciplinary approach to the science and application of light.

The papers included in this volume were part of the technical conference cited on the cover and title page. Papers were selected and subject to review by the editors and conference program committee. Some conference presentations may not be available for publication. The papers published in these proceedings reflect the work and thoughts of the authors and are published herein as submitted. The publisher is not responsible for the validity of the information or for any outcomes resulting from reliance thereon.

Please use the following format to cite material from this book:
   Author(s), "Title of Paper," in *Ground-based and Airborne Telescopes IV*, edited by Larry M. Stepp, Roberto Gilmozzi, Helen J. Hall, Proceedings of SPIE Vol. 8444 (SPIE, Bellingham, WA, 2012) Article CID Number.

ISSN: 0277-786X
ISBN: 9780819491459

Published by
SPIE
P.O. Box 10, Bellingham, Washington 98227-0010 USA
Telephone +1 360 676 3290 (Pacific Time) · Fax +1 360 647 1445
SPIE.org

Copyright © 2012, Society of Photo-Optical Instrumentation Engineers.

Copying of material in this book for internal or personal use, or for the internal or personal use of specific clients, beyond the fair use provisions granted by the U.S. Copyright Law is authorized by SPIE subject to payment of copying fees. The Transactional Reporting Service base fee for this volume is $18.00 per article (or portion thereof), which should be paid directly to the Copyright Clearance Center (CCC), 222 Rosewood Drive, Danvers, MA 01923. Payment may also be made electronically through CCC Online at copyright.com. Other copying for republication, resale, advertising or promotion, or any form of systematic or multiple reproduction of any material in this book is prohibited except with permission in writing from the publisher. The CCC fee code is 0277-786X/12/$18.00.

Printed in the United States of America.

Publication of record for individual papers is online in the SPIE Digital Library.

SPIEDigitalLibrary.org

Paper Numbering: Proceedings of SPIE follow an e-First publication model, with papers published first online and then in print and on CD-ROM. Papers are published as they are submitted and meet publication criteria. A unique, consistent, permanent citation identifier (CID) number is assigned to each article at the time of the first publication. Utilization of CIDs allows articles to be fully citable as soon as they are published online, and connects the same identifier to all online, print, and electronic versions of the publication. SPIE uses a six-digit CID article numbering system in which:
  ▪ The first four digits correspond to the SPIE volume number.
  ▪ The last two digits indicate publication order within the volume using a Base 36 numbering system employing both numerals and letters. These two-number sets start with 00, 01, 02, 03, 04, 05, 06, 07, 08, 09, 0A, 0B ... 0Z, followed by 10-1Z, 20-2Z, etc.
The CID Number appears on each page of the manuscript. The complete citation is used on the first page, and an abbreviated version on subsequent pages. Numbers in the index correspond to the last two digits of the six-digit CID Number.

Contents

xxxi *Conference Committee*

xxxvii *Introduction*

xxxix *The cosmic microwave background: observing directly the early universe (Plenary Paper)* [8442-506]
P. de Bernardis, S. Masi, Univ. degli Studi di Roma La Sapienza (Italy)

Part 1

SOLAR TELESCOPES I

8444 03 **The 1.6 m off-axis New Solar Telescope (NST) in Big Bear (Invited Paper)** [8444-1]
P. R. Goode, W. Cao, Ctr. for Solar-Terrestrial Research, New Jersey Institute of Technology (United States) and Big Bear Solar Observatory (United States)

8444 04 **Applications of infrared techniques in solar telescopes NVST** [8444-141]
Y. Li, F. Xu, S. Huang, G. Liu, Yunnan Astronomical Observatory (China)

8444 05 **Introduction to the Chinese Giant Solar Telescope** [8444-3]
Z. Liu, Yunnan Astronomical Observatory (China); Y. Deng, National Astronomical Observatories (China); Z. Jin, Yunnan Astronomical Observatory (China); H. Ji, Purple Mountain Observatory (China)

8444 06 **Large-field high-resolution mosaic movies** [8444-4]
R. H. Hammerschlag, Utrecht Univ. (Netherlands); G. Sliepen, Institute for Solar Physics, Royal Swedish Academy of Sciences (Sweden); F. C. M. Bettonvil, ASTRON (Netherlands) and Leiden Observatory (Netherlands); A. P. L. Jägers, Utrecht Univ. (Netherlands); P. Sütterlin, Institute for Solar Physics, Royal Swedish Academy of Sciences (Sweden); S. F. Martin, Helio Research (United States)

SOLAR TELESCOPES II

8444 07 **The Advanced Technology Solar Telescope: design and early construction (Invited Paper)** [8444-5]
J. P. McMullin, T. R. Rimmele, S. L. Keil, M. Warner, S. Barden, S. Bulau, S. Craig, B. Goodrich, E. Hansen, S. Hegwer, R. Hubbard, W. McBride, S. Shimko, F. Woeger, J. Ditsler, National Solar Observatory (United States)

8444 08 **ATST Enclosure final design and construction plans** [8444-6]
G. Murga, IDOM (United States); H. Marshall, National Solar Observatory (United States); J. Ariño, T. Lorentz, IDOM (United States)

8444 09 **Progress making the top end optical assembly (TEOA) for the 4-meter Advanced Technology Solar Telescope** [8444-7]
B. Canzian, L-3 Integrated Optical Systems Brashear (United States); J. Barentine, L-3 Integrated Optical Systems Tinsley (United States); J. Arendt, S. Bader, G. Danyo, C. Heller, L-3 Integrated Optical Systems Brashear (United States)

8444 0A **The azimuth axes mechanisms for the ATST telescope mount assembly** [8444-8]
H. J. Kärcher, U. Weis, O. Dreyer, MT Mechatronics GmbH (Germany); P. Jeffers, National Solar Observatory (United States); G. Bonomi, Ingersoll Machine Tools, Inc. (United States)

TELESCOPES FOR SYNOPTIC AND SURVEY OBSERVATIONS I

8444 0C **Manufacturing and testing of the large lenses for Dark Energy Survey (DES) at THALES SESO** [8444-10]
D. Fappani, J. Fourez, THALES SESO (France); P. Doel, D. Brooks, Univ. College London (United Kingdom); B. Flaugher, Fermilab Chicago (United States)

8444 0D **The Transneptunian Automated Occultation Survey (TAOS II)** [8444-11]
M. J. Lehner, Institute of Astronomy and Astrophysics (Taiwan) and Harvard-Smithsonian Ctr. for Astrophysics (United States) and Univ. of Pennsylvania (United States); S.-Y. Wang, Institute of Astronomy and Astrophysics (Taiwan); C. A. Alcock, Harvard-Smithsonian Ctr. for Astrophysics (United States); K. H. Cook, Institute of Astronomy and Astrophysics (Taiwan); G. Furesz, J. C. Geary, Harvard-Smithsonian Ctr. for Astrophysics (United States); D. Hiriart, Instituto de Astronomía, Univ. Nacional Autónoma de México (Mexico); P. T. Ho, Institute of Astronomy and Astrophysics (Taiwan); W. H. Lee, Instituto de Astronomía, Univ. Nacional Autónoma de México (Mexico); F. Melsheimer, DFM Engineering, Inc. (United States); T. Norton, Harvard-Smithsonian Ctr. for Astrophysics (United States); M. Reyes-Ruiz, M. Richer, Instituto de Astronomía, Univ. Nacional Autónoma de México (Mexico); A. Szentgyorgyi, Harvard-Smithsonian Ctr. for Astrophysics (United States); W.-L. Yen, Z.-W. Zhang, Institute of Astronomy and Astrophysics (Taiwan)

8444 0E **NGTS: a robotic transit survey to detect Neptune and super-Earth mass planets** [8444-12]
B. Chazelas, Observatoire de l'Univ. de Genève (Switzerland); D. Pollacco, Queen's Univ. Belfast (United Kingdom) and The Univ. of Warwick (United Kingdom); D. Queloz, Observatoire de l'Univ. de Genève (Switzerland); H. Rauer, Deutsches Zentrum für Luft- und Raumfahrt (Germany) and Technische Univ. Berlin (Germany); P. J. Wheatley, The Univ. of Warwick (United Kingdom); R. West, Univ. of Leicester (United Kingdom); J. Da Silva Bento, The Univ. of Warwick (United Kingdom); M. Burlegih, Leicester Univ. (United Kingdom); J. McCormac, Queen's Univ. Belfast (United Kingdom); P. Eigmüller, A. Erikson, Deutsches Zentrum für Luft- und Raumfahrt (Germany); L. Genolet, Observatoire de l'Univ. de Genève (Switzerland); M. Goad, Univ. of Leicester (United Kingdom); A. Jordán, Pontificia Univ. Católica de Chile (Chile); M. Neveu, Observatoire de l'Univ. de Genève (Switzerland); S. Walker, The Univ. of Warwick (United Kingdom)

8444 0F **Design of a compact wide field telescope for space situational awareness** [8444-13]
D. Lee, A. Born, P. Parr-Burman, P. Hastings, B. Stobie, N. Bezawada, UK Astronomy Technology Ctr., Royal Observatory (United Kingdom)

TELESCOPES FOR SYNOPTIC AND SURVEY OBSERVATIONS II

8444 0G **OAJ: 2.6m wide field survey telescope** [8444-14]
O. Pirnay, V. Moreau, G. Lousberg, Advanced Mechanical and Optical Systems S.A. (Belgium)

8444 0H **Design differences between the Pan-STARRS PS1 and PS2 telescopes** [8444-15]
J. S. Morgan, N. Kaiser, Institute for Astronomy, Univ. of Hawai'i (United States); V. Moreau, AMOS Ltd. (Belgium); D. Anderson, Rayleigh Optical Corp. (United States); W. Burgett, Institute for Astronomy, Univ. of Hawai'i (United States)

8444 0I **Ground-based search for the brightest transiting planets with the Multi-site All-Sky CAmeRA: MASCARA** [8444-16]
I. A. G. Snellen, R. Stuik, Leiden Observatory, Leiden Univ. (Netherlands); R. Navarro, F. Bettonvil, ASTRON (Netherlands); M. Kenworthy, Leiden Observatory, Leiden Univ. (Netherlands); E. de Mooij, Univ. of Toronto (Canada); G. Otten, Leiden Observatory, Leiden Univ. (Netherlands); R. ter Horst, ASTRON (Netherlands); R. le Poole, Leiden Observatory, Leiden Univ. (Netherlands)

8444 0J **LSST secondary mirror assembly baseline design** [8444-17]
D. R. Neill, W. J. Gressler, J. Sebag, O. Wiecha, National Optical Astronomy Observatory (United States); M. Warner, National Optical Astronomy Observatory (Chile); B. Schoening, J. DeVries, J. Andrew, National Optical Astronomy Observatory (United States); G. Schumacher, National Optical Astronomy Observatory (Chile); E. Hileman, National Optical Astronomy Observatory (United States)

UPGRADES TO EXISTING OBSERVATORIES

8444 0K **Current status of the Hobby-Eberly Telescope wide field upgrade** [8444-19]
G. J. Hill, J. A. Booth, M. E. Cornell, J. M. Good, McDonald Observatory, The Univ. of Texas at Austin (United States); K. Gebhardt, The Univ. of Texas at Austin (United States); H. J. Kriel, H. Lee, R. Leck, W. Moriera, P. J. MacQueen, D. M. Perry, M. D. Rafal, T. H. Rafferty, C. Ramiller, R. D. Savage, C. A. Taylor, B. L. Vattiat, McDonald Observatory, The Univ. of Texas at Austin (United States); L. W. Ramsey, The Pennsylvania State Univ. (United States); J. H. Beno, T. A. Beets, J. D. Esguerra, Ctr. for Electromechanics, The Univ. of Texas at Austin (United States); M. Häuser, Univ.-Sternwarte Munich (Germany); R. J. Hayes, J. T. Heisler, I. M. Soukup, J. J. Zierer, M. S. Worthington, N. T. Mollison, D. R. Wardell, G. A. Wedeking, Ctr. for Electromechanics, The Univ. of Texas at Austin (United States)

TELESCOPE MOUNTS AND ENCLOSURES

8444 0L **ATST telescope pier** [8444-20]
P. Jeffers, National Solar Observatory (United States); E. Manuel, M3 Engineering & Technology Corp. (United States); O. Dreyer, H. Kärcher, MT Mechatronics GmbH (Germany)

8444 0M **Design concepts for the EST mount** [8444-21]
H. J. Kärcher, M. Süß, D. Fischer, MT Mechatronics GmbH (Germany)

8444 0N **Progress on the structural and mechanical design of the Giant Magellan Telescope** [8444-22]
M. Sheehan, Giant Magellan Telescope Organization Corp. (United States); S. Gunnels, Paragon Engineering (United States); C. Hull, J. Kern, C. Smith, M. Johns, S. Shectman, Giant Magellan Telescope Organization Corp. (United States)

8444 0P **E-ELT dome for modified baseline design** [8444-24]
A. Bilbao, G. Murga, C. Gómez, IDOM (Spain)

8444 0Q **The E-ELT project: the dome detailed design study** [8444-25]
G. Marchiori, S. De Lorenzi, A. Busatta, European Industrial Engineering s.r.l. (Italy)

DESIGN TO WITHSTAND EARTHQUAKES

8444 0R **Seismic design accelerations for the LSST telescope** [8444-113]
D. R. Neill, National Optical Astronomy Observatory (United States); M. Warner, Cerro Tololo Inter-American Observatory (Chile); J. Sebag, National Optical Astronomy Observatory (United States)

8444 0T **Seismic analysis of the LSST telescope** [8444-26]
D. R. Neill, National Optical Astronomy Observatory (United States)

MODELING, MEASUREMENT, AND CONTROL OF WIND BUFFETING

8444 0U **GMT enclosure wind and thermal study** [8444-29]
A. Farahani, Giant Magellan Telescope Organization Corp. (United States); A. Kolesnikov, L. Cochran, CPP, Inc. (United States); C. Hull, M. Johns, Giant Magellan Telescope Organization Corp. (United States)

8444 0V **Vibration mitigation for wind-induced jitter for the Giant Magellan Telescope** [8444-30]
R. M. Glaese, Moog-CSA Engineering (United States); M. Sheehan, Giant Magellan Telescope Organization (United States)

CONCEPTS FOR FUTURE TELESCOPES

8444 0W **Feasibility studies to upgrade the Canada-France-Hawaii Telescope site for the next generation Canada-France-Hawaii Telescope** [8444-31]
K. Szeto, NRC Herzberg Institute of Astrophysics (Canada); M. Angers, C. Breckenridge, Dynamic Structures Ltd. (Canada); S. Bauman, Canada-France-Hawaii Telescope (United States); N. Loewen, Dynamic Structures Ltd. (Canada); D. Loop, A. McConnachie, J. Pazder, NRC Herzberg Institute of Astrophysics (Canada); D. Salmon, Canada-France-Hawaii Telescope Corp. (United States); P. Spano, NRC Herzberg Institute of Astrophysics (Canada); S. Stiemer, The Univ. of British Columbia (Canada); C. Veillet, Canada-France-Hawaii Telescope Corp. (United States)

8444 0X **The Astronomical Telescope of New York: a new 12-meter astronomical telescope** [8444-32]
T. Sebring, Xoptx LLC (United States); R. Junquist, Optical Consultant (United States); C. Stutzki, Stutzki Engineering, Inc. (United States); P. Sebring, Sebring Mechanical Design (United States); S. Baum, Rochester Institute of Technology (United States)

8444 0Y **Reviewing off-axis telescope concepts: a quest for highest possible dynamic range for photometry and angular resolution** [8444-107]
G. Moretto, Institut de Physique Nucléaire de Lyon (France); J. R. Kuhn, Institute for Astronomy, Univ. of Hawai'i (United States); P. R. Goode, Big Bear Solar Observatory (United States)

AIRBORNE TELESCOPES I

8444 10 **Early science results from SOFIA (Invited Paper)** [8444-35]
E. T. Young, SOFIA Science Ctr., NASA Ames Research Ctr. (United States); T. L. Herter, Cornell Univ. (United States); R. Güsten, Max-Planck-Institut für Radioastronomie (Germany); E. W. Dunham, Lowell Observatory (United States); E. E. Becklin, P. M. Marcum, SOFIA Science Ctr., NASA Ames Research Ctr. (United States); A. Krabbe, Deutsches SOFIA Institut, Univ. Stuttgart (Germany); B. Andersson, W. T. Reach, SOFIA Science Ctr., NASA Ames Research Ctr. (United States); H. Zinnecker, SOFIA Science Ctr., NASA Ames Research Ctr. (United States) and Deutsches SOFIA Institut (Germany)

8444 11 **Active damping of the SOFIA Telescope assembly** [8444-36]
P. J. Keas, Moog-CSA Engineering (United States); E. Dunham, Lowell Observatory (United States); U. Lampater, Deutsches SOFIA Institut, Univ. Stuttgart (Germany); E. Pfüller, SOFIA Science Ctr., NASA Ames Research Ctr. (United States) and Univ. of Stuttgart (Germany); S. Teufel, Deutsches SOFIA Institut, Univ. Stuttgart (Germany); H.-P. Roeser, Univ. of Stuttgart (Germany); M. Wiedemann, J. Wolf, SOFIA Science Ctr., NASA Ames Research Ctr. (United States) and Univ. of Stuttgart (Germany)

8444 12 **Evaluation of the aero-optical properties of the SOFIA cavity by means of computational fluid dynamics and a super fast diagnostic camera** [8444-37]
C. Engfer, E. Pfüller, M. Wiedemann, J. Wolf, Deutsches SOFIA Institut, Univ. of Stuttgart (Germany) and SOFIA Airborne Systems Operations Ctr., NASA Dryden Flight Research Ctr. (United States); T. Lutz, E. Krämer, Institute of Aerodynamics and Gas Dynamics, Univ. of Stuttgart (Germany); H.-P. Röser, Institute of Space Systems, Univ. Stuttgart (Germany)

8444 13 **Optical characterization of the SOFIA telescope using fast EM-CCD cameras** [8444-38]
E. Pfüller, J. Wolf, Deutsches SOFIA Institut, Univ. of Stuttgart (Germany) and SOFIA Science Ctr., NASA Ames Research Ctr. (United States); H. Hall, SOFIA Science Ctr., NASA Ames Research Ctr. (United States); H.-P. Röser, Institute of Space Systems, Univ. of Stuttgart (Germany)

AIRBORNE TELESCOPES II

8444 14 **SOFIA observatory performance and characterization** [8444-39]
P. Temi, P. M. Marcum, NASA Ames Research Ctr. (United States); W. E. Miller, Orbital Science Corp. (United States); E. W. Dunham, Lowell Observatory (United States); I. S. McLean, Univ. of California, Los Angeles (United States); J. Wolf, Deutsches SOFIA Institut, Univ. of Stuttgart (Germany); E. E. Becklin, SOFIA Science Ctr, NASA Ames Research Ctr. (United States); T. A. Bida, Lowell Observatory (United States); R. Brewster, Orbital Science Corp. (United States); S. C. Casey, SOFIA Science Ctr, NASA Ames Research Ctr. (United States); P. L. Collins, Lowell Observatory (United States); S. D. Horner, NASA Ames Research Ctr. (United States); H. Jakob, Deutsches SOFIA Institut, Univ. of Stuttgart (Germany); S. C. Jensen, NASA Dryden Flight Research Ctr. (United States); J. L. Killebrew, NASA Marshall Space Flight Ctr. (United States); U. Lampater, NASA Ames Research Ctr. (United States); G. I. Mandushev, Lowell Observatory (United States); A. W. Meyer, SOFIA Science Ctr, NASA Ames Research Ctr. (United States); E. Pfueller, A. Reinacher, Deutsches SOFIA Institut, Univ. of Stuttgart (Germany); J. Rho, SOFIA Science Ctr, NASA Ames Research Ctr. (United States); T. L. Roellig, NASA Ames Research Ctr. (United States); M. L. Savage, SOFIA Science Ctr, NASA Ames Research Ctr. (United States); E. C. Smith, NASA Ames Research Ctr. (United States); S. Teufel, M. Wiedemann, Deutsches SOFIA Institut, Univ. of Stuttgart (Germany)

8444 15 **The balloon-borne large-aperture submillimeter telescope for polarimetry-BLASTPol: performance and results from the 2010 Antarctic flight** [8444-40]
E. Pascale, P. A. R. Ade, Cardiff Univ. (United Kingdom); F. E. Angilè, Univ. of Pennsylvania (United States); S. J. Benton, Univ. of Toronto (Canada); M. J. Devlin, B. Dober, Univ. of Pennsylvania (United States); L. M. Fissel, Univ. of Toronto (Canada); Y. Fukui, Nagoya Univ. (Japan); N. N. Gandilo, Univ. of Toronto (Canada); J. O. Gundersen, Univ. of Miami (United States); P. C. Hargrave, Cardiff Univ. (United Kingdom); J. Klein, Univ. of Pennsylvania (United States); A. L. Korotkov, Brown Univ. (United States); T. G. Matthews, Northwestern Univ. (United States); L. Moncelsi, California Institute of Technology (United States) and Cardiff Univ. (United Kingdom); T. K. Mroczkowski, California Institute of Technology (United States); C. B. Netterfield, Cardiff Univ. (United Kingdom) and Univ. of Toronto (Canada); G. Novak, Northwestern Univ. (United States); D. Nutter, Cardiff Univ. (United Kingdom); L. Olmi, Univ. of Puerto Rico (United States) and INAF-Osservatorio Astrofisico di Arcetri (Italy); F. Poidevin, G. Savini, Univ. College London (United Kingdom); D. Scott, Univ. of British Columbia (Canada); J. A. Shariff, J. Soler, Univ. of Pennsylvania (United States); N. E. Thomas, Univ. of Miami (United States); M. D. Truch, Univ. of Pennsylvania (United States); C. E. Tucker, Cardiff Univ. (United Kingdom); G. S. Tucker, Brown Univ. (United States); D. Ward-Thompson, Cardiff Univ. (United Kingdom)

GAMMA RAY TELESCOPES

8444 17 **Optical design and calibration of a medium size telescope prototype for the CTA** [8444-42]
B. Behera, J. Bähr, Deutsches Elektronen-Synchrotron (Germany); S. Grünewald, Humboldt Univ. (Germany); G. Hughes, Deutsches Elektronen-Synchrotron (Germany); I. Oya, Humboldt Univ. (Germany); D. Melkumyan, S. Schlenstedt, Deutsches Elektronen-Synchrotron (Germany); U. Schwanke, Humboldt Univ. (Germany)

8444 18   Development of a mid-sized Schwarzschild-Couder Telescope for the Cherenkov
          Telescope Array [8444-43]
          R. A. Cameron, SLAC National Accelerator Lab. (United States)

ASSEMBLY, INTEGRATION, VERIFICATION, AND COMMISSIONING

8444 19   Status and performance of the Discovery Channel Telescope during commissioning
          (Invited Paper) [8444-44]
          S. E. Levine, T. A. Bida, T. Chylek, P. L. Collins, W. T. DeGroff, E. W. Dunham, P. J. Lotz,
          A. J. Venetiou, S. Zoonemat Kermani, Lowell Observatory (United States)

8444 1A   The Large Binocular Telescope [8444-45]
          J. M. Hill, R. F. Green, D. S. Ashby, J. G. Brynnel, N. J. Cushing, J. K. Little, J. H. Slagle,
          R. M. Wagner, Large Binocular Telescope Observatory, Univ. of Arizona (United States)

8444 1B   New Fraunhofer Telescope Wendelstein: assembly, installation, and current status [8444-46]
          H. Thiele, N. Ageorges, D. Kampf, M. Hartl, S. Egner, Kayser-Threde GmbH (Germany);
          P. Aniol, Astelco Systems GmbH (Germany); M. Ruder, Tautec (Germany); C. Abfalter,
          Astelco Systems GmbH (Germany); U. Hopp, R. Bender, C. Gössl, F. Grupp, F. Lang-Bardl,
          W. Mitsch, Univ.-Sternwarte München (Germany)

8444 1C   VST: from commissioning to science [8444-47]
          P. Schipani, INAF - Osservatorio Astronomico di Capodimonte (Italy); M. Capaccioli, INAF -
          Osservatorio Astronomico di Capodimonte (Italy) and Univ. Federico II of Naples (Italy);
          C. Arcidiacono, INAF - Osservatorio Astronomico di Bologna (Italy) and INAF - Osservatorio
          Astrofisico di Arcetri (Italy); J. Argomedo, European Southern Observatory (Germany);
          M. Dall'Ora, S. D'Orsi, INAF - Osservatorio Astronomico di Capodimonte (Italy); J. Farinato,
          D. Magrin, INAF - Osservatorio Astronomico di Padova (Italy); L. Marty, INAF - Osservatorio
          Astronomico di Capodimonte (Italy); R. Ragazzoni, INAF - Osservatorio Astronomico di
          Padova (Italy); G. Umbriaco, Univ. of Padua (Italy)

8444 1D   Commissioning results from the Large Binocular Telescope [8444-48]
          J. G. Brynnel, N. J. Cushing, R. F. Green, J. M. Hill, D. L. Miller, A. Rakich, K. Boutsia, Large
          Binocular Telescope Observatory, Univ. of Arizona (United States)

8444 1E   Discovery Channel Telescope active optics system early integration and test [8444-49]
          A. J. Venetiou, T. A. Bida, Lowell Observatory (United States)

EXTREMELY LARGE TELESCOPES

8444 1F   E-ELT update of project and effect of change to 39m design (Invited Paper) [8444-50]
          A. McPherson, J. Spyromilio, M. Kissler-Patig, S. Ramsay, E. Brunetto, P. Dierickx, M. Cassali,
          European Southern Observatory (Germany)

8444 1G   Thirty Meter Telescope project update (Invited Paper) [8444-51]
          L. Stepp, Thirty Meter Telescope Observatory Corp. (United States)

8444 1H **Giant Magellan Telescope: overview (Invited Paper)** [8444-52]
M. Johns, P. McCarthy, K. Raybould, A. Bouchez, A. Farahani, J. Filgueira, G. Jacoby, S. Shectman, M. Sheehan, Giant Magellan Telescope Organization Corp. (United States)

8444 1I **Science with the re-baselined European Extremely Large Telescope** [8444-54]
J. Liske, P. Padovani, M. Kissler-Patig, European Southern Observatory (Germany)

SITE CHARACTERIZATION, TESTING, AND DEVELOPMENT

8444 1J **Opacity measurements at Summit Camp on Greenland and PEARL in northern Canada with a 225-GHz tipping radiometer** [8444-55]
K. Asada, P. L. Martin-Cocher, C.-P. Chen, S. Matsushita, M.-T. Chen, Y.-D. Huang, M. Inoue, Institute of Astronomy and Astrophysics (Taiwan); P. T. P. Ho, Academia Sinica Institute of Astronomy and Astrophysics (Taiwan) and Harvard-Smithsonian Ctr. for Astrophyics (United States); S. N. Paine, Harvard-Smithsonian Ctr. for Astrophysics (United States); E. Steinbring, National Research Council Canada (Canada)

8444 1K **Site characterization studies in high plateau of Tibet** [8444-56]
Y. Yao, H. Wang, L. Liu, Y. Wang, X. Qian, J. Yin, National Astronomical Observatories (China)

8444 1L **New instruments to calibrate atmospheric transmission** [8444-57]
P. Zimmer, J. T. McGraw, D. C. Zirzow, The Univ. of New Mexico (United States); C. Cramer, K. Lykke, J. T. Woodward IV, National Institute of Standards and Technology (United States)

DESIGN OF TELESCOPES FOR EXTREME ENVIRONMENTS

8444 1N **The Greenland Telescope** [8444-59]
P. Grimes, R. Blundell, Smithsonian Astrophysical Observatory (United States)

8444 1O **Status of the first Antarctic survey telescopes for Dome A** [8444-60]
Z. Li, Nanjing Institute of Astronomical Optics & Technology (China) and Graduate Univ. of Chinese Academy of Sciences (China); X. Yuan, X. Cui, Nanjing Institute of Astronomical Optics & Technology (China) and Chinese Ctr. for Antarctic Astronomy (China); D. Wang, Nanjing Institute of Astronomical Optics & Technology (China); X. Gong, Nanjing Institute of Astronomical Optics & Technology (China) and Chinese Ctr. for Antarctic Astronomy (China); F. Du, Y. Zhang, Nanjing Institute of Astronomical Optics & Technology (China); Y. Hu, National Astronomical Observatories (China); H. Wen, X. Li, L. Xu, Nanjing Institute of Astronomical Optics & Technology (China); Z. Shang, National Astronomical Observatories (China) and Chinese Ctr. for Antarctic Astronomy (China); L. Wang, Purple Mountain Observatory (China) and Chinese Ctr. for Antarctic Astronomy (China)

8444 1P **Ukpik: testbed for a miniaturized robotic astronomical observatory on a high Arctic mountain** [8444-61]
E. Steinbring, B. Leckie, T. Hardy, K. Caputa, M. Fletcher, National Research Council Canada (Canada)

8444 1Q **The Gattini South Pole UV experiment** [8444-62]
A. M. Moore, Caltech Optical Observatories (United States); S. Ahmed, California Institute of Technology (United States); M. C. Ashley, The Univ. of New South Wales (Australia);

E. Croner, A. Delacroix, Caltech Optical Observatories (United States); Y. Ebihara, Research Institute for Sustainable Humanosphere, Kyoto Univ. (Japan); J. Fucik, Caltech Optical Observatories (United States); D. Martin, California Institute of Technology (United States); V. Velur, Caltech Optical Observatories (United States); A. Weatherwax, Siena College (United States)

8444 1R   **PLATO-R: a new concept for Antarctic science** [8444-63]
M. C. B. Ashley, Y. Augarten, C. S. Bonner, M. G. Burton, L. Bycroft, The Univ. of New South Wales (Australia); J. S. Lawrence, Australian Astronomical Observatory (Australia); D. M. Luong-Van, S. McDaid, C. McLaren, G. Sims, J. V. Storey, The Univ. of New South Wales (Australia)

CONTROL OF THERMAL ENVIRONMENT

8444 1S   **Canada-France-Hawaii Telescope image quality improvement initiative: thermal assay of the observing environment** [8444-64]
K. Thanjavur, K. Ho, S. Gajadhar, M. Baril, T. Benedict, S. Bauman, D. Salmon, Canada-France-Hawaii Telescope (United States)

PROJECT REVIEWS

8444 1T   **New optical telescope projects at Devasthal Observatory (Invited Paper)** [8444-65]
R. Sagar, B. Kumar, A. Omar, A. K. Pandey, Aryabhatta Research Institute of Observational Sciences (India)

8444 1U   **Towards a national astronomy observatory for the United Arab Emirates** [8444-66]
S. Els, Gaia Data Processing & Analysis Consortium (Spain); J. Maree, S. Al Marri, Y. Al Muqbel, A. Yousif, Emirates Institution for Advanced Science and Technology (United Arab Emirates); H. Al Naimiy, Univ. of Sharjah (United Arab Emirates)

8444 1V   **The 3,6 m Indo-Belgian Devasthal Optical Telescope: general description** [8444-67]
N. Ninane, C. Flebus, AMOS Ltd. (Belgium); B. Kumar, Aryabhatta Research Institute of Observational Sciences (India)

8444 1W   **Manufacturing optics of a 2.5m telescope** [8444-68]
F. Poutriquet, P. Plainchamp, J. Billet, B. Pernet, T. Lagrange, C. Cavadore, J. Carel, H. Leplan, E. Ruch, R. Geyl, J. Jouve, Sagem Défense Sécurité (France)

ENABLING TECHNOLOGIES FOR EXTREMELY LARGE TELESCOPES I

8444 1X   **E-ELT optomechanics: overview** [8444-69]
M. Cayrel, European Southern Observatory (Germany)

8444 1Y   **E-ELT M1 test facility** [8444-70]
M. Dimmler, J. Marrero, S. Leveque, P. Barriga, B. Sedghi, M. Mueller, European Southern Observatory (Germany)

Part 2

8444 1Z **Active damping strategies for control of the E-ELT field stabilization mirror** [8444-71]
B. Sedghi, M. Dimmler, M. Müller, European Southern Observatory (Germany)

8444 20 **Development of a fast steering secondary mirror prototype for the Giant Magellan Telescope** [8444-117]
M. Cho, National Optical Astronomy Observatory (United States); A. Corredor, C. Dribusch, The Univ. of Arizona (United States); K. Park, Y.-S. Kim, Korea Astronomy and Space Science Institute (Korea, Republic of); I.-K. Moon, Korea Research Institute of Standards and Science (Korea, Republic of); W.-H. Park, College of Optical Sciences, The Univ. of Arizona (United States)

8444 21 **Repairing stress induced cracks in the Keck primary mirror segments** [8444-73]
D. McBride, J. S. Hudek, S. Panteleev, W.M. Keck Observatory (United States)

ENABLING TECHNOLOGIES FOR EXTREMELY LARGE TELESCOPES II

8444 22 **Alignment algorithms for the Thirty Meter Telescope** [8444-74]
G. Chanan, Univ. of California, Irvine (United States)

8444 23 **Phasing metrology system for the GMT** [8444-75]
D. S. Acton, Ball Aerospace & Technologies Corp. (United States); A. Bouchez, Giant Magellan Telescope Project (United States)

8444 24 **Performance prediction of the fast steering secondary mirror for the Giant Magellan Telescope** [8444-76]
M. Cho, National Optical Astronomy Observatory (United States); A. Corredor, C. Dribusch, The Univ. of Arizona (United States); W.-H. Park, College of Optical Sciences, The Univ. of Arizona (United States); M. Sheehan, M. Johns, S. Shectman, J. Kern, C. Hull, Giant Magellan Telescope Project (United States); Y.-S. Kim, Korea Astronomy and Space Science Institute (Korea, Republic of); J. Bagnasco, Naval Postgraduate School (United States)

8444 25 **Dynamics, active optics, and scale effects in future extremely large telescopes** [8444-77]
R. Bastaits, B. Mokrani, Active Structures Lab., Univ. Libre de Bruxelles (Belgium); G. Rodrigues, European Space Agency (Netherlands); A. Preumont, Active Structures Lab., Univ. Libre de Bruxelles (Belgium)

SEGMENTED MIRROR ALIGNMENT, PHASING, AND WAVEFRONT CONTROL

8444 26 **The development of the actuator prototypes for the active reflector of FAST** [8444-78]
Q. Wang, M. Wu, M. Zhu, J. Xue, Q. Zhao, X. Gu, National Astronomical Observatories (China)

OBSERVATORY FACILITIES

8444 28   **Design, development, and manufacturing of highly advanced and cost effective aluminium sputtering plant for large area telescopic mirrors.** [8444-80]
R. R. Pillai, S. K. K., K. Mohanachandran, N. Sakhamuri, Hind High Vacuum Co. Pvt. Ltd. (India); V. Shukla, A. Gupta, Aryabhatta Research Institute of Observational Sciences (India)

SQUARE KILOMETER ARRAY AND SKA PATHFINDERS

8444 2A   **The Australian SKA Pathfinder (Invited Paper)** [8444-82]
A. E. T. Schinckel, J. D. Bunton, T. J. Cornwell, I. Feain, S. G. Hay, Commonwealth Scientific and Industrial Research Organisation (Australia)

8444 2B   **LOFAR, the low frequency array (Invited Paper)** [8444-83]
R. C. Vermeulen, ASTRON (Netherlands)

RADIO TELESCOPES

8444 2D   **The RAEGE VLBI 2010 radiotelescope design** [8444-85]
E. Sust, MT Mechatronics GmbH (Germany); J. López Fernández, Instituto Geográfico Nacional (Spain)

8444 2E   **Architecture of the metrology for the SRT** [8444-86]
T. Pisanu, F. Buffa, G. L. Deiana, P. Marongiu, INAF - Osservatorio Astronomico di Cagliari (Italy); M. Morsiani, INAF - Istituto di Radioastronomia (Italy); C. Pernechele, INAF - Osservatorio Astronomico di Padova (Italy); S. Poppi, G. Serra, G. Vargiu, INAF - Osservatorio Astronomico di Cagliari (Italy)

8444 2F   **Requirements and considerations of the surface error control for the active reflector of FAST** [8444-87]
M. Wu, Q. Wang, X. Gu, B. Zhao, National Astronomical Observatories (China)

8444 2G   **The Sardinia Radio Telescope (SRT) optical alignment** [8444-88]
M. Süß, D. Koch, MT Mechatronics GmbH (Germany); H. Paluszek, Sigma3D GmbH (Germany)

MILLIMETER AND SUBMILLIMETER WAVELENGTH TELESCOPES I

8444 2J   **Final tests and performances verification of the European ALMA antennas** [8444-91]
G. Marchiori, F. Rampini, European Industrial Engineering s.r.l. (Italy)

8444 2K   **ALMA system verification** [8444-92]
R. Sramek, K.-I. Morita, M. Sugimoto, P. Napier, M. Miccolis, Joint ALMA Observatory (Chile); P. Yagoubov, European Southern Observatory (Germany); D. Barkats, W. Dent, Joint ALMA Observatory (Chile); S. Matsushita, Academia Sinica Institute of Astronomy and Astrophysics (Taiwan); N. Whyborn, S. Asayama, Joint ALMA Observatory (Chile); J. Marti Canales, European Southern Observatory (Chile); R. Bhatia, E. DuVall, S. Blair, Joint ALMA Observatory (Chile)

MILLIMETER AND SUBMILLIMETER WAVELENGTH TELESCOPES II

8444 2M **The CCAT 25m diameter submillimeter-wave telescope** [8444-94]
D. Woody, Owens Valley Radio Observatory (United States); S. Padin, California Institute of Technology (United States); E. Chauvin, B. Clavel, Eric Chauvin Consulting (United States); G. Cortes, Cornell Univ. (United States); A. Kissil, J. Lou, Jet Propulsion Lab. (United States); P. Rasmussen, Owens Valley Radio Observatory (United States); D. Redding, Jet Propulsion Lab. (United States); J. Zolwoker, Cornell Univ. (United States)

8444 2N **High performance holography mapping with the LMT** [8444-95]
D. R. Smith, MERLAB, P.C. (United States); K. Souccar, Large Millimeter Telescope, Univ. of Massachusetts Amherst (United States)

8444 2O **Photonic local oscillator technics for large-scale interferometers** [8444-96]
H. Kiuchi, M. Saito, S. Iguchi, National Astronomical Observatory of Japan (Japan)

POSTER SESSION: AIRBORNE TELESCOPES

8444 2Q **First technological steps toward opening a near-IR window at stratospheric altitudes** [8444-98]
F. Pedichini, M. Centrone, D. Lorenzetti, M. Mattioli, M. Ricci, F. Vitali, INAF - Osservatorio Astronomico di Roma (Italy)

8444 2R **SOFIA in operation: telescope performance during the basic science flights** [8444-99]
H. J. Kärcher, MT Mechatronics GmbH (Germany); J. Wagner, A. Krabbe, Deutsches SOFIA Institut, Univ. Stuttgart (Germany); U. Lampater, Deutsches SOFIA Institut, NASA Dryden Flight Research Ctr (United States); T. Keilig, Deutsches SOFIA Institut, Univ. Stuttgart (Germany); J. Wolf, SOFIA Science Ctr., NASA Ames Research Ctr. (United States)

8444 2S **A new backup secondary mirror for SOFIA** [8444-100]
M. Lachenmann, M. J. Burgdorf, J. Wolf, Deutsches SOFIA Institut, Univ. Stuttgart (Germany) and SOFIA Science Ctr., NASA Ames Research Ctr. (United States); R. Brewster, Orbital Sciences Corp., NASA Ames Research Ctr. (United States)

8444 2T **Upgrade of the SOFIA target acquisition and tracking cameras** [8444-101]
M. Wiedemann, J. Wolf, Deutsches SOFIA Institut, Univ. Stuttgart (Germany) and SOFIA Science Ctr., NASA Ames Research Ctr. (United States); H. Roeser, Institute of Space Systems, Univ. Stuttgart (Germany)

POSTER SESSION: ASSEMBLY, INTEGRATION, VERIFICATION, AND COMMISSIONING

8444 2U **The 3,6 m Indo-Belgian Devasthal Optical Telescope: assembly, integration and tests at AMOS** [8444-102]
N. Ninane, C. Bastin, J. de Ville, F. Michel, M. Piérard, E. Gabriel, C. Flebus, AMOS Ltd. (Belgium); A. Omar, Aryabhatta Research Institute of Observational Sciences (India)

8444 2V **First tests of the compact low scattered-light 2m-Wendelstein Fraunhofer Telescope** [8444-103]
U. Hopp, R. Bender, F. Grupp, Univ.-Sternwarte München (Germany) and Max-Planck-Institut für extraterrestrische Physik (Germany); H. Thiele, N. Ageorges, Kayser-Threde GmbH (Germany); P. Aniol, Astelco Systems GmbH (Germany); H. Barwig, C. Gössl, F. Lang-Bardl, W. Mitsch, Univ.-Sternwarte München (Germany); M. Ruder, Astelco Systems GmbH (Germany)

8444 2W **SALT's transition to science operations** [8444-104]
D. A. H. Buckley, J. C. Coetzee, S. M. Crawford, South African Astronomical Observatory (South Africa); K. H. Nordsieck, Space Astronomy Lab., Univ. of Wisconsin-Madison (United States); D. O'Donoghue, South African Astronomical Observatory (South Africa); T. B. Williams, Rutgers, The State Univ. of New Jersey (United States)

POSTER SESSION: CONCEPTS FOR FUTURE TELESCOPES

8444 2Y **The QUIJOTE-CMB experiment: studying the polarisation of the galactic and cosmological microwave emissions** [8444-106]
J. A. Rubiño-Martín, Instituto de Astrofísica de Canarias (Spain) and Univ. de La Laguna (Spain); R. Rebolo, Instituto de Astrofísica de Canarias (Spain) and Univ. de La Laguna (Spain) and Consejo Superior de Investigaciones Científicas (Spain); M. Aguiar, Instituto de Astrofísica de Canarias (Spain); R. Génova-Santos, Instituto de Astrofísica de Canarias (Spain) and Univ. de La Laguna (Spain); F. Gómez-Reñasco, J. M. Herreros, R. J. Hoyland, Instituto de Astrofísica de Canarias (Spain); C. López-Caraballo, A. E. Pelaez Santos, Instituto de Astrofísica de Canarias (Spain) and Univ. de La Laguna (Spain); V. Sanchez de la Rosa, A. Vega-Moreno, T. Viera-Curbelo, Instituto de Astrofísica de Canarias (Spain); E. Martínez-Gonzalez, R. B. Barreiro, F. J. Casas, J. M. Diego, R. Fernández-Cobos, D. Herranz, M. López-Caniego, D. Ortiz, P. Vielva, Instituto de Fisica de Cantabria, Univ. de Cantabria (Spain); E. Artal, B. Aja, J. Cagigas, J. L. Cano, L. de la Fuente, A. Mediavilla, J. V. Terán, E. Villa, DICOM spol. s.r.o. (Spain); L. Piccirillo, R. Battye, E. Blackhurst, M. Brown, R. D. Davies, R. J. Davis, C. Dickinson, S. Harper, B. Maffei, M. McCulloch, S. Melhuish, G. Pisano, R. A. Watson, Jodrell Bank Ctr. for Astrophysics, The Univ. of Manchester (United Kingdom); M. Hobson, K. Grainge, Cavendish Lab., Univ. of Cambridge (United Kingdom); A. Lasenby, Cavendish Lab., Univ. of Cambridge (United Kingdom) and Kavli Institute for Cosmology, Univ. of Cambridge (United States); R. Saunders, P. Scott, Cavendish Lab., Univ. of Cambridge (United Kingdom)

8444 2Z **The next generation of the Canada-France-Hawaii Telescope: science requirements and survey strategies** [8444-108]
A. McConnachie, P. Côté, D. Crampton, NRC Herzberg Institute of Astrophysics (Canada); D. Devost, D. Simons, Canada-France-Hawaii Telescope Corp. (United States); K. Szeto, NRC Herzberg Institute of Astrophysics (Canada)

8444 30 **The optics and detector-simulation of the air fluorescence telescope FAMOUS for the detection of cosmic rays** [8444-109]
T. Niggemann, T. Hebbeker, M. Lauscher, C. Meurer, L. Middendorf, J. Schumacher, M. Stephan, RWTH Aachen Univ. (Germany)

POSTER SESSION: CONTROL OF THERMAL ENVIRONMENT

8444 31 **Experimental characterization of the turbulence inside the dome and in the surface layer** [8444-110]
A. Ziad, D.-A. Wassila, J. Borgnino, Observatoire de la Côte d'Azur, Univ. de Nice Sophia Antipolis, CNRS (France); M. Sarazin, European Southern Observatory (Germany)

8444 32 **Seeing trends from deployable Shack-Hartmann wavefront sensors, MMT Observatory, Arizona, USA** [8444-111]
J. D. Gibson, G. G. Williams, T. Trebisky, MMT Observatory, Univ. of Arizona (United States)

8444 33 **An updated T-series thermocouple measurement system for high-accuracy temperature measurements of the MMT primary mirror** [8444-112]
D. Clark, J. D. Gibson, MMT Observatory, Univ. of Arizona (United States)

POSTER SESSION: ENABLING TECHNOLOGIES FOR EXTREMELY LARGE TELESCOPES

8444 35 **A spectropolarimetric focal station for the ESO E-ELT** [8444-115]
K. G. Strassmeier, I. Di Varano, I. Ilyin, M. Woche, Leibniz-Institut für Astrophysik Potsdam (Germany); U. Laux, Thüringer Landessternwarte Tautenburg (Germany)

8444 37 **Performance of industrial scale production of ZERODUR mirrors with diameter of 1.5 m proves readiness for the ELT M1 segments** [8444-119]
T. Westerhoff, P. Hartmann, R. Jedamzik, A. Werz, SCHOTT AG (Germany)

POSTER SESSION: EXTREMELY LARGE TELESCOPES

8444 38 **E-ELT project: geotechnical investigation at Cerro Armazones** [8444-120]
P. Ghiretti, V. Heinz, European Southern Observatory (Germany); D. Pollak, J. Lagos, ARCADIS Chile S.A. (Chile)

POSTER SESSION: GAMMA RAY TELESCOPES

8444 39 **Technological developments toward the small size telescopes of the Cherenkov Telescope Array** [8444-121]
R. Canestrari, INAF - Osservatorio Astronomico di Brera (Italy); T. Greenshaw, Univ. of Liverpool (United Kingdom); G. Pareschi, INAF - Osservatorio Astronomico di Brera (Italy); R. White, Univ. of Leicester (United Kingdom)

8444 3A **SST-GATE: an innovative telescope for very high energy astronomy** [8444-254]
P. Laporte, J.-L. Dournaux, H. Sol, Observatoire de Paris, CNRS, Univ. Paris Diderot (France); S. Blake, Durham Univ. (United Kingdom); C. Boisson, P. Chadwick, D. Dumas, G. Fasola, F. de Frondat, Observatoire de Paris, CNRS, Univ. Paris Diderot (France); T. Greenshaw, Univ. of Liverpool (United Kingdom); O. Hervet, Observatoire de Paris, CNRS, Univ. Paris Diderot (France); J. Hinton, Univ. of Leicester (United Kingdom); D. Horville, J.-M. Huet, I. Jégouzo, Observatoire de Paris, CNRS, Univ. Paris Diderot (France); J. Schmoll, Durham Univ. (United Kingdom); R. White, Univ. of Leicester (United Kingdom); A. Zech, Observatoire de Paris, CNRS, Univ. Paris Diderot (France)

POSTER SESSION: INDUSTRIAL PERSPECTIVES

8444 3B **A new era for the 2-4 meters class observatories: an innovative integrated system telescope-dome** [8444-122]
G. Marchiori, A. Busatta, S. De Lorenzi, F. Rampini, European Industrial Engineering s.r.l. (Italy); C. Perna, G. Vettolani, Istituto Nazionale di Astrofisica (Italy)

POSTER SESSION: MEASUREMENT AND CONTROL OF TELESCOPE VIBRATION

8444 3C **Low-frequency high-sensitivity horizontal monolithic folded-pendulum as sensor in the automatic control of ground-based and space telescopes** [8444-123]
F. Acernese, Univ. degli Studi di Salerno (Italy) and Istituto Nazionale di Fisica Nucleare (Italy); R. De Rosa, Istituto Nazionale di Fisica Nucleare (Italy) and Univ. degli Studi di Napoli Federico II (Italy); G. Giordano, Univ. degli Studi di Salerno (Italy); R. Romano, F. Barone, Univ. degli Studi di Salerno (Italy) and Istituto Nazionale di Fisica Nucleare (Italy)

8444 3D **Herzberg Institute of Astrophysics' vibration measurement capabilities with applications to astronomical instrumentation** [8444-124]
P. W. G. Byrnes, NRC Herzberg Institute of Astrophysics (Canada)

POSTER SESSION: MILLIMETER AND SUBMILLIMETER WAVELENGTH TELESCOPES II

8444 3F **ALMA array element astronomical verification** [8444-126]
S. Asayama, Joint ALMA Observatory (Chile) and National Astronomical Observatory of Japan (Japan); L. B. G. Knee, Joint ALMA Observatory (Chile) and NRC Herzberg Institute of Astrophysics (Canada); P. G. Calisse, Joint ALMA Observatory (Chile) and European Southern Observatory (Chile); P. C. Cortés, Joint ALMA Observatory (Chile) and National Radio Astronomy Observatory (United States); R. Jager, Joint ALMA Observatory (Chile) and European Southern Observatory (Chile); B. López, C. López, Joint ALMA Observatory (Chile); T. Nakos, N. Phillips, Joint ALMA Observatory (Chile) and European Southern Observatory (Chile); M. Radiszcz, Joint ALMA Observatory (Chile); R. Simon, Joint ALMA Observatory (Chile) and National Radio Astronomy Observatory (United States); I. Toledo, Joint ALMA Observatory (Chile); N. Whyborn, Joint ALMA Observatory (Chile) and European Southern Observatory (Chile); H. Yatagai, Joint ALMA Observatory (Chile) and National Astronomical Observatory of Japan (Japan); J. P. McMullin, National Solar Observatory (United States); P. Planesas, Observatorio Astronómico Nacional (Spain)

8444 3G **Trajectory generation for parametric rotating scan patterns at the LMT** [8444-127]
D. R. Smith, MERLAB, P.C. (United States); K. Souccar, Large Millimeter Telescope, Univ. of Massachusetts Amherst (United States)

8444 3H **Atacama compact array antennas** [8444-128]
M. Saito, National Astronomical Observatory of Japan (Japan) and Joint ALMA Observatory (Chile); J. Inatani, National Astronomical Observatory of Japan (Japan); K. Nakanishi, National Astronomical Observatory of Japan (Japan) and Joint ALMA Observatory (Chile); H. Saito, S. Iguchi, National Astronomical Observatory of Japan (Japan)

8444 3I **Very large millimeter/submillimeter array toward search for 2nd Earth** [8444-129]
S. Iguchi, National Astronomical Observatory of Japan (Japan); M. Saito, National Astronomical Observatory of Japan (Japan) and Joint ALMA Observatory (Chile)

8444 3K **ACA phase calibration scheme with the ALMA water vapor radiometers** [8444-253]
Y. Asaki, Institute of Space and Astronautical Science (Japan) and The Graduate Univ. for Advanced Studies (Japan); S. Matsushita, Academia Sinica Institute of Astronomy and Astrophysics (Taiwan) and Joint ALMA Observatory (Chile); K.-I. Morita, National Astronomical Observatory of Japan (Japan) and Joint ALMA Observatory (Chile); B. Nikolic, Univ. of Cambridge (United Kingdom)

POSTER SESSION: SOLAR TELESCOPES

8444 3L **Functional safety for the Advanced Technology Solar Telescope** [8444-132]
S. Bulau, T. R. Williams, National Solar Observatory (United States)

8444 3M **Facility level thermal systems for the Advanced Technology Solar Telescope** [8444-133]
L. Phelps, National Solar Observatory (United States); G. Murga, AEC IDOM (United States); M. Fraser, M3 Engineering & Technology Corp. (United States); T. Climent, AEC IDOM (United States)

8444 3N **Stray light and polarimetry considerations for the COSMO K-Coronagraph** [8444-134]
A. G. de Wijn, J. T. Burkepile, S. Tomczyk, National Ctr. for Atmospheric Research (United States); P. G. Nelson, Sierra Scientific Solutions LLC (United States); P. Huang, Consultant (United States); D. Gallagher, National Ctr. for Atmospheric Research (United States)

8444 3O **Quasi-static wavefront control for the Advanced Technology Solar Telescope** [8444-135]
L. C. Johnson, National Solar Observatory (United States); R. Upton, Sigma Space Corp. (United States); T. Rimmele, S. Barden, National Solar Observatory (United States)

8444 3P **Optical design of the COSMO large coronagraph** [8444-136]
D. Gallagher, S. Tomczyk, National Ctr. for Atmospheric Research (United States); H. Zhang, Nanjing Institute of Astronomical Optics & Tech. (China); P. G. Nelson, Sierra Scientific Solutions LLC (United States)

8444 3S **Behavior of a horizontal air curtain subjected to a vertical pressure gradient** [8444-140]
J. Linden, L. Phelps, National Solar Observatory (United States)

POSTER SESSION: TELESCOPE MOUNTS AND ENCLOSURES

8444 3T **ATST telescope mount: machine tool or telescope** [8444-143]
P. Jeffers, National Solar Observatory (United States); G. Stolz, G. Bonomi, Ingersoll Machine Tools, Inc. (United States); O. Dreyer, H. Kärcher, MT Mechatronics GmbH (Germany)

8444 3U **Performance introduction of a 2.5m telescope mount** [8444-144]
G. Wang, B. Gu, S. Yang, X. Jiang, Z. Zhang, Y. Ye, J. Xu, Nanjing Institute of Astronomical Optics & Technology (China)

8444 3V **Installation and verification of high precision mechanics in concrete structures at the example of ALMA antenna interfaces** [8444-145]
V. Heinz, M. Kraus, European Southern Observatory (Germany); E. Orellana, Bautek S.A. (Chile)

8444 3W **E-ELT telescope main structure** [8444-146]
A. Orden Martínez, A. Dilla Martínez, N. Ballesteros Pérez, M. Alcantud Abellán, Empresarios Agrupados (Spain)

8444 3X **Testing, characterization, and control of a multi-axis, high precision drive system for the Hobby-Eberly Telescope Wide Field Upgrade** [8444-147]
I. M. Soukup, J. H. Beno, The Univ. of Texas Ctr. for Electromechanics (United States); G. J. Hill, J. M. Good, The Univ. of Texas McDonald Observatory (United States); C. E. Penney, T. A. Beets, J. D. Esguerra, R. J. Hayes, J. T. Heisler, J. J. Zierer, G. A. Wedeking, M. S. Worthington, D. R. Wardell, The Univ. of Texas Ctr. for Electromechanics (United States); J. A. Booth, M. E. Cornell, M. D. Rafal, The Univ. of Texas McDonald Observatory (United States)

8444 3Y **Enclosure rotation on the Large Binocular Telescope** [8444-148]
J. Howard, R. Meeks, D. Ashby, Large Binocular Telescope Observatory (United States); W. Davison, Steward Observatory, Univ. of Arizona (United States); J. Wiese, J. Urban, R. Hansen, J. Schuh, Large Binocular Telescope Observatory (United States)

8444 3Z **The 3,6 m Indo-Belgian Devasthal Optical Telescope: the hydrostatic azimuth bearing** [8444-150]
J. de Ville, M. Piérard, C. Bastin, AMOS Ltd. (Belgium)

8444 40 **Telescope positioning and drive system based on magnetic bearings, technical challenges and possible applications in optical stellar interferometry** [8444-151]
R. Lemke, Ruhr-Univ. Bochum (Germany); H. J. Kärcher, MT Mechatronics GmbH (Germany); L. Noethe, European Southern Observatory (Germany)

8444 41 **Enclosure design for the ARIES 3.6m optical telescope** [8444-152]
A. K. Pandey, V. Shukla, T. Bangia, Aryabhatta Research Institute of Observational Sciences (India); R. D. Raskar, R. R. Kulkarni, A. S. Ghanti, Precision Precast Solutions Pvt. Ltd. (India)

8444 42 **An innovative alt-alt telescope for small observatories and amateur astronomers** [8444-153]
M. Riva, S. Basso, R. Canestrari, P. Conconi, D. Fugazza, M. Ghigo, M. Landoni, G. Pareschi, P. Spanó, INAF - Osservatorio Astronomico di Brera (Italy); R. Tomelleri, Tomelleri s.r.l. (Italy); F. M. Zerbi, INAF - Osservatorio Astronomico di Brera (Italy)

8444 43 **Prototype enclosure design for the Korea Microlensing Telescope Network (KMTNet)** [8444-154]
N. Kappler, L. Kappler, TBR Construction & Engineering (United States); W. M. Poteet, H. K. Cauthen, CP Systems, Inc. (United States); B.-G. Park, C.-U. Lee, S.-L. Kim, S.-M. Cha, Korea Astronomy and Space Science Institute (Korea, Republic of)

POSTER SESSION: TELESCOPES FOR SYNOPTIC AND SURVEY OBSERVATIONS

8444 44 **Initial alignment and commissioning plan for the LSST** [8444-18]
W. J. Gressler, J. Sebag, National Optical Astronomy Observatory (United States); C. Claver, LSST Corp. (United States)

8444 45 **Dark energy camera installation at CTIO: overview** [8444-155]
T. M. C. Abbott, F. Muñoz, A. R. Walker, R. C. Smith, A. Montane, B. Gregory, R. Tighe, P. Schurter, N. S. van der Bliek, G. Schumacher, Cerro Tololo Inter-American Observatory (Chile)

8444 46 **Dark Energy Camera installation at CTIO: technical challenges** [8444-156]
F. Muñoz A., A. Montane, R. Tighe, M. Warner, T. M. C. Abbott, Cerro Tololo Inter-American Observatory (Chile)

8444 47 **Korea Microlensing Telescope Network: science cases** [8444-157]
B.-G. Park, S.-L. Kim, J.-W. Lee, B.-C. Lee, C.-U. Lee, Korea Astronomy and Space Science Institute (Korea, Republic of); C. Han, Chungbuk National Univ. (Korea, Republic of); M. Kim, Korea Astronomy and Space Science Institute (Korea, Republic of) and The Observatories of the Carnegie Institution for Science (United States); D.-S. Moon, Univ. of Toronto (Canada); H.-K. Moon, Korea Astronomy and Space Science Institute (Korea, Republic of); S.-C. Rey, Chungnam National Univ. (Korea, Republic of); E.-C. Sung, Korea Astronomy and Space Science Institute (Korea, Republic of); H. Sung, Sejong Univ. (Korea, Republic of)

8444 48 **Design and development of a wide field telescope** [8444-158]
I. Moon, Korea Research Institute of Standards and Science (Korea, Republic of); S. Lee, Korea Research Institute of Standards and Science (Korea, Republic of) and Hannam Univ. (Korea, Republic of); J. Lim, Korea Research Institute of Standards and Science (Korea, Republic of) and Kyung Hee Univ. (Korea, Republic of); H.-S. Yang, H.-G. Rhee, J.-B. Song, Y.-W. Lee, Korea Research Institute of Standards and Science (Korea, Republic of); J.-U. Lee, Cheongju Univ. (Korea, Republic of); H. Jin, Kyung Hee Univ. (Korea, Republic of)

8444 4A **Achieving high precision photometry for transiting exoplanets with a low cost robotic DSLR-based imaging system** [8444-160]
O. Guyon, Subaru Telescope, National Astronomical Observatory of Japan (United States) and Steward Observatory, Univ. of Arizona (United States); F. Martinache, Subaru Telescope, National Astronomical Observatory of Japan (United States)

POSTER SESSION: UPGRADES TO EXISTING OBSERVATORIES

8444 4B **An active surface upgrade for the Delingha 13.7-m Radio Telescope** [8444-163]
D. Yang, Y. Zhang, G. Zhou, National Astronomical Observatories (China) and Nanjing Institute of Astronomical Optics & Technology (China); A. Li, National Astronomical Observatories (China) and Nanjing Institute of Astronomical Optics & Technology (China) and Graduate Univ. of Chinese Academy of Sciences (China); K. Chen, Z. Zhang, G. Li, National Astronomical Observatories (China) and Nanjing Institute of Astronomical Optics & Technology (China); Y. Zuo, Y. Xu, Graduate Univ. of Chinese Academy of Sciences (China)

Part 3

8444 4D   **Development of a compact precision linear actuator for the active surface upgrade of the Delingha 13.7-m radio telescope** [8444-165]
G. Zhou, Nanjing Institute of Astronomical Optics & Technology (China); A. Li, Nanjing Institute of Astronomical Optics & Technology (China) and Graduate Univ. of Chinese Academy of Sciences (China); D. Yang, Z. Zhang, G. Li, Nanjing Institute of Astronomical Optics & Technology (China)

8444 4E   **Upgrading the TNT Telescope: remote observing and future perspectives** [8444-166]
G. Di Rico, INAF - Osservatorio Astronomico di Teramo (Italy); M. Fiaschi, MFC Elettronica (Italy); G. Valentini, A. Di Cianno, A. Valentini, INAF - Osservatorio Astronomico di Teramo (Italy)

8444 4F   **ESPRESSO: design and analysis of a Coudé-train for a stable and efficient simultaneous optical feeding from the four VLT unit telescopes** [8444-167]
A. Cabral, A. Moitinho, J. Coelho, J. Lima, Univ. de Lisboa (Portugal); G. Ávila, B.-A. Delabre, European Southern Observatory (Germany); R. Gomes, Univ. de Lisboa (Portugal); D. Mégevand, Observatoire de l'Univ. de Genève (Switzerland); F. Zerbi, INAF - Osservatorio Astronomico di Brera (Italy); P. Di Marcantonio, INAF - Osservatorio Astronomico di Trieste (Italy); C. Lovis, Observatoire de l'Univ. de Genève (Switzerland); N. C. Santos, Ctr. de Astrofísica, Univ. do Porto (Portugal) and Univ. do Porto (Portugal)

8444 4G   **Recent performance improvements for the Large Binocular Telescope primary mirror system** [8444-169]
R. L. Meeks, D. Ashby, C. Biddick, A. Chatila, M. Gusick, Large Binocular Telescope Observatory, Univ. of Arizona (United States)

8444 4H   **Modernization of the 1 meter Swope and 2.5 meter Du Pont telescopes at Las Campanas Observatory** [8444-170]
F. Perez, A. Bagish, Carnegie Observatories (United States); G. Bredthauer, Semiconductor Technology Associates (United States); J. Espoz, P. Jones, P. Pinto, Carnegie Observatories (United States)

8444 4I   **A happy conclusion to the SALT image quality saga** [8444-171]
L. A. Crause, South African Astronomical Observatory (South Africa); D. E. O'Donoghue, Southern African Large Telescope (South Africa); J. E. O'Connor, F. Strümpfer, O. J. Strydom, C. Sass, South African Astronomical Observatory (South Africa); C. du Plessis, E. Wiid, J. Love, Southern African Large Telescope (South Africa); J. D. Brink, South African Astronomical Observatory (South Africa); M. Wilkinson, C. Coetzee, Southern African Large Telescope (South Africa)

8444 4J   **Facility calibration unit of Hobby Eberly Telescope wide field upgrade** [8444-172]
H. Lee, G. J. Hill, B. Vattiat, McDonald Observatory, The Univ. of Texas at Austin (United States); M. P. Smith, Univ. of Wisconsin-Madison (United States); M. Häuser, Univ. Observatory Munich, Univ. of Munich (Germany)

8444 4K   **Solid telescopes for interferometric enhancement of existing telescopes** [8444-173]
A. Riva, M. Gai, INAF - Osservatorio Astrofisico di Torino (Italy)

POSTER SESSION: ACTIVE OPTICS AND PRECISION POSITION CONTROL MECHANISMS

8444 4L **Optics and the mechanical system of the 62-cm telescope at the Severo Díaz Galindo Observatory in Guadalajara, Jalisco, México** [8444-168]
E. de la Fuente, Univ. de Guadalajara (Mexico); J. M. Nuñez, S. Zazueta, Observatorio Astronómico Nacional, Univ. Nacional Autónoma de México (Mexico); S. E. Ibarra, Univ. de Guadalajara (Mexico); B. García, Observatorio Astronómico Nacional, Univ. Nacional Autónoma de México (Mexico); B. Martínez, Univ. de Guadalajara (Mexico); J. L. Ochoa, G. Sierra, F. Lazo, D. Hirart, Observatorio Astronómico Nacional, Univ. Nacional Autónoma de México (Mexico); L. Corral, J. L. Flores, J. Almaguer, S. Kemp, S. G. Navarro, A. Nigoche-Netro, G. Ramos-Larios, J. P. Phillips, A. Chávez, G. García-Torales, O. Blanco Alonso, T. Oceguera-Becerra, D. de Alba, R. Bautista, Univ. de Guadalajara (Mexico)

8444 4N **Folded Cassegrain sets of the Gran Telescopio Canarias (GTC)** [8444-175]
A. Gomez, R. Sanquirce, G. Murga, B. Etxeita, A. Vizcargüenaga, A. San Vicente, E. Fernandez, O. Vega, IDOM (Spain); B. Siegel, GRANTECAN S.A. (Spain)

8444 4O **Design, testing, and installation of a high-precision hexapod for the Hobby-Eberly Telescope dark energy experiment (HETDEX)** [8444-176]
J. J. Zierer, J. H. Beno, D. A. Weeks, I. M. Soukup, The Univ. of Texas at Austin, Ctr. for Electromechanics (United States); J. M. Good, J. A. Booth, G. J. Hill, M. D. Rafal, The Univ. of Texas at Austin, McDonald Observatory (United States)

8444 4P **Prototype pipeline for LSST wavefront sensing and reconstruction** [8444-177]
C. Claver, S. Chandrasekharan, M. Liang, National Optical Astronomy Observatory (United States); B. Xin, E. Alagoz, K. Arndt, I. P. Shipsey, Purdue Univ. (United States)

8444 4Q **Active optics in Large Synoptic Survey Telescope** [8444-178]
M. Liang, V. Krabbendam, C. F. Claver, S. Chandrasekharan, National Optical Astronomy Observatory (United States); B. Xin, Purdue Univ. (United States)

8444 4R **Keck 1 deployable tertiary mirror (K1DM3)** [8444-179]
J. X. Prochaska, C. Pistor, G. Cabak, D. J. Cowley, J. Nelson, Univ. of California Observatories (United States)

8444 4S **Metrology systems of Hobby-Eberly Telescope wide field upgrade** [8444-181]
H. Lee, G. J. Hill, M. E. Cornell, B. Vattiat, D. Perry, T. Rafferty, T. Taylor, McDonald Observatory, The Univ. of Texas at Austin (United States); M. Hart, Hart Scientific Consulting International L.L.C. (United States); M. D. Rafal, R. D. Savage, McDonald Observatory, The Univ. of Texas at Austin (United States)

8444 4T **Optics derotator servo control system for SONG Telescope** [8444-183]
J. Xu, C. Ren, Y. Ye, Nanjing Institute of Astronomical Optics & Technology (China)

8444 4U **Active optical control system design of the SONG-China Telescope** [8444-185]
Y. Ye, Nanjing Institute of Astronomical Optics & Technology (China) and Graduate Univ. of Chinese Academy of Sciences (China); S. Kou, D. Niu, Nanjing Institute of Astronomical Optics & Technology (China); C. Li, Nanjing Institute of Astronomical Optics & Technology (China) and Graduate Univ. of Chinese Academy of Sciences (China); G. Wang, Nanjing Institute of Astronomical Optics & Technology (China)

8444 4V   **The 3,6m Indo-Belgian Devasthal Optical: the active M1 mirror support** [8444-186]
M. Piérard, C. Flebus, N. Ninane, AMOS Ltd. (Belgium)

8444 4W   **Synchronous redundant control algorithm in the telescope drive system** [8444-187]
C. Ren, Nanjing Institute of Astronomical Optics & Technology (China); Y. Niu, X. Song, Nanjing Institute of Astronomical Optics & Technology (China) and Graduate Univ. of Chinese Academy of Sciences (China); J. Xu, X. Li, Nanjing Institute of Astronomical Optics & Technology (China)

8444 4X   **The M_2 & M_3 positioning control systems of a 2.5m telescope** [8444-188]
Y. Ye, C. Pei, Nanjing Institute of Astronomical Optics & Technology (China) and Graduate Univ. of Chinese Academy of Sciences (China); Z. Zhang, B. Gu, Nanjing Institute of Astronomical Optics & Technology (China)

8444 4Y   **Progress of the active reflector antenna using laser angle metrology system** [8444-189]
Y. Zhang, Nanjing Institute of Astronomical Optics & Technology (China) and National Astronomical Observatories (China); J. Zhang, Nanjing Institute of Astronomical Optics & Technology (China) and National Astronomical Observatories (China) and Graduate Univ. of Chinese Academy of Sciences (China); D. Yang, G. Zhou, A. Li, G. Li, National Astronomical Observatories (China)

8444 4Z   **The active optics system of the VST: concepts and results** [8444-190]
P. Schipani, INAF - Osservatorio Astronomico di Capodimonte (Italy); D. Magrin, INAF - Osservatorio Astronomico di Padova (Italy); L. Noethe, European Southern Observatory (Germany); C. Arcidiacono, INAF - Osservatorio Astronomico di Bologna (Italy) and INAF - Osservatorio Astrofisico di Arcetri (Italy); J. Argomedo, European Southern Observatory (Germany); M. Dall'Ora, S. D'Orsi, INAF - Osservatorio Astronomico di Capodimonte (Italy); J. Farinato, INAF - Osservatorio Astronomico di Padova (Italy); L. Marty, INAF - Osservatorio Astronomico di Capodimonte (Italy); R. Ragazzoni, INAF - Osservatorio Astronomico di Padova (Italy); G. Umbriaco, Univ. of Padua (Italy)

8444 50   **Performance comparison between two active support schemes for 1-m primary mirror** [8444-191]
D. Niu, Nanjing Institute of Astronomical Optics & Technology (China) and Graduate Univ. of Chinese Academy of Sciences (China); G. Wang, B. Gu, Nanjing Institute of Astronomical Optics & Technology (China)

8444 51   **Design, development, and testing of the DCT Cassegrain instrument support assembly** [8444-192]
T. A. Bida, E. W. Dunham, R. A. Nye, T. Chylek, R. C. Oliver, Lowell Observatory (United States)

POSTER SESSION: ALIGNMENT OF TELESCOPE OPTICS

8444 53   **Experience of primary surface alignment for the LMT using a laser tracker in a non-metrology environment** [8444-195]
D. M. Gale, Lab. de Superficies Asféricas, Instituto Nacional de Astrofísica, Óptica y Electrónica (Mexico)

8444 54   **Using a laser tracker for active alignment on the Large Binocular Telescope** [8444-196]
A. Rakich, Large Binocular Telescope Observatory (United States) and European Southern Observatory (Germany)

8444 55   **Generic misalignment aberration patterns and the subspace of benign misalignment** [8444-197]
P. L. Schechter, R. S. Levinson, Kavli Institute for Astrophysics and Space Research (United States) and Massachusetts Institute of Technology (United States)

8444 56   **The VST alignment: strategy and results** [8444-198]
P. Schipani, INAF - Osservatorio Astronomico di Capodimonte (Italy); L. Noethe, European Southern Observatory (Germany); K. Kuijken, Leiden Univ. (Netherlands); C. Arcidiacono, INAF - Osservatorio Astronomico di Bologna (Italy) and INAF - Osservatorio Astrofisico di Arcetri (Italy); J. Argomedo, European Southern Observatory (Germany); M. Dall'Ora, S. D'Orsi, INAF - Osservatorio Astronomico di Capodimonte (Italy); J. Farinato, D. Magrin, INAF - Osservatorio Astronomico di Padova (Italy); L. Marty, INAF - Osservatorio Astronomico di Capodimonte (Italy); R. Ragazzoni, INAF - Osservatorio Astronomico di Padova (Italy); G. Umbriaco, Univ. of Padua (Italy)

8444 58   **Test system for a Shack-Hartmann sensor based telescope alignment demonstrated at the 40cm Wendelstein Telescope** [8444-200]
S. Bogner, M. Becker, Ernst-Abbe Fachhochschule (Germany); F. Grupp, Max-Planck-Institut für extraterrestrische Physik (Germany) and Univ.-Sternwarte München (Germany); F. Lang-Bardl, Univ.-Sternwarte München (Germany); S.-M. Hu, Shandong Univ. at Weihai (China); M. Beyerlein, J. Lamprecht, J. Pfund, OPTOCRAFT GmbH (Germany); U. Hopp, Univ.-Sternwarte München (Germany); R. Bender, Univ.-Sternwarte München (Germany) and Max-Planck-Institut für extraterrestrische Physik (Germany); B. Fleck, Ernst-Abbe Fachhochschule (Germany)

8444 59   **An improved collimation algorithm for the Large Binocular Telescope using source extractor and an on-the-fly reconstructor** [8444-201]
D. L. Miller, A. Rakich, T. Leibold, Large Binocular Telescope Observatory (United States)

8444 5A   **Features of a laser metrology subsystem for astrometric telescopes** [8444-202]
A. Riva, M. Gai, M. G. Lattanzi, INAF - Osservatorio Astrofisico di Torino (Italy)

POSTER SESSION: DESIGN OF TELESCOPES FOR EXTREME ENVIRONMENTS

8444 5B   **Conceptual design of a 5-m terahertz telescope at Dome A** [8444-203]
D. Yang, H. Wang, Y. Zhang, Y. Chen, G. Zhou, Nanjing Institute of Astronomical Optics & Technology (China); J. Cheng, National Radio Astronomy Observatory (United States); G. Li, Nanjing Institute of Astronomical Optics & Technology (China)

8444 5C   **New Exoplanet Surveys in the Canadian High Arctic at 80 Degrees North** [8444-204]
N. M. Law, S. Sivanandam, Dunlap Institute for Astronomy & Astrophysics, Univ. of Toronto (Canada); R. Murowinski, National Research Council Canada (Canada); R. Carlberg, W. Ngan, Univ. of Toronto (Canada); P. Salbi, Dunlap Institute for Astronomy & Astrophysics, Univ. of Toronto (Canada); A. Ahmadi, Univ. of Calgary (Canada); E. Steinbring, M. Halman, National Research Council Canada (Canada); J. Graham, Dunlap Institute for Astronomy & Astrophysics, Univ. of Toronto (Canada)

8444 5E **An off-axis telescope concept for Antarctic astronomy** [8444-206]
G. Moretto, Lyon Institute of Origins, Institute of Nuclear Physics of Lyon, CNRS (France); N. Epchtein, Lab. J.L. Lagrange, CNRS, Univ. of Nice Sophia-Antipolis (France); M. Langlois, I. Vauglin, Ctr. de Recherche Astronomique de Lyon, CNRS (France)

8444 5F **The package cushioning design of the first AST3 and its dynamics analysis** [8444-207]
H. Wen, X. Gong, R. Zhang, Nanjing Institute of Astronomical Optics & Technology (China)

8444 5G **Nonlinear disturbance to Large Optical Antarctic Telescope** [8444-208]
S. Yang, Nanjing Institute of Astronomical Optics & Technology (China) and Graduate Univ. of Chinese Academy of Sciences (China)

8444 5H **Where is Ridge A?** [8444-209]
G. Sims, The Univ. of New South Wales (Australia); C. Kulesa, Univ. of Arizona (United States); M. C. B. Ashley, The Univ. of New South Wales (Australia); J. S. Lawrence, Macquarie Univ. (Australia) and Australian Astronomical Observatory (Australia); W. Saunders, Australian Astronomical Observatory (Australia); J. W. V. Storey, The Univ. of New South Wales (Australia)

8444 5I **Two years of polar winter observations with the ASTEP400 telescope** [8444-210]
L. Abe, J. Rivet, A. Agabi, E. Aristidi, D. Mekarnia, I. Goncalves, T. Guillot, Lab. J.L. Lagrange, CNRS, Univ. of Nice Sophia-Antipolis (France); M. Barbieri, Lab. J.L. Lagrange, CNRS, Univ. of Nice Sophia-Antipolis (France) and Univ. of Padova (Italy); N. Crouzet, Space Telescope Science Institute (United States); F. Fressin, Harvard-Smithsonian Ctr. for Astrophysics (United States); F. Schmider, Y. Fantei-Caujolle, J. Daban, C. Gouvret, S. Peron, P. Petit, A. Robini, M. Dugue, E. Bondoux, Lab. J.L. Lagrange, CNRS, Univ. of Nice Sophia-Antipolis (France); T. Fruth, A. Erikson, H. Rauer, DLR (Germany); F. Pont, A. Alapini, Univ. of Exeter (United Kingdom); S. Aigrain, Univ. of Oxford (United Kingdom); J. Szulagyi, Konkoly Observatory, Research Ctr. for Astronomy and Earth Sciences (Hungary); P. Blanc, A. Le Van Suu, Observatoire de Haute-Provence (France)

POSTER SESSION: OBSERVATORY CONTROL SYSTEMS

8444 5J **HETDEX tracker control system design and implementation** [8444-211]
J. Beno, R. Hayes, Ctr. for Electromechanics, The Univ. of Texas at Austin (United States); R. Leck, McDonald Observatory, The Univ. of Texas at Austin (United States); C. E. Penney, I. M. Soukup, Ctr. for Electromechanics, The Univ. of Texas at Austin (United States)

8444 5K **An upgrade to the telescope control system (TCS) for the Canada-France-Hawaii Telescope** [8444-212]
K. K. Y. Ho, W. Cruise, J. Thomas, Canada-France-Hawaii Telescope (United States)

8444 5L **Automation of the OAN/SPM 1.5-meter Johnson telescope for operations with RATIR** [8444-214]
A. M. Watson, M. G. Richer, Univ. Nacional Autónoma de México (Mexico); J. S. Bloom, Univ. of California, Berkeley (United States); N. R. Butler, Arizona State Univ. (United States); U. Ceseña, D. Clark, E. Colorado, A. Córdova, A. Farah, L. Fox-Machado, Univ. Nacional Autónoma de México (Mexico); O. D. Fox, NASA Goddard Space Flight Ctr. (United States); B. García, L. N. Georgiev, J. J. González, G. Guisa, L. Gutiérrez, J. Herrera, Univ. Nacional Autónoma de México (Mexico); C. R. Klein, Univ. of California, Berkeley (United States); A. S. Kutyrev, NASA Goddard Space Flight Ctr. (United States) and Univ. of

California Observatories/Lick Observatory, Univ. of California (United States); F. Lazo, W. H. Lee, E. López, E. Luna, B. Martínez, F. Murillo, J. M. Murillo, J. M. Núñez, Univ. Nacional Autónoma de México (Mexico); J. Prochaska, Univ. of Maryland, College Park (United States); J. Ochoa, F. Quirós, Univ. Nacional Autónoma de México (Mexico); D. A. Rapchun, NASA Goddard Space Flight Ctr. (United States) and Global Science & Technology (United States); C. Román-Zúñiga, G. Valyavin, Univ. Nacional Autónoma de México (Mexico)

8444 5M **Control system for the first three Antarctic Survey Telescopes (AST3-1)** [8444-216]
X. Li, D. Wang, L. Xu, J. Zhao, F. Du, Y. Zhang, Nanjing Institute of Astronomical Optics & Technology (China)

8444 5N **Development of an EtherCAT enabled digital servo controller for the Green Bank Telescope** [8444-217]
P. G. Whiteis, M. J. Mello, National Radio Astronomy Observatory (United States)

8444 5O **Design and development of telescope control system and software for the 50/80 cm Schmidt telescope** [8444-218]
T. S. Kumar, Indian Institute of Technology Bombay (India) and Aryabhatta Research Institute of Observational Sciences (India); R. N. Banavar, Indian Institute of Technology Bombay (India)

8444 5P **Upgrading the MMT primary mirror actuator test stand: a unique vehicle for evaluating EtherCAT as a future I/O standard for systems** [8444-219]
D. Clark, S. Schaller, MMT Observatory, Univ. of Arizona (United States)

8444 5Q **MMT nightly tracking logs: a web-enabled database for continuous evaluation of tracking performance** [8444-220]
D. Clark, J. D. Gibson, D. Porter, T. Trebisky, MMT Observatory, Univ. of Arizona (United States)

8444 5R **Pointing and tracking results of the VST telescope** [8444-221]
P. Schipani, INAF - Osservatorio Astronomico di Capodimonte (Italy); C. Arcidiacono, INAF - Osservatorio Astronomico di Bologna (Italy) and INAF - Osservatorio Astrofisico di Arcetri (Italy); J. Argomedo, European Southern Observatory (Germany); M. Dall'Ora, S. D'Orsi, INAF - Osservatorio Astronomico di Capodimonte (Italy); J. Farinato, D. Magrin, INAF - Osservatorio Astronomico di Padova (Italy); L. Marty, INAF - Osservatorio Astronomico di Capodimonte (Italy); R. Ragazzoni, INAF - Osservatorio Astronomico di Padova (Italy); G. Umbriaco, Univ. of Padua (Italy)

POSTER SESSION: PROJECT REVIEWS

8444 5S **Design and fabrication of three 1.6-meter telescopes for the Korea Microlensing Telescope Network (KMTNet)** [8444-223]
W. M. Poteet, H. K. Cauthen, CP Systems, Inc. (United States); N. Kappler, L. G. Kappler, TBR Construction & Engineering (United States); B.-G. Park, C.-U. Lee, S.-L. Kim, S.-M. Cha, Korea Astronomy and Space Science Institute (Korea, Republic of)

8444 5T **Introduction of Chinese SONG Telescope** [8444-224]
G. Wang, S. Kou, D. Niu, Z. Zhang, X. Jiang, C. Ren, Nanjing Institute of Astronomical Optics & Technology (China)

8444 5U   **Perspectives of astronomy in Kazakhstan: from new ground-based telescopes to space ones** [8444-225]
Ch. T. Omarov, Fessenkov Astrophysical Institute (Kazakhstan); Zh. Sh. Zhantayev, National Ctr. of Space Research and Technology (Kazakhstan)

8444 5V   **Deployment status of the Las Cumbres Observatory Global Telescope** [8444-226]
A. J. Pickles, W. Rosing, J. Martinez, B. J. Fulton, D. Sand, Las Cumbres Observatory Global Telescope Network (United States)

POSTER SESSION: RADIO TELESCOPES

8444 5W   **The microwave holography system for the Sardinia Radio Telescope** [8444-227]
G. Serra, P. Bolli, INAF - Osservatorio Astronomico di Cagliari (Italy); G. Busonera, CRS4 (Italy); T. Pisanu, S. Poppi, F. Gaudiomonte, INAF - Osservatorio Astronomico di Cagliari (Italy); G. Zacchiroli, J. Roda, M. Morsiani, INAF - Istituto di Radioastronomia (Italy); J. A. López-Pérez, Ctr. Astronómico de Yebes (Spain)

8444 5X   **Structural optimization of the outer ring of FAST Telescope** [8444-228]
X. Zhang, National Astronomical Observatories (China) and Graduate Univ. of Chinese Academy of Sciences (China); H. Li, S. Yang, National Astronomical Observatories (China)

8444 5Y   **Experimental study on the damping of FAST cabin suspension system** [8444-229]
H. Li, J. Sun, National Astronomical Observatories (China); X. Zhang, National Astronomical Observatories (China) and Graduate Univ. of Chinese Academy of Sciences (China); W. Zhu, G. Pan, Q. Yang, National Astronomical Observatories (China)

POSTER SESSION: SEGMENTED MIRROR ALIGNMENT, PHASING, AND WAVEFRONT CONTROL

8444 5Z   **Control algorithm for the petal-shape segmented-mirror telescope with 18 mirrors** [8444-230]
A. Shimono, The Univ. of Tokyo (Japan); F. Iwamuro, M. Kurita, Kyoto Univ. (Japan); Y. Moritani, Kyoto Univ. (Japan) and Hiroshima Univ. (Japan); M. Kino, Nagoya Univ. (Japan); T. Maihara, Nano-Optonics Energy Inc. (Japan); H. Izumiura, National Astronomical Observatory of Japan (Japan); M. Yoshida, Hiroshima Univ. (Japan)

8444 60   **How to calibrate edge sensors on segmented mirror telescopes** [8444-231]
C. Shelton, L. C. Roberts, Jet Propulsion Lab. (United States)

8444 61   **Outdoors phasing progress of dispersed fringe sensing technology in NIAOT, China** [8444-232]
Y. Zhang, X. Cui, G. Liu, Y. Wang, J. Ni, H. Li, Y. Zeng, A. Li, Y. Li, Nanjing Institute of Astronomical Optics & Technology (China); Z. Wu, Nanjing Institute of Astronomical Optics & Technology (China) and Graduate Univ. of Chinese Academy of Sciences (China)

POSTER SESSION: SITE CHARACTERIZATION, TESTING, AND DEVELOPMENT

8444 62   **The new TNG-DIMM: calibrations and first data analysis** [8444-233]
E. Molinari, A. G. de Gurtubai, Telescopio Nazionale Galileo (Spain); A. della Valle, INAF - Osservatorio Astronomico di Bologna (Italy); S. Ortolani, Univ. degli Studi di Padova (Italy); J. San Juan, A. F. Martinez Fiorenzano, Telescopio Nazionale Galileo (Spain); V. Zitelli, INAF - Osservatorio Astronomico di Bologna (Italy)

8444 64   **Atmospheric turbulence measurements at Ali Observatory, Tibet** [8444-235]
L. Liu, National Astronomical Observatories (China) and Lab. Lagrange, CNRS, Univ. de Nice Sophia-Antipolis (France); Y. Yao, National Astronomical Observatories (China); J. Vernin, M. Chadid, Lab. J.L. Lagrange, CNRS, Univ. de Nice Sophia-Antipolis (France); Y. Wang, H. Wang, J. Yin, National Astronomical Observatories (China); C. Giordano, Lab. J.L. Lagrange, CNRS, Univ. de Nice Sophia-Antipolis (France); X. Qian, National Astronomical Observatories (China)

8444 68   **Dust concentration and soil properties at the TMT candidate sites** [8444-239]
S. G. Els, Gaia Data Processing & Analysis Consortium (Spain) and Cerro Tololo Inter-American Observatory (Chile) and TMT Observatory Corp. (United States); R. Riddle, Thirty Meter Telescope Observatory Corp. (United States) and Caltech Optical Observatories (United States); M. Schöck, W. Skidmore, T. Travouillon, Thirty Meter Telescope Observatory Corp. (United States)

8444 69   **Surface layer turbulence measurements on the LSST site El Peñon using microthermal sensors and the lunar scintillometer LuSci** [8444-240]
J. Sebag, National Optical Astronomy Observatory (United States); P. Zimmer, J. Turner, J. McGraw, The Univ. of New Mexico (United States); V. Krabbendam, National Optical Astronomy Observatory (United States); A. Tokovinin, E. Bustos, M. Warner, CTIO (Chile); O. Wiecha, National Optical Astronomy Observatory (United States)

8444 6A   **Overview of site monitoring at the SAAO** [8444-241]
T. Pickering, South African Astronomical Observatory (South Africa) and Southern African Large Telescope (South Africa); S. M. Crawford, South African Astronomical Observatory (South Africa); L. Catala, South African Astronomical Observatory (South Africa) and Univ. of Cape Town (South Africa); D. Buckley, South African Astronomical Observatory (South Africa); A. Ziad, Observatoire de la Côte d'Azur, CNRS, Univ. de Nice (France); R. Wilson, Univ. of Durham (United Kingdom)

8444 6B   **Evaluations of new atmospheric windows at thirty micron wavelengths for astronomy** [8444-242]
T. Miyata, S. Sako, T. Kamizuka, Institute of Astronomy, The Univ. of Tokyo (Japan); T. Nakamura, Institute of Astronomy, The Univ. of Tokyo (Japan) and The Univ. of Tokyo (Japan); K. Asano, M. Uchiyama, M. Konishi, Institute of Astronomy, The Univ. of Tokyo (Japan); M. Yoneda, Planetary Plasma and Atmospheric Research Ctr., Tohoku Univ. (Japan); N. Takato, Subaru Telescope, National Astronomical Observatory of Japan (United States); Y. Yoshii, M. Doi, K. Kohno, K. Kawara, M. Tanaka, K. Motohara, T. Minezaki, T. Tanabe, T. Morokuma, Y. Tamura, Institute of Astronomy, The Univ. of Tokyo (Japan); T. Aoki, T. Soyano, K. Tarusawa, Kiso Observatory, Institute of Astronomy, The Univ. of Tokyo (Japan); H. Takahashi, S. Koshida, N. M. Kato, Institute of Astronomy, The Univ. of Tokyo (Japan)

8444 6C **Atmospheric seeing measurements obtained with MISOLFA in the framework of the PICARD Mission** [8444-243]
R. Ikhlef, Observatoire de la Côte d'Azur, CNRS, Univ. de Nice Sophia Antipolis (France) and Observatoire d'Alger, Ctr. de Recherche en Astronomie, Astrophysique et Géophysique (Algeria); T. Corbard, Observatoire de la Côte d'Azur, CNRS, Univ. de Nice Sophia Antipolis (France); A. Irbah, Lab. Atmosphères, Milieux, Observations Spatiales, CNRS, Univ. Versailles St-Quentin (France); F. Morand, Observatoire de la Côte d'Azur, CNRS, Univ. de Nice Sophia Antipolis (France); M. Fodil, Observatoire d'Alger, Ctr. de Recherche en Astronomie, Astrophysique et Géophysique (Algeria); B. Chauvineau, P. Assus, C. Renaud, Observatoire de la Côte d'Azur, CNRS, Univ. de Nice Sophia Antipolis (France); M. Meftah, S. Abbaki, Lab. Atmosphères, Milieux, Observations Spatiales, CNRS, Univ. Versailles St-Quentin (France); J. Borgnino, Observatoire de la Côte d'Azur, CNRS, Univ. de Nice Sophia Antipolis (France); E. M. Cissé, E. D'Almeida, A. Hauchecorne, Lab. Atmosphères, Milieux, Observations Spatiales, CNRS, Univ. Versailles St-Quentin (France); F. Lalcare, Observatoire de la Côte d'Azur, CNRS, Univ. de Nice Sophia Antipolis (France); P. Lesueur, M. Lin, Lab. Atmosphères, Milieux, Observations Spatiales, CNRS, Univ. Versailles St-Quentin (France); F. Martin, Observatoire de la Côte d'Azur, CNRS, Univ. de Nice Sophia Antipolis (France); G. Poiet, Lab. Atmosphères, Milieux, Observations Spatiales, CNRS, Univ. Versailles St-Quentin (France); M. Rouzé, Ctr. National d'Études Spatiales (France); G. Thuillier, Lab. Atmosphères, Milieux, Observations Spatiales, CNRS, Univ. Versailles St-Quentin (France); A. Ziad, Observatoire de la Côte d'Azur, CNRS, Univ. de Nice Sophia Antipolis (France)

POSTER SESSION: TELESCOPE OPTICAL DESIGNS

8444 6D **Optical system of Chinese SONG Telescope** [8444-244]
S. Kou, G. Liu, G. Wang, Nanjing Institute of Astronomical Optics & Technology (China)

8444 6E **Design of an off-axis optical reflecting system** [8444-245]
Y. V. Bazhanov, V. B. Vlahko, Precision Systems and Instruments Corp. (Russian Federation)

8444 6G **Dome flat-field system for 1.3-m Araki Telescope** [8444-248]
T. Yoshikawa, Y. Ikeda, N. Fujishiro, Koyama Astronomical Observatory, Kyoto Sangyo Univ. (Japan); S. Ichizawa, Cybernet Systems Co., Ltd. (Japan); A. Arai, M. Isogai, A. Yonehara, H. Kawakita, Koyama Astronomical Observatory, Kyoto Sangyo Univ. (Japan)

8444 6H **Fast and compact wide-field Gregorian telescope** [8444-249]
M. Bahrami, A. V. Goncharov, National Univ. of Ireland, Galway (Ireland)

8444 6I **Optical design for amateur reflecting telescopes based on tilted axial-symmetrical planoidal mirror** [8444-250]
S. A. Chuprakov, Institute of Solar-Terrestrial Physics (Russian Federation)

8444 6J **Preliminary optical design for the WEAVE two-degree prime focus corrector** [8444-251]
T. Agócs, ASTRON (Netherlands); D. C. Abrams, D. Cano Infantes, N. O'Mahony, Isaac Newton Group of Telescopes (Spain); K. Dee, Engineering & Project Solutions Ltd. (United Kingdom); J.-B. Daban, C. Gouvret, S. Ottogalli, Observatoire de la Côte d'Azur, Lab. Lagrange, CNRS, Univ. de Nice Sophia Antipolis (France)

Author Index

Active damping strategies for control of the E-ELT field stabilization mirror

B. Sedghi[a] and M. Dimmler[a] and M. Müller[a]

[a]European Southern Observatory (ESO), Karl-Schwarzschild-Strasse. 2, Garching, Germany

ABSTRACT

The fifth mirror unit (M5) of the E-ELT is a field stabilization unit responsible to correct for the dynamical tip and tilt caused mainly due to the wind load on the telescope. The unit is composed of: i) an electromechanical subunit, and ii) an elliptical mirror with a size of approximately 2.4 by 3-m. The M5 unit has been designed and prototyped using a three point support for the mirror actuated by piezo actuators without the need of a counter weight system. To be able to meet the requirements of the telescope, i.e. sufficient wavefront rejection capability, the unit shall exhibit a sufficient bandwidth for tip/tilt reference commands. In the presence of the low damped mechanical resonant modes, such a bandwidth can be guaranteed thanks to an active damping loop. In this paper, different active damping strategies for the M5 unit are presented. The efficiency of the approaches are analyzed using a detailed model of the unit. On a scale-one prototype active damping was implemented and the efficiency was demonstrated.

Keywords: E-ELT, field stabilization unit, active damping, control, robust stability

1. INTRODUCTION

The European Extremely Large Telescope (E–ELT) is a project led by ESO for a next generation optical and near–infrared, ground–based telescope. Its optical design is based on a three–mirror anastigmat with two folding flat mirrors sending the beam to either of the two Nasmyth foci along the elevation axis of the telescope[1],[2]. The elliptical primary mirror consists of almost 800 off-axis aspherical segments, each 1.45-m in size and 50-mm thick. The secondary and tertiary mirrors are designed as convex and concave aspherical mirrors, respectively, providing active position and shape control. The quarternary mirror is adaptive aiming at the compensation of fast wavefront distortions which are mainly due to atmospheric turbulence. The main purpose of the ultra–lightweight fifth mirror is to provide the compensation of image motion. The opto-mechanical mirror units (M1– M5)[3] are held by the main structure, which also supports the instruments at the Nasmyth platforms, all handling tools and all equipment necessary for the altitude azimuth kinematics. The main structure also holds the pre–focal stations, which contain the on–sky metrology for wavefront control.

The M5 unit of the E-ELT together with the adaptive M4 are responsible for correcting the tip/tilt errors of the telescope mainly due to the atmospheric and wind load perturbations. The entire unit is inclined at 53.5-degrees and mounted on a rotating stage provided by the telescope main structure. The unit is composed of: i) an electromechanical subunit, and ii) a mirror with a size of approximately 2.4 by 3-m. The electromechanical unit is composed of a support unit based on a fixed frame - main structure - that carries the base frame upon which the actuators and the central restraint are located. The M5 electromechanical unit has been designed and prototyped by NTE/CSEM[4] on the basis of a three point actuated support of the mirror without a counter weight system. The actuator is based on a CEDRAT APA[5] design custom built for the E-ELT. A preloaded elliptical steel ring is forced open by the action of a piezo stack running along the major axis. In the relaxed state the actuator is at its maximum extent and expanding the piezo compresses the actuator.

The intrinsic stiffness of the actuator is high (45-N/μm). A position sensor is used to provide absolute calibration and feedback to the actuator. It is mounted on the same flange as the actuator. The position sensor selected is an eddy current device that provides a 15-nm resolution over a 1-mm range. The actuator provides interfaces for accelerometers to be mounted at different locations, e.g. ends of piezo stack, or on top near the

---

Further author information: (Send correspondence to B. Sedghi)
B. Sedghi: E-mail: bsedghi@eso.org, Telephone: +49 89 32006529

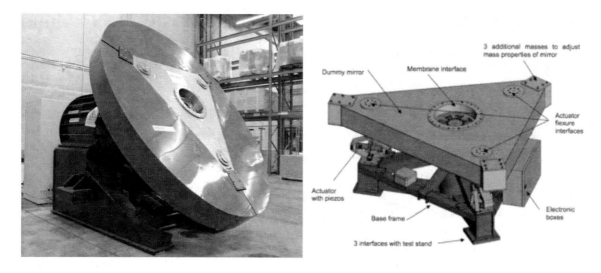

Figure 1. M5 scale-one prototype (left) and the overview (right)

position sensor. The total stroke of each actuator is 700-μm, a large portion of which is used for fine alignment and to compensate for gravity deformations and the remaining for the atmospheric and wind rejection.

The M5 unit scale-one prototype as designed by the contractor assumes a mirror with characteristics similar to a closed-back Ultra Low Expansion (ULE) mirror where the actuators are connected to the mirror without the need for an axial support system (whiffle tree). The mirror is restrained laterally using a central membrane. In the prototype a cut-out of an optical table with the same mass but lower eigen-frequencies has been used as dummy mirror. For tip-tilt control this does not affect the performance of the system which is dominated by the modes of the electro-mechanical part. If a open-back Zerodur mirror option is considered, a whiffle tree system with nine support points is necessary. Figure 2 compares the FEM representation of the M5 unit assuming a closed-back ULE with an open-back Zerodur mirror option.

Table 1. M5 mirror option main characteristics

Zerodur	495 kg ($90 kg/m^2$)	open-back	first mode (free): 139Hz	9 point support
ULE	390 kg ($70 kg/m^2$)	closed-back	first mode (free): 300Hz	3 point support

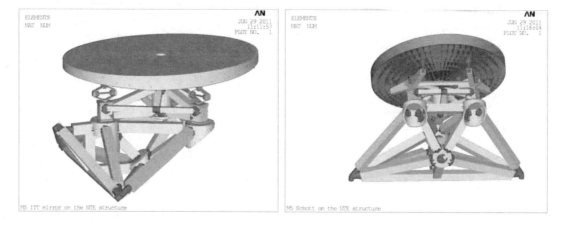

Figure 2. M5 FEM: closed-back ULE mirror (left), open-back Zerodur mirror with whiffle tree support (right)

To be able to meet the telescope requirements the unit from any tip/tilt command input to output shall exhibit sufficient bandwidth[6]. The required bandwidth can be guaranteed thanks to the local control loops of the electromechanical unit: i) an active damping loop, ii) a position control loop. A trade-off study and analysis was performed to answer the following main questions: Is the damping strategy efficient for all mirror options? Which modes limit the stability and robustness? Can the desired unit bandwidth of 10Hz be achieved robustly for all mirror options? What are the main active damping strategies and algorithms which can be considered? How do they compare? What are the main factors limiting the efficiency of the active damping of the unit? How does the mechanical interface of the unit to the telescope structure affect the control and performance of M5?

In this paper, a summary of the important results and conclusions of the analysis is presented. In Section 2 different damping strategies are introduced. The active damping strategies are tested on the dynamical model extracted from FE analysis and the results are compared in terms of efficiency and stability/robustness for two different unit mirror designs, i.e. closed-back ULE vs. open-back Zerodur. In Section 3 some measurement results on the scale-one prototype are presented, and the conclusions are given in Section 4.

2. ACTIVE DAMPING STRATEGIES AND MAIN RESULTS

2.1 FE Models for two mirror designs and comparison of the open-loop frequency responses

From a detailed FEM (capturing the mechanical modes up to 600Hz) of the M5 unit for the two mirror options, closed-back ULE (no whiffle tree) vs. open-back Zerodur (with three tripods for a nine-point support), the frequency responses of the actuator input command to a collocated sensor measuring the stroke of the actuator are derived. It is assumed that the unit is attached to an infinite stiff structure. The effect of the telescope structure was as well analyzed but the results are out of scope of this document. Figure 3 compares the open-loop responses for three actuators for ULE and Zerodur options respectively.

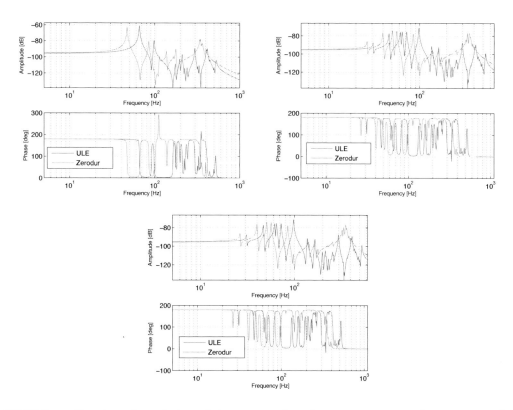

Figure 3. M5 ULE vs. Zerodur mirror design: Actuator open-loop frequency responses (input command force to local sensor), Act #1 (top left), Act #2 (top right), Act #3 (bottom)

Due to a non symmetric installation of actuators on the M5 unit one actuator response differs from the other two. The mechanical modes of the Zerodur mirror design option are lower in frequency. The first large amplitude mechanical mode seen by the sensor on Actuator #1 of ULE and Zerodur option designs are at 66Hz and 47Hz, respectively. Actuator #2 and #3 exhibit the same response with first visible mechanical mode (though small) at 31Hz and 26Hz for ULE and Zerodur designs, respectively. For ULE mirror the first largest amplitude mode is at 46Hz while for the Zerodur this mode is at 40Hz. For the ULE design a mode at 98Hz has an important amplitude while the similar mode for the other design option has a smaller amplitude. However, the Zerodur mirror option exhibits a large amplitude mode at 340Hz which is the local mode of the actuator ellipses and the whiffle tree tripods.

The requirement on M5 unit closed-loop bandwidth is 10Hz. For both mirror options the required robust stable closed-loop bandwidth cannot be achieved unless the first mechanical modes are damped or notched. The approach based on notch filters is disadvantageous since a good knowledge of the location of the problematic modes is required and in general is non-robust. In addition, while solving the problem of control it does not solve the problem of response to any perturbation excitation. In case of using an active damping strategy the location of the sensing system and control strategy plays an important role in damping of the harmful mechanical modes. In the next section different strategies and their outcome on the two mirror options are discussed. The models have the necessary inputs and outputs to perform the required trade-off analysis and control design: the inputs are the actuator forces and the outputs are considered to be the absolute or differential acceleration of actuators, the elongation of three actuators (position sensor), force at the interface of actuator and mirror (force sensors) together with 100 mechanical modes of the electromechanical unit.

2.2 Strategies and Results

The active damping strategies are classified based on the type of the sensors, their physical location and the type of the implemented algorithm.[7] In this work three main schemes on the basis of the sensing system for active damping are assumed: i) accelerometer, ii) force sensor, and iii) position sensor. In the case of accelerometer based active damping approach two possibilities were investigated: ia) accelerometers are located on the top of each actuator measuring the absolute signal, ib) two accelerometers are installed, on the top and bottom of each actuator, and the differential signal value is used for control.

One objective of the work is to evaluate the efficiency of the active damping schemes in face of problematic resonant modes of the unit. Therefore, to be able to compare mirror options, identical control parameters were implemented for different mirror options and for the same strategy. For each mirror design and damping strategy a Multi-Input Multi-Output (MIMO) robust stability criterion is verified *. In order to distinguish the effect of the cross-coupling and the control structure interaction on the robustness, the Single-Input Single-Output (SISO) sensitivity transfer function, e.g. for the tip response, is compared with that of the MIMO criterion. The control of the position loops are implemented in the piston/tip/tilt (PTT) space. Hence, a geometrical transformation projects the signals of actuator position sensors to the piston/tip and tilt of the M5 unit. The frequency responses from PTT input commands to the PTT generated from the local position sensors are used as a basis for design of three identical integral position controllers, $C_p = \dfrac{k_i}{s}$, for the piston/tip and tilt signals constructed from the position sensors. Assuming enough damping is introduced by the active damping the gain of $k_i = 56$ leads to a tip/tilt reference to output closed-loop bandwidth (-3dB) of 10Hz.

2.2.1 Acceleration Feedback

The strategy consists on either to integrate the acceleration signal a to obtain the velocity and then by a direct gain controller K_a introduce damping, i.e.

$$u = C_{damp} * a = -\frac{K_a}{s} * \frac{s}{s+\omega_b} * a = -\frac{K_a}{s+\omega_b} * a \tag{1}$$

---

*using the characteristic transfer functions CTFs, i.e. eigenvalues of the MIMO loop transfer matrix which take into account the dynamical cross-coupling and control and structure interactions,[8] and the infinity norm of the MIMO sensitivity transfer function

or by passing the acceleration signal through a second order filter and by generating a force proportional to the output of the filter, i.e.

$$u = C_{damp} * a = -\frac{K_a}{s^2 + 2\xi_f \omega_f s + \omega_f^2} * \frac{s}{s + \omega_b} * a \qquad (2)$$

where damping factor ξ_f at a selected frequency ω_f are chosen by the designer.

In practice, acceleration signals at low frequencies are noisy and thus a high-pass filters are included in the controllers. The high-pass filter corner frequency is presented by ω_b.

Closed-back ULE mirror option: Figure 4 compares the tilt frequency response constructed from position sensors before and after implementing the active damping using the direct velocity Eq.(1) approach for both cases of accelerometer on the top of an actuator and the differential acceleration signal and for the closed-back ULE mirror option.

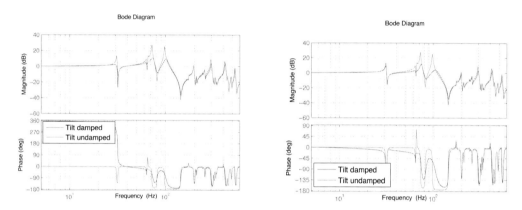

Figure 4. Closed-back ULE mirror: tilt frequency response before and after damping. Accelerometer on top of each actuator (left), Accelerometers on top and bottom of each actuator (right). Note: The large phase transition of 360deg is an artifact.

In the case of accelerometer on top of each actuator, the first 3 important modes are damped. Both the active damping and the position loops are MIMO robust stable (see Figure 5). Further investigation showed that mechanical modes at 100Hz could lead to stability issues for the damping loop and the non-damped mode at 31Hz could be problematic for the outer-loop (position) loop.

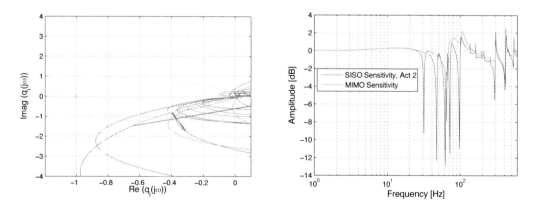

Figure 5. Closed-back ULE mirror active damping with accelerometer on top of each actuator: MIMO Nyquist curves (left) and Sensitivity transfer function $||S||_\infty$ (right) of the active damping loop.

When the accelerometer on top and bottom of each actuator are used (differential signal), one expects a perfect collocation and simply by increasing the damping loop gain better results can be obtained. Although this was verified, for the sake of comparison the results for the same control gains are presented here. In this case the important modes except the mode at 31Hz are damped. The active damping loop is robust stable. The position loop is stable, however due to the undamped mechanical mode at 31Hz and the cross-coupling effects the position loop is not robust, i.e. $||S||_\infty > 6$dB (see Figure 6).

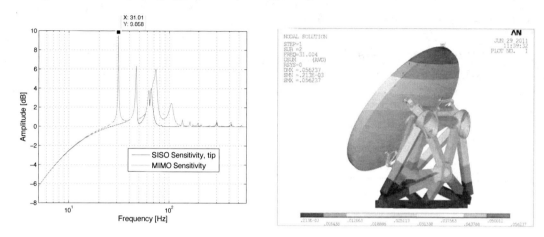

Figure 6. Closed-back ULE mirror active damping with accelerometer on top and bottom of each actuator: Sensitivity transfer function $||S||_\infty$ of the PTT position loop (left), Graphical representation of the mode at 31Hz (right)

The mode at 31Hz is mainly related to the tilt motion of the mirror together with the back structure. This is the main reason that the mode is not seen on the differential measurement and consequently cannot be damped by this strategy. The graphical representation of this mode is depicted in Figure 6. The results for the second order filter algorithm are similar and to avoid a repetition they are not presented here.

Open-back Zerodur mirror option: Figure 7 compares the piston frequency response constructed from position sensors before and after implementing the active damping using the direct velocity Eq.(1) approach for both cases of accelerometer on top of an actuator and the differential acceleration signal for the open-back Zerodur mirror option.

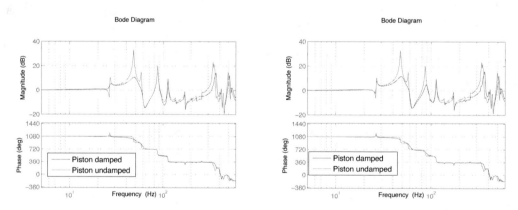

Figure 7. Open-back Zerodur mirror : piston frequency response before and after damping. Accelerometer on top of each actuator (left), Accelerometers on top and bottom of each actuator (right)

For this case and for accelerometer on the top of each actuator the first 3 important modes are damped. However, it can be seen from Figure 8 that the high frequency modes, e.g. mode at 366Hz, cause robustness issue for the damping loop. It was observed that in case of the accelerometer on the top of each actuator the mode at 86Hz (piston mode) is not damped and the system with the selected position loop gain becomes unstable. If the differential accelerometer signal is used instead, this mode is damped and hence no stability issue for the position loop is observed. However, the lowest mode at 26Hz is a global mirror support mode and is not seen by the differential accelerometer and is consequently not damped in this approach. As a results the position loop for the selected control gains and the desired bandwidth is not robust stable at this frequency. Similar results and observations were obtained in the case of the second order filter algorithm. To avoid the repetition they are not presented in this paper.

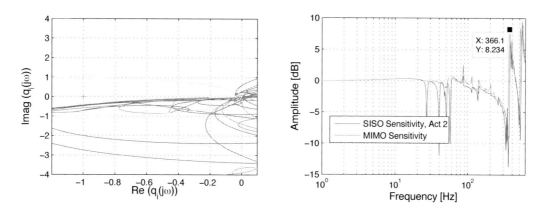

Figure 8. Open-back Zerodur mirror option: accelerometer on top of each actuator: MIMO Nyquist curves (left) and Sensitivity transfer function $||S||_\infty$ (right) of the active damping loop

2.2.2 Force feedback

The approach is more or less equivalent to the control strategies based on acceleration feedback. There are however some advantages with force feedback, i) force sensors have often better sensitivity than accelerometers, ii) the stability properties of the force feedback are often better than the acceleration feedback.[7] Let K_{act} be the axial stiffness of each actuator, then the force sensor y measures the force at the interface of the actuator and the mirror, i.e. $y = -K_{act}z + u_{act}$, where u_{act} is the active control force of an actuator. The controller is the integral of the measured force, i.e.

$$u = C_{damp} * y = -\frac{K_f}{s} * y \qquad (3)$$

In practice often high-pass filters are added to the control to remove possible offsets and low frequency perturbations.

Closed-back ULE mirror option: Figure 9 compares the piston and tilt frequency responses constructed from position sensors before and after implementing the active damping using force feedback Eq.(3).

The active damping loop is stable and robust. However, due to non collocation the mode 410Hz limiting the efficiency of the damping loop and reduced the robustness (see Figures 9 and 10). The position loop is robust stable.

Open-back Zerodur mirror option: Figure 11 compares the piston and tilt frequency responses constructed from position sensors before and after implementing the active damping using force feedback Eq.(3). The damping is efficient for three of the important mechanical modes.

The active damping loop is stable and robust. However, due to non collocation the mode 355Hz limiting the efficiency of the damping loop and reduced the robustness . The position loop is consequently robust stable where the modes at 340Hz and 355Hz are the limiting modes. In comparison to the case of closed-back ULE mirror the high amplitude of the non damped high frequency parasitic modes are limiting the robustness of the position loop with the desired bandwidth of 10Hz.

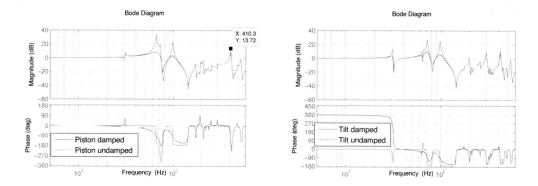

Figure 9. Closed-back ULE mirror with force feedback active damping: piston (left) and tilt (right) frequency responses before and after damping

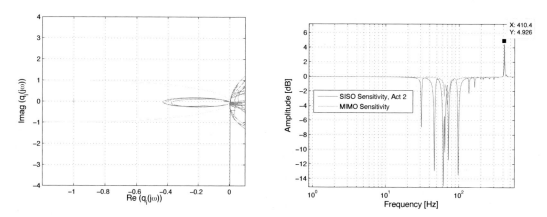

Figure 10. Closed-back ULE mirror with force feedback active damping: MIMO Nyquist curves (left) and Sensitivity transfer function $||S||_\infty$ (right) of the active damping loop.

2.2.3 Positive position feedback

The approach consist of filtering the position signal (from the position sensor) with a second order filter and feed it back to generate a force in a positive feedback constellation. This is not an intuitive approach while in general most of the control loops are negative feedback loops. The effect and damping capabilities are more or less like the acceleration based active damping strategies with an additional property that the control adds -40dB/decade roll-off at higher frequencies.[7] The filter has the frequency ω_f and the damping ξ_f as design parameters:

$$u = C_{damp} * z = + \frac{K_z}{s^2 + 2\xi_f \omega_f s + \omega_f^2} * z \qquad (4)$$

where z is the position signal.

Closed-back ULE mirror option: Figure 12 compares the piston and tilt frequency responses constructed from position sensors before and after implementing the active damping using the positive position feedback (Eq. 4) for the closed-back ULE mirror option.

The damping is efficient for three of the important mechanical modes. The scheme changes the static gain of the inner-loop (see Figure 12). Therefore, to maintain the required position closed-loop bandwidth at 10Hz and in order to keep the position loop control gain identical for this case as for the other schemes, a scaling gain factor in addition should be applied .

The active damping loop is stable and robust. Note that due to the specific shape of the frequency response of this approach the range of suitable steady state gains is limited. The damping loop controller gain was adjusted

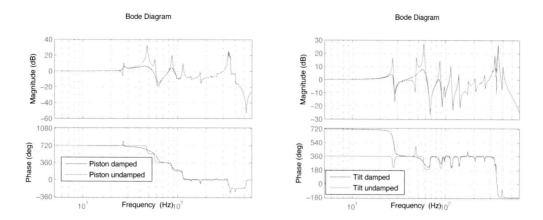

Figure 11. Open-back Zerodur mirror with force feedback active damping: piston (left) and tilt (right) frequency responses before and after damping

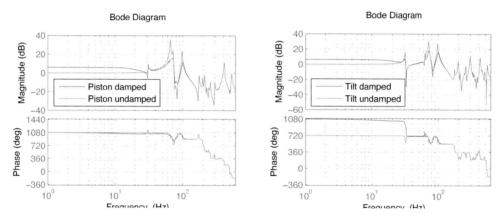

Figure 12. Closed-back ULE mirror with positive position feedback active damping: piston (left) and tilt (right) frequency responses before and after damping

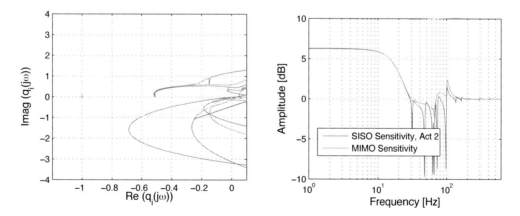

Figure 13. Closed-back ULE mirror with positive position feedback active damping: MIMO Nyquist curves (left) and Sensitivity transfer function $||S||_\infty$ (right) of the active damping loop.

such that the sensitivity gain remains under 6dB. The limiting robustness gains are mainly at low frequencies (see Figure 13). In general, at low frequency there are less uncertainties on the amplitude and phase of the system. Therefore, one could expect the damping loop gain can be increased without an important consequence on the stability of the design.

Open-back Zerodur mirror option: Figure 14 compares the piston and tilt frequency responses constructed from position sensors before and after implementing the active damping using the positive position feedback (Eq. 4) for the open-back Zerodur mirror option.

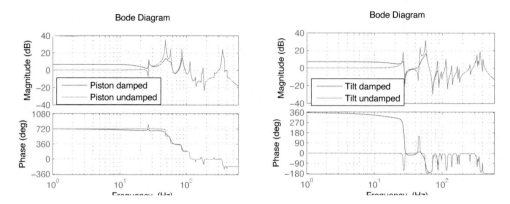

Figure 14. Open-back Zerodur mirror with with positive position feedback active damping: piston (left) and tilt (right) frequency responses before and after damping

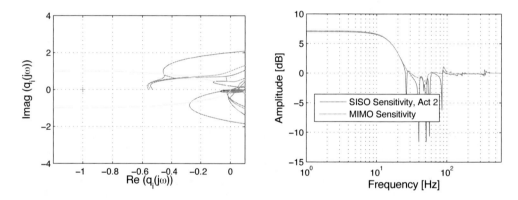

Figure 15. Open-back Zerodur mirror with positive position feedback active damping: MIMO Nyquist curves (left) and Sensitivity transfer function $||S||_\infty$ (right) of the active damping loop

The active damping loop is robust stable. The limiting robustness margins are mainly at low frequencies (see Figure 15).

3. MEASUREMENT RESULTS ON THE M5 PROTOTYPE UNIT

During a test period on M5 unit prototype the active damping strategy based on the velocity feedback using the accelerometer sensor was investigated. Different type of accelerometers and their location on the actuator were tested as well. After extensive measurements and comparisons it turned out that the most favorable accelerometer location to be the one with the accelerometer on top of the actuator on the position sensor support. Here, the results related to this case is presented. The test procedure consists of measuring the open loop frequency of the accelerometer and actuator position sensor before and after the implementation of the active damping loop:

Actuators under test were excited with band-limited random noise signals. Open-loop frequency responses with frequency range of 0-200Hz and 0-500Hz were measured. Additionally, the cross-power spectra (coherence) were measured in order to verify the measurement quality. The configurable high-pass filters in the accelerometer preamplifiers were set to 1Hz. From the measured frequency response of each actuator input command to the accelerometer a parametric model was identified which in turn was used as a basis of controller parameter tuning. The reference response was fit by an 18th order polynomial model using a least squares method. As shown in Figure 16 the model represents well all modes until 200Hz and it is minimum-phase.

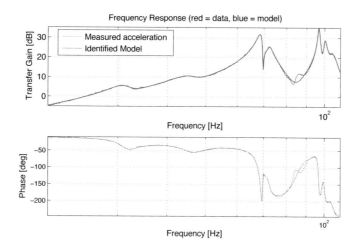

Figure 16. Measured accelerometer frequency response vs. identified model response

The tuning parameters were the gain and filter cut-off frequency K_a and ω_b as in Eq.(1). As a compromise between sufficient stability and robustness margins and sufficient damping (min. 10 dB for first resonant mode) a setting of $\omega_b = 2*\pi*10$Hz cut-off frequency and a gain of $K_a = 30$ was selected. To verify the stability and effect of higher order mechanical modes the system components (controller and controlled system) are represented by their frequency responses and the closed-loop frequency responses are calculated point-by-point over the complete frequency range of interest (here 0-500Hz).

The controller was implemented on the real time controller and the frequency response of the actuator input voltage to the position sensor was measured and compared with that of the open loop and expected response from the design. Figure 17 shows these measurement responses. In closed-loop response the 70Hz and 100Hz resonance modes were both damped to the extend predicted by analysis.

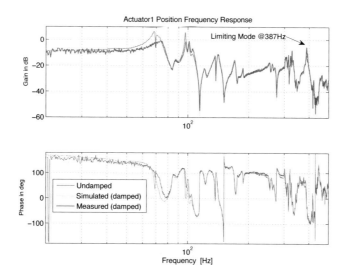

Figure 17. The effect of the active damping loop on the mechanical modes seen by actuator position sensor. Measured response before and after implementation of the active damping loop.

In order to explore the gain margin and the limitations due to model mismatch the feedback gain was increased until the system started to show tendency to oscillate. As shown in Figure 18, the increase in gain further improved the damping of the 70Hz and 100Hz modes. Further increase of gain led strong oscillations with a frequency of 387Hz.

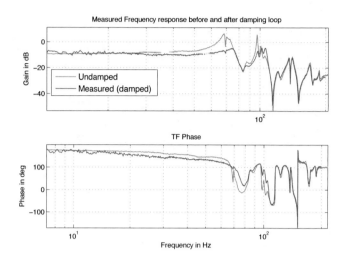

Figure 18. Strong damping on the main modes but at the limit of the stability

The nominal setting was also tested with the two other actuators. For both the active damping was stable and the first mode was damped like on actuator the first actuator.

4. CONCLUSION

From the analysis of different damping strategies and the two mirror options the following points can be concluded:

- All damping strategies discussed in the paper introduce damping to the main mechanical modes in the direction of sensing/control. Their efficiency, i.e. amount of the introduced damping, depends on the structural mechanical modes which are not damped (parasitic modes) or on the control/structure interaction combined with the effect of latency in the control system limiting the damping loop gain.

- Depending on the location of the sensor, e.g. absolute acceleration or differential, the parasitic modes affecting the efficiency of the damping loop differ.

- Due to the presence of higher number of parasitic modes with a higher amplitude (mostly high frequency) at the actuator and whiffle-tree interface, the damping loop for the open-back Zerodur mirror option is less robust and consequently less effective. Specifically, the approach with accelerometers are shown to be not efficient as in the case of closed-back ULE mirror.

- The difference between direct velocity feedback Eq.(1), and the second order filter of acceleration signal Eq.(2), are mainly algorithmic: the direct velocity feedback has one gain to tune while the second order filter scheme requires a prior knowledge on the frequency or frequency ranges where the damping should be efficient. The direct velocity feedback approach using accelerometer located of the top of the actuators was demonstrated successfully on the scale-one prototype. In the next measurement campaign the effect of the MIMO implementation (simultaneous operation of three damping loops) will be tested.

- The analysis showed that integral force feedback Eq.(3) is promising for both mirror options. The approach requires a force sensor located properly at actuator/mirror (whiffle-tree) interface. If the sensor is well collocated with the actuator the damping loop is well efficient. One advantage could be that the sensitivity of the available commercial force sensors are often higher than the accelerometers. This approach is not foreseen to be tested on this prototype because it requires a major design modification to include the force sensors. It will be tested in an alternative test setup.

- The analysis showed that the positive feedback strategy Eq.(4) is promising for both mirror options. The loss of robustness at low frequencies is a drawback. However, in general the uncertainties at low frequencies are lower so lower stability margins can be tolerated. Since no addition sensor or modification to the actual system is required, it will be tested on the scale-one prototype.

REFERENCES

1. J. Spyromilio, "E-ELT telescope: the status at the end of detailed design," in *Ground-based and Airborne Telescopes III*, L. M. Stepp and R. Gilmozzi, eds., *Proc. SPIE* **7733**, 2010.
2. A. M. McPherson, E. T. Brunetto, P. Dierickx, M. M. Casali, and M. Kissler-Patig, "E-ELT update of project and effect of change to 39m design," in *Ground-based and Airborne Telescopes IV*, L. M. Stepp, R. Gilmozzi, and H. J. Hall, eds., *Proc. SPIE* **8444**, 2012.
3. M. Cayrel, "E-ELT optomechanics: overview," in *Ground-based and Airborne Telescopes IV*, L. M. Stepp, R. Gilmozzi, and H. J. Hall, eds., *Proc. SPIE* **8444**, 2012.
4. J. M. Casalta, J. Barriga, J. Arino, J. Mercader, M. S. Andrés, J. Serra, I. Kjelberg, N. Hubin, L. Jochum, E. Vernet, M. Dimmler, and M. Müller, "E-ELT M5 field stabilisation unit scale-1 demonstrator design and performances evaluation," in *Adaptive Optics Systems II*, B. L. Ellerbroek and et al, eds., *Proc. SPIE* **7736**, 2010.
5. P. Bouchilloux, F. Claeyssen, and R. L. Letty, "Amplified piezoelectric actuators: from aerospace to underwater applications," in *Smart Structures and Materials 2004*, E. H. Anderson, ed., *Proc. SPIE* **5388**, 2004.
6. B. Sedghi, M. Müller, H. Bonnet, and B. Bauvir, "Field stabilization (tip/tilt control) of E-ELT," in *Ground-based and Airborne Telescopes III*, L. M. Stepp and R. Gilmozzi, eds., *Proc. SPIE* **7733**, 2010.
7. A. Preumont, *Vibration Conrtrol of active structures, an introduction*, Kluwer academic publishers, 2nd Edition, 2002.
8. O. N. Gasparyan, *Linear and Nonlinear Multivariable Feedback Control*, John Wiley and Sons Ltd, 2008.

Development of a Fast Steering Secondary Mirror Prototype for the Giant Magellan Telescope

Myung Cho[a], Andrew Corredor[b], Christoph Dribusch[b], Kwijong Park[c], Young-Soo Kim[c], Il-Kweon Moon[d], Won Hyun Park[e]

[a]GSMT Program Office, National Optical Astronomy Observatory, USA
[b]Aerospace and Mechanical Engineering Department, University of Arizona, USA
[c]Korea Astronomy and Space Science Institute, Korea
[d]Korea Research Institute of Standards and Science, Korea
[e]College of Optical Sciences, University of Arizona, USA

ABSTRACT

The Giant Magellan Telescope (GMT) will be a 25m class telescope currently in the design and development phase. The GMT will be a Gregorian telescope and equipped with a fast-steering secondary mirror (FSM). This secondary mirror is 3.2 m in diameter and built as seven 1.1 m diameter circular segments conjugated 1:1 to the seven 8.4m segments of the primary. The prototype of FSM (FSMP) development effort is led by the Korea Astronomy and Space Science Institute (KASI) with several collaborators in Korea, and the National Optical Astronomy Observatory (NOAO) in USA. The FSM has a tip-tilt feature to compensate image motions from the telescope structure jitters and the wind buffeting. For its dynamic performance, each of the FSM segments is designed in a lightweight mirror. Support system of the lightweight mirror consists of three axial actuators, one lateral support at the center, and a vacuum system. A parametric design study to optimize the FSM mirror configuration was performed. In this trade study, the optical image qualities and structure functions for the axial and lateral gravity print-through cases, thermal gradient effects, and dynamic performances will be discussed.

Keywords: Extremely large telescope, GMT Secondary Mirror Prototype, mirror support, off-axis

1. INTRODUCTION

The Fast Steering Secondary Mirror (FSM) of the Giant Magellan Telescope (GMT) is one of the next generation extremely large telescope projects under the design and development phase. The FSM consists of seven separated circular segment mirrors which are conjugated 1:1 to the segments of the primary mirror. An artesian CAD model of the GMT telescope and the optical system configurations are shown in Figure 1. The FSM system has a tip-tilt feature to correct wind vibration and telescope jitters. The FSM system will mainly serve as a commissioning system and a backup system when the adaptive secondary mirror (ASM) is not in operation.

In order to develop a full-size off-axis mirror and a tip-tilt test-bed, the Korea Astronomy and Space Science Institute (KASI) organized a consortium. The consortium consists of institutions in Korea - KASI, Korea Research Institute of Standards and Science (KRISS), Institute for Advanced Engineering (IAE), Gwangju Institute of Science and Technology (GIST) – and National Optical Astronomy Observatory (NOAO) in the US. The collaboration efforts are involved in the design and the development of the FSM Prototype (FSMP). Main purpose of FSMP development is the creation of a technical infra structure for the high precision optical telescopes for astronomy in Korea. The FSMP design and development activities will promote the optical technology in high precision optics in Research institutions, Universities, and industries in Korea.

---

* mcho@noao.edu; phone 1 520 318-8544; fax 1 520 318-8424; www.noao.edu

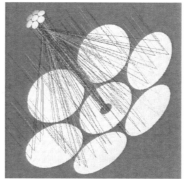

Fig. 1. The GMT telescope artesian model (left) and GMT optics: f/8 Gregorian beams with 1:1 conjugated Primary and Secondary mirrors (right).

For GMT FSM design and development, a conservative engineering approach is taken; utilizing concepts established from the F/11 M2 of Magellan telescope. The FSM system was optimized to meet the requirements defined in "GMT Image Size and Wavefront Error (WFE) Budgets"[3]. Preliminary WFE budget allocation for the FSM mirror figure, from mirror supports, are as follows: 40nm "root mean squared" (RMS) at Zenith and 60nm RMS at 60 degrees elevation. These WFEs are equivalent to the RMS surface errors of 20nm and 30nm at Zenith and at 60 degrees elevation, respectively. The WFE allocations were set as a design goal based on the experiences and opto-mechanical analysis with the similar class of mirrors designed at NOAO. In a seeing limited operation, the FSM optical surface figure errors can be corrected by a fast steering tip-tilt system. The FSM tip-tilt feature will accommodate the dynamic disturbances from wind vibrations and the telescope jitters. The tip-tilt system requires a fast tip-tilt of ±20 arc-seconds at a bandwidth of 10 Hz (goal of 20 Hz). To provide the tip-tilt motions, the FSM cell assembly requires an excellent dynamic stiffness and the stiffness goal was set as 60 Hz.

2. FSM MIRROR CONFIGURATION

The FSM is a lightweight concave mirror formed by seven separate, circular segments. The FSM converts the beam reflected from the f/0.8 Primary Mirror into an f/8 Gregorian beam for the science instruments as shown in Figure 1. Each lightweight mirror segment has a nominal diameter of 1.06 m and a thickness of 140 mm. Each of the segments has a three-point axial support and a single lateral support. The axial supports are mounted on the back surface of the mirror and oriented parallel to the optical axis (z-axis) and the lateral support consists of a single flexure located at the mirror's center position. A design concept of the FSM Assembly was developed by NOAO in a collaborative effort with KASI. The FSM Assembly consists of the FSM mirror, support system, and a mirror cell. The off-axis segment solid model is shown in Figure 2. The solid model of the FSM Assembly is shown in Figure 3. As a baseline mirror material, Zerodur™ glass material and common metal material properties were assumed in the FE mirror models.

2.1. FSM Mirror Trade Study

Several different finite element (FE) models were created to serve various calculations. Typical FE mirror model of FSM is composed with several layers of elements with a total of 298,900 solid elements and 504,300 nodes as shown in Figure 4 (a). This FE model assumes a solid concave lightweight mirror with a diameter of 1.06m, 140mm thick, and a radius of curvature of 4.2m (best fit sphere). With this specific model, the FSM mass was estimated as 105kg from the FE model.

For a specific case, FE models with 2D thin shell elements were utilized to facilitate a fast turn-around time in solutions during a parametric trade study as shown in Figure 4(b) and (c). A local coordinate system in the FE model was assumed as follows: (1) the positive Z-axis corresponds to the line which connects the vertex of the primary mirror segment to the vertex of the secondary mirror segment which is conjugated to primary segment; (2) the positive X-axis corresponds to the telescope's mechanical elevation axis; (3) the positive Y-axis is defined by the right hand rule. Based on this coordinate system, detailed FE results are presented.

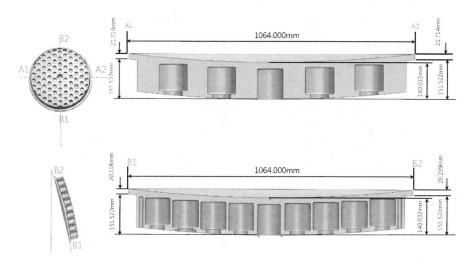

Fig. 2. FSMP mirror off-axis segments baseline configurations and nominal dimension.

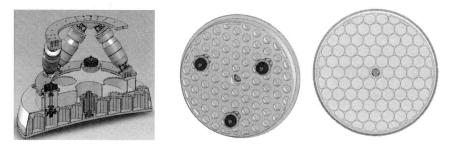

Fig. 3. A design concept of the FSM Cell Assembly consisting of FSM, Support system, and Cell, as shown in a half model at a cutaway view (left); center view shows FSM, support systems, and hidden Cell. Three axial supports are shown in dark blue, and lateral support mounted at the center of the FSM mirror; for visibility of lightweight pattern, axial actuators and the back surface was hidden (right).

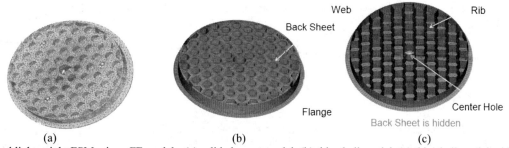

Fig. 4. Typical lightweight FSM mirror FE models, (a) solid element model; (b) thin shell model; (c) thin shell model with back sheet hidden.

In order to achieve an optimized baseline mirror blank configuration, a parametric trade study was performed. Several mirror configurations were considered, and five mirror configurations as shown in Figure 5 were intensively examined and their performances were evaluated. Through the extensive optimization and parametric processes in gravity print-through and natural frequency, a baseline configuration was selected based on stiffness and strength merits. Most favorable baseline configuration is a depth of 140mm, face sheet thickness of 20mm, and a mass of 105kg. Lightweight wall thicknesses of the baseline mirror configuration are: Rib wall of 5 mm, outer web wall of 10mm, and wall of 10mm at the center hole. To this baseline configuration (Figure 2), detailed mechanical and optical performance analyses were further investigated using I-DEAS finite element analysis program and the PCFRINGE optical program.

Fig. 5. Five candidate lightweight mirror configurations employed in trade study. Typical mirror wall thickness used in analysis: face sheet thickness=15mm, back sheet=15mm, flange thickness=19mm, rib wall thinness=5mm, outer web wall thickness=10mm; and center hole thickness=10mm. Baseline configuration was selected through optimization process based on stiffness and strength merits.

2.2. FSM On-axis Mirror Natural Frequency

Natural frequencies of the mirror were calculated by using a solid full FE mirror model with a free-free boundary condition. These frequency modes are characteristic mirror bending shapes and were obtained after removing rigid body motions (piston and tilts). The natural frequencies, up to 22 modes, were calculated and the corresponding characteristic mode shapes were examined. The lowest mode was found at 717 Hz, as an astigmatic shape. The rest low mode shapes up to 10 modes are at 1099, 1411, 1854, 1947, and 2150hz. These low frequency modes are similar to low order Zernike polynomials, but not in the same order.

3. FSMP SUPPORT SYSTEMS

3.1. FSMP Axial support system

The baseline mirror support system developed by NOAO contains three axial supports with a tip-tilt capability and a lateral support flexure mounted at the center of the mirror. This FSM support system was optimized for minimum gravity induced errors. The axial support was optimized for the telescope at Zenith pointing. Parametric modeling iterations were conducted for the support system optimization. These iterative calculations utilize an optimization scheme for a minimum global surface deformation over the optical surface. In order to achieve the optical performance goal of 20nm RMS surface, extensive parametric calculations were made for an optimum axial support system.

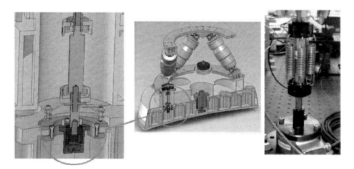

Fig. 6: Sectional view of FSMP mirror blank with FSM mirror cell assembly in a Solidworks model, and localized view of an axial actuator mounted at the mirror, and an axial actuator (PI) on a test table at IAE.

Optimization was performed based on the maximum stress and displacement over the mirror surface at the axial support. Localized FE models were created to implement variations of the back surface thickness at the mount and the wall thicknesses around the support mount. Figure below shows a sectional view of FSMP mirror blank, FSM mirror cell assembly in a Solidworks model, and localized view of an axial actuator mounted at the mirror. Detailed analysis for the Von-Mises stresses and displacement calculations were performed by linear FE simulations and the results were addressed in Chapter 4 with Nominal operational mode and Survival mode.

3.2. FSMP Lateral support system

The design concept of the lateral support system was mainly based on the heritage of the Magellan telescope[4]. The FSM baseline lateral support configuration developed by NOAO is a flexure diaphragm mounted at the center hole of the FSM. Two different lateral support center flexure concepts have been analyzed and tested in order to identify a design that provides an optimum lateral performance in terms of stiffness, strength and elastic stability. The two diaphragm configurations utilized in such analyses are shown in the Figure 7. As shown in the figure, configuration 1 is a flat diaphragm with slotted rim and configuration 3 is a modified star-shaped diaphragm sandwiched between two solid rings. Moreover, it must be noted that the material used for configuration 1 is Invar 36 and the materials utilized for configuration 3 are RH 950 Steel (star-shape) and Invar 36 (Solid Rings). These two lateral flexures were manufactured by KASI, provided from IAE. For both flexures a few modifications from the initial design were made to account for lateral tests at NOAO described in Chapter 8. Current baseline flexure Configuration is a diameter of 100mm, a central hub of 32mm with a thin flexure blade thickness of 0.4mm as shown in Figure 7 (configuration 1). Extensive trade studies were conducted for the performance, cost, and risk.

Fig. 7 Lateral Flexures configurations, configuration 1 (left) is a flat diaphragm with slotted rim, configuration 3 (right) is a star-shape sandwiched between two solid rings (manufactured by KASI/IAE).

4. MIRROR SENSITIVITY

The FSM segment will be mounted on a positioner, or a hexapod. Rigid body motions of the segment such as decenter and clocking will degrade the image quality. We estimated aberrations due to a misalignment of 0.1 mm decenter along x axis and y axis, and a 0.01° clocking of the segment. In misalignment sensitivity calculations, we introduced an induced aberration by the perturbations from the hexapod as a demonstration purpose. This implies that this level of perturbations can be compensated by only piston and tip-tilt as adjustments of the hexapod. The hexapod behavior is summarized in Table 1 for the decenter and clocking cases.

Table 1. Values of piston and tip-tilt to compensate induced aberration by the misalignments.

	Piston	x tilt	y tilt
0.1mm decenter along x axis	8.537μm	0°	0.00134° (4.8")
0.1mm decenter along y axis	0.265μm	0.00129° (4.6")	0°
0.01° clocking about z axis	14.657μm	<< 1"	<< 1"

For the perturbations, effects of the position sensitivity were calculated. For the case of a 0.1mm decenter along x axis, the optical wavefront map along with the image spot is shown in Figure 8. It shows that the wavefront PV of 1.09waves and RMS of 0.2 waves at a wavelength of 500nm, and a geometrical radius of the spots at 36.1μm RMS. Similarly, the optical impacts for the y-decenter case are a PV wavefront error of 1.68 waves and a RMS of 0.35 waves, and a geometrical radius of the spots at 89.1μm RMS. For the z-clocking, a PV wavefront error of 2.17 waves and a RMS of 0.39 waves, and a geometrical radius of the spots at 71.5μm RMS.

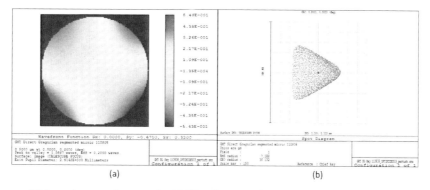

(a) (b)

Fig. 8. Results for the case of 0.1mm decenter along x axis. (a) Wavefront map in Zemax. This map shows only the perturbed segment. Unit is waves at 500nm light. PV is 1.09 waves, and RMS is 0.2 waves. (b) Spot diagram shows light dispersion by the misalignment after adjustments by the hexapod. Geometrical radius of the spots is 36.1μm. RMS radius is biased for this analysis because spots formed by the rest 6 segments are concentrated in the center.

5. AXIAL SUPPORT STRENGTH STUDY

Level of stress and its distribution at the optical components are always of interest. The stress distributions in the mirror around the axial support were calculated. Optimization was performed based on the maximum stress and displacement over the mirror surface at the axial support. Localized FE models were created to implement variations of the back surface thickness at the mount and the wall thicknesses around the support mount. Figure 9 shows a sectional view of FSMP mirror blank, FSM mirror cell assembly in a Solidworks model, and localized view of an axial actuator mounted at the mirror.

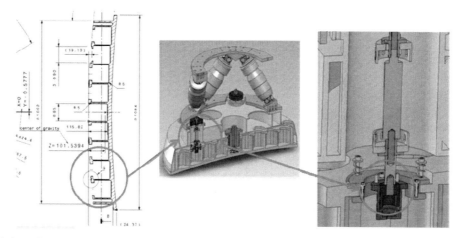

Fig. 9: Sectional view of FSMP mirror blank, FSM mirror cell assembly in a Solidworks model, and localized view of an axial actuator mounted at the mirror.

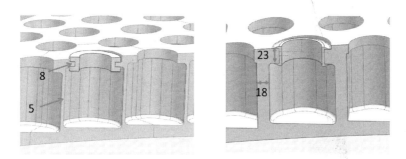

Fig. 10. Two design alternatives: Bracket thickness and wall thickness shown in Original Design and Optimized Design.

The Von-Mises stresses and displacements were calculated by linear FE simulations and the results were summarized. To demonstrate the stress effects, two modes are considered as: (1) Nominal mode; (2) Survival mode. In each mode, Von-Mises stresses and displacements were calculated and compared between the Original Design and an Optimized Design. These two design alternatives are shown in Figure 10.

4.1. Normal Mode

The design concept of the axial support system is that the mirror cell is pressurized to balance the mirror gravity during the operation. Therefore, the balancing between gravity and pressure will not produce an axial force at the axial support. During the initial installation and calibration, the full gravity of the mirror will be held by the axial supports. As a nominal mode, it was assumed that the mirror gravity is equally distributed over the three axial supports. For the axial gravity, Von Mises stresses were calculated for Original Design and Optimized Design. Maximum Von-Mises stress is 1.7Mpa and 0.3Mpa, respectively. The stress distributions are shown in Fig. 11. The displacements for both designs are small.

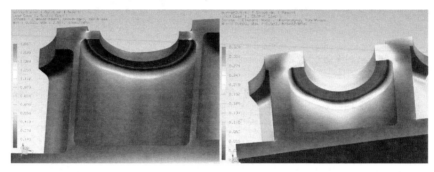

Fig. 11. Von-Mises stress distributions in Original Design (left) and Optimized Design (right). For Original design, the maximum Von-Mises stress is 1.7Mpa, whereas 0.3Mpa in Optimized Design.

4.2. Survival Mode

In order to secure the mirror from excessive loadings and extreme conditions, considered are survival and failure modes during handling or maintenance operations. As a demonstration purpose, considered is a case when the mirror gravity load is carried by only two axial supports. In this scenario, each of the two axial supports will carry excessive loads. As a survival mode, Von Mises stresses were calculated for Original Design and Optimized Design. Maximum Von-Mises stress is 10.5Mpa and 3.6Mpa, respectively. The stress distributions are shown in Fig. 12. The nominal working stress of Zerodur is known as 20Mpa; however, a conservative approach was taken into the design of FSM.

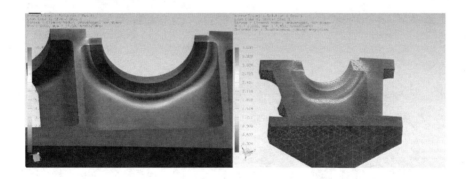

Fig. 12. As a survival mode, Von-Mises stress distributions were calculated for Original Design (left) and Optimized Design (right). For Original design, the maximum Von-Mises stress is 10.5Mpa, whereas 3.6Mpa in Optimized Design.

6. LATERAL FLEXURE TRADE STUDY

Several lateral support center flexure concepts have been studied and the results are evaluated to find a preferable design that provides an optimum structural and thermal performance. This study includes flexure shape merits and material trades with stainless steel and invar flexures. Typical flexure configurations utilized in such analyses are shown in Figure 13. Material used in flexure Configuration-1 is invar, Configuration-2 and Configuration-3 are stainless steel for a inner blade and invar for a mounting ring to glass.

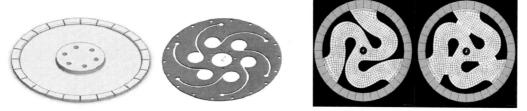

Fig. 13. Typical flexure configurations. Figures from left to right show Configuration-1 (V1), Configuration-2 (V2), and Configuration-3 (V3 - flexure used in the Magellan M2, and V3A).

Static and dynamic analyses including lateral gravity loading, linear buckling, geometric nonlinear - large deflections, and natural frequency, have been performed using each of the three different configurations. Such analyses were performed in order to identify which concept provides the optimal amount of axial compliance, while retaining sufficient lateral stiffness, strength, and elastic stability. Summary of some of the results for such analyses were reported in the reference [1]. Currently, more extensive trade studies are being performed, and the detailed analysis will continue as the design evolves in the future.

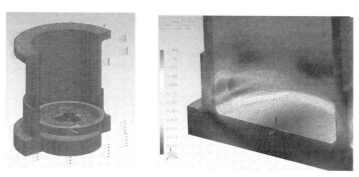

Fig. 14. Localized FE model at the lateral support (left), Von-Mises stress in the mirror at the lateral support with the maximum value of 1.25 MPa (right)

For the strength of the mirror, a localized FE model was employed. As shown in Figure 14, lateral forces in the y-direction are applied on two outside diameters. The forces are selected such that they sum to 1000N and their moments with respect to the plane of the flexure cancel. This model simulates the case which the entire mirror gravity acts onto the local model. The applied force on the top diameter far from the flexure is 200.92N. The force applied on the bottom diameter close to the flexure is 799.08N. The center of the flexure is fixed. This boundary condition is the "shadow" at the center of close to the coordinate system. Von-Mises stress in the mirror at the lateral support was observed at the maximum value of 1.25 MPa.

7. MIRROR CELL ASSEBLY

Baseline design concept of the FSM Cell Assembly is to provide appropriate stiffness against the static loads and for dynamic operations. The FSM cell assembly consists of FSM mirror, Support system, and Cell. In a seeing limited operation, the FSM optical surface figure errors can be corrected by a fast steering tip-tilt system. The FSM tip-tilt feature will accommodate the dynamic disturbances from wind vibrations and the telescope jitters. The tip-tilt system requires a fast tip-tilt of ±20 arc-seconds at a bandwidth of 10 Hz with a goal of 20 Hz. To capture the tip-tilt motions, a sample simulation as a combination of unit cases was considered. As a unit case, unit axial displacement of 1mm was applied at each of the three axial actuators for a tip-tilt motion. Typical illustration is shown in Figure 15.

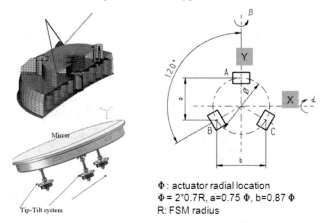

Fig.15. FE model of the FSM assembly (half model, top left), FSM mirror held by 3 actuators (bottom left), model of axial actuators to simulate a tip-tilt motion (right).

The FSM mirror cell must be stiff enough to sustain static gravity and dynamic operations. Current mirror cell design assumes thin walled aluminum plates for most parts and reinforced sections at the parts under direct load paths along the hexapod interface. Total estimated mass in the FE model is approximately 190kg excluding a reaction mass system for the tip-tilt mechanism which is under development. In order to achieve the design goals, static analyses with gravity induced deformation and thermal distortions were performed for the mirror cell assembly. The static deformations and stress levels are relatively small and favorably met the design goals. Current design goal in the cell structural deformation is an overall displacement of 100 microns including the effects from gravity and thermal loadings. The goal will be refined and adjusted as the cell assembly design evolves and the tip-tilt system becomes mature. For a dynamic stiffness requirement, natural frequencies of the mirror cell were calculated. The first 10 significant modes were obtained excluding the lowest mode, torsion mirror clocking mode shape. The fundamental structural mode (2nd mode) occurred at 82Hz which is in an astigmatic shape of the cell. This result indicated that the cell has an excellent dynamic stiffness for the tip-tilt motion at the goal of 60Hz.

8. LATERAL FLEXURE TEST

In order to evaluate the performances of the lateral flexure, we have performed extensive finite element analyses. The analyses include the static gravity, nonlinear behaviors, thermal effects, and buckling analysis. Two optimized flexures (V1 and V3 in Figure 13) installed at test mirrors (diameter of 200mm with a radius of curvature of 1m) to validate the lateral flexure design. The two test mirror setup configurations are: (1) a membrane flexure (V1) bonded to a test mirror using epoxy with a bondline thickness of 250μm; (2) a flexure used in the Magellan M2 (V2) bonded to a mirror using RTV with a bondline thickness of 1mm. These two test mirrors are shown in Figure 16. The main purpose of this test is to validate the design concept for lateral stiffness and its axial stiffness.

For the stiffness of the lateral flexures, we arranged the test setups with two loading configurations, that is, the lateral loading test and the axial loading test. In the lateral test, a maximum force of 1000N was applied to simulate the FSM mass of 100kg at an increment of 10N. This lateral test is intended to verify the mechanical behaviors from the entire mirror mass. Figure 17 illustrates the test setup which includes air compressor and heavy duty piston, a thick aluminum plate to mount the test mirror with lateral flexure, a linear gauge and a gauge tip located near to the plane of the lateral flexure. Since the test setup has an offset between the lateral loading point and line of action in the flexure, we anticipated a moment due to the offset. Mechanical measurements for the lateral and axial stiffness of lateral flexures were summarized and the results were compared with FE models. Figure 18 shows a lateral stiffness plot (displacements to loads) for the Epoxy flexure, and results indicate a good agreement between the measurements and the off-plane loading FE model (maximum displacement of 0.009mm vs. 0.0075nm, 20% variation).

 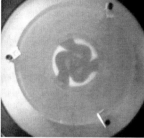

(a) V1 flexure boned with epoxy (b) V3 flexure boned with RTV

Fig. 16. (a) Lateral flexure boned to a test mirror with epoxy (V1 model). Grey areas forming a circle are epoxy layer of which thickness is about 250μm. (b) Lateral lexure boned to a mirror with RTV (V3 model). The RTV layer is 1mm thick between flexures outer rim and mirror the side of inner hole.

Fig. 17. Experiment set-up to measure lateral stiffness; (1) air compressor and heavy duty piston up to 1000N. (2) a thick aluminum plate, (3) A linear gauge, (4) mirror with lateral flexure, (5) a gauge tip contacting very lower part of the central hub.

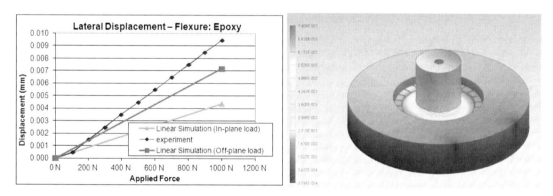

Fig. 18. Lateral Stiffness of the Epoxy flexure, results indicated a good agreement with FE model (20% variation with FE off-plane load, max Displacement = 0.0075mm)

For the axial test of the lateral flexures, a maximum force of 100N was applied. In the design of the axial stiffness in the lateral flexures, the axial force range of interest is approximately 10N. However, we conducted the axial test for an extended axial force up to 100N with an increment of 10N to monitor for any extraordinary behavior. Figure 19 illustrates the axial test setup including the test mirror, gimbal mount, and an interferometer. Total of three LVDT were installed; two LVDT indicators measure displacement of the indicator plate at the very edge and one LVDT catches up-down displacement of the plate.

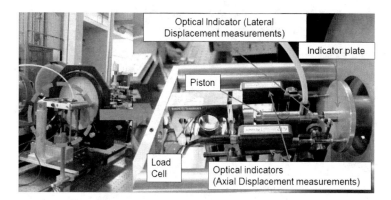

Fig. 19. Experiment set-up to measure axial stiffness of lateral flexures; Axial test setup including target mirror, gimbal mount, interferometer and air baffle, a load frame - mechanical structures on the back of the gimbal mount with force controlling components and measuring instruments. Two LVDT indicator measure displacement of the indicator plate at the very edge and one LVDT for up-down displacement of the plate. The small piston worked by air pressure provides varying forces regulated by a control software which monitors Load cell readings.

Fig. 20. Experiment set-up for the axial stiffness test including target mirror, gimbal mount, interferometer for the optical tests, and air baffle (right).

For the axial test, we established test setups for mechanical and optical measurements in order to evaluate the flexure behaviors and the optical surface deformations. The mechanical measurements for the axial stiffness of lateral flexures were recorded and the results were compared with FE models. For the optical measurements, we installed an interferometer with a light baffle to minimize air turbulence effects as shown in Figure 20. We further conducted the optical tests early in the morning to avoid potential noise sources from the surrounding environment at the NOAO optics shop.

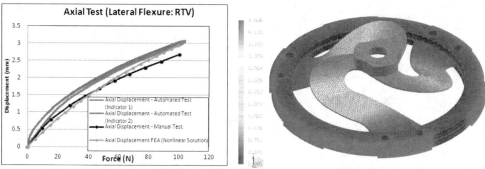

Fig. 21. Axial Stiffness (force to displacement) of the RTV flexure, test results indicated a good agreement with non-linear FE model.

Figure 21 shows the axial stiffness (force to displacement) of the RTV flexure. The axial stiff measurement results indicate a good agreement with the non-linear FE model (less than 10% variation). We are currently collecting optical measurements, processing the data, and the results will soon be available for discussions.

9. FSMP DEVELOPMENT STATUS AT KASI

In order to develop a full-size off-axis mirror and a tip-tilt test-bed, KASI organized a consortium. The consortium consists of institutions in Korea - KASI, Korea Research Institute of Standards and Science (KRISS), Institute for Advanced Engineering (IAE), Gwangju Institute of Science and Technology (GIST) – and National Optical Astronomy Observatory (NOAO) in the US. The collaboration efforts are involved in the design and the development of the FSM Prototype. The design and development of the FSM prototype includes the efforts in a prototype mirror, mirror cell, supports, and a vacuum system. KASI is leading the efforts and working on the FSMP systems engineering, KRISS for off-axis mirror fabrication, IAE for tip-tilt test-bed and the development, GIST for controls, and NOAO for the design and development of opto-mechanics.

Fig. 22.Mirror fabrication activities at KASI/KRISS, light weighting from a solid mirror blank (top left), checking dimensions of lightweight cell and wall thickness (top right), lifting the mirror for polishing (bottom left), working support for polishing (bottom center), and mirror mounted on the support.

KASI has hosted several milestone review meetings for the FSMP development. Such meetings are Preliminary Design Review (PDR) in 2010, Critical Design Review (CDR) in March 2011, Manufacturing Readiness Review (MRR) in September 2011, and Test Readiness Review (TRR) in April 2012. KASI plans to host Final Test Review (FTR) in later this year. At KRISS, the off-axis FSMP mirror is being fabricated. Figure 22 shows snapshots of some of the major activities at KRISS for the mirror fabrication. The design works produced for the FSM prototype. The major efforts are: Light weighting from a solid mirror blank (top left), checking dimensions of lightweight cell and wall thickness (top right), lifting the mirror for polishing (bottom left), working support for polishing (bottom center), and mirror mounted on the support. A full-size mirror segment with off-axis aspheric surface shape will be fabricated and its quality will be demonstrated. Fabricating mirrors of 1.06m in diameter is within the KRISS's capability. KRISS has equipments for light-weighting, polishing, and figuring mirrors up to 2m in diameter. Fabricating off-axis mirrors with the fast focal ratio of 0.65 is a challenging task; however we anticipate this task can be achievable. Figuring and testing of the off-axis is also challenging. Currently, we plan to test the mirror by using two different computer generating hologram methods, and other optical testing schemes.

In order to demonstrate capability of tip-tilt controls, a tip-tilt test-bed has been manufactured and assembled by the KASI team at IAE. The test-bed consists of a dummy mirror, a cell with axial and lateral supports, and a test-bed frame. Those are shown in Figure 23. Top left in the figure shows a half model of the mirror cell assembly design consisting of a FSM mirror with a cell and hexapod. At the back surface of the mirror, axial supports and central lateral support are installed. The mirror cell model and its physical hardware are shown in top center and right, respectively. A dummy aluminum mirror is a 1.06m full-size center segment light-weighted, and its weight is 113.5 Kg. Three axial supports are assembled, each of which includes a piezo actuator for tip-tilt and a load cell for vacuum control. The figure illustrates the dummy mirror mounted on the cell (bottom left), and the dummy mirror assembly for a vacuum test (bottom center), the cell attached onto the test-bed frame for a preparation for off-axis test (bottom right). The mirror cell together with a vacuum control system is under design. After the completion of the assembly tests, the test-bed would provide valuable data for the tip-tilt control, such as reaction forces according to frequencies as well as resolution and range of tilt angles, and optimal vacuum pressure.

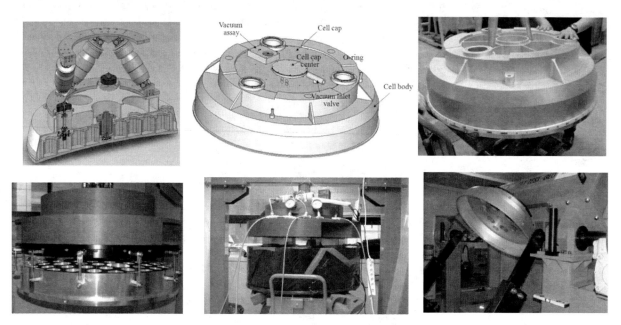

Fig. 23. Mirror cell test-bed activities at KASI/IAE; a half model of the cell assembly design consisting of a FSM mirror with a cell and hexapod (top left). The mirror cell model and its physical hardware (top center and right); the dummy mirror mounted on the cell (bottom left), and the dummy mirror assembly for a vacuum test (bottom center), the cell attached onto the test-bed frame for a preparation for off-axis test (bottom right).

10. SUMMARY

We have performed extensive finite element analyses and optical calculations for an optimized Fast Steering Secondary Mirror (FSM) and its support system. The performance prediction of the FSM Prototype mirror was based on a mirror configuration with a diameter of 1.064m (off-axis), depth of 140mm, face plate thickness of 20mm, and mass of approximately 105kg. The optical surface deformations for various Zenith angles were evaluated by combining cases of the effects from axial and lateral gravities. The results indicated that the current GMT FSM mirror and its support system adequately met the optical performance goal of 20nm surface RMS, and also satisfies the FSM surface figure accuracy requirement defined at EE80.

With the prototype of FSM, KASI is currently engaged in validations and verification work activities. Two optimized lateral flexures were modeled and tested to validate the lateral flexure design. Through the lateral and the axial loading tests, we verified the test measurements with the FE predictions. The results adequately validated the lateral flexure design concepts. The off-axis mirror fabrication is in progress, and a test-bed of the tip-tilt system advances rapidly for practical simulations. Several trade studies in a few design aspects are planned to be performed for the merits among the performance, cost, and risk. Majority of the efforts and activities for the prototype and the test-bed are expected to be completed in the end of this year. Final completion of the FSMP activities will be in the first part of the next year.

ACKNOWLEDGMENTS

This research was carried out at the National Optical Astronomy Observatory, and was sponsored in part by KASI. The authors gratefully acknowledge the support of the GMT Project Office. This work was partially contributed by the scientists, engineers, and students from the KASI team (KASI, KRISS, IAE, GIST) in Korea which is one of the key partners of the GMT Project.

REFERENCES

[1] Cho, M., Corredor, A., Dribusch, C., A., Park, K., Kim, Y., Moon, I., "Design and Development of a Fast Steering Secondary Mirror for the Giant Magellan Telescope," Proc. SPIE 8125, (2011).

[2] Cho, M. and Richard, R.M., "PCFRINGE Program – Optical Performance Analysis using Structural Deflections and Optical Test Data," the Optical Sciences Center, University of Arizona, 35, (1990).

[3] GMT Office, "GMT Image Size and Wavefront Error Budgets," GMTO, August, (2007).

[4] GMT Office, "The F/11 Secondary Mirror of Magellan telescope," GMTO, August, (2002).

[5] Cho, M., "Performance Prediction of the TMT Secondary Mirror Support System," Proc. SPIE 7018-65, (2008).

[6] Cho, M., Corredor, A., Park, K., "K-GMT FSMP CDR," KASI, Korea, (2011)

[7] Blanco, D., Cho, M., Daggert, L., Daly, P., DeVries, J., Elias, J., Fitz-Patrick, B., Hileman, E., Hunten, M., Liang, Nickerson, M., Pearson, E., Rosin, D., Sirota, M. and Stepp, L., "Control and support of 4-meter class secondary and tertiary mirrors for the Thirty Meter Telescope," Proc. SPIE 6273-65, (2006).

[8] Williams, Eric, et. al., "Advancement of the Segment Support System for the Thirty Meter Telescope Primary Mirror," Proc. SPIE 7018-37, (2008).

Repairing stress induced cracks in the Keck primary mirror segments

Dennis McBride, John S. Hudek, Sergey Panteleev

W.M. Keck Observatory, 65-1120 Mamalahoa Hwy., Kamuela HI, USA 96743

ABSTRACT

Stress induced cracks have developed in the Zerodur glass at bonded supports for the primary mirror segments of the W.M. Keck Observatory telescopes. This has been a slow process that has advanced over the 20 year life of the telescopes. All mirror segments exhibit cracks to varying degrees. The number and severity of cracks has now reached a stage at which repairs are mandatory. A project is under way to determine the root causes of the cracks, and to develop a repair strategy. New supports and bonding methods are being designed and tested that will replace all of the original supports.

Keywords: Keck, mirror segment, stress, crack, axial insert, radial pad, Zerodur

1. INTRODUCTION

The W.M. Keck Observatory (WMKO) operates two optical / infrared telescopes at an elevation of 4145 meters on Mauna Kea on the island of Hawaii. The primary mirrors are 10 meters in diameter, made up of 36 hexagonal segments which are 1.8 meters vertex to opposite vertex, and approximately 75 mm thick. The mirrors are Zerodur low expansion glass ceramic by Schott AG, having coefficient of thermal expansion (CTE) of 0 ± 10^{-7} °C^{-1} as measured at production [1]. The nighttime temperature on Mauna Kea is normally 0 ± 8°C, and the telescope domes are refrigerated during the daytime.

The telescopes are a Richey-Chrétien design. The primary mirror surface is hyperbolic with a 34.974 meter radius of curvature and a conic constant of -1.003683[2]. There are six different segment types which are formed as off-axis hyperbolic sections to make up the primary mirror, as shown in Figure 1. There are two spare segments of each type, which are rotated into the telescopes during segment re-coatings.

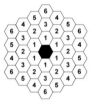

Figure 1. Keck primary mirror segment types.

The back of a segment is polished spherically convex with 35 meter radius in order to maintain an approximately constant thickness across the glass surface. Each primary mirror segment is supported by two kinematically decoupled support systems for the radial (in plane) and axial (normal) loads.

For the radial support, a central hole is machined into the center of the back surface. The hole is 254mm diameter, and 55mm deep. Six equally spaced support pads of Invar 36 material are bonded to the circumference of the hole, with the center line of the pads at the mid-plane of the segment. The adhesive used is Hysol EA-9313 epoxy with bond line thickness of approximately 0.76mm. The six radial pads are connected by flexures to a support ring and the radial support post. A diaphragm in the radial post provides isolation between the radial and axial supports.

Axial support is provided by 36 axial inserts of Invar 36 material, which are bonded at the bottom of 18mm diameter by 39mm deep holes in the back of the segment. The adhesive is Hysol EA-9313 with 0.25mm bond line. The bond line is positioned at the mid-plane of the segment in order to minimize front surface deflections due to bending moments that are applied by the bonded inserts. Subsequent analysis has shown that this is less of a concern than originally expected. The axial supports for newer segmented mirror designs are bonded on the back surface of the segments.

The axial inserts are connected by flexible rods to three whiffletree structures which distribute the axial load. A system of adjustable warping beams is attached to the whiffletree structures to allow fine correction of the front surface figure. The axial inserts carry a maximum axial force of ± 310 N for gravity and warping loads. The total weight of a mirror segment and whiffletree assemblies is 4493 N.

The radial post and whiffletrees ultimately are connected to a sub-cell assembly. A diagram of a mirror segment is shown in Figure 2, and a cross-section through the radial post in Figure 3.

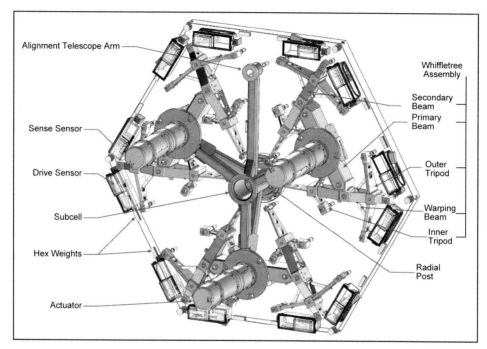

Figure 2. Keck primary mirror segment.

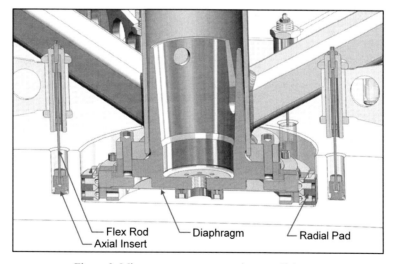

Figure 3. Mirror segment cross section at radial post.

In addition to the 36 axial inserts which support the axial segment load, there are 12 axial inserts which support the position sensors near the segment vertices. These adjust the phasing of each segment relative to its neighbors. (Dummy weights are installed in positions where there are no neighbors.) One additional insert is used as a reference for clocking the segment in the mirror cell. Overall, there are 49 axial inserts in each mirror segment.

2. PROJECT HISTORY

Mirror segments were fabricated between 1988 and 1992. Blanks for the mirror segments were manufactured by Schott AG. Polishing was performed by Itek and Tinsley. Machining and assembly were performed by Itek. Ion figuring was performed by Kodak.

The initial specifications for the segment optics indicated that there should be no "hairline or conchoidal fractures on any surface."[3] It is believed that the segments were originally manufactured with sufficient care that there were no visible cracks. From 1994 to 2007, cracks were found and repaired in radial pads of nine segments. The first of these was caused by a handling error when a segment was being removed from the telescope for re-coating. This repair was performed by a team from Itek. Subsequent repairs were performed in house by WMKO staff. However, the specific cause of the cracks and justification for their repair is not documented.

To repair the radial pad cracks the radial post and pad were removed. The damaged glass surface was ground out to a larger radius with a custom fixture, then etched. A new radial pad with a matching radius was machined and bonded in place. A photo of the radial pad support ring is in Figure 4, a radial pad in Figure 5, and a repaired pad in Figure 6.

Figure 4

Figure 5

Figure 6

An investigation of the mirror crack problem started in 2008 by a team of WMKO personnel. The extent and severity of cracks led to the initiation of the Mirror Repair Project in December 2008. The initial focus was on the radial pads, since there was a history of repairs, and they had the most obvious damage. Because the cause of the cracks was unknown and it was felt that handling the segments could result in further damage, the segment exchange and re-coating process was put on hold while the problem was investigated.

A finite element analysis (FEA) program was started in March 2009 to understand the stresses the at radial pad bonds. This analysis was performed by Ozen Engineering in Sunnyvale, California. Analysis results are discussed in Section 6.

In August 2009 a review of the segment handling procedures was conducted. Steps in the procedures where accidental mishandling could result in overturning loads and large stresses placed on the radial pads were identified. Engineering and procedural changes were put in place to reduce these risks.

A photographic survey of the mirror segments was started in August 2009. Evaluation of the photos from the first few segments surveyed indicated that there were numerous cracks in the axial inserts as well. These are small and difficult to see by eye. They can only be viewed through the front surface of the mirror when the aluminum coating has been removed. It was noted that there was one segment on which an axial insert had completely failed. In this condition the whiffletree cannot properly distribute the load among the axial inserts, and it is not possible to apply warping loads to the segment. It was at this point that the focus shifted to the axial inserts. As the photo survey progressed, it became obvious that there were extensive cracks at the axial insert bonds.

Radial pad cracks, although still being monitored carefully, are fairly stable at this time. The improvement in handling procedures has reduced the chance of a major radial pad failure.

Since 2009, a new axial insert has been designed. Numerous tests have been performed to evaluate the performance of the new insert, adhesives, and the effects of surface finish and etching on the surface strength Zerodur. An external review of the new insert design was held in October 2011. The design was approved, with recommendations for additional testing. Construction of a repair facility at the Keck headquarters is currently under way.

3. STATUS

The current status is that two segments are out of service due to axial insert cracks, and one segment is out of service due to radial pad cracks. There are four segments in the Keck I telescope, and one in Keck II telescope that we do not want to remove for re-coating until the repair procedures are in place. One segment in Keck II has a failed sensor insert.

The complete failure of an axial insert does not present a danger of major damage to a segment, but a failed insert prevents the whiffletree from distributing the load properly, and the segment cannot be warped. This affects the ability to re-coat segments, due to the lack of spares.

The radial pad cracks are more likely to propagate to such an extent that a large piece of glass could be dislodged which would be difficult or impossible to repair. The state of the radial pads is being carefully monitored until repair procedures are in place. Any segment deemed to be at risk is taken out of service.

Examples from the photo survey of axial insert cracks and how they develop are shown in Figures 7, 8, and 9. The view is looking through the glass from the front surface towards the bottom of the axial insert.

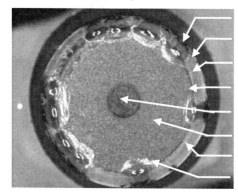

Glass cracks start at the fillet and grow inward
Glass fillet radius = .04"
Unsupported adhesive generates high stress when cold
Edge of axial insert
Adhesive nub spacer
Air bubbles
Cracks in adhesive due to aging
Glass cracks grow conchoidally

Figure 7

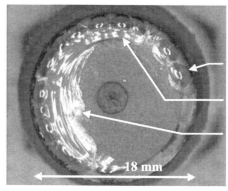

Cracks start here
Cracks merge as they grow
Eventually cracks breach the bottom of the insert

Figure 8

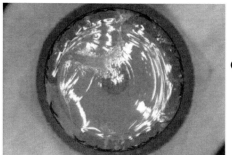

Complete failure

Figure 9

Typical radial pad cracks are shown in Figures 10 and 11.

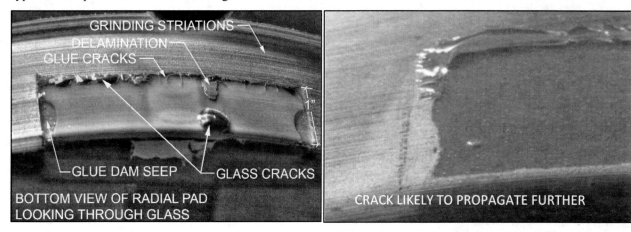

Figure 10   Figure 11

The radial pads were originally installed by holding them in place with an alignment fixture and injecting epoxy beneath the pads. A dam was made around three sides of each pad by using a soft adhesive to secure small Teflon tubes along the edges of the pads. The tubes were removed after the epoxy set.[4] Note in Figure 10 that some of the adhesive for the tubes flowed under the pad.

The low viscosity epoxy adhesive EA-9313 was injected into two slots in the face of the pad until adhesive flowed out of the top edge of the pad. This left a thick fillet of epoxy along the top edge. The fillet generates high stresses in the glass when cooled. Also, epoxy remains in the injection slots. The large round bubble crack seen in Figure 10 is at the bottom of one of the slots. The crack in this location is commonly seen on the radial pads. Many of the radial pad cracks appear to be self-terminating. However, the crack seen in Figure 11 extends into the glass and may propagate further.

There are also cracks in the epoxy itself, due to aging. In some locations the epoxy has de-bonded from the glass.

All of the radial pads and axial inserts are photographed through the front surface when the segment has been stripped for re-coating. At this writing, 65 of the total 84 segments have been surveyed.[5] Of these, 71% of the axial inserts have cracks, and 77% have radial pads have cracks. Overall, the severity of the axial insert cracks is substantially greater than for radial pads, and ongoing monitoring indicates that the axial inserts cracks are growing more rapidly than the radial pad cracks.

Classification schemes have been developed for the axial inserts and radial pads to categorize the severity of cracks based on the number, size and location of the cracks. A numerical value from 0 (no cracks) to 5 (failed or near failure) is assigned to each axial insert and radial pad. Summary results are shown in Table 1.

Axial Insert Crack Classification		Crack Class					
		0	1	2	3	4	5
Segments w/ cracks	Qty (%)	0 (0%)	65 (100%)	65 (100%)	56 (86%)	34 (52%)	9 (14%)
Inserts w/ cracks	Qty (%)	939 (29%)	1071 (34%)	598 (19%)	431 (14%)	133 (4%)	13 (0.4%)
Radial Pad Crack Classification		Crack Class					
		0	1	2	3	4	5
Segments w/ cracks	Qty (%)	2 (3%)	63 (97%)	54 (83%)	22 (34%)	2 (3%)	1 (2%)
Pads w/ cracks	Qty (%)	89 (23%)	102 (26%)	150 (38%)	45 (12%)	4 (1%)	1 (0.3%)

Table 1

Statistical analysis was performed to determine if cracks occur systematically. There is no significant difference in the crack distribution based on segment type for axial inserts or radial pads. ($\chi^2 = 4.9$, $\chi^2_{p=.05} = 11.1$).

There is a significant difference in the crack distribution by axial insert location ($\chi^2 = 26.7$, $\chi^2_{p=.05} = 6.0$). Axial inserts supporting the whiffletrees are more likely to have cracks than inserts supporting the sensors. Since the load on the whiffletree inserts is cyclical and greater than the sensor inserts, it is inferred that loading and cyclic fatigue has an influence on the likelihood of cracking. Also, although the humidity on Mauna Kea is generally low, high humidity conditions occur as clouds blow in and during storms. There is evidence of moisture accumulation in the axial insert holes. It is well known that water vapor causes crack growth through the process of stress corrosion cracking.[6]

There is a significant difference in the axial insert crack distribution by telescope fabrication ($\chi^2 = 5.1$, $\chi^2_{p=.05} = 3.8$). An analysis of the number of axial insert cracks versus fabrication date shows the number of cracks in the first 42 mirror segments produced for the Keck I telescope is consistent. However, the number of cracks in segments produced Keck II telescope increased over time. The results for radial pads show a uniform number of pads with cracks by serial number.

4. ROOT CAUSE

Several factors that contribute to the development of cracks have been identified. These are summarized in Figures 12 and 13, and the most significant are discussed below.

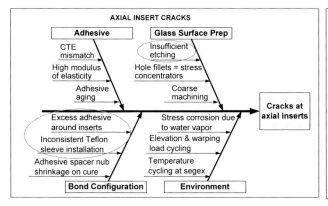

Figure 12

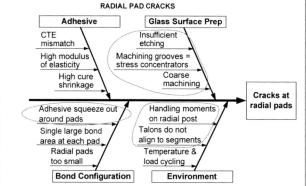

Figure 13

Two key elements for both axial inserts and radial pads are glass surface preparation and the excess unsupported adhesive around the bonds. Subsurface damage (SSD) left from machining is the starting point for crack growth. Schott has shown that the surface strength of Zerodur can be increased greatly by etching sufficiently to remove SSD[7,8]. In general, to produce a high strength surface for bonding, a minimum amount of material must be removed by etching.[9] Mirror segment specifications called for "Acid etching to remove stresses." However, there is no indication how much material was to be removed by etching, or what the final surface profile should be.

For the central radial hole, machining is seen to be very coarse, and there are striations left by the cutting tool. Measurements of a typical surface profile with a Mahr PS1 profilometer, and processed by the NIST Surface Metrology System[10] shows the very rough surface in Figure 14. The mean surface roughness parameter Ra = 8.6 μm. By comparison, Zerodur ground with 240 grit diamond (FEPA D64) typically has a surface roughness parameter Ra less than 2 μm (depending on tool speed and load). Finite element studies have shown that the striations produce stress concentrators that increase stresses by a factor of up to three.

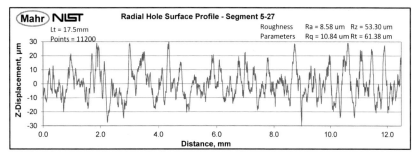

Figure 14

The original axial inserts were bonded at the bottoms of the holes, as in Figure 15. The procedure was to insert a Teflon sleeve into the hole, pour adhesive into the hole, and then push the insert into place. A previously cured nub of adhesive machined to 0.25mm thickness served as a spacer. The Teflon sleeve was removed after the adhesive cured. This method left a layer of unsupported adhesive about 0.5mm to 1 mm thick surrounding the insert which intersects the glass at the fillet, as in Figure 16. The thickness of the adhesive is determined by how far the Teflon sleeve was pushed into the hole. The coefficient of thermal expansion (CTE) of the adhesive is about 83 ppm/°C.

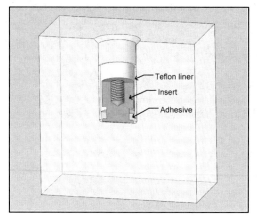 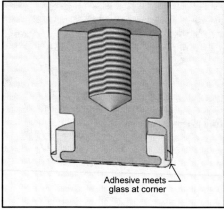

Figure 15 Figure 16

Figure 17 shows a close up of an insert installed into a test block that was made at the time the original segments were fabricated. The thick layer of unsupported adhesive is seen in the right side photo. Notice that the edge of the adhesive meets the glass approximately at the center of the glass fillet, which acts as a stress concentrator.

Figure 17

Seen through a polariscope in Figure 18, the high stress generated at low temperature is evident.

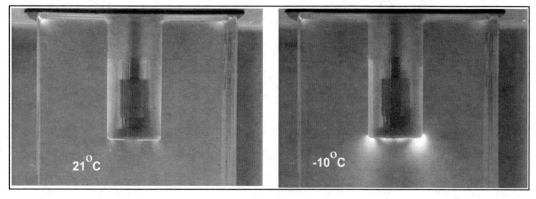

Figure 18

5. MIRROR SEGMENT REPAIR

Evaluation of the extent and growth of axial insert and radial pad cracks led to the conclusion that it is necessary to rebuild all 84 mirror segments with new supports. Due to a number of factors, including space, worker efficiency, and logistics it was decided not to do the repairs at the observatory. A repair facility is currently under construction at the WMKO headquarters in Waimea with workstations for four mirror segments. A protective cradle is being designed to transport the mirror segments safely between the observatory and the repair facility.

The required steps in the repair process have been identified, and detailed procedures are being developed.[11] The original positions of all support hardware (radial post, radial pads, whiffletrees, and axial inserts) will be measured with a high accuracy laser tracker. All of the support hardware and edge sensors will be removed. The original radial pads and axial inserts will be removed and the residual adhesive dissolved by solvents. The glass surface will be machined as necessary to remove cracks, and etched to remove all SSD. New radial and axial supports will be installed. Radial post, whiffletrees, and sensors will be re-installed and re-aligned to the original tolerances with the laser tracker. This process will take about one month per segment. Overall, it is anticipated that repair of all mirror segments can be completed in about four years from the start of production.

Axial Insert Repair

Since the glass at the bottoms of the axial holes is damaged, it is not possible to replace the inserts bonded at the same locations. Several concepts were considered for the repair of axial inserts. These included injecting a filler adhesive into the existing bonds, boring out the holes to bond a new insert at the bottom, and bonding to the back side of the segment. All of these have associated problems that made them unacceptable. The solution that was adopted is to bond new inserts to the circumference of the holes.

A simple method of removing the original axial inserts was developed using a spring loaded tool and induction heater to quickly heat the inserts. This softens the adhesive without locally heating the glass above 70°C. (Heating Zerodur above 120°C can affect the CTE of the material.) This same technique has been applied to removal of the radial pads.

The damaged glass at the bottoms of the holes will be ground out. Currently, the options of developing a small portable machine to do this work versus using a large glass capable CNC machine are being considered. The glass will be etched about 100μm in order to insure that all SSD has been removed.

New axial inserts have been designed as shown in Figure 19. These have six "fingers" which are axially stiff but radially compliant. This allows the bond thickness to be maintained at 0.25mm by a shim within the tolerance range of the hole diameter. Also, the fingers can flex radially to adjust for shrinkage of the adhesive on cure and at cold temperature. A tool holds the insert in place while precise amounts of adhesive are injected through small holes in the fingers. There is no adhesive "squeeze out" past the edge of the bonding surfaces. A bonded insert is shown in Figure 20, and a cross-section is in Figure 21.

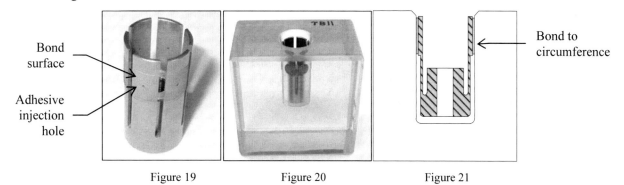

Figure 19 Figure 20 Figure 21

Radial Pad Repair

Repair of the radial pad cracks presents more difficult challenges than that for the axial inserts. At present, 77% of radial pads have some degree of cracking. Whereas the morphology of cracks in the axial inserts is fairly uniform, radial pads exhibit a variety of cracks, some of which can propagate further into the glass.

Also, the striations in the machined glass surface need to be removed to eliminate stress concentrators. The proposed repair process is to remove the original radial pads, manually grind away the damaged glass, machine the surface to remove the striations, etch the surface to remove SSD, and bond new radial pads clocked 30 degrees from the original positions, as shown in Figure 22.

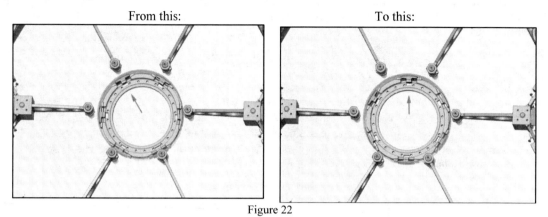

Figure 22

As for the axial inserts, the options of using a small portable tool or CNC for machining are being evaluated. The glass surface will be etched about 100μm to remove all SSD.

A new radial pad and adhesive bonding system are being developed that will provide a larger bonding surface and eliminate the problem of adhesive squeeze out around the pad. Adhesive will be injected through holes in the pads similarly to the new axial inserts.

A segment is removed from the telescope by jacking it up with a tool that attaches to the radial post and pushes it out of the mirror cell. This action applies an axial load on the radial pads, which normally have only in-plane loads. The segment is transferred to a crane that supports the segments around the edges. During the load transfer, and at other points in the handling process, the radial pads are susceptible to excessive overturning loads that can be applied accidentally due to equipment misalignment. Such loads are capable of initiating new cracks in the radial pads, even with the new design. Procedures are in place to reduce the likelihood of such an event. Nevertheless, another project being considered is to redesign the segment removal system so that high loads cannot be placed on the radial pads inadvertently.

6. TESTING AND ANALYSIS

A number of tests have been performed to assess the stresses generated at the mirror segment bonds, and to evaluate the proposed repairs. Some of the significant results are described below.

Finite Element Analysis of radial pad stresses

The conservative nominal bending strength of Zerodur is 10 MPa[12]. The realized bending strength is dependent on the speed of application of stress, and on the surface finish. Bending strength in excess of 100 MPa has been achieved with adequate machining and etching[7].

A large scale finite element analysis (FEA) of the radial pads was performed to evaluate static structural effects of gravity and thermal loading, and dynamic shock effects. The stress produced by overturn loading at the radial pads that would be caused by mishandling during segment exchange was also studied. From the large scale model, sub-models were developed to evaluate the effects of bond line thickness, fillets around the pads due to adhesive squeeze out, and striations in the glass substrate due to machining. Some of the key results of the analysis are:

- The maximum first principal stress at the radial pad bonds due to gravity loading is about 5 MPa. However, stress due to temperature change to the nominal minimum operating temperature of -10°C can exceed 25 MPa at the edges of the adhesive bonds.

- The stress concentration produced by machining striations under the fillets of unsupported adhesive along the top edge of the radial pads can produce stresses above 50 MPa, depending on the striation depth and period.

- Assuming one edge of a segment is lifted 10mm by the segment handling crane to produce an overturning load on the radial post, the first principal stress at the radial pads exceeds 35 MPa. If the radial post is jammed, the stress can exceed 100 MPa.

- Glass stress at the radial pads increases slightly with decreasing bond line thickness. Bond line thickness variations from 0.6mm to 0.15mm produce an increase in stress of only about 1.5 MPa.

Fracture mechanics analyzes sudden fracture for a single flaw in terms of a stress intensity factor

$$K_I \approx Y \sigma_o a^{1/2}$$

where Y is a geometry factor, σ_o is the stress normal to the crack plane, and a is the depth of the crack. A crack in a thick plate will result in fracture if

$$K_I \geq K_{IC}$$

where K_{IC} is the critical stress intensity factor for mode I (tensile force normal to the crack plane), and is a material constant determined experimentally.

For mode I fracture $K_{IC} = 0.9$ MPa m$^{1/2}$ for Zerodur.[13] With Y=2, a normal stress of 10 MPa will produce mode I fracture for an initial crack length of 2mm. With a stress of 50 MPa, fracture will occur for an initial crack length of just 130μm.

Crack growth in Keck mirror segments has been a slow process. Slow (subcritical) crack growth in brittle materials can occur for $K_I \ll K_{IC}$ due to static loads where moisture is present (stress corrosion cracking), and under cyclic loading (cyclic fatigue). These effects have been studied with finite element fracture mechanics studies. It is clear that the very coarse surfaces of the ground radial and axial holes left subsurface damage that led to the generation of subcritical crack growth, and eventual fracture.

FEA of axial insert stresses

Finite element models of the original and new axial inserts were produced to evaluate the stresses placed on the glass by thermal and mechanical loads[14]. The models show that very high stresses were produced at low temperature by the excess unsupported adhesive in the original installation. The new insert design substantially reduces the glass stress. The conclusions of the FEA studies indicate that for the original axial inserts:

- Where adhesive is applied in between an Invar 36 substrate and Zerodur, the Invar acts to stabilize the adhesive and prevent it from shrinking at low temperature to the extent it would by itself. The unsupported adhesive layer at the bottoms of the axial holes generates very large stress in the glass at low temperature, and the stress generated is approximately proportional to the thickness of the adhesive layer. Thermally induced stress by the unsupported adhesive is much greater than the stress induced by gravity or warping loads.

- The morphology of cracks at axial insert bonds is consistent. However, there is a wide range in the severity of cracks. This varies from insert to insert and from segment to segment. There are a number of variables at production such as ambient temperature, humidity, adhesive mix ratio, handling time, and others which might have some effect. However, the finite element models have identified a single factor which produces substantial variation in thermally induced stress in the unsupported adhesive at the bottoms of the axial holes. As mentioned above, Teflon sleeves were installed into the holes prior to bonding. The thickness of the unsupported adhesive layers at the bottoms of the holes is dependent on how far the Teflon sleeves were pushed into the holes. The thickness of this layer has a substantial effect on the thermally induced stress in the glass. For a 0.5mm layer at -10°C, the average maximum principal stress at the adhesive edge is 20 MPa. For a 1mm layer, this increases to about 45 MPa. This wide variation serves to explain the difference in crack severity.

For the new inserts analysis was performed to evaluate the effect of bond line thickness, mechanical loading, and adhesive elastic modulus at temperature down to -10°C. Bond line thickness has minimal effect for bond lines between 0.127mm and 0.381mm. A bond line of .254mm was chosen for use with the new inserts.

The maximum applied load on the axial inserts for both gravity and warping loads is ±310 N. The maximum stress in the glass at the adhesive bonds for both thermal and mechanical loads is less than 10 MPa at -10°C. For variations in elastic modulus of the adhesive between 1000 MPa to 2500 MPa, the variation in peak stress applied to the glass is roughly proportional to $0.7*(E/E0)$.

Etching

Etching is a critical step in preparation for bonding to glass to remove SSD. Hydrofluoric acid (HF) is commonly used for etching glass. Due to the risks inherent in handling HF, tests were performed to compare the performance of ammonium bifluoride (NH_4HF_2) as an alternative. Ammonium bifluoride is the active ingredient in chemicals such as ArmourEtch and EtchAll, which are sold in craft stores for frosting glass. Liquid and paste versions are available containing approximately 20% to 30% ammonium bifluoride. Although still releasing fluoride ions, these compounds are less aggressive than HF, and have the perception of being safer to handle.

Glass ceramics such as Zerodur are difficult to etch because of the formation of insoluble fluorides, such as AlF_3, MgF_3, and CaF_2. These form precipitates on the surface during etching which slow the etch rate and result in the formation of uneven surfaces. The addition of hydrochloric acid (HCl) can dissolve the insoluble precipitates and produce a higher quality and more consistent etch.

Schott has recommended an etch solution of HF and HCl in preparation for bonding Zerodur[15]. The mixture used for our tests consists of HF(49%) + HCl(36%) + H_2O in the proportions 2:1:1.5. For ammonium bifluoride, we found that the highest practical concentration at room temperature to avoid super saturation is 50%. Also, the etch rate for ammonium bifluoride slows after about 30 minutes, so multiple etch cycles were performed.

Several test samples were prepared to compare the etch depth and bonding strength of Zerodur etched with HF solution versus ammonium bifluoride solution[16]. The samples were ground to give an initial surface profile Ra = 0.9μm. The etch depth was measured, and block shear tests per ASTM D4501 were performed on 12.7mm diameter coupons bonded with Loctite E-120HP adhesive to evaluate the relative strength of the bonded surfaces. The mean values of test results are shown in Table 2.

Sample Preparation	Etch Depth um	Shear Strength N	Shear stress MPa
Unetched surface	-	1477	11.7
Etch 3 x 30 minutes with NH_4HF_2(50%)	27	2210	17.5
Etch 3 x 30 minutes with [3] NH_4HF_2(50%) + [1]HCl(36%)	35	2349	18.5
Etch 1 x 30 minutes with [2]HF(49%) + [1]HCl(36%) + [1.5]H_2O	86	3149	24.9

Table 2

Etching one time for 30 minutes with HF solution produces a much stronger bond surface than etching three times with ammonium bifluoride solution. Based on these results it was decided that the HF + HCl solution will be used to prepare the Zerodur surfaces for both axial insert and radial pad repairs.

Adhesive selection

Several adhesives were evaluated to use in the mirror segment repairs. Four candidates were selected for testing[17]. These are Armstrong A-12, Hysol Loctite E-120HP, 3M EC-2216 B/A Gray, and Summers Milbond. As a baseline, the original Hysol EA-3913 adhesive was tested. It was decided not to use the EA-9313 adhesive to repair segments for several reasons, including the difficulty of mixing and de-gassing, as well as the low viscosity making it difficult to control the quantity of adhesive applied.

Initial tests have been based on the new axial insert which has been designed. It is anticipated that the same adhesive will be used for radial pad repairs, but tests specific for the radial pads will be conducted. Samples were tested for shear strength, glass stress generated at low temperature, load versus displacement, and creep.

One of the most popular adhesives for structural bonding to glass is 3M EC-2216. It has relatively high strength and low elastic modulus, which minimizes stress induced in glass at low temperature. It has been tested and used extensively by NASA and others. However, it has a low glass transition temperature of 29°C which makes it questionable for use at room temperature. Mirror segments are at room temperature (~20°C) during the recoating process. The creep performance tests indicate that it may not be suitable for the mirror segment repairs. Based on the tests, the current candidate adhesive selected for repair of the mirror segments is Loctite E-120HP. Some of the tests that have been performed are described below. Additional testing with larger sample sizes is planned to validate the results.

Strength and adhesion

The new axial inserts will be bonded in shear. Block shear tests were performed to determine the shear strengths of the adhesive bonds. Invar coupons were bonded to etched Zerodur glass blocks and tested to ASTM D4501 standard. Tests were performed with and without adhesive primers. Milbond comes with its own primer, and 3M 3901 silane primer was used for the others. On average, the use of primer increased bond shear strength by about 20%. All of the adhesives proved strong enough to give a strength safety factor of at least 5X.

Stress evaluation

Shrinkage on cure of the adhesive produces some initial stress at glass bonds. At low temperature and with no mechanical loads, the adhesive properties that generate stress are the CTE and elastic modulus. An adhesive may have a high CTE, but if it is soft and stretches easily it does not generate high stresses. In general, the range of CTE values available for adhesives is much smaller than the elastic modulus values. For the adhesives that were studied, the CTE values range from $62 \times 10^{-6}/°C$ to $102 \times 10^{-6}/°C$. However, the elastic modulus values vary from 70 MPa to 2800 MPa. To evaluate the stress generated by the adhesives, a combination of FEA models and tests with a polariscope were performed. As expected, the higher modulus adhesives generated higher stresses in the glass.

Load versus displacement

It is important that the mirror segment supports remain stable and do not allow for motion of the segments with variations in gravity load as telescopes change elevation. Load versus displacement curves were measured for the adhesives at various temperatures. The low elastic modulus adhesives EC-2216 and Milbond exhibited much larger displacements under load than the stiffer adhesives.

Creep

Polymer materials such as adhesives have time dependent viscoelastic properties under load. When a constant load is applied the first response is elastic deformation. If the adhesive were purely elastic, it would reach an equilibrium state for a given load and no additional deformation would occur. However, for viscoelastic materials a sustained load will cause polymer chains to unwind and untangle, allowing them to slide internally. This disentanglement results in additional deformation over time. Once polymer chains have stretched as far as they can, further motion comes from viscous slippage between non-crosslinked chains. This is more prevalent at higher temperatures when the adhesive is near or above its glass transition temperature. Permanent deformation or rupture can occur if the stress, temperature and length of time under load are sufficient. This deformation is termed creep.

Samples of the new axial inserts with the test adhesives were evaluated for creep performance using the standard ASTM D2990 test. Samples were loaded to 150% of the nominal maximum working load at 15°C. The samples were held at the test load for 24 hours, and the displacement was recorded. At the end of the test, the load was ramped down to no load and the displacement was measured for one hour. The Hysol EA-9313, Armstrong A-12, and Loctite E-120HP adhesives did not show any creep response throughout the test, and there was no indication of permanent plastic deformation. However, the 3M EC-2216 and Milbond adhesives continued to deform throughout the test period and did not return to their initial lengths when unloaded. The permanent plastic deformation of EC-2216 was 20μm and the Milbond deformation was 8μm.

Cyclic loading

Cyclic loading of an adhesive bond can lead to changes in the properties of the adhesive, including changes in elastic modulus and CTE. In order to determine if there are changes in the adhesive properties with large load variations, tests were performed over 5400 cycles at -10°C on a new axial insert bonded to Zerodur with the candidate adhesive Loctite E-120HP. For this test, the sample was loaded in tension in a sinusoidal pattern with a period of 2 minutes, from 71 N to 468 N, which is 23% to 150% of the maximum in-service load. In this test, dynamic stiffness and energy dissipated per cycle remained constant, indicating that there were no changes in the internal structure of the adhesive.

Evaluation of front surface aberrations

A major concern with replacing the radial pads and axial inserts is that changing the support positions will affect the front surface figure of the mirrors. A full segment finite element model is being constructed to evaluate the effect of changing the support positions. This calculation is expected to give some confidence in the outcome.

To validate the FEA models for the axial inserts, tests were performed on the effects of thermal and mechanical loads in an environmental chamber.[18] Both original and new axial inserts were bonded into 178mm diameter Zerodur flats, and front surface deformations were measured with a Zygo interferometer under varying axial loads and temperatures. The results matched well with the finite element models. The local front surface deformation produced by the original insert at -10°C is about 20 times greater than that for the new insert. The local deformation due to axial loading on the original insert is about twice the deformation produced by the new insert. This is because the old insert is bonded closer to the front surface, so the load translated to the front is distributed over a smaller area.

Due to the 35 meter focal length of the mirror segments, it is cost prohibitive to perform figure measurements directly with an interferometer or autocollimator. On sky tests with a Shack-Hartmann camera will be used to evaluate the results as segments are rebuilt, using the procedure that is currently applied to warp segments after they are recoated and reinstalled in the telescopes. This test has an accuracy of 15nm rms, which is sufficient to evaluate the optical surface after repair.

Extended life study

One of the questions to be answered is how long will these repairs last. Predicting the lifetime of an adhesive bonded joint is difficult because there are multiple factors such as temperature, moisture and loading that have an effect. Cracks in brittle materials develop over time at lower stress levels than for short term tests.[19] In order to get an estimate of the expected life of the repairs and extended life study will be conducted. The general procedure is to load samples at different percentages of the short term strength (e.g., 80%, 70%, 60%, 50%). The time to failure for each sample is recorded. The data can be extrapolated to estimate the lifetime under actual service conditions. Such tests may be conducted at elevated temperature to accelerate the aging process. An Arrhenius relationship (log failure against reciprocal temperature) is used to calculate the failure rate under normal conditions. Data variance for such tests is high, so a large sample set is needed, and the test may last for more than a year for the lowest stress levels.

7. CONCLUSION

The cracks that have developed at supports in the primary mirror segments over the 20 year life of the Keck telescopes have progressed to the point that repairs are now necessary. Over the last 3 years, extensive surveys and analysis have been performed to evaluate the status of the cracks, determine the underlying causes, and develop a repair strategy. It has been determined that both the axial and radial bonded supports should be replaced.

A new design for the axial inserts has been completed. Preliminary tests indicate that it will perform adequately and will reduce the stress generated in the glass such that new cracks will not form in the future. Additional testing will be performed to insure this result.

A conceptual design for repair of the segment radial pads by clocking the support locations by 30 degrees has been proposed. Finite element models of the mirror segments are being developed to determine if this change will have an effect on the figure of the mirrors. Ultimately, on sky tests will be used to measure the result of the repairs. New handling procedures have been put in place to mitigate the excessive loads that can be placed on the radial pads by misalignment of the lifting systems.

The project is now at the testing and implementation stage. It is anticipated that testing and development of the tools and procedures to perform the repairs will be completed at the end of 2013. The actual repair process is anticipated to take about 4 years.

The exact lifetime of the Keck telescopes is not known. The Mirror Repair Project team is driven by the challenge to insure that the Keck telescopes can continue to perform breakthrough astronomy for many decades in the future.

8. ACKNOWLEDGEMENTS

The W. M. Keck Observatory is operated as a scientific partnership among the California Institute of Technology, the University of California, and the National Aeronautics and Space Administration. The Observatory was made possible by the generous financial support of the W. M. Keck Foundation. The authors wish to recognize and acknowledge the very significant cultural role and reverence that the summit of Mauna Kea has always had within the indigenous Hawaiian community. We are most fortunate to have the opportunity to conduct observations from this mountain.

REFERENCES

1. Schott AG, "TIE-37: Thermal expansion of ZERODUR", Schott Technical Paper (2006)
2. Nelson, J., "KOTN 163: Optical Parameters of the Keck Telescope", Keck Observatory Technical Note (1986)
3. Mast, T., Nelson, J., Magner, J, "KOTN 89: Specifications for Generating the Primary Mirror Segments", Keck Observatory Technical Note (1985)
4. Lawrence Berkeley Laboratory, "Procedures for Segment Assembly at Itek", Keck Observatory Procedure, TMD-765-Rev G (1989)
5. McBride, D., "KOTN 591: Axial Insert Crack Statistics", Keck Observatory Technical Note (2011)
6. Wiederhorn, S., Bolz, L., "Stress Corrosion and Static Fatigue of Glass", J. Am. Ceramic Society, Vol. 53, No. 10 (1970)
7. Hartmann, P., Nattermann, K., Dohring, T., Kuhr, M., Thomas, P., Kling, G., Gath, P., Lucarelli, S., "ZERODUR Glass Ceramics - Strength Data for the Design of Structures with High Mechanical Stresses", Proc. SPIE Vol. 7018 70180P (2008)
8. Nattermann, K., Hartmann, P., Kling, G., Gath, P., Lucarelli, S., Sesserschmidt, B., "ZERODUR Glass Ceramics – Design of Structures with High Mechanical Stresses", Proc. SPIE Vo. 7018 70180Q (2008)
9. Hartmann, P., Nattermann, K., Dohring, T., Jedamzik, R., Kuhr, M., Thomas, P., Kling, G., Lucarelli, S., "ZERODUR Glass Ceramics for High Stress Applications", Proc. SPIE Vol. 7425-22 (2009)
10. National Institute of Standards and Technology, Internet Based Surface Metrology Algorithm Testing System
11. McBride, D., "KOTN 581: Mirror Repair Process Sequence", Keck Observatory Technical Note (2011)
12. Schott AG, "TIE-33: Design Strength of Optical Glass and ZERODUR", Schott Technical Paper (2004)
13. Viens, M., "Fracture toughness and crack growth parameters of ZERODUR", NASA Technical Memo 4185 (1990)
14. McBride, D., "KOTN 587: Finite Element Analysis of Axial Insert Stresses", Keck Observatory Technical Note (2011)
15. Schott AG, "TIE-45: ZERODUR Adhesive Bonding Recommendation", (2009)
16. McBride, D., Hudek, J.S., "KOTN 579: Etching Zerodur with Hydrofluoric Acid Vs. Ammonium Bifluoride", Keck Observatory Technical Note (2011)
17. McBride, D., Hudek, J.S., "KOTN 580: Adhesive Selection for Axial Insert Repairs", Keck Observatory Technical Note (2011)
18. Panteleev, S., McBride, D., "KOTN 590: Interferometric Tests on 7-Inch Glass Samples", Keck Observatory Technical Note (2011)
19. Beevers, A., "Durability testing and life prediction of adhesive joints", International Journal of Materials and Product Technology, Vol. 14, No. 5/6 (1999)

Alignment Algorithms for the Thirty Meter Telescope

Gary Chanan

Department of Physics and Astronomy
University of California, Irvine
Irvine, CA 92697

ABSTRACT

A variety of algorithms utilized in the alignment of segmented telescopes were developed for and implemented on the Keck telescopes in the 1990s. The algorithms associated with the Keck segmented primary mirrors are very similar to those that will be used for the Thirty Meter Telescope (TMT). However, there are alignment or related wavefront measurement tasks associated with the TMT secondary and tertiary mirrors for which the corresponding Keck algorithms either did not exist or are not adequate for TMT. We discuss two particular algorithms associated with the TMT secondary and tertiary mirrors.

1. INTRODUCTION

The primary mirror (M1) of the Thirty Meter Telescope will have 492 segments, compared to 36 segments for each of the Keck Telescopes. Nevertheless, the algorithms for aligning and phasing[1,2] the Keck segments are scalable, and thus the Keck primary mirror algorithms can be utilized at TMT with few modifications. [The phasing of the TMT segments will likely involve Fresnel diffraction,[3] whereas Keck phasing involved Fraunhofer diffraction, but this makes little difference for the relevant algorithms.] However, an improvement to the Keck algorithm for aligning the secondary mirror (M2) is needed for TMT, and an algorithm to verify the flatness of the tertiary (M3) did not exist for Keck, but is desirable for TMT.

As was the case for Keck, the surfaces of M2 and M3 for TMT will have no active control. In general, focus and astigmatism errors associated with M2 can be corrected at all field points by correcting M1 alone.[4] We expect that for TMT the higher order M2 aberrations can also be well-corrected on-axis with M1, since its segments have a total of $492 \times 3 = 1476$ rigid body degrees of freedom. This cancellation will no longer be perfect off-axis, but since the telescope beam only moves on M2 by about 3% of its diameter over the full telescope field of view, the cancellation should still be good for the modest spatial frequencies that one could hope to attain with a reasonable active control system. This argument breaks down for very high order aberrations, but for a reasonable aberration spectrum, such high order residuals should be small. This leaves the problem of aligning the rigid body degrees of freedom of M2, which we discuss in the following section. Although the secondary mirrors of the Keck Telescopes are also controlled in rigid body degrees of freedom only, the associated Keck algorithm was not entirely satisfactory and we propose to adopt a different approach for TMT, as discussed in Section 2.

Since the TMT M3 is a flat, it should be possible to fabricate and support it so that no active control, other than that for rigid body degrees of freedom, is necessary. Nevertheless, the mirror is large enough that one would like to be able to verify its flatness when it is in the telescope, and we describe an algorithm for doing so in Section 3. The Keck tertiary, also a flat, was small enough that it was not considered cost-effective to provide this capability.

Email: gachanan@uci.edu

2. RIGID BODY ALIGNMENT OF M2

At Keck the M2 rigid body alignment algorithm was based on analytical equations, e.g. for the overall wavefront coma associated with M2 tip/tilt.[5] Although this approach worked reasonably well for Keck, it is impractical for TMT because (1) many, if not most, of the results available in the literature contain unevaluated approximations, so that the accuracy of the analysis is formally unknown. In addition, it is conceivable that the errors associated with these approximations may increase as the F-ratio of the primary mirror decreases, and so may be significant for TMT with its F/1 primary. (2) It is easy to make a mistake with even relatively simple analytical results; in fact this happened with Keck. [Fortunately the magnitude of the associated error was small.]

We propose to determine the TMT M2 rigid body misalignment by an analytical (not numerical) raytrace algorithm. Because the M1 and M2 surfaces have simple analytical representations (we can ignore the flat M3 for the purposes of this calculation), the relevant raytrace calculations are simple and straightforward to implement. We outline the calculation below.

A ray is specified by the Cartesian coordinates (x_0, y_0, z_0) of one point on the ray (typically its intersection with a surface) and its three direction cosines $(\alpha_1, \alpha_2, \alpha_3)$. It is useful to represent the ray in parametric form:

$$x = x_0 + \alpha_1 t, \qquad y = y_0 + \alpha_2 t, \qquad z = z_0 + \alpha_3 t \tag{1}$$

where t is (to within a sign) the distance between the points (x_0, y_0, z_0) and (x, y, z). The surfaces of M1 and M2 are conic sections:

$$(1 + K)z^2 - 2Rz + (x^2 + y^2) = 0 \tag{2}$$

where K is the conic constant and R is the radius of curvature; R may be positive or negative. Note that given the (x, y) coordinates of a point on the conic section, one can find z simply by solving a quadratic equation, although care must be taken to choose the appropriate root. One can similarly find the intersection of an arbitrary ray (in parametric form) and a given conic section by solving a quadratic equation for the parameter t. From the radius of curvature and conic constant, we can find the normal to the surface at a given point by finding the direction cosines $(\beta_1, \beta_2, \beta_3)$:

$$\beta_1 = x/b, \qquad \beta_2 = y/b, \qquad \beta_3 = [-R + (1+K)z]/b \tag{3}$$

where

$$b^2 = x^2 + y^2 + [-R + (1+K)z]^2 \tag{4}$$

Finally one can construct the direction cosines α'_i ($i = 1, 2, 3$) of the reflected ray[6] by:

$$\alpha'_i = \alpha_i - 2(\boldsymbol{\alpha} \cdot \boldsymbol{\beta})\beta_i \tag{5}$$

Once the raytrace program exists it is a straightforward matter to generate either a 5-column control matrix for M2 piston, tip, tilt and x- and y-translation or a 3-column control matrix for M2 piston, tip, and tilt. The argument for controlling only three degrees of freedom is that M2 x- and y-translation are nearly degenerate with tip and tilt. This fact is well-known,[5] but the analytical raytrace quantifies this approximation. Figure 1 shows the spot diagrams from the output of the raytrace program for the perfect TMT, for a 1 mm decenter of M2 along x, for a 1 arcminute tilt of M2 about y, and for a 1 mm decenter of M2 along x combined with the value of y-tilt that most nearly cancels the decentering aberration. Note that the residual aberrations in the last figure are significantly larger than for the perfect TMT, but are nevertheless only at the milliarcsecond level.

The piston column of the above control matrix consists of a concatenation of the theoretical x- and y-offsets from a (Shack-Hartmann [SH]) raytrace for unit piston error of M2, and similarly for the other columns. The error multipliers for the piston and tip/tilt measurements can be obtained directly via Singular Value Decomposition

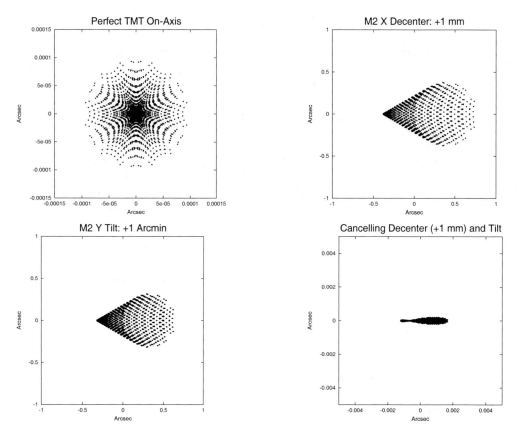

Figure 1. Results of the analytical raytracing code described in the text for the perfect TMT, for a 1 mm x-decenter of M2, for a 1 arcminute tilt of M2 about y, and for a 1 mm decenter of M2 along x combined with the y-tilt that most nearly cancels the decentering aberration. Note that the residual aberrations in the last figure are significantly larger than for the perfect TMT, but are nevertheless only at the milliarcsecond level.

of the control matrix. The control law and the associated error multipliers can be varied by conditioning the SH data in different ways.

We assume that each M1 segment has multiple Shack-Hartmann subapertures. A simple and robust control law can then be obtained by conditioning the data by subtracting from every subimage centroid offset associated with a given segment the mean of the offsets for that segment (so that the new mean offset for each segment is zero). The virtue of this control law, essentially equivalent to what was done at Keck, is that the results are independent of segment tip/tilt errors, and M2 can be aligned before one begins to align the M1 segments. For Keck, the error multipliers are 69 μm of piston per arcsecond uncertainty (1-dim) in the location of the SH centroids, and 40 arcseconds (each) for M2 tip and tilt per arcsecond. For TMT the multipliers are smaller because of the larger number of segments and the larger number of lenslets per segment (at least 19 for TMT vs. 13 for Keck): 35 μm of piston per arcsecond uncertainty, and 12 arcseconds per arcsecond for tip and tilt. Strictly speaking this analysis only applies when the errors are Gaussian, which is not the case for the usual atmospheric-dominated SH uncertainties, but it is nevertheless reasonable to estimate the TMT uncertainties by scaling from Keck. For Keck, the typical uncertainty in the M2 piston, obtained empirically from the fluctuations in a series of measurements, is 3.6 μm; the uncertainties in the M2 tip and tilt are 2.6 arcseconds. [The effective 1-dim uncertainties in the Keck centroids are therefore 0.05 to 0.06 arcseconds.] Scaling to TMT yields a piston uncertainty of 2.1 μm and a tip/tilt uncertainty of 0.7 arcseconds. Although one could in principle obtain much smaller error multipliers by forcing only the global mean (not each segment mean) offset to be zero, this is probably not a good strategy since the results would no longer be independent of segment tip/tilt errors (a significant disadvantage) and the above uncertainties are already almost certainly sufficiently small.

3. DETERMINATION OF THE M3 FIGURE ERROR: A STITCHLESS ALGORITHM

As noted in Section 1, since the TMT M3 is a flat, it should be possible to fabricate and support it so that no active control, other than what is required for rigid body degrees of freedom, is necessary. However, we would still like to be able to verify its flatness when it is in the telescope. Although the telescope beam moves very little on M2 as one moves about the field of view, the same is not true for M3: there the beamprint moves by 100% of its diameter between the center and the edge of the field. Characterization of the M3 surface therefore requires measurements at multiple positions around the telescope field of view.

As was the case for M2, the second order Zernike aberrations of M3 cannot be detected by tip/tilt-removed wavefront measurements, so we are interested here in higher order aberrations. These can be estimated for M3 by the following method. First we actively align the system on-axis, i.e. we force the sum of the on-axis aberrations of the three mirrors to be approximately zero by adjusting the M1 actuators. We then make measurements about the field of view. The analysis proceeds on the basis of two approximations: (1) the motion of the beam on M2 as we move around the field is neglected, and (2) the compensation of M2 and M3 by M1 on-axis is taken to be perfect. Because of the first approximation, the characterization of the surface of M3 is limited to an expansion in terms of Zernike polynomials (or similar functions) of order about 8 or less.

Once the on-axis alignment is carried out, the sum of the M1 plus M2 aberrations will be equal and opposite to the aberrations in the center of the M3 mirror. If we now move to an off-axis position and neglect the motion of the beam on M2 (as described above), the shear of M1 plus M2 with respect to M3 will give rise to aberrations in the overall wavefront. In the following analysis it is convenient to imagine an extended primary mirror of diameter d/α, where d is the true diameter of M1 (nominally 30 meters) and α is the ratio of the beam diameter at M3 to the full diameter of M3. [For TMT the nominal value of α is $\frac{1}{2}$, so that $d/\alpha \sim 60$ meters.] We consider dimensionless radial coordinates on both mirrors, where $r_3 = 1$ corresponds to the edge of M3 and $r_1 = 1$ corresponds to the edge of the imaginary (extended) M1, i.e. $r_1 = \alpha$ corresponds to the physical edge of the true M1. Thus the Zernike expansion of the surface of M1 is only meaningful for $r_1 \leq \alpha$. It is essential to the algorithm below to define the coordinates in this way; if we were instead to use $r_1 = 1$ to define the physical edge of M1, then there would be no simple relationship between the Zernike expansions of the two mirrors, even though their aberrations were equal and opposite.

Consider an observation of a star located at an off-axis position in the telescope field of view. Suppose a ray from this star that strikes the center of M1 ultimately strikes M3 at a point specified by the vector position $\mathbf{b} = (b_x, b_y)$ on M3. [Note that the maximum possible length of $\mathbf{b}$ for an unvignetted observation is $1 - \alpha$.] Then a ray which strikes M1 at the point $\mathbf{r}$ will ultimately strike M3 at the point $\mathbf{r} + \mathbf{b}$. If the overall M3 surface is described by a single Zernike polynomial Z_j of unit amplitude, then the wavefront error at this ray will be twice the "sheared Zernike polynomial":

$$\Delta Z_j(\mathbf{r}; \mathbf{b}) = Z_j(\mathbf{r} + \mathbf{b}) - Z_j(\mathbf{r}) \qquad (6)$$

It follows that if we want to characterize M3 in terms of Zernike polynomials, we should fit the Shack-Hartmann offsets for off-axis measurements to the gradients of sheared Zernike polynomials, rather than to the gradients of the Zernike polynomials themselves.

Since sheared Zernike polynomials are the appropriate basis functions for the current algorithm, it is useful to have a fast and efficient method for generating them. To this end consider Taylor's theorem applied to a polynomial function of two variables. This yields a finite sum:

$$Z_j(\mathbf{r} + \mathbf{b}) = \sum_{\ell=0}^{n(j)} \frac{1}{\ell!} (\mathbf{b} \cdot \nabla)^\ell Z_j(\mathbf{r}) \qquad (7)$$

where $n(j)$ is the radial order of the Zernike polynomial identified by the single running index j. We follow the single indexing scheme of Noll.[7] To convert this from an equation for $Z_j(\mathbf{r} + \mathbf{b})$ to an equation for $\Delta Z_j(\mathbf{r}; \mathbf{b})$ simply delete the $\ell = 0$ term:

$$\Delta Z_j(\mathbf{r}; \mathbf{b}) = \sum_{\ell=1}^{n(j)} \frac{1}{\ell!} (\mathbf{b} \cdot \nabla)^\ell Z_j(\mathbf{r}) \tag{8}$$

Noll also provides an expansion for the gradients of Zernike polynomials:

$$\frac{\partial}{\partial x} Z_j(\mathbf{r}) = \sum_{k=1}^{j^*} G_{jk}^x Z_k(\mathbf{r}) \tag{9}$$

$$\frac{\partial}{\partial y} Z_j(\mathbf{r}) = \sum_{k=1}^{j^*} G_{jk}^y Z_k(\mathbf{r}) \tag{10}$$

where Z_{j^*} is the highest index polynomial whose radial order is one less than the radial order of Z_j, G_{jk}^x is an element of a matrix of constant coefficients tabulated by Noll, and similarly for G_{jk}^y. Eqs. (8) through (10) can then be combined to yield a convenient expression for the sheared Zernike polynomials:

$$\Delta Z_j(\mathbf{r}; \mathbf{b}) = \sum_{k=1}^{j^*} H_{jk}(\mathbf{b}) Z_k(\mathbf{r}) \tag{11}$$

where $H_{jk}(\mathbf{b})$ is an element of the matrix $\mathbf{H}(\mathbf{b})$:

$$\mathbf{H}(\mathbf{b}) = \sum_{\ell=1}^{n} \frac{1}{\ell!} (b_x \mathbf{G}^x + b_y \mathbf{G}^y)^\ell \tag{12}$$

In the last equation we have written n rather than $n(j)$ as the upper limit of the sum so that the dimensions of $\mathbf{H}$ are independent of j. In this case n is the highest radial order of the finite set of Zernike polynomials under consideration. This replacement reflects the practical programming matter that it is easier to fill out $\mathbf{H}$ with zeros as appropriate rather than to keep track of the dimensions.

For the actual data analysis, we require not the sheared Zernike polynomials but their gradients. However, these can readily be obtained from the polynomials themselves with an additional matrix multiply, as in Eqs. (9) and (10).

Note that in order to implement this scheme, one must be able to compare the off-axis Shack-Hartmann measurements with their ideal or desired values, which are significantly different from zero because the off-axis aberrations of TMT are significant. However, the analytical raytrace program presented in the previous section provides a fast and efficient solution to this problem. Figure 2 shows typical off-axis spot diagrams (readily converted to Shack-Hartmann patterns) obtained from the analytical raytrace program described above.

In many wavefront sensing applications where only a fraction of the wavefront is sampled in any given measurement, the overall (full aperture) wavefront is reconstructed by "stitching" the individual measurements together. There are many algorithms for doing this, but they are not trivial to implement and are subject to a variety of systematic errors. By contrast, the proposed algorithm is virtually "stitchless" - the stitching problem is circumvented by using from the beginning basis functions (the sheared Zernike polynomials) that are appropriate to the sampling of the overall wavefront. Note that because these functions can be generated very efficiently, the sampling pattern does not need to be specified in advance, rather the required theoretical Shack-Hartmann patterns can be generated "on the fly" and there is no need to store the very large matrices appropriate to highly sampled wavefronts corresponding to multiple field points.

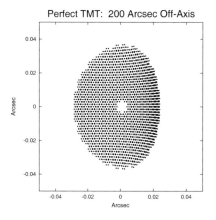 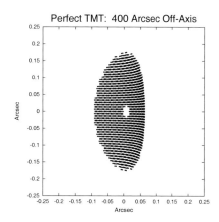

Figure 2. Spot diagram for a perfect TMT at two off-axis field points as calculated using the raytrace program described in Section 2. The ability to generate these diagrams quickly is essential to the scheme for verifying the figure of M3.

Figure 3 illustrates the difference between a conventional stitching-dependent algorithm applied to the current problem and the proposed stitchless algorithm. This is essentially the difference between the complicated sum of terms on the right hand side of Eq. (8) (different for each different field point) and the single term on the left hand side. Suppose for example that the aberration on M3 consists of 1 micron of the single 5th-order Zernike corresponding to $j = 19$. The left hand panel shows simulated spectra with respect to a conventional Zernike basis resulting from Shack-Hartmann measurements at five randomly chosen points around the field of view; these data would then need to be stitched together to produce the overall wavefront map. The right hand panel shows the much simpler and in effect already stitched spectra (now indistinguishable from field point to field point) with respect to a basis of sheared Zernike polynomials.

Since, as noted above, focus and astigmatism on M3 can be corrected at all field points by correcting M1 alone, the proposed algorithm is necessarily blind to these second order aberrations. The figure corresponding to Figure 3 for these Zernikes ($j = 4, 5, 6$) would show only piston and tip/tilt errors in the left-hand panel and therefore the piston- and tip/tilt-removed wavefronts would produce a perfect surface in the right-hand panel. This reflects the fact that these aberrations are benign for a properly aligned telescope. However, if the tertiary were then rotated to access a different focal station on TMT, the astigmatic errors on M1, M2, and M3 would no longer cancel, and the telescope would have to be re-aligned or new corrections would have to be applied to M1 via a look-up table.

4. ACKNOWLEDGMENTS

We gratefully acknowledge the support of the TMT partner institutions. They are the Association of Canadian Universities for Research in Astronomy (ACURA), the California Institute of Technology and the University of California. This work was supported as well by the Gordon and Betty Moore Foundation, the Canada Foundation for Innovation, the Ontario Ministry of Research and Innovation, the National Research Council of Canada, and the U.S. National Science Foundation.

REFERENCES

[1] G. Chanan, M. Troy, F. Dekens, S. Michaels, J. Nelson, T. Mast, and D. Kirkman, "Phasing the Mirror Segments of the Keck Telescopes: The Broadband Phasing Algorithm," Appl. Opt. **37**, 140-155 (1998).

[2] G. Chanan, C. Ohara, and M. Troy, "Phasing the Mirror Segments of the Keck Telescopes II: The Narrowband Phasing Algorithm," Appl. Opt. **39**, 4706-4714 (2000).

[3] G. Chanan, M. Troy, I. Surdej, G. Gutt, and L. C. Roberts, Jr., "Fresnel phasing of segmented mirror telescopes," Appl. Opt. **50**, 6283-6293 (2011).

[4] P. Piatrou and G. Chanan, "Tomographic alignment algorithm for an extremely large three-mirror telescope: invisible modes," Appl. Opt. **49**, 6395-6401 (2010).

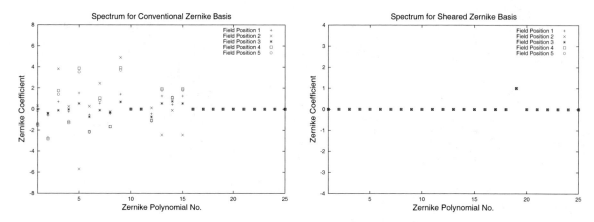

Figure 3. Left panel: Simulated M3 aberration spectra with respect to a conventional basis of Zernike polynomials corresponding to wavefront measurements from five randomly chosen field points, for an aberration equal to 1 micron of the 5th order polynomial Z_{19}. Note that each field point produces a different spectrum. Right panel: Simulated spectra for the same field points and the same aberration determined by the proposed algorithm. Each field point now produces the same spectrum.

[5] D. J. Schroeder, "Astronomical Optics," (Academic, New York, 1987).
[6] B. de Greve, "Reflections and Refractions in Raytracing,"
  http://users.skynet.be/bdegreve/writings/reflection_transmission.pdf, November, 2006.
[7] R. J. Noll, "Zernike polynomials and atmospheric turbulence," J. Opt. Soc. Am. **66**, 207 (1976).

Phasing metrology system for the GMT

D. Scott Acton[1], Antonin Bouchez[2]

[1]Ball Aerospace and Technologies Corporation, 1600 Commerce Street, Boulder, CO 80501
[2]Giant Magellan Telescope, 251 Lake Street, Suite 300, Pasadena, CA 91101

ABSTRACT

The Giant Magellan Telescope (GMT) is a 25.4 m diameter ground-based segmented Gregorian telescope, composed of 7 8.4 meter diameter primary mirror segments, and 7 1 meter diameter adaptive secondary mirror segments. Co-phasing of the integrated optical system will be partially achieved by making real-time measurements of the wavefront of an off-axis guide star. However, slowly varying aberrations due to thermal and gravitational effects, as well as wind buffeting, will make it difficult to maintain alignment using real-time optical measurements alone. Consequently, we are proposing internal metrology systems to maintain the relative alignment of the optical elements. In this paper we describe a differential capacitive edge sensing system to maintain the relative alignment of the adaptive secondary mirror reference bodies. We also propose an interferometric system for sensing of the relative displacements of primary mirror segments.

Keywords: Extremely Large Telescopes, Wavefront Sensing and Controls

1. INTRODUCTION

The GMT is an extremely large, segmented Gregorian telescope, which will be constructed on Cerro Las Campanas in Chile. A notional drawing of the GMT is shown in Figure 1. The primary mirror (M1) consists of 7 8.4-meter diameter circular segments (shown on the left). The secondary mirror (M2) is identically segmented (shown on the right), consisting of 7 1 meter round mirrors. Each of the primary and secondary mirror segments are mounted on hexapod positioners, capable of controlling the 6 rigid body motions to micron-level accuracy. The primary mirror segments also contain multiple force actuators for active control of the mirror figure. Likewise, each secondary mirror segment has hundreds of actuators for real-time adaptive correction of the atmospheric turbulence.

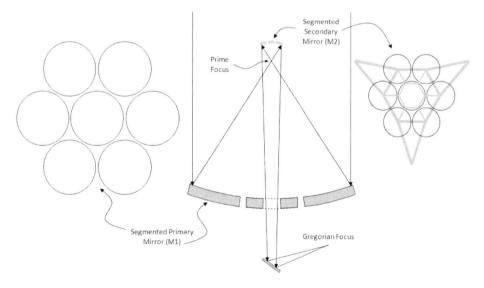

Figure 1. The Giant Magellan Telescope.

Closed-loop operation of the GMT will involve interactions between the adaptive optics, which correct the atmospheric turbulence, the active optics, which keep the telescope aligned, and optical sensors which establish and maintain the phasing of the telescope [1]. A detailed description of this interaction is beyond the scope of this paper. However, one of the key elements is the internal metrology systems which detect rigid-body motions of the outer mirror segments relative to center segments in both M1 and M2.

The metrology system for M1 must meet the following requirements:
- It must permit up to 3 cm of motion of any M1 segment, without creating a collision in the sensor components.
- It must facilitate rapid recovery of M1 segment positions, to within a few microns (absolute), after being shut down, or after an off-nominal event.
- It must be able to sense segment-level piston fluctuations to within about 10 nm (wavefront) which result from vibration and wind buffeting, at about 50-100 Hz.
- It must be stable in piston to better than the capture range of the AO wavefront sensors (200-300 nm, wavefront) for several minutes at a time, to support slewing to and acquiring new science objects.

Because the secondary mirror (M2) is concave, it is possible to place a calibration source at the prime focus, to stimulate a wavefront sensor at the Gregorian focus. Consequently, it should be relatively straightforward to co-align the M2 segments prior to initiating an observing session. The goal of the M2 metrology system will be to maintain this alignment throughout the observing session. Because the segments in M1 are also adjustable, one can conceive of scenarios where an optical error in M2 is compensated by a similar "error" in M1. For many optical aberrations, this is acceptable. Segment-level tilt, however is problematic because even a small amount will create a field-dependent piston term in the wavefront. In some observing modes, the nominal segment-level piston will be monitored with an off-axis piston sensor, resulting in an uncorrected piston error within the science field. Consequently, it is vital that the M2 metrology system be capable of maintaining relative tilt between the segments.

Specific requirements for the M2 metrology system are:
- Must fit in the small space afforded by the M2 structure.
- Must be located within the shadow of the M2 support structure (gray bars, right of Figure 1) as seen by M1 (to minimize thermal emissivity).
- Must not prevent the removal of an M2 segment.
- Should measure segment-level piston accurately to about 10 nm (wavefront) at speeds of 50-100 Hz.
- Must sense segment-level tilt to about 1.0 microradians stably over 12 hours.

The conceptual designs for the metrology systems are presented in the balance of this paper.

2. THE SENSORS

One of the most difficult challenges in building a metrology system for M1 is the need to allow for up to 3 cm motion between segments. Any sensor on an M1 segment must therefore be located at least 6 cm from hardware on an adjacent segment. Given that we must measure piston values on the order of 10 nm in the wavefront, we are motivated to use an interferometric type sensor, of which there are several commercial options. These sensors measure displacement changes by counting interference fringes in quadrature. If the beam is interrupted, the distance information is largely lost. Therefore, whenever lock is lost on a sensor/retro pair, the zero point has to be reestablished.

We propose a hybrid metrology system that supplements the interference based sensors with a more coarse sensor with larger capture range. The coarse sensors will make it possible to quickly establish wavefront sensing capture range (about 10 microns of piston error). Once the optical phasing has been achieved, we can then rely on the interferometric metrology system to maintain the phasing between optical sensor updates.

The space constraints of the M2 sensors, however, argue for capacitive edge sensors, which have considerable heritage for segmented telescopes [2].

2.1 M1 fine sensors

Our proposed fine sensor for the M1 metrology system is the Renishaw Distance Measuring Interferometer (DMI). Each sensor consists of a frequency-stabilized laser coupled to a 3-meter single-mode polarization-maintaining fiber. At the end of the fiber is a small interferometer head that projects the laser to a cube and reads the return signal. This is illustrated in Figure 2. The heads would be mounted to the side-walls of the M1 segments, with the electronics and lasers mounted in the cell below the M1 segments. If properly implemented, the DMI's should be able to sense ~1 nm length changes.

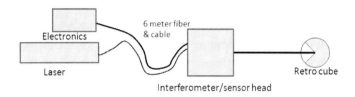

Figure 2. DMI.

2.2 M1 coarse sensors

For the M1 metrology system coarse sensor, we propose an absolute imaging encoder described in references [3] and [4], and illustrated in Figure 3. A special target is attached to the side-wall of an M1 segment. An image of the target is formed via a lens onto a simple 8-bit camera, mounted to an adjoining segment. The target is designed in such a way so that the camera (and its associated software) can know in an absolute sense where it is pointing, while still giving sub-micron sensitivity. As such, this type of encoder would have a capture range of several cm. The short-term sensitivity to translation is on the order of 0.2 microns, The long term sensitivity should be about 2 microns, limited by the stability of the detector/optics/segment interface.

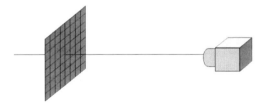

Figure 3. Absolute Imaging Encoder concept.

2.3 M2 sensors

For M2, we propose a sensor that uses a pair of differential capacitive plates, as illustrated in Figure 4. This sensor will permit the measuring of 6 different capacitances, which we denote with the letter "C":

- C(1,1a)
- C(1,1b)
- C(1,1a+1b)
- C(2,2a)
- C(2,2b)
- C(2,2a+2b)

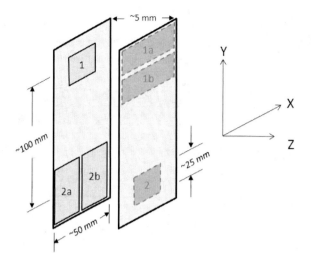

Figure 4. Differential capacitive edge sensor concept.

When the right-most half of the sensor is moved relative to the left-most half, the capacitance values change. Some modes can be sensed strongly; others not at all:

- ΔX derived from $C(2,2a) - C(2,2b)$
- ΔY derived from $C(1,1a) - C(1,1b)$
- ΔZ derived from $C(1,1a+1b) + C(2,2a+2b)$
- θX derived from $C(1,1a+1b) - C(2,2a+2b)$
- θY not observed
- θZ not observed

(Note that θ refers to rotation about the specified axis.) We can estimate the sensitivity of the differential sensor by first establishing the sensitivity of a hypothetical capacitor with 25 X 25 mm plates, separated by 5 mm. We could appeal to basic physics to establish the initial sensitivity. For this conceptual design, however, it will be sufficient to estimate the sensitivity from commercially available products, noting that in a rough sense,

- The sensitivity increases linearly with the plate area
- The sensitivity decreases linearly with separation
- The sensitivity decreases with the square-root of the temporal bandpass

In Table 1, we have picked the closest matching product from 3 different manufacturers, and estimated the performance of our hypothetical capacitor. (Note that this involves a bit of extrapolation, so these predictions should only be considered estimates.)

	Queensgate	Physik-Instrumente	Micro-Epsilon
Model	NXD	D-100.00	CS-05
Plate area (mm^2)	282	113.1	125
Spacing (mm)	1.25	0.3	5.0
Estimated Noise at 50 Hz	1.3 nm RMS	0.2 nm RMS	0.6 nm RMS
Implied sensitivity of GMT capacitor	2.3 nm RMS	0.6 nm RMS	0.125 nm RMS

Table 4. Comparison of three commercial products with the hypothetical capacitor.

We will take the middle value of 0.6 nm RMS at 50 Hz for the assumed sensitivity of our hypothetical capacitive sensor. We can use this value to estimate the sensitivity for measuring each degree of freedom of our proposed sensor in Fig. 4.

ΔX, ΔY

ΔX and ΔY are actually differential measurements obtained by splitting the small plate in half, and comparing the change in capacitance on one half to that of the other (Figure 5). This cuts the sensitivity in half as well. However, the final translation is determined by combining two uncorrelated signals, which gives us an *increase* of sqrt(2) in sensitivity, for a net decrease of sqrt(2). The sensitivity for these measurements is therefore 0.85 nm rms.

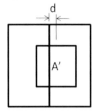

Figure 5. Differential measurements.

ΔZ

Displacement along the Z axis is derived from 4 uncorrelated capacitance measurements. This should give us an *improvement* of 2x in the sensitivity. However, they are made with plates cut in half, so we lose this sensitivity and we are back to 0.6 nm rms.

θX

This is very similar to the ΔZ measurement, except we need to subtract the top result from the bottom. Consequently, we are differencing two results instead of averaging, and we lose a factor of sqrt(2) in sensitivity instead of gaining it. The result is 1.2 nm acting over a distance of about 100 mm, which in turn gives us an angular sensitivity of about 12 nanoradians.

3. SENSOR LOCATIONS

Our proposed layout of sensors for M1 is illustrated in Figures 6, 7 and 8. Suites of sensors are mounted at 24 locations. The separation between segments at this location is ~600 mm. Two fine sensors and 1 coarse sensor are required at each location on M1. As shown in Figure 7, two fine sensors form a criss-cross pattern across an M1 segment boundary. A coarse sensor, however, simply looks at its associated target on an adjoining segment. The two cameras at each segment interface are mounted with a 20-degree angle between them, so that absolute displacement between the segments can be sensed.

The proposed sensor layout for M2 is shown on the left of Figure 9. Note that only 1 sensor is used per segment interface, for a total of 12 sensors.

4. SYSTEM MATRICES

The sensors have been arranged so that each outer segment has the same geometry, relative to segment-local coordinates which rotate with the segment position (Figures 5, 6, and 9). As a result, we only need to calculate how the sensed values will change on a single segment in response to the usual 6 degrees of freedom of motion, relative to the center segment. On M1, sensors were assumed to be located at the natural segment boundaries, with a 20 degree angle between each half. For the fine sensors in M1, we are simply measuring the change of length between the retro cubes and the sensor head in response to these motions. Once the locations are known, therefore, simple geometry can be used to calculate the change length in response to the applied mode change. The resulting response is shown for the fine sensors in Table 1. The units are microns of length change per micron or microradian of motion of the segment.

In a similar manner, the image motions in response to a pose change can be worked out for the coarse sensor geometry. In this case, we adopt a right-hand coordinate system on the target, as seen from the perspective of the camera. The Y axis will always be out of the page. We will define the image motion as the motion of the chief ray of the camera optics as it hits the target. The results are shown in Table 2. R is the radius of a segment, which is 4.2 meters.

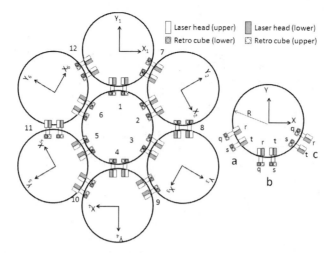

Figure 6. Locations of the M1 fine sensors. A total of 48 DMIs are used.

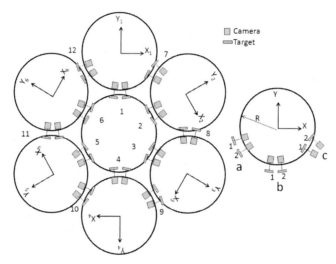

Figure 7. Locations of the M1 coarse sensors. A total of 24 cameras are used.

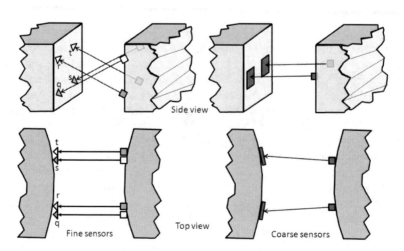

Figure 8. Coarse and fine sensor geometry for M1.

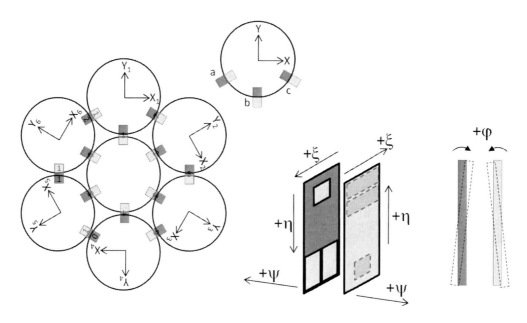

Figure 9. Left: M2 sensor location. Right: local sensor coordinate system.

	a				b				c			
Mode	q	r	s	t	q	r	s	t	q	r	s	t
X-tran	0.67	0.67	0.61	0.61	-0.01	-0.01	0.01	0.01	-0.61	-0.61	-0.66	-0.66
Y-tran	0.36	0.36	0.44	0.44	0.79	0.79	0.79	0.79	0.44	0.44	0.36	0.36
Piston	0.66	-0.66	0.66	-0.66	0.61	-0.61	0.61	-0.61	0.66	-0.66	0.66	-0.66
Theta-X	-0.93	0.93	-1.79	1.79	-2.50	2.52	-2.50	2.52	-1.79	1.79	-0.93	0.93
Theta-Y	2.60	-2.59	2.11	-2.10	0.45	-0.45	-0.45	0.45	-2.10	2.11	-2.59	2.60
Theta-Z	0.49	0.49	-0.26	-0.26	0.65	0.65	-0.63	-0.63	0.29	0.29	-0.46	-0.46

Table 1: Response of a pose change of a local segment, relative to its neighbors. Units are microns of length change per micron (or microradian) of pose change.

mode	a				b				c			
	1		2		1		2		1		2	
	X	Y	X	Y	X	Y	X	Y	X	Y	X	Y
X	-sin(20)	0	-sin(40)	0	-sin(80)	0	-sin(80)	0	-sin(40)	0	-sin(20)	0
Y	cos(20)	0	cos(40)	0	cos(80)	0	-cos(80)	0	-cos(40)	0	-cos(20)	0
Z	0	1	0	1	0	1	0	1	0	1	0	1
θX	0	-Rsin(20)	0	-Rsin(40)	0	-Rsin(80)	0	-Rsin(80)	0	Rsin(40)	0	Rsin(20)
θY	0	Rcos(20)	0	Rcos(40)	0	Rcos(80)	0	-Rcos(80)	0	Rcos(40)	0	Rcos(20)
θZ	-R	0	-R	0	-R	0	-R	0	-R	0	-R	0

Table 2: Image motion on each of the 6 coarse sensors in response to motion of the segment.

The geometry of the M2 sensor is shown on the right of Figure 9. To avoid confusion, Greek letters are used to denote the 4 degrees of freedom that can be sensed in a single sensor assembly. Moving an M2 segment in 6 DOF will only affect the 3 nearest sensors. If a segment is associated with sensors a, b, and c (Figure 9), then making unit perturbations on the segment will create the signals shown in Table 3. The numbers in Table 3 assume that the spacing between the M2 segment centers is 1 meter. The units are microns (or microradian) per micron (or microradian).

	ξ_a	η_a	ψ_a	ϕ_a	ξ_b	η_b	ψ_b	ϕ_b	ξ_c	η_c	ψ_c	ϕ_c
X	-0.5		+0.866		-1				-0.5		-0.866	
Y	+0.866		+0.5			+1			-0.866		0.5	
Z		+1				-1				-1		
θX		-0.25		+0.5	+0.5			+1		+0.25		+0.5
θY		+0.433		-0.866						+0.433		0.866
θZ	-0.5				-0.5				-0.5			

Table 3: Changes in sensor displacements and rotations that result from unit perturbations of an M2 segment.

A "system matrix" is a representation of the expected sensor readings in response to a unit perturbation of each controlled degree of freedom. We are controlling 6 degrees of freedom of each outer segment, relative to the center segment, so there are 36 controlled modes in both M1 and M2. Each of the 3 metrology systems measures 48 values. (In the case of the M1 coarse system, each of the 24 cameras measures two image translation parameters.) As a result, we end up with 48 X 36 element system matrices. Since the response to movement of a generic outer segment is known, it is easy to build up system matrices simply by replicating these values, keeping track of the segment location and its neighbors. As an example, the system matrix for the M2 metrology system is shown in Figure 10, with the matrix value mapped to image intensity.

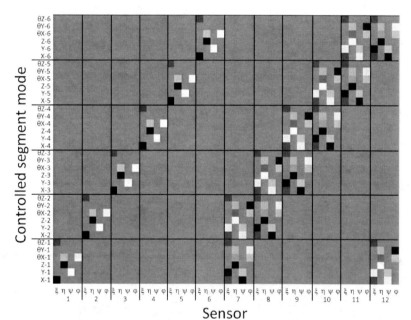

Figure 10. The system matrix for the M2 metrology system.

A singular-valued decomposition routine was applied to the 3 system matrices to analyze the natural modes of each system. None of the system matrices leads to unobservable modes. The Eigen values for the systems are plotted in Figure 11. As an example, the 36 natural modes are displayed for the M2 system in Figure 12, with the mode value mapped to intensity. The first mode is shown in the lower left corner, the last mode in the upper right. The most crucial modes for the M2 system are piston, tip, and tilt. Fortunately, these terms are represented mostly by the stronger Eigen modes.

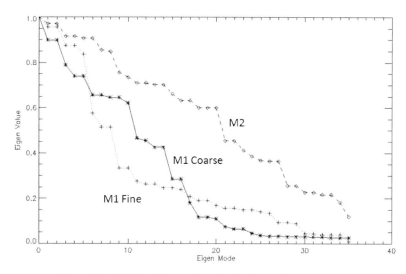

Figure 11. Relative Eigen values for the 3 metrology systems.

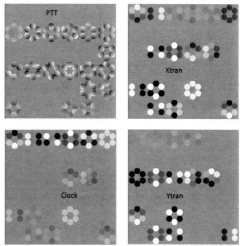

Figure 12. Eigen modes for the M2 metrology system. Each mode is the combination of the represented Piston/Tip/Tilt, X-translation, Y-translation, and clocking terms, which are mapped to intensity in the display.

5. NOISE PROPAGATION

Taking the pseudo inverse of the system matrix leads to the "control matrix" which can be used to calculate the modes associated with a given set of sensor measurements. By generating random sensor values for an input, we can use the control matrix to determine the associated segment positions that would be inferred from those values. Keeping track of the statistics in a Monte-Carlo simulation will give us a good estimate of the "noise floor" of each control system.

Recall that we are assuming the following sensitivities:
- M1 Coarse: 2 microns (long-term stability)
- M1 Fine: 1 nm
- M2: ΔX, ΔY: 0.85 nm, ΔZ: 0.6 nm, θX: 12 nRad

Using a simple Monte Carlo analysis involving 1000 random sensor realizations, the sensitivity predictions in Table 6 were generated. The only result in Table 6 that is a bit concerning is the piston sensitivity for M2 of 5.9 nm. This is slightly over the stated goal of 5 nm (or 10 nm in the wavefront). We will need to watch this term carefully as the preliminary design is completed. As stated in Section 1, we need a long term (12 hour) stability of about 1 microradian for the tip/tilt of M2. The predicted noise floor of the M2 sensors is about 100X better than this, which is encouraging. The real issue, however, is the stability of the M2 sensors. The results shown in Table 6 for M2 were derived from the assumption that the hypothetical capacitor described in Section 2.3 has a sensitivity of 0.6 nm. Therefore, the same capacitor would need to be stable to better than 60 nm for 12 hours. Consequently, this will also be a chief consideration for the preliminary design.

Mode	M1 Coarse	M1 Fine	M2
X translation	4.1 µm	0.7 nm	0.4 nm
Y translation	3.2 µm	0.6 nm	0.5 nm
Piston	4.1 µm	2.3 nm	5.9 nm
Theta-X	1.0 µRad	0.6 nRad	11.1 nRad
Theta-Y	1.4 µRad	0.5 nRad	9.5 nRad
Theta-Z	0.9 µRad	0.6 nRad	0.2 nRad

Table 3: Predicted coarse and fine sensitivity.

6. CONCLUSION

We have proposed a conceptual design of metrology systems for maintaining optical alignment of the M1 and M2 mirror segments. A first-order analysis indicates that these metrology systems will meet or exceed our alignment requirements. Upcoming design, analysis, and prototyping will refine our understanding of the expected performance.

7. ACKNOLEDGEMENTS

This work has been supported by the GMTO Corporation, a non-profit organization operated on behalf of an international consortium of universities and institutions: Astronomy Australia Ltd, the Australian National University, the Carnegie Institution for Science, Harvard University, the Korea Astronomy and Space Science Institute, the Smithsonian Institution, The University of Texas at Austin, Texas A&M University, University of Arizona and University of Chicago.

REFERENCES

[1] A. Bouchez, et al., "The Giant Magellan Telescope phasing system," SPIE 8447-138, 2012. These proceedings.
[2] T. Mast, G. Chanan, J. Nelson, R. Minor, R. Jared, "Edge sensor design for the TMT," Proc. SPIE 6267, 62672S, 2006.
[3] Douglas B. Leviton, Jeff Kirk and Luke Lobsinger, "Ultrahigh-resolution Cartesian absolute optical encoder", Proc. SPIE 5190, 111 (2003)
[4] Douglas B. Leviton, "Ultrahigh resolution absolute Cartesian electronic autocollimator", Proc. SPIE 5190, 468 (2003)

Performance Prediction of the Fast Steering Secondary Mirror for the Giant Magellan Telescope

Myung Cho[1], Andrew Corredor[2], Christoph Dribusch[2], Won-Hyun Park[3], Michael Sheehan[4], Matt Johns[4], Stephen Shectman[4], Jonathan Kern[4], Charlie Hull[4], Young-Soo Kim[5], John Bagnasco[6]

[1]GSMT Program Office, National Optical Astronomy Observatory
950 N. Cherry Ave., Tucson, AZ 85719, USA
[2]Aerospace and Mechanical Engineering Department, University of Arizona, USA
[3]College of Optical Sciences, University of Arizona, USA
[4] GMT Project Office, Pasadena, USA
[5]Korea Astronomy and Space Science Institute, Daejeon, South Korea
[6]Department of Mechanical and Astronautical Engineering Naval Postgraduate School, USA

ABSTRACT

The Giant Magellan Telescope (GMT) Fast-steering secondary mirror (FSM) is one of the GMT two Gregorian secondary mirrors. The FSM is 3.2 m in diameter and built as seven 1.1 m diameter circular segments conjugated 1:1 to the seven 8.4m segments of the primary. A parametric study and optimization of the FSM mirror blank and central lateral flexure design were performed. For the optimized FSM configuration, the optical image qualities and structure functions for the axial and lateral gravity print-through cases, thermal gradient effects, and dynamic performances will be discussed. This paper reports performance predictions of the optimized FSM. To validate our lateral flexure design concept, mechanical and optical tests were conducted on test mirrors installed with two different lateral flexures.

Keywords: Extremely large telescope, GMT Secondary Mirror, optimum configuration, mirror support, image quality

1. INTRODUCTION

The Fast Steering Secondary Mirror (FSM) of the Giant Magellan Telescope (GMT) is one of the next generation extremely large telescope projects under the design and development phase. The FSM consists of seven separated circular segment mirrors which are conjugated 1:1 to the segments of the primary mirror. An artesian CAD model of the GMT telescope and the optical system configurations are shown in Figure 1. The FSM system has a tip-tilt feature to correct wind vibration and telescope jitters. The FSM system will mainly serve as a commissioning system and a backup system when the adaptive secondary mirror (ASM) is not in operation.

High fidelity mechanical and optical modeling and analysis of FSM were employed for the performance predictions and evaluations of the FSM systems. The segmented FSM is a lightweight mirror with a diameter of 1.058 m (on-axis segment) and 1.048m (off-axis segments), and a nominal thickness of 120 mm. This lightweight segment is held by three equally spaced axial supports and a single lateral support at the center of the mirror. The axial supports are mounted on the back surface of the mirror and oriented parallel to the optical axis, and the lateral support consists of a single flexure located at the mirror's center position. For the optical and mechanical analysis of the GMT Fast Steering Secondary Mirror, the design features of the Magellan Secondary Mirror and their functions were extensively studied. Several finite element models of the Magellan M2 were created and the performances were evaluated. Modeling and analysis for the FSM was performed using finite element analysis in NX 8.0 and optical analyses with PCFRINGE™.

* mcho@noao.edu; phone: 1 520 318-8544; fax: 1 520 318-8424; www.noao.edu

Fig. 1. The GMT telescope artesian model (left), and Top end structure and optics of FSM (right).

For an initial design and development of the FSM, a baseline FSM design was created. The baseline was a lightweight mirror with a diameter of 1.064 m and a nominal thickness of 140 mm. Detail descriptions of the baseline design and its performance predictions were reported [1]. The baseline design of FSM mirror met the functional requirements of the gravity, thermal, and dynamics. Since the telescope and the optical design evolve, extended modeling efforts were required to accommodate the design change and to implement for the high fidelity design.

The FSM system was optimized to meet the requirements defined in "GMT Image Size and Wavefront Error (WFE) Budgets"[4]. The FSM error budget in terms of arc-seconds of the encircled energy at an 80% diameter (EE80) is listed in Table 1. These error budgets are given in two extreme telescope orientations and a figure error. Preliminary WFE budget allocation for the FSM mirror figure, from mirror supports, are as follows: 40nm "root mean squared" (RMS) at Zenith and 60nm RMS at 60 degrees elevation. These WFEs are equivalent to the RMS surface errors of 20nm and 30nm at Zenith and at 60 degrees elevation, respectively. The WFE allocations were set as a design goal based on the experiences and opto-mechanical analysis with the similar class of mirrors designed at National Optical Astronomy Observatory (NOAO). To fulfill the optical and mechanical performance requirements, extensive finite element analyses using NX 8.0 and optical analyses with PCFRINGE™ have been conducted. Mechanical and optical analyses performed include static gravity induced deformations, thermal responses, natural frequency calculations, and support system sensitivity evaluations. For GMT FSM design and development, NOAO took a conservative engineering approach and utilized concepts established from the f/11 Secondary Mirror of Magellan telescope, where f/11 is a f-number which is the focal length divided by the diameter of aperture of FSM mirror.

Table 1. FSM error Budget.

FSM Error budget: Encircled Energy diameters at 80% (EE80)	
Orientation	80% EE Specifications
Zenith	0.020" (arc-seconds)
Horizon	0.030" (arc-seconds)
Figure error	0.039" (arc-seconds)

In a seeing limited operation, the FSM optical surface figure errors can be corrected by a fast steering tip-tilt system. The FSM tip-tilt feature will accommodate the dynamic disturbances from wind vibrations and the telescope jitters. The tip-tilt system requires a fast tip-tilt of ±20 arc-seconds at a bandwidth of 10 Hz (goal of 20 Hz). To provide the tip-tilt motions, the FSM cell assembly requires an excellent dynamic stiffness and the stiffness goal was set as 60 Hz. The GMT Gregorian optical system and its latest optical prescription are summarized and listed in Table 2. In this study, the FSM was modeled based on the optical parameters in the prescription.

2. FSM MIRROR CONFIGURATION

The FSM is a meniscus concave mirror formed by seven separate, circular segments. The FSM converts the beam reflected from the f/0.8 Primary Mirror into an f/8 Gregorian beam for the science instruments. Each lightweight mirror segment has a nominal diameter of 1.058 m (on-axis) and a thickness of 120 mm. These design parameters were slightly modified from the previous baseline design of FSM mirror. Since the telescope and the optical design evolved, extended modeling efforts were required to accommodate the design change and to implement for the high fidelity design. The new physical segment diameters are slightly under sized, change in the back focal length, and other optical parameters are modified.

Table 2. GMT Gregorian optical system layout and the optical prescription as of May 2012

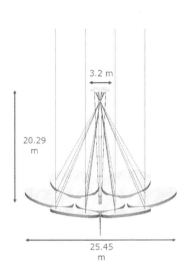

	Primary Mirror (M1)		
Diameter (D1)	25.448 m	Non-circular aperture.	
Radius of curvature (R1)	36.000 m		
Conic constant (K1)	-0.99829	Ellipsoid	
Segment diameter (D_c1)	8.365 m	Circular clear aperture. Off-axis segments tilted 13.522°	
	FSM Secondary Mirror (M2)		
Diameter (D2)	3.2 m	Pupil stop. Non-circular aperture.	
Radius of curvature (R2)	4.1639 m	Concave	
Conic constant (K2)	-0.716927	Ellipsoid	
Segment diameter (D_c2)	1.058 m		
	System		
Effective Area	368 m²		
Effective Focal Ratio	8.16	Full aperture	
Effective Focal Length	207.5888 m		
Back Focal Distance	5.830 m		
Field Curvature	2193 mm	Towards instrument	

We have performed an extensive study to quantify the effects of the shading or blocking caused by the depth of the segments (mainly from off-axis segments). For the previous baseline segment depth of 140mm, the optical performance results indicated that the shading effect may cause an image degradation. In order to overcome the effect, we have looked at the following four options: (1) extended flange of the front sheet, (2) tapered back, (3) chamfered edge, (4) combined. With the trade off study, we concluded that the new depth of 120mm provides with a superb optical performance and merits in optical fabrications.

Fig. 2. FSM mirror segment on-axis configuration.

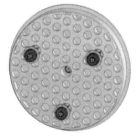

Fig. 3. A design concept of the FSM Cell Assembly consisting of FSM, Support system, and Cell, as shown in a half model at a cut-away view (left); right view shows FSM, support systems, and hidden Cell. Three axial supports are shown in dark blue, and lateral support mounted at the center of the FSM mirror.

Each of the segments has a three-point axial support and a single lateral support. The axial supports are mounted on the back surface of the mirror and oriented parallel to the optical axis (z-axis) and the lateral support consists of a single flexure located at the mirror's center position. The FSM Assembly consists of the FSM mirror, support system, and a mirror cell. The latest optical prescription defines on-axis segment clear aperture of 1.0583m and the off-axis segment of 1.0495m. The clear aperture sizes are as follows: the on-axis segment of 1.0503m and the off-axis segment of 1.0415m. The optical analysis and the image quality calculations in this paper are based on this prescription. The on-axis segment solid model is shown in Figure 2. The solid model of the FSM Assembly is shown in Figure 3.

As a baseline mirror material, Zerodur™ glass material was assumed. Zerodur properties and other material properties used in the FE mirror models are summarized in Table 3. The Coefficient of Thermal Expansion (CTE) of Zerodur varies among the manufacture's Class; however, the baseline FSM assumes a CTE of Class-0, 20×10^{-9} m/m/°C, unless specified differently.

Table 3. Material properties used in FE mirror models.

Material	Invar 36 (Flexure)	3M 2216 (Adhesive)	Zerodur (Mirror)
Young's modulus	1.47×10^{11} N/m^2	9.92×10^{8} N/m^2	9.2×10^{10} N/m^2
Poisson's ratio	0.29	0.45	0.24
Density	8050 kg/m^3	1330 kg/m^3	2530 kg/m^3
Conductivity	10.5 W/m°C	0.4 W/m°C	1.46 W/m°C
Specific Heat	515 J/kg°C	1884 J/kg°C	800 J/kg°C

2.1. FSM On-axis Mirror Modeling

A finite element (FE) model of the On-axis mirror was created in NX to serve various calculations. Such model consists of 195,840 elements and 236,753 nodes as shown in Figures 4 (a) and (b). This FE model assumes a solid concave lightweight mirror with a diameter of 1.058m, 120mm thick, and a radius of curvature of 4.2m (best fit sphere).

A local coordinate system in the FE model was assumed as follows: (1) the negative Z-axis corresponds to the line which connects the vertex of the primary mirror segment to the vertex of the secondary mirror segment which is conjugated to primary segment; (2) the positive X-axis corresponds to the telescope's mechanical elevation axis; (3) the positive Y-axis is defined by the right hand rule. Based on this coordinate system, detailed FE results are presented.

Fig. 4. FSM On-axis mirror FE model, (a) Isometric front view; (b) Isometric back view.

2.2. FSM On-axis Mirror Natural Frequency

Natural frequencies of the mirror were calculated by using the solid full FE mirror model with a free-free boundary condition. These frequency modes are characteristic mirror bending shapes and were obtained after removing rigid body motions (piston and tilts). The natural frequencies, up to 10 modes, were calculated and the corresponding characteristic mode shapes were examined. The first 6 modes are localized modes considered as rigid body motion. Mode shapes 7 thru 10 have frequencies of 568, 939, and 1159Hz respectively. Such characteristic shapes are shown in Figure 5.

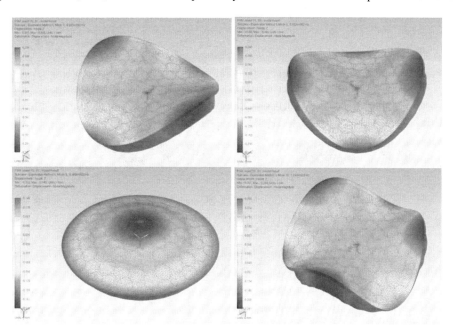

Fig. 5 Mode shapes shown (7-10) with natural mirror mode shapes (free-free).

2.3. Asphericity and Position Sensitivity

The FSM consists of seven separated circular segment mirrors, and they are arranged as: one on-axis segment at the center and six off-axis segments around the periphery. Current optical prescription defines that each off-axis FSM segment has a diameter of 1.0495 m and a nominal thickness of 120 mm, and each of which has the same optical surface configuration at the segment local coordinate system. Since the off-axis segment is an ellipse, the aspheric departure from the sphere, asphericity, is of interest. The asphericity was calculated and further decomposed in terms of lower order Zernike polynomials. For the asphericity calculation, an off-axis distance of 1089mm, a nominal radius of curvature of 4167mm, and conic of -0.7154 were assumed. As shown in Figure 6, the aspheric departure of the off-axis FSM segment is a P-V of 3.1mm and a RMS of 0.62mm as a raw data. After correcting piston, tilts, and focus from the raw data, the P-V reduced to 1.8mm and the RMS to 0.32mm. The asphericity calculations before and after correction are shown in Figure 6 (a) and (b), respectively.

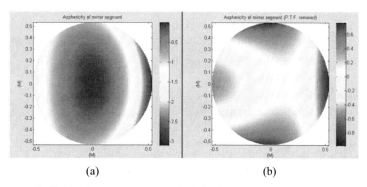

(a)   (b)

Fig. 6. Aspheric departure of off-axis FSM segment. (a) asphericity from raw data, P-V=3.1mm, RMS=0.62mm; (b) asphericity after piston, tilts, and focus corrected from the raw data, P-V=1.8mm, RMS=0.32mm.

As shown in Figure 6, the aspheric departure of the off-axis FSM segment is not rotationally symmetric; therefore, the position of the segment commonly requires a tight tolerance. Two types of position sensitivity cases were considered; (1) segment mis-located by 1mm along the radial direction with respect to its parent coordinate (delta r), and (2) segment is rotated by 1mm at the edge about its local coordinate (clocking). The effects were calculated and further decomposed in terms of Zernike polynomials. For the "delta r" case, a P-V of 4470nm and a RMS of 960nm were obtained, whereas a P-V of 7670nm and a RMS of 1260nm were obtained for the "clocking". These position errors are very significant so that the segment position should be tightly controlled. The optical surface maps for both cases are shown in Figure 7 (a) and (b), respectively. The optical surface figures of the aspheric departure and the position sensitivity cases were calculated in the first 10 Zernike terms as the FRINGE Zernike description without normalization[3]. The results were summarized and listed in Table 4.

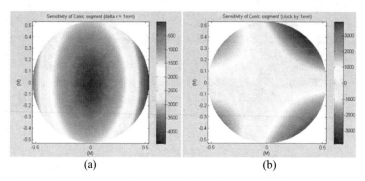

(a) (b)

Fig. 7. Position sensitivity of off-axis FSM segment. (a) sensitivity of conic segment (delta r = 1 mm); (b) sensitivity of conic segment (clock by 1mm)

Table 4. Aspericity and position sensitivities in the first 10 Zernike terms (FRINGE Zernike)

Zernike	Zernike Coefficients (mm)			description
	Asphericity	delta r=1mm	clocking 1mm	
0	-0.817711	-0.001366	0.000000	Piston
1	-0.471133	-0.000313	0.000001	Tilt-x
2	0.000000	0.000000	-0.000886	Tilt-y
3	-0.832891	-0.001364	0.000000	Focus
4	-0.769098	-0.001315	0.000005	0-Astigmatism
5	0.000000	0.000000	-0.002891	45-Astigmatism
6	-0.236015	-0.000157	0.000000	0-Coma
7	0.000000	0.000000	-0.000444	90-Coma
8	-0.015227	0.000002	0.000000	Spherical
9	0.008494	0.000021	0.000000	0-Trifoil
10	0.000000	0.000000	0.000048	30-Trifoil

3. FSM PERFORMANCE EVALUATIONS

3.1. FSM On-axis Mirror

In a seeing limited operation, the FSM optical surface figure errors can be corrected by a fast steering tip-tilt system. For the thermal distortions due to temperature variations on the FSM, an FE model was created with a unit thermal gradient of 1°C along the z-direction (Optical axis). The linear gradient of 1°C along the thickness, 1°C/0.12m indicates the top surface is 1°C warmer than the back surface. Furthermore, a CTE of 50×10^{-9}/°C was assumed in this thermal gradient case. For this particular case, a P-V surface error of 50 nm and RMS surface error of 14.5 nm were calculated, after removing piston and tilt. Mechanical deformation and the optical surface error maps are shown in Figure 8. Other thermal cases, such as a unit thermal soak, unit gradient along the X axis and Y axis, and radial thermal gradient cases were not examined. From previous analyses in similar models, it was observed that these effects are much smaller than that of the gradient case along the thickness (less than 10%).

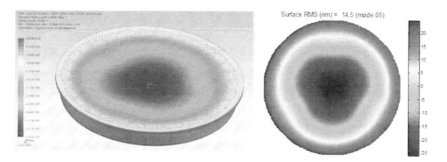

Fig. 8. Thermal deformation due to thermal gradient through the thickness; delta T of 1°C over a thickness of 0.12m; (a) overall mirror deformation, (b) the optical surface deformation with RMS surface error of 14.5 nm.

The mechanical deformation of the on-axis mirror was obtained when axial and lateral gravity loads were applied to the model. For the axial gravity case, the model was constrained in the z-direction at the three axial supports, gravity was applied parallel to the optical axis, and atmospheric pressure was applied on the optical surface. From the analysis, the minimum and maximum displacements calculated along the z-direction were -13 and 25 nm (Figure 9), and the corresponding surface RMS error obtained was 9.4nm.

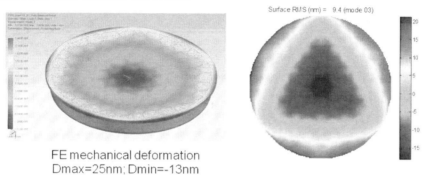

Fig. 9. Fully Compensated model for Axial gravity (a) Mechanical Deformation (z-axis), (b) Optical surface deformation.

For the lateral gravity case, the model was constrained in the x and y-directions at the lateral support and he gravitational load was applied perpendicular to the optical axis. In this particular case, the minimum and maximum displacements obtained along the z-direction were -48 and 48 nm, respectively (Figure 10). Moreover, the resulting RMS surface error calculated was 7.7 nm.

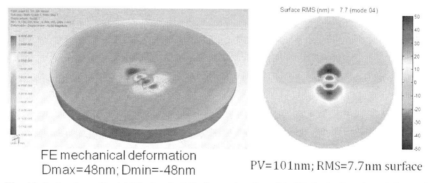

Fig. 10. Lateral gravity (a) Mechanical Deformation (z-axis), (b) Optical surface deformation.

The image quality analysis was performed for the axial and lateral support print-through from gravity. For axial support gravity print-through at Zenith when axial gravity is fully balanced with vacuum, the Spot diagram and Encircled Energy distribution plot are shown in Figure 11. The Spot diagram is shown in diameter (arc-seconds). Its Encircled Energy distribution at an 80% (EE80) diameter shows 0.012 arc-seconds. Similarly, the image quality from the lateral support gravity effects is shown in Figure 12. We demonstrated these two extreme gravity cases to simulate the effect of the image quality for an optical system consisting of the on-axis M1 segment and the conjugated M2 segment. The optical analysis for a fully populated system with seven segments is in progress.

If the secondary mirror was polished, figured, and tested at its face up position, then no gravity support error would exist at the face up position. After the FSM is installed in the telescope, the FSM would be in a -2g axial gravity (reversed gravity impact from null figuring at faced up position) at the telescope in Zenith position. At a 90 degrees Zenith angle (horizon position), the support gravity print-through would be a combination of the axial gravity and lateral gravity errors. Therefore, the resulting surface error becomes 13.1nm RMS with a quadratic sum of errors from -1g axial (reversed gravity impact) and 1g lateral gravity cases.

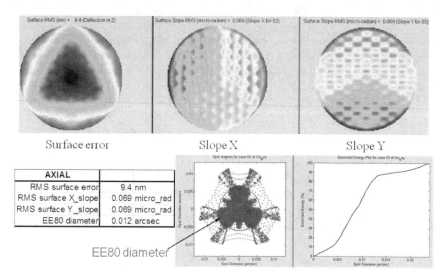

Fig. 11. Image quality analysis of the axial gravity print-through. Top figures show the optical surface map of the gravity support print-through (9.4nm RMS surface) and the corresponding slope errors. Bottom figures show the spot diagram in diameter (arc-seconds) and its encircled energy distribution. EE80 diameter shows 0.012 arc-seconds.

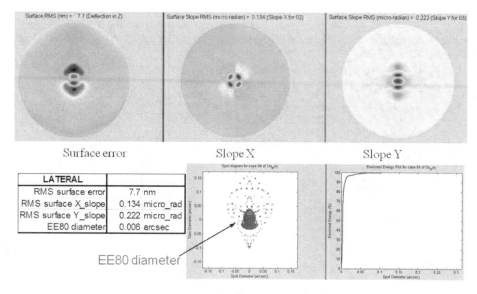

Fig. 12. Image quality analysis at Zenith angle of 90 degrees. Top figures show the optical surface map of the gravity support print-through (7.7nm RMS) and the corresponding slope errors. Bottom figures show the spot diagram in diameter (arc-seconds) and its encircled energy distribution. EE80 diameter shows 0.006 arc-seconds.

In order to control the amplitude of surface figure errors as a function of their spatial frequency, the FSM System Design Requirements Document (DRD) specifies the requirement for surface figure accuracy in terms of an Encircled Energy and a Structure Function (SF). The value of the SF for each separation distance is calculated in terms of the optical path difference (OPD) for each pair of points on the OPD map. SF is defined as:

$$D(r) = \langle [\Phi(x+r) - \Phi(x)]^2 \rangle$$

where Φ is the OPD at a position x. Structure functions for the FSM at several Zenith angles (ZA) were calculated. A sample SF calculated for the FSM at ZA of 0 degrees at all spatial scales is shown in Figure 13. In the SF calculation, a scale factor of two (2) was applied to convert the surface error to the wavefront error for the OPD. The structure function for the gravity print through was compared to that of the SF requirement in the DRD.

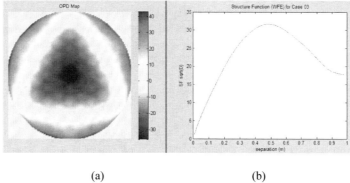

(a) (b)

Fig. 13. (a) OPD map of the gravity support print-through at Zenith angle of 0 degrees with a 9.4nm RMS surface, or 18nm RMS WFE on OPD; (b) the square root of the structure function, sqrt(D(r)), calculated from the OPD map.

3.2. FSM Off-axis Mirror

Similarly, performance evaluations were made for a FSM off-axis mirror. The mechanical deformation of the mirror due to axial gravity was obtained by applying the gravitational load at a 13 degree angle to represent the actual orientation of the mirror, as shown in Figure 14(a). Atmospheric pressure was applied on the optical surface to balance the axial mirror gravity. In this off-axis mirror at Zenith pointing, the mirror is exerted by the lateral component of its gravity. The boundary conditions imposed on the model were the following: (1) displacements in the z-direction were constrained at the axial supports and (2) displacements in the x and y-directions were constrained at the lateral support. From the analysis, the minimum and maximum displacements calculated along the z-direction were -21.5 and 27 nm as shown in Figure 14(b), and the corresponding surface errors obtained were PV of 21nm and RSM of 7.5nm (Figure 14(c)).

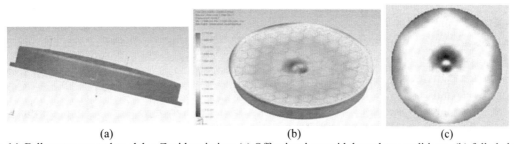

(a) (b) (c)

Fig. 14. Fully compensated model at Zenith pointing; (a) Off-axis mirror with boundary conditions, (b) fully balanced case: Mechanical Deformation magnitude in z-axis, (c) the optical surface deformation with RMS surface of 7.5nm.

Moreover, the structure function calculated for the FSM off-axis mirror at ZA of 0 degrees at all spatial scales is shown in Figure 15. In the SF calculation, a scale factor of two (2) was applied to convert the surface error to the wavefront error for the OPD. The structure function for the gravity print through was compared to that of the SF requirement in the DRD.

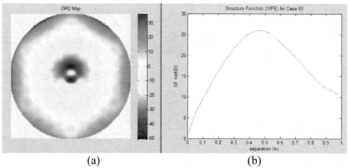

Fig. 15. (a) OPD map of the gravity support print-through at Zenith angle of 0 degrees with a 7.5nm RMS surface, or 15nm RMS WFE on OPD; (b) the square root of the structure function, sqrt(D(r)), calculated from the OPD map.

4. MIRROR CELL ASSEBLY

The design concept of the FSM Cell Assembly is to provide appropriate stiffness against the static loads and for dynamic operations. The FSM cell assembly consists of FSM mirror, Support system, and Cell, as described in Chapter 2. We are currently enhancing the previous baseline design of the FSM Cell Assembly. The new cell design provides with a higher stiffness and compactness by arranging the axial actuators and hexapod legs at the common locations for a direct load path. Such structural integrity would offer appropriate stiffness against the static loads and for dynamic operations.

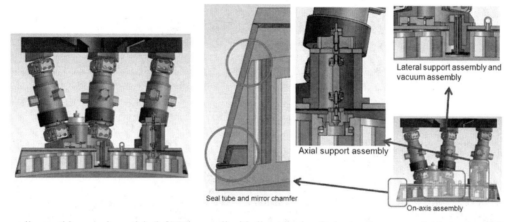

Fig. 16. Mirror cell assembly on-axis model; (left) Mirror cell with direct load path sharing the same point between axial supports and hexapod legs; (right) design details for the seal, axial support and lateral support

In a seeing limited operation, the FSM optical surface figure errors can be corrected by a fast steering tip-tilt system. The FSM tip-tilt feature will accommodate the dynamic disturbances from wind vibrations and the telescope jitters. The tip-tilt system requires a fast tip-tilt of ±20 arc-seconds at a bandwidth of 10 Hz with a goal of 20 Hz. To capture the tip-tilt motions, a sample simulation as a combination of unit cases was considered. As a unit case, unit axial displacement of 1mm was applied at each of the three axial actuators for a tip-tilt motion. The results were discussed in the previous paper [1].

The FSM mirror cell must be stiff enough to sustain static gravity and dynamic operations. Current mirror cell design assumes thin walled aluminum plates for most parts and reinforced sections at the parts under direct load paths along the hexapod interface. Total estimated mass in the FE model is approximately 190kg excluding a reaction mass system for the tip-tilt mechanism which is under development. In order to achieve the design goals, static analyses with gravity induced deformation and thermal distortions were performed for the mirror cell assembly. The static deformations and stress levels are relatively small and favorably met the design goals. Current design goal in the cell structural deformation is an overall displacement of 100 microns including the effects from gravity and thermal loadings. The goal will be refined and adjusted as the cell assembly design evolves and the tip-tilt system becomes mature. Since we are now engaged in a new mirror cell design for robustness with a direct load path and compactness, this work is in progress and its performance will be evaluated later. With the previous cell design, natural frequencies of the mirror cell were calculated. The first 10 significant modes were obtained excluding the lowest mode, torsion mirror clocking mode shape. The fundamental structural mode (2nd mode) occurred at 82Hz which is in an astigmatic shape of the cell. This result indicated that the cell has an excellent dynamic stiffness for the tip-tilt motion at the goal of 60Hz.

5. LATERAL FLEXURE TEST

In order to evaluate the performance of the lateral flexures, we have performed finite element analyses on several flexure options. These analyses include the static gravity, nonlinear behaviors, thermal effects, and buckling analysis. In this Chapter, two optimized flexures were tested to validate the lateral flexure design. The two test mirror setups are: (1) a membrane flexure (diaphragm) bonded to a test mirror using epoxy with a bondline thickness of 250μm; (2) a flexure used in the Magellan M2 bonded to a mirror using RTV with a bondline thickness of 1mm. For the nature of these tests conducting at the optical lab in NOAO, we did not intend to capture the thermal characteristics of bonding because of the test setup at room temperature. These two test mirrors are shown in Figure 17. The main purpose of this test is to validate the design concept for lateral stiffness and its axial stiffness.

 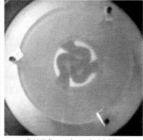

Fig. 17. (a) Lateral flexure bonded to a test mirror with epoxy (V1 model). Grey areas forming a circle are epoxy layer of which thickness is about 250μm. (b) Lateral lexure bonded to a mirror with RTV (V3 model). The RTV layer is 1mm thick between flexures outer rim and mirror the side of inner hole.

 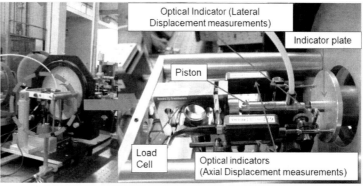

Fig. 18. (a) Experiment set-up to measure lateral stiffness; (1) air compressor and heavy duty piston up to 1000N. (2) a thick aluminum plate, (3) a linear digital gauge, (4) mirror with lateral flexure, (5) a gauge tip contacting very lower part of the central hub. (b) Axial test setup including target mirror, gimbal mount, interferometer and air baffle, a load frame - mechanical structures on the back of the gimbal mount with force controlling components and measuring instruments. Two LVDT indicator measure displacement of the indicator plate at the very edge and one LVDT catches up-down displacement of the plate. The small piston worked by air pressure provides varying forces regulated by a control software which monitors Load cell readings.

Two test configurations were prepared, which are the lateral test and the axial test. For the lateral test, a maximum force of 1000N was applied to simulate the FSM mass of 100kg at an increment of 10N. This lateral test is intended to verify the mechanical behaviors from the entire mirror mass (Figure 18 (a)). For the axial test of the lateral flexures, a maximum force of 100N was applied. In the design of axial stiffness in the lateral flexures, the force range of interest is approximately 10N. However, we conducted the axial test for an extended axial force up to 100N with an increment of 10N to monitor for any extraordinary behavior. Figure 18(b) illustrates the axial test setup including the test mirror, gimbal mount, and interferometer. Total of three LVDT were installed; two LVDT indicators measure displacement of the indicator plate at the very edge and one LVDT catches up-down displacement of the plate.

We established test setups for mechanical and optical measurements in order to evaluate the flexure behaviors and the optical surface errors resulting from the loads. The mechanical measurements for the lateral and axial stiffness of lateral flexures were summarized and the results were compared with FE models. Figure 19(a) shows the axial Stiffness (force to displacement) of the RTV flexure. The axial stiffness results are in good agreement with the non-linear FE model (less than 10% variation). In addition, the lateral stiffness of the Epoxy flexure was recorded, the results indicate a good agreement between the measurements and the off-plane (offset) loading FE model (20% variation), as shown in Figure 19(b). Since the test setup has an offset between the lateral loading point and line of action in the flexure, we anticipated a moment due to the offset.

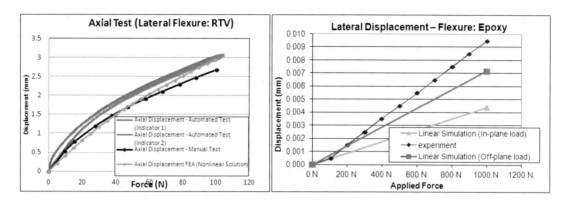

Fig. 19. (a) Axial Stiffness (force to displacement) of the RTV flexure, results indicated a good agreement with non-linear FE model; (b) Lateral Stiffness of the Epoxy flexure, results indicated a good agreement with FE model (20% variation with FE off-plane load).

6. SUMMARY

We have performed extensive finite element analyses and optical calculations for the optimized Fast Steering Secondary Mirror (FSM) and its support system. The performance predictions of the FSM were based on a FSM mirror configuration with a diameter of 1.058m (on-axis), 1.048m (off-axis); depth of 120mm; face plate thickness of 20mm; and mass of approximately 90kg. In this report, the depth of FSM of 120mm was chosen in order to minimize interference with the incoming beam path. The optical surface deformations for various Zenith angles were evaluated by combining cases of the effects from axial and lateral gravities. The results indicated that the current GMT FSM mirror and its support system adequately met the optical performance goal of 20nm surface RMS, and also satisfies the FSM surface figure accuracy requirement defined at EE80. That is, for Axial EE80 = 0.012 arc-seconds (< 0.020) and Lateral EE80 = 0.006 arc-seconds (< 0.020). The image quality calculations at a fully populated system with seven segments are in progress.

With the latest GMT optical prescriptions and changes in FSM configurations, currently validation and verification work activities are in progress. Sensitivity analyses will be conducted with several sample cases to quantify the optical surface deformations affected by uncertainties in design and potential errors involved in polishing, assembly and system integrations. Tip-tilt simulations will be performed to capture the impacts from tip-tilt motions. The mirror cell assembly indicated an excellent dynamic stiffness for the tip-tilt motion at the goal of 60 Hz. Integrated FE models with the mirror, supports, and mirror cell structure need to be established for further optimizations to refine design parameters of the mirror cell and support systems. A high fidelity finite element model will be required to evaluate more extensive sensitivity cases, structural interaction effects, thermal mismatches, and other opto-mechanical effects. The FE model may include features of support pads, mounting blocks, linkage, and other detail hardware parts which may contribute to mechanical and optical performance degradation. Upon the previous study, the results from the earlier baseline model indicated that no significant impacts exist on the mechanical performance between different flexure configurations. Trade study between the lateral diaphragm flexure and Magellan type configuration should be continued for the merits among the performance, cost, and risk.

ACKNOWLEDGMENTS

This research was carried out at the National Optical Astronomy Observatory, and was sponsored in part by the GMT. The authors gratefully acknowledge the support of the GMT Project Office. This work was partially contributed by the scientists, engineers, and students from KASI (KASI, KRISS, IAE, GIST) in Korea which is one of the key partners of the GMT.

REFERENCES

[1] Cho, M., Corredor, A., Dribusch, C., A., Park, K., Kim, Y., Moon, I., "Design and Development of a Fast Steering Secondary Mirror for the Giant Magellan Telescope," Proc. SPIE 8125, (2011).

[2] Wilson, R., "Fourier Series and Optical Transform Techniques in Contemporary Optics," Willey Intersc., (1995).

[3] Cho, M. and Richard, R.M., "PCFRINGE Program – Optical Performance Analysis using Structural Deflections and Optical Test Data," the Optical Sciences Center, University of Arizona, 35, (1990).

[4] GMT Office, "GMT Image Size and Wavefront Error Budgets," GMTO, August, (2007).

[5] GMT Office, "The F/11 Secondary Mirror of Magellan telescope," GMTO, August, (2002).

[6] Cho, M., "Performance Prediction of the TMT Secondary Mirror Support System," Proc. SPIE 7018-65, (2008).

[7] Cho, M., Corredor, A., Park, K., "K-GMT FSMP CDR," KASI, Korea, (2011)

[8] Cho, M., Corredor, A., Vogiatzis, K., and Angeli, G., "Thermal Performance Prediction of the TMT Optics," Proc. SPIE 7017-43, (2008).

[9] Blanco, D., Cho, M., Daggert, L., Daly, P., DeVries, J., Elias, J., Fitz-Patrick, B., Hileman, E., Hunten, M., Liang, Nickerson, M., Pearson, E., Rosin, D., Sirota, M. and Stepp, L., "Control and support of 4-meter class secondary and tertiary mirrors for the Thirty Meter Telescope," Proc. SPIE 6273-65, (2006).

[10] Williams, Eric, et. al., "Advancement of the Segment Support System for the Thirty Meter Telescope Primary Mirror," Proc. SPIE 7018-37, (2008).

Dynamics, Active Optics and Scale Effects in Future Extremely Large Telescopes

Renaud Bastaits[a], Bilal Mokrani[a], Gonçalo Rodrigues[b], André Preumont[a]

[a]Active Structures Laboratory, Université Libre de Bruxelles, Brussels, Belgium
[b]ESA-ESTEC, Structures Section, Noordwijk, The Netherlands

ABSTRACT

This paper examines the active optics of future large segmented telescopes from the point of view of dynamic simulation and control. The first part of the paper is devoted to the modelling of the mirror. The model has a moderate size and separates the quasi-static behavior of the mirror (primary response) from the dynamic response (secondary or residual response). The second part of the paper is devoted to control. The control strategy considers explicitly the primary response of the telescope through a singular value controller. The control-structure interaction is addressed with the general robustness theory of multivariable feedback systems, where the secondary response is considered as uncertainty. This approach is very fast and allows extensive parametric studies. The study is illustrated with an example involving 90 segments, 270 inputs, and 654 outputs.

Keywords: ELTs, Active Optics, Scale effects, Control-Structure interaction

1. INTRODUCTION

Keck (located in Hawaï) and the GTC (located in the Cannary Islands) are the largest optical telescopes in operation, with a diameter of the primary mirror (M1) a little larger than 10 m. There are plans to build Extremely Large Telescopes (ELTs), Fig.1, with a primary mirror of diameter up to 42 m (E-ELT), and even larger ones are foreseen in a more distant future, with a diameter up to 100 m (OWL). Table 1 compares the main characteristics of the Keck and the European E-ELT telescopes. Note that the collecting area (and also the area exposed to wind) of E-ELT is 15 times larger than that of Keck, and that the natural frequency of the first mode of the primary mirror is 4 times smaller; this has an important impact on the structural response of the telescopes. As their size increases, large telescopes become increasingly sensitive to external disturbances such as thermal gradients, gravity and wind, and also to internal disturbances from support equipments such as pumps, cryocoolers, fans,... Because even minute vibrations can deteriorate significantly the image quality, there are legitimate concerns about the feasibility of extremely large telescopes, in spite of several active control layers known as *active optics* (which handles the large amplitude, low frequency disturbances coming from the thermal gradients and gravity, and also part of the wind response) and *adaptive optics* (which uses a smaller, deformable mirror to compensate the low amplitude, high frequency disturbances to the wave front, mostly due to atmospheric turbulence).

Monolithic mirrors of diameter larger than 8 m are difficult to manufacture.[1] As a result, all future large telescopes will be segmented. Currently, telescopes of 30 m and more are under design,[2,3] involving several hundreds of segments. Figure 2 shows the primary mirror (M1) of the European extremely large telescope (E-ELT); it has a diameter of 42 m and consists of 984 aspherical segments; every segment is equipped with three two-stage position actuators, controlling the piston and the two tilts, and six edge sensors measuring the position of the segment with respect to its six neighbors. The motion of the position actuators is transmitted to the segment through a whiffletree. Overall, there are 2952 position actuators **a** and 5604 edge sensors y_1.

Further author information:
A. Preumont: E-mail: apreumont@ulb.ac.be
R. Bastaits: E-mail: rbastait@ulb.ac.be

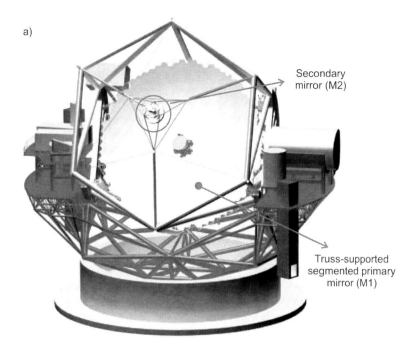

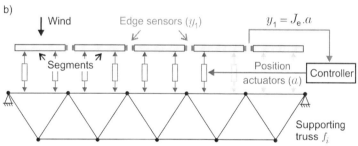

FIGURE 1. (a) View of an extremely large telescope (the one shown is the future TMT). (b) Schematic view of the segmented primary mirror, the supporting truss and the actuators and sensors involved in the active optics. The quasi-static behavior of the reflector follows $\mathbf{y}_1 = \mathbf{J}_e \cdot \mathbf{a}$ where $\mathbf{a}$ is the control input (position actuators), $\mathbf{y}_1$ is the edge sensor output, and $\mathbf{J}_e$ is the Jacobian of the segmented mirror.

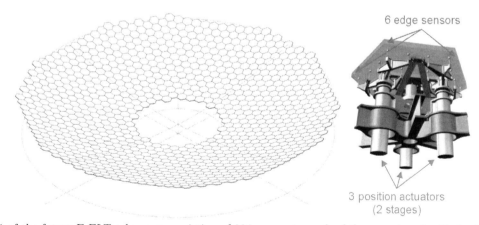

FIGURE 2. M1 of the future E-ELT telescope consisting of 984 segments, each of them equipped with 6 edge sensors and 3 two-stage position actuators. The segments are connected to the actuators by a whiffletree.[4]

TABLE 1. Main characteristics of the Keck and of the future E-ELT telescopes.

	Keck	E-ELT
M1 diameter : D	10 m	42 m
Segment size	1.8 m	1.4 m
Collecting Area	76 m^2	1250 m^2
# Segments : N	36	984
# Actuators	108	2952
# Edge Sensors	168	5604
$f_{segment}$ (+ whiffletree)	25 Hz	$\sim$ 60 Hz
f_1 (M_1)	$\sim$ 10 Hz	$\sim$ 2.5 Hz
f_2 (M_2)	$\sim$ 5 Hz	$\sim$ 1-2 Hz
Adaptive Optics # d.o.f.	$\sim$ 350	$\sim$ 8000
Tube and mount mass	$\sim$ 110 t	$\sim$ 2000 t

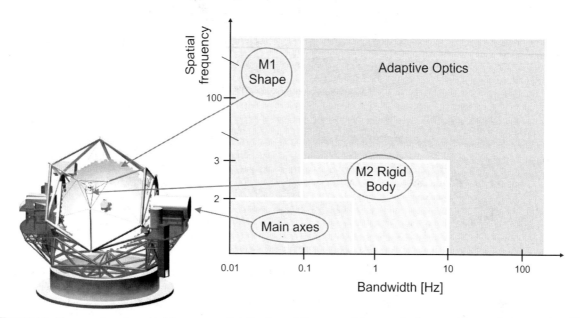

FIGURE 3. Temporal and spatial frequency distribution of the control layers of a large telescope (adapted from[5]).

Figure 3 describes the temporal and spatial frequency distribution of the various layers of the control system involved in the wave front correction of a large telescope[5]; the spatial frequency is expressed in terms of Zernike modes. In general, adaptive optics operates on a smaller deformable mirror and the amplitudes are small, typically a few microns. Our discussion is focused on M1; the amplitudes to be corrected by the active optics are typically several hundred microns.[6]

Modelling for control and controlling large complex active structures such as telescopes poses several challenges : (1) Finite element techniques classically used in structural modelling tend to use a very large number of degrees of freedom that reflect the complex geometry of the structural components. The size of the control model (used in the control design and to evaluate the control-structure interaction) must be drastically reduced while preserving the main features of the system, statically and dynamically. The control model must give access to all actuator inputs, sensor outputs, and optical performance metrics, and it is essential to preserve the kinematic relationship between the position actuators and the quasi-static position of the segmented mirror (primary response). (2) The control algorithm must be simple enough to be implemented in real time (in spite of the large number of inputs and outputs), provide enough gain in low frequency to achieve performance, and enough roll off outside the bandwidth to reduce spillover. A reliable lower bound for the stability margin should be evaluated from the control model, and, if possible, the critical mode(s) for stability should be identified. The control model should allow extensive parametric studies and sensitivity analysis. These challenges have only been partially covered by the existing literature.[4,7–11]

The objective of this study is threefold : (1) To develop a representative numerical model of the segmented mirror and its supporting truss that can be used for control design and robustness evaluation; this model should have a minimum complexity to allow extensive parametric studies and control-structure interaction. In a later stage, we intend to also use it for wind response calculation and optical performance evaluation. (2) To implement a control strategy based on the primary response of the mirror and two sensor arrays, the edge sensors measuring the relative displacements of adjacent mirrors and the normal to the segments. (3) To examine the interaction between the controller and the structural dynamics with various multivariable robustness tests in the frequency domain, where the secondary response is considered as uncertainty, and evaluate the requirements in terms of frequency and damping for the supporting truss to guarantee the performance, control bandwidth, and an appropriate stability margin for the control system.

2. QUASI-STATIC CONTROL

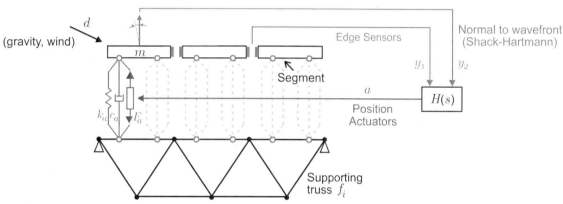

FIGURE 4. Active optics control flow for large segmented mirrors. Only the axial d.o.f. at both ends of the actuators are kept after a Craig-Bampton reduction is performed.

The numerical model of the primary mirror and its control system is represented schematically in Fig.4. The shape control of M1 is based on position actuators of displacement vector **a** and on two sets of sensors : The edge sensors $\mathbf{y_1}$ measuring the relative height between adjacent segments, and the normal sensors $\mathbf{y_2}$ measuring the orientation of the segments (measured by an optical wavefront sensor such as a Shack Hartmann). Each position actuator is represented by a force $\mathbf{F}_a$ acting on a spring k_a that is taken as the stiffness of the whiffletree; the force is related to the unconstrained displacement by $\mathbf{a} = \mathbf{F}_a/k_a$. In this way, the local flexibility of the whiffletree is well represented in the system, which is very important if one wants to simulate their effect in the control-structure interaction (this turned out to be essential in Keck that experienced control-structure interaction with its segments, with a local resonance at 25 Hz; later designs use stiffer whiffletrees). The position actuators rest on a supporting truss carrying the whole mirror. The disturbances **d** applied to the system come from thermal gradients, changing gravity vector with the elevation of the telescope, and wind.

Currently, similarly to the Keck telescope,[14] the control strategy envisaged by ESO for co-phasing the segments of the future E-ELT[4] assumes that the supporting truss is rigid and that the natural frequency of the whiffletree (connecting the actuators to the segments) is well above the bandwidth of the control system (which is realistic). In this case, the behavior of the segmented mirror is assumed quasi-static (the influence of the residual response is supposed negligible) and the relationship between the actuator displacements **a** and the sensors output **y** depends only on the geometry of the actuator and sensor arrays :

$$\mathbf{y} = \begin{pmatrix} \mathbf{y_1} \\ \mathbf{y_2} \end{pmatrix} = \begin{pmatrix} \mathbf{J}_e \\ \mathbf{J}_n \end{pmatrix} \mathbf{a} = \mathbf{J}\mathbf{a} \qquad (1)$$

where $\mathbf{J}_e$ is the Jacobian of the edge sensors, $\mathbf{J}_n$ the Jacobian of the normal to the segments and $\mathbf{J}$ the extended Jacobian. The pseudo-inverse of the Jacobian $\mathbf{J}^+$ is best obtained by singular value decomposition (SVD) :

$$\mathbf{J} = \mathbf{U\Sigma V}^T \qquad (2)$$

where the column of **U** are the orthonormalized sensor modes, the column of **V** are the orthonormalized actuator modes, and $\mathbf{\Sigma}$ is a rectangular matrix which contains the singular values σ_i on its diagonal. The control system works according to Fig.5, called SVD controller. The diagonal matrix $\mathbf{\Sigma}^{-1}$ contains the inverse of the singular values, σ_i^{-1}, on its diagonal; it provides an equal authority on all singular value modes. Only the modes with non-zero singular values are considered in this control block. The set of filters $\mathbf{H}(s)$ provide adequate disturbance rejection and stability margins (we will come back to this). $\mathbf{H}(s)$ may be a scalar function if the same loop shaping is applied to all SVD modes; in this case, all the loops have essentially the same control gain.

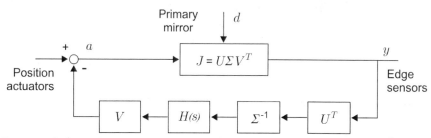

FIGURE 5. Block diagram of the cophasing control system (SVD controller). The block $\mathbf{\Sigma}^{-1}$ is limited to the non-zero singular values.[8]

Typical distributions of the singular values of the Jacobian of the edge sensor $\mathbf{J}_e$ and the extended Jacobian **J**, ranked by increasing order, are compared in Fig.6. The lowest singular values of $\mathbf{J}_e$ correspond to the lowest optical modes, piston, tilt and defocus, which are unobservable from the edge sensors, and the modes with the lowest singular values next to them are astigmatism, trefoil and coma. This is not very good from an optical control viewpoint and illustrates the requirement for the full control system to include another set of sensors, $\mathbf{y}_2$, measuring the normal to the segments. One also sees that, except for the global piston mode which is not observable (in practice, it can be solved by locking any group of three position actuators to define a reference plane), the singular values of the extended Jacobian **J** are quite well conditioned and ready for implementation according to Fig.5.

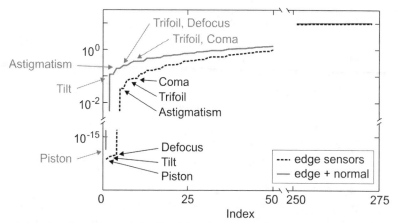

FIGURE 6. Comparison of the singular values of the Jacobian $\mathbf{J}_e$ of the edge sensors with the extended Jacobian **J** (edge sensor measurements and displacement actuators are expressed in meters, tilt angles in radians).

3. STRUCTURAL DYNAMICS

The dynamics of the mirror consists of global modes involving the supporting truss and the segments, and local modes involving the segments alone. The global modes are critical for the control-structure interaction;

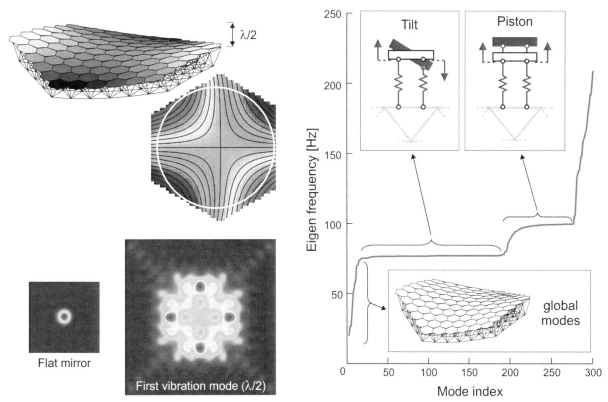

FIGURE 7. Segmented mirror consisting of 91 segments supported by a truss. Left : Effect of a low frequency vibration mode on the PSF produced by an equivalent mirror. The diameter of the segments is $d = 2$ m. The PSF corresponds to the largest circular aperture inside the mirror, when it is flat and when the vibration amplitude is $\lambda/2$ (peak to valley) Right : Eigen frequency distribution of the full FE model.

the segments are designed in such a way that their local modes have frequencies far above the critical frequency range, but their quasi-static response (to the actuator as well as to gravity and wind disturbances) must be dealt with accurately. In order to handle large optical configurations, it is important to reduce the model as much as possible, without losing the features mentioned before.

The starting point is a standard finite element model undergoing a Craig-Bampton reduction.[12,13] We choose to retain the axial DOF at both ends of the position actuators (Fig.4). For a segmented mirror with N segments, the size of $\mathbf{x}_1$ is thus $\sim 6N$. Because of the way they have been selected, the reduced coordinates $\mathbf{x}_1$ describe fully the rigid body motion of the segments and the sensor output can be expressed by

$$\mathbf{y}_1 = \mathbf{S}_{y_1}\mathbf{x}_1 \qquad (3)$$

$$\mathbf{y}_2 = \mathbf{S}_{y_2}\mathbf{x}_1 \qquad (4)$$

where the matrices $\mathbf{S}_{y_1}$ and $\mathbf{S}_{y_2}$ describe the topology of the sensor arrays.

3.1 Modal analysis

In order to illustrate our analysis, consider the segmented mirror of Fig.7, consisting of 91 segments supported by a truss. The mass is assumed to be distributed equally between the segments, 50 kg each, and the truss, 4550 kg. The actuator stiffness k_a has been selected in order that the first piston mode of the segments is 100 Hz. The truss stiffness has been chosen so that the first global mode is $f_1 = 20$ Hz. The modal damping is assumed uniformly $\xi_i = 0.01$. To illustrate how a tiny vibration can deteriorate the image quality, Fig.7 (left) shows how a vibration mode (which is essentially astigmatism) affects the optical performance of the mirror; the Point

Spread Function (PSF) corresponds to the largest circular aperture inside the mirror, when it is flat and when the vibration amplitude is $\lambda/2$ (peak to valley). Figure 7 also shows the eigen frequency distribution of the full FE model; the first 20 modes or so are global modes; their mode shapes are a combination of optical aberration modes of low order. Next follow the local modes of the segments (tilt near 75 Hz and piston near 100 Hz). For a large segmented mirror, even a reduced model is far too complex to be included in full in a control-structure interaction analysis, and only the low frequency modes (which can potentially jeopardize the stability) really matter. The reduced model is thus truncated, but the static behavior is not altered by the truncation.

3.2 Static response

The static response of the system is[15]

$$\mathbf{x}_1 = \hat{\mathbf{K}}_{11}^{-1}[\mathbf{S}_a \mathbf{F}_a + \mathbf{d}] \tag{5}$$

$$\mathbf{y}_1 = \mathbf{S}_{y_1}\mathbf{x}_1 = \mathbf{S}_{y_1}\hat{\mathbf{K}}_{11}^{-1}[\mathbf{S}_a\mathbf{F}_a + \mathbf{d}] = \mathbf{S}_{y_1}\hat{\mathbf{K}}_{11}^{-1}\mathbf{S}_a k_a \mathbf{a} + \mathbf{S}_{y_1}\hat{\mathbf{K}}_{11}^{-1}\mathbf{d} \tag{6}$$

Comparing this equation with Eq.(1), one gets

$$\mathbf{J}_e = \mathbf{S}_{y_1}\hat{\mathbf{K}}_{11}^{-1}\mathbf{S}_a k_a \tag{7}$$

Similarly,

$$\mathbf{y}_2 = \mathbf{S}_{y_2}\mathbf{x}_2 = \mathbf{S}_{y_2}\hat{\mathbf{K}}_{11}^{-1}[\mathbf{S}_a\mathbf{F}_a + \mathbf{d}] = \mathbf{S}_{y_2}\hat{\mathbf{K}}_{11}^{-1}\mathbf{S}_a k_a \mathbf{a} + \mathbf{S}_{y_2}\hat{\mathbf{K}}_{11}^{-1}\mathbf{d} \tag{8}$$

$$\mathbf{J}_n = \mathbf{S}_{y_2}\hat{\mathbf{K}}_{11}^{-1}\mathbf{S}_a k_a \tag{9}$$

Both $\mathbf{J}_e$ and $\mathbf{J}_n$ can be obtained geometrically without resorting to the FE model.

3.3 Dynamic response in modal coordinates

Assume that the eigenvalue problem has been solved for the reduced system and that the eigen modes have frequencies ω_i and have been normalized to a unit modal mass. Let ϕ_i be the partition of the eigen modes corresponding to the boundary d.o.f. $\mathbf{x}_1$. Because the control force and the disturbance are only applied to those d.o.f., the equation governing the dynamic response of mode i is

$$\ddot{z}_i + 2\xi_i\omega_i\dot{z}_i + \omega_i^2 z_i = \phi_i^T \mathbf{S}_a \mathbf{F}_a + \phi_i^T \mathbf{d} \quad (i = 1, \ldots, m) \tag{10}$$

However, only the m lowest frequency modes, within or close to the bandwidth of the disturbance respond dynamically; the higher ones ($i > m$) respond in a quasi-static manner and may be regarded as a singular perturbation. The static response of the previous section includes all modes, and if a flexible modes is accounted for dynamically, one must remove its contribution from the flexibility matrix; this is obtained by subtracting from $\hat{\mathbf{K}}_{11}^{-1}$ the contribution $\mathbf{F}_m$ of the m modes which respond dynamically according to Eq.(10):

$$\mathbf{F}_m = \sum_{i=1}^{m} \frac{\phi_i \phi_i^T}{\omega_i^2} \tag{11}$$

Thus, the flexibility matrix of the high frequency modes is $\hat{\mathbf{K}}_{11}^{-1} - \mathbf{F}_m$ (depending on m) [,[16] p.22] and the overall dynamic response at the d.o.f. $\mathbf{x}_1$ reads

$$\mathbf{x}_1 = \mathbf{\Phi}_m \mathbf{z} + [\hat{\mathbf{K}}_{11}^{-1} - \mathbf{F}_m][\mathbf{S}_a \mathbf{F}_a + \mathbf{d}] \tag{12}$$

where the components of **z** are solutions of Eq.(10) (dynamic response), $\mathbf{\Phi}_m = (\phi_1, \ldots, \phi_m)$ is the matrix of mode shapes at the boundary d.o.f., and the second term is the quasi-static response (singular perturbation) of all modes beyond m. From Eq.(3), the edge sensor output $\mathbf{y}_1$ is

$$\mathbf{y}_1 = \mathbf{S}_{y_1}\mathbf{\Phi}_m\mathbf{z} + [\mathbf{J}_e - \mathbf{S}_{y_1}\mathbf{F}_m\mathbf{S}_a k_a]\mathbf{a} + \mathbf{S}_{y_1}[\hat{\mathbf{K}}_{11}^{-1} - \mathbf{F}_m]\mathbf{d} \qquad (13)$$

where Eq.(7) has been used. Similarly, from Eq.(4), the normal sensor output $\mathbf{y}_2$ reads

$$\mathbf{y}_2 = \mathbf{S}_{y_2}\mathbf{\Phi}_m\mathbf{z} + [\mathbf{J}_n - \mathbf{S}_{y_2}\mathbf{F}_m\mathbf{S}_a k_a]\mathbf{a} + \mathbf{S}_{y_2}[\hat{\mathbf{K}}_{11}^{-1} - \mathbf{F}_m]\mathbf{d} \qquad (14)$$

after using Eq.(9). In these equations, $\mathbf{S}_{y_1}\mathbf{\Phi}_m$ and $\mathbf{S}_{y_2}\mathbf{\Phi}_m$ are the modal components, for the first m modes of the edge sensor output and the normal sensor output, respectively. Equations (13) and (14) are reduced to (6) and (8) if none of the modes respond dynamically ($m = 0$), which confirms that the reduced model in modal coordinates does convey the full kinematic relationship between the position actuators and the two sets of sensors (primary response). The matrices describing the structural response can be reorganized as shown in Fig.8 to decompose the input-output relationship in terms of the primary response $\mathbf{G}_0(s) = \mathbf{J}$ and the residual response $\mathbf{G}_R(s)$; the latter is not taken into account in the controller structure and is regarded as an uncertainty.

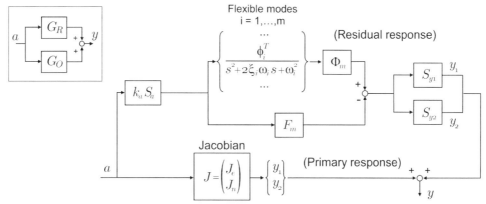

FIGURE 8. Input-output relationship of the segmented mirror. The nominal plant $\mathbf{G}_0(s) = \mathbf{J}$ accounts for the quasi-static response (primary response) and the dynamic deviation $\mathbf{G}_R(s)$ is regarded as an additive uncertainty (residual response).

4. LOOP SHAPING OF THE SVD CONTROLLER

The SVD controller has the form of Fig.5 and, if one assumes the same loop shape for all singular value modes, $\mathbf{H}(s) = \mathbf{I}h(s)$, the controller

$$\mathbf{K}(s) = h(s)\mathbf{V}\mathbf{\Sigma}^{-1}\mathbf{U}^T \qquad (15)$$

essentially inverts the Jacobian of the mirror, leading to $\mathbf{G}_0(s)\mathbf{K}(s)$ being diagonal, with all non-zero singular values being equal to

$$\sigma(\mathbf{G}_0\mathbf{K}) = |h(j\omega)| \qquad (16)$$

Thus, the loop shaping can be done as a SISO controller, according to classical techniques (,[17] p.129) or frequency shaping using the Bode integrals (,[18] p.74). The control objective is to maximize the loop gain in the frequency band where the disturbance has a significant energy content while keeping the decay slow enough near crossover to achieve a good phase margin. An integral component is necessary to eliminate the static error in the mirror shape and a large gain at the earth rotation frequency to compensate for the gravity deformations.

Fig.9 shows the Bode plots and the Nichols chart of the controller used in this study; the compensator $h(s)$ consists of an integrator, a lag filter, followed by a lead and a second order Butterworth filter. The crossover is $f_c = 0.25$ Hz and the attenuation at the earth rotation frequency is 125 dB. The robustness margins (of the quasi-static controller) are clearly visible on the Nichols chart [the exclusion zone around the critical point $(-180^0, 0$ dB) corresponds to a phase margin $PM = \pm 45^0$ and a gain margin $GM = \pm 10$ dB]. However, the Nichols chart does not say anything about the control-structure interaction since the flexibility of the supporting truss was ignored in the controller design.

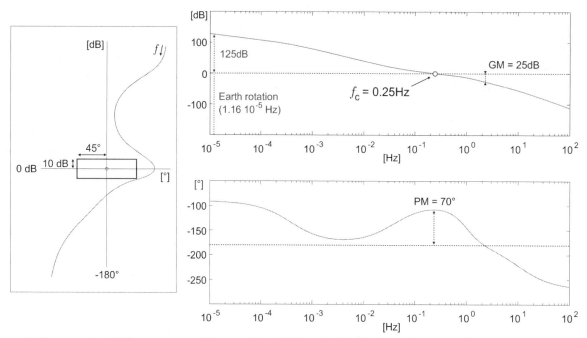

FIGURE 9. Compensator $h(s)$ common to all loops of the SVD controller. Left : Nichols chart [the exclusion zone around the critical point $(-180^0, 0$ dB) corresponds to $(PM = \pm 45^0, GM = \pm 10$ dB)]. Right : Bode plots showing the gain of 125 dB at the earth rotation frequency and a cut-off frequency of $f_c = 0.25$ Hz. The Nichols chart is invariant with respect to a shift of the Bode amplitude and phase diagrams along the frequency axis.

5. CONTROL-STRUCTURE INTERACTION

From Fig.5, if the mirror would respond in a quasi-static kinematic manner, the controller transfer matrix would essentially invert that of the mirror. However, because the response of the mirror includes a dynamic contribution at the frequency of the lowest structural modes and above, the system behaves according to Fig.8 and the robustness with respect to control-structure interaction must be examined with care.[19,20] The structure of the control system is that of Fig.10.a, where the primary response $\mathbf{G}_0(s)$ corresponds to the quasi-static response described earlier and the residual response $\mathbf{G}_R(s)$ is the deviation resulting from the dynamic amplification of the flexible modes; $\mathbf{K}(s)$ is the controller.

The control-structure interaction may be addressed with the general robustness theory of multivariable feedback systems,[21–23] with the residual response being considered as uncertainty, either multiplicative or additive.

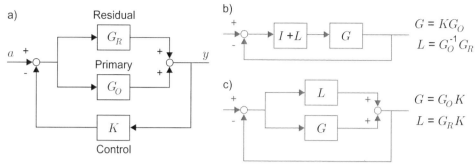

FIGURE 10. Block diagram of the control system (a) Mirror represented by its primary and residual dynamics. (b) multiplicative uncertainty, and (c) additive uncertainty.

5.1 Multiplicative uncertainty

For a multiplicative uncertainty, the standard structure of Fig.10.b applies with $\mathbf{G}(s) = \mathbf{K}(s)\mathbf{G}_0(s)$ and $\mathbf{L}(s) = \mathbf{G}_0^{-1}(s)\mathbf{G}_R(s)$. The general theory of MIMO systems shows that a sufficient condition for stability is that [*]

$$\bar{\sigma}[\mathbf{L}(j\omega)] < \underline{\sigma}[I + \mathbf{G}^{-1}(j\omega)] \quad , \quad \forall \, \omega > 0 \tag{17}$$

($\bar{\sigma}$ and $\underline{\sigma}$ stand respectively for the maximum and the minimum singular value), which is transformed here into

$$\bar{\sigma}[\mathbf{G}_0^{-1}\mathbf{G}_R(j\omega)] < \underline{\sigma}[I + (\mathbf{K}\mathbf{G}_0)^{-1}(j\omega)] \quad , \quad \forall \, \omega > 0 \tag{18}$$

This test is quite meaningful and is illustrated in Fig.11.a. The left hand side is independent of the controller; it starts from 0 at low frequency where the residual dynamics is negligible and increases gradually when the frequency approaches the flexible modes of the mirror structure, which are not included in the nominal model $\mathbf{G}_0$; the amplitude is maximum at the resonance frequencies where it is only limited by the structural damping. The right hand side starts from unity at low frequency where $|\mathbf{K}\mathbf{G}_0| \gg 1$ ($\mathbf{K}\mathbf{G}_0$ controls the performance of the control system) and grows larger than 1 outside the bandwidth of the control system where the system rolls off ($|\mathbf{K}\mathbf{G}_0| \ll 1$).

The critical point A corresponds to the closest distance between these curves. The vertical distance between A and the upper curve should not be smaller than the requested gain margin GM. When the natural frequency of the structure changes from f_1 to f_1^*, point A moves horizontally according to the ratio f_1^*/f_1 (increasing the frequency will move A to the right). Similarly, changing the damping ratio from ξ_1 to ξ_1^* will change the amplitude according to ξ_1/ξ_1^* (increasing the damping will decrease the amplitude of A).

5.2 Additive uncertainty

For an additive uncertainty, the standard structure of Fig.10.c applies with $\mathbf{G}(s) = \mathbf{G}_0(s)\mathbf{K}(s)$ and $\mathbf{L}(s) = \mathbf{G}_R(s)\mathbf{K}(s)$; a sufficient condition for stability is that

$$\bar{\sigma}[\mathbf{L}(j\omega)] < \underline{\sigma}[I + \mathbf{G}(j\omega)] \quad , \quad \forall \, \omega > 0 \tag{19}$$

which is translated into

$$\bar{\sigma}[\mathbf{G}_R\mathbf{K}(j\omega)] < \underline{\sigma}[I + \mathbf{G}_0\mathbf{K}(j\omega)] \quad , \quad \forall \, \omega > 0 \tag{20}$$

A typical result for this robustness test is depicted in Fig.11.b; in this case, both terms depend on the controller. The robustness conditions in Eq.(18) and (20) come from small-gain theorem; being sufficient conditions, they are both conservative and one may be more conservative than the other.

---

*. The inverse of rectangular matrices should be understood in the sense of pseudo-inverse.

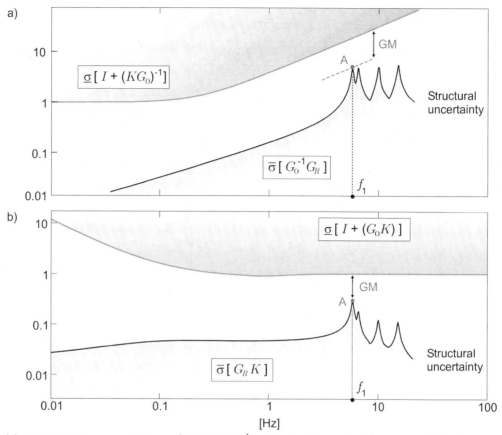

FIGURE 11. (a) Multiplicative uncertainty : $\underline{\sigma}[I + (\mathbf{KG}_0)^{-1}]$ refers to the nominal system used in the controller design. $\bar{\sigma}[\mathbf{G}_0^{-1}\mathbf{G}_R]$ is an upper bound to the relative magnitude of the residual dynamics - (b) Additive uncertainty : $\underline{\sigma}[I + \mathbf{G}_0\mathbf{K}]$ refers to the nominal system used in the controller design. $\bar{\sigma}[\mathbf{G}_R\mathbf{K}]$ is an upper bound to the effect of the controller on the residual dynamics. In both tests, the critical point A corresponds to the closest distance between the nominal system and the structural uncertainty. The vertical distance between A and the upper curve is a lower bound to the gain stability margin.

6. DISCUSSION

To illustrate the foregoing discussion, consider again the segmented mirror of Fig.7. Although this example does not correspond to a particular existing telescope, it is representative of the current generation of large telescopes (Keck, GTC); the parameters have been chosen so that $f_1 = 20$ Hz and $\xi_i = 0.01$ uniformly. The SVD controller is that of Fig.9 with a crossover frequency $f_c = 0.25$ Hz (we do not distinguish between the bandwidth and the crossover frequency).

Figure 12.a shows the robustness test (18); with a ratio $f_1/f_c \simeq 80$ the system exhibits a substantial gain margin of 45 dB with respect to control structure interaction. Figure 12.b shows the stability limit when A touches the upper curve; $f_1 = 2.2$ Hz in this case. The same procedure is followed for the robustness test (20) shown in Fig.13.a, for $f_1 = 20$ Hz. The system exhibits a lower but still substantial gain margin of 36 dB with respect to control structure interaction. Figure 13.b shows the stability limit when A touches the upper curve, for $f_1 = 3.7$ Hz. Therefore, the test based on multiplicative uncertainty appears to be less conservative than that based on additive uncertainty and is therefore more relevant in this context.

Figure 14 shows the evolution of the gain margin with the frequency ratio f_1/f_c for various values of the damping ratio ξ in the case of the multiplicative uncertainty. This curve is obtained by gradually softening the Young's modulus of the supporting structure. If the first structural mode of the supporting structure has a damping ratio of 1%, a gain margin $GM = 10$ requires a frequency separation f_1/f_c significantly larger than one decade.

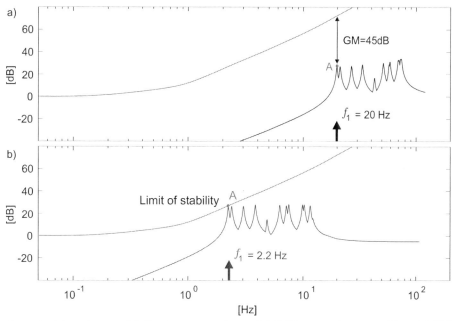

FIGURE 12. Robustness test based on multiplicative uncertainty : (a) Stiff supporting truss ($f_1 = 20$ Hz) ; the gain margin is $GM = 45$ dB - (b) Soft supporting truss at the stability limit.

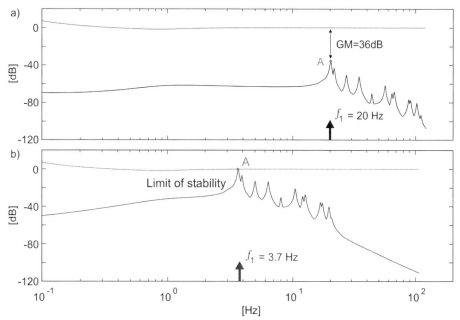

FIGURE 13. Robustness test based on additive uncertainty : (a) Stiff supporting truss ($f_1 = 20$ Hz) ; the gain margin is $GM = 36$ dB - (b) Soft supporting truss at the stability limit.

7. CONCLUSIONS

The numerical modelling of large segmented mirrors has been addressed. This model gives a full account of the quasi-static behavior of the segmented mirror and has minimum complexity to account for control-structure interaction. Two control strategies have been discussed, both based on a quasi-static behavior of the mirror. The control-structure interaction has been addressed through a frequency domain robustness test for MIMO systems ; this test separates the dynamic response (residual) from the quasi-static response (primary) of the mirror. The

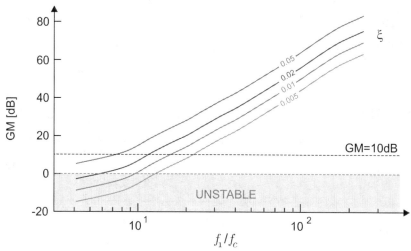

FIGURE 14. Robustness test based on multiplicative uncertainty : Evolution of the gain margin with the frequency ratio f_1/f_c for various values of the damping ratio ξ_i.

evolution of the robustness test with the dominant frequency f_1 of the supporting truss and the damping ratio ξ has been analyzed. The study has been illustrated with an example involving 90 segments, 270 inputs (excluding the central segment) and 654 outputs (edge sensors and normal sensors), and a SVD controller; the procedure is very fast and allows extensive parametric studies. An important conclusion of this study is that, for a structural damping ratio of $\xi = 1\%$, the frequency gap between the crossover frequency of the controller, f_c, and the critical flexible mode, f_1, must be at least one decade. This condition may be more and more difficult to fulfill as the size of the telescope grows.[24–26]

ACKNOWLEDGMENTS

The work of Renaud Bastaits was supported by Fonds National de la Recherche Scientifique (FNRS), Belgium, under FRIA PhD grant FC76554.

REFERENCES

[1] Bely P.Y., *The Design and Construction of Large Optical Telescopes*, Springer, 2003.
[2] Gilmozzi, R., Spyromilio, J., The 42m European ELT : status, In *Ground-based and Airborne Telescopes II* - SPIE 7012 (2008), Larry M. Stepp, Ed.
[3] Nelson, J., Sanders, G., The Status of the Thirty Meter Telescope Project, In *Ground-based and Airborne Telescopes II* - SPIE 7012 (2008), Larry M. Stepp, Ed.
[4] Dimmler M., Erm T., Bauvir B., Sedghi B., Bonnet H., Müller M., Wallander A., E-ELT primary mirror control system, In *Ground-based and Airborne Telescopes II* - SPIE 7012 (2008), Larry M. Stepp, Ed.
[5] Angeli, G.Z., Cho, M.K., Whorton, M.S., Active optics and control architecture for a giant segmented mirror telescope, in *Future Giant Telescopes* (Angel, and Gilmozzi, eds.), 129-139, 2002.
[6] European Southern Observatory, *The VLT White Book*, ESO, 1998.
[7] Angeli, G.Z., Upton, R., Serguson, A., and Ellerbroek, B., Active Optics Challenges of a Thirty meter segmented mirror telescope, In *Second Backaskog Workshop on Extremely Large Telescopes* - SPIE 5382 (2004), Ardeberg A. L. and Andersen T., Ed.
[8] Chanan, G. MacMartin, D.G., Nelson, J. & Mast, T., Control and alignment of segmented-mirror telescopes : matrices, modes, and error propagation, *Applied Optics*, Vol.43, No 6, 1223-1232, Feb. 2004.
[9] MacMynowski, D.G., Thompson, P.M., and Sirota, M.J., Analysis of TMT Primary Mirror Control-Structure Interaction, In *Modeling, Systems Engineering and Project Management for Astronomy III* - SPIE 7017, Angeli G.Z. and Cullum M.J., Ed.
[10] MacMynowski, D.G., Thompson, P.M., and Sirota, M.J., Control of Many-Coupled Oscillators and Application to Segmented-Mirror Telescopes, *AIAA Guidance, Navigation and Control Conference*, AIAA Paper 2008-6638, Aug. 2008.

[11] Jiang, S., Voulgaris, P.G., Holloway, L.E., and Thompson, L.A., H2 Control of Large Segmented Telescopes, *Journal of Vibration and Control*, Vol. 15, No. 6, 2009, pp. 923-949.

[12] Geradin, M., Rixen, D., Mechanical Vibrations, Wiley, 1994.

[13] Craig, R.R., Bampton, M.C.C, Coupling of Substructures for Dynamic Analyses, AIAA Journal, Vol.6(7), 1313-1319, 1968.

[14] Cohen R.W., Mast T.S., Nelson J.E., Performance of the W.M. Keck telescope active mirror control system, In *Advanced Technology Optical Telescopes V* - SPIE 2199, Stepp L.M., Ed.

[15] Bastaits R., Rodrigues G., Mokrani B., Preumont A., Active Optics of Large Segmented Mirrors : Dynamics and Control, *AIAA Journal of Guidance, Dynamics and Control*, Vol. 32(6), 1795-1803, Nov.-Dec., 2009.

[16] Preumont A., *Vibration Control of Active Structures, An Introduction*, 2nd edition, Kluwer, 2002.

[17] Franklin, G.F., Powell, J.D., Emani-Naemi, A., *Feedback Control of Dynamic Systems*, Addison-Wesley, 1986.

[18] Lurie, B.J. and Enright, P.J., *Classical Feedback Control*, Marcel Dekker, 2000.

[19] Aubrun, J.N., Lorell, K.R., Mast, T.S. & Nelson, J.E., Dynamic Analysis of the Actively Controlled Segmented Mirror of the W.M. Keck Ten-Meter Telescope, *IEEE Control Systems Magazine*, 3-10, December 1987.

[20] Aubrun, J.N., Lorell, K.R., Havas & T.W., Henninger, W.C., Performance Analysis of the Segment Alignment Control System for the Ten-Meter Telescope, *Automatica*, Vol.24, No 4, 437-453, 1988.

[21] Doyle, J.C. and Stein, G., Multivariate Feedback Design : Concepts for a Modern/classical Synthesis, *IEEE Trans. on Automatic Control*, Vol. AC-26, pp.4-17, Feb.1981.

[22] Maciejowski, *Multivariable Feedback Design*, Addison-Wesley, 1989.

[23] Kosut, R.L., Salzwedel, H., Emami-Naeini, A., Robust Control of Flexible Spacecraft, *AIAA J. of Guidance*, Vol.6, No 2, 104-111, March-April 1983.

[24] Preumont A., Bastaits R., Rodrigues G., Scale Effects in Active Optics of Large Segmented Mirrors, *Mechatronics*, Vol. 19, No 8, 1286-1293, December, 2009.

[25] Bastaits R., Preumont A., On the Structural Response of Extremely Large Telescopes, *AIAA Journal of Guidance, Dynamics and Control*, accepted for publication.

[26] Bastaits, R., *Extremely Large Segmented Mirrors : Dynamics, Control and Scale Effects*, PhD thesis, Active Structures Laboratory - Université Libre de Bruxelles, June 2010.

The development of the actuator prototypes for the active reflector of FAST

QiMing Wang*, MingChang Wu, Ming Zhu, JianXing Xue, Qing Zhao, XueDong Gu
National Astronomical Observatories, CAS, 20A, Datun Road, Chaoyang District, Beijing, CHINA 100012;

ABSTRACT

Upon its completion, the Five-hundred-meter Aperture Spherical radio Telescope (FAST) will be the largest single dish radio telescope ever in the world. The construction has been initiated in March 2011 in Guizhou province of China. The whole construction process is expected to be completed in September 2016, with duration of 5.5 years.

With an aperture of 500 meters and an illumination aperture of 300 meters, the active reflector is one of the most important parts of FAST. The reflector is composed of a ring beam, a cable net and thousands of panels, tie-down cables, actuators and anchors. For the observation process of source switching and source tracking, the parabola shape of the reflector is achieved by drawing back of the tie-down cables by the actuators. The motion performance and the reliability of the actuators are of great importance to the telescope.

In this paper, the motion models of the actuators are analyzed for the observation process of source switching and source tracking. Several design schemes are proposed, including mechanical and hydraulic design. The electric, mechanical and hydraulic characteristics of these designs are discussed. Related experimental studies are performed to investigate the electric and mechanical performances of these actuator prototypes. Based on the analysis and test results, a final type of actuator will be optimally concluded to meet the requirements of the reflector of FAST.

Keywords: Radio Telescope, Active Reflector, Actuator, Reliability, Prototype

1. INTRODUCTION

The large radio telescope FAST is the largest single dish radio telescope ever planned. The concept was presented by Chinese astronomers in 1994[1]. With a spherical dish as its primary reflector, it is of Arecibo type. Underneath the reflector, over two thousand actuators are applied to realize the active deformation of the dish to transiently form the 300-meter aperture paraboloid of different optical axis for astronomical observation (Fig. 1). The focus ratio of the parabola is 0.467. The frequency range of the telescope covers from 70MHz to 3GHz. It is located in the Karst depression in GuiZhou province of south-west of China. The project was commenced in March 2011, expected to be completed in September of 2016.

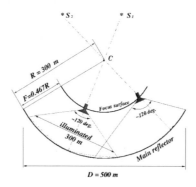

Figure 1. Optical design of FAST

*qmwang@bao.ac.cn; phone 86 10 64877293; fax 86 10 64877278

The whole FAST project can be divided into six systems: The site reconnaissance and excavation system, the observatory, the active reflector system, the feed support system, the measurement and control system, the receiver and backend system. Also a science department is formed to study the early science and observation program.

The active reflector is one of the key components of FAST[2]. It is composed of ground anchors, actuators, tie-down cables, ring beam(500m in diameter), ring beam pillars, cable net, reflector panels, back frames and health monitoring system[3], as shown in table 1. Among all these components, the only driving component is the actuator. The actuator is used to realize the active deformation of the reflector. The remaining components are all passively moved. The development of the actuator and the test of its prototype are of great importance for the construction, operation and maintenance of the FAST telescope.

Table 1. Amount of parts of the active reflector

Parts	Amount
Ground anchors	2235
Actuators	2235
Tie-down cables	2235
Ring beam pillars	50
Cable net segments	~7000
Reflector panels	~4400
Back frames	~4400

2. TECHNICAL REQUIREMENTS AND TYPE SELECTION ANALYSIS OF THE ACTUATORS

2.1 Technical requirements

Results of structural analysis show that the maximal operation load of the actuator is about 50kN when operated in the active deformation mode. The maximal working stroke is 950mm. The error budget of the actuator itself is 0.25mm. Based on the observational mode, the actuators have two types of operation modes: source switching mode and source tracking mode. When working in source switching mode, the related actuators are required to complete the process in 10 minutes, which means a maximal speed of 1.6mm/s. For the source tracking mode, the related actuators will operate with a varied speed ranged from 0 to 0.6mm/s. as shown in Figure 2. Due to the difference in maximal speed in the two modes, the actuators must have good speed control capability. The mechanical drive and hydraulic drive are studied for type selection of the actuators.

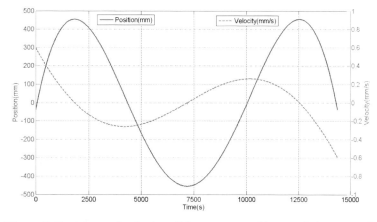

Figure 2. Position and velocity of the source tracking mode in one cycle

From figure 2 it can seen that the maximal speed in the source tracking process is about 0.6mm/s, and for most of the time in the process, the speed is slower than 0.2mm/s. Considering the precision requirement of 0.25mm, we can conservatively and reasonably think that if an actuator can operation at the speed of 0.1mm/s for enough long time, the actuator will be functional in the source tracking process. The reason is that for the position with relating speed below 0.1mm/s, the position control can be achieved by on-offs of the motor in the worst case, without lowering the position control precision of 0.25mm.

2.2 Type selection analysis of the actuators

Based on the motor types, the mechanical actuators may be classified to servo mode drive, variable frequency motor drive, stepping motor drive and three phase AC squirrel cage motor drive etc. Based on the mechanical transmission components, the actuators can also be classified to worm reducer(Self-locking) with ball screw transmission, or gear reducer with T type screw transmission (Self-locking).

For the hydraulic actuators, speed control can be realized more continuously with less mechanical friction. In the case of overload, the piston can be moved with the cable more freely in the absence of self-locking components. The shortage is that its position sustainability is worse than mechanical one. Another shortage is that there are more requirements in valve components reliability.

The first step in the type selection process is the selection of motor type.

(1)The three phase AC squirrel cage motor has no capability in speed control. The actuator may be designed to meet the requirement of maximal speed of 1.6mm/s. However, the motor can only work in frequently on-off mode to realize the speed curve in source tracking mode as shown in figure 2.

For instance, if a 4-pole motor with nominal rotational speed of 1440rpm is used, the motor must have about ten on-off switches in one minute to work in source tracking mode with a position precision better than 1mm. This mode of frequently on-offs will lead to heat generation in the coil of the motor, the aging of the insulation between the coils, the shock, and wear of the mechanical components. All these will lead to shorter life and less reliability of the actuators.

(2) In common case, the stepping motor is used in case of small torque. For large load, the stepping motor will lost steps and generate strong noise which is not favorable for the synchronous operation of about one thousand motors in observational process.

(3)The variable frequency motor can be used in speed control with less cost than servo motor. However, when working at low frequency mode (0~2Hz) , the forced cooling fan is necessary, which will lower the reliability of the actuators. Another disadvantage is that the variable frequency inverter will induce RFI(Radio Frequency Interference) problems to the telescope, which must be handled with care.

(4)The servo motor is best to meet the speed control requirement of the actuators. The disadvantages are higher price and RFI problems similar to the variable frequency inverter.

The second step is the selection of the mechanical transmission components.

(1) One choice is using planetary gear reducer or cycloid reducer, connecting to the T type screw through a coupling joint. The planetary gear reducer or cycloid reducer has high mechanical efficiency. The T type screw is a self-locking component, which can ensure the position sustainability of the actuator. T type screw is suitable for high load and low speed application, which is similar to the requirements of FAST actuators. However, the T type screw is of sliding friction and is normally lubricated with grease. To avoid poor lubrication induced wear of the screw pair, the lubrication system must be specially designed.

(2) Another choice is using worm reducer, connecting to the ball screw through a coupling joint. The ball screw mechanism has high mechanical efficiency. The worm reducer is a self-locking component. The ball screw is more expensive than the T type screw. Similar to the T type screw, the worm gear pair is of sliding friction, which will lead to low efficiency. The worm reducer is located in a box, which is easier for valid lubrication using lubricating oil.

Based on the former analysis of the motor type and the mechanical transmission system, both the mechanical and the hydraulic design have been performed for the development and test of prototype actuators.

3. PROTOTYPE DESIGN

3.1 Prototype design of a mechanical actuator[4,5]

Here we present an actuator design of variable frequency inverter driven, with mechanical transmission of T type screw and tumbler gear. The maximal working load F_{max} is 50kN. The maximal speed V_{max} is 1.6mm/s. The net output power W_0 can be given by:

$$W_0 = F_{max} \times V_{max} = 80 \text{ W} \tag{1}$$

For a factor of safety of S=1.5, the raw output power W_1 should be:

$$W_1 = W_0 \times S = 120 \text{ W} \tag{2}$$

For the transmission efficiency, we choose η_1=0.3 for T type screw, η_2=0.96 for ear, η_3=0.98 for bearing.

The input power should be:

$$W = W_1/(\eta_1 \times \eta_2 \times \eta_3) = 425.17 \text{ W} \tag{3}$$

If the lead of the T type screw is P=6, the driving force per 1 Nm input torque is F_1, F_1 can be calculated as:

$$F_1 = 1 \times \eta_1 \times 2\pi/P = 314.15 \text{N} \tag{4}$$

The input torque should be:

$$M_s = F_{max}/F_1/(\eta_2 \times \eta_3) = 169.17 \text{ Nm} \tag{5}$$

Considering the friction between the components, an effective margin of 50% is used to ensure the valid output of the actuator. The actual input torque M_s' should be:

$$M_s' = (1+0.5) \times M_s = 254 \text{Nm} \tag{3}$$

Based on these specifications, the motor and reducer assembly of SEW R47F DT is chosen. The output power is 550W, the efficiency is 0.86, the nominal output torque is 330Nm, and the output speed is 16rpm.

For this case, the effective out power of the motor is 550×0.86=473W, which is larger than the designed input power of 425.17W. The nominal output torque of the motor reducer assembly is 330Nm which is larger than the designed input torque of 254Nm. The nominal output speed is 16rpm. The lead of the T type screw is designed as P=6, the output speed of the T type screw is 16/60×6=1.6mm/s. So the selection of the motor reducer assembly of SEW can satisfy the design requirements. The design drawing is shown in figure 3.

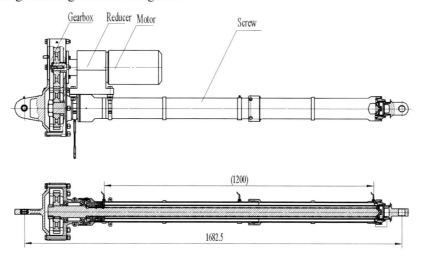

Figure 3. Design drawing of a mechanical actuator

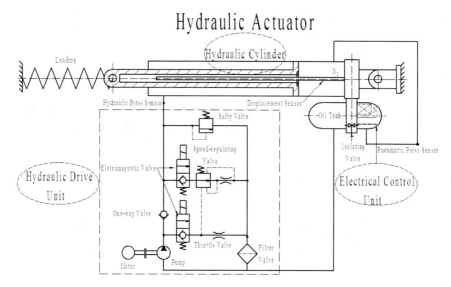

Figure 4. Design of a hydraulic actuator

3.2 Prototype design of a hydraulic actuator

As can be seen from figure 4, the hydraulic actuator is composed of hydraulic driving unit, electrical control unit and hydraulic cylinder. The hydraulic driving unit is totally immersed in the oil in the tank. This special, compact design is of help to the heat dissipation and lubrication of rotating parts.

The electrical control unit is composed of driving PCB board, controlling PCB board and an inverter. The unit is settled in the Nitrogen gas tank behind the oil tank, which is full of dry and low pressure Nitrogen gas (A little higher than the atmosphere pressure). The Nitrogen gas is used to provide a friendly environment for the electrical parts. The corrosion resistance capability of the inner surface of the tank is enhanced at the same time.

The design is of single action. The displacement sensor is used to measure the absolute position of the piston. The oil pressure sensor and the gas pressure sensor can provide instant data for operation state of the hydraulic tank. To improve the system reliability, the low pressure Nitrogen gas is also connected to the oil tank and the hydraulic cylinder without piston to reduce the corrosion possibility. A stop valve is design to stop the oil loop when in transportation to reduce the possibility of oil flowing to the gas tank. The stop value will be switched on upon the installation of the actuator.

The primary mechanism of the hydraulic actuator is using the variable frequency inverter to drive the three-phase asynchronous motor to pump out high pressure oil. In drawing back phase, the variable frequency inverter controls the output oil flow rate according to the instruction. The oil flows into the hydraulic cylinder without piston through the one-way valve. The piston will be drawn back at a speed of 0~1.6mm/s. In pulling out state, the flow valve with pressure compensation will combine with electromagnetic valve into an oil flow channel. The oil in the hydraulic cylinder without piston will flow back to the oil tank through the oil flow channel, the filter valve under the action of external force. The bottom electromagnetic valve loop will control the upper oil flow channel by oil pressure, enabling the piston to be pulled out at a speed from 0~1.6mm/s.

4. THE EXPERIMENTAL STUDY OF THE PROTOTYPE ACTUATORS

To verify the feasibility of the two designs, one mechanical prototype and a hydraulic prototype are produced and tested according to the operation mode of FAST. Both variable load and constant load test for the actuators are performed as shown in figure 5. It is worth noting that, the actuators pull the cable net of the active reflector through tie-down cables. The tie-down cable is a kind of flexible steel wire rope, which will function only in pulled state. For this reason, the actuator will always bear pulling force. The T type screw and the nut are contacted in one side, which will lead to poor lubrication between the T type screw and the nut. The test results proved the analysis. After a long period of operation, the outer cylinder of the actuator has significant temperature rise and loud friction noise. Significant wear can be found on the nut when the actuator was disassembled.

(a) Mechanical actuator(variable load test) (b) Hydraulic actuator(variable load test)

(c) Mechanical actuator(constant load test)

Figure 5. Actuator and the test platforms

To finish a source tracking cycle, at least 4 hours is necessary. One method to expedite this is to test the actuator using to and fro cycles at low speed about 0.1mm/s, as explained earlier.

For the mechanical actuator, continuous operation in to and fro mode of fifty days have been finished. The actuator ran twenty to and fro cycles within 60mm at low speed(about 0.1mm/s, corresponding to 2Hz), with load varying from 1.8t to 4.5t. The actuator then ran two to and fro cycles within 400mm at high speed(about 1.7mm/s, corresponding to 50Hz) to lubricate the nut and the T type screw, and to simulate the source switching process. Figure 6 shows a sample data of the operation.

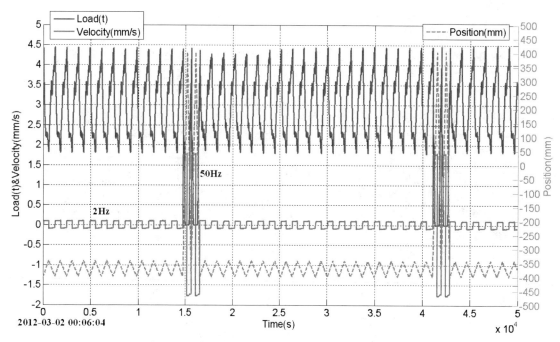

Figure 6. Sample test data for the mechanical actuator

For the hydraulic actuator, continuous operation in to and fro mode of about five days have been finished. The actuator ran ten to and fro cycles within 100mm at low speed(about 0.15mm/s), with load varying from 2.0t to 5.2t. The actuator then ran five to and fro cycles within 500mm at high speed(about 2.0mm/s) to simulate the source switching process. Figure 7 shows a sample data of the operation.

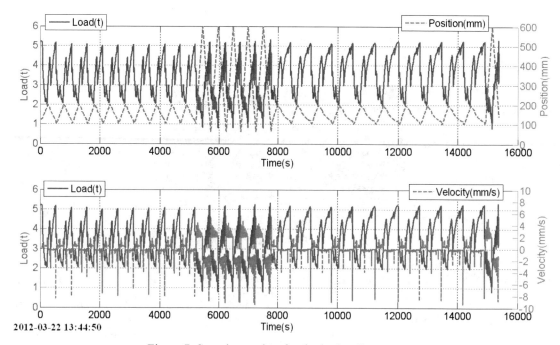

Figure 7. Sample test data for the hydraulic actuator

The variable load test platform uses chain mechanism, which brings load shock to the actuators. The mechanical actuator can keep position control fairly well, as can be seen from figure 6. However, under the load shock, the hydraulic actuator can not control the position steadily, which can be seen from the velocity peaks in figure 7.

5. CONCLUSION

The paper discussed the operation requirements and the prototype design of the actuators of FAST. The development and test of a mechanical actuator and a hydraulic actuator are conducted. The work is important for the type selection for the FAST actuators. Based on these work, tentative actuator designs will be analyzed and tested to acquire an optimized actuator prototype both in performance and reliability.

ACKNOWLEDGEMENTS

This work was supported by the National Natural Science Foundation of China (Grant No. 11173035). I would like to thank all my colleagues for their contributions to the study of the actuators for FAST.

REFERENCES

[1] Nan R.D., et al., "The Five-Hundred Aperture Spherical Radio Telescope (fast) Project", IJMPD, 20, 989(2011)
[2] Qiu Y. H., "A novel design for a giant Arecibo-type spherical radio telescope with an active main reflector", MNRAS, 301, 827-830 (1998).
[3] Qi-ming Wang, Lichun Zhu, Rendong Nan, "Development of Active Reflector of the 30-Meter FAST Model", Proc. of SPIE, Vol. 7018 70184A-8 (2008).
[4] Robert, L. Norton, [Design of Machinery: An Introduction to the Synthesis and Analysis of Mechanisms and Machines], McGraw-Hill, New York (2012).
[5] Jiang Huang, ZhongXing Guo, [Applied Motor Design: Calculation Manual], ShangHai Science and Technology Press, ShangHai (2010).

Design, development and manufacturing of highly advanced and cost effective aluminum sputtering plant for large area telescopic mirrors.

Rajeev R Pillai*[a], Sanjth K K[a], K Mohanchandran[a], Nagarjun Sakhamuri[a],
Vishal Shukla[b], Alok Gupta[b]

[a]Hind High Vacuum Co. Pvt. Ltd., Bangalore, India-560058; [b]Aryabhatta Research Institute of Observational Sciences, Nainital, India-263129

ABSTRACT

The design, development and manufacturing of a fully automated and cost effective aluminum sputtering unit for the deposition of aluminum on large area telescopic mirrors (maximum diameter of 3600mm) is presented here. The unit employs DC planar magnetron sputtering for the deposition process. A large area glow discharge unit is also designed for the pre-cleaning of the mirrors prior to aluminum coating. A special kinematic support structure with rotation is designed to support heavy mirrors of large area to minimize the deflection of the mirrors during deposition process. A custom designed 'mask' is employed in the magnetron system to improve the thickness uniformity within $<\pm 3\%$. The adhesion, thickness uniformity and reflectivity properties are studied in detail to validate the sputtering plant. Special fixtures have been designed for the system to accommodate smaller mirrors and studies have been conducted for the coatings and reported in the paper. The unit was successfully tested at HHV facility in Bangalore and will be installed at the ARIES Facility, Nainital.

Keywords: Telescopic mirror, magnetron sputtering, sputtering mask, glow discharge, whiffletree, reflectivity.

1. INTRODUCTION

The modern day very large telescope (VLT) mirrors are a marvel of engineering and glassmaking, having a very precise curved shape and surfaces polished to within a few wavelengths of light. These large mirror surfaces are covered with a thin layer of highly reflecting material (normally aluminum or silver), ensures approximately 90% of the photons from stars impinging on the mirror to be registered by the sensitive telescope instruments. The optical telescope mirror coatings used in observatories get damaged and tarnished over time due to oxidation, humidity, dust, impact of charged particles from space etc. In order to maintain the accuracy of observations these mirrors must be re-coated frequently. It is preferable to have the coating unit near the telescope as the mirrors are heavy, have large diameters and have to be handled delicately. The recoating process is usually carried out under precisely controlled high vacuum conditions using either magnetron sputtering or thermal evaporation technique. The films deposited using this technique should be spotlessly clean, and extremely uniform. In addition to that, in the case of large telescopes, the re-coating equipment needs to employ specially designed, whiffletree support system which distributes the weight of the mirrors across a series of points to minimize deformation.

HHV is a pioneering high vacuum equipment manufacturer in India, established in 1965 and is one of the leading players in designing and developing VLT mirror coating units. HHV has demonstrated its technological strength by commissioning telescope re-coating equipments in various parts of India. They include the 2.1diameter telescope mirror coating unit, which uses thermal evaporation technique, installed at the Hanle observatory located in the Ladakh region of the Himalayas, at a height of 4570 meters. In 2006, a re-coater is commissioned with sputtering process for a telescope mirror of 2.2 metre diameter at Girawali near Mumbai at an altitude of 1000 metres above sea level.

The present paper discusses the design, manufacturing and testing of re-coating equipment for the 3.6m mirror for the telescope, which is to be installed at 'The Aryabhatta Research Institute of Observational Sciences (ARIES) facility, Nainital, India at an altitude of 2500 meters above sea level.

*rajeevpillai@hhv.in; phone +91-080-22633740; fax +91-080-28394874; www.hhv.in

2. THE COATING UNIT

Figure 1. ARIES telescopic mirror coating unit.

The coating unit consists of a vacuum chamber fabricated out of stainless steel, with 2 halves, torrispherical dished ends welded with flanges and clamped together. The chamber has an overall dimension of 4000mm dia x 1700mm (ht.) (i.e. from centre of the bottom dish to the centre of the top dish). The bottom chamber is supported on a 4 -tubular support structure in order to support of the total weight of the chamber. A separate support structure made of mild steel tubes are provided for bearing housing of the rotary drive mechanism to transfer the total load to the ground.

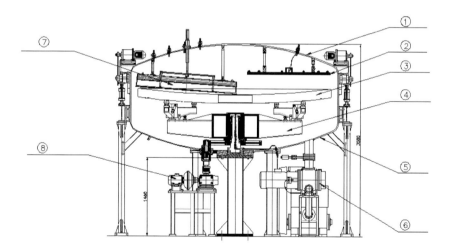

Figure 2. Main components of ARIES telescopic mirror coating unit.(1)Top chamber,(2)Glow discharge gadget,(3)Primary mirror,(4)Mirror support system,(5)Bottom chamber,(6)Roots-rotary combination ,(7)magnetron, shield ,shutter,(8)work holder rotary mechanism.

It has a custom built mechanism for opening and closing the chamber's top lid. During opening of the chamber, the top lid is lifted up by a pair of hydraulic cylinders, and then it is moved horizontally on the channels with wheels provided to the lid located on the railings, by means of a drive mechanism. The vacuum pumping systems consists of a combination of rough pumping system with (EDWARDS 412J) direct drive mechanical booster pump (EDWARDS EH 2600) and high vacuum pumping system consisting of two cryo-pumps (CVI TM 450) regenerated with rotary pump (EDWARDS E2M40).

Rotary work holder for 3.6m diameter mirror is designed to hold the large thin mirrors within the deflection threshold. The mirror stability is evaluated using finite element analysis. The work holder consists of a whiffletree structure comprising of a central hub attached to a thick hollow tube. This hub on its periphery is welded with three stainless steel arms 120° apart. Each arm is provided with a three point kinematic support with soft pads on the top. The entire assembly is supported on thrust bearings and roller bearings enclosed in a bearing housing to take care of the total load of the mirror which is 4.5 tons. This hollow shaft will pass through a specially designed vacuum shaft seal using Viton lip seal for vacuum tightness during substrate rotation.

Figure 3. Rotatable Whiffle tree.

The drive mechanism for the entire assembly is achieved by means of reduction gear box with an AC motor and AC drive mechanism attached to the drive shaft through gear arrangement. The rotary mechanism is kept between 0-1 rpm. Sensors are used for sensing and counting of the work holder rotation, enabling precise timing of start and stop of deposition process. It helps in the starting and stopping of the work holder rotation at the same point. The whole work holder structure is designed and evaluated by finite element analysis method prior to manufacturing. The manufactured structure has proven the evaluation correct by its functionality and reliability.

The coating unit is capable of accommodating individual separate substrate holders for smaller mirrors of size 1.313m, 0.980m and 0.610m in diameter. These substrate holders were designed taking into consideration the radius of curvature of the mirrors. These substrate holders are located on the arms of the main rotary work holder. At any given point of time, one substrate can be used depending upon the mirror on which aluminum deposition is carried out.

The present system employs DC magnetron sputtering technology for depositing the aluminum film on the mirror. Water cooled rectangular Magnetron Source (HHV), with split shutters, is used to sputter aluminum target (99.999%) of size 180mm (W) x 2000mm (L) x 6mm (T). The magnetron is mounted on the inner side of the top lid with suitable supports and adjusting mechanisms to vary the distance and angle for downward sputtering of aluminum on to the telescope mirrors. A glow discharge gadget, powered with an HT power supply is provided inside the chamber for the ion bombardment cleaning of the mirror, prior to sputtering deposition.

A specially designed stainless steel mask (trim shield) is provided below the magnetron source to trim the deposition of the sputtered aluminum and to achieve a uniform thickness of aluminum on the mirror. The mask is custom designed so that the same mask can be used for the uniform deposition of mirrors of sizes 1.313m, 0.980m and 0.610m in diameter respectively. This attractive feature of the mask helps the user in avoiding the hustles of mask changing during the coating of each type of mirrors and also reduces the cost of designing and developing individual masks for each mirror, in-turn reducing the cost of the coating unit as a whole.

3. EXPERIMENTAL PROCEDURE

The system can be operated both in auto and manual mode. In auto mode the recipe for the whole process of pumping, glow discharge cleaning and deposition can be programmed. In order to optimize the deposition process, coating trials were conducted by fabricating a special dummy work holder (test platform) with specified radius of curvature, which exactly replicates the surface of the mirrors. The glass test coupons (50x50 mm, 1 mm thick) were then mounted onto the dummy work holder surface at intervals of 50 mm along the mirror radius. The position of the slides is recorded on the back of the slides, together with the coating run number. Once the sample loading is completed, the chamber is sealed and is pumped down to achieve the base vacuum level of 1×10^{-6} mbar.

The glow discharge cleaning of the mirror glass surface is done by introducing argon gas (99.999%) to the chamber and a process pressure of 1×10^{-2} mbar is maintained by throttling. During the glow discharge cleaning process, mirror is rotated at a speed of 1rpm and the glow discharge is carried out at 3 kV, 500mA for 45 minutes. On the completion of the glow discharge cleaning process, the pumping is throttled to attain a process pressure of 1×10^{-3} mbar. Then the magnetron is switched on and the aluminum target (99.99%) was pre sputtered with a DC power (Huttinger) of 7kW, until the target surface is clean. During pre-sputtering the shutters of the magnetron is closed in order to prevent any coating of the mirror surface. Once the pre-sputtering is completed the split shutter is opened and the coating is carried out for a predetermined time, after which the magnetron is switched off in such a way that a joint free film is deposited on the mirror. Once the coating is over the chamber is opened. The test coupons are unloaded and are taken for various measurements. The thickness of the test coupons are measured by stylus profiler (Veeco-Dektak 6M), the reflectance study is performed using spectrophotometer (Ocean Optics UV-VIS-NIR DH 2000) and adhesion test is carried out by scotch tape pull test.

4. RESULTS & DISCUSSION

Finite element (FE) analysis of the mirror was performed using the parameters of glass i.e. Young's modulus and density. Using this technique, maximum stress, maximum sag and points of minimal strain have been identified. Considering several parameters, the design has been finalized to a nine point load distribution mechanism as shown in figure 3.

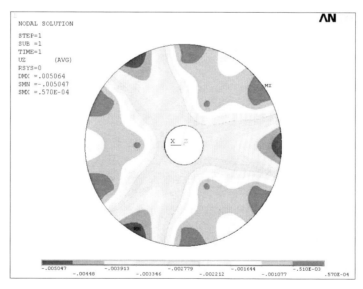

Figure 4. Displacement contours by FE analysis.

Figure 4 shows the displacement contours due to gravitational deformation of the 3.6m mirror with nine point support. The maximum sag of the glass is found to be only 0.005047 mm. The load distribution of the 3.6m mirror over the nine point contact is measured using load cell and the data is tabulated in table 1.

Table 1. Load distribution studies.

Load distribution study of the mirror using load cell	
Load cell position	load (kg)
A1	304.5
A2	304.9
A3	325.7
B1	305.4
B2	304.6
B3	323.3
C1	303.5
C2	303.1
C3	321.3

During the magnetron sputtering process the temperature of the substrate is found to increase to a maximum of 15°C. For the optimized conditions, a typical aluminum sputter rate of 15Å/min is obtained. The deposited aluminum films deposited is found to be uniform and shows good adhesion. The film thickness of the sputtered aluminum is investigated using a stylus profiler by selectively removing coating from the test coupons. The optimization goal was obtain a film thickness of 1000 Å±50Å. For a given test coupon, thickness is measured at least three places, in order to confirm the accuracy. The thickness uniformity was calculated from standard deviation of the thickness measurements and a graph between thickness and radial distance is shown in figure 5 and fig 6.

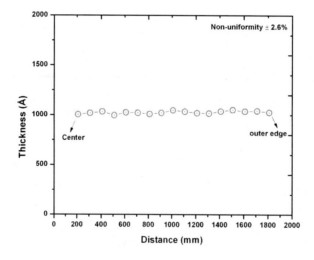

Figure 5. Thickness profiles of 3.6m mirror with a thickness uniformity of ±2.6%.

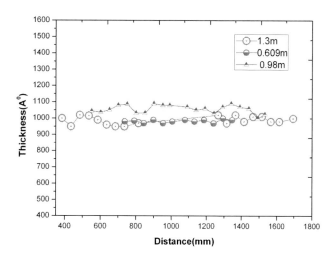

Figure 6. Thickness profiles of 1.3 m, 0.98m, 0.609m mirrors with thickness uniformities of ±2.5%, 2.36% & 0.97%.

The reflectance study of the test coupons are investigated using spectrophotometer and the reflectance measurement was measured relative to a STAN-SSH-NIST standard reference material (front surface protected aluminum mirror on fused silica substrate), whose reflectance was known. The spectrum is taken over the required wavelength region of 400-900nm. Test coupons from different positions of the same trial are measured for reflectance and the graphs are plotted and are shown in figure 7.

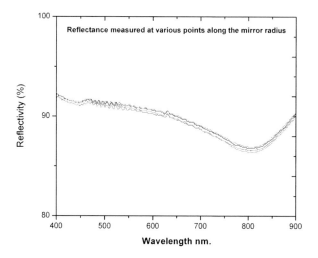

Figure 7. Reflectance measured along 3.6m mirror radius.

5. CONCLUSIONS

HHV has successfully designed and manufactured a fully automated and cost effective sputtering unit for the deposition of aluminum on large area astronomical telescope mirrors. The custom designed whiffletree arrangement is found to distribute the whole weight of the mirror over nine points with minimum deformation of the mirror. The aluminum coating on the mirror was optimized by investigating various power pressure combinations and optimizing magnetron position and the profile of the magnetron mask. This custom built mask has removed the complexity of using different masks for different mirrors, which in turn has enabled a uniform deposition thickness, (within ±3%) with good adhesion on all substrates. The reflectance values of the mirrors in the visible and NIR regions are found to be very close to the expected value.

REFERENCES

[1] David.A.Glocker., Ismuth shah.s.,[Handbook of thin film process technology.vol1& 2],Institute of physics (IoP)publishing,Bristol& Philadelphia(1995).
[2] Albrecht., Bodo.,W.D. (translator)., [The New Cosmos: An Introduction to Astronomy and Astrophysics], Springer, Berlin, New York: (2001).
[3] Kasturi.L.Chopra[Thin film phenomena], McGraw-Hill,New York(1969).
[4] Marcel Dekker., R. R., [Practical Design and Production of Optical Thin Films], Willey, New York, 294-298 (2002).
[5] http://www.eso.org/sci/publications/messenger/archive/no.97-sep99/messenger-no97-4-8.pdf.

The Australian SKA Pathfinder

Antony E.T. Schinckel*[a], John D. Bunton[b], Tim J. Cornwell[a], Ilana Feain[a], Stuart G. Hay[b]
[a]CSIRO Astronomy and Space Science, PO Box 76, Epping, NSW, 1710 Australia;
[b]CSIRO ICT, PO Box 76, Epping, NSW, 1710, Australia;

ABSTRACT

The Australian Square Kilometre Array Pathfinder (ASKAP) will be the fastest cm-wave survey radio-telescope and is under construction on the new Murchison Radio-astronomy Observatory (MRO) in Western Australia. ASKAP consists of 36 12-meter 3-axis antennas, each with a large chequerboard phased array feed (PAF) operating from 0.7 to 1.8 GHz, and digital beamformer preceding the correlator. The PAF has 94 dual-polarization elements (188 receivers) and the beamformer will provide about 36 beams (at 1.4 GHz) to produce a 30 square degree field of view, allowing rapid, deep surveys of the entire visible sky. As well as a large field of view ASKAP has high spectral resolution across the 304 MHz of bandwidth processed at any one time generating a large data-rate (30Gb/sec in to the imaging system) that requires real-time processing of the data. To minimise this processing and maximise the field of view for long observations the antenna incorporates a third axis, which keeps the PAF field of view and sidelobes fixed relative to the sky. This largely eliminates time varying artefact in the data that is processed.

The MRO is 315 kilometres north-east of Geraldton, in Western Australia's Mid West region. The primary infrastructure construction for ASKAP and other telescopes hosted at the Murchison Radio-astronomy Observatory has now been completed by CSIRO, the MRO manager, including installation of the fibre connection from the MRO site to Perth via Geraldton. The radio-quietness of the region is protected by the Mid West Radio Quiet Zone, implemented by the Australian Federal Government, out to a radius of 260km surrounding the MRO.

Keywords: ASKAP, SKA, survey, phased array feed, PAF, interferometer, radio astronomy

1. INTRODUCTION

In the early to mid-1990's it became apparent to the world's radio-astronomy community that to make significant progress in a range of areas traditionally studied by cm and metre radio astronomy, a significant new telescope was required, now called the Square Kilometre Array (SKA)[1] [2] [3]. In particular, studying the earliest epoch of the universe's formation would require the ability to detect much fainter sources, and to examine large scale structures and fluctuations.

To achieve this, a combination of both a major improvement in the sensitivity of the telescope, and the ability to map large areas of the sky quickly was required. The ability to map the sky quickly requires a large field of view and in the early days of the SKA gave rise to concepts that used Luneburg lens [4], cylindrical reflectors [5] and phased arrays. The first two concepts have now been abandoned for the SKA. Initial phased array concepts used a phased array placed on the ground to give an aperture array. Each element of the aperture array has a field of view that encompasses the whole visible sky. At lower frequencies (below 400MHz) this concept is demonstrated by telescopes such as LOFAR [6], MWA[7], LWA[8] and PAPER [9]. A higher frequency demonstrator, EMRACE [10], extends operation to 1.55GHz.

Another approach to the use of phased arrays is to place them at the focus of a dish. The field of view is reduced compared to the aperture array but it is still more than an order of magnitude greater than a dish with a single feed. This concept, a dish array with a Phased Array Feed (PAF), exchanges field of view for a reduction in the cost of electronics for the same sensitivity. For a given sensitivity and field of view the electronics cost is approximately proportional to the square of frequency. Hence for a given electronics cost, PAFs are more cost-effective to a higher frequency than aperture arrays. Demonstrators for PAFs are APERTIF [11], PHAD [12], NRAO/BYU L-band array [13] and the subject of this paper, the Australian SKA Pathfinder (ASKAP) [14].

Australian has a long history in radio astronomy dating back to World War II, with the main players being a number of universities and CSIRO. The CSIRO Division of Radiophysics (now CASS – CSIRO Astronomy and Space Science) built the Parkes 64 meter telescope as well as the 6 antenna Australia Telescope Compact Array, and has a strong skill base in radio-telescope instrumentation.

CSIRO and the Australian Federal Government determined that Australia could be an ideal site for the SKA. Australia has many areas which are sufficiently sparsely populated to give them the low-RFI background required for the SKA. To demonstrate this, a decision was made for Australia to build a pathfinder (or precursor) instrument, and CSIRO acquired a suitable site in Western Australia, the Boolardy Pastoral Station, and the Murchison Radio Observatory (_MRO_) was established within it. Appropriate legislation has been enacted to protect the site from incursions of Radio Frequency Interference (RFI).

The pathfinder would not only develop SKA technologies demonstrating wide field-of-view high-dynamic range performance on a world-class array, but would demonstrate the radio-quiet nature of the site and be a world class radio telescope in its own right. Initially conceived as a cylindrical reflector array (HYFAR) [15] it rapidly became ASKAP which is a dish array with phased array feeds.

2. SCIENCE WITH ASKAP

ASKAP was conceived as a fast survey instrument that could map the entire sky quickly, with an angular resolution of up to 10 arcsecond with excellent polarisation purity. The 36 dishes in ASKAP give a high quality instantaneous point spread function. This coupled with a three axis dish mount to keep the sidelobes fixed on the sky and the extremely low RFI background will provide very high (over 50 dB) signal to noise ratio in images. The sky coverage can also be exchanged for deep integrations on smaller areas of the sky to allow detection of extremely faint sources.

The correlator output data rate is 30Gb/s and this requires that the imaging be done in real time. This opens up possibilities for rapid follow-up of transient events. Transient observing is further enhanced by having 36 tied array beams that combine signals from all antennas for maximum sensitivity on compact sources.

The initial science drivers for ASKAP [16] [17] come from a subset of the SKA science [1]. With this as a basis, observing proposals were solicited for ASKAP based on the Science User policy completed in October 2009. The notion of a Survey Science Project was introduced and defined to be large (>1500 hrs) and coherent survey projects that utilise ASKAP's wide field-of-view and fast survey speed to enable major science outcomes. In line with the ASKAP guiding principles, the ASKAP Survey Science Projects, which in total will utilise at least 75% of observing time in the first five years, were selected following an open international call for proposals with competitive peer-review. The proposals have now been reduced to ten ASKAP survey science projects:

- EMU[18] will observe 75% of the sky in continuum with a sensitivity of 10μJy/beam rms. It will typically probe star forming galaxies to a redshift of 1.

- WALLABY[19] is an extra galactic neutral hydrogen survey over 75% of the sky. It will detect up to 500,00 galaxies up to a redshift of 0.26

- FLASH[20] will use the spectral line capabilities to perform a blind HI absorption survey. It focuses on the neutral gas content of galaxies in the redshift range 0.5 to 1.

- GASKAP[21] is a galactic spectral line survey of HI and OH. It will achieve an order of magnitude improvement in brightness sensitivity and resolution compared to previous surveys.

- POSSUM[22] is a polarization survey and will yield a fine grid of polarization rotation measures (RM) over 75% of the sky

- DINGO[23] studies the evolution of HI out to redshift of about 0.5. It will make measurements of key cosmological distributions.

There are two ASKAP Science Projects that explore transients:

- VAST[24] uses the ability of ASKAP to generate an image for each 5 second correlation dump. This allows transients with time scales as short as 5 seconds to be detected in real time.

- CRAFT[25] will detect fast transient with time scales of 1ms to 5 sec. These are transients associated with the most energetic single events in the universe.

A further two projects were selected as Strategic projects for SKA development

- COAST[26] will undertake a pulsar timing program to investigate gravitational waves and the limits of general relativity. COAST will also undertake a blind search for pulsars.
- VLBI[27] will work in conjunction with existing radio telescopes and the high speed data links to them. ASKAP is ideally placed to enhance VLBI capabilities in the southern hemisphere. ASKAP together with dishes in New Zealand increase the maximum baseline from 2000 to 5000km. Science includes the parallax and proper motion of pulsars and OH masers, and high resolution mapping of Active Galactic Nuclei.

3. SYSTEM ARCHITECTURE

The architecture of the ASKAP system is that of a synthesis radio-telescope where the feed element is a phased array. Examination of the science drivers for ASKAP resulted in the requirement for a reasonable number (25 – 50) of antennas distributed over a configuration of approximately 5 – 10 km to provide the combination of good u-v coverage, and angular resolution of up to 10 arcsec. Most of the science requires a 30 arcsec angular resolution and the final antenna configuration, **Figure 1**, has 30 antennas in a compact core about 2km across and six outlier antenna bringing the extent of the array to 6km. The inner 30 antennas generate u-v coverage that is approximately Gaussian. This results in an instantaneous point spread function with very low sidelobe levels. Good snapshot imaging is possible with this array. With the outer six antennas added the u-v coverage is less complete and high dynamic range imaging requires longer observations.

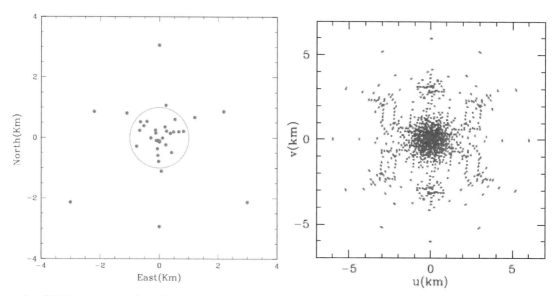

Figure 1. ASKAP antenna configuration on the left and the resulting instantaneous u-v coverage on the right.

The data path from the phased array feed element to the imaging computer is shown in Figure 2. The RF signal from the 188 element phased array feed is transported to the MRO Control and Correlator building over optical fibre. There it is coupled into a frequency domain digital beamformer that generates 36 dual polarization beams with high aperture efficiency (this is the Mark II implementation). The first six antennas are built to an earlier (Mark I) design [14] with downconversion and digitization in the antenna pedestal and digital links to the Control and Correlator building. The development of economical DFB lasers over the past few years has allowed the move to this "RF over Fiber" (RFoF) design. This significantly simplifies the requirements for electronics at each antenna. The Mark II correlator will process outputs from the Mark I and II beamformers. The correlator is an FX correlator based on polyphase filterbanks[28]. The correlator processes the whole 304MHz instantaneous bandwidth at a spectral resolution of

18.5kHz. Lower spectral resolutions are obtained by averaging across frequency channels. The high spectral resolution is achieved with cascaded polyphase filterbank[29]. The first (coarse) polyphase filterbank precedes the beamformer. This allows a simple weight-and-add approach to be used for beamforming.

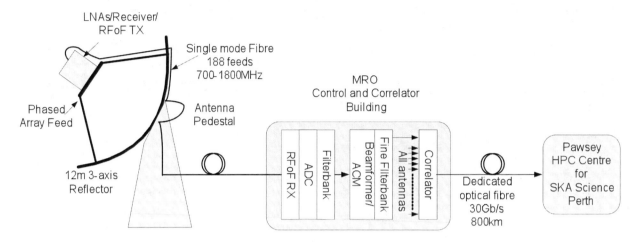

Figure 2. Simplified System diagram of ASKAP using Mark II PAFs

With a 5 second dump time and 16,416 frequency channels the correlator generates 30Gb/s of data. This is transported on a fibre based network to the Pawsey HPC (High Performance Computer) Centre for SKA Science in Perth where it will be reduced to images in real time. This 800 kilometre link has an extra 10Gb/s to provide for other users of the MRO site. In Perth the data is processed by a number of data pipelines. In one data pipeline a continuum image is generated for each 5 second integration. The self calibration parameters from this imaging are fed-back to the MRO and forms part of the continuous calibration of the telescope. Source finding on these images provides a key input into VAST. There is also a spectral line imaging pipeline and a continuum imaging data pipeline. Both these can handle long track observations. The performance of ASKAP for these last two types of imaging is shown in Table 1.

Table 1 Indicative survey speeds and sensitivities as a function of angular resolution for ASKAP assuming 50K system temperature and an aperture efficiency of 0.8 at an observing frequency of 1.4 GHz.

	10"	18"	30"	90"	180"
Survey Speed (deg^2 hr^{-1})					
Continuum (300MHz, 100µJy beam^{-1})	154	253	187	38	12
Spectral Line (100 kHz, 5mJy beam^{-1})	129	211	156	32	10
Surface Brightness (5 kHz, 1K)	-	-	0.77	13	66
Sensitivity 1 hr					
Continuum (300 MHz, µJy beam^{-1})	37	29	34	74	132
Spectral Line (100 kHz, mJy beam^{-1})	2.1	1.6	1.9	4.1	7.3
Surface Brightness (5 kHz, K)	-	-	5.2	1.3	0.56

WALLABY, FLASH, GASKAP and DINGO use the spectral images and EMU, VAST and POSSOM the continuum images. VAST, COAST and VLBI require tied array beams. For each of the 36 antenna beams a single tied array beam is generated. Calibration for this beamforming comes from the feed-back loop of calibration data from the 5 second integrations. CRAFT takes as its primary data 1MHz power spectra generated in the beamformer. These are provided with a 1ms cadence. CRAFT takes this data and de-chirps it over a range of Dispersion Measures (DMs) to detect transients in real time. A number of seconds of real time data is stored in the beamformer. This data can be retrieved for post processing if a transient is detected.

4. ANTENNAS

Analysis of the basic system requirements, and a detailed examination of the international antenna market place, resulted in the decision that antennas of approximately 12 metre diameter would best fit ASKAP's requirements.

The requirement for very high signal to noise ratio and high polarization purity dictated a "sky mount" antenna design that maintains a fixed orientation of the focal plane with respect to the celestial sphere during long observations (i.e. fixed parallactic angle). This also keeps off-axis beams pointing at a fixed sky position during an observation. Without this feature the imaging software, which is already taxing current techniques and resources, would need to handle beams (and sidelobes) that rotate on the sky. The alternative was to have the beamformer dynamically reposition the beams so that they remain fixed on the sky. As the technology for fixed-beam beamforming is yet to demonstrate high dynamic range imaging, the added risk was unacceptable. Keeping sidelobes fixed relative to sky also greatly enhances the dynamic range possible.

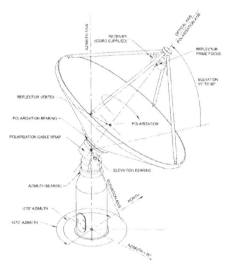

Figure 4. Generic "sky mount" antenna based on a standard alt-az mount with a third polarisation bearing that allows the whole reflector to rotate about a "polarisation" axis.

An equatorially mounted antenna is the traditional way of achieving a "sky mount". Generally alt-az mount antennas are known to be cheaper for a given size. It is also easier to achieve acceptable sky coverage. Optical telescope also achieve a fixed parallactic angle by rotating the imaging system in alt-az mount designs. In a radio telescope this function is implemented with a feed rotator. With this approach the beams are fixed but sidelobes still rotate on the sky. In ASKAP another possibility is explored: rotate the whole feed and reflector structure with a third axis of rotation. When compared to the cost of a feed rotator it was found that the three axis design was competitive on prices, and it achieved a true "sky mount". The third axis allows the antenna to exactly mimic the motion of an equatorially mounted antenna.

In 2008, an international tender process for the delivery of 36, 3-axis 12 metre antennas meeting a number of key requirements (performance such as surface RMS, pointing and tracking allowing operations to 10 GHz, etc) resulted in a contract with the 54th Research Institute of the China Electronics Technology Group (known as CETC54 from Shijiazhuang, in Hebei Province, China) for the design and development of these antennas. The first antenna successfully passed the factory acceptance test in and was delivered and assembled on site and five more were installed by October 2010 and the final 30 were completed between June 2011 and June 2012. They are currently undergoing final acceptance testing.

5. RECEIVERS

A key requirement for ASKAP was for large areas of sky to be surveyed quickly and deeply. This can only be done if it is possible to image a significant area of the sky with multiple individual beams simultaneously, with sensitivity not significantly lower than traditional single pixel receivers.

Multiple feed receivers systems are not new [30]. These systems used multiple feed horns at the focus of a dish. The spacing between the feed horns is such that it is only possible to capture part of the energy at the focus. Also the off-axis feed horns cannot compensate for aberration. An alternative approach with a number of advantages is to place a phased array feed (PAF) at the focus of the dish (Figure 4) [31]. The PAF captures most of the incident energy and by weighting and summing the signals of the individual PAF elements, multiple beams with high sensitivity can be produced. Such beamforming is performed in the digital domain, allowing calibration of the element responses and the flexibility to tailor beam properties such as spacing, sidelobes and cross polarization to best suit the astronomy requirements.

CSIRO has developed a novel PAF based on a planar connected array[32]. In this structure, neighbouring elements of a chequerboard array of conducting patches are connected by two-wire transmission lines that divert the array signals through low-noise amplifiers (LNAs) situated behind a ground-plane. Connecting the array elements in this way allows operation over a wide frequency range with the array element spacing less than half the minimum wavelength. The approach requires accurate analysis and design of the array [33] in concert with LNAs that have differential impedances in the order of 300ohm [34] [35]. A PAF testing facility developed at Parkes has been vital to this development [36]. Preliminary measurements of noise temperatures less than 60K on an enhanced array and LNA design indicate that very good low-noise performance has been obtained over the full 0.7-1.8GHz frequency range required for ASKAP.

Figure 5. Three ASKAP antennas at the MRO with Phased Array Feeds installed.

The ASKAP PAF development also includes a package of enhancements to the RF and digital signal processing chains [37]. In the Mark I PAFs the LNAs are mounted directly on the back of the ground-plane. A frame is then constructed over them to hold gain cards that also provide sub-octave band filtering, gain levelling and monitoring. In the Mark II PAFs all these functions occur in a "Domino" that is removable without deconstructing any of the PAF mechanical structure. The Mark II design has also seen the change from chilled water temperature stabilisation to Peltier cooling, simplifying the support infrastructure requirements. In the Mark II design heat is conducted to the edge of the PAF through heat pipes embedded in the ground plane. Here they connect to Peltier cooling elements that conduct the heat to the exterior of the PAF where it is dissipated to the air.

6. DATA TRANSPORT

The result of using area type receivers in the focal plane, even ones with only a few hundred pixels, on a 36 antenna interferometer is that raw data rates are reaching magnitudes that have not been seen before in radio-astronomy. Each PAF produces approximately 100GHz of bandwidth. In the Mark II design this is digitized by 12-bit ADCs to generate 3.4 Terabits of data per second (Tb/sec). After first stage processing, the output for all antennas into the beamformer is up to 103Tbs on 10,368 fibres.

For the Mark I PAF [14], the RF signal is transported from the final stage amplifier output on 3.15 mm coaxial cables. These run from the rear of the PAF, down the quadrupod support legs, through the cable wraps to the pedestal of the antenna. Here the signal is down-converted, digitized, and passed through a 1MHz resolution filterbank. A total of 304 of the 1MHz channels are then transmitted on fibre to the Control and Correlator building. This fibre is Prysmian Fusionlink, an economical loose sheath, ribbon fibre design which has 12 G652 fibres in each ribbon, with 18 ribbons in a single cable. These ribbons can be terminated in a single connector or via a ribbon to single fibre interconnect, allowing a range of single fibre connectors to be used and the signals to be routed as required.

As a result of the significant advances in technology over the last few years, and the resultant cost reduction in optical fibre components, it is now possible to economically modulate the raw RF onto the fibre directly from the PAF using uncooled distributed feedback DFB lasers at 1310nm. This is the basis of the Mark II PAF which will be installed on all 36 antennas, and it significantly reduces the quantity of both cables and electronics required within each antenna, as the fibres can be installed in a continuous run all the way from the PAF to the Control and Correlator building.

7. DIGITAL SIGNAL PROCESSING

For the Mark II PAF, digitizing and the first filterbank occurs in the Correlator room. Instead of the dual heterodyne receiver used on the Mark I PAFs, the RF signal is directly digitized in the second or third Nyquist zones of the Analog-to-Digital Convertor (ADC). The ADC is operated at either 5/6 or 2 times the sample rate of the Mark I digitizer and the size of the filterbank is increased accordingly. Thus the 1MHz channels of the Mark II PAF can be made to exactly align with the 1MHz channels of the Mark I PAF. This feature allows the two generations of PAF systems to connect to a common correlator.

Digitising in the Mark II system occurs in a CSIRO developed DragonFly-3 system, Figure 5. Each DragonFly-3 system is a standard 1U server chassis and power supply which houses:

- 8 dual channel RF-over-Fibre receivers. The receivers incorporate filters to select the correct Nyquist zone and attenuators for level setting.
- A DragonFly-3 board which has 8 dual channel 12 bit ADCs that can clock at 1.6GS/s, four XC7K325T FPGAs to implement the filterbank, the largest of which is a 1536 real input filterbank that clocks at 1.536GHz, and a pair of Avago MiniPOD multimode fibre ribbon transmitter (total output 240Gb/s)
- An ADC clock synthesizer that can generate the 8 ADC clock with a timing jitter of 0.18ps
- A Bullant-3 control board that performs all command and control function including control of fans and power to the DragonFly-3

The RF data is digitized on the DragonFly-3 board and passed through the filterbanks in the FPGAs. 1MHz channel data is then exchanged between the four FPGAs so that each has one quarter of the bandwidth data for all 16 RF inputs. This data is then packetized and transported out the system on two 12 fibre multimode ribbons.

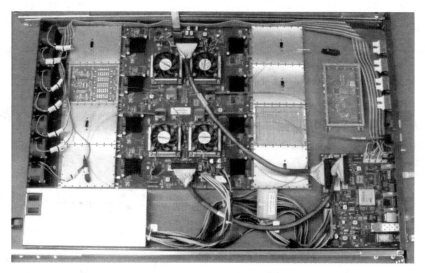

Figure 5. DragonFly-3 system under development by CSIRO Astronomy and Space Science (CASS)

There are 12 DragonFly-3 systems per antenna and the fiber ribbons from corresponding optical transmitters are cross connected passively. This then connects to a Redback-3 processing system. The Redback-3 system uses the same infrastructure as the DragonFly-3 system but omits the ADC synthesizer and replaces the DragonFly-3 board with a Redback-3 processing board. This board has six Xilinx Kintex FPGAs, either XC7K480T, 420T or 355T. Each FPGA has two SODIMM DRAMs and is fully connected to all other FPGAs. Data I/O is through three 12-fibre transmitters and three 12-fibre receivers giving a data bandwidth of 432Gb/s.

After the passive optical cross connect each Redback-3 system receives three 12-fibre ribbons from the DragonFly-3 systems, which transport 48MHz of bandwidth for all 192 RFoF inputs. The Redback-3 system implements four major functions on this data:

- beamforming of 36 dual polarization beams, with each beam being the weighted sum of 63 PAF elements;
- a fine filter bank on the 1MHz channels to give a final frequency resolution of 18.5kHz;
- calculation of an Array Covariance Matrix, and
- storage of real-time beam data to DRAM to allow the capture of full data sets of transient events

The Array covariance matrix is a correlation between all ports from the PAF as well as 4 extra ports for other signals. The processing of an ACM for all 304 1MHz channels for all time samples has compute requirements that exceed that of the correlator that processes the beam data. To make it manageable, only 1 in 4 frequency channels are processed at any one time and then only every fourth time sample is used. The ACM is used for the calculation of beam weights. It includes correlations between PAF ports and a calibration signal, which are used to track gain and phase variations in the PAF analogue electronics.

For ASKAP, the beamforming results in data compression and a single 12-fibre ribbon is sufficient to transport data out of the Redback-3 beamformers. A second passive optical cross-connect is used to aggregate beam data from 12 antennas onto individual 12-fibre ribbons and three of these go to Redback-3 systems that are used in the correlator. The three ribbons carry data for all 36 antennas. Each Redback-3 correlator processes 48 MHz of bandwidth and implements a full Stokes correlator on all 36 antenna beams. In addition the Reback-3 correlator will generate a single tied array beam for each antenna. This gives a total of 36 dual polarization tied array beams that used for transient process such a pulsar timing and short time spectral analysis of variables.

Raw correlations are dumped to the Telescope Operating System (TOS) every 5 seconds. TOS formats the data and adds metadata before transmitting the data over the 40Gb/s link to the Pawsey HPC Centre in Perth

8. SOFTWARE AND COMPUTING

The ASKAP Computing System architecture has been described previously [38]. Here we concentrate on the science data processing.

The requirement of high survey speed inevitably leads to high computational loads, no matter how the survey speed is achieved. For ASKAP, one can think of each separate beam as giving a separate (large) radio interferometric array. Thus, computationally, ASKAP requires roughly 36 times the processing power of a single beam system. Since each single beam has 16,416 channels and 36 antennas, the processing rate can become very large. A sequence of estimates and benchmarks has led to a specification of 100TFlops for the entire processing load, including all pipelines and the transient detection. This will be realised in a dedicated Real Time Computer (RTC) at the Pawsey Centre, consisting of roughly 10,000 x86 cores.

Controlling the data processing interactively will not be advisable or even possible. Consequently, the processing will be automatic, being triggered when all relevant data are available. The steps in processing are:

- Ingest of data from the correlator and meta-data from the Telescope Operating System
- Re-ordering of the data as needed for subsequent processing
- Editing of radio frequency interference (RFI) and other bad data
- Estimation of calibration parameters using a previously obtained Sky Model for the region being observed
- Application of calibration parameters
- Construction of images from the measured and calibrated visibilities
- Source finding and fitting

Construction of the images is by far the most time consuming portion of the processing, and the gridding/de-gridding dominates. Consequently, we have investigated the gridding operation in considerable detail [39]. During the gridding operation, we correct for two effects: first the w term [40] of the baseline vector, and second the antenna/beam illumination pattern. There is no globally optimal way to construct the image but for ASKAP, the best approach is to use snapshot imaging to collapse all observations at one parallactic angle onto a nearly two-dimensional space, and then use convolution to apply the net effect, Cornwell, Voronkov, Humphreys (in prep).

Broadband continuum images require several cycles of deconvolution and self-calibration. In addition, spectral index effects must be estimated and corrected [41]. Finally, we expect that time-variable sources must be estimated and removed from the data on a few minutes time-scale.

Spectral line imaging is simpler but similarly expensive. The Sky Model must be removed from the observed data, after estimation and correction of the antenna/beam bandpasses. Then each of the 16,416 channel images must be calculates in turn. We anticipate that no deconvolution will be necessary except for a moderate number of nearby galaxies.

The efficacy of this automated processing relies upon a number of factors:

- The Sky at these frequencies and resolutions is relatively simple, consisting largely of unresolved or partially resolved sources
- ASKAP has excellent Fourier plane coverage, especially once multi-frequency synthesis is accounted for
- The atmosphere (troposphere and ionosphere) is relatively benign and isoplanatic, requiring only estimation every few minutes at most
- The Phased Array Feed calibration upstream corrects most of the short-term calibration drifts.

All of these, save the last, are well-established, and the last remains to be demonstrated when observing commences.

Once processed, the data products will be placed in the ASKAP Science Archive Facility (ASDAF) for access by the Survey Science Team members. The Survey Science Teams are responsible for evaluating science data quality, and authorising general release. This quality control process is on top of the normal QC process followed by ASKAP.

9. THE MURCHISON RADIO-ASTRONOMY OBSERVATORY

The Murchison Radio-astronomy Observatory (MRO) is Australia's site for the core of the SKA and was described previously [14]. The infrastructure construction (antenna foundations, roads, power and fibre distribution system, Control and Correlator building) was designed by Aurecon Pty Ltd and constructed by McConnell Dowell Constructors (Australia) Pty Ltd during 2011 – 2012. The power station will be built during 2013, and will be a hybrid system, using diesel generators for the base load, initially with photovoltaic solar cells providing up to 500 kW of power during the day, but with an expansion of the renewable aspects of the power plant planned for 2014.

The MRO also supports a number of other projects, including the Murchison Widefield Array (MWA) [7] [42]. MWA operates in the frequency range 80-300MHz. The antenna MWA uses is a tile composed of a 4x4 grid of dipoles. The tiles are RF beamformed to produce a single beam on the sky. MWA currently has a 32 tile (32T) array operating at the MRO and is currently upgrading to 128 tiles (128T). In 128T the correlator is housed in the Control and Correlator building and all cabling to the building from the MWA site is in place.

The Mid West Radio Quiet Zone has been in force since 2005 and is providing effective protection for radio astronomy in the region including and surrounding the MRO. The radio-quietness is primarily protected by the Mid West Radio Quiet Zone Frequency Band Plan, a legislative instrument implemented by the Australian Communications and Media Authority, Australia's spectrum regulator. Band Plans designate how spectrum can be used in Australia. Within 70km radius from the MRO, all other services are secondary to radio astronomy. The Band Plan extends to 150km radius from the MRO and stipulates that other potential spectrum users should consult with the managers of the MRO, with the aim of ensuring that their proposed activity does not cause detrimental interference to the radio astronomy activities in the region. A further ACMA instrument extends protection to 260km radius from the MRO centre. In addition, the Government of Western Australia has implemented measures to protect the radio astronomy activities in the MRO and surrounding areas from incidental emissions from mining.

10. CONCLUSION

ASKAP has passed several key milestones, including meeting Phased Array Feed performance requirements (driven by the science goals). With the completion of the core MRO infrastructure, and the installation of the first PAFs on antennas, ASKAP has reached the early commissioning phase. During 2012 and 2013, we will begin the process of commissioning ASKAP and understanding how to do astronomy with Phased Array Feeds.

11. ACKNOWLEDGEMENTS

ASKAP is the result of a large team at CSIRO – over 80 people have contributed directly to the development and construction of ASKAP, and the authors thank the entire team for their work in developing and building ASKAP.

This scientific work uses data obtained from the Murchison Radio-astronomy Observatory (MRO), which is jointly funded by the Commonwealth Government of Australia and State Government of Western Australia. The MRO is managed by the CSIRO, who also provide operational support to ASKAP. We acknowledge the Wajarri Yamatji people as the traditional owners of the Observatory site.

12. REFERENCES

[1] Carilli, R., and Rawlings, S., Eds, "Science with the Square Kilometer Array," Elsevier, (2004) Amsterdam
[2] Taylor, A.R and Braun, R., Eds, "Science with the Square Kilometer Array : a next generation world radio observatory," Netherlands Foundation for Research in Astronomy (1999)
[3] M.A. Garrett, J.M. Cordes, D. de Boer, J.L. Jonas, S. Rawlings, R.T. Schilizzi "A Concept for SKA Phase 1 (SKA1)," SKA Memo 125, (2010)
[4] Kot, J.S., Donelson, R., a Nikolic, N., Hayman, D., O'shea, M. & Peeters, G., "A Spherical Lens For The SKA," Experimental Astronomy 17, 141–148, (2004)

[5] Bunton, J.D., "Cylindrical Reflectors," Experimental Astronomy 17, 185-189, (2004)
[6] de Vos, M., Gunst, A.W., Nijboer, R., "The LOFAR Telescope: System Architecture and Signal Processing," Proceedings of the IEEE , 97(8), 1431-1437, (2009)
[7] Lonsdale, C.J. et al, "The Murchison Widefield Array: Design Overview" IEEE Proceedings, 97(8), 1497 - 1506 (2009)
[8] Ellingson, S.W.; Clarke, T.E.; Cohen, A.; Craig, J.; Kassim, N.E.; Pihlstrom, Y.; Rickard, L.J.; Taylor, G.B.; , "The Long Wavelength Array," *Proceedings of the IEEE* , 97(8), 1421-1430 (2009)
[9] Parsons, A.R. et al, "The Precision Array For Probing The Epoch Of Re-Ionization: Eight Station Results" The Astronomical Journal 139, 1468–1480, (2010)
[10] Ardenne, A., Wilkinson, P., Patel, p., bij de Vaate, J.G., "Electronic Multi-beam Radio Astronomy Concept: Embrace a Demonstrator for the European SKA Program, "Experimental Astronomy, 17 (1-3), 65-77, (2005)
[11] Verheijen, M.A.W., Oosterloo, T. A., van Cappellen, W. A., Bakker, L., Ivashina, M.V., and van der Hulst, J.M., "Apertif, a focal plane array for the WSRT" AIP Conf. Proc. 1035, 265-271, (2008)
[12] Veidt, B. and Dewdney P., "A phased-array feed demonstrator for radio telescopes," Proc. XXVIIth URSI General Assembly, Vigyan Bhavan, New Delhi, India, (2005)
[13] Warnick, K.F.; Jeffs, B.D.; Landon, J.; Waldron, J.; Jones, D.; Fisher, J.R.; Norrod, R., "Beamforming and imaging with the BYU/NRAO L-band 19-element phased array feed," Antenna Technology and Applied Electromagnetics and the Canadian Radio Science Meeting. ANTEM/URSI 2009. 13th International Symposium on, 15-18 Feb. (2009)
[14] DeBoer, D.R. et al "Australian SKA Pathfinder: A High-Dynamic Range Wide-Field of View Survey Telescope Array," IEEE Proceedings, 97(8) pp1507-1521 Aug 2009
[15] Bunton, J.D., Briggs, F.H. and Blake, C.A., " HYFAR: An Array of Cylindrical Reflectors for Precision Cosmology," WARS'04 Conference, Hobart (2004)
[16] Johnston, S., et al, "Science with the Australian Square Kilometre Array Pathfinder," PASA 24, 174-188, (2007)
[17] Johnston, S et al, " Science with ASKAP," Exp Astron 22, 151-273, (2008)
[18] Norris, R.P., et al, "EMU: Evolutionary Map of the Universe," PASA, vol 28, pp 215 – 248, (2011)
[19] Duffy, A.R., Moss, A., Staverley-Smith, L., "Cosmological surveys with the Australian Square Kilometre Array Pathfinder" Accepted for publication PASA
[20] Allison, J. R., Curran, S. J., Emonts, B. H. C., Geréb, K., Mahony, E. K., Reeves, S., Sadler, E. M., Tanna, A., Whiting, M. T. and Zwaan, M. A. "A search for 21 cm HI absorption in AT20G compact radio galaxies." Monthly Notices of the Royal Astronomical Society, 423: 2601–2616 (2012),
[21] Stanimirovič, S., Dickey, J.M., Gibson, S.J., Gómez, J.F., Imai, H., Jones, P.A. and van Loon, J.T. "GASKAP: The Galactic ASKAP Survey," Proceedings of the International Astronomical Union, 5, 819-819 (2009).
[22] Gaensler, B.M., Landecker, T.L., Taylor, A.R., "Survey Science with ASKAP: polarization Sky Survey of the Universe's Magnetism (POSSUM)," Bulletin of the American Astronomical Society, 42, 515 (2010)
[23] Meyer, M. et al, "Exploring the HI Universe with ASKAP," Proceedings of Panoramic Radio Astronomy: Wide-field 1-2 GHz research on galaxy evolution. June 2-5. Groningen, the Netherlands. Edited by G. Heald and P. Serra.(2009)
[24] Chatterjee, S., Murphy, T., VAST collaboration, "Survey Science with ASKAP: Variables and Slow Transients (VAST)," Bulletin of the American Astronomical Society, 42, 515 (2010)
[25] Marquart, J-P., "The Commensal Real-Time ASKAP Fast-Transients (CRAFT) Survey," PASA, 27(3) 272 – 282
[26] Stairs, I. H., "Pulsars wit the Australian Square Kilometre Array Pathfinder," AIP Conference Proceedings, 1357, (2011)
[27] Leonid, P. et al, "First Geodetic Observations Using New VLBI Stations ASKAP-29 and WARK12M," Publications of the Astronomical Society of Australia 28, 107–116(2011)
[28] Bunton, J.D., "SKA Correlator Advances," Experimental Astronomy, 17(1-3)251-259, (2004)
[29] Tuthill, J., Hampson, G., Bunton, J., Brown, A., Neuhold, S., Bateman, T., Jayasri, J. & de Souza, L. "Development of Multi-Stage Filter Banks for ASKAP," ICEAA 2012, Cape Town, South Africa, 2-7 September 2012
[30] Staveley-Smith, L. et al "The Parkes 21cm multibeam receiver," Publications Astronomical Society of Australia, 13(3) 243-248 (1996)

[31] Fisher, J.R., and R.F. Bradley, "Full-sampling array feeds for radio telescopes," Proceedings of the SPIE, 4015, 308-319, 2000.
[32] Hay, S.G., and O'Sullivan, J.D., "Analysis of Common-Mode Effects in a Dual-Polarized Planar Connected-Array Antenna," Radio Science, 43, RS4S06, 2008.
[33] Hay, S.G., O'Sullivan J.D. and Mittra, R "Connected patch array analysis using the Characteristic Basis Function Method," *IEEE Trans. Ant. Prop.*, 59(6) 1828 - 1837 (2011)
[34] Hay, S.G., "Maximum-sensitivity matching of connected-array antennas subject to Lange noise constants," *International Journal of Microwave and Optical Technology*, 5(6) 375-383 (2010)
[35] Shaw, R.D, Hay, S.G and Ranga, Y. "Development of a low-noise active balun for a dual-polarized planar connected-array antenna for ASKAP," *International Conference on Electromagnetics in Advanced Applications*, Cape Town, South Africa, September 2012
[36] Chippendale, A., O'Sullivan, J.D., Reynolds, J., Gough, R., Hayman, D., and Hay, S., "Phased Array Feed Testing for Astronomy with ASKAP," *Phased Array Systems and Technology*, 2010 IEEE International Symposium on, (2010)
[37] G. Hampson, A. Macleod, R. Beresford, M. Brothers, A. Brown, J. Bunton, C. Cantrall, R. Chekkala, W. Cheng, R. Forsyth, R. Gough, S. Hay, J. Kanapathippillai, D. Kiral1,M. Leach, N. Morison, S. Neuhold, P. Roberts, R. Shaw, A. Schinckel, M. Shields, and J. Tuthill "ASKAP PAF ADE - advancing an L-band PAF design towards SKA," *International Conference on Electromagnetics in Advanced Applications*, Cape Town, South Africa, September 2012
[38] Guzman, J.C. and Humphreys, B.. "The Australian SKA Pathfinder (ASKAP) software architecture," In Proceedings of SPIE, 7740, page 77401J, (2010)
[39] Humphreys, B. and Cornwell, T.J., "Analysis of convolutional resampling algorithm performance,"SKA memo 132. Technical report, SKA Telescope, (2011).
[40] Cornwell, T.J. and Perley, R.A., "Radio-interferometric imaging of very large fields - The problem of non-coplanar arrays," Astron. & Astrophys., 261 353–364 (1992)
[41] Rau U., and Cornwell, T.J., "A multi-scale multi-frequency deconvolution algorithm for synthesis imaging in radio interferometry," Astron. & Astrophys 532, article A71 (2011)
[42] Tingay, S.J. et al., "The Murchison Widefield Array: the Square Kilometre Array Precursor at low frequencies," PASA submitted, arXiv.1206.6945, (2012)

LOFAR, the low frequency array

R.C. Vermeulen[a], on behalf of the LOFAR collaboration
[a]ASTRON Netherlands Institute for Radio Astronomy,
P.O. Box 2, 7990 AA Dwingeloo, The Netherlands

ABSTRACT

LOFAR, the Low Frequency Array, is a next-generation radio telescope designed by ASTRON, with antenna stations concentrated in the north of the Netherlands and currently spread into Germany, France, Sweden and the United Kingdom; plans for more LOFAR stations exist in several other countries. Utilizing a novel, phased-array design, LOFAR is optimized for the largely unexplored low frequency range between 30 and 240 MHz. Digital beam-forming techniques make the LOFAR system agile and allow for rapid re-pointing of the telescopes as well as the potential for multiple simultaneous observations. Processing (e.g. cross-correlation) takes place in the LOFAR BlueGene/P supercomputer, and associated post-processing facilities. With its dense core (inner few km) array and long (more than 1000 km) interferometric baselines, LOFAR reaches unparalleled sensitivity and resolution in the low frequency radio regime.

The International LOFAR Telescope (ILT) is now issuing its first call for observing projects that will be peer reviewed and selected for observing starting in December. Part of the allocations will be made on the basis of a fully Open Skies policy; there are also reserved fractions assigned by national consortia in return for contributions from their country to the ILT. In this invited talk, the gradually expanding complement of operationally verified observing modes and capabilities are reviewed, and some of the exciting first astronomical results are presented.

Keywords: Radio astronomy, radio telescopes, radio surveys, LOFAR

1. INTRODUCTION; TECHNICAL DESIGN

1.1 Dipole receptors and phased array beam-forming

LOFAR, the Low Frequency Array, was designed and built by ASTRON for innovative science and in-depth exploration of the spectral window below 240 MHz. The International LOFAR Telescope (ILT) is an array of dipole antenna stations distributed across parts of the Netherlands and into other European countries with central processing facilities for interferometry and other applications. The stations have no moving parts: dipole low-band antennas (LBA) function down to 10 MHz and are optimized for 30 – 80 MHz, while the separate upper range 120 – 240 MHz (avoiding the FM radio broadcasting band) is covered with high-band antenna tiles (HBA) of 4x4 dipoles with built-in amplifiers and an analogue beam-former. Both receptor types are shown in Figure 1.

Figure 1. Low Band Antenna dipoles (left) and High Band Antenna tiles of dipoles (right) give very wide field-of-view.

Each LOFAR station in the Netherlands has 96 LBA dipoles (selectable at any time as either 48 Inner or 48 Outer) and 48 HBA antenna tiles; international stations have both 96 LBA antennas and 96 HBA tiles. This architecture, with effectively all-sky coverage using many low-cost dipoles, gives LOFAR an unprecedentedly large field-of-view (many degrees, depending on frequency). The signals from individual dipoles or tiles are combined digitally into phased array stations. Layered electronic beam-forming techniques at the tile, station, and multi-station levels not only allow for rapid re-pointing of the telescope but even for simultaneous observing of multiple, independent areas of the sky, as illustrated in Figure 2. A unique feature of the LOFAR stations are the Transient Buffer Boards, that are soon to be extended to allow the recovery at any moment of the last 5.2 sec of full bandwidth un-averaged data from the individual dipoles or tiles. This allows post-facto beam-forming in any arbitrary direction, in response to an externally or internally generated trigger concerning a transient event anywhere on the sky.

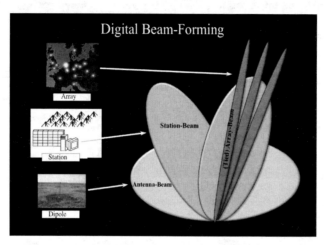

Figure 2. Electronic beam-forming at the station and multi-station levels allows the wide field of view of the individual LOFAR dipoles to be used for simultaneous observations of multiple, independent areas of the sky.

1.2 Station and array topology

The core area in the Netherlands was chosen to be near the town of Exloo in the northeastern province of Drenthe, which has relatively low population density and a comparatively benign radio frequency interference environment, which is reinforced through collaboration with a nature conservancy agency to establish a wetlands area. The distribution of the LOFAR core stations is optimized for high fidelity imaging of spatially extended sources: there are 24 stations within the inner 2 km wide area, of which 6 are densely packed onto a 300 m diameter area called the Superterp, that is slightly raised above the surrounding wetlands. Most of the planned 16 remote stations out to distances up to 80 km in the Netherlands have now also been rolled out; these allow good higher resolution imaging. The array in the Netherlands is currently further complemented with 5 stations in Germany, and 1 each in France, Sweden, and the United Kingdom; these give baselines up to 1000 km and allow sub-arcsecond resolution in the high band. Further European stations are in advanced stages of planning. The international stations, while individually owned, are jointly operated within the International LOFAR Telescope (ILT). The inner and outer topologies of the ILT are illustrated in Figure 3.

1.3 Central processing and storage

All ILT stations can be centrally controlled from the operations room of ASTRON in Dwingeloo, the Netherlands. Each station can also be operated locally in stand-alone mode; some international station owners have installed local data processing equipment for private observations in addition to participating in ILT observing. All stations have dedicated (owned or leased) 3 Gb/s or 10 Gb/s data connections to the LOFAR central processing facilities at the computing centre of the University of Groningen in the Netherlands. The entire array, or arbitrary subsets of stations, can operate in interferometric mode for imaging. The core can be coherently or incoherently summed to give tied array. Any and all single station data streams flowing into Groningen can also be used for time series or dynamic spectra.

Figure 3. Left: The LOFAR Superterp is the innermost area, densely packed with the hardware of 6 stations. It is surrounded by several of the innermost core stations, visible as separate patches. Right: the locations of the current 8 international LOFAR stations in relation to the concentration of stations in the Netherlands.

The data processing capacity in Groningen currently consists of a BlueGene/P supercomputer offering 42 Tflops and 640 Gbit/s I/O, connected to a multi-node CPU cluster with 20 Tflops and 2 Pbyte temporary storage for post-processing operations. The ILT Long Term Archive is expected to be expanding at a rate of 2 – 4 Pbyte per year, based on collaborations involving the computing centers of Amsterdam, Groningen, and Jülich with other partners. While the raw visibility datasets are too large to store, the long-term archive will contain averaged visibilities as well as processed data such as images, time series, source and event lists, etc. The data in this e-archive will be open to the general public after a proprietary period (by default one year), and are expected to play an important part in the overall science yield obtained with this revolutionary new instrument.

2. THE INTERNATIONAL LOFAR TELESCOPE FOUNDATION

As part of the formal opening of LOFAR by Her Majesty queen Beatrix of the Netherlands, on 12 June 2010, a memorandum of understanding was signed by ASTRON and LOFAR astronomy consortia in Germany, France, the Netherlands, Sweden, and the United Kingdom. This paved the way to the formation in the Netherlands of the International LOFAR Telescope (ILT) Foundation, on 9 November 2010, as the vehicle for the international collaboration between the owners of LOFAR infrastructure (stations, computing, archives) and the astronomical stakeholders. The ILT seeks to maximize the science yield for its stakeholders and the full astronomical community, through the joint exploitation under a common science policy of the LOFAR infrastructure owned by the partners.

The LOFAR astronomical consortia bundle the interests of the parties involved in LOFAR (universities and other research institutes, station owners, etc.) in their country. The detailed setup of the consortia varies with local needs. ASTRON is the central operational organization within the ILT; all other station owners share in the cash and in-kind costs of central operations. Since 21 June 2011, the ILT Board, which sets the overall science policies, has representatives of ASTRON and all of the consortia in the countries where LOFAR stations are currently located.

3. FIRST PROPOSAL CALL WITH OPEN SKIES

After a period of concerted commissioning work, the ILT is now making its first call for Regular Proposals. This call will for the first time allow participation in LOFAR from the worldwide community. During Cycle 0, running from 01 December 2012 to 31 May 2013, there will be 241 hours, corresponding to 10% of the available operational time, available under a fully Open Skies policy based purely on scientific merit as well as technical feasibility. After uniform review of all Regular proposals, other parts of the observing time will be allocated in tandem between the independent Programme Committee and the individual LOFAR astronomy consortia listed above. They will generally also take into account relevance in their respective local contexts. This mechanism also allows a number of large Key Science Projects to get under way; details of all Reserved Access proposals that have already received umbrella allocations are publicly available online. The fully Open Skies time will expand later.

For Cycle 0, there is a limited but diverse set of standard observing and data processing capabilities. LOFAR version 1 delivers correlated visibility data, and in/coherently added single and multiple station data (several beam-formed modes), and transient buffer read-out. Synthesis images can be produced, and there is also a pipeline for known pulsar data processing. Proposers should take careful note of the current limitations, such as on field sizes, resolutions, and noise levels of initial images if produced from an automatic pipeline. For many science goals further processing steps requiring expertise and/or specific tooling; some recent results are presented below. Computing and storage for data processing will be a limiting resource for many projects. All projects require a careful technical justification, referring, where appropriate, to expertise, tools, manpower, and computing capacity required from the ILT or supported within the proposing team. Expertise from the worldwide community to participate in projects selected under Open Skies Central is welcomed. Full details are available from http://www.astron.nl/observatory/lofar.

LOFAR Version 1 already allows novel and cutting-edge science projects, while considerable development is ongoing to reach the full potential of the ILT. The online information about LOFAR capabilities for Cycle 0 will be supplemented with further details until 31 August 2012, two weeks before the proposal submission deadline of 15 September 2012.

4. COMMISSIONING AND FIRST SCIENCE RESULTS

The past two years have been a period of intensive commissioning work, in which the software developers and astronomers at ASTRON have collaborated closely with the Key Science Project teams. Below, the main goals of each of these Key Science Project teams and some of the most promising first results are briefly presented. The joint commissioning work also extends to the Multi-Snapshot Sky Survey (MSSS), which is designed to yield an all-sky catalogue of calibrator sources, to be used for detailed calibration of the deeper imaging observations that are to follow. The MSSS catalog itself will also be used for population studies and other astrophysical applications. MSSS was started in late 2011; the first pass of observing, covering the 30-80 MHz (LBA) band, is currently being completed, and HBA observing will follow. The catalog and images will be put into the public archive. The requirements of MSSS have been a prime driver both to develop the full, robust end-to-end LOFAR observing system, as well as to develop an initial automatic imaging pipeline, which is currently being extended and improved.

There are six Key Science Projects, each in fact covering a range of broad astrophysical topics that are enabled by a particular observing technique or feature of the LOFAR functionality. The design specifications and the development of specific hardware and software for LOFAR have been driven in part by the requirements of these Key Science Projects, worked out in long-standing collaborations between the design team at ASTRON and the astronomers involved.

Arguably the most challenging Key Science Project is aimed at the detection of the Epoch of Reionisation via its spectral and spatial radio emission/absorption signature in the redshifted HI 21 line. Brightness temperature steps of about 30 mK are expected to occur over a range in redshift that overlaps the LOFAR HBA frequency range ($z=6-11$). The team intends to observe a small number of fields for several hundreds of hours each. The cardinal issue facing them is not so much to accumulate raw sensitivity, but rather to achieve very high dynamic range and precise subtraction of numerous foreground components, which places stringent requirements on calibration and repeatability. It is most gratifying, therefore, that for a single 6-hr observing run the team has already been able to reach a thermal-noise limited image (0.15 mJy/beam) with a 600000:1 dynamic range, for their primary chosen field, and after a lot of dedicated processing. A recent summary of this project was given by Zaroubi et al. (2012, MNRAS in press; arXiv:1205.3449).

Another group is preparing for a three-tiered large-area sky survey in multiple LOFAR radio continuum bands. The main science goals of the surveys Key Science Project team are to study the formation and evolution of the first massive black holes, galaxies, and clusters of galaxies at redshifts $z=6$ and higher, to chart in detail the grand era of star formation at redshifts $z\sim3$, and to make in depth studies of the gas in galaxy clusters at $z\sim1$. Their plans are described in detail by Röttgering (2010, http://pos.sisa.it/cgi-bin/reader/conf.cgi?confid=112,p50). The commissioning efforts with the team at ASTRON to date have focused on understanding the challenges of extremely wide-field imaging, that requires handling of the non-planar and non-isoplanatic effects, and understanding the detailed response function of the antennas, antenna tiles, and antenna stations. The effects the brightest sources need to be removed from the data regardless of their separation on the sky form the central field of interest. The effort to reach the ultimate, noise-limited image quality using procedures that can be automated is still proceeding; handling the ionospheric phase screen under current Solar maximum conditions is a particularly difficult challenge. The range of images that have already been produced, however, is already impressive; a gallery of examples is in Figure 4.

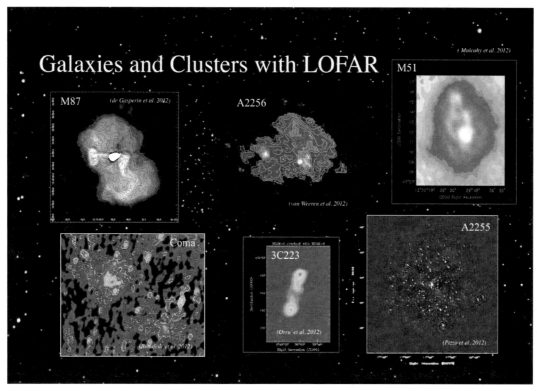

Figure 4. Gallery of early LOFAR images; some show the full field-of-view, others depict only the central source targeted by the observation. Credits to the main investigator involved are on each image.

There are also already impressive demonstrations of the use of the international LOFAR stations to achieve low frequency radio images with sub-arcsecond resolution, such as the pilot observation of Cas-A shown in Figure 5.

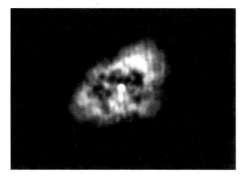

Figure 5. LOFAR high-band image of the Crab Nebula by Wucknitz et al.; in this complex field, the central pulsar is detected at sub-arcsecond resolution using the international LOFAR baselines.

A separate Key Science Project group focuses on achieving high quality full polarization imaging with LOFAR, in order to study the magnetic universe from near to far. Their main topics of interest include the origin of cosmic magnetism and the existence of seed fields and cosmic dynamos, magnetic collimation and acceleration models for mass flows, the physics of synchrotron radiation and interaction with particles in radio hot spots, and studies of planetary atmospheres. A recent overview of this project was given by Anderson et al. (2012, arXiv:1203.2467).

A further Key Science Project group is dedicated to Solar and interplanetary medium studies. It will use both imaging and dynamic spectrum modes. Detailed goals have recently been described by Mann, Vocks & Breitling (2011, PRE VII, p. 507). Observing the Sun – a strong, variable, and extended source – is obviously rather more challenging than imaging a more typical astronomical field. Preliminary images of the Sun have already been produced and verified against images produced by other telescopes; this work is being extended to achieve routine snapshot Solar imaging at higher resolution, as well as the production of dynamic spectra.

LOFAR is also eminently suited for studies of variable sources, including exoplanets, variable stars and pulsars, as well as supermassive black holes and gamma-ray burst sources; see the recent summary by Fender (2012, IAU Symp. 285, p.11). Pulsar observations, both targeted to known pulsars, as well as surveys, have already started, based on the excellent sensitivity and high time resolution of LOFAR through its coherent beam-formed mode for the core stations, as described extensively by Stappers et al. (2011, A&A, 530, 80). Variable source surveys can already occur by analyzing series of images made at intervals ranging from months down to hours. This is being extended to snapshots in a cadence as rapid as minutes and even seconds. Software also is being developed to allow patrols for variable sources to occur in parallel on the data for any targeted imaging observation. Transient studies will also benefit tremendously from LOFAR's unique Transient Buffer Boards. These revolving buffers can hold the last 5 sec of data, and longer at reduced bandwidth. Upon receipt of an internally or externally generated trigger, and interruption of the running observation, this data can be read back to form an *a posteriori* image anywhere in the field of the individual dipole antennas or tiles; in the low frequency band, this means most of the instantaneously visible sky!

The LOFAR Transient Buffer Boards are also essential for the final Key Science Project, on the detection of high energy cosmic rays. As described and demonstrated by Corstanje et al. (2011, arXiv:1109.5805), such particles lead to airshowers when they impact the upper atmosphere, of which the short flash of radio emission can be detected by LOFAR, with the precise timing, spectrum, and footprint giving both localization as well as a measure of the energy of the particle. An even more ambitious project is in preparation, that aims to detect radio flashes from ultra-high energy cosmic rays impact the surface of the Moon.

In short, with first call for Regular Proposals, heralding the start of both the Key Science Projects as well as the Open Skies access, that should draw both new expertise and new ideas from the wider community, LOFAR is poised to make a major impact in many areas of astronomy, with revolutionary observing capabilities that will be steadily expanding in the next few years.

THE RAEGE VLBI2010 RADIOTELESCOPES

Eberhard Sust, MT Mechatronics, Germany
José Antonio López Fernández, Instututo Geografico National, Spain

Keywords: VLBI, Geogesy, Radiotelescope, RAEGE

1. INTRODUCTION

The goal of the RAEGE (Red Atlantica Estaciones Geodinamicas Espaciales) project is the establishment of a Spanish-Portuguese network of geodynamical and spatial geodesy stations by the installation and operation of four fundamental geodetic / astronomical stations provided with radio telescopes located at
- Yebes, close to Madrid / Spain
- Tenerife, Canary Islands / Spain
- Santa Maria, Azores Islands / Portugal

Figure 1. Location of the RAEGE project radio telescopes.

VLBI 2010 radiotelescopes are belonging to a new generation of radiotelescopes suitable for high precision geodetical earth observation and measurements, that shall allow to built up a high precision global reference system.

The design of the radiotelescopes has been finished by MT Mechatronics in summer 2011 and currently three radiotelescopes are being manufactured. The first one is scheduled for installation in summer 2012 at Yebes Observatory close to Madrid.

2. THE RADIOTELESCOPE DESIGN

The radiotelescope is designed as elevation/azimuth turning head telescope. The optical design is based on a 13,2m ring focus reflector. In the basic configuration, the observation frequency is in the range from 2 – 40 GHz. It can be enhanced up to 100 GHZ by using additional options like insulation and air conditioning.

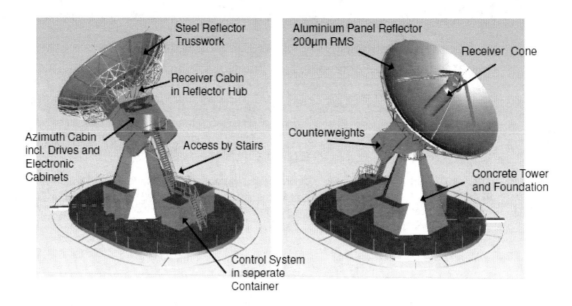

Figure 2. Design overview on the RAEGE project radio telescopes.

2.1 Basic Design Considerations

The most important Design requirements are:

- Use of wide band receiving systems working in the range of S-band to Ka-band (2-40 GHz)
- Fast slewing velocities of 12 deg/s in azimuth and 6 deg/s in elevation
- Low settlind times for minimizing the slewing times
- Low pathlenght error of less than 0,26 mm

Essential for geodetical telescopes is the possibility of measuring the position of the intersection between azimuth and elevation axis. Therefore a concrete pillar is installed at the center of the telescope tower, allowing the installation of a measurement system, located at the axis intersection, being visible from the outside through openings.

2.2 Telescope Optical Design

The telescope will be equipped with a symmetrical ringfocus reflector design as shown in the following picture:

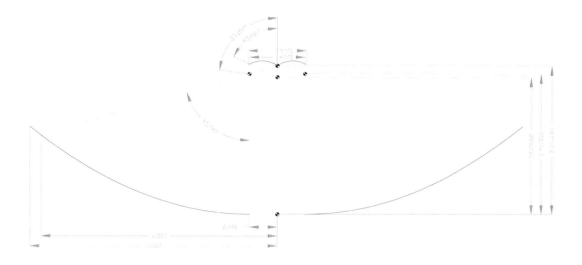

Figure 3. Optical Design of the VLBI 2010 radiotelescopes.

The optics will have the following approximate dimensions:

Diameter primary mirror: D = 13,2 m
Focal length pm   fp = 3,7 m
Primary paraboloid definition   fp/D = 0,28
Precision of surface single panel: < 0,1 mm RMS
Separation between panels: 2mm +/- 1mm at 20°C
Secondary Mirror Diameter: d = 1,5m
Precision surface Secondary Mirror: < 0,05 mm RMS
Final f/D: f/D = 0,4m

The operating frequency of the telescope will be 1 GHz to 40 GHz standard. The telescope can as an option be delivered with a surface accuracy of better than 200 μm that allows an enlarged of the operating frequency up to 100 GHz when some changes at the telescope will be performed.

The basic system design values can be seen in the following table. Elevation depending values are given for a elevation angle of 37 deg that will be the alignment angle of the telescopes.

	S-band	X-band	Ka-band
Aperture Efficiency (%)	71	75.6	80.4
Antenna Noise Temperature (K)	11,3	3,4	0,9
Antenna Gain (dB)	49,1	59,9	71,5
G/T	33,8	44,3	53,8

Figure 4. Antenna optical system design values.

2.3 Telescope structural and mechanical Design

The influence of the mechanical & servo system of the VLBI 13m antenna on the overall performance are determined by the surface accuracy of the reflecting surfaces, by the pointing accuracy of the overall antenna system mainly influenced by the servo system in cooperation with the mechanical positioning system as well as the pathlength error. These errors are described on mechanical and servo system level in the form of error budgets.

The errors contributors can be categorized into influences of the environment (mainly into gravity, wind and temperature influences), into influences of manufacturing tolerances, into influences of measuring and alignment tolerances, and into influences of the servo system components.

For the surface error contributors and the tracking error contributors, statistical distribution of the errors can be assumed, the errors are understood as "rms" values (3σ), the error contributions are superposed in rss (root-summed-square) method. For the pointing errors, repeatable errors (e.g. repeatable axis misalignment) will be corrected by the servo system via look-up tables.

The following analysis is based on finite element model (ANSYS) for the complete antenna in five selected elevation positions (EL = $0°, 30°, 50°, 70°, 90°$).

The models are used for calculation the gravity, wind and thermal deformations and for the dynamic analysis. The figures show the locked rotor frequencies in the two extreme positions (horizon, zenith). The frequencies are above 4 Hz.

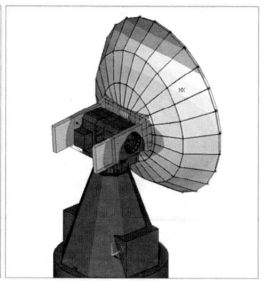

Figure 5. RAEGE Telescopes first Eigenfrequencies.

The overall **surface accuracy** will be in a range of 200μm RMS, taking into account the contributions from

- Gravity deformation: The adjustment of the reflector will be done at 37° elevation angle
- Operational wind of 40km/h
- Temperature deformation assuming a temperature difference of 4 K in the reflector
- Main Reflectopr and Subreflector manufacturing deformation

Source and Type of Error (all values in µm rms)	40GHz
Environmental Influences	
Gravity Deformations	
Backup Structure	140
Panels	20
Subreflector	20
Wind Deformations 40km/h	
Backup Structure	30
Panels	15
Subreflector	15
Thermal Deformations 4K	
Backup Structure	50
Panels	20
Subreflector	20
Reflector Manufacturing	
Panels	100
Subreflector	70
Mechanical Alignment	
Panels	70
Subreflector Position	5
Margin	50
Overall Surface Error	218

Figure 6. RAEGE Telescopes Surface Budget.

The Basic theory of the **pathlength error** is shown in following sketch. The budget has been elaborated from the FE calculations.

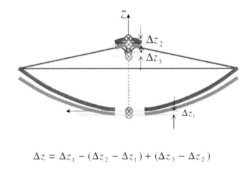

$$\Delta z = \Delta z_1 - (\Delta z_2 - \Delta z_1) + (\Delta z_3 - \Delta z_2)$$

Source and Type of Error (all values in µm)	40GHz	
	repeatable	non-rep
Environmental Influences (steady-state)		
Gravity Deformations	600	50
Wind Deformations 40km/h	200	50
FBC System		100
Thermal Deformations 4K	500	50
ThM System		100
Mechanical Alignment		
Reflectors, Feed	200	100
Main axis	200	100
Margin		100
Overall Longitudinal Elongation Error		240

Figure 7. RAEGE Telescopes Pathlength Budget.

The Basic theory of the **pointing error** is shown in following:

$$\Delta \varphi = k_1 \cdot \Delta \varphi_1 + k_2 \cdot \Delta y_1 + k_3 \cdot \Delta \varphi_2 + k_4 \cdot \Delta y_2 + k_5 \cdot \Delta y_3$$

description		parameter	[]
main reflector tilt	K_1	1.700	mrad/mrad
main reflector translation	K_2	-0.169	mrad/mm
sub reflector tilt	K_3	-0.067	mrad/mrad
sub reflector translation	K_4	+0.090	mrad/mm
feed translation	K_5	+0.085	mrad/mm

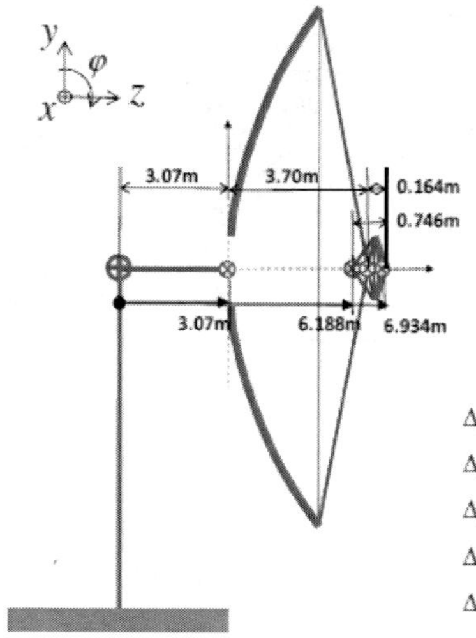

$$\Delta \overline{y}_1 = \Delta y_1 - 3{,}07 * \Delta \varphi_{enc}$$
$$\Delta \overline{y}_2 = \Delta y_1 - 6{,}934 * \Delta \varphi_{enc}$$
$$\Delta \overline{y}_3 = \Delta y_3 - 6{,}188 * \Delta \varphi_{enc}$$
$$\Delta \overline{\varphi}_1 = \Delta \varphi_1 + \Delta \varphi_{enc}$$
$$\Delta \overline{\varphi}_2 = \Delta \varphi_2 + \Delta \varphi_{enc}$$

Figure 8. Pointing Theory.

The following budget has been elaborated from the FE calculations:

Source and Type of Error (all values in arcsec rms)	40GHz repeatable	40GHz non-rep
Environmental Influences (steady-state)		
Gravity Deformations	35.0	5.0
Wind Deformations 40km/h	15.0	3.0
FBC System		3.0
Thermal Deformations 4K	7.0	7.0
Mechanical Alignment		
Reflectors, Feed	10.0	5.0
Main axis	10.0	5.0
Environmental Influences (dynamic)		
Gusts on Servo		1.0
Gusts on Structure		1.0
Servo		
Sensors		1.0
Actuators		1.0
Controllers		1.0
Servo Commands		1.0
Margin		7.0
Overall Pointing Error		14.0

Figure 9. RAEGE Telescopes Pathlength Budget.

Reference Point Accuracy

The reference point is defined as intersection between azimuth and elevation axis. The accuracy has been measured for the first antenna with less than 0,1 mm.

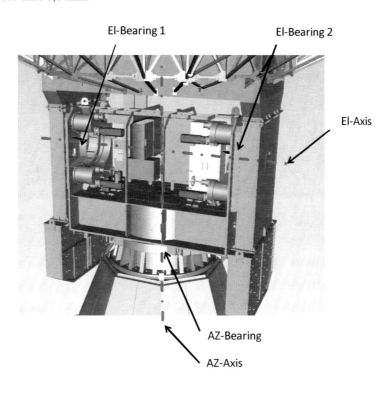

Figure 10. Telescopes Reference Point.

Reference Point Stability

The reference point stability has been calculated by FE Analysis. It will be less than 0.6 mm
The following reference point stability budget shows the maximum RSS sum values. All values are in mm:

Elevation position = 0 deg	X-direction	Y-direction	Z-direction
Elevation movement 0 – 90 deg	0	0	0
Temperature difference 10 deg 1)	0	0	0,24
Wind 36 km/h	0,005	0,136	0,004
Solar radiation impact (+3 deg)1)	0,012	0,387	0,111
Sum RSS	0,013	0,410	0,265
Total RSS	0,488		

Figure 11. RAEGE Telescopes reference point stability.

For Achieving the values in the table the following measures have been taken into account:

The height of the invariant point from ground (app 8,5m) requires a thermal insulation of the telescope concrete tower as well as an inside air conditioning to minimize the impact of thermal elongation during temperature change and solar radiation, so that the remaining movement of the invariant point will be within the specified value. The height of the tower will be constantly measured by an invar wire, that enables the permanent compensation of the thermal movements up to the azimuth cabin floor.

Settling Time

For optimization of the slewing time, the telescope is following a trajectory curve during slewing and positioning. The trajectory is calculated in the ACU considering the distance to final position, current velocity, maximum velocity. The trajectory is calculated in such a way that no overshoot does occur. A classical settling time caused by the servo and control system does not appear. The following features contribute to an optimum dynamic behavior:
Modern state-of-the-art digital servo amplifiers

The settlingtime of the telescope slewing with 1 deg/sec will be less than 1 sec.

2.4 Telescope Basic Kinematics

The telescope will be mounted on a concrete tower. The mount or pedestal is made from conventional steel structure, suitable for quick on-site assembly. It is designed for a lifetime of 20 years with automated and continuous operation at minimized downtimes.
The azimuth mount structure contains interfaces to the azimuth as well as to the elevation bearings. Its size allows the complete machining in one piece. This guarantees the position from elevation axis to the azimuth axis within machining accuracy.
The elevation portion as well as the counterweights are connected to the azimuth mount structure by two elevation bearings.
The ballast arm is a hollow box design, also acting as part of the elevation cabin. It is screwed to the reflector center hub and burries the counterweight steel plates, screwed to the rear side of the ballast arm.
The reflector hub is designed rectangular to maximize the space inside the cabin.
Weight of the structure (excluding supporting tower) is < 80,000 kg

2.5 Main Axis Drive System

The main mechanical components are the bearings and drives in elevation and azimuth. The elevation axis is defined by two roller bearings mounted to the sidewalls of the azimuth cabin

The elevation drive system consists of two toothed gear rims, part of the elevation bearings. Two motor/gear units arelocated on each side. The azimuth axis is defined by one cross roller bearing on top of the tower interface ring. The azimuth drive system consists of four mechanical reducers, which mesh with their pinions into the toothed rim of the azimuth bearing. Backlash is eliminated by biasing the drive units of each axis. The drive units are equipped with the necessary safety devices as stow pins, breaks, limit switches etc.

The mechanical components of the drives are designed for the following design parameters:

	Elevation	Azimuth
Travel range	-0 – 100 deg	± 275 deg
Max. traveling velocity	6 deg/sec	12 deg/sec
Max acceleration	3 deg/sec^2	3 deg/sec^2

Figure 12. Performance data Telescope Main Axis

Hand cranks make sure that the telescope can be moved manually in case of current failure.
Both axis are equipped with
- stow pins, electrically driven
- automatic lubrication systems
- limit switches
- end stops

All equipment sensitive to environments are located inside the azimuth cabin. This makes the telescope operable in harsh environments like arctic or tropic conditions.

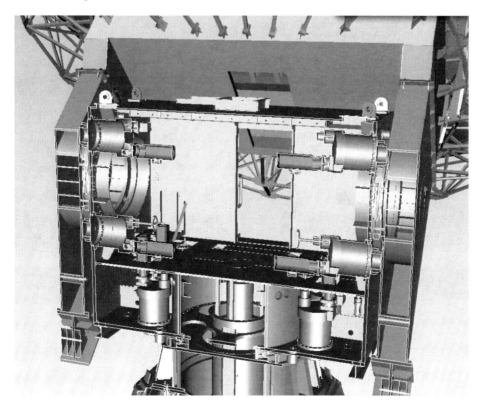

Figure 13. Mechanical components protected inside the azimuth cabin

Accuracies:

- The separation of the AZ / EL axes is < 0.3 mm
- The stability of the invariant point is < 0,3 mm
- The Orthogonality of the AZ / EL axes is < 10 arcsec
- The separation of the axis of elevation and that of the reflector is < 0.3 mm
- The Orthogonality of the axes of elevation and that of reflector is < 10 arcsec
- Deviation between the axis of azimuth and that of gravity (measured at the level of azimuth bearings) is < 5 arcsec
- The deviation between the axis of elevation and that of gravity (measured at the level of azimuth bearing) is < 5 arcsec
- The concentricity error of azimuth bearing is < ± 150 µm
- The tilt of elevation axis is < ± 50 µm

2.6 Subreflector Positioner Design

The sub-reflector is moved with a positioner of type hexapod. The positioner is a complete unit which will be assembled, commissioned and tested in the workshop. On site the unit will be mounted to the quadripod structure and electrical connected with plugs.
The unit consists of high precise actuators which are fixed to the supporting structure with backlash-free joints.
The mechanical and electronic parts are located in a weather protected area. An integrated heating and ventilation system provides climatic conditions which are necessary for accurate and safe operation.
The payload of the positioner is 100 kg

Figure 14. Subreflector positioner with and without weather protection

	X-direction	Y-Direction	Z-direction
Velocities of displacement	±1 mm/s	±1 mm/s	±2 mm/s

Ranges of displacement	±25 mm	±25 mm	±75 mm
Repeat precision in the position	±30 μm	±30 μm	±30 μm
Repeat precision in inclination	±10 arcsec	±10 arcsec	±10 arcsec

Figure 15. Performance data Subreflector positioner

2.7 Servo and Control System

In the standard configuration, the control System is installed in a separate container and the azimuth cabin. For EMC reasons, the elevation cabin is foreseen only for HF equipment.

A hand held panel with touch screen is used for driving the telescope in maintenance mode. Connections for the HHP are located in the azimuth cabin as well as the control container.

An easy upgrade to 100 GHz is possible since sensors for the 100 GHz option are already installed in baseline antenna.

The tracking system is state-of-the-art and based on servo drives that are controlled electronically.

- The sensor technology is on the bases of high-resolution angle measuring devices.
- The mounting of the primary encoders are on axis
- Control technology is based on commercially available, reliable and proven components, which comply the EMC and grounding standards.
- The modules used for the control show an interference level – in the range between 1 GHz and 40 GHz – that cannot be detected by the receivers of the radio telescope
- Software modules are of proven technology in accordance with the standard EN 61131-3.
- Safety technology is with parameterizable safety terminals and state of the art.
- The operation concept is flexible with redundant design, compatible control interfaces, and possibility of remote diagnostics

The control system is designed for low operation costs and low energy. This includes anenergy saving mode by means of optimized application of velocity and acceleration and the possibility of regenerating braking energy when the drives are in generatoric mode.

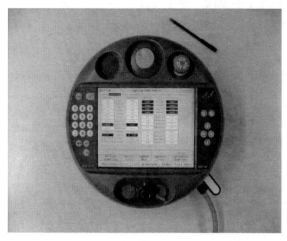

Figure 16. Control Cabinet and Hand Held Panel for Maintenance Operation

Encoders:

The performance of the telescope is reached by using high precision encoders as well as two axis inclinometers for observations above the 40 GHz range.

The encoder requirements are
- Resolution: 27 bit
- Precision: < 0.35 arcsec

Due to the mechanical design and the experience on recent antenna and telescope projects a tape encoder system has been selected for the Azimuth axis. The tape encoder system consists of a distance coded (absolute) tape from Heidenhain mounted on the inner diameter of the AZ bearing.

Figure 17. Tape encoder system used for azimuth and elevation axis

2.8 Tower Design

The foundation, on which the antenna and pedestal is designed to support the antenna deadweight and the calculated survival loads including seismic loads. The design of the foundation has to be tailored to the ground conditions found at site.It contains earthing and grounding lines as well as lightning arresters to protect the system from overvoltage.
To assure the required thermal stability, the outside walls will be isolated in such a way that the asymmetric thermal warm up due to sun radiation will be minimized.

An independent central tower made from concrete is installed in the center of the telescope tower. Its upper diameter is 30 cm. The height of the tower is until 25 cm below the Azimuth Cabin floor level. A free volume on top of the central tower is foreseen for installation of an independent measurement system detecting the movement of the invariant point of the telescope.

The central concrete tower and the telescope supporting tower have a common foundation, but are independent above the foundation.

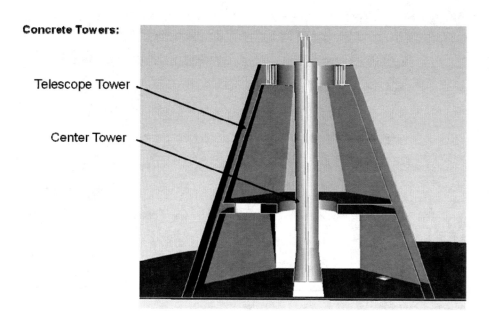

Figure 18. Telescope Towers

2.9 Metrology:

Measurement of invariant point located on the intersection of azimuth and elevation axis can be performed by an Instrument support on separate pier on separate foundation.
- For height measurement of the invariant point an interface for fixing a height measurement system using invar wire through the center of the azimuth axes is foreseen.
- Targets are located on elevation rotating structure for measurement of elevation axis
- A fix target on ground is installed in the center of azimuth rotation
- Fix targets outside, visible through doors and openings are located on separate pilars
- Targets are located in the Az-Cabin for measurement of azimuth axis
- Targets are located in the El-Cabin to be seen in different elevation positions

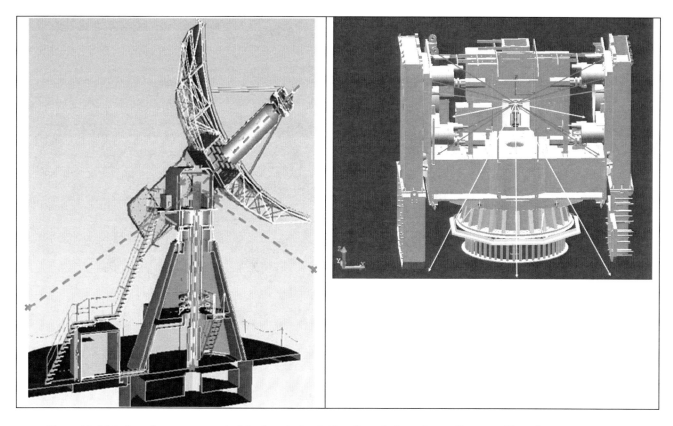

Figure 19. Metrology for measurement of the invariant point location relative to inner pilar or outside references

2.10 Workshop Assembly and Testing

The Telescope is designed in such a way that many requirements can be verified already during workshop assemblies and tests:

- Test Setup 1: Azimuth Cabin including complete Mechanical and Servo Equipment
- Test Setup 2: Reflector Backup Structure
- Test Setup 3: Subreflector Mechanism including Subreflector Control Mechanism

The assembly, alignment and commissioning of the complete mechanical system will be done in the manufacturing workshop. Since the system will not be dismounted for transport, the commissioning does not need to be repeated on site.

The reflector including quadrupod and cone will be assembled, aligned, and bolted in such a way that reassembly on site will be done without any additional alignments. Only the reflector panels will have to be aligned on site by photogrammetry.

Figure 20. Workshop assembly and testing

2.11 Transportability

The antenna main structural and mechanical components will be disassembled in the workshop and transported in large but good transportable pieces to the site. All parts will be transported in 12 standard containers, partly open top, partly flats. No oversize transports are necessary.

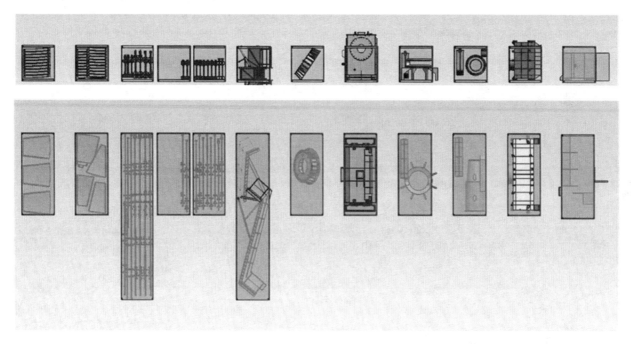

Figure 21. Transport Configuration

Architecture of the Metrology for the SRT

Tonino Pisanu[a], Franco Buffa[a], Gian Luigi Deiana[a], Pasqualino Marongiu[a], Marco Morsiani[b], Claudio Pernechele[c], Sergio Poppi[a], Giampaolo Serra[a], Giampaolo Vargiu[a]

[a]INAF - Cagliari Astronomical Observatory, loc. Poggio dei Pini, Capoterra (CA) Italy
[b]INAF - Institute of Radio Astronomy Via P. Gobetti, 101 40129 Bologna, Italy
[c]INAF – Astronomical Observatory of Padova, Vicolo dell'Osservatorio 5, Padova, Italy

ABSTRACT

The Sardinia Radio Telescope (SRT) Metrology team is planning to install an initial group of devices on the new 64 meters radio-telescope. These devices will be devoted for the realization of the antenna deformation control system: an electronic inclinometer able to monitor the alidade deformations and a Position Sensing Device (PSD) able to map the antenna secondary mirror (M2) displacements and tilts. The inclinometer will be used to map the rail conditions, the azimuthal axis inclination and the thermal effects on the alidade structure. The PSD will be used to measure the secondary mirror displacements induced by the gravity and by the thermal deformations that produce shifts and tilts with respect to its ideal optical alignment. The PSD will be traced by diode laser installed on a mechanically stable position inside the elevation equipment room. The inclinometer has been tested in laboratory with the aim to compare its performances with a reference measurement system. The PSD and the laser have been characterized by a long-term tests to assess their stability and accuracy, thus simulating the open air conditions that will be experienced by the device during its operative life. M2 may move freely in space thanks to a six axis actuator system (hexapod). The PSD measurements are processed by a hexapod kinematic model (HKM) to evaluate the correct actuator elongations, thus closing the control loop. The sensors will be acquired and recorded by a dedicated PC installed in the Alidade equipment room and connected to the sensors via the Ethernet network.

1. INTRODUCTION

The Metrology team of the SRT has the aim of projecting, building and testing all the needed measuring systems to maintain as high as possible in whatever climatic condition the scheduled antenna efficiency and pointing performances. The SRT has an homology configuration and has been projected to operate up to 22 GHz frequency with a total RMS surface error of around 600 µm and a repeatable pointing error of around 5 arcmin, without any measurement or correction at "Normal" condition. For Normal we mean, clear sky condition, air Temperature range between -10°C to +40°C, and mean wind speed of around 8 m/sec. At higher frequencies (the antenna has been designed to operate up to 100 GHz), active measuring and correcting systems are needed.

The SRT has a Gregorian configuration, with a quasi parabolic shaped primary mirror with an f/D of 0.329 and an elliptical shape secondary mirror with an f/D of 2.3523. The secondary reflector sensitivities in terms of pointing error factors for tilt ($\Delta\theta Beam/\Delta\theta tilt$) and translation ($\Delta\theta Beam/\Delta S$) of the sub-reflector are:

- Subreflector rotation: 0.837 arcsec/mdeg
- Subreflector translation 7.898 arcsec/mm

Where $\Delta\theta Beam$ is the pointing error of the beam, $\Delta\theta tilt$ is the secondary reflector tilt angle with respect to reflector vertex and ΔS is the lateral secondary reflector displacement [1].

In the following sections the first set of metrological devices we planned to install and the related characterization procedures will be described.

2. INCLINOMETER

The two inclinometers that we want to install on SRT are from Wyler©, the ZEROMATIC 2/2 models, each one with two ZEROTRONIC sensors, that are able to measure the inclination of a plane around 2 axis with a measuring range of ± 1 degree and with an accuracy of ± 1 Arcsec.

The inclinometers will be mounted on the N1 and N1X nodes of the alidade structure, as close as possible to the elevation bearings and encoders, as indicated on the next figures. They are installed inside a box as shown in figure, in which we have installed also a polyethylene cylinder to stabilize a little bit the internal temperature.

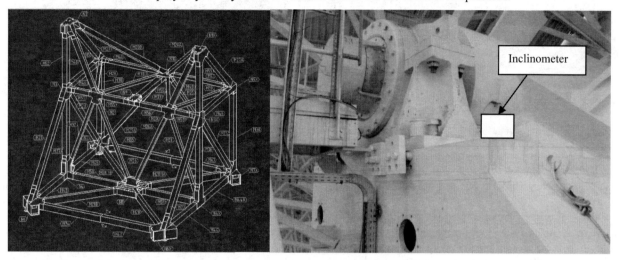

Figure 1: On the left, the design of the Alidade structure with the indication of nodes and trusses, on the right a picture of the particular of the N1 node where we want to install the first inclinometer

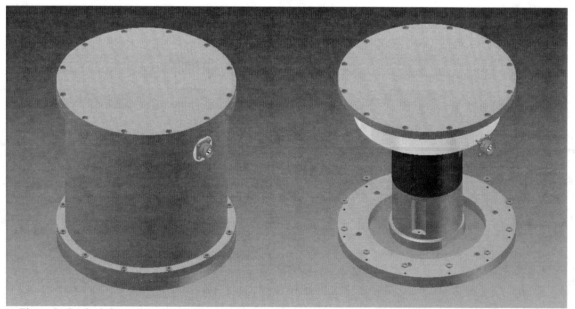

Figura 2: On the left we show the design on the enclosure for protecting and isolating the inclinometer which is the cylinder in red shown on the left

Each inclinometer will be aligned in such a way that it can measure the alidade inclination, around the elevation axis (hereafter LivEL) and perpendicularly to the elevation axis (hereafter LivXEL). LivEL describe the effects related to the elevation pointing errors while LivXEL may be used to reduce the azimuth errors [2]. The LivEl measures directly the elevation errors ΔEL=δLivEL, while LivXEL measures the tilt along the elevation axis allowing to infer the azimuth error ΔAZ=-δLivXEL*tan(θel), where θel is the elevation angle.

It should be noted that our inclinometer sensors, are not able to perform measures when the antenna is moving quickly, therefore it is possible to collect reliable data only when the antenna is still or moving slowly.

Before to install the inclinometers on the antenna, we tested them in laboratory to characterize their resolution and accuracy. To do so we prepared a setup like the one showed in the next figure in which is visible an aluminum bar 1,130 mm long, mounted at one side on a hinge that permit its rotation and on the other side supported from a mechanical micrometer. First of all we zeroed the inclinometer and then we moved it up and down in an angular range θ of ± 5 mm/m which correspond to an angular range of ± 0.28° with a ΔZ of 1 micron which correspond to a Δθ of around 0.2 arcsec. The result of the linearity and accuracy of the inclinometers are shown in the next figures.

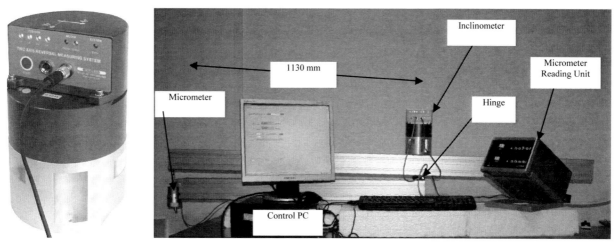

Figure 3. On the left the ZEROMATIC 2/2 inclinometer device, on the left the setup we used to characterize the instrument performances.

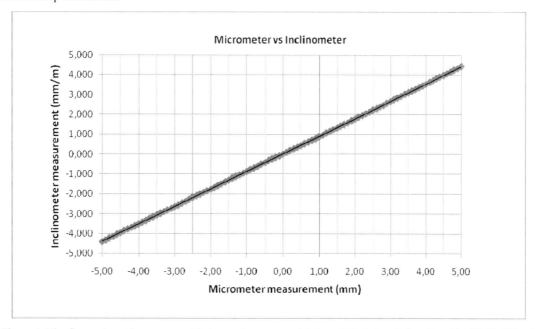

Figure 4. The figure show the agreement between the measured data and the interpolation line. The RMS of the data is of the order of few microns as expected from the datasheet of the inclinometer.

3. SUBREFLECTOR KINEMATIC MODEL

The SRT subreflector (M2) movements are obtained by means of six linear actuators (hexapod) that confer six degrees of freedom to the system. The actuator arrangement is depicted in Fig. 1, the actuator's positions and the M2 position in space are referred to the geometrical position of the subreflector vertex.

The "real" position of M2 in space depends on external factors, such as gravity, wind and thermal gradients. The M2 misalignment affects the pointing and efficiency error budget.

The position corrections which will be imposed on M2 will be measured by a position sensing device (PSD) described elsewhere in this paper. In order to close the control loop, a kinematic model is needed to provide the functional transformation tying the actuator elongations to the spatial position of M2.

In synthesis the subreflector kinematic study should achieve two goals:

1. Define the *direct* transformation from the M2 position, where the latter is described by a roto-translation defined by a position vector and by three Eulerian angles, to the corresponding actuator elongations;
2. Define the *inverse* transformation from the six actuator elongation as measured by the actuator linear encoders to the corresponding roto-translation.

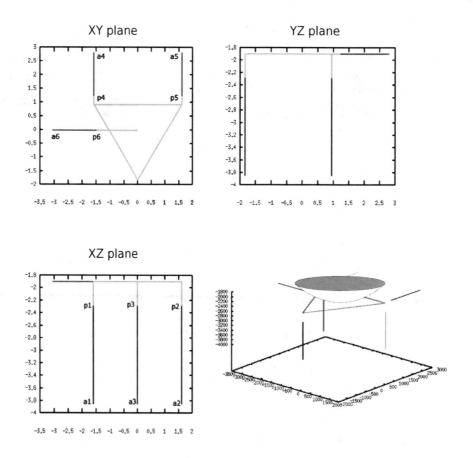

Figure 5: Schematic representation of the SRT hexapod. Each actuator is represented as a segment $a_i p_i$. The actuators are constrained to the support structure in a_i, while p_i may move in space providing six DOF to the subreflector. The subreflector sketch is not in scale.

The direct problem is easily solvable taking into account the geometry of the system, while the inverse transformation is a typical non linear problem. Assuming the reference system described in Fig. 5, it can be seen that if we move the subreflector, the a_i points are constrained to the structure while p_i points assume a new position p'_i in space described by the transformation:

$$p'_i = [x'_i, y'_i, z'_i]^T = R_z R_y R_x [x_i, y_i, z_i]^T + [B_x, B_y, B_z]^T \qquad i = 1, 2, \ldots 6 \qquad (1)$$

where R_j is the rotation matrix around the j axis (j = x, y, z) and B_j is the displacement vector. If R_j and B_j are known, the corresponding actuator elongations are simply:

$$d_i = [(a_i p'_i)(a_i p'_i)^T]^{1/2} \qquad i = 1, 2, \ldots 6 \qquad (2)$$

The inverse problem may be expressed in form of a set of non linear equations:

$$F_i = d^*_i - [(a_i p'_i)(a_i p'_i)^T]^{1/2} = 0 \qquad i = 1, 2, \ldots 6 \qquad (3)$$

or in compact form:

$$F(r) = 0 \qquad (4)$$

where the elongations are known parameters and the roto-translation vector r (rotation angles and translation components) is the unknown to be determined.

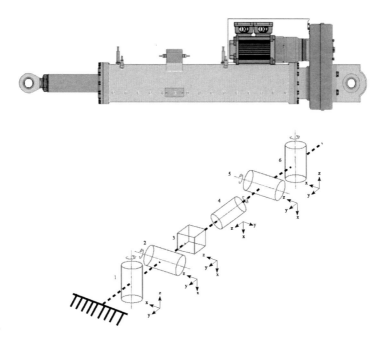

Figure 6. Technical draw of a M2 actuator (top) and relative decomposition in elementary movements (bottom): revolution (cylinders) and prismatic (cube) joints.

The ordinary Newton method could be applied to solve the problem linearizing the equation set defined in (4):

$$Jdr = -F(r) \tag{5}$$

Where J is the Jacobian of F.

Simulation tests we carried out to check the algorithm reliability shown convergence problem of Newton's algorithm, especially if the initial parameter estimate is poor. The Powell's Hybrid method fortunately overcomes this limitation combining the standard Newton and steepest descending methods [3]. We successfully implemented this algorithm in a C library which has been integrated in the hexapod control routines.

The kinematic chain of each actuator has been decomposed following the Denavit-Hartenberg representation (DHR) [4]. Formally this procedure is not needed to properly actuate M2, whereas such approach is useful to predict the proper gimbal working range and, in extreme cases, to prevent gimbal locks. Following DHR the spherical and linear joints of each actuator has been represented in terms of elementary movements, i.e. revolution and prismatic joints (Fig. 6). In such a way a complex kinematic chain may be easily represented and studied. A new library implementing the DHR for M2 actuators has been written in Octave [5] language in order to map the whole working range of each elementary joint. In fact a great number (10^5) of actuator elongations has been generate randomly, for each configuration set, the DHR has been applied allowing us to map a representative working range of each joint.

4. SUBREFLECTOR METROLOGY

In order to measure and correct the position of the secondary mirror of the SRT which is subjected to gravitational and thermal deformations, we are planning to install an optical measurement system. The system is mainly composed from a laser diode installed in a stable position inside the vertex room of the antenna, which emits a laser beam pointed towards a PSD installed on the subreflector structure. The laser diode is a FERMION 1 fiber optic laser from Micro Laser Systems, Inc. with a wavelength of 658 nm, a power of 10 mW, an high stability and coupled with a single-mode optical fiber at an FC connector optical collimator, with a divergence of < 0.25 mrad.

The PSD [6] is a Silicon photodetector mod. AlignMeter LA from DUMA Optronics ltd, with a circular sensitive area of 22.5 mm in diameter, an operational spectral range of 350-1100 nm and a power range from 100μW to 5 mW. It can measure 4 degrees of freedom, the xy position and two tilts around the x and y axis. The detection position resolution is better than +/- 2 μm, the accuracy is ± 50 μm over the calibrated area, the angle measurement range is ± 2° and the angle resolution is ± 1 arcsec.

In order to properly actuate the subreflector corrections, the PSD measurements should be included in the telemetry flux of the antenna control system (ACS). To do that we integrated the ACTIVEX controls provided by DUMA on a TCP/IP server that is able to poll the PSD and, at the same time, to handle the ACS requests [7].

The accurate measurement of the spatial position of the subreflector is a key factor to close the M2 control loop. Rotations (tip and tilt) and translations (decenter) may affect both focus and pointing, then a suitable metrological system is needed to mitigate such effects. The M2 misalignments may be induced by repeatable effects (gravity) and non-repeatable effects (thermal gradients, wind). A reliable mapping of repeatable effects may be realized by means of specific campaigns of photogrammetric measurements or by means of finite element model (FEM) simulations. On the other hands, after that the correction of repeatable errors has been applied, non-predictable effects have to be monitored in real-time. The effects of M2 tip-tilt and decenter on the pointing may be estimated in terms of 0.837 arcsec/mdeg for subreflector rotations and 7.898 arcsec/mm for the translations [1]. This means that in the case of decenter effects the M2 spatial position should be estimated with sub-millimeter accuracy.

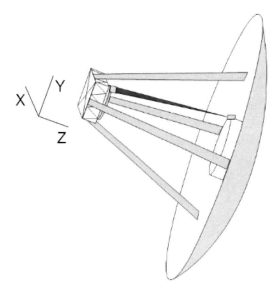

Figure 7. PSD arrangement on board the radiotelescope. The laser diode is placed near the reference structure of SRT, while the PSD is hold on the subreflector support structure. The beam path is about 20 m.

In this frame the SRT Metrological team defined a measurement system based on the DUMA Optronics position sensing device (or laser position sensitive detector) which is able to measure both X-Y displacements and tilts.

The metrological system will be arranged as described in Fig. 7. The laser diode will be placed in Vertex room, while the PSD will be hold on a mechanical frame linked to M2.

The spatial measurement range of the PSD is limited to ± 10 mm. This range is fortunately wider than the predicted range of misalignments induced by non repeatable causes. Anyway, in some cases a wider measurement range could be necessary, for example during alignments or if the predicted FEM corrections for repeatable errors are not sufficiently accurate.

We are studying a photogrammetric alignment system able to provide a wider detection range to be used as ancillary system for the PSD. The system is based on the subpixel image registration technique (IRT) [8]. The classic photogrammetry is based on calibrated cameras acquiring the object images. The object displacement is detected recognizing and tracking control points materialized as photogrammetric targets placed on the object surface. The displacements are measured solving the non linear problem related to the evaluation of the 3D camera external orientation and comparing the obtained solution with a reference configuration. IRT may drastically reduce the complexity of the classic photogrammetric approach if a 2D rigid translations retrieval is sufficient. In this case the IRT technique concerns with the computation of the fast Fourier transform (FFT) cross correlation of the image to be registered with the reference one, and locating its peak that correspond to the 2D displacement. The IRT approach takes into account all the picture textured information, while in the classic photogrammetric case just the (usually few) pixels representing the targets could be considered.

Subpixel accuracy may be achieved upsampling the images, but the inverse FFT arising from high resolution images, requires computational power and memory that are far from the nowadays personal computer capabilities. The IRT techniques overcome this problem considering just a small region about an initial estimate of the cross-correlation peak.

We implemented the IRT algorithm proposed in [8] on an Octave framework. In our test the registration of a 10 Mpx image takes just 3 seconds on an Linux PC.

In order to test both PSD and IRT techniques at the same time, we realized a simple framework based on a XY micrometric table, on which we hold the PSD and a target with a suitable texture for IRT. The laser diode and the camera where placed about 20 meters away from the measure table as shown in Fig. 8, in such a way the calibration system was able to mimic the real operative configuration. Obviously in the real case the textured image will be replaced by a part or by the whole subreflector surface.

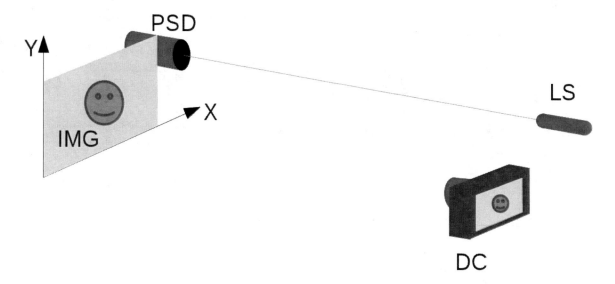

Figure 8. PSD calibration setup, LD = laser diode, IMG = target image, DC = digital camera. The laser diode and the camera where fixed and placed about 20 meters away from the XY translating table, the PSD and the target image rigidly translate with the table.

The plot in Fig. 9 shows - for X axis - the residuals obtained as difference between measured displacements and the micrometer readings. It is clearly shown as the PSD tends to deviate from the linearity when the laser beam reach the periphery of the detector.

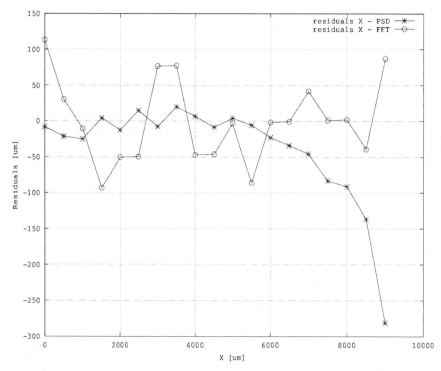

Figure 9. PSD and IRT (FFT) X axis residuals with respect to the table displacement. The quantities are expressed in µm.

The measured accuracy of the photogrammetric system is about 50 μm for X and Y axis, while the X and Y accuracy reached by the PSD is about 15 μm (in the linearity interval). Out from the linearity interval, the PSD accuracy downgrades to 100 μm or more. Of course the photogrammetric system doesn't suffer for this problem: in any case a suitable objective focal length (500 mm in our case) may be chosen in order to keep the whole frame in the field.

5. CONCLUSIONS

The SRT control system will be made by many different metrological apparatus. We described the results of the characterization of the first devices we intend to install on-board the antenna: an off-the-shelf electronic inclinometer and an optical PSD able to measure respectively the alidade and subreflector deformations. These results made us confident that their accuracy are able to guarantee a good antenna alignment at higher observation frequencies.

The secondary mirror is moved by six linear actuators (hexapod) that allow to control M2 reaching the best antenna optical alignment. We presented the SRT subreflector hexapod kinematic model (HKM) which is necessary to evaluate the subreflector position in space, and consequently to close properly the control loop.

REFERENCES

[1] Vertex RSI – A TriPoint Global Company "64-METER Sardinia Radio Telescope final design report optics and RF design"
[2] Pisanu T., Buffa F., Morsiani M., Pernechele C., Poppi S., "Thermal behavior of the Medicina 32 meter Radio Telescope" Modern Technologies in Space- and Ground-based Telescopes and Instrumentation, Proc. of SPIE Vol. 7739, 773935 · © 2010 SPIE
[3] Galassi, M. et al., "GNU Scientific Library Reference Manual (3rd Ed.)", Network Theory Ltd, (2009).
[4] Denavit, J., Hartenberg, R. S., "Kinematic Synthesis of Linkages", New York, McGraw-Hill (1964).
[5] Eaton, J. W., Bateam, D. and Hauberg, S., Gnu Octave Manual. Network Theory Ltd. (2008)
[6] http://www.duma.co.il/align-models.html#alignmeterLA
[7] Orlati A., Buttu M., Melis A., Migoni C., Poppi S., Righini S., "The control software for the Sardinia Radio Telescope", SPIE 8541, 8451-101 (2012)
[8] Guizar-Sicairos, M., Thurman, S. T. and Fienup, J. R., "Efficient Subpixel Image Registration Algorithms", Opt. Lett. 33, 156-158 (2008).

Requirements and considerations of the surface error control for the active reflector of FAST

MingChang Wu*, QiMing Wang, XueDong Gu, BaoQing Zhao
National Astronomical Observatories, CAS, 20A, DaTun Road, ChaoYang District, Beijing, CHINA 100012;

ABSTRACT

The Five-hundred-meter Aperture Spherical radio Telescope (FAST) is currently under construction at a Karst depression in the Guizhou province of China. The active reflector of the telescope is composed of 4395 triangular panels laid on a cable-net structure. The aperture of the spherical surface is 500 meters, with open angle of about 110~120 degrees. Acting as the nodes of the reflector, the joint of these panels are adjusted by 2235 down-tie cables drawn by actuators. The RMS error of the parabola reflector is expected to be 5mm.

To form the parabola shape of the reflector, for each of the actuators, a minimal working stroke of 950mm is required, with maximal speed of 1.6mm/s at the load of 50kN. Considering the elastic deformation of the down-tie cable and other factors, a positioning error within 0.25mm is required for the actuators.

In this paper, the base formula for the motion of a general actuator at a typical observation time is studied analytically. The results are used to estimate the control error of the actuators and the pointing error of the whole reflector. Based on the designed error budgets, a statistical method is employed to estimate the overall surface error of the parabola reflector. The overall surface error is a comprehensive result of the panel design error, panel fabrication error, thermal deformation error, panel wind load induced error, cable-net error, installation error, measurement and control error etc. The results may be used as a reference in the measurement and control of the active reflector when in operation.

Keywords: Surface error, Radio Telescope, Active Reflector, Actuator, Cable, Panel, Measurement, Control

1. INTRODUCTION

Using cable supported light feed cabin and active reflector, the large radio telescope FAST initiates a new mode to construct giant radio telescopes, which breaks the engineering limits of constructing ground based giant telescopes[1,2]. It is the fast development of modern electro-mechanical technology and measurement and control technology that makes this possible.

One of the main characteristics of FAST is to form a paraboloid of aperture 300m along the observation direction using active control[3,4]. Using integrated optical-mechanical-electrical cable-suspended light feed cabin and secondary adjusting facility in the cabin, high precision pointing is achieved without rigid connection between the feed cabin and the reflector.

The base surface of the active reflector is a spherical cap of radius 300m with aperture 500m. The cap is divided into 4395 triangular panels, based on the precision requirements of fitting to a paraboloid. Each of the panels is pieced together using multiple aluminum plates of transmission 50%. The panels are supported by a back frame of rigid space structure, and are connected through cable-net nodes. For each node of the cable-net, there is a down-tie cable drawn by an actuator. The position of the vertexes of the triangular panels is adjusted by the controlling of the node position of the nodes of the main cables, the surface profile of the reflector is adjusted in this way. A scaled model of 30m of FAST was constructed in Beijing for technology verification[5].

*mcwu@bao.ac.cn; phone 86 10 64863369-5111; fax 86 10 64877278

2. THE SURFACE PRECISION REQUIREMENTS OF THE ACTIVE REFLECTOR

The active reflector is consisted of 4395 reflector panels which will reflect the incident radio waves to the feed. The panels are fixed to the nodes of the cable net at three corner points. Each of these nodes will be controlled by a down-tie cable. The base spherical surface will be changed to paraboloid of different observational directions.

Each of the reflector panels are consisted of triangular panels and back frames of side length 10~12m. The fit RMS of the supporting nodes will be better than 2.5mm. The related error budgets are described in table 1.

Table 1. Error budgets for the active reflector elements

Item	Error Budget
Panel design error	2.2mm
Panel manufacture error	2.5mm
Normal error of thermal effect on panel	1mm
Wind load error on panels	1mm
Cable net error	2mm
Installation error	1mm
Reflector measurement and control error	2mm

3. DESIGN OF THE SUPPORTING CABLE NET

3.1 The supporting cable net of FAST

The main structure of the active reflector is a cable net structure with diameter of 500m. The cable net is used to support the reflector panels. The down-tie cables are controlled to change the spherical reflector to a paraboloid reflector of aperture 300m and focus ratio of 0.467. The cable net structure is consisted of main cable net, down-tie cables, cable nodes and ring beam, as shown in Figure 1.

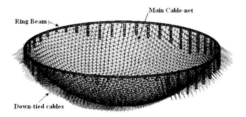

Figure 1. The cable-net hung on the ring beam

The spherical cable net is divided using geodesic method. The triangular grid formed with geodesic method has the advantage of better stability, better uniformity in stress, less grid number, less grid types, and easier to fix back frames. The grids are shown in figure 2.

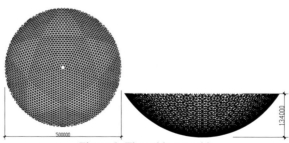

Figure 2. The cable net grids

There are 6725 cable segments of length 10~12m. The 4395 reflector panels are fixed to the cable net at 2235 nodes. Each of the nodes is tied to a down-tie cable which will be controlled by an actuator. The down-tie cables act both as stabling cables and controlling cables. Each of the 4395 reflector panels is fixed to the cable net which is hanged on the ring beam. The beam is supported with 50 pillars in the Karst depression. The inner diameter of the ring beam is 500m.

3.2 Installation of the cable net

The installation of the cable net is a continuous process. The stress state of the structure change correspondingly for each stage. The stress and displacement of the structure in the former stage will surely affect the state of the next stage. Numerical analysis will be performed for each stage to get the stress and displacement state, to calculate the accumulating effect of each stage. For each stage, the construction state will be verified to limit the errors to be acceptable, to avoid the errors being brought to the next stage, which may result to unacceptable structure internal force and displacement.

4. DESIGN OF THE ACTIVE REFLECTOR PANELS

The theoretical side length of the reflector panels is from 10.379m to 12.401m. The structure is of relative large stiffness. It is consisted of panel elements, purlins, and back frame[6]. The reflector panel is the basic element of FAST. Adjustable structure is used at the connection point between the back frame and the panel elements for easy adjustment of the surface error of the reflector panel to meet observation requirements. Each of the triangular panels is connected to the nodes of the cable net at the three corner points through displacement adjustable nodes. This enables the auto compensation of the changes in length and angle which are induced by the deformation of the spherical reflector surface to parabola surface. The design specifications are shown in table 2. The main loads are shown in table 3.

Table 2. Design specifications of the reflector panels

Items	Specifications
Radius of curvature	315m
Open area	≥ 50%
Maximum deflection under weight	1mm
Gap width between the panels	5cm
Fitting precision (RMS) at the adjusting points(For single panel)	RMS≤2.5mm

Table 3. Main loads on the reflector panels

Items	Loads value
Constant load: Self weight of the nodes, supportings, panels, rivets, purlins, bolts etc.	≤8.5kg/m²
Wind load	working speed=4m/s, maximal speed=14m/s
Maintenance load:	1.0kN point load at any node
Temperature load:	-10°C ~ 40°C
Variable load(Leaves etc.)	0.01kN/m²

4.1 Type selection of the back frame

Figure 3. The prototype back frame of the triangular pyramid, bolt ball space truss with double layers

The reflector panels are supported by the back frames, which will provide the necessary strength and stiffness to bear the self weight, wind load, snow load, maintenance load etc. The back frames shown in figure 3 are connected to the cable net nodes at three corner points. The back frames are of triangular pyramid, bolt ball space truss with double layers. To lower the self weight of the structure, and to provide enough corrosion resistance to the atmospheric environment, most of the parts are made of aluminum alloy, except for the high strength bolts and fastening screws, which are made of stainless steel. Each of the 4395 space trusses are fixed to nodes of the cable net. The average surface density of the back frame is 8.5kg/m².

The reflector of FAST is controlled by the position adjustment of the cable net nodes. Therefore, the nodes should have enough strength, stiffness and durability. Furthermore, when the positions of the nodes are adjusted by the actuators, the reflector panels should move with the nodes smoothly. One design of the nodes is shown in figure 4[6].

Figure 4. Joint nodes for the connection of the back frames and the cable net

4.2 Type selection and design of the panel elements

As the main functional parts of FAST, the panel elements are attached to the related back frames. Each of the panel elements is composed of perforated aluminum plates, supporting aluminum purlins, fastening rivets, supporting and its adjusting structures etc. The adjusting structures are used to adjust the surface shape of the reflector panels. Of the perforated aluminum plates and the stretched aluminum screen, the perforated plates are selected for better flatness, stiffness, as shown in figure 5 and figure 6.

Figure 5. The perforated aluminum plate

Figure 6. Structure of the reflector panel

4.3 Selection of materials

To acquire larger strength and stiffness for the reflector panels with lower weight, and to meet the requirements of corrosion resistance, most parts of the back frames, and all of the perforated plates, are made of aluminum, as shown in table 4. On the other hand, instead of welding, patented mechanical connection is widely used in the back frames to improve the structure strength and stiffness. Fastening and loose proof measures are applied as shown in figure 7.

Table 4. Material for different parts

No.	Parts	Material
1	Rods, ball joints, seal plates of the rods, fastening screws, Rod bolts, panel elements	Aluminum alloy (6061-T6, 2A12, A1-50, A2-70, 3003)
2	Supporting joints to the cable net	Stainless steel(304)

(a) Loose proof bolts

(b) Loose proof washers

Figure 7. Loose proof measures

5. SOURCE SEARCHING AND SOURCE TRACKING METHODOLOGY

5.1 Source searching and source tracking methodology

High precision contactless position measurement system is required for FAST to acquire the three dimension coordinates of the cable net nodes at the frequency of nearly 1Hz.

For the source searching, the measurement results are used to adjust the nodes in the illuminated area to position on the target parabola surface by controlling related actuators.

For the source tracking, the measurement results are used to adjust the nodes in the transient illuminated area to position on the transient target parabola surface by controlling related actuators. For the nodes beyond the current illuminated area, the nodes should be restored to base state by corresponding actuators. The source tracking is a slow (The maximal radial speed of the nodes is about 1.6mm/s) dynamic process which can be thought as quasi static. Figure 8 shows the methodology. Figure 9 shows two typical deformation effects simulated in ANSYS software.

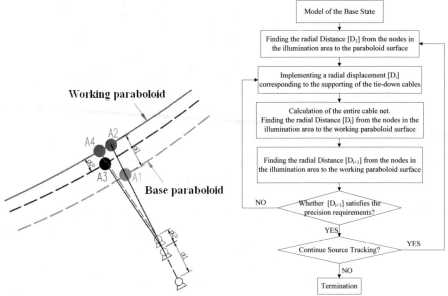

Figure 8. Source searching and source tracking methodology

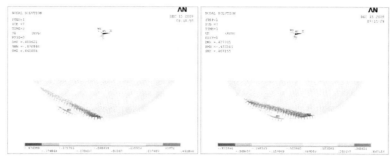

Figure 9. The deformation effect of the cable-net of FAST

5.2 Requirements for the actuators

The actuator is a mechanism which is used to control the position of the nodes through a down-tie cable. Each of the actuators is fixed on one end to its ground anchor, with the other end hinged to a cable net node. The actuator operates at the instruction of the upper computer to generate tension in the down-tie cable. As the actuator pulls out or draws back, the distance between the ground anchor and the cable net nodes changes correspondingly. The overall result of the coordination operation of the actuators within the illumination aperture is a formation of a segmental parabola surface.

Structural analysis results show that the maximal operation load of the actuator is about 50kN when operated in the active deformation mode. Considering the elastic deformation of the down-tie cables, the position control precision of the actuator is limited to be better than 0.25mm. Based on the observational mode, the actuator should function in two types of operation modes: source switching mode and source tracking mode. When working in source switching mode, the related actuators are required to complete the process in 10 minutes, which means a maximal speed of 1.6mm/s. For the source tracking mode, the related actuators will operate with a varied speed ranged from 0 to 0.6mm/s, which will be shown in Figure 11. Due to the difference in maximal speed in the two modes, the actuators must have a good speed control capability. Currently, both mechanical drive and hydraulic drive are studied for the type selection of the actuators.

5.3 Control formula for the actuators

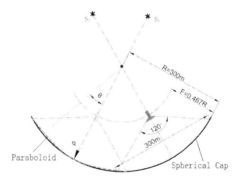

Figure 10. Geometry of the paraboloid and the spherical cap

As shown in figure 10, the coordinate origin is set at the center of the spherical cap. The polar axis is the axis of the paraboloid. The illuminated paraboloid can be described as equation (1):

$$r^2 \sin^2 \theta + 4Fr\cos\theta - 4FR = 0 \qquad (1)$$

Where r is the distance between an interested point and the origin. θ is the angle from the optical axis to the line connecting the center and the point. R=300m, F=0.467R, θ =-30°~30°.

Correspondingly, the actuator should operate as equation (2).

$$r_a = \frac{-2F\cos\theta + 2\sqrt{F^2 \cos^2\theta + FR\sin^2\theta}}{\sin^2\theta} \qquad (2)$$

Where for a source tracking process, θ varies with time t (in seconds):

$$\theta = \theta_0 + \varpi t \qquad (3)$$

Where θ_0 =-30°, ϖ is the angular speed of the earth rotation, ϖ =4.167e-3(degree/s).

The position (Zero point modified to the stroke center of the actuator) and the velocity of a typical actuator is shown in figure 11. It can seen that the maximal speed in the source tracking process is about 0.6mm/s, and for most of the time in the process, the speed is slower than 0.2mm/s. Considering the precision requirement of 0.25mm, we can conservatively and reasonably think that if an actuator can operation at the speed of 0.1mm/s for enough long time, the actuator will be functional in the source tracking process. The reason is that for the position with relating speed below 0.1mm/s, the position control can be achieved by on-offs of the motor in the worst case, without lowering the position control precision of 0.25mm.

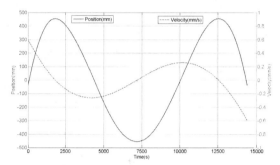

Figure 11. Control position and velocity curve of a typical actuator in source tracking mode

5.4 Monte Carlo simulation for the RMS of the illuminated paraboloid

For an illuminated paraboloid, about 1000 actuators will be in operation. Currently, the position control error budget of a single actuator is 0.25mm. When in operation, the 1000 actuators will not necessarily operate with the same precision. And the precision of the cable net nodes will surely be worse than a single actuator due to the elastic deformation, the measurement and control error, the wind load induced error, and other unseen errors. For the final position control precision of the cable net nodes, a statistical Monte Carlo method is used to estimate the overall RMS of the illuminated aperture. The results are shown in figure 12, with the 50000,100000, and 200000 simulations.

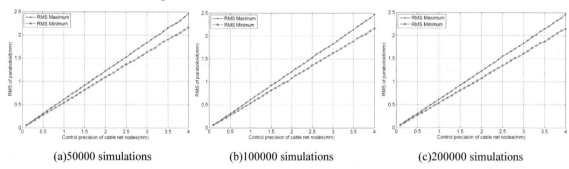

(a)50000 simulations　　　　　　　　(b)100000 simulations　　　　　　　　(c)200000 simulations

Figure 12. RMS maximum and minimum for different control error budgets of cable net nodes

The results show little differences for simulations more than 50000. The results also show that, to achieve a RMS value of 2.5mm within the illuminated area, the control precision of the cable net nodes should be better than 4mm. to achieve a RMS value of 5mm within the illuminated area, the control precision of the cable net nodes should be better than 8mm, as can be seen in figure 13.

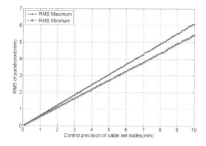

Figure 13. 50000 simulations for different control precisions(up to 10mm) of the cable net nodes

6. THE CONTROL OF THE ACTIVE REFLECTOR

Since there are no rigid connections between the feed cabin and the reflector, time synchronization and precise position control is critical for the operation of the telescope. The main measurement and control tasks of FAST are shown in table 5.

Table 5. Main measurement and control tasks of FAST

No.	Tasks
1	High precision topographic mapping
2	High precision(better than 1mm) ground base control net and time reference
3	Position measurement in the production and installation process of large scale structures
4	Real time control and surface test of the surface nodes of the active reflector
5	Real time measurement and control of the position and pointing of the feed cabin
6	System monitoring and diagnostics in the operation process of the telescope

To meet the specifications required by scientific goals, all these will be performed on a distance scale of 1000m, real time, and with enough sample frequency. The measurement and control flow of the active reflector of FAST is shown in figure 14.

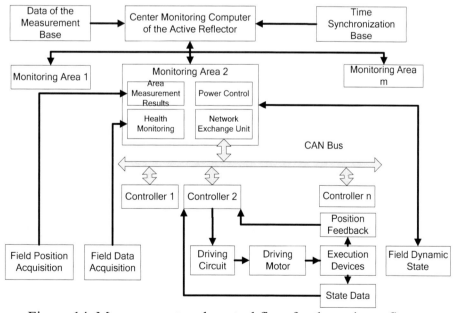

Figure 14. Measurement and control flow for the active reflector

In the preliminary design of FAST, 3GHz observation frequency is a must with possible upgrade to 8GHz. Correspondingly, precision requirements related to the measurement and control of the active reflector are shown in table 6.

Table 6. Precision requirements related to the measurement and control of the active reflector

Base measurement	Measurement and control of the reflector
Absolute positioning precision of the base points: 2cm	Positioning precision of the ground anchors: 2cm
Absolute pointing precision of the base points: 0.5"	Positioning precision of the fixed point of the cable net on the ring beam: 2cm
Baseline length precision: 1mm	Positioning precision of the nodes on the neutral plane of the cable net: 1.5mm
Relative coordinates precision of base points: 1mm	Dynamic measurement and control precision of the nodes on the cable net: 2mm
Relative coordinates precision of control points: 1mm	Dynamic measurement frequency of the nodes on the cable net: 0.02Hz
Time precision: 1ms	

7. CONCLUSION

The paper summarized the requirements of the surface precision of FAST active reflector. The related cable net design and reflector panel design are presented. The two basic operation modes of source searching and source tracking are also given. The requirements and the control formula for the actuators are deduced. Monte Carlo simulation is used to estimate the restriction of surface RMS of the paraboloid to the control precision of the cable net nodes. The results show that, to achieve a RMS of 5mm for the illuminated paraboloid, node control precision of 8mm is necessary. Precision requirements of the measurement and control process are also presented. For the current design, the surface RMS of FAST will be satisfied.

ACKNOWLEDGEMENTS

This work was supported by the National Natural Science Foundation of China (Grant No. 11173035). Thanks also go to my colleagues for their contributions to the work on the active reflector for FAST.

REFERENCES

[1] Nan R.D. et al., "The Five-Hundred Aperture Spherical Radio Telescope (fast) Project", IJMPD, 20, 989(2011)
[2] Qiu Y. H., "A novel design for a giant Arecibo-type spherical radio telescope with an active main reflector", MNRAS, 301, 827-830 (1998).
[3] FAST Group, [Preliminary Design of FAST], National Astronomical Observatories, Beijing (2008).
[4] FAST Group, [Preliminary Design of FAST-Active Reflector Volume], National Astronomical Observatories, Beijing (2008).
[5] Qi-ming Wang, Lichun Zhu, Rendong Nan, "Development of Active Reflector of the 30-Meter FAST Model", Proc. of SPIE, Vol. 7018 70184A-8 (2008).
[6] RongWei Tang et al., [Optimization Design Report for the FAST reflector panels], China Academy of Building Research, Beijing (2012).

The Sardinia Radio Telescope (SRT) optical alignment

Martin Süß [a], Dietmar Koch [a], Heiko Paluszek [b]

[a] MT Mechatronics GmbH, Weberstrasse 21, D-55130 Mainz, Germany
[b] Sigma3D GmbH, Max-Hufschmidt-Strasse 4a, D-55130 Mainz, Germany

ABSTRACT

The Sardinia Radio Telescope (SRT) is the largest radio telescope recently built in Europe – a 64m Radio Telescope designed to operate in a wavelength regime down to 1mm.

The SRT is designed in a classical Gregorian configuration, allowing access to the primary mirror focus (F1), the Gregorian focus (F2) as well as a further translation to different F3 using a beam waveguide system and an automated change between different F3 receiver positions.

The primary mirror M1, 64m in diameter, is composed by 1008 individual panels. The surface can be actively controlled. It's surface, as well as the one of the 8 m Gregorian subreflector, needed to be adjusted after panel mounting at the Sardinia site. The measurement technique used is photogrammetry. In case of the large scale M1 a dedicated combination of a large scale and a small scale approach was developed to achieve extremely high accuracy on the large scale dimension. The measurement/alignment efforts were carried out in 2010 and 2011, with a final completion in spring 2012. The results obtained are presented and discussed.

The overall alignment approach also included the absolute adjustments of M2 to M1 and the alignments of M3, M4 and M5. M3 is a rotating mirror guiding the RF beam to M4 or M5, depending on the operational scenario. These adjustments are based on Lasertracker measurements and have been carried out in an integrated approach.

Keywords: SRT, Sardinia Radio Telescope, Optical Alignment, Mirrors, Beam Waveguide (BWG), Laser Tracker, Photogrammetry

1. THE SRT OPTICAL CONFIGURATION

The SRT optical configuration is a classical Gregorian system with a shaped parabola primary mirror M1, an elliptical shape of the secondary mirror M2. It is followed by additional beam waveguide mirrors M3-M5. The M3 can be rotated along an axis parallel to the line of sight of the primary, and can thereby direct the beam to either M4 or M5. In practice, there are two more mirror positions available for future extended use [1],[2].

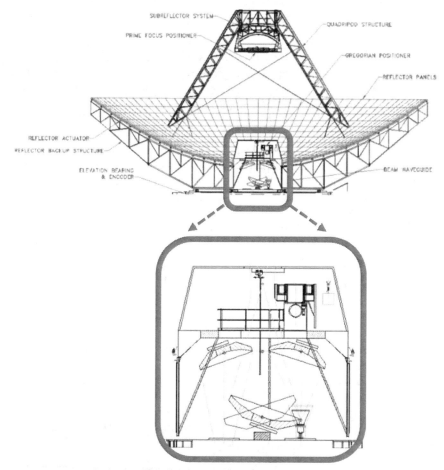

Figure 01: The SRT mirror arrangement

The drawback of the Gregorian configuration is higher effort on the mechanical side (since the M2 is further away from the M1). This is accepted to use the advantage of the Gregorian principle (compared to a cassegrain configuration with a hyperbolic secondary mirror): the access to the focal point F1 of the primary mirror: At SRT, a set of receivers can be placed in the F1 in front of the secondary mirror by the use of the "Prime Focus Positioner", essentially a swing mechanism, Furthermore at F2 there is a revolver-type receiver mechanism, and finally the M3 is capable of rotating and can thereby redirect the beam to M4 or M5 to make use of the final focus F3 at different positions / receivers. This system allows a large variety of receivers and a fast change between them.

2. REQUIREMENTS AND MEASUREMENT APPROACHES

2.1 Primary Mirror Surface

The surface of the primary is composed by individual panels made from aluminum. They are arranged in 14 rows with an increasing number of Panels per row, starting from 24 at the inner ring to 96 at the outermost ring. In total the reflector is formed by 1008 panels. The panels have a 4-point mount, with one support point at each corner. An actuator is attached to the intersection of 4 Panels, so this actuator equally drives the corners of 4 different panels at a time. A total of 1116 actuators are installed on a steel backup structure and are individually aligned prior to panel installation. Unlike in other concepts [3], the active surface will be controlled in an open-loop control, based on surface measurements in different elevation angles.

Each individual panel has been measured and verified in-shop by a photogrammetry measurement, using a target projection technique suitable for serial verification measurements. 4 holes, 3.2mm diameter, are referencing the panel geometry. The overall RMS of the panel manufacturing process was better than 65μm.

For the primary mirror surface, two different kinds of specifications apply:

1.) The typical surface accuracy requirement asks for an overall RMS of 500μm, with the goal to achieve 300μm. This relates to the passive system, the active surface system can be used for further improvements.

2.) A relative accuracy, measured in the direction of the local surface normal, of ±100μm relative between the 4 panel corners.

Both measurements have to be done at 45° elevation angle.

In addition, a surface measurement in different elevation angles (15°, 30°, 45°, 60°, 75° and 90°) is requested to provide input for a lookup table for the active surface actuators as a function of elevation angle.

2.2 M2 Surface

For the subreflector surface, the requirement is defined in terms of an overall RMS value. This is 50 μm RMS on the 8m subreflector. The measurement shall be taken in a position representing 45° elevation angle.

Compared to the M1 requirements, this is somewhat easier to achieve due to the scaling. Also practical works require much less effort.

2.3 Alignment of M2 with reference to M1

The M2 Unit is supported by a steel framework structure connecting the M2 with 4 legs to the Primary mirror backup structure (we call it "Quadropod"). The steelworks in this dimension (one of the legs is >25 m, weighting approx. 30 metric tons) underlies manufacturing and assembly tolerances which are one to two orders of magnitude higher compared to the ones applying to the subreflector position.

The final alignment of the M2 unit with respect to the M1 is specified with the following ranges:

± 2.5mm distance between the actual and the theoretical subreflector vertex.

<= 250" (arcsec) tilt between the M2 and M1 LOS axes.

The measurement shall be done at 45° elevation. Repeated measurements at 5°, 15°, 30°, 75° and 90° also have to be performed. The alignment is performed by variation of shim thickness at the Hexapod actuator mounting pints.

2.4 M3 – M5 mirrors

The mirrors M3-M5, also called the beam-waveguide mirrors (BWG) are formed by one piece cast aluminum. The mirrors have been measured in-shop and have 4 reference points each (in form of 3.2mm holes) to define the mirror position for the alignment process.

To the mirror position itself the following requirements apply:

1.) The mirror positions shall be in nominal position within ±0.5 mm in X, Y and Z-direction.

2.) A maximum tilt of ± 60" (arcsec) from nominal direction is accepted.

The M3 mirror redirects the beam to M4 or M5. To do this, it is rotated along an axis which shall be in parallel with the line-of-sight- (LOS-) axis of the M1. The tolerances on this are limited to be within ± 0.5mm with the M1 axis, with a maximum tilt of 30" (arcsec).

Again, the measurement works have to be performed at 45° elevation. Mirror positions in 5° and 90° also need to be measured and documented.

3. MEASUREMENT CONCEPTS

Several common practices for mirror alignments have been performed in the past. Next to the conventional techniques like direct measurements using gauges, angular intersection methods (based on theodolite measurements), become more and more enhanced by accurate distance measurements in modern "total stations" or Laser-Trackers. A different concept is the photogrammetric technique, used in many fields of science and technology like digital land modeling, architectural and archeological data acquisition etc. and in many fields of industrial measurements, like deformation analysis, surface shapes for documentation and quality assurance tasks [7]. It is also frequently used for the measurements of RF reflecting surfaces, since they somehow form an ideal object for this technique, both in terms of accuracy and practicability. On the other end there are techniques based on wave propagation measurements, from simple gain measurements up to microwave holography. This is nicely compared in [3].

In the optical alignment tasks of the SRT there are two aspects which need to be considered: obviously the technique selected must be capable of delivering the required accuracy to perform the alignment and verification within the tolerances given. Additionally the alignment needs to be performed in a reasonable timeframe. This puts additional requirements on the measurement practice. Ideally there is a direct readout and feedback of the actual results, allowing effective adjustments. This is especially important in our case: the majority of the tasks have to be performed in 45° elevation, where accessibility and safe working (and adjustment) conditions are other serious boundary conditions.

The surfaces of M1 and M2 are measured using photogrammetry. While for the M2 this is quite straightforward, for the M1 this needed to be carefully analyzed due to the different requirements: While the overall RMS requirement is also clearly a candidate for photogrammetry, the relative adjustment of the panel corners requires much higher measurement accuracy. Typically, this relative adjustment could easily be done by locally applied dial gauges. However, this is not easily possible in a 45°-position, where a manlift (capable to operate up to approx. 70m and with noticeable overhang) would be required to apply the tooling and to perform the adjustments. Multiplying this by 4032 adjustment points and considering the difficulties of finding the local normal direction, this becomes a significant effort in terms of cost and schedule.

Figure 02: The SRT in 45° Elevation

Finally it was chosen to use photogrammetry also for this task: targets needed to be applied anyway for the global RMS measurement and the 45° Elevation is not a serious problem for taking the pictures. Only the predicted measurement accuracy of approx. 100μm confidence range (1σ) on global scale is not sufficient to do the adjustments against a specification of the same amount. Therefore the approach was to do a combination of "global" and "local" photogrammetry: The global set is used to accurately determine point coordinates within the mirror reference frame. The local set is combined with these results and analyzed with focus on the differential local deviations.

For the alignment tasks of M3 – M5, as well as for the positioning of M2 with respect to M1, Laser-tracker measurements have been chosen. The measurement itself is quite straightforward; the effort was to prepare safe working platforms for the measurement and adjustment team.

4. PRIMARY MIRROR MEASUREMENT AND ALIGNMENT

4.1 The Photogrammetric Approach

Since the requirements are pushing the limits of available measurement techniques, several estimates and simulation runs have been performed to evaluate the best possible set-up of targets. Using a state-of the art camera (Nikon D3X x 28mm, 24Mpixles, 36x24mm sensor) we were experimenting with different kinds of sizes of retro-reflecting targets, 50 and 70mm diameter. A large number of coded targets are also required to allow the orientation of pictures on a local scale; both targets have been specially produced for this task.

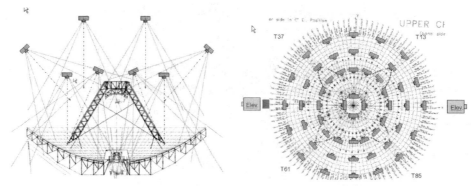

Figure 03: Ideal Camera Positions (here for 90° elevation)

The reference for the global coordinate system and the scaling of the data set is formed by a set of 6 fixed holders for CCR targets. These do have the advantage that they can be used one time with a corner cube reflector (CCR) to be accurately determined in the reference coordinate frame using a total station. Afterwards, the same holder is used with a hemisphere and a 60mm retro-reflector on the flat side to appear in the photogrammetry pictures. In this way, a relation between the photogrammetry data against the reference frame can be achieved easily. The reference points are positioned on the 4 quadropod legs (1 each) and 6 on the cabin in the M1 center, which forms the most stable basis in this region.

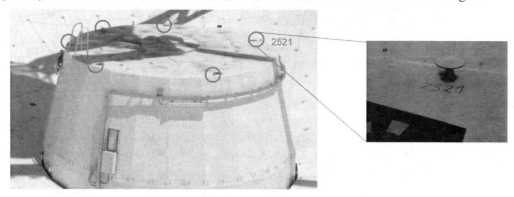

Figure 04: Reference coordinate system

The practical problem in taking the photogrammetry pictures is the accessibility of the required camera positions in terms of a fast, continuous process. Several possibilities have been investigated, starting from a balloon born camera (blimp), remote controlled airships (quadrocopter, etc., [5]) to manlifts and cranes. Finally the decision was taken to use a mobile crane, since it promised the best ratio of reliability, controllability and cost. However, a suitable crane still is a major cost contribution in the overall process.

Figure 05: Crane setup for photogrammetry measurement

The actual measurements were carried out during nighttime, trying to meet the phase of most stable thermal environment. Also other environmental conditions are critical for successful measurements: the humidity shall be low to avoid condensation on the panel surface (and the targets). With condensing humidity, the retro-targets do not work properly. In this case the points are either detected on a wrong position or cannot be detected at all. For the same reason rain is also a cause to stop a campaign. Another restriction is the wind situation: the crane itself can be operated up to 8 m/s wind speed, not mentioning the difficulties to achieve good perspectives in regions close to the surface >= 60m above ground!

Figure 06: Photogrammetric measurement campaign

In good conditions, a typical measurement is a combination of two sets of measurements: a first set, taken at the beginning of the night for the global approach and a second set, where pictures are taken from a closer distance. These are later on evaluated with respect to the local deviations. Each set typically contains approx. 400 – 550 pictures and it took approximately 1 – 1.5 hours to take them, using the setup described above, depending on the environmental conditions.

Analysis of the photogrammetric measurements is essentially a post-processing of the sets of pictures. This is done in 4 steps:

a) measurement of the 2D picture-coordinates in each picture

b) pre-orientation of the image in the global coordinate system

c) point mapping of identical data points

d) bundle adjustment, delivering 3D coordinates

In a) the center of the target is measured in each picture in a picture coordinate system. The analysis software allows the automated detection of coded and uncoded points, see Figure 07. The centroiding algorithm delivers sub-pixel accuracy in the center detection.

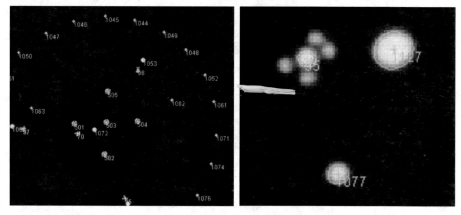

Figure 07: point detection and centroiding during image processing

In b) and c) for each picture the camera position and orientation is identified, resulting in the information which points are visible in the pictures (Figure 08, left) or for a selected point in which and how many pictures it appears (Figure 08, right). The latter one already allows estimating whether a selected point is sufficiently covered: If the number of pictures is small and/or if the camera positions are not sufficiently distributed (all from one direction), the 3D coordinate evaluation will be inaccurate.

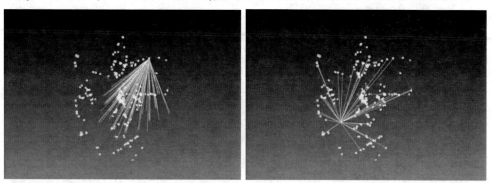

Figure 08: Points in a selected picture (left) and camera positions of pictures showing a selected point (right)

Finally in d) the 3-D coordinates of each point are calculated. In this "bundle adjustment", the highly over-constrained set of equations is solved, using least square error algorithms. This finally delivers the extreme accuracy which can be achieved by the photogrammetric approach.

Comparing the 3D Data set with the specified surface contour, adjustment values for the surface alignment process can be determined.

4.2 Primary Mirror Alignment

For purpose of alignment, the results are printed into lists, sorted after a certain numbering scheme: the reflector is subdivided into "trusses" (the radial girders of the backup structure, where the surface actuators are mounted) and "hoops", the circumferential connections between the trusses. So for every individual adjustment point, a dedicated offset is derived. This process has been varied throughout the proceeding: while at the beginning the results from the global photogrammetry have been used to define the adjustment offsets, later on the local deviations are used. This is because the global values are less accurate, but improve the overall RMS figure. Also there are more points to be adjusted. Once we had reached a status where global RMS was better than the goal (and far below spec), only the local deviations have been used to define adjustment offsets.

The adjustments have been performed in zenith position (but based on values derived at 45°) using a dedicated tool set of adjustment wrenches. Additionally there is a holder for dial gauges which is attached to the actuator through an access hole, formed by the chamfers of 4 adjacent panels. The dial gauges are placed on the panel surface close to the reflective target and set to zero. From there on, the values given in the adjustment tables are directly adjusted (including sign convention).

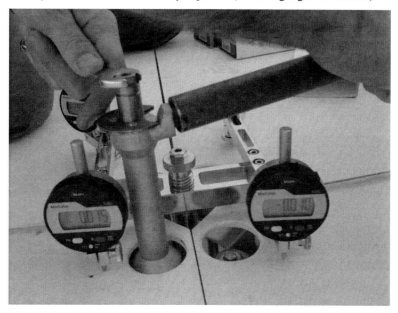

Figure 09: Alignment process

4.3 Results

A first measurement has been conducted in August 2011, followed by 3 iterations during the second half of September 2011. After a period of bad weather conditions, it was decided to postpone activities towards late spring 2012. So the activities continued end of May 2012 and were completed mid of June 2012. Altogether 11 measurements were completed.

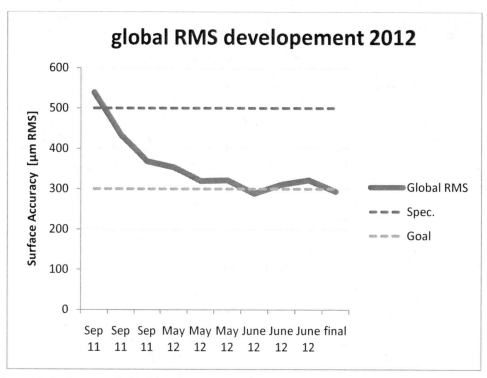

Figure 10: development of global RMS

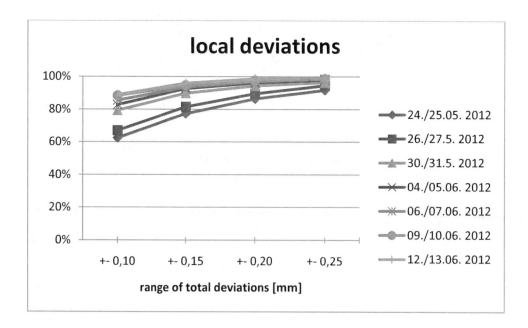

Figure 11: developement of local deviations

In the final nights we achieved measurement accuracies on local scale in the range of 31 μm. The global scale photogrammetry was in the range of 100 μm. The global RMS is below 300μm, in the final set at 293μm.

For the local deviations, the efforts have settled slightly out of spec. While in terms of specification all (100%) of the points have to be within ±100 μm, in practice we have distribution as shown above in Figure 11, and a remaining number of points which are out. A detailed error analysis, including not only the measurement itself, but also error sources like adjustment accuracy and environmental influences, theoretically confirms this result.

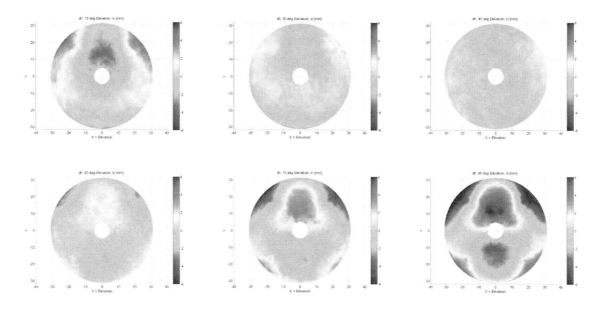

Figure 13: surface deflection measurement results in varying elevation angles

5. SECONDARY MIRROR ALIGNMENT

As for M1, the M2 is formed by individual panels which need to be aligned. Due to the size, the effort is less, anyway the requirement to achieve a 50μm RMS is still challenging. The works have been performed prior to mounting the M2 unit to the quadropod, in a dedicated assembly mock-up on ground. The measurement is again performed in a 45° angle, as shown in figure 14.

Figure 14: The SRT secondary mirror in the test / alignment mount at the Sardinia site

The adjustment of panels is done from the backside in horizontal mirror position, therefore a change between measurement and alignment position is required.

In May 2010, a total of 14 iterations from an initial setup to the finally achieved result have been performed. It needs to be noticed that the adjustment metrics did cause some difficulties. At the beginning, having large offsets, a significant influence from second order effects was observed. These have been considered in later adjustments, bringing down the overall RMS figure to a final number of 57μm rms.

In terms of measurement technique, challenges are mainly from environmental conditions. Analysis effort as well as accuracy requirements have been well within sufficient range.

Nr.	7	8	9	10	11	12	13	14
Messun	Messun 04, 12.05.2010	Messun 01, 13.05.2010	Messun 02, 13.05.2010	Messun 03, 13.05.2010	Messun 03, 13.05.2010, efittet	Messun 01, 15.05.2010, efittet	Messun 02, 15.05.2010, efittet	Messun 02, 16.05.2010, efittet
Mittelwert mm	0,010	0,004	-0,006	-0,018	-0,023	-0,02	-0,032	-0,024
RMS [mm]	0,259	0,312	0,138	0,187	0,111	0,066	0,080	0,057

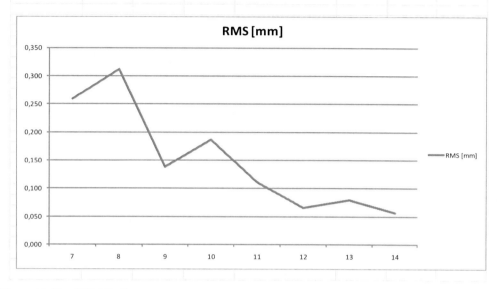

Figure 15: The SRT M2 alignment progress

6. M1 TO M2 ALIGNMENT

The subreflector best fit axis and the primary reflector best fit axis are to be aligned to their theoretical position within the specified tolerances. The theoretical position needed to be verified in the 45°-Elevation position. Corrections to the M2 position at this stage are done by varying the thickness of shims, which are mounted between the apex structure and the SR positioner actuators.

Following the alignment position measurement have been done in different elevation angles (5°, 15°, 30°,60°, 75° and 90°) in order to create a correction matrix as input for the subreflector positioner lookup tables.

The measurements have been performed using a Lasertracker system. Due to the required variation of elevation angles, the setup needed to be carefully prepared. The Laser tracker is located on a specially designed mount in the center of the primary mirror, see figure 16.

Figure 16: Mount of Laser Tracker for 45° measurement

While this allows the placement of the Laser Tracker in a well-defined position with respect to the reference frame, the operation cannot be done directly. The Laser Tracker in use is equipped with a remote control, so that the operator only needs to be within the operational range of the remote. Thanks to a previous teach-in procedure of reference- and measurement targets, the actual measurement itself is quite simple: the expected positions are acquired, and the final readout has been found within the search range. The scenario is shown in figure 17.

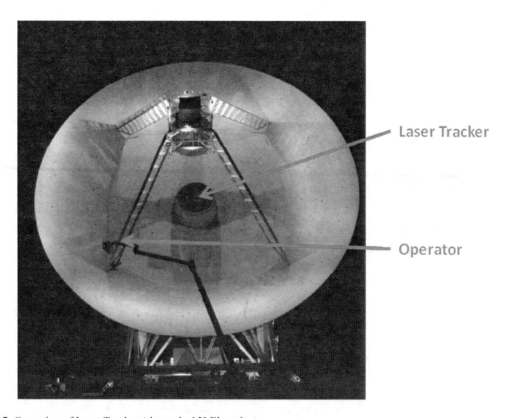

Figure 17: Operation of Laser Tracker (shown in 15° Elevation)

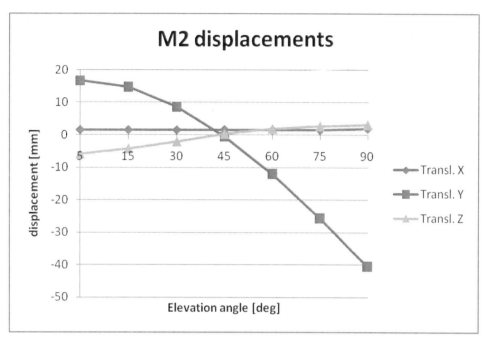

Figure 18: M2 displacement vs. elevation angle

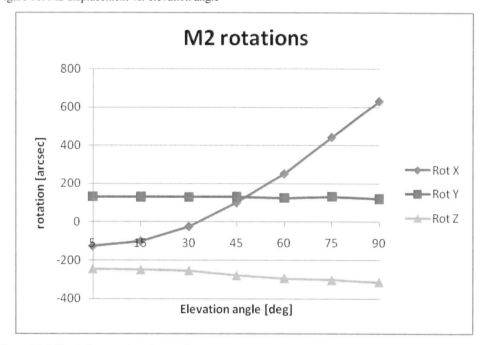

Figure 19: M2 rotation vs. elevation angle

As expected, deviations occur in y- and z- directions. Y is the direction of gravity in horizontal, z in zenith position. The different values are due to the different stiffness of the structure when loaded lateral (bending) or perpendicular (compression).

In rotational direction the predominant effect is the tilt along the elevation axis. Somehow noticeable is also some rotational movement in z, the line of sight. It is not yet clarified whether this is an artifact within the measurement or an actual second order effect.

7. M3 – M5 ALIGNMENT

For the alignment of the M3 axis, a corner cube reflector is placed on a (randomly chosen) point at the M3, which is visible from the laser tracker position. The M3 is then rotated along its axis, and the position of the target is recorded. The laser tracker is oriented in the M1 reference coordinate frame, so the readout directly delivers a sinusoidal curve. While the amplitude of this readout delivers the tilt of the M3 axis, the phase shift gives the direction of the tilt axis. Adjustments are done by shimming the mounting flange of the M3, and a repeated measurement proved the correct adjustment.

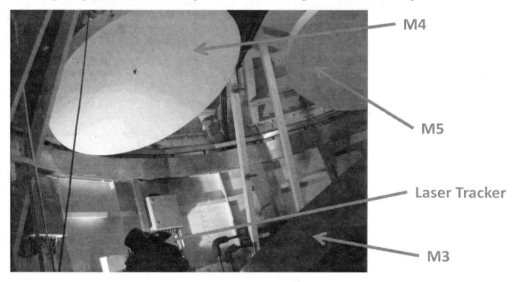

Figure 20: Laser Tracker assembly for 45 deg measurement at M3 – M5

Afterwards, the measurement and alignment of the beam waveguide mirror was performed as a classical Lasertracker measurement. The mirrors are monolithic cast aluminum mirrors. Their contours are defined and manufactured with reference to 4 reference holes, which are used again for the mirror alignment. CCRs are used as targets, the loop between adjustment and readout is in real-time. From the RF analysis the nominal mirror positions are given. Therefore, the adjustment itself was straightforward: The operator of the Lasertracker was observing the actual readout and the adjustment people are doing the correction of the positions. Finally the achieved positions have been measured 45 and 90 deg to determine possible pointing corrections terms for the M3-M5 mirrors. It was found that deviations are in fractions of millimeter between the two positions.

8. REFERENCES

[1] Orfei, A., Morsiani, M., Zacchiroli, G., Maccaferri, G., Roda, J., Fiocchi, F., "Active surface system for the new Sardinia Radio Tele-scope", Astronomical Structures and Mechanisms Technology, Proc. SPIE 5495, 2004

[2] Olmi, L. & Grueff, G. "SRT: design and technical specifications" , Memorie della Società Astronomica Italiana Supplement, v.10, p.19 (2006)

[3] Parker, D.H., Payne, J.M., "Active Surface Architectures of Large Radio Telescopes", Proceedings of the XXVIIth URSI General Assembly in Maastricht, 2002

[4] Jodoin, S.J., "The calibration of a parabolic antenna with the aid of close range photogrammetry and surveying", Geodesy and Geomatics Engineering UNB, Technical Report No. 130 (1987).

[5] Mayr, W., Schroth, R., "Autonomous Aerial Sensing - Fast Response and Personalized", FIG Congress 2010, Sydney, Australia, 11-16 April 2010

[6] http://www.srt.inaf.it/

[7] Luhmann, T., "Photogrammetrische Verfahren in der industriellen Messtechnik", DGPF, Band 9, 2001

Final tests and performances verification of the European ALMA Antenna

Gianpietro Marchiori[*a] and Francesco Rampini[a],
[a] European Industrial Engineering (EIE GROUP), via Torino 151A – 30172 VENEZIA–MESTRE (Italy)

ABSTRACT

The Atacama Large Millimeter Array (ALMA) is under erection in Northern Chile. The array consists of a large number (up to 64) of 12 m diameter antennas and a number of smaller antennas, to be operated on the Chajnantor plateau at 5000 m altitude. The antennas will operate up to 950 GHz so that their mechanical performances, in terms of surface accuracy, pointing precision and dimensional stability, are very tight.

The AEM consortium constituted by Thales Alenia Space France, Thales Alenia Space Italy, European Industrial Engineering (EIE GROUP), and MT Mechatronics is assembling and testing the 25 antennas. As of today, the first set of antennas have been delivered to ALMA for science.

During the test phase with ESO and ALMA, the European antennas have shown excellent performances ensuring the specification requirements widely.

The purpose of this paper is to present the different results obtained during the test campaign: surface accuracy, pointing error, fast motion capability and residual delay. Very important was also the test phases that led to the validation of the FE model showing that the antenna is working with a good margin than predicted at design level thanks also to the assembly and integration techniques.

Keywords: ALMA, surface accuracy, pointing error, fast motion, residual delay, FE model, System Engineering

1. INTRODUCTION

The European Antennas are now in an advanced phase of delivery, and some of them are already being used for scientific activities.

All commissioning and test phases have been performer by the AEM Consortium in close cooperation with ESO Team. In particular, the first two antennas have been subject of very accurate tests with the goal of evaluating in detail all the performances of the antenna.

During this phase, all the specification requirements have been tested, with a particular attention to: General inspection and dimension, Kinematic and Dynamics performance, antenna power distribution, front-end and back-end installation, Computing and Antenna control unit, Axes position, reflector system and surface accuracy, pointing performance, metrology, path-length delay, thermal control, electrical performance and electromagnetic compatibility, maintainability, safety.

The following chapters report only the tests that were aimed at validating the scientific performances of the antenna. The first step was to evaluate in detail the correspondence between the results of the FEM Model and the final structure. The FEM Model has in fact been used afterwards to evaluate all

---

[*] gmarchiori@eie.it, phone +390415317906, fax +390415317757

the structural performances that cannot be evaluated with tests in the sky or with other instruments. We have therefore evaluated the surface accuracy of the antenna, the pointing and path length performances.

Today eight antennas are doing science at 5000m, other two antennas have been delivered to ALMA for the completion of the acceptance tests, two antennas are under acceptance at OSF camp by ESO and other five antenna are under integration at OSF camp at different stage level, all the other antennas have been manufactured in Europe and the different parts are ready for final assembly.

Figure 1 OSF camp. Antenna during assembly phase

2. MAIN STRUCTURE VALIDATION AND FE MODEL

Not all the performances of the antenna can be evaluate during the observation operations, or anyway it is not easy to obtain in a short time results that can give result to structural characteristics. To be sure of the behavior of the antenna in all operational conditions with respect to the specification requirements we have therefore performed a series of structural tests that demonstrate that the FEM Model, with which the entire evaluation of the antenna behavior during the design phases has been carried out, is reliable and it corresponds to reality.

The tests, started during the design phase with the qualification of the materials, of the gluing, of the bearing etc., have involved all the main components of the antenna. Hereafter, some results of the tests:

- Elevation structure behavior;
- Steel structure behavior;
- Eigenfrequency evaluation.

Due to the uncertain of the measurements (measurement systems error, boundary conditions, test performed not in laboratory but in the external environment), the goal was to have for each test a deviation lower than the 20% between the FEM prediction and real measurement.

2.1 Elevation structure behavior

This test has been performer to evaluate the stiffness of the elevation structure at different elevation positions. A Laser Tracker was located in the central hole of the main reflector. 4 targets have been placed at the level of the subreflector and 8 targets have been fixed on the rim of the main reflector. Then the antenna has been rotated from 90° to 2° in steps of 15°.

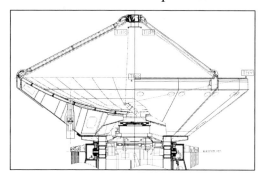

Figure 2 Elevation structure stability and deviation measurement setup

The measurement results have been compared with the results obtained with the FE model. In figure 3 the cyan curves represent the measured value of the subreflector deviation test, while the yellow curves represent the value expected by FEM analysis.

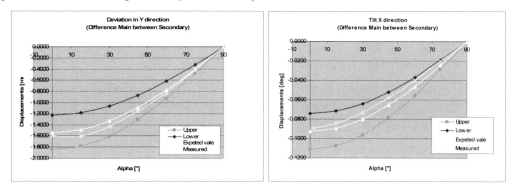

Figure 3 Comparison between test and FE odel: UY and RotX

The maximum measured differences between 90° elevation and 0° elevation of the translation UY and rotation rotX of the Subreflector are :

- TY=1.63mm
- ROTX=0.097deg

While the FEM results are:

- UY=1.54mm
- ROTX=0.093deg

The results show that the difference between the measured displacement ante FEM prediction is well inside the band of ±20%.

2.2 Steel structure behavior

A pull test of the yoke arms from the ground has been done in order to verify the stiffness of the steel structure. For the execution of these tests two additional inclinometers have been assembled at the level of the top arm.

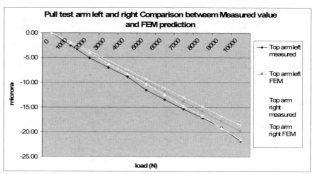

Figure 4 Alidade structure pull test measurement setup and results

In figure 4 it is clear that the real steel structure behavior is very similar to the predicted FE analyses.

2.3 Eigen frequency evaluation.

An additional important test that gives us a feedback about the antenna attitude was the evaluation of the Azimuth and Elevation close loop and open loop transfer function. With these tests we have evaluated the main Eigen-frequencies of the antenna.

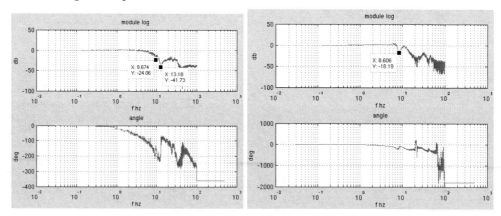

Figure 5 Azimuth close loop and elevation close loop transfer function

The visible eigen-frequencies are at 8.6 Hz in elevation and 9.67 Hz and 13.18 Hz in Azimuth.
From the FE analysis the first eigen-frequency around the elevation axis is 9,29 Hz when the antenna is at 0° elevation and 9.21Hz when the antenna is at 90° elevation. The tests show a first eigen frequency in elevation of 8.6 Hz. With a difference of -7%.
The first eigen mode that can be detected by the azimuth axis is about 9,45 Hz with the antenna at 0° and 9,35 Hz with the antenna at 90°. The test show a first eigenmode around the azimuth axis of 9.67, with a difference of +3%.

3. ANTENNA SURFACE ACCURACY

The Antenna surface accuracy includes contributions of the panels system (manufacturing error, gravity wind and thermal deformation), panels adjuster (accuracy, thermal deformation), backup structure (gravity wind and thermal deformation, ageing) and subreflector (manufacturing error, gravity, wind and thermal deformation, coma error).

To evaluate the Antenna surface accuracy many of these contributions have been measured or tested during the manufacturing of the single components or in the antenna on site.

One of the main test was the holography measurement of the primary mirror surface accuracy and its thermal behavior under primary condition.

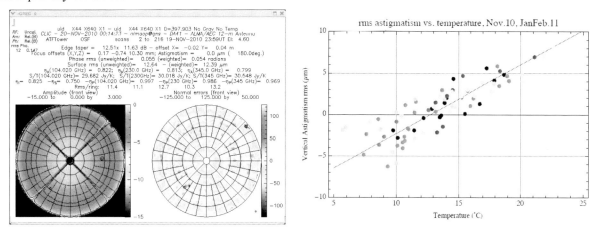

Figure 6 Holography map and Surface accuracy behaviour with respect ambient temperature variation

In the first antenna the surface accuracy of the primary mirror was 12.4 microns RMS.

4. POINTING ERROR AND TRACKING TEST

During the acceptance phases, the ESO team performed the below-reported tests on sky, and derived the pointing capability of each antenna. Up to now, all the antennas have shown very good results with a good margin. It was also surprising that the antennas almost immediately met the pointing specification requirement without any particular debugging. Actually we are working with ESO scientists because there are some margins to improve the pointing capability.

The main pointing requirements are:

All sky pointing error or Absolute Pointing Errors
These errors shall not exceed in the whole sky travel range 2.0 arcsec RSS.

Offset Pointing Errors
The offset pointing and tracking error shall be limited to the error when the antenna is pointed within 2 degree from any starting position, tracking over a 15 minute period and it shall not exceed 0.6 arcsec RSS.

Step response analysis or fast switching phase calibration
In the fast switching cycle, the antenna shall perform steps of 1.5 degrees on the sky and settle to within 3 arcsec peak pointing error, all in 1.5 seconds of time. The antenna shall then track and integrate on a calibration source for few seconds.

4.1 All-sky pointing error

The evaluation of all-sky pointing error has been done with an Optical Telescope installed in the backup structure. The first step has been the evaluation of the pointing model. In the first antenna ESO made 49 individual tests (6274 measurements).

The predicted all-sky pointing performance of the antenna at AOS, was 1.38 arcsec rms at night and 1.50 arcsec rms during the day, all the tests were performed with the thermal pointing correction predicted by the metrology thermal matrix. The pointing model stability was checked with excellent results and thermal metrology correction significantly improves the stability of the pointing model during the time.

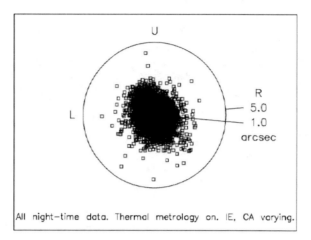

Figure 7 Pointing error with metrology enabled. The rms of 6274 individual measurement performed during 49 pointing runs is 1.00 arcsec.

Table 1 reports as example some day time pointing runs. Column1: UT date. Col. 2: UT start time. Cols 3 – 5: temperature, temperature gradient and wind speed. Col. 6: wind speed from the weather station close to the antenna. Col. 7: number of observations. Col. 8: population SD derived from TPOINT. Col. 9: predicted rms at AOS.

UT date	UT hhmmss	T C	Meteo-3 dT/dt deg/hr	w ms^{-1}	AEM w ms^{-1}	N	σ arcsec	σ_{aosmw} arcsec
05152011	122609	10.09	−0.08	1.36	1.81	76	1.22	1.26
05152011	135431	10.84	1.19	1.47	1.80	65	1.54	1.57
05152011	150548	12.20	0.71	2.27	2.33	71	1.65	1.67
05152011	162245	13.68	1.24	2.62	2.56	84	1.50	1.52
05152011	172742	14.98	1.25	2.89	2.58	99	1.39	1.41
05152011	184820	16.27	1.37	1.99	1.91	109	1.28	1.31
05152011	203013	16.78	−1.13	2.72	2.48	133	1.33	1.36
05162011	101209	9.71	−0.09	3.15	3.94	120	1.26	1.26
05162011	120939	10.59	0.64	0.98	1.12	113	1.40	1.44
05162011	134051	11.45	0.72	1.64	1.94	61	1.79	1.81
05162011	143413	11.88	0.92	2.25	2.42	66	1.65	1.67
05272011	101845	10.15	−0.41	2.13	3.08	128	1.67	1.68

Table 1 Summary of day-time pointing observations.

4.2 Offset Pointing and Tracking Error

The offset pointing and tracking performances evaluation have been done with the thermal and dynamic metrology enabled. The rms error at OSF is 0.28 arcsec, with a worst case of 0.77 arcsec, compared with a specification of 0.6 arcsec. The related predicted performance at AOS is 0.36 arcsec rms with a worst case of 0.80 arcsec.

In order to reproduce the service operational model the thermal metrology was active and dynamic metrology was reset and enabled at the start of each pointing section and disabled at the end. The whole sky was covered both in azimuth and elevation position. We note that a number of tests were performed at wind speeds exceeding 5ms^{-1}.

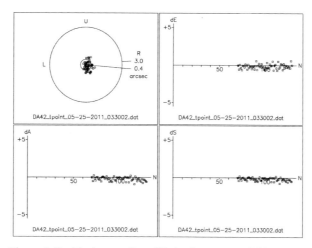

Figure 8: Residuals error from TPoint (σ aosmw = 0.24 arcsec.)

4.3 Fast Motion behavior

One of the main specification requirement of the Alma antenna is the fast motion capabilities. The specification of the step response is: for movements of 1.5° on the sky the antenna position as measured by the encoders must settles to 3 arcsec peak error within 1.5 s and stays within that limit and the rms error in the time period 2 – 4 s following the start of the switch is <0.6 arcsec. Table 2 reports the result of the test campaign performed at different azimuth and elevation positions.

A grid of 21 positions in Azimuth and Elevation at four different position angles on the sky has been tested. For each position has been done 30 individual. All the test passed. The average position of all the test after 1.5sec was 0.344 arcsec. The maximum position error was 0.99 arcsec. The rms tracking error in the period 2 – 4 s has a mean value of 0.037 arcsec and a worst-case value of 0.108 arcsec.

The test were performed with dynamic metrology on and off, there were no visible difference with the Metrology dynamic correction On or Off. Some tests have been performed also in windy conditions (wind speed 10m/sec). A degradation of the performance has been observed, but still all passed, and well in specification.

Elevation (deg)	Azimuth deg							PA deg
	−180	−120	−60	0	60	120	180	
30	0.83	0.52	0.47	0.59	0.59	0.71	0.73	0
	0.14	0.14	0.15	0.15	0.15	0.15	0.15	90
	0.19	0.17	0.19	0.21	0.21	0.17	0.20	225
	0.13	0.10	0.19	0.14	0.14	0.12	0.13	−45
45	0.72	0.60	0.59	0.66	0.66	0.63	0.62	0
	0.15	0.13	0.15	0.16	0.16	0.14	0.15	90
	0.22	0.26	0.27	0.29	0.29	0.30	0.22	225
	0.13	0.08	0.17	0.11	0.11	0.37	0.07	−45
60	0.78	0.80	0.80	0.88	0.88	0.71	0.80	0
	0.15	0.13	0.15	0.15	0.15	0.15	0.15	90
	0.47	0.51	0.44	0.51	0.51	0.53	0.45	225
	0.41	0.40	0.38	0.40	0.40	0.42	0.39	−45

Table 2 Fast Switching results.

5. PATH LENGHT ERROR

Another important requirement is the residual delay evaluation. This is subdivided into repeatable residual delay and non-repeatable residual delay.

The repeatable delay is caused by difference in gravity deformation between an antenna and the nominal antenna. The main contributors to this error are: axis alignment errors, bearing run out, bearing alignment, etc.
The repeatable residual delay for an antenna must not change by more than 20 microns when the antenna moves between any two point 2 degree apart in the sky. During the manufacturing/ qualification, commissioning and acceptance phases all these contribution were measured. In the below table the evaluation of the repeatable residual delay.

Contributors	Value recorded
Azimuth Axis run-out	4 microns
Azimuth bearing axial run-out	2 microns
Elevation Axis run-out	2 microns
Subreflector gravity displacement	5 microns
Primary reflector change of focus length with gravity	1.2 microns
Feed support gravity deviations from theoretical	2,2 microns
	16.4 microns

Table 3 Repeatable residual delay

The non-repeatable residual delay is the delay component that varies with time or is not repeatable as function of the antenna position. According to technical specifications, the contribution to the residual delay budget, due to slowly varying sources such as thermal effects, may be limited to the differential residual delay over a solid angle of 2° radius on sky and over a 3 minute period when tracking at the sidereal rate. Also, this error has been checked during the manufacturing/ qualification, commissioning and acceptance phases of the antenna.

Contributors	Value recorded
Non repeatable component of the Azimuth Axis run-out	0.4 microns
Non repeatable component of the azimuth bearing.	0.2 microns
Non repeatable component of the Elevation Axis run-out	0.2 microns
Subreflector wind displacement	7.3 microns
Primary reflector change of focus length with wind	
Component along the boresight axis direction of the elevation axis displacement with respect to its theoretical position due to wind	
Subreflector thermal displacement	3 microns
Primary reflector change of focus length due to temperature change	4.1 microns
Component along the boresight axis direction of the elevation axis displacement with respect to its theoretical position due to temperature change. (Steel Structure)	3,67 microns
	9.63 microns

Table 4 Non-repeatable residual delay

While the gravity and wind contributions have been evaluated with the FE (see chapter 2), a specific attention has been done on the evaluation of the error induced by temperature variation.

5.1 Antenna Thermal Characterization

The first step has been the evaluation of the of the steel structure elongation due to temperature change. The goal of this test was to establish is in 3 minutes period the thermal insulation is capable to maintain stable enough the antenna of is the thermal metrology correction must be applied to the antenna.

The test has been done with an API laser system and the measurement has been compared with the thermal matrix prediction coming from the 86 thermal sensor installed in the antenna. The test took place during the day and the night and it has been performed first in the antenna base then in the Yoke arm.

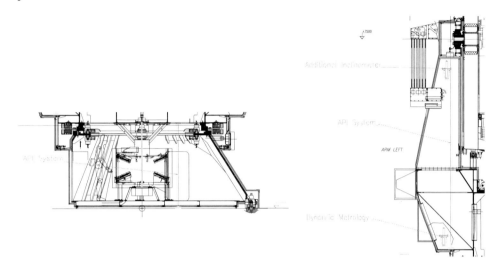

Figure 9: API system inside the base and yoke arm.

Base:
Here are presented the measurement deformation of the Base structure during sunset and sunrise. Blue line metrology prediction Orange line API measurement.

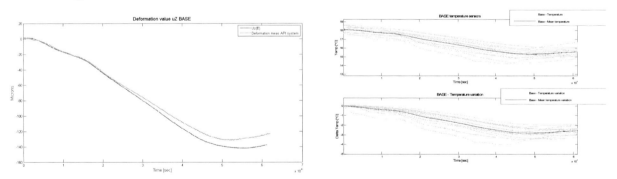

Figure 10: Base measurement deformation during sunset and thermal sensor acquisition.

Yoke Structure:
The same measurement has been performed also in the yoke base for a full day. Blue line metrology prediction Orange line API measurement.

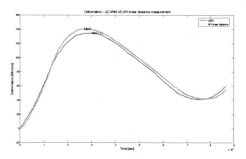 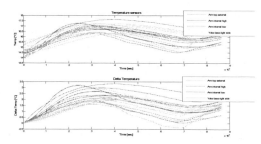

Figure 11: Yoke measurement deformation during one day and thermal sensor acquisition

Many tests have been done for three week, at the end we did a normalization of all the measurements with respect to the delta air temperature and the operational conditions at 5000m. The wind during the test was frequently lower that 5 m/sec and that the solar flux measured at OSF was really similar to the high site.

Base Structure			
delta delay measured	Delta T measured in 30'	Delta T from Spec	Delta Delay
0,962	1,65	1,8	1,05
1,55	1,09	1,8	2,56
1,27	1,2	1,8	1,91
		average	**1,84**

Yoke Structure			
delta delay measured	Delta T measured in 30'	Delta T from Spec	Delta Delay
1,43	2,42	1,8	1,06
2,27	1,23	1,8	3,32
1,72	1,41	1,8	2,20
1,3	1	1,8	2,34
0,08	0,68	1,8	0,21
		average	**1,83**

Table 5 Temperature variation: Path length error

On the base of the above table we can consider that the foreseen path length delay for the steel structure in 3 min period was 3.67 microns, as reported in table 4.

Also the reflectors change of focus length due to temperature variation has been done with a laser tracker positioned in the center of the antenna. Same test setup used for the elevation structure behavior evaluation figure 2.
The distance from the parabola vertex (created using 16 distant targets around the BUS) to the subreflector (materialized by 4 targets) was measured every 1.5 minute.

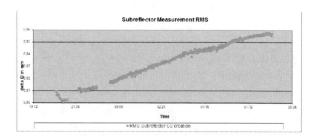

Figure 12: Subreflector measurement during temperature variation

The following graph represents the delta distance variation of the Subreflector with respect to the parabola vertex in a period of 12 hours.

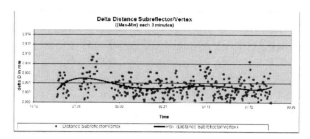

Figure 13: Distance variation between subreflector and parabola vertex.

The mean value of the elevation structure focus length change was 0,003mm with a standard deviation = 0,002mm and a maximum value of 0,005mm.

6. CONCLUSION

A lot of additional tests and measurements have been done during the manufacturing, commissioning and acceptance activities in order to check the real mechanical behavior of the structure and the antenna capabilities. Also the maintainability and reliability that are important tasks for these antennas at 5000m were successfully tested. Anyway, in this paper we presented only the most important ones linked to observation requirements. At the end of the works we were well satisfied because the antennas demonstrated to meet the scientific requirements with a good margin.
Also the feedback obtained during the antenna test phase was very important and it underlined that all performances were in agreement with the prediction done during the design phase.

ACKNOWLEDGEMENTS

We wish to thank ESO (in particular Mr R. Laing, Mr S. Stanghellini, Mr. P. Martinez that performed a lot of work during the tests campaign) and the staff of the AEM Consortium (TAS-F, MTM, TAS-I, EIE) who provided the support for the manufacturing, erection and tests of the antennas.

REFERENCE DOCUMENTS

[1] Robert Laing, Krister Wirenstrand, Pascal Martinez, Silvio Rossi, Reynald Boutembourg, AEM Production Antenna DA41: All-Sky Pointing and Thermal Metrology Analysis, 2011-04-14
[2] Robert Laing, Krister Wirenstrand, Pascal Martinez, Silvio Rossi, Reynald Boutembourg, Gianni Marconi, AEM Production Antenna DA41: Offset Pointing Analysis.
[3] Robert Laing, Massimiliano Marchesi, AEM Production Antenna DA41: Step-response Analysis

ALMA system verification

Richard Sramek*[a], Koh-Ichiro Morita[a], Masahiro Sugimoto[a], Peter Napier[a], Maurizio Miccolis[a], Pavel Yagoubov[b], Denis Barkats[a], William Dent[a], Satoki Matsushita[c], Nicholas Whyborn[a], Shin'ichiro Asayama[a], Javier Marti-Canales[d], Ravinder Bhatia[a], Eugene DuVall[a], Samantha Blair[a]

[a]Joint ALMA Observatory (JAO), Alonso de Cordova 3107, Vitacura, Santiago, Chile; [b]ESO, Karl-Schwarzschild-Strasse 2, D-85748 Garching bei Muenchen, Germany; [c] Institute of Astronomy & Astrophysics, Academia Sinica, P.O.Box 23-141, Taipei 106, Taiwan, R.O.C.; [d]ESO, E-ELT Project, Alonso de Cordova 3107, Vitacura, Santiago, Chile

ABSTRACT

The ALMA aperture synthesis radio telescope is under construction in northern Chile. This paper presents the organization and process of ALMA System Verification. The purpose of System Verification is to measure the performance of the integrated instrument with respect to the ALMA System Technical Requirements. The System Technical Requirements flow down from the Science Requirements of the telescope and are intended to guide the design of the array and set the standards for technical performance. The process of System Verification will help determine how well the ALMA telescope meets its science goals. Some verification results are discussed.

Keywords: ALMA, System Engineering, System Verification

1. INTRODUCTION

ALMA, the Atacama Large Millimeter/submillimeter Array, is an international astronomy facility and synthesis radio telescope currently under construction in the Atacama Desert of northern Chile at an altitude of 5,000 meters above sea level. With completion in 2013, ALMA will consist of a main array of 50 12-meter diameter antennas plus the Atacama Compact Array (ACA) consisting of four 12-meter antennas and twelve 7-meter diameter antennas. When fully outfitted, the 66 antennas will be able to observe in ten bands in the frequency range 31 GHz to 950 GHz.

(For more information on the ALMA telescope presented at these 2012 SPIE Conferences, see Plenary Session presentation "ALMA construction and early science" by Thijs de Graauw, Thursday 5-Jul-2012, and paper 8444-89.)

This paper presents the organization, process and some results of the top level performance verification of the telescope, System Verification (SV). The purpose of SV, which is a component of project System Engineering (SE), is to measure the performance of the integrated instrument with respect to the ALMA System Technical Requirements[1] (see paper 8449-6 for more discussion of ALMA System Engineering).

SV is different from system Assembly, Integration and Verification (AIV) testing which is largely done through testing after integration and unit testing of each outfitted antenna as it is being accepted into the array. It is also different from Commissioning and Science Verification (CSV), which starts with the assembled & integrated array and carries it forward to its final state as a functioning scientific instrument. CSV tasks include developing observing, calibration and test procedures, developing user interfaces, plus debugging these procedures and interfaces. For example, the CSV group develops measurement procedures for system temperature, antenna pointing, antenna focus, instrumental delays, antenna baseline measurements, etc. It also develops and implements calibration techniques such as interferometric amplitude and phase calibration, bandpass calibration and polarization calibration. All of these efforts are critical to SV. (See papers 8449-24, 8444-126 and 8444-90 concerning AIV and CSV).

Although the purpose of System Verification is different from AIV and CSV, there is large overlap in the methods and procedures used and SV draws heavily upon the techniques and results of CSV and AIV. Where possible, synergies are exploited such that tests conducted by these groups for their own needs, are also used as verification of design performance with respect to the system requirements.

*rsramek@alma.cl

2. ALMA SYSTEM REQUIREMENTS

The ALMA System Technical Requirements[1] flow down from the Science Requirements[2] of the telescope and are intended to guide the design of the array and set the standards for technical performance.

There are a total of 108 system requirements. Of these, 65 need to be verified by system level tests, usually on-the-sky astronomical observations, utilizing at least a subset of the full array. Eleven of the system requirements can best be verified by combining tests done at the sub-system level. For example, the phase drift of the complete LO chain is difficult to separate from tropospheric delay variations during observing tests, but can be estimated from the RSS sum of the drift of LO chain components. Finally, thirty two system requirements can be verified by review of the design and construction reports; these are more like design guidelines rather that system performance requirements.

Significant parts of the ALMA System Technical Requirements [1] are the notes and explanatory text. The short phrase of the text of a requirement might leave questions of interpretation. The notes can provide more detail on the meaning and intent of a requirement, the reasoning or analysis behind the value of a performance requirement and possibly suggestions for a verification procedure; the notes are important in defining the requirement.

Some principal areas of the System Requirements and their impact on achieving the performance goals of ALMA are:

- antenna surface accuracy - array sensitivity;
- antenna pointing accuracy - array sensitivity and image quality (including mosaic images);
- primary beam characteristics and their stability - mosaic image quality and target intensity accuracy;
- receiver temperatures - array sensitivity;
- cross-polarization levels - polarization intensity accuracy;
- cross-polarization stability with time, angle and frequency - calibration needed to achieve polarization intensity accuracy;
- radiometric gain stability with time - single dish autocorrelation sensitivity;
- interferometric phase and amplitude stability with time and angle - calibration needed to achieve array sensitivity and target position and amplitude accuracy;
- spectral bandpass stability - spectral dynamic range in interferometric and autocorrelation modes;
- spurious signal levels - spectral dynamic range and array sensitivity and image quality;
- spurious signal suppression - spectral dynamic range and array sensitivity and image quality;
- return to phase after frequency change and/or bandpass change - phase calibration needed to achieve required array phase accuracy;
- array switching time between target/calibrator sources and between observing frequencies - phase calibration needed to achieve required array phase accuracy.

3. SYSTEM VERIFICATION PLAN AND STATUS

The SV tests are carried out as a multidisciplinary effort by scientists, engineers and control software developers. A plan for conducting ALMA SV[3] and a report on the current SV status[4] was prepared by Koh-Ichiro Morita.

Of the sixty-five system requirements that need verifications by test, about forty tests are currently ongoing. Five tests are being done by the AIV group as part of their regular acceptance of outfitted antennas (e.g., receiver temperature measurements, antenna surface accuracy, gain stability measurements, etc.). Based on these results, a summary report will be prepared by the SV team verifying the compliance of the design with the system requirements.

Most verification tests are being conducted as a joint effort of the SV team and the CSV diagnostic group. For those tests which need astronomical measurements, the SV team will prepare measurement procedures and control scripts in collaboration with the CSV group. In general, astronomical measurements at the array high site greatly depend on the weather conditions, telescope conditions, priority of CSV projects, and the Early Science schedule. Therefore, the SV team leaves execution of these measurements to CSV. The analysis of the measurements is done by the SV team or CSV diagnostics group. The final reporting is done by the SV team. In the same manner, some of the verification tests are also done in collaboration with the Array Systems Group (the ASG is an element of ALMA observatory operations).

The permanent SV staff of 4 people (2.7 FTE) will soon be augmented by two additional staff to be transferred from the ALMA system engineering group. However, the group is quite small and progress on the SV tasks is only possible with major contributions from people in the ASG, project SE, sub-system SE groups and a large effort for the CSV staff.

Currently about 30% of SV work has been accomplished with emphasis on those requirements that impact a broad range of potential observations, where there may be a serious deficiency in meeting the performance requirement, where it is suspected that there is a design flaw that needs to be understood and/or addressed, where the requirement is critical to ALMA performance, or where very little is known about system performance regarding the requirement..

As the SV tasks are being carried out, at times problems are uncovered of unknown origin, either in performance or in actually conducting the tests. The SV effort then turns into a diagnostic effort, isolating the problem, assisting with forming a mitigation plan and, if needed, suggesting a design change. About a third of the SV effort is consumed in such diagnostic effort.

4. EXAMPLES OF REQUIREMENT VERIFCIATION

In is not practical to discuss in this short paper the current verification status of all of the ALMA requirements. However, a few examples presented here will serve to illustrate typical verification tests and results.

4.1 Return to Phase

It is important that an interferometer array like ALMA return to the same phase as the array is switched from one observing frequency to another and then back again to the original frequency. The interferometer phases at the two frequencies do not need to be the same, but they do need to remain constant, at least to within the permitted variations with time, on the time scale of the external calibration period.

Return to phase is most important for fast switching phase calibration where the calibration band and target observing band might be different. The phase at the two frequencies can be determined by external calibration using an unresolved calibrator source, but for reasons of observing efficiency, this period is typically about 300 seconds.

Achieving return to phase is done through hardware and control software design of the entire LO chain, so it is a rather encompassing system level requirement and test. An example of return to phase testing is show in Figure 1, taken from a CSV report by Ed Fomalont [5].

Figure 1 shows the phase (top plot) and phase difference (bottom plot) versus time at two ALMA bands; the Band 3 data is observed at 86.3 GHz and Band 6 data at 230.6 GHz. The observations at the two bands are interleaved, switching at an 85 second interval, thus returning to the same frequency after 170 seconds. The top plot alone can be used to demonstrate return to phase and can be used for system verification. The requirement is that return to phase be achieved to better than 22 fsec, which is 0.7 deg at 86 GHz (Band 3). In this example, the variations are typically a few times greater than, this and, even at the short baseline used here of 70 meters, are likely due to atmospheric delay changes. Repeating the test under the best conditions of atmospheric stability is needed.

Demonstrating LO performance to the extremely high accuracy required by ALMA is very difficult outside a controlled laboratory environment. As mentioned earlier, a series of sub-system lab tests may in the end be the best verification. However, it should be noted, that any software/hardware failure that undermines the return to phase is likely to result in wildly large errors, so even the result given here is very encouraging.

The bottom plot of Figure 1 shows how well phase transfer between frequencies can be used to calibrate phase at a high frequency using a calibrator at a lower frequency. The stability shown is adequate for imaging and will be greatly improved when software upgrades bring the switching period down from 170 second to 10 or 20 seconds.

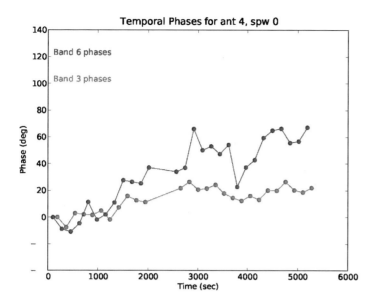

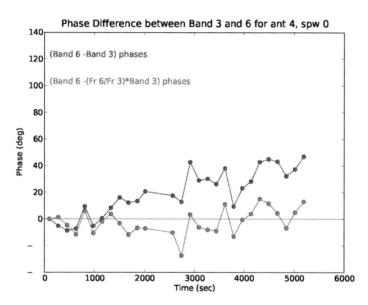

Figure 1- (top) Phase vrs time at two ALMA bands; (bottom) phase difference vrs time; the upper series is simply the phase differences and the lower series has Band 3 phase corrected by the observing frequency ratio.

4.2 Spurious Signals

The requirements on spurious signals apply to self signals originating within the ALMA array electronic and computing systems, plus unintended astronomical signals that enter the system through indirect means such as alias responses, harmonics or direct feed through. These requirements do not address external RFI such as that arising from satellites, aircraft, radio communications, radars, digital controllers, etc. However, these requirements may be used as a guide to establish harmful levels of emissions from such man-made devices.

The most damaging effect of spurious signals (or spurs) in the telescope system is the creation false spectral features that may be mistaken for astronomical emission lines. Although there are techniques for identifying and suppressing self generated spurious lines, these techniques fail if the spurious lines are too strong and/or numerous.

There are two basic types of spurious signals, those that are incoherent among antennas and those that are coherent. The incoherent spurious signals are a problem for autocorrelation observations, but not for interferometry which uses cross-correlations between antennas. The coherent spurious signals are a problem for both.

The first step in verifying the system level requirement on spurious signals is to perform a survey across the full RF bandpass of all of the ALMA receiving bands. This need only be a moderate sensitivity survey since it is designed to detect only the strong lines that might not be amenable to spur suppression.

One example of a 2 GHz baseband spectrum from the spur survey is shown in Figure 2 [6]. This observation was done at 230 GHz in interferometer mode. The array was pointed at the South Celestial Pole so that there would be no spur suppression due to fringe rotation; this left the array most open and vulnerable to spurious signals. The spur suppression techniques of phase switching and LO-offsetting were turned off. The spectrum shows numerous spurs which are coherent between antennas. Many of these spur are exact multiples of 7.8125 MHz and are stationary in the baseband, i.e., they originate after the 1st and 2nd LOs. The origin of these spurs is not yet known. These spurs are not necessarily a significant problem depending on how well they can be removed by spur suppression. However, the survey reveals where to do an intensive examination of the effectiveness of spur suppression.

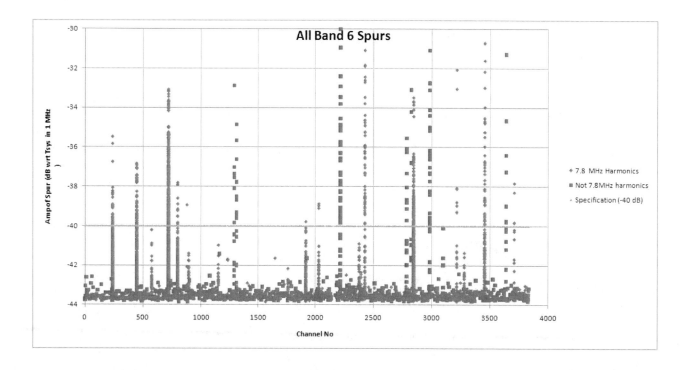

Figure 2 - This plot shows the spurs detected in one baseband of the interferometric spur survey at 230 GHz and identifies those which are in the correct channel to be harmonics of 7.8125 MHz.

An example of an incoherent spurious signal is shown in Figure 3. This feature is known to be incoherent between antennas because it appears in auto-correlation spectra but not in cross-correlation spectra. Narrow spurs like this can arise from unlocked LO signals that cross-couple between sub-systems. Examples in the ALMA Front End are the harmonics of the 1st LO chain for receiver bands that are powered up but in standby mode. These LOs are not phase locked to the LO Reference, so they are incoherent. The mitigation for these spurs is to identify for LOs in standby mode safe parking frequencies which do not have harmonics in the input RF of the active band. This is a mitigation to be implemented in software.

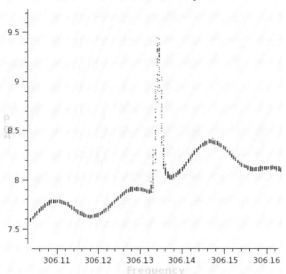

Figure 3 - Spur seen in auto-correlation mode due to leakage from an unlocked LO of a receiving band in standby mode or the WVR.

4.3 Walsh Switching to Suppress spurs

One effective way to suppress spurious signals in the cross-correlation spectral output is to insert a 180 degree phase shift in the 1st LO and then remove this phase shift by sign reversal in the digitized baseband signal. The sequence of phase reversals used are Walsh functions, with a different order of Walsh function applied at each antenna. In this way, astronomical signals entering the receiver will see two 180-degree phase shifts and be thus unchanged, but spurs introduced after the 1st LO will see only one 180-degree phase shift. The cross-correlation of these spurious signals will also be a Walsh function with an equal number of 0-degree and 180-degree phase shifts; the average cross-correlation of the spur will therefore tend to zero. The degree of suppression will depend on the accuracy and timing of the phase shifts introduced at the two antennas.

The system requirement which must be verified is the implementation of Walsh switching; the desired degree of spur suppression is 30 dB. Like return to phase, spur suppression is a rather encompassing system level requirement and test. Meeting this requirement depends on the carely timed interaction of hardware and software among many sub-systems in different parts of the array.

An example of a verification test for Walsh switching is given in a test report by Bill Dent [7]. Figure 3 taken from that report shows a baseband spectrun of with and without Walsh switching turned on. In this test 26 dB of spur suppression was demonstrated, which is encouraging since the accuracy of this test was limited by thermal noise and probably does not represent the true limit of spur suppression. More work is needed.

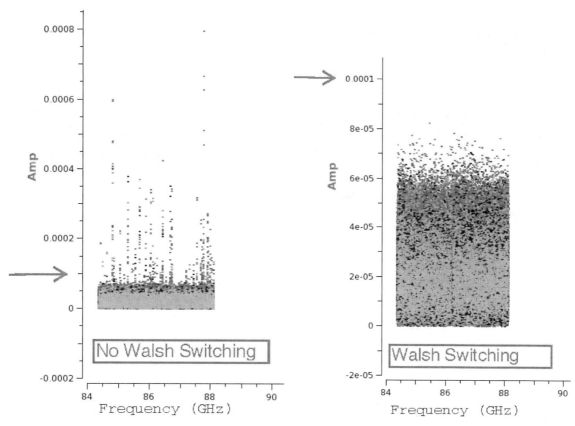

Figure 3 – Cross-correlation spectra with and without Walsh 180-degree phase switching (note the change in scale in the two plots). When Walsh Switching is turned on the strongest spurious signals, which are off scale on the left hand plot, are suppressed by >26 dB.

4.4 Antenna Location (Array Baseline) Stability

The ability of the array to produce precise radio images depends on knowing the vector separation of the antennas to high accuracy. These measurements are made astronomically, but can take about an hour of array time, so such "baseline" determinations would ideally be done only once every couple weeks. Therefore there is a system requirement that the antenna locations be stable to better than 65 microns over a time scale of 14 days.

Verification of this requirement is done by repeatedly measuring the array baselines and looking for variations. Figure 4 show the results of one such study by Masahiro Sugimoto [8]. Baselines were measured every two to three hours for a 24 hour period. This series of measurements shows shifts of baseline vector components of many tenths of a millimeter, far beyond the specification. It is not clear if these are true shifts in the location of the antenna axis intersection or if the results are due to possible errors in applying atmosphere delay corrections. This is currently being investigated.

ACKNOWLEDGEMENTS

It is not possible to acknowledge all of the ALMA staff members that are contributing to System Verification. The listed authors are just a sampling of the many people in Chile, Europe, North America and Asia who are aiding this effort.

This paper is dedicated to the memory of Koh-Ichiro Morita, recently deceased, who until 7-May-2012 lead the ALMA System Verification effort.

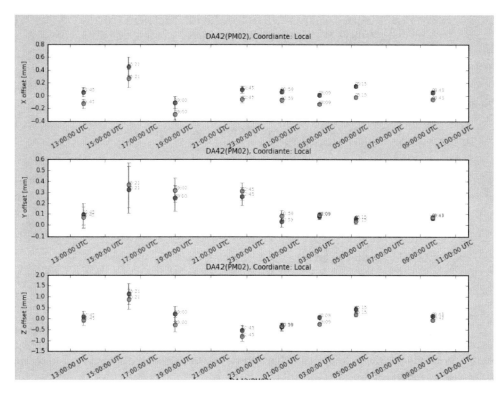

Figure 3 – The components of the vector separation of two antennas; the measurements were repeated eight times in a 24 hour period. The two data points for each measurement represent two different analysis methods (delay and phase) for determining the baseline vector. The changes in the vector during this period are more than allowed, but measurement errors like improper atmospheric delay corrections have not been ruled out.

REFERENCES

[1] Sramek, R. and Haupt, C., "ALMA System Technical Requirements for 12m Array", 21-Sep-2006, ALMA-80.04.00.00-005-B-SPE.
[2] Wootten, A. and Wilson, T., "ALMA Scientific Specifications and Requirements", 28-Jul-2006, ALMA-90.00.00.00-001-A- SPE.
[3] Morita, K., "ALMA System Verification Progress Plan", 29-Apr-2012 Draft, SYSE-88.00.00.00-0002-A-PLA.
[4] Morita, K., "ALMA System Verification Progress Report", 29-Apr-2012 Draft, SYSE-88.00.00.00-0003-A-REP.
[5] Fomalont, E.,"Band 3 to Band 6 Phase Transfer", 24-Mar-2011, ALMA JIRA issue CSV-413
[6] Napier, P., "Spur survey Band 6", 31-May-2011, ALMA JIRA issue AIV-1475.
[7] Dent, W., "Effect of 180-deg Walsh switching on spurious signals", 25-Aug-2011, ALMA JIRA issue AIV-1475.
[8] Sugimoto, M., "An interim report of analysis of 24hrs baseline measurement", 01-Feb-2012, ALMA JIRA issue CSV-457.

The CCAT 25 m diameter submillimeter-wave telescope

David Woody[*a], Steve Padin[b], Eric Chauvin[c], Bruno Clavel[c], German Cortes[d], Andy Kissil[e], John Lou[e], Paul Rasmussen[a], David Redding[e] and Jeff Zolkower[d],

[a]Owens Valley Radio Observatory, Caltech, 100 Leighton Lane, Big Pine, CA, USA 93513-0968; [b]Caltech, Pasadena, CA, USA; [c]Eric Chauvin Consulting, Pasadena, CA, USA; [d]Cornell, Ithaca, NY, USA; [e]JPL, Pasadena, CA, USA

ABSTRACT

CCAT will be a 25 m diameter telescope operating in the 2 to 0.2 mm wavelength range. It will be located at an altitude of 5600 m on Cerro Chajnantor in Northern Chile. The telescope will be equipped with wide-field, multi-color cameras for surveys and multi-object spectrometers for spectroscopic follow up. Several innovations have been developed to meet the <0.5 arcsec pointing error and 10 μm surface error requirements while keeping within the modest budget appropriate for radio telescopes.

Keywords: Telescopes, optics, telescope structures, telescope mounts, telescope surface control

1. INTRODUCTION

CCAT will be a 25 m diameter telescope, operating in the 0.2 to 2 mm wavelength range. It will be located at an altitude of 5600 m on Cerro Chajnantor in Northern Chile, near ALMA. This is one of the best submillimeter sites on Earth.[1,2] This paper describes the design features of the telescope developed to meet the demanding requirements for a radio telescope operating at short submillimeter wavelengths.

CCAT will be equipped with wide-field, multi-color cameras for surveys, and multi-object spectrometers for spectroscopic follow up. It will observe structures on all scales ranging from Kuiper Belt objects to clusters of galaxies, but its immediate impact is likely to be in tracing the process of galaxy formation from the first giant starbursts. The large 1° field of view (FoV) will be covered using sub-field cameras. This allows the first light instrument to be modest sized (few x10k pixels at 350 μm) with later additions to fill the full FoV. CCAT will be sensitive with the first light instruments reaching the 5σ confusion limit in a few hundred hours of observing time. The 1 deg^2 image will contain tens of thousands of submillimeter galaxies at redshift ~2, and a few strongly lensed galaxies at redshift >6.3.[3,4] Photometry in a few bands will yield rough estimates of the redshifts of these objects, but a multi-object spectrometer (MOS) will clearly be needed for follow up. CCAT will eventually be equipped with cameras covering all the atmospheric windows in the λ = 0.2 to 2 mm range, with a FoV up to 1°, and an MOS that can select tens of objects from the 1° FoV.

The optical design developed to provide high efficiency over the full 1° deg FoV is described in section 2. The primary reflector and support structure are described in section 3. This section also describes the sensors and control system for maintaining the surface figure between infrequent wave front measurements. The secondary and tertiary are described in section 4. The mechanical design of the mount and drive system and pointing are presented in section 5. A summary of the design features and expected performance are given in section 6.

2. OPTICAL DESIGN

CCAT is a wide-field Gregory telescope with cameras and spectrometers at both Nasmyth foci.[5,6] All the CCAT instruments will be available all the time, switching between instruments will be done by changing the pointing and rotating the tertiary mirror. This approach minimizes instrument changes, which is an advantage for a remote site.

---

[*] dwoody@caltech.edu; phone 1 760-938-2075x111; fax 1 760-938-2075; http://www.ccatobservatory.org

At short submillimeter wavelengths, the transmission of the atmosphere is only ~50%, so there is little to be gained by building a telescope with very low emissivity. CCAT is therefore an on-axis telescope with a total emissivity of ~9% (3% from the reflectivity of the mirror surfaces, 2% from the secondary, 2% from the secondary support, and 2% from gaps between reflector tiles).

CCAT has a primary focal ratio of f/0.4, so the volume swept out by the telescope is limited by the diameter of the primary. This minimizes the size and cost of the enclosure, which is important because the cost of the enclosure is large (~1/4 of the cost of the telescope) and it scales at least with surface area.

CCAT will spend much of its time on surveys, so survey speed is an important consideration. The optical design of CCAT is therefore optimized for wide FoV. At mm wavelengths, the FoV is limited to ~1deg by the strong curvature of the focal surface. 1° FoV is a fairly hard limit and it sets the final focal ratio of the telescope in order to pass the beam from the secondary through a hole in the primary that is no larger than the secondary as shown in the upper left panel in Figure 1. For CCAT, the final focal ratio is f/6, which is considerably faster than many existing telescopes. The fast final focus allows a fairly simple relay to generate an ~f/3 beam to feed the detectors, but it moves the instruments close to the tertiary[6]. Table 1 gives the basic optical parameters for CCAT.

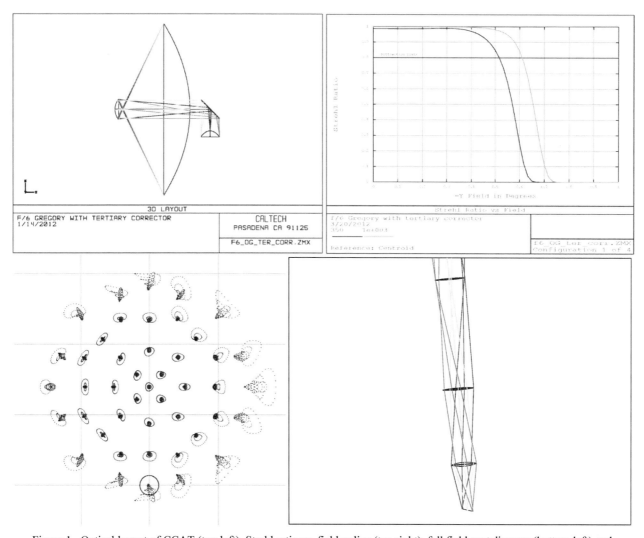

Figure 1. Optical layout of CCAT (top left), Strehl ratio vs. field radius (top right), full field spot diagram (bottom left) and relay details for a sub field camera with 5' fov (bottom right). In the full field spot diagram, the circle is the Airy disc at lambda=350um and the spots are magnified 50x. The sub field camera uses high density polyethylene lenses which are small in order to keep the total loss below ~20%. Much larger sub fields can be used if the lenses are made of Si.

Table 1. Optical parameters for CCAT.

Parameter	Value
Primary diameter	25 m
Primary focal ratio	0.4
Final focal ratio	6
Back focal distance	6.1 m
Primary to secondary separation	11.15 m
Secondary to tertiary separation	14.15 m
Field of view	1°

The secondary is 3 m diameter, strongly curved, and its surface must be set within ~2 μm RMS. This presents a significant challenge because direct mechanical measurements, e.g., with a coordinate measuring machine (CMM), cannot achieve the required accuracy. An optical measurement must be used, and this is much easier if the secondary is concave. CCAT is therefore a Gregory telescope in order to allow measurements of the surface profile of the secondary using millimeter wave holography[7,8].

At the shorter submillimeter wavelengths, aberrations are important, so the FoV can be increased by adding a corrector. In CCAT, the correction is done at the tertiary. This not ideal, because the tertiary is not a constant distance from the focal surface, but Figure 1 shows that good wide field performance can be obtained, and since the tertiary is made of machined aluminum tiles, it does not cost much to implement a complicated profile. To maximize the Strehl ratio over the field, the conic constants of the primary and secondary are also adjusted slightly from the nominal values for an aplanatic design. The CCAT primary will be cut as a paraboloid and the tiles set to the required profile when the segments are assembled.

It is impractical to build relay optics for the full 1° FoV of CCAT, so the field must be broken up into smaller pieces. The field selection must be done at a focus, where beams from adjacent fields do not overlap. This leads to the sub-field camera approach shown in Figure 1, where the refractive relay fits within the cylinder defined by the field lens. Reflective sub-field camera schemes are also possible, but it is not easy to close-pack the mirrors.

3. PRIMARY

The accuracy requirement for the primary surface is <7 μm RMS under all operating conditions. The telescope will be inside an enclosure which will greatly reduce the wind loading and keep direct sunlight off the primary and its support structure. Even with this shielding, the required surface is better than can be achieved for a passive 25 m diameter radio telescope structure utilizing the best carbon fiber reinforced plastic (CFRP) material. CCAT will have an active primary surface consisting of 162 reflector segments mounted on three computer controlled actuators to give independent piston, tip and tilt control of each segment. A control system based on an innovative imaging displacement sensor (IDS) that measures the six degrees of freedom (DoF) displacements between neighboring segments will command the actuators to maintain the primary surface figure. To achieve the best passive performance and to improve the thermal stability, the support structure for the reflector segments and secondary mirror will be a carbon fiber reinforced plastic (CFRP) truss.

3.1 Segments

Figure 2 shows a schematic diagram of a primary segment. A segment consists of aluminum reflector tiles mounted on CFRP sub-frames. This compound segment design was the result of a study looking at the options for light weight segments appropriate for a large submillimeter-wave telescope.[9] The precision machining of the reflecting surface occurs on manageable ½ m x ½ m aluminum tiles. Machining accuracy of <2 μm RMS can be achieved on tiles this size with five point support. The tiles can be very light weight with an areal density of ~10 kg/m^2. The sub-frames are fabricated from CFRP providing good thermal stability and high stiffness. The coefficient of thermal expansion (CTE) is <0.5 ppm/C with uniformity better than ± 0.1 ppm/C. The sub-frames don't have to be manufactured with high accuracy which is a large cost savings. The overall segment accuracy is achieved by factory setting of the five adjusters supporting each tile using a CMM. This concept increases the complexity of the segments but overcomes many of the difficulties and risks associated with monolithic segments. It also allows the fabrication of segments as large as 2 m x 2 m that meet the CCAT specifications of less than 4 μm RMS segment error under all operating conditions. The increased size reduces the number of computer controlled actuators and the node density in the reflector support structure.

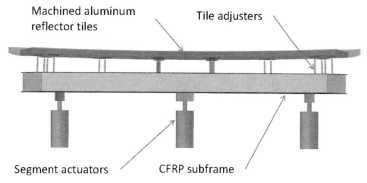

Figure 2. Schematic diagram of a primary segment.

The CCAT segmentation plan is shown in Figure 3. There are 162 segments in six rings with the segments sized to fit into a standard 2 m x 2 m x 1 m CMM. The layout avoids continuous radial lines which helps minimize the side lobes in the point spread function while also removing some of the degeneracy in the IDS configuration, improving the controllability of the surface.

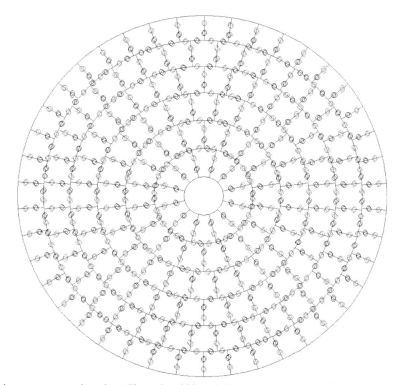

Figure 3. CCAT primary segmentation plan. The red and blue circles are the locations of IDSs.

The actuators and connection of the actuators to the CFRP sub-frames are designed to carry the full load of the segment in all orientations and under the -30 to +20 C temperature ranges with minimal distortion of the segment. The actuators will have an absolute precision of the better than ±1 µm over their 10 mm stroke and 0.1 µm readout and setting resolution.

3.2 Support truss

The primary truss is constructed from struts with CFRP tubes and steel and Invar end-fittings and nodes. Figure 4 shows the truss structure including the secondary support structure. The truss has five layers of nodes with nine different classes of struts, one class for each node layer and a different class for struts interconnecting successive node layers. The large FoV and large instrument packages are accommodated by providing a 3 m diameter unobstructed cylinder through the structure along the elevation axis. The secondary support system is designed to have very low blockage while not compromising stiffness. The inner section of the support legs (green in Figure 4) are above the convergent cone of rays that meet at the prime focus and hence only block the plane wave incident on the aperture. The lower tripods (orange in Figure 4) connect to relatively stiff nodes in the primary support truss. This avoids having excessively long support legs with low net stiffness and resonant frequencies at the cost of a small amount of blockage of the rays converging towards prime focus.

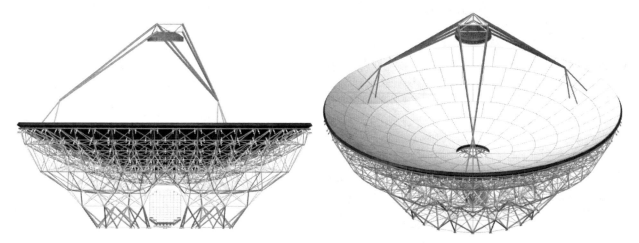

Figure 4. CCAT primary, secondary and tertiary. Left: side view along the elevation axis. Right: top front view.

One of the major challenges for CFRP telescope truss is the interface to the steel tipping structure which must allow for the differential thermal expansion of the two materials without distorting the primary surface or compromising the stiffness of the total structure. This is accomplished using the concept of thermal homology[10] which defines how multiple connections can be made between the CFRP and steel structures throughout a 3D volume without inducing thermal deformation stresses. The truss shown in Figure 4 has two concentric rings of blade flexures that connect the bottom set of nodes to the steel tipping structure plus a single rigid pillar at the center of the bottom layer. The blade flexures plus the radial CFRP struts connecting to the central pillar effectively constrain the bottom nodes in three DoF while allowing the steel tipping platform to expand relative to the central pillar. These added radial connections greatly improve the stiffness and resonant frequency of the primary support structure.

The nodes and end-fitting play a critical role in determining the performance of the CFRP truss. Figure 5 shows one end of a strut assembly. The metallic end-fittings and nodes dominate the effective CTE for the structure and can dominate the structural mass. In addition the ultimate strength of the truss will depend upon the bonding of the metallic fittings to the CFRP tubes. The design shown in Figure 5 has a few basic components which can be tailored for each strut end to accommodate the close packing at each node and are optimized to give the best stiffness to mass ratio. The CFRP tubes will have a negative CTE of ~ -0.3 ppm/C, depending upon the exact choice of material and layup, and the amount of Invar and steel in the end-fittings is adjusted to yield an effective CTE of 0.2 ppm/C from node-to-node for the various strut lengths which range from 0.8 to 3.8 m. Although the tailoring and optimization of each type of strut and node interface adds complexity to the design, the overall performance as measured by the mass and resonant frequency of the whole structure is significantly better than can be achieved using a common uniform end-fitting. The custom tailoring only requires adjusting the thickness of the steel spacers and setting the length of the threaded Invar studs.

The frequency of the lowest resonant mode was used as the evaluation parameter for the optimization of the structural design. The primary segments and secondary were treated as passive loads on a distributed structural spring. Following the formulation for a single mass on a distributed spring the resonant frequency can be written as

$$f_0 = \frac{\beta}{\sqrt{1 + \frac{C_1 M_1 + C_2 M_2}{M_S}}}, \qquad (1)$$

where M_1 and M_2 are the masses of the primary segments and the secondary reflector respectively and M_S is the structural or spring mass. β and the coefficients C_1 and C_2 are determined by fitting the FEA results for several configurations with different masses for the major components. β is effectively the self-resonant frequency of the structure in the absence of any load from the primary segments or secondary and is a measure of the quality or stiffness-to-mass ratio for the structure. Knowing C_1 and C_2, Equation 1 can be used to calculate β from the f_0 determine from the finite element analysis (FEA) of a given design. The design is optimized by varying the geometry and strut cross-sections to maximize β. Once the design has been optimized Equation 1 can be used to calculate the structural mass, M_S, required to meet a given f_0 requirement or to accommodate changes in the load masses while preserving f_0.

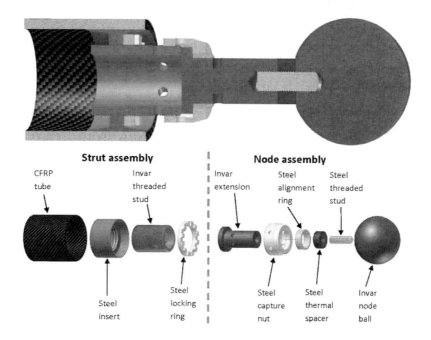

Figure 5. CCAT primary truss strut end assembly.

CFRP is an engineered material with a complex trade space for the fabricated tube elastic modulus, Y, CTE and cost. It can be a difficult task to optimize the performance per \$. The self-resonant frequency β of the structure will scale as $Y^{1/2}$ and this can be used in combination with Equation 1 to estimate the amount of CFRP required to achieve a given f_0 as a function of Y. This is then used in estimating the costs for fabricating the truss from different types of CFRP tubes.

The fast mapping requirement and high pointing accuracy requires a high f_0. The design goal for CCAT is $f_0 \sim 10$ Hz. Table 2 shows a sample set of material properties and masses for the primary, secondary and support truss. Table 3 summarizes the FEA results for this version of the CCAT primary and secondary under a variety of load conditions.

3.3 Surface control

The reflector segments will be mounted on computer controlled actuators. Although the primary structure is designed to have very small thermal distortions and perform well using lookup tables for gravity and large temperature changes, it is anticipated that active control of the surface based on a sensor system will be required to achieve the best performance at short submillimeter wavelengths. The sensor based active control strategy can also correct unanticipated or poorly characterized deformations in the mount, primary support structure, actuators and segment mounting.

Table 2. Sample material and mass table for the CCAT primary and secondary.

Number of nodes	517			
Number of struts	2880			
Structural material	Modulus [GPa]	Density [kg/m^3]	CTE [ppm/C]	Mass [kg]
steel	210	7,850	11.4	1,500
Invar	141	8,000	0.9	1,200
CFRP tubes	140	1,850	-0.5	7,000
Strut assembly (average)	141	2,472	0.2	**9,700**
Load masses	[kg/m^2]			
Segments + actuators	30			**16,000**
Secondary	80			**600**
total				**26,300**

Table 3. FEA results.

Analysis case	FEA results
Lowest resonance	10 [Hz]
	RMS distortion
Z-gravity*	112 [μm]
Y-gravity*	218 [μm]
$\Delta T = 10$ C temperature change	0.1 [μm]
0.1 ppm/C CTE random variation & 10C temperature change	5.9 [μm]
$\Delta T = 1$ C front-to-back , before focus correction	3.1 [μm]
$\Delta T = 1$ C front-to-back*	0.2 [μm]
$\Delta T = 1$ C top-to-bottom, before pointing correction	1.5 [μm]
$\Delta T = 1$ C top-to-bottom*	0.1 [μm]
$\Delta T = \pm 0.5$ C random in CFRP truss	0.6 [μm]
$\Delta T = \pm 0.5$ C random in steel axle structure	6.7 [μm]

*RMS distortion after correcting for focus and pointing

At millimeter and submillimeter wavelengths only the planets provide enough flux for a reliable wave front measurement with ~1 μm accuracy and even then it will take several hours to measure the surface at the ~1/2 m scale size required to accurately set each segment. Since the planets of the appropriate angular size and position in the sky will not always be available, the primary structure and any sensor or metrology system must have excellent stability over timescales of several months. An imaging displacement sensor (IDS) has been developed which utilizes a simple light emitting diode (LED) and pinhole collimator on one segment and a charge coupled device (CCD) camera on the neighboring segment to measure displacements in the plane normal to the collimator beam. By suitable placement of four such collimators and CCD cameras along an edge all six degrees of freedom motion between two segments can be measured with better than 0.1 mm accuracy. This sensor system is described in a companion paper at this conference.[11]

A state-control system has been developed and simulated to study the performance of the IDS sensor system for the full 162 segment CCAT surface. It uses the response matrix of IDS readings for all six DoF motion of each segment to estimate the system state from the sensor readings. The estimated state is projected to wavefront space using the response matrix to wavefront system state change. Actuator motions are computed and applied to minimize the wavefront error. The details of the control system are described in Redding et. al.[12,13] with the modeling and simulation results given in Lou et. al.[14] The simulations include expected gravity distortions of the primary support truss as well as applying random piston, tip and tilt displacements of the segments. After removing the six rigid body motions for the whole primary all 6x162-6 = 966 Eigen modes are measured with high fidelity. As expected, the weakest or poorest sensed Eigen modes mimic low order Zernike modes.

Simulations of the performance of the control system have been analyzed under a variety of surface distortions. The simulations used the full model of the IDSs, including their locations and orientations on the segments, with a range of sensor noise levels. A simple characterization of the performance is the EMF (error multiplying factor), ratio of the RMS wave front error after applying control to the sensor noise. The expected EMF based on the Eigen values for all modes is ~10 which is also consistent with all of the simulations. Figure 6 shows the simulations using the gravity distortions predicted from FEA for the primary structure. The IDS system accurately predicts the large initial wavefront error of 489 μm RMS and the RMS wavefront after control of the three actuators on each segment is only 2 μm. Simulations using random motions in the six DoF of each segment give similar control performance.

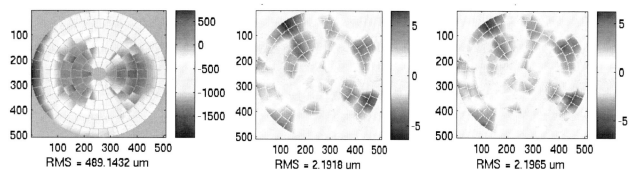

Figure 6. Control system simulation for FEA predicted gravity distortion at 0 deg elevation and 0.35 μm RMS sensor noise. Left: wave front for the gravity distortion. Center: difference between the initial wave front and the state estimated wave front determined from the IDS system. Right: resulting wave front after applying control to the segment actuators. The RMS values below each panel are the full wave front error.

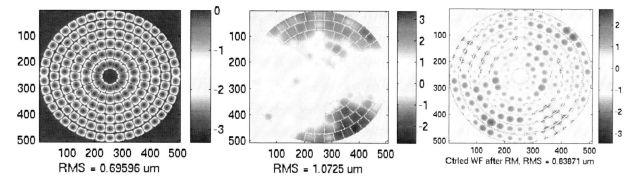

Figure 7. Control system simulation with 1 μm cupping on all segments and 0.07 μm RMS sensor noise. Left: wave front for the applied cupping distortion. Center: difference between the initial wave front and the state estimated wave front determined from the IDS system. Right: resulting wave front after applying control to the segment actuators. The RMS values below each panel are the full wave front error.

The more challenging case is a coherent or uniform distortion of all segments. Temperature gradients through the segments or sub-frames with slightly different CTE for the front and back face sheets can produce a change in curvature or cupping of the segments. To analyze this case a segment figure distortion proportional to X^2+Y^2 in the local segment frame was added as a seventh DoF for each segment in the IDS response matrix. The IDS system is able to measure the amount of cupping in each segment. Figure 7 shows the simulation for 1 μm of cupping applied to all segments. In this simulation the sensor noise was 0.07 μm (similar to the measured noise for the prototype IDS sensor in a single CCD image frame[11]). The state estimator accurately determined the cupping from the IDS readings consistent with the EMF and sensor noise. The control system successfully adjusted the piston of each segment to minimize the effect of the cupping despite the actuators not having the ability to change the curvature of the segments. Edge sensor systems that do not measure all six DoF displacements of a segment relative to its neighbor can actually amplify the cupping distortion across the surface, greatly increasing the wave front error.[9] The IDS and state control system also has good performance for random six DoF rigid body displacement of the segments along with random segment cupping distortions.

4. SECONDARY AND TERTIARY

The secondary and tertiary mirrors use the same machined aluminum tile and sub-frame technology as the primary segments. Both mirrors are challenging, the tertiary because it is large (2.8 x 3.8 m), and the secondary because it is large and strongly curved (3 m diameter and 2.16 m radius of curvature at the apex). The secondary and tertiary sub-frames must have a CTE of a few x0.1±0.1 ppm/C in order to maintain its profile within a few μm rms with diurnal and annual temperature changes of ~20 C.

4.1 Secondary

The secondary has 3 rings of keystone-shaped tiles on a CFRP or Invar sub-frame. The tiling pattern is shown in the left panel of Figure 8. A CFRP sub-frame offers low mass, but maintaining low and uniform CTE for a strongly curved structure may be difficult. With a CFRP sub-frame, gravitational deformations of ~2 um rms require a ~0.3 m thick sub-frame. An Invar sub-frame may give a more uniform CTE, but Invar honeycomb material is only available up to 0.1 m thick, so a space frame support structure is needed. The secondary sub-frame is supported on a hexapod that is continuously adjusted to compensate gravitational deflection of the secondary support.

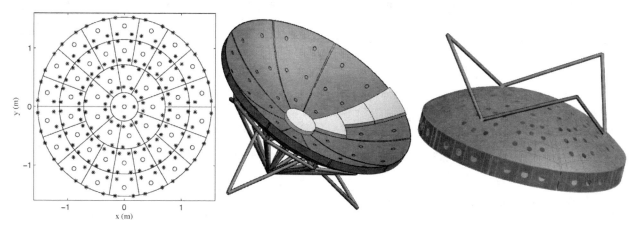

Figure 8. Secondary tiling pattern (left), Invar sub-frame (center) and CFRP sub-frame (right). Crosses and circles show the locations of tile adjusters on the sub-frame. Both subframes have 9 identical pie-shaped segments. The CFRP sub-frame has 6 mm thick face sheets and 3 mm thick ribs. The holes in the face sheets allow access to interior glue joints and to the tile adjusters.

4.2 Tertiary

The 1° field of view will project on the tertiary mirror as a 2.727 m x 3.856 m ellipse. The reflecting surface will be constructed of 105, ~0.29 m square aluminum tiles as shown in Figure 9. Each tile's surface will be uniquely machined to provide aberration correction. Each tile will be mounted on the CFRP sub-frame through a 5-point mounting arrangement on differential screw adjusters. The CFRP sub-frame will be ~0.3 m thick and will be constructed by a regular pattern of slotted intersecting ribs. The 3 mm thick ribs will be bonded to each other and to 3 mm top and bottom face sheets. The structure of the sub-frame core is closely coupled to the pattern of reflecting tiles on the mirror because the tiles must be attached at points where the core is stiff. The tiles have a radial support at the center, so this point is placed over the intersection of 2 ribs in the sub-frame core. The tile corner supports carry mainly axial loads, so these can be placed over a single rib, but close to an intersection of 2 ribs. Invar pads will be bonded to the rib core at all mounting points.

The sub-frame is supported on a steel space frame structure. Differential expansion between the low CTE, CRFP sub-frame and the steel space frame will be taken up by a set of blade flexures. The steel space frame will be attached to a rotator that will provide 180° of rotation in order to direct the optical beam to either Nasymth focus.

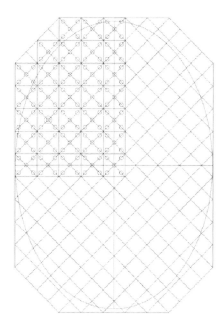
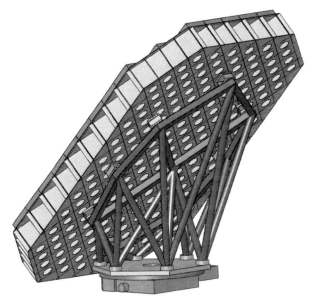

Figure 9. Left: Tertiary Mirror aluminum reflecting tile and sub-frame rib pattern. Right: Tertiary Mirror Assembly and Support Structure.

5. MOUNT AND POINTING

The critical specifications driving the mount design are the pointing, tracking accuracy and slew speed for wide field mapping.

5.1 Mount

The mount is an elevation over azimuth system designed to support and drive the primary, secondary, tertiary and scientific instruments as shown in Figure 10. The mount consists of an azimuth structure sitting on a stationary structure and supporting the elevation structure. The best pointing and tracking performance is achieved by using hydrostatic bearings (HSB), tape encoders and linear drives for both the elevation and azimuth axes.

Because of the size of the mount and the need for a large circular track for the azimuth HSB system, the azimuth structure is a traditional steel space-frame. This concept allows not only for a direct load path from the elevation structure to the foundation with beams working in tension and compression, but it also results in smaller and more manageable parts to ship to the site. A pintle bearing at the center of the track maintains the azimuth rotating structure on axis and accommodates the shear loads that may occur during seismic events. Similarly, a combination of a shell-design and a more traditional space-frame made of steel material and built as an extension of the truss of the primary appears the most appropriate for the elevation structure. This concept is compact enough to keep the mount naturally balanced around the elevation-axis despite the light weight of the primary.

The nominal diameter of the azimuth track is 20,225 mm, the height of the elevation-axis from the top of the azimuth track is 12,500 mm, the nominal distance between the elevation journals is 16,000 mm, the nominal diameter of the elevation journals is 5,400 mm, the inner diameter of the instrument rotators is 2,800 mm and the swept radius of the structures is 14,500 mm. The total mass of the telescope is 500,000 kg, the total mass of the azimuth rotating structure is 450,000 kg and the total mass of the elevation tipping mass is 200,000 kg, including the 35,000 kg primary, secondary and tertiary, 10,000 kg of instruments and 20,000 kg of equipment.

The HSB system uses master and slave pads. The master pads are fixed in position and arranged to create an iso-static system and the slave pads are free in position and are used to support the loads and to improve the dynamic performance without over constraining the system. There are six HSB pads for the azimuth track, four masters and two slaves. The elevation HSB system consists of two large elevation journals, one at each side of the mount, and four master HSB pads to carry the elevation rotating structure. To maintain the system in position and to accommodate the shear loads that may occur during seismic events, eight HSB pads are mounted laterally to the elevation journals; two master pads on one side and six slave pads on the other side, as shown in Figure 11. The operating pressure of the oil supply system is 100 bars with a total flow of 195 l/min at the highest temperature.

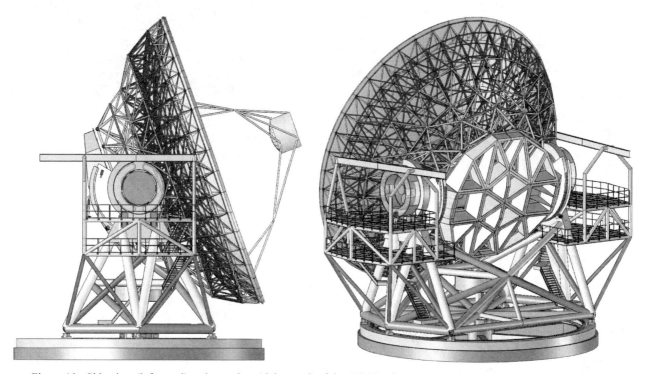

Figure 10. Side view (left panel) and rear view (right panel) of the CCAT telescope at 20 deg elevation limit. The primary truss and secondary are an earlier version compared to what is shown in Figure 4.

Two different styles of linear drives are used. The azimuth linear drive uses fixed passive magnetic on the inside of the track with 48 forcers on the azimuth structure. The flux lines are parallel to the azimuth axis. This allows the azimuth rotating structure to float laterally with a constant air gap between the magnets and the forcers without risking physical interferences that could occur during seismic events or under a difference of thermal expansion between the structures. The azimuth drive is capable of developing 1,150 kNm of nominal torque and to accelerate the mount at 2 deg/s^2. The elevation drive system consists of two large sectors of magnets mounted at both sides of the elevation structure and 18 radial flux forcers distributed between the elevation HSB pads at each side of the azimuth structure. In this configuration the elevation rotating structure can float laterally along the elevation-axis with a constant air gap between the magnets and the forcers without risking physical interferences during seismic events or thermal expansion between the structures. The elevation drive is capable of developing 200 kNm of nominal torque and to accelerate the mount at 2 deg/s^2. Both azimuth and elevation configurations allow the mounting surfaces of the HSB pads, magnets and forcers to be precisely machined, proof assembled and tested at the shop before shipping, like a single motor, so there are fewer alignment issues during installation at the site.

The mount design also includes the structures necessary for the installation and operation of the large instruments that will be used on CCAT. There are two Nasmyth platforms mounted at each side of the azimuth structure to provide access to the instruments inside the elevation tube and support ancillary equipment. The platforms are equipped with cranes designed to lift the instruments from the floor of the observatory to the top of the platforms. Motorized rotators, coaxial to the elevation-axis provide supports for the instruments at each side of the elevation structure. The rotators can be controlled to track the parallactic angle on the sky or to keep the instrument in a fixed gravity frame relative to the ground. The clear volume for instruments inside each rotator is 2.8 m diameter by 5 m long and can support as many as seven separate cryogenically cooled instruments. Essentially all of the machinery will be mounted on the azimuth structure with air and liquid heat exchangers to expel the heat to outside of the telescope enclosure. All rotating axes, azimuth, elevation and rotator will be equipped with chain style cable wraps to handle the necessary power cables, fluid lines, gas lines, vacuum lines and signal and control cables.

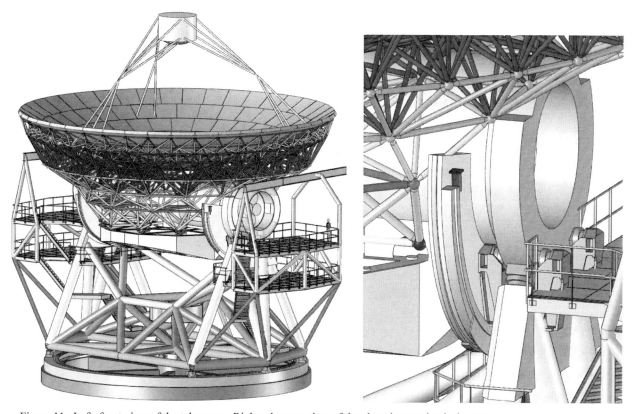

Figure 11. Left: front view of the telescope. Right: close up view of the elevation mechanical system.

Preliminary FEA has been performed to pre-optimize the mount and to determine the static and dynamic performance of the telescope. These analyses show that the azimuth structure is the main contributor to the lowest resonance frequencies of the telescope. The first resonance mode is a side-to-side sway at 4.8 Hz with most of the bending at the top of the azimuth structure that supports the elevation HSB pads. The second resonance mode is a fore-aft bending of the azimuth structure at 6.0 Hz and the third resonance mode is a twist around the azimuth-axis at 10.0 Hz. These modes are not directly sensed by the azimuth and elevation encoders. The analysis also shows that the elevation structure is very stiff with ±0.4 mm deviation from a plane at the interface between the mount and the primary truss as a function of elevation. The peak stresses of full structure are below the yield limit of the various materials even for a static 1 G of seismic acceleration in any direction.

5.2 Pointing

The pointing performance of CCAT is limited mainly by deflection of the CFRP primary truss when the telescope scans, thermal deformation of the steel mount, and thermal deformation of the CFRP secondary support. The pointing error (PE) due to deflection of the truss is PE ~ $\Omega/(2\pi f_0)^2$, where Ω is the angular acceleration of the truss and f_0 is the natural frequency. For CCAT, f_0 ~ 10 Hz and Ω = 0.3*(λ/350 µm)deg/s^2, so at λ = 350 µm, PE ~ 0.3". This is essentially the entire 0.35"x(λ /350 um) pointing error budget, so a pointing correction must be applied based on the commanded or actual acceleration of the structure.

The ~10 K p-p diurnal air temperature variations at the CCAT site, combined with the different conductivities and heat capacities of the mount parts, cause ~0.1 C/hr changes in temperature gradient across the mount. The aspect ratio of the mount is ~1, so 0.1 C/hr change in temperature gradient causes pointing changes of ~0.2"/hr. A pointing measurement every ~1/2 hr is therefore required to meet the 0.35" pointing requirement at λ =350 µm.

The secondary support is made of CFRP with CTE ~ 0.5 ppm/C, so a typical 1 C temperature gradient across the structure causes a secondary decenter of ~6 μm. This corresponds to a pointing error of ~0.2". Temperature gradients inside the enclosure change on timescales of an hour, so a pointing measurement is required every ~1/2 hr.

Pointing errors due to thermal deformation of the mount and secondary support can be estimated based on measurements of the temperature of the structure. Such measurements should not be needed to meet the pointing requirements, but ~100 temperature sensors will be installed roughly uniformly over the CCAT structure.

6. SUMMARY

The basic environment parameters and observing parameters for CCAT are listed in Table 4. The telescope system design and various components are designed to meet the overall wavefront and pointing requirements for these operating conditions.

Table 4. Observatory parameters.

Parameter	Value	Units	Notes
Wavelength	350	μm	
Mean outside wind speed	6	m/s	3rd quartile (from submm.org)
Wind speed for pressure	6.77	m/s	$2^{1/2}(2/n)^{1/2} v_{outside}$, see CCAT-TM-56
Density of air	0.7	kg m^{-3}	At 5600 m altitude
Scan acceleration	0.3	deg s^{-2}	0.3°s^{-2} × λ/350μm
rms temp gradient in dome	1	K	From TMT CFD
Soak temp change	20	K	Diurnal & longer (from submm.org)
Relative humidity change	30	%	Diurnal & longer (from submm.org)
Flux density of pointing source	0.1	Jy	For >1 source/deg^2, S<0.3Jy at λ=350μm and S<40mJy at λ=850μm, see Fig. 4.9 in feasibility study
Pointing integration time	120	s	< a few min for reasonable observing efficiency
Field angle	0.30	deg	

The reflecting surfaces for the primary, secondary and tertiary will be precision machined light weight Al ~0.5 m x 0.5 m tiles. The reflector tiles are attached to CFRP frames by five manual adjusters. The frames provide a stiff and thermally stable structure for supporting the optical surface. The frames for the secondary and tertiary mirrors are kinematically attached to the telescope structure. In the case of the primary there are 162 ~2 m x ~2 m sub-frames which are mounted on three computer controlled actuators. The reflector tiles on the frames or sub-frames will be set in the factory or laboratory using a CMM or other metrology system.

The primary support structure is a truss consisting of CFRP tubes with Invar nodes and steel and Invar end-fittings. The metallic fittings are tailored for each strut end to produce an effective node-to-node CTE of 0.2±0.03 ppm/C while also maximizing the stiffness to mass ratio for the strut assemblies. The truss and secondary support structure are structurally very efficient, using only ~10,000 kg of structural mass to support ~16,000 kg of optics payload with a resonant frequency of ~10 Hz for the 25 m diameter telescope.

The active control of the primary surface uses a new type of IDS system that measures the full six DoF position of two segments relative to each other with an accuracy of better than 0.1 μm. The six DoF motion of the segments as well as a change in the radius of curvature of the segments are well sensed and the control system can maintain an RMS wave front error that is only ~10 times larger than sensor noise.

The mount will utilize hydrostatic bearings for both the azimuth and elevation axes. These axes will also use linear drives and tape encoders. This combination has very low friction and stiction with a very stiff drive closely coupled to the encoders. This will allow high bandwidth drive control and provide precision pointing with small tracking errors.

Table 5 gives the high level itemized error budget for CCAT. Each item in Table 5 is an RMS sum of many smaller component level errors. Note that three different options for the primary are given in Table 5; CFRP truss with open-loop control using lookup tables to control the actuators, steel truss using the IDS system for closed-loop control of the actuators and CFRP truss using the IDS system for closed-loop control. A steel truss running open loop misses the half wave front error (HWFE) by a large margin and is not included. The CFRP truss with open-loop control satisfies the basic HWFE requirement but with only a small margin while adding closed-loop control gives a significant margin and will produce significantly better Strehl ratios for observations through the 200 μm wavelength atmospheric window that is available from the Cerro Chajnantor site at 5600 m altitude.

Table 5. HWFE budget for reflector, pointing errors and emissity.

Contribution	HWFE (μm rms)	PE (arcsec rms)	EM	Notes
Aberrations	2.70	0.00		Gregory with tertiary corrector
Primary open-loop, CFRP truss	8.83	0.05	0.019	
Primary closed-loop, steel truss	7.52	0.55		
Primary closed-loop, CFRP truss	6.49	0.04		
Secondary	6.17	0.31	0.053	
Tertiary	5.08	0.10	0.017	
Instrument	0.05	0.04		
Mount	0.00	0.32		No HWFE from mount
Alignment	2.19	0.10		Regular pointing with science camera, occasional wavefront measurements with WFS
Telescope total open loop, CFRP	12.41	0.47	0.089	
Telescope total closed-loop, steel	11.51	0.72		
Telescope total closed-loop, CFRP	10.87	0.47		
Telescope requirement	12.50	0.35	0.100	HWFE for <50% increase in integration time at λ=350μm, PE<1/10th beam, EM for <25% increase in integration time, from CCAT-TM-48
Atmosphere	5.74	0.23		1st quartile

ACKNOWLEDGEMENTS

This work was supported by the John B. and Nelly Kilroy Foundation.

REFERENCES

[1] Radford, S.J.E., et al., Submillimeter observing conditions on Cerro Chajnantor. Proc. SPIE. 7012 (2008).
[2] Giovanelli, R., et al., The Optical/Infrared Astronomical Quality of High Atacama Sites. II. Infrared Characteristics. PASP. 113: p. 10 (2001).
[3] Blain, A.W., et al., Submillimeter Galaxies. Phys. Rep. 369: p. 65 (2002).
[4] Negrello, M., et al., Astrophysical and cosmological information from large-scale submillimetre surveys of extragalactic sources. Mon. Not. R. Astron. Soc. 377: p. 11 (2007).
[5] Stacey, G.L., et al., Instrumentation for the CCAT telescope. Proc. Soc. Photo-Opt. Instrum. Eng. (2006).
[6] Padin, S., et al., CCAT Optics. Proc. SPIE. 7733: p. 77334Y-1-77334Y-11 (2010).
[7] Bennett, J.C., et al., Microwave holographic metrology of large reflector antennas. IEEE Transactions on Antennas and Propagation. AP-24(3): p. 295-303 (1976).
[8] Scott, P.F. and M. Ryle, A rapid method for measuring the figure of a radio telescope reflector. Monthly Notices of the Royal Astronomical Society. 178: p. 539-545 (1977).
[9] Woody, D.P., et al., Panel options for large precision radio telescopes. Proc. SPIE. 7018: p. 70180T1-70180T11 (2008).
[10] Woody, D.P., S. Padin, and T.A. Sebring, CFRP truss for the CCAT 25m diameter submillimeter-wave telescope Proc. SPIE. 7733: p. 77332B-1-77332B-10 (2010).
[11] Woody, D.P. and D.C. Redding, An Imaging Displacement Sensor with Nanometer Accuracy. Proc. SPIE. 8450 (2012).
[12] Redding, D.C., et al., Wavefront controls for a large submillimeter-wave observatory Proc. SPIE. 7733: p. 773329-1-773329-11 (2010).
[13] Redding, D.C. and e. al., Model-based Wavefront Control for CCAT. Proc. SPIE. 8339 (2011).
[14] Lou, J.Z., et al., Modeling a large submillimeter-wave observatory Proc. SPIE. 7733: p. 773326-1-773326-13 (2010).

High performance holography mapping with the LMT

David R. Smith*

MERLAB, P.C., 357 S. Candler St., Decatur, GA 30030, USA

Kamal Souccar

Large Millimeter Telescope, Astronomy Department,
University of Massachusetts, Amherst, MA 01003, USA

ABSTRACT

When making holography measurements on a large telescope, there are many factors that make it difficult to obtain a consistent map. Two of these factors are variations of the satellite and temperature-induced deformations of the reflecting surface. The former requires frequent returns to the center of the map to check the satellite and the latter requires that the map be completed rapidly and at night. While holography mapping has traditionally been performed using point-by-point or raster scanning, these methods involve substantial overhead in the frequent movements back to the center of the map.

In performing holography maps of the Large Millimeter Telescope (LMT) for the first light campaign, the observing team proposed a radial scanning approach. This strategy has the advantage that every scan passes through the center of the map. However, such a scan results in a disproportionate amount of telescope time near the center region. To achieve more uniform coverage, the team proposed a velocity profile that is inversely proportional to the distance from the center of the map. Because the velocity profile is defined with respect to position rather than time, this new approach required an extension of the existing parametric scanning capabilities at the LMT. The high axis rates resulting from this velocity profile present additional challenges.

This paper describes the implementation and performance results for holography maps that use a radial scan pattern with a position-dependent velocity profile at the LMT. Both theoretical and experimental results are presented.

Keywords: Large telescopes, Main axis control, Holography, LMT, Parametric scanning

1. INTRODUCTION

In a companion paper[1], it has been shown how to incorporate a time parametric tracking profile into the LMT main axis control. The formulation works by calculating the position, velocity, and acceleration of the total motion and passing it to the LMT trajectory generator[2]. The trajectory generator offers the usual advantages of smooth motions that are within the axis motion limits while providing the necessary information for feedforward control to improve the servo accuracy.

Unfortunately, it is not always possible to parameterize a desired scan pattern in terms of time. A clear example of this arose during the LMT millimeter-wave first light holography measurements in 2011[3]. Obtaining a holography map using a satellite reference requires mapping around the satellite position. This can be accomplished using conventional raster scanning or through time parametric mapping. Unfortunately, holography mapping also requires frequent returns to the center of the map to check the reference signal. The overhead

*David_Smith@merlab.com; Phone 404-378-2138

introduced by these frequent interruptions of the map results in increased elapsed time. Not only does the increased time per map reduce the number of maps that can be made in a single night of observation, but it also allows more time for the structure to change (*e.g.*, due to temperature variations).

To address this, an alternate mapping strategy for holography was proposed, in which the map would consist of a series of straight lines that cut through the center of the map. After each cut, the position would be rotated about the map center and another cut performed. This approach can be implemented using the time parametric approach described in the companion paper[1]. As a simple approach, the offset can be parameterized as

$$x_l = A\cos(at)$$
$$y_l = 0. \tag{1}$$

This pattern can then be rotated discretely or continuously. Unfortunately, this approach results in a disproportionate amount of time spent in the center of the map. The center pixel received about a factor of twelve times more integration time than points that were far off-axis.

To improve this, the holography team requested a cut for which the velocity varied with respect to the distance from the center of the map. Specifically, they requested a velocity variation of the form

$$\dot{x}_l = \frac{vdir * A}{x_l}$$
$$\dot{y}_l = 0. \tag{2}$$

where *vdir* is the direction sign of the motion and A is a constant. The maximum velocity is limited by the drive system near the very center of the map (x_l near zero) so that within the central area the velocity is smoothly blended to the maximum value. In the limit as the telescope velocity could reach infinity, the time spent at each point would be equal. Figure 1 shows three cases for comparison. The first case is the sinusoidal scan defined in equation (1). The other two cases have the $1/x_l$ velocity pattern in the outer area and velocity-limited motion in a central section. For a map of radius *xmax*, one case assumed that the motion would be velocity limited within 0.2*xmax* and the other within 0.05*xmax*.

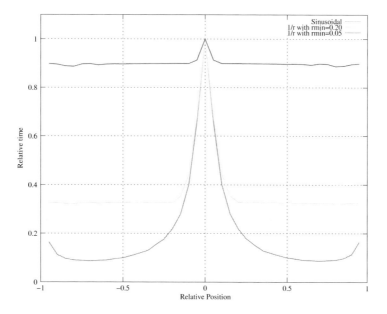

Figure 1: Comparison of time distribution across the map

The benefits of the $1/x_l$ velocity profile are immediately evident. Even if the motion is velocity limited within $0.2x\mathit{max}$, the variation in integration time drops to about a factor of three between the center and elsewhere on the map. If the velocity-limited area can be held to the central $0.05x\mathit{max}$, the variation is only about 12%. In both cases, the rest of the map is also more uniform than in the simple sinusoidal case.

Because of the benefits in distribution of integration time across the map, the holography team requested implementation of this mapping strategy. However, such a piecewise parametrization of velocity with respect to position cannot be accomplished using the time parametric approach already developed. This paper expands on the development of the time parametric case to allow position-dependent parameterization of the map. The method even allows a piecewise-continuous parameterization, and led to significantly faster holography mapping.

2. THE LMT TRAJECTORY GENERATOR

The LMT trajectory generator[2] guarantees that the commanded path always remains within specified velocity, acceleration, and jerk limits. Further, it provides feedforward information, which enables improved controller tracking performance. The trajectory generator is called separately in each axis with a function call of the form:

$$\text{coeffs} = f(s_0, v_0, a_0, s_1, v_1, a_1, v_{\lim}, guess_v, guess_a, dt). \tag{3}$$

The initial conditions of the commanded position, velocity, and acceleration of the path (s_0, v_0, a_0) are always known from the previous call to the function and the state of the telescope. As described in the companion paper[1], the scan pattern must be defined in such a way that it provides the next state of the path (s_1, v_1, a_1) for the next dt time step of the astronomical tracking update date. In the case of the LMT, this is 40 ms (25 Hz update rate). The trajectory generator polynomial is then used for calculating the controller commands at the much faster (250 Hz) servo rate.

3. SCAN PATTERNS WITH POSITION DEPENDENCE

As with the rotating parametric pattern[1], it is helpful to consider the system as the sum of the target position, velocity, and acceleration and the added rotating pattern. The resulting vectors are

$$\begin{aligned}\mathbf{s} &= \mathbf{s}_{\text{track}} + \mathbf{s}_{\text{rotatingpattern}} = \mathbf{s}_t + \mathbf{s}_r \\ \mathbf{v} &= \mathbf{v}_{\text{track}} + \mathbf{v}_{\text{rotatingpattern}} = \mathbf{v}_t + \mathbf{v}_r \\ \mathbf{a} &= \mathbf{a}_{\text{track}} + \mathbf{a}_{\text{rotatingpattern}} = \mathbf{a}_t + \mathbf{a}_r.\end{aligned} \tag{4}$$

Each of these have an azimuth and elevation component. For this paper, x is the azimuth angle and y is the elevation angle. The track in azimuth and elevation (*i.e.*, x and y), as well as all of its derivatives are assumed to be known from the astronomical calculations or interpolated from the positions. This leaves the task of determining the effect of the additional pattern on the target position, velocity, and acceleration in azimuth $(x, \dot{x}, \ddot{x})$ and elevation $(y, \dot{y}, \ddot{y})$.

3.1 Calculating the scan pattern

Before adding the effects of rotating the pattern, it is first necessary to calculate the nominal position, velocity, and acceleration of the desired scan pattern. The development below applies to any scan pattern that is parameterized in terms of position and thus has position-dependent velocity and acceleration. However, to clarify the method, a specific case is also presented as an illustrative example. For this definition, the scan angles are left in joint-space coordinates. That is, they are driving the azimuth and elevation angle of the telescope rather than trying to produce the pattern on the sky. The required correction in azimuth due to the elevation angle must be applied after rotating the pattern. By keeping the scan pattern separate from the rotation of the pattern, there is maximum commonality between this type of scan and the time parametric version.

An unusual feature of this approach is that there is, in general, no definition of the scan pattern position as a function of time. However, since the velocity depends on position, an estimate must be constructed. Thus,

$$\begin{aligned} x_l &= x_l(t+dt) = x_l(t) + \dot{x}_l(x_l, y_l)dt \\ y_l &= y_l(t+dt) = y_l(t) + \dot{y}_l(x_l, y_l)dt. \end{aligned} \quad (5)$$

It is worth noting that since $x_l(t)$ is not known explicitly, it is not guaranteed that the path will receive a particular position command (*e.g.*, that it will ever be commanded to exactly a zero offset). However, this approach does guarantee that the commanded velocity will have the desired relationship with the position.

The example case is for a straight line motion with a velocity that is inversely proportional to the distance from the center. This means that the example scan pattern has the form:

$$\begin{aligned} x_l &= x_l(t+dt) = x_l(t) + \dot{x}_l(x_l, y_l)dt \\ y_l &= 0. \end{aligned} \quad (6)$$

The velocity is defined as a function of the position in the scan pattern. Thus,

$$\begin{aligned} \dot{x}_l &= f(x_l, y_l) \\ \dot{y}_l &= g(x_l, y_l). \end{aligned} \quad (7)$$

Because the telescope trajectory generator will smooth out any discontinuities, this function can even have a piece-wise definition, as long as the derivatives are evaluated correctly. However, the telescope will track the trajectory generator path more accurately if the prescribed function is continuous in position, velocity, and acceleration.

Again, the desired example case is for a straight line motion with a velocity that is inversely proportional to the distance from the center. However, to avoid a singularity at the center, it is also necessary to limit the velocity when it is within a distance $x_{\min}$ from the center of the map, and we do so in such a way that the acceleration is continuous. So, if $x_l > x_{\min}$, the velocity is defined as

$$\begin{aligned} \dot{x}_l &= \frac{\text{vdir} * A}{x_l} \\ \dot{y}_l &= 0. \end{aligned} \quad (8)$$

If $x_l < x_{\min}$, the definition is

$$\begin{aligned} \dot{x}_l &= \text{vdir} * (v_{\max} - \beta x_l^2) \\ \dot{y}_l &= 0. \end{aligned} \quad (9)$$

The constant β is chosen so that the acceleration matches at $x_{\min}$.

$$\beta = \frac{A}{2x_{\min}^3}. \tag{10}$$

Choosing the constant A so that the velocity also matches at $x_{\min}$ results in

$$A = \frac{2}{3} v_{\max} x_{\min}. \tag{11}$$

Substituting back into the previous equations,

$$\beta = \frac{(2/3) v_{\max} x_{\min}}{2 x_{\min}^3} = \frac{v_{\max}}{3 x_{\min}^2}.$$

The resulting velocity for $x_l \leq x_{\min}$ is then

$$\begin{aligned}
\dot{x}_l &= \text{vdir} * (v_{\max} - \frac{v_{\max}}{3 x_{\min}^2} x_l^2) \\
&= \text{vdir} * v_{\max} \left(1 - \frac{1}{3}(\frac{x_l}{x_{\min}})^2\right).
\end{aligned} \tag{12}$$

Similarly, the velocity for $x_l > x_{\min}$ is

$$\dot{x}_l = \text{vdir} * \frac{2 * v_{\max} x_{\min}}{3 x_l}. \tag{13}$$

Finally, when x_l reaches the desired maximum value, xmax, the pattern must reverse, so vdir is set to $-\text{vdir}$. When this happens, the trajectory generator will calculate a smooth path for reversal, but there will be some settling time for each direction change at the outer edge of the map. An additional advantage of this velocity profile is that the reversal happens when the telescope is already at lower velocity, so the demands on the drive system are lower and the settling time is shorter.

Because the velocities are defined, the azimuth and elevation scan accelerations are simply the time derivatives of the position. However, since the velocities are given as a function of position, it is necessary to use the chain rule. Thus,

$$\begin{aligned}
\ddot{x}_l &= \frac{\partial \dot{x}_l}{\partial x} \dot{x}_l + \frac{\partial \dot{x}_l}{\partial y} \dot{y}_l \\
\ddot{y}_l &= \frac{\partial \dot{y}_l}{\partial x} \dot{x}_l + \frac{\partial \dot{y}_l}{\partial y} \dot{y}_l.
\end{aligned} \tag{14}$$

For the desired example, there are two cases. If $x_l > x_{\min}$, the acceleration is defined as

$$\ddot{x}_l = \frac{-2 * \text{vdir} * v_{\max} x_{\min} \dot{x}_l}{3 x_l^2}$$
$$\ddot{y}_l = 0. \tag{15}$$

If $x_l \leq x_{\min}$, the definition is

$$\ddot{x}_l = -\frac{2 * \text{vdir} * v_{\max} x_l \dot{x}_l}{3 x_{\min}^2}.$$
$$\ddot{y}_l = 0. \tag{16}$$

3.2 Calculating the rotated scan pattern

Since some observing techniques, including this holography mapping approach, require rotating a given pattern on the sky, either step-wise or at a constant rate, it is convenient to separate this part of the calculation from the parametric scan pattern defined in the previous section. That way, whether the pattern is a time parametric pattern such as a Lissajous figure or if is a position-dependent velocity pattern such as the one considered here, the rotation section is applied in the same manner. The rotation rate Ω is assumed to be constant. If the pattern does not rotate during the scan, this section must still be applied, but with $\Omega = 0$. The principal difference between this development and the time parametric case is that here the rotations are defined by the angle $\Omega t + \phi$, where the angle ϕ has been introduced to allow for step-wise rotation of the map. In this example case, $\Omega = 0$, but whenever the scan position x_l reaches the end of the map and the velocity is reversed, ϕ is incremented and again held fixed during the next scan. Because the rotation equations have already been treated in detail in the companion paper[1], they are presented here only in summary form.

The general equations for rotating the pattern in telescope joint space are simply

$$x_r = (x_l \cos(\Omega t + \phi) - y_l \sin(\Omega t + \phi))/\cos\theta_{el}$$
$$y_r = x_l \sin(\Omega t + \phi) + y_l \cos(\Omega t + \phi), \quad (17)$$

where $\theta_{el} = y_t + y_r$ and the azimuth angle x_r has been corrected by $1/\cos\theta_{el}$ to provide the proper position on the sky. For step-wise rotation, these equations can be written

$$x_r = (x_l \cos\phi - y_l \sin\phi)/\cos\theta_{el}$$
$$y_r = x_l \sin\phi + y_l \cos\phi. \quad (18)$$

As shown in the companion paper[1], the velocity for the continuously-rotating case is as follows:

$$\dot{x}_r = [(\dot{x}_l - y_l\Omega + x_l\dot{\theta}_{el}\tan\theta_{el})\cos(\Omega t + \phi) - (\dot{y}_l + x_l\Omega + y_l\dot{\theta}_{el}\tan\theta_{el})\sin(\Omega t + \phi)]/\cos\theta_{el}$$
$$\dot{y}_r = (\dot{x}_l - y_l\Omega)\sin(\Omega t + \phi) + (\dot{y}_l + x_l\Omega)\cos(\Omega t + \phi). \quad (19)$$

Where $\dot{\theta}_{el} = \dot{y}_t + \dot{y}_r$. For step-wise rotation, these equations can be written

$$\dot{x}_r = [(\dot{x}_l + x_l\dot{\theta}_{el}\tan\theta_{el})\cos\phi - (\dot{y}_l + y_l\dot{\theta}_{el}\tan\theta_{el})\sin\phi]/\cos\theta_{el}$$
$$\dot{y}_r = \dot{x}_l \sin\phi + \dot{y}_l \cos\phi. \quad (20)$$

Again referring to the companion paper[1], the equations for the acceleration for the continuously-rotating case are as follows:

$$\ddot{x}_r = [(\ddot{x}_l - 2\dot{y}_l\Omega - x_l\Omega^2 + x_l\ddot{\theta}_{el}\tan\theta_{el} + x_l\dot{\theta}_{el}^2(1 + 2\tan^2\theta_{el}) + 2(\dot{x}_l - y_l\Omega)\dot{\theta}_{el}\tan\theta_{el})\cos(\Omega t + \phi)$$
$$- (\ddot{y}_l + 2\dot{x}_l\Omega - y_l\Omega^2 + y_l\ddot{\theta}_{el}\tan\theta_{el} + y_l\dot{\theta}_{el}^2(1 + 2\tan^2\theta_{el}) + 2(\dot{y}_l + x_l\Omega)\dot{\theta}_{el}\tan\theta_{el})\sin(\Omega t + \phi)]/\cos\theta_{el} \quad (21)$$
$$\ddot{y}_r = (\ddot{x}_l - 2\dot{y}_l\Omega - x_l\Omega^2)\sin(\Omega t + \phi) + (\ddot{y}_l + 2\dot{x}_l\Omega - y_l\Omega^2)\cos(\Omega t + \phi).$$

In all cases, $\theta_{el} = y_t + y_r$, $\dot{\theta}_{el} = \dot{y}_t + \dot{y}_r$, and $\ddot{\theta}_{el} = \ddot{y}_t + \ddot{y}_r = \ddot{y}_r$. Finally, for step-wise rotation, the acceleration equations can be written

$$\ddot{x}_r = [(\ddot{x}_l + x_l\ddot{\theta}_{el}\tan\theta_{el} + x_l\dot{\theta}_{el}^2(1 + 2\tan^2\theta_{el}) + 2\dot{x}_l\dot{\theta}_{el}\tan\theta_{el})\cos\phi$$
$$- (\ddot{y}_l + y_l\ddot{\theta}_{el}\tan\theta_{el} + y_l\dot{\theta}_{el}^2(1 + 2\tan^2\theta_{el}) + 2\dot{y}_l\dot{\theta}_{el}\tan\theta_{el})\sin\phi]/\cos\theta_{el} \quad (22)$$
$$\ddot{y}_r = \ddot{x}_l \sin\phi + \ddot{y}_l \cos\phi.$$

3.3 Pointing corrections

As with the time-parametric case, pointing corrections are slowly-varying, and can be added at the end of the calculation. These are assumed to be position-only corrections, with all derivatives equal to zero.

4. SOFTWARE IMPLEMENTATION

To implement this in the software requires calculating the tracking profile, calculating the pattern, rotating the pattern, and including the pointing corrections. Most of this is identical to the time-parametric case[1]. However, an important difference lies in the fact that the position and velocity calculation are interdependent.

4.1 Calculating the scan pattern

Throughout this section, the time variable t is the current time t_0 plus dt. That is, the purpose of the calculation is to determine the desired position, velocity, and acceleration at the next time step.

1. The first step is to calculate and save the values of x_l and y_l. Since there are no parametric equations as a function of time, it is necessary to estimate the next position based on the current value and the derivative. In the simplest case, the estimates can be calculated from the equations

$$x_l = x_l(t) = x_l(t_0) + \dot{x}_l(x_l(t_0), y_l(t_0))dt$$
$$y_l = y_l(t) = y_l(t_0) + \dot{y}_l(x_l(t_0), y_l(t_0))dt. \qquad (23)$$

However, this is just a coupled system of two first order differential equations that are already in standard form. As a result, a much better estimate can be obtained with only slight additional computation using a low-order differential equation solving method (*e.g.*, 4th order Runge-Kutta).

2. Once the values of t, x_l, and y_l have been determined, calculate and save $\dot{x}_l$, $\dot{y}_l$, $\ddot{x}_l$, and $\ddot{y}_l$ at time t.

4.2 Calculate the rotated scan pattern

Calculating the rotated scan pattern requires the most computation, so it is worthwhile to look for values that appear repeatedly. Note that once the scan pattern (x_l, y_l) and its derivatives have been calculated, the approach is the same, whether the values came from time-parametric or position-parametric definitions. The calculation steps are as follows:

1. Calculate $omt = \Omega t + \phi$. Again, throughout this section, remember that the value of t in all equations is the current time t_0 plus the time step dt. If the pattern does not continuously rotate, such as in the holography case, set $\Omega = 0$ and ϕ to the desired sky rotation angle. As a further simplification, for a straight-line path such as the holography mapping strategy employed at the LMT, $y_l = \dot{y}_l = \ddot{y}_l = 0$, which eliminates many terms.
2. Save the value of $somt = \sin(\Omega t + \phi)$ and $comt = \cos(\Omega t + \phi)$. For the stepwise rotation case, this is the same as $somt = \sin\phi$ and $comt = \cos\phi$.
3. Calculate $y_r = x_l * somt + y_l * comt$.
4. Calculate $\theta_{\text{el}} = y_t + y_r$.
5. Calculate and save the value of $sthel = \sin\theta_{\text{el}}$, $cthel = \cos\theta_{\text{el}}$, and $tthel = \tan\theta_{\text{el}}$.
6. Calculate $x_r = (x_l * comt - y_l * somt)/cthel$.
7. Calculate and save $tmp1 = \dot{x}_l - y_l\Omega$ and $tmp2 = \dot{y}_l + x_l\Omega$. For stepwise rotation, this step is unnecessary.
8. Calculate $\dot{y}_r = tmp1 * somt + tmp2 * comt$. For the stepwise rotation case, $\dot{y}_r = \dot{x}_l * somt + \dot{y}_l * comt$.
9. Calculate and save $\dot{\theta}_{\text{el}} = \dot{y}_t + \dot{y}_r$.
10. Calculate and save $theldottan = \dot{\theta}_{\text{el}} * tthel$.

11. Calculate $\dot{x}_r = ((tmp1 + x_l * theldottan) * comt - (tmp2 + y_l * theldottan) * somt)/cthel$.
12. Calculate $tmp3 = \ddot{x}_l - 2\dot{y}_l\Omega - x_l\Omega^2$ and $tmp4 = \ddot{y}_l + 2\dot{x}_l\Omega - y_l\Omega^2$. For the stepwise rotation case, this step is unnecessary.
13. Calculate $\ddot{y}_r = tmp3 * somt + tmp4 * comt$, or, for the step-wise rotation case, $\dot{y}_r = \ddot{x}_l * somt + \ddot{y}_l * comt$.
14. Calculate and save $\ddot{\theta}_{el} = \ddot{y}_t + \ddot{y}_r$.
15. Calculate and save $thelddottan = \ddot{\theta}_{el} * tthel$, $theldotsq = \dot{\theta}_{el}^2$, $tanval = 1 + 2 * tthel * tthel$, and $tmp5 = thelddottan + theldotsq * tanval$.
16. Calculate $\ddot{x}_r = [(tmp3 + x_l * tmp5 + 2 * tmp1 * theldottan) * comt - (tmp4 + y_l * tmp5 + 2 * tmp2 * theldottan) * somt]/cthel$.

4.3 Trajectory generator commands

As before, the pointing corrections (x_p, y_p) are added only to the positions. These corrections result in a final set of commands for the trajectory generator as follows:

$$\begin{aligned} s_1 &= (x_t + x_r + x_p, y_t + y_r + y_p) \\ v_1 &= (\dot{x}_t + \dot{x}_r, \dot{y}_t + \dot{y}_r) \\ a_1 &= (\ddot{x}_t + \ddot{x}_r, \ddot{y}_t + \ddot{y}_r). \end{aligned} \quad (24)$$

5. EXPERIMENTAL RESULTS

The improved scan pattern was tested at the LMT during the 2011 initial holography campaign. Unfortunately, the telescope test data is only available from a time range where only three of the four elevation motors were operational. While the system still provided sufficient performance to perform the map, there is some acceleration lag in the servo tracking that becomes evident during the transition area where the velocities and accelerations are both high. Figure 2 shows the elevation position versus time for a few scans that are nearly in the pure elevation direction ($\Omega t \approx 90°$).

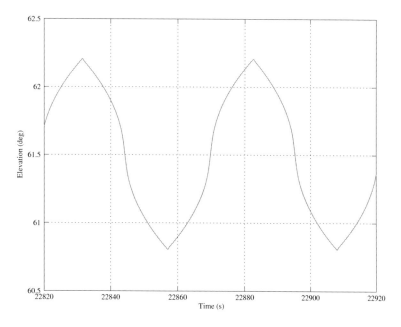

Figure 2: Scan position vs time

The corresponding profile of velocity versus time is shown in Figure 3. Finally, Figure 4 shows the commanded and actual velocity versus position. Due to the acceleration lag, the velocity overshoots noticeably, and, in the worst case, the resulting position errors are of order 40". This was within the tolerances for the relatively low frequencies of the holography observations, but better results are expected using all of the axis drives and by correcting for more of the acceleration lag using the feedforward part of the controller.

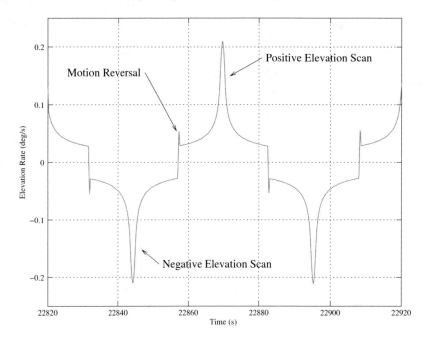

Figure 3: Scan velocity vs time

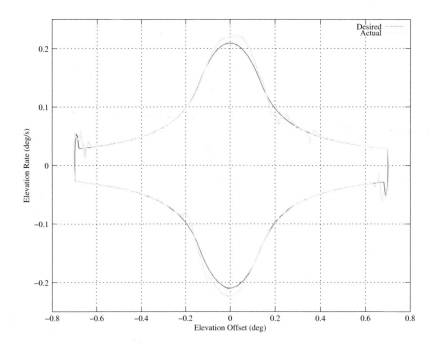

Figure 4: Commanded and actual velocity vs position

6. CONCLUSIONS

We have expanded the capabilities of the LMT time-parametric rotating scan patterns so that it is also possible to include patterns that are parametric with position. This has been implemented in an improved high-performance holography mapping strategy that provides more uniform coverage of the mapped area. While the mapping approach was implemented successfully, the only experimental data available are for a degraded operation in which the telescope was missing a drive motor. Even in this case, the drive system was able to maintain pointing tolerances within the requirements for holography observations. However, further improvements are certainly possible with a fully-operational system and some additional tuning of the servo feedforward gains.

ACKNOWLEDGMENTS

The authors wish to express appreciation to the LMT site crew for their technical and logistical support in carrying out this work and to the holography team for proposing this observing strategy. This material is based in part upon work supported by National Science Foundation Grant Number AST-0838222 and/or Contract Number NNX09AR37G from the National Aeronautics and Space Administration.

REFERENCES

[1] Smith, D.R., and Souccar, K., "Trajectory generation for parametric rotating scan patterns at the LMT," *Proc. SPIE Astronomical Telescopes and Instrumentation*, Amsterdam, 2012.
[2] Smith, D.R., and Souccar, K., "A Polynomial-based trajectory generator for improved telescope control," *Proc. SPIE*, **7019**, 701909 (2008).
[3] Hughes, D.H., *et al.*, "The Large Millimeter Telescope (LMT): current status and preparations for early science observations," *Proc. SPIE Astronomical Telescopes and Instrumentation*, Amsterdam, 2012.

Photonic local oscillator technics for large-scale interferometers

Kiuchi H.[a], Saito M.[a] and Iguchi S.[a]

[a]National Astronomical Observatory of Japan, 2-21-1 Osawa, Mitaka, Tokyo, Japan

ABSTRACT

In signal transmission through optical fiber, cable length delay fluctuation accompanied by chromatic and polarization-mode dispersion affects the coherence of distributed signals. To maintain signal coherence, it is very important to generate very-high-frequency signals with minimum phase noise and transmission loss. In a photonic local signal generation/distribution system with a microwave-photonic signal generator and a real-time microwave-photonic signal phase stabilizer that we developed as an alternative photonic LO system for ALMA (Atacama Large Millimeter/sub-millimeter Array), signals are transmitted in the form of frequency difference between two coherent light waves, effectively maintaining the coherence of distributed reference signals. Through the development of the real-time phase stabilizer, we discovered that the system would be further improved with the introduction of a post-processing scheme phase stabilizer and confirmed its effectiveness by experiments.

Keywords: Photonic Local, Microwave photonic signal generation, Microwave photonic signal transmission, Allan variance, Phase stability

1. INTRODUCTION

The photonic signal generation technique at millimeter and submillimeter wavelengths can be widely used to various applications. Using this technique, we generated very-high-frequency signals with stable phase noise and reduced transmission loss, which cannot be attained by current electronics-based techniques. The use of optical fiber cables to transmit high-frequency signals provides us with unique system solutions. In this paper, we will discuss the phase stability of the generated/transmitted microwave-photonic local signal. It is well known that the transmission signal phase fluctuates when the fiber swings. This is not due to fiber length change but due to refractive index change, which is easily generated by the temperature gradient and antenna motion. Chromatic dispersion is specific to single mode fibers where lights are transmitted at different wavelengths and at different speeds. In optical fiber transmission, the refraction index is a function of the wavelength of the transmitted light. The value of the refraction index will increase when a very high frequency signal is transmitted over very long distances.

2. MICROWAVE-PHOTONIC SIGNAL GENERATOR

We have developed a microwave-photonic signal generator using an optical modulator as an alternative photonic LO system for ALMA (Atacama Large Millimeter/submillimeter Array). A microwave-photonic signal is generated by a high-extinction ratio lithium niobate ($LiNbO_3$) Mach-Zehnder intensity modulator (LN-MZM). The transmission microwave frequency is transmitted in the form of the optical frequency difference between the two optical signals. Compared to the optical phase lock scheme,[1,2] the LN-MZM has significant advantages in terms of stability (free from the influence of the input laser stability), robustness to mechanical vibration and acoustic noise, and capability of maintaining polarization state of the input laser.

Further author information: (Send correspondence to H. Kiuchi)
Hitoshi Kiuchi: E-mail: hitoshi.kiuchi@nao.ac.jp, Telephone: +81 422 34 3761
Masao Saito: E-mail: masao.saito@nao.ac.jp, Telephone: +81 422 34 3633
Satoru Iguchi: E-mail: S.Iguchi@nao.ac.jp, Telephone: +81 422 34 3762

2.1 Scheme of the microwave-photonic signal generator

In the high-extinction ratio LN-MZM,[3],[4] the optical frequency difference between the two optical signals is exactly twice (or four times) the modulation frequency, while the output signal is equivalent to the frequency shift keying (FSK) spectrum. The output spectrum depends on the DC bias voltage applied to the electrodes of the Mach-Zehnder structure.

The LN-MZM has two operation modes: null-bias point operation mode and full-bias point operation mode. When the bias of the LN-MZM is set to a minimum transmission point (null-bias point) of the input laser signal, the first-order upper side band (USB) and lower side band (LSB) components are intensified, and the input laser signal is suppressed. The frequency difference between the two spectral components is twice the modulation sinusoidal signal frequency. On the other hand, in case of using a conventional optical filter with a fiber bragg grating (FBG), when the bias is set to a maximum transmission point (full-bias point), the optical frequency of even-order (zero- and second-order) components would remain, while the zero-order component would be eliminated. After eliminating the zero-order component (same as the input laser wave-length), there remains a two-tone optical spectrum whose frequency is four times the modulation frequency. A block diagram of the photonic millimeter-wave generator for a 20-GHz external synthesizer is shown in Fig. 1. The photonic millimeter-wave generator is composed of an LN-modulator (Sumitomo-Osaka cement T.FSX1.5-10-P-O ALMA), an FBG (Athermal type: Bandwidth is 50 GHz) with an isolator, RF-drive amplifiers and some optical components with polarization maintaining capability. The null-bias mode is driven by the 13-17 GHz amplifier, and the full-bias mode is driven by the 17-31 GHz amplifier. Each mode can be selected by the optical switch. If there is a 40 GHz external synthesizer available, a wide-bandwidth amplifier (1 dB compression level is more than 25 dBm: e.g. Centellax TA2U50HA) would be acceptable. In this case, the microwave drive circuits in Fig. 1 can be replaced by the wide bandwidth amplifier. Measured phase stability in Allan standard deviation is 2×10^{-14} in white-PM noise (Fig. 2).

The photonic technique has been applied to the holography receiver as the direct photonic LO signal for ALMA antenna surface measurement. The measured RMS surface error is better than $4.4 \mu m$, the repeatability is better than $2 \mu m$, which provide evidence of the Photonic LO effectiveness.

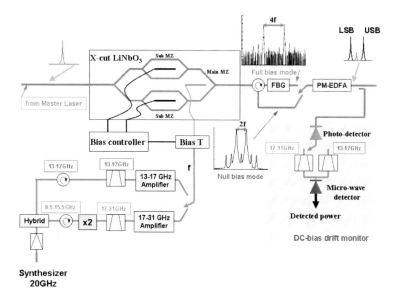

Figure 1. Block diagram of the photonic millimeter-wave generator for an external 20-GHz synthesizer (equivalent to the ALMA synthesizer)

3. MICROWAVE PHOTONIC SIGNAL PHASE STABILIZER

The optical signal phase stabilizer is classified into two types according to the transmission signal. One is a phase stabilizer for a single light-wave signal,[5],[6] and the other is for a transmission microwave signal which

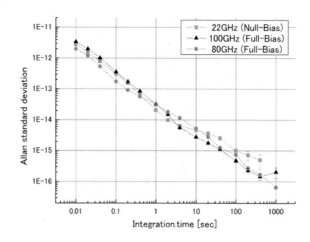

Figure 2. Measured phase stability of the photonic millimeter-wave generator.

is transmitted in the form of frequency difference between two coherent light waves. Since the ALMA phase stabilizer[7] belongs to the latter, it is necessary to consider chromatic dispersion,[8],[9,10] between two light waves. In a real-time phase stabilizer that we developed,[11],[12] signals generated by the microwave photonic signal generator (LN-MZM) are sent to the antennas via a long single mode fiber. During the signal transmission through the fiber optics, the cable length delay fluctuation occurs along with chromatic dispersion, which affects the performance of coherent signal distribution. The real-time phase stabilizer uses a dual difference round-trip phase measurement method with Michelson's interferometer. The roundtrip phase measurement is performed on each light-wave signal separately. When a microwave signal is transmitted in the form of light, two lights at different wavelengths (wavelengths λ_1 and λ_2) are used. A block diagram of the real-time phase stabilizer is shown in Fig. 3. In this composition, the two lights (wavelengths λ_1 and λ_2), are commonly subject to external influence during the fiber transmission, and thus the phase detector detects only the differential phase between the two lights (wavelengths λ_1 and λ_2).

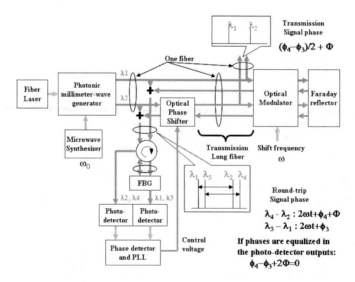

Figure 3. Block diagram of the real-time phase stabilizer.

In the real-time phase stabilizer process, the effect of the chromatic dispersion can be reduced by using dual difference phases of two optical signals, each of which are obtained from a round-trip measurement. As a result, we can measure the instrumental delay phase (twice of the cable delay phase). Moreover, this method does not require the transmission of the modulation signal (ω), which means we do not have to consider any phase delay of

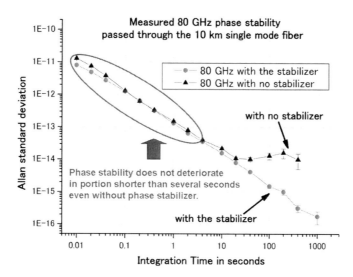

Figure 4. Measured typical phase stability of the real-time phase stabilizer. When the optical signal (at 80 GHz) is transmitted through the single mode fiber cable (10 km), the phase stability decreases around 10 seconds of integration time. With the high-frequency phase stabilizer, the decrease of the phase stability is staved off. The measured phase noise is the white phase noise.

the modulation signal (ω). The measured double differential phase (($\lambda_1 - \lambda_3$) - ($\lambda_2 - \lambda_4$)) is used to compensate the instrumental delay / phase change (($\lambda_3 - \lambda_4$) - ($\lambda_1 - \lambda_2$)).

3.1 Possibility of the post-processing phase stabilizer

Figure 4 shows a typical phase stability measurement result of the real-time phase stabilizer. The figure demonstrates that the phase stability was effectively improved with the real-time phase stabilizer by transmitting a signal via an optical line in the form of a relationship between the measured phase stability and time (at 80 GHz and with a transmission path length of 10 km). In the case of using only a fiber cable (triangle marks), flicker frequency noise appears for 10 seconds or longer, and the phase becomes unstable (the Allan standard deviation value is constant regardless of the lapse of time with a characteristic line being substantially horizontal). On the other hand, in the case of using the real-time phase stabilizer (circle marks) together with a fiber cable, it is clear that flicker frequency noise is suppressed and white phase noise is detected with a stable phase for a longer time (the Allan standard deviation value is in inverse proportion to time).

For example, in astronomical observation, obtained signals, which are extremely faint, are time-integrated to improve its signal-to-noise ratio, and then necessary data is acquired. As shown in Fig. 4, it is clear that there is not much difference in the phase stability between use and non-use of a real-time phase stabilizer in an integration time shorter than several seconds. In other words, optical fiber cable transmission does not have much influence on the short-term phase stability.

Therefore, in a case where an integration processing is performed at the signal transmission destination for a time shorter than several seconds (approximately three seconds), there is no need to use a microwave signal phase shifter or a phase-locked loop circuit to perform real-time round-trip transmission phase compensation, because these devices make the system configuration complicate, thereby increasing signal loss and decreasing phase stability of a transmitted signal. Furthermore, it gives negative impact on digitization and larger packaging densities, which would be an impediment to multipoint transmission compensation.

3.2 Scheme of the post-processing phase stabilizer

The post-processing phase stabilizer scheme has been designed to improve the long term stability with simplified optical transmission system. Exclusion of real-time operation contributes to the reduction of signal loss and the improvement of phase stability of the transmitted signal, which facilitates digitization and larger packaging

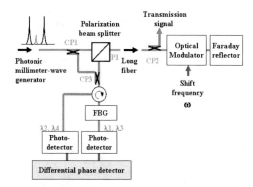

Figure 5. Post-processing phase stabilizer. The photo-detectors detect 2ω (50 MHz) signals.

densities as well as multipoint transmission compensation.

In the optical transmission system, phase compensation data for round-trip transmission is acquired at time intervals synchronized with the integration timings. In the post-processing scheme, the transmission phase compensation is carried out when the integration processing is performed at the signal transmission destination for a time shorter than several seconds (approximately three seconds in experiments).

At the signal transmission destination, off-line transmission path phase compensation is made for the time integration data acquired by a user by utilizing the round-trip transmission phase compensation data acquired at a signal transmission source. Naturally, in this process, phase shifting devices become unnecessary because phase swinging compensation is not performed within a time shorter than an integration time. In our experiments, the integration processing was performed within a time shorter than three seconds to confirm its effectiveness. The optical transmission system includes a round-trip transmission function using the return light that has been shifted in frequency in the high-stability optical signal transmission based on the frequency difference between two light signals. When the integration processing is performed within a time shorter than several seconds (approximately three seconds) at the signal transmission destination, the transmission light and round-trip return light are detected according to the principle of the Michelson interferometer to distinguish round-trip transmission and reception signals by setting a polarization state in which transmission light and reception light are made orthogonal to each other. And then, the transmission light and the return light are separated by optical signals or microwave signals, and in the optical transmission system, the round-trip transmission phase compensation data is acquired at time intervals synchronized with signal integration timings at the signal transmission destination to perform the transmission phase compensation in the post-processing.

A simple block diagram of the phase stabilizer is shown in Fig. 5. Transmission delay on the fiber is measured as the differential phase of the optical round-trip delay of each light-wave signal. At first, the two coherent-optical-signals generated by the photonic millimeter-wave generator have a vertical and high-extinction ratio polarization. In a series of processing in the ground unit, the polarization is maintained.

The signal in Fig. 5, passing through the optical coupler (CP1) and polarization beam splitter (P1), is sent to the antenna. The signal is divided into two signals at the optical coupler (CP2) after passing through a long single-mode fiber. One of the divided signals is converted to a millimeter wave by a photo-mixer, and the other is reflected by a Faraday reflector after the frequency shift by an optical frequency shifter (Acous-Optics frequency shifter). The reflected signal is converted into a 90-degree different optical polarization signal by the Faraday reflector. The signal, after passing through the frequency shifter again, is returned back to the polarization beam splitter (P1) in the ground unit. While the signal is going through the optical reciprocal process, the received signal maintains a horizontal (90-degree different polarization angle to the transmission signal) polarization state. After this, the signal is finally recombined with the divided transmission signal at the optical coupler (CP3).

After passing through the beam splitter (P1), the signal is divided into two wavelengths (λ_1 and λ_2). Wavelength λ_1 signal is reflected by the fiber bragg grating (FBG) and returned to the circulator. The differential phases on the angular frequency 2ω between transmission and round-trip signals on each light-wave signal are detected by low-frequency photo-mixers after wavelength separation by the FBG optical filter. These measured phases are equivalent to the round-trip phases on both lightwave signals.

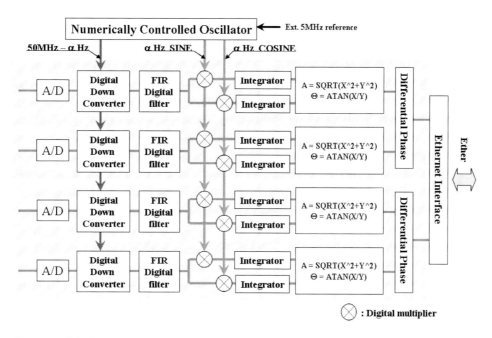

Figure 6. Block diagram of the differential phase measurement board for two antennas.

3.3 Differential phase detector

Figure is a block diagram illustrating a configuration of the phase difference detector to support digitization in the optical transmission system.. The input signals are sent to the phase difference detector which detects the frequency twice as high as the shift frequency ω, 50 MHz. The two input signals are A/D-converted separately by 8-bit 108-Msps analog-digital converters, and are digitally processed sequentially by all the other components. First, a reference signal of F-Hz (approximately several α Hz), whose frequency is adjusted to be twice as high as the shifted frequency, and a reference clock signal of F-Hz (49.96 MHz) are created by a 32-bit numerically controlled oscillator, and then multiplied by digital mixers to perform digital frequency-conversion. Unnecessary waves are removed by digital filters at this point. A phase difference detection circuit performs a correlation integration processing on output signals from the digital filters by an orthogonal wave detection method, and calculates the phases and amplitudes of the signal. The phase difference between the two waves is calculated and output by the phase difference detection circuit. For orthogonal wave detection, a Hilbert transform can be used but the description is given by a digital averaging phase detector, which is more suitable for a digital circuit. A sine wave signal of F-Hz and a cosine wave signal of F-Hz corresponding to the reference clock signal are created by the numerically controlled oscillator. The phase difference detection circuit multiplies the sine wave signal and the cosine wave signal of F-Hz respectively by the output from the digital filter to obtain four correlations. The phase difference detection circuit averages the four correlation outputs separately (four averaged signals are "Real(X)" (cosine), "Imag(X)" (sine), "Real(Y)" (cosine), and "Imag(Y)" (sine) in the stated order) and obtains amplitudes (root sum square: Amplitude $A1 = \sqrt{Real(X)^2 + Imag(X)^2}$ and Amplitude $A2 = \sqrt{Real(Y)^2 + Imag(Y)^2}$ and phases (arc tangent: $Phase1 = \tan^{-1}[Imag(X)/Real(X)]$ and $Phase2 = \tan^{-1}[Imag(Y)/Real(Y)]$. The phase difference during the round trip transmission through the optical fiber is derived from the difference between the two obtained phases (1 and 2).

3.4 Phase Linearity

We conducted a linearity testing of the digital differential phase detector using two AD9854 (Analog Devices) digital synthesizers generating 50 MHz signals, which are transmitted to the differential phase measurement board. Measured linearity error is shown in Fig. 7.

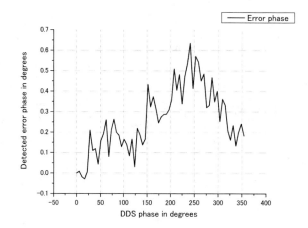

Figure 7. Measured linearity phase error of the differential phase detector.

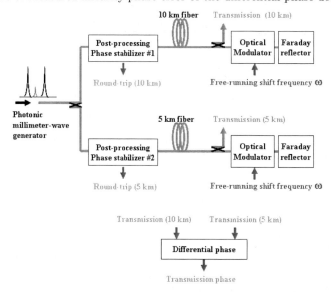

Figure 8. Block diagram of the phase consistency measurement.

3.5 Performance

3.5.1 Phase consistency between transmitted and round-tripped signals

We carried out a phase consistency measurement between a transmitted signal and a round-tripped signal in a configuration as shown in Fig. 8. The transmission frequency generated by a photonic millimeter-wave generator is 80 GHz. The measurement was performed on two different fiber cables under a condition close to their actual radio interferometer operation. The cable lengths are 10 km (ch1) and 5 km (ch2). The differential round-trip signal phase is double the differential transmission signal phase via the two cables (10km and 5km). The differential transmission signal phase and half of differential roundtrip signal phase are shown in Fig. 9. Both phases are well aligned.

3.5.2 Phase stability

We measured the phase stability of the transmitted signal (80 GHz) through the single mode fiber cable spool (10 km) by the time-domain Allan standard deviation method,[13],[14],[15],[16],[17],[18] The measurements were conducted with/without the phase stabilizer. Figure 10 shows that the compensated phase stability is effectively improved with the optical transmission system. The integration time was 0.1 seconds. In our experiment without the

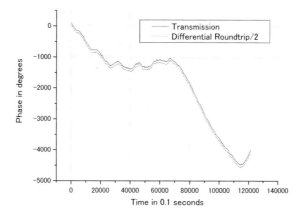

Figure 9. Phase difference between transmission phase and round-trip phase

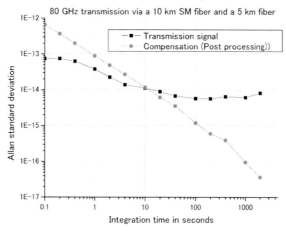

Figure 10. Compensated phase stability.

stabilizer, we detected no White-PM noise in the range shorter than 10 seconds integration time, although normally White-PM noise is generated as shown in Fig. 4. Also, it is clear that there is no flicker frequency noise as in the case of real-time transmission path compensation. Another advantage is that the entire system configuration is simplified, which contributes to the reduction of signal loss and to the improvement of phase stability of the transmitted signal.

4. CONCLUSION

The result of our experiment clearly shows that the photonic local signal generation and distribution system would be further improved with the introduction of a post-processing scheme phase stabilizer instead of a current real-time stabilizer. We confirmed by experiments that there is no frequency limit on transmitted signals in the post-processing scheme and it is effectively used for large-scale radio interferometers with multi baselines that would require terahertz transmission.

As a future application of the photonic techniques, we will place a small low-powered photonic millimeter-wave source (CW and noise) on one of the mountain peaks overlooking the ALMA operations site for ALMA antenna/receiver system calibration.

REFERENCES

[1] J. Cliche, and B. Shillue, "Precision timing control for radioastronomy, maintaining femtosecond synchronization in Atacama Large Millimeter Array", *IEEE control system magazine*, pp. 19–26, 2006.

[2] S. Ayotte, P. Poulin, N. Poulin, A. Jeanneau, M. J. Picard, D. Poulin, C. A., Davidson, M. Aube, I. Alexandre, F. Costin, F. Pelletier, J. F. Clich?, M. Tetu and B. Shillue, "Laser synthesizer of the ALMAtelescope: design and performance", *Microwave Photonics (MWP)*, pp. 249–252, 2010.

[3] T. Kawanishi, T. Sakamoto, and M. Izutsu, "High-speed control of lightwave amplitude, phase, and frequency by use of electro optic effect", *IEEE Journal of selected topics in quantum electronics*, **13**, 2007.

[4] H. Kiuchi, T. Kawanishi, M. Yamada, T. Sakamoto, M. Tsuchiya, J. Amagai, and M. Izutsu, "High Extinction Ratio Mach-Zehnder Modulator Applied to a Highly Stable Optical Signal Generator," *IEEE Trans. Microwave Theory and Techniques*, **55**, pp. 1964–1972, 2007.

[5] C. Daussy, O. Lopez, A. Amy-Klein, A. Goncharov, M. Guinet, C. Chardonnet, F. Narbonneau, M. Lours, D. Chambon, S. Bize, A. Clairon, and G. Santarelli ,"Long-Distance Frequency Dissemination with a Resolution of 10^{-17}," *Physical Review Letters*, **94**, 2005.

[6] S. M. Foreman, A. D. Ludlow, M. H. G. de Miranda, J. E. Stalnaker, S. A. Diddams, and J. Ye ,"Coherent optical phase transfer over a 32-km fiber with 1-s instability at 10^{-17}," *LEOS Summer Topical Meetings, 2007 Digest of the IEEE*, pp. 184–185, 2007.

[7] B. Shillue, "Atacama large millimeter array photonic local oscillator: femtosecond-level synchronization for radio astronomy", *Frequency control symposium (FCS)*, pp. 569–571, 2010.

[8] P. Ciprut, B. Gisin, N. Gisin, R. Passy, J. P. Von der Weid, F. Prieto, and C. W. Zimmer, "Second-order polarization mode dispersion: Impact on analog and digital transmissions," *IEEE Journal of lightwave technology*, **16**, pp. 757–771, 1998.

[9] G. P. Agrawal, *Fiber optic communication system, third edition*, John Wiley & Sons Inc., 2002.

[10] D. Derickson, *Fiber optics test and measurement*, Prentice Hall PTR, 1998.

[11] H. Kiuchi, "Highly stable millimeter-wave signal distribution with an optical round-trip phase stabilizer," *IEEE Trans. Microwave Theory and Techniques*, **56**, pp. 1493–1500, 2008.

[12] H. Kiuchi, "Optical transmission signal phase compensation method using an image rejection mixer," *IEEE Photonics Journal*, **3**, pp. 89–99, 2011.

[13] D. W. Allan, "Statistics of Atomic Frequency Standards," *Proc. IEEE*, **54**, 1966.

[14] D. W. Allan,"Report on NBS dual mixer time difference system (DMTD) built for time domain measurements associated with phase 1 of GPS," *NBS IR*, **75**, 1976.

[15] A. E. E. Rogers and J. M. Moran, "Coherence limits for very-long-baseline interferometry," *IEEE Trans. Instrum. Meas.*, **30**, pp. 283–286, 1981.

[16] A. E. E. Rogers, A. T. Moffet, D. C. Backer and J. M. Moran,"Coherence limits in VLBI observation at 3-millimeter wavelength," *Radio Science*, **19**, pp. 1552–1560, 1984.

[17] D. J. Healey III, "Flicker of frequency and phase and white frequency and phase; Fluctuations in frequency sources", *Proc. 25th Annu. Sympo. On Frequency Control (ASFC)*, pp. 29–42, 1972.

[18] N. Kawaguchi, "Coherence loss and delay observation error in Very-Long-Baseline Interferometry," *J. of Radio Research Labs.*, **30**, pp. 59–87, 1983.

First Technological Steps Toward Opening a near-IR Window at Stratospheric Altitudes

Fernando Pedichini[a], Mauro Centrone[a], Dario Lorenzetti[a],
Massimiliano Mattioli[a], Massimo Ricci[a], Fabrizio Vitali[a]

[a] INAF – Osservatorio Astronomico di Roma, Via Frascati 33, 00040 Monte Porzio Catone, Italy

ABSTRACT

The possibility to open a near-IR window at stratospheric altitude is crucial for a large variety of astronomical issues, from cosmology to the star formation processes. Up to now, one of the main issue is the role of the OH and thermal sky emission that are rising the sky background level when such observations are performed through ground based telescopes. We present the results of our technological activity aimed at affording some critical aspects typical of balloon flights. In particular, the obtained performances of prototype systems for rough and fine tracking will be illustrated. Both these systems constitute a high precision device ($\leq$ 1 arcsec) for pointing and tracking light telescopes on board stratospheric balloons. We give the details concerning the optical and mechanical layout, as well as the detector and the control system. We demonstrate how such devices, when used at the focal plane of enough large telescopes(2-4m, F/10), may be capable to provide diffraction limited images in the near infrared bands. We have also developed a prototypal single channel photometer *NISBA* (Near Infrared Sky Background at Arctic pole), working in the H band (1.65 µm), able to evaluate, during a high-latitude balloon flight, how OH emission affects the sky background during the arctic night. The laboratory tests and performance on sky are presented and analyzed.

Keywords: Astronomical instrumentation, stratospheric balloons, tracking systems, infrared

1. INTRODUCTION

The excellent conditions present in the high stratosphere (25-35km) especially for near-infrared (NIR) observations allow balloon-borne telescopes to approach space telescopes performances. The stratosphere is very dry, which minimizes atmospheric opacity, an advantage magnified by the low pressure which reduces line broadening. Moreover, the ambient temperature is also low (T=220-240 K), which minimizes thermal emission from the telescope optics and reduces atmospheric radiance. Finally, low turbulence reduces seeing to negligible effects. On the other hand, observations from aircraft suffer from image degradation due to air turbulence and vibration. Observations from space require the development of new technological aspects and are intrinsically very expensive. Balloon flights, although offer an observing time limited by the flight duration, provide conditions close approaching to space ones, but at much more reasonable costs. Well known problems related to the realization of a large balloon-borne NIR telescope mainly concern: (*i*) the use of lightweight mirrors to keep the total weight of the experiment below 1000-1500 kg; (*ii*) the realization of long duration balloons which can remain at 30 km of altitude for several weeks;(*iii*) accurate attitude control to keep pointing stability at the level of 0.05 arcsec; (*iv*) recovery of the payload and telescope for next flights. New positive results are being obtained in some of the technological aspects mentioned above. SiC mirrors up to 1m have been realized with optical quality and larger mirrors are coming soon [6,7]: their surface weight is of the order of 20 kg/m^2 implying, for a 4m diameter mirror, a total weight of about only 400 kg. Long duration flights up to 40 days have been recently obtained in cosmic ray experiments in Antarctica (CREAM experiment [2]). Development of Adaptive Optics (AO) facilities for large ground-based telescopes allows the realization of innovative lightweight optical systems able also to keep an excellent pointing stability. Finally, airbag technology for the soft landing of experiments on the surface of external planets could allow the realization of a safe recovery of the telescope(eg. http://www.lockheedmartin.com/). Remarkably, many of these issues were afforded (and successfully solved) more than 40 years ago (Stratoscope II Balloon-Borne Telescope [4, 8]). Aiming at finding low-cost solutions of some critical aspects typical of the balloon flights, we have undertaken a series of technological problems. In the present paper we present our preliminary results obtained in the framework of the attitude control to keep pointing stability within acceptable levels. This paper is structured as follows: in Sect.2 the modalities of the gondola oscillations are described, while in Sections 3, 4 and 5 the adopted solutions for rough and fine tracking, are presented, respectively. Sections 6 is dedicated to describe both the opto-mechanical characteristic and the performances of a prototypal IR photometer (dubbed *NISBA*) we have developed for a test balloon flight.

2. STRATOSPHERIC GONDOLA OSCILLATIONS

The correct pointing of a telescope on board a stratospheric balloon is mainly hampered by the payload pendular motion which is essentially due to the residual atmospheric turbulence still present at a stratospheric altitude. The most evident effect of such turbulence is to provoke balloon rotation and, to a lesser extent, to make the roll and pitch angles oscillating. Usually, azimuth rotation is reduced by some decoupling devices located in between the gondola and its suspension cable: they aim at both nulling the torsion of this latter and keeping the payload in a state of rest. Residual oscillations typical of this state are characterized by a less than 1 arcmin amplitude and by a sub-Hz frequency. Roll and pitch angles present oscillations of small amplitude at low frequency, as well; typical values of the balloon and gondola frequencies are 0.1 and 1 Hz, respectively [5]: our fine tracking system aims at correcting such residual oscillations. It is based on three different control levels:

1. A dedicated Inertial Measurement Unit (IMU), based on gyroscopes, computes instant angular velocities and provides the telescope servo-motors with correction signals to keep its pointing well within 1 arcmin.
2. A star sensor camera computes the absolute pointing with enough accuracy(< 1 arcmin); provides the signal for IMU long-term drift compensation, and improves the telescope coarse pointing until the target field is reached.
3. Once on target, with the attitude pre-stabilized by using the IMU signals a fine optical tracking system shall lock on one (or more) reference star(s) with a 0.1 pixel accuracy by using a predictive algorithm. Field rotation of the focal plane instrumentation is supposed to be provided by a standard mechanical system (if needed, gyroscope assisted).

3. THE INERTIAL MEASUREMENT UNIT

The Inertial Measurement Unit provides angular velocities and acceleration measurements for the basic attitude calculation. These data are useful to stabilize the telescope axes within a fraction of arcmin by using the fine-slewing motors. Gyros measurements are affected by Gaussian noise and bias error. The bias error grows with time with non known law, so we have to determine the instant bias due to other sensors. We will use accelerometer measurements to do this, in fact we can obtain pitch and roll angle by comparing them with the *g* vector. For the 2d problem, with a double-pendulum model of the payload-balloon system, a Kalman filter provides the attitude angles together with the gyro's biases having gyros and accelerometers measurements as input. For the real case with a much more complex dynamical model (triple-3d-pendulum) we used other digital filters like Butterworth and Chebyshev [12] that do not need to know the dynamical model of the phenomena. Then, to estimate the attitude, we compared it with the angles provided by the accelerometers to calculate the bias error under the approximation of negligible apparent forces (this approximation is valid on a balloon gondola where centripetal forces are very low with respect to *g*-force).

GYROSCOPES		ACCELEROMETERS	
Dynamical Range [°/s]	±75	Dynamical Range [g]	±5
Initial Sensitivity [°/s/LSB]	0.003125	Initial Sensitivity [mg/LSB]	0.25
Output noise [°/s rms]	X,Y axes 0.75 Z axis 0.25	Output noise [mg rms]	3.3

Table 1

We used a Micro Electro-Mechanical Systems (MEMS) sensor for our tests: it is a low cost device we used to test algorithms and give an idea of the required performances for our scopes. The specific model we used is an Analog Devices ADIS16385 with its characteristic listed in the Table 1. As a first analysis, we modeled the sensor to make a numeric simulation of the attitude determination performances. Simulations were made using a numerical model of the IMU performing a simple sinusoidal motion. It was imposed to oscillates with amplitude of 2÷3 arc-min and period of 5 s. The results gave a good attitude evaluation, with error < 1 arcmin. Then we put the real IMU onto an oscillating table in a strap-down configuration. Tests showed that we can measure the platform attitude with an error of about 1 arcmin slightly above the requirements for the lock of the fine tracking system(< 1 arcmin). Moreover the typical MEMS bias error allows a good tracking for only a few minutes. These results provide essential information about the required performances of the sensors for the IMU. For our scopes we should have to use a Fiber Optical Gyroscope (FOG) yielding performances one hundred better than MEMS gyroscopes. With these devices we could achieve attitude pre-stabilization well below 1 arcmin for a long time because of their low drift.

4. THE STAR SENSOR

The star sensor provides the equatorial coordinates of the observed field center, in order to recognize the sky area pointed by the telescope. It uses a CMOS camera whose field of view (about 50 deg) is enough to allow us observing, at least, three bright stars (visual magnitude < 3). An IDL dedicated software, running in real time on a laptop, provides the astrometric solution in equatorial coordinates of the field center in less than 100 ms after the last image is acquired.

4.1 Recognition of the stellar field

In order to recognize the stellar field the following procedure is adopted. From any image, the coordinates (in pixels) of the bright stars are derived, and the angular distances between them are computed (see Figure 1) as a function of the parameters associated to the optical system. The software procedure is able to identify stars in any sky image by comparing their angular distances with those ones given by a stellar catalog, assuming that the obtained images are gnomonic projection of the celestial sphere. Such a comparison is done through the relationship:

$$cos(ang.distance) = sin(\delta 1) sin(\delta 2) + cos(\delta 1) cos(\delta 2) cos(\alpha 1 - \alpha 2) \qquad (1)$$

where $\alpha 1, \alpha 2, \delta 1, \delta 2$ are the Right Ascension (RA) and the Declination (Dec) of a couple of stars. In this way, the three stars selected for providing the astrometric solution are identified. The center field RA and Dec are found by means of a numerical and analytical solution, respectively.

4.2 Test with real images

A star sensor prototype has been assembled by using a 1.3 Megapixel CMOS detector (e2v EOS-AN012) with a pixel size of 5.3×5.3 µm, equipped with a 12mm F/3.0 objective (Sunsex DSL901C). The field of view corresponds to about 50×50 deg^2 with an angular resolution of 3 arcmin/pixel and our calibration tests did not evidenced any significant optical distortion. Figure 2 shows a pair of the images obtained with the prototype CMOS board camera during real sky tests. Results are fully compatible with camera resolution and indicate an absolute pointing error below one pixel (i.e. < 3 arcmin) when comparing the real time data with standard astronomical off-line software packages (i.e. sextractor).

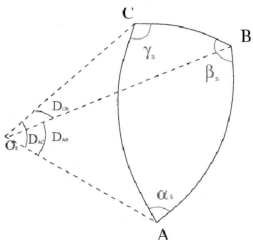

Fig. 1. Sketch of three generic stars named A,B, and C on the celestial sphere. The corresponding angles are α_s, β_s, and γ_s, while their relative angular distances are DAB, DAC, and DCB. These latter are needed for the stars identification and for the comparison with the stellar catalog. OS indicates the center of the celestial sphere.

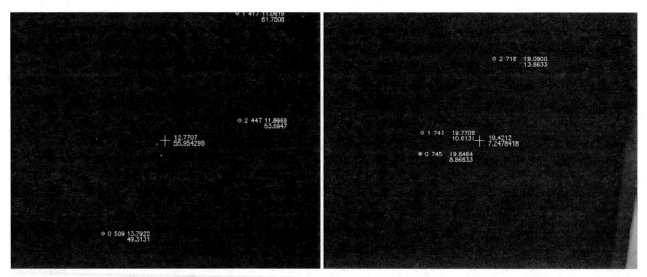

Fig. 2. Images of the Ursa Mayor constellation (left panel) and Aquila (right panel) obtained with our camera (see text). The stars selected to obtain the astrometric solution are indicated with both a progressive number (0, 1, 2,..) assigned by our software and the corresponding catalog (J2000 NEW FK5) identification. RA (in hours) and Dec (in degrees) complete the label of any recognized star. A cross indicates the field center (in equatorial coordinates).

5. THE STAR TRACKER

The star tracker performs a real time correction of the image smearing due to the payload oscillations. At variance with other tracking systems based on PID (Proportional Integral Derivative) correction algorithms, the one described here presents a novelty, namely it is conceptually based on predicting the position that the reference source centroid is expected to have. This is accomplished through the mathematical analysis of a stationary time series.

5.1 Predicting the baloon trajectory

For predicting the trajectory, an auto-regressive model is used: it allows us to write the predicted value x_t of a time series $X = [x_0, x_1, x_2, ..., x_{t-1}]$ as a linear combination of p values, already measured, to which a random error z_t is added:

$$xt = a1xt-1 + a2xt-2 + ... + apxt-p + zt \qquad (2)$$

It works in a way similar to a model of linear regression, in which, however, x_t is not defined by independent variables, but by its already measured values. The coefficients $a_1, a_2, ..., a_p$ are computed in order to minimize the uncorrelated term zt. Noticeably, the order p of the auto-regressive process that best represents a given time series, is hard to be derived: a viable approach consists in increasing progressively the order of the process until the sum of the residuals reaches a required value [3]. In our case, the time series representing an oscillating gondola is well described by an auto-regressive process of order 4.

5.2 Testing the predictive software

The experimental test bench of the fine tracking predictive software simply consists in a 1.3 Mpixels CMOS camera MAGZERO MZ-5m, with a lens of 100mm focal length (corresponding to a plate scale of 10 arcsec/pixel). Moreover, to simulate the payload pendular motion, the camera has been stiffly mounted over an oscillating optical bench (at about 2 Hz). The reference source is a light spot from an optical fiber. About 100 images have been taken (each 10 msec integrated, repeated every 50 msec). The predictive software analyzes these images and provides the predicted position to be compared with the real one. Thus, the goodness of our prediction can be evaluated. The final star image FWHM resulted just 10% worse than that corresponding to individual images (see Figure 3). The plots in Figure 4 depicts real and predicted trajectories, together with their difference. Such difference presents a standard deviation of 0.18 and 0.05 pixel in Y and X directions, respectively or a reduction of a factor greater than 100 of the peak to valley amplitude of the oscillations. It is worthwhile noting that our processing is accomplished in real time (i.e. results are delivered in a time much shorter than the sampling time).

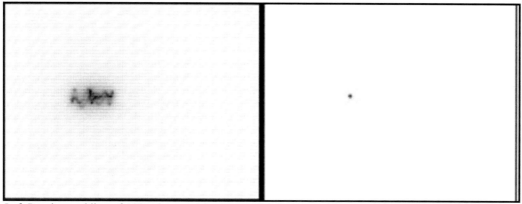

Fig. 3. Left Panel: co-adding of 100 subsequent images, each taken every 50 msec, (10msec exposure) highlights the real trajectory of the oscillating bench. Right Panel: re-centered co-adding operated by the predictive software. Evidently, the effects of the bench oscillations are negligible here.

5.3 Closed Loop Experimental Hardware for predictive fine tracking

The capability of our system to work in a closed loop configuration has been tested by exploiting the experimental layout shown in Figures 5 and 6 where a movable mirror, commanded trough the predictive algorithm, is trying to null the bench oscillations. To verify our predicting model, just one axis has been corrected by using only one camera. Indeed, two cameras should be used, one dedicated to the acquisition of the final image, and the other one as a sensor for the tracking system. Instead, our simplified configuration forced us to feed the predictive algorithm with the re-constructed position of the star, and then to accept a larger error induced by the backlash present in the steerable mirror. Figure 7 shows some test results: while the re-centering via software can be considered as an excellent performance; the results obtained with the movable mirror are not at the same level of goodness since the used mechanics is not precise enough. However, we note that oscillations larger than 300 pixels were reduced down to about 30 pixels, which represents the limit imposed by the backlash of the used hardware. To get our final goal (to have error tracking less than 1 pixel) we plan two future actions: (i) to improve the quality of the steering mirror; (ii) to employ a two camera system (imager and star tracker) to avoid any systematic error. The performances offered by the currently available tip-tilt optical modules used in adaptive optics, when working in closed loop, are more than enough for our scope.

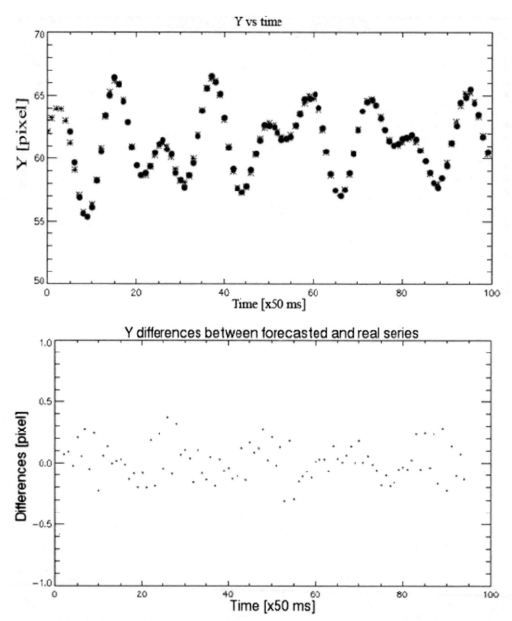

Fig. 4. Upper Panel: Y coordinate (of the source centroid) corresponding to the 100 frames depicted in Figure 3 Left Panel: asterisks represent the positions of the real trajectory and dots positions predicted by means of our algorithm. Lower Panel: Residuals (in pixel) between the two trajectories. It is worthwhile noticing the performance of the predictive algorithm; even when working on complex trajectories it is converging after a few computational steps.

5.4 Application to real star fields

The experimental tests of our predictive algorithm have demonstrated it works fine. A new tip-tilt system of higher performance represents a next step of implementation. A rough analysis of the S/N ratio of a stellar source at the focal plane of a class 2m telescope, shows that a centroid of 0.1 arcsec is easily reachable for a 16.5 mag star with an integration of just 0.03 sec. Hence, the fine tracking of the slow gondola oscillations is possible on a wide sky area up to high galactic latitudes, providing an available field of at least 50 arcmin^2. In fact, the model of Bahcall & Soneira [1] indicates a density of 220 stars / deg^2 brighter than 15.5 mag at a latitude of 90 deg (one of the sky area less populated of stars). Therefore, in a field of 7×7 arcmin^2, about 3 stars are always present and suitable for the balloon fine tracking.

At the light of the presented considerations, the fine tracking topic appears as a solved problem towards the flight of an optical-IR telescope on board of a stratospheric balloon.

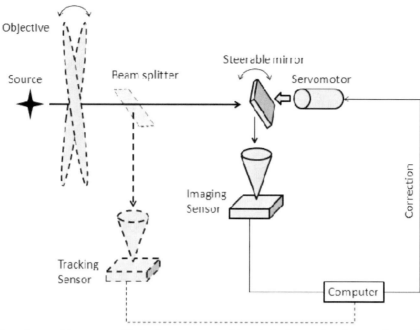

Fig. 5. Layout of our fine tracking system. Dashed components, although fundamental to improve its performance, have not been used during the original laboratory tests.

Fig. 6. Left: the arrow indicates the star tracker device on our oscillating optical bench. Right: the correction system composed by the CMOS camera (A), the movable mirror (B), and the Faulhaber servo-motor (C).

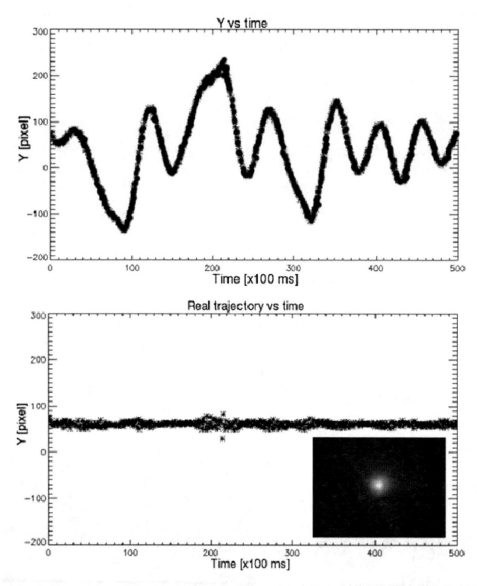

Fig. 7. Upper Panel: Y coordinate of the source centroid images vs. time with the servo mirror in open loop configuration (i.e. servo OFF) Bottom Panel: asterisks represent the positions of the real trajectory when the servo is working in closed loop using the predictive software. Bottom right: the final image of the oscillating point like source. The FWHM is about 10 times the original one.

6. NISBA: THE H BAND WIDE FIELD PHOTOMETER

NISBA (Near-Infrared Sky Background at Arctic pole) is a single channel photometer working in the near-IR (H band) and capable of covering a very wide field of view (Figure 7). It is built to be used both as a piggy back experiment on board of stratospheric balloon and as a ground probe to measure the day and night sky background. The peculiar thermal shielding of NISBA, showed in Figures 8, was developed for the cold stratosphere (200K), thus making it well suited to be used also on polar regions during the arctic night.

6.1 Technical description

NISBA is essentially based on an Hamamatsu G5851-23 1-D photodiode: a semiconductor detector with a good linear response with respect to the light flux. Its active area (Ø=3mm) can be thermo-electrically cooled down to -20 °C for lab tests; during the balloon flight, it shall be passive cooled due to a cooling strap connected to the cold balloon gondola. The lenses of the front optics (Figure 9 Left) conjugates onto the detector sensitive area a circular field of view with a diameter of 10 deg equivalent to a sky area of about 80 deg^2. Signals well below the dark current noise level are detected by means of a lock-in amplifier combined with an optical chopper driven by a brushless motor.

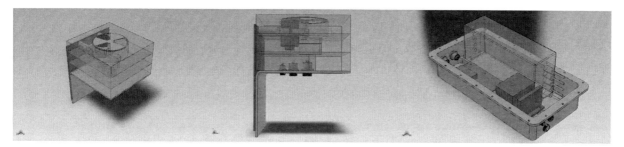

Fig. 7. Left: Optical head CAD rendering. Center: Optical head side view. Right: Electronic box open view.

The chopper is directly connected to the motor shaft allowing a simple readout of its position trough the motor's Hall sensors. The special firmware loaded inside the motor driver electronic processes the Hall sensors signal to generate a square wave phased to the shaft rotation with a frequency that is 3 times the shaft revolution frequency. The chopper has 3 slots and 3 blanks doing a light modulation well matching with the previously described signal that is routed to the lock-in reference input.

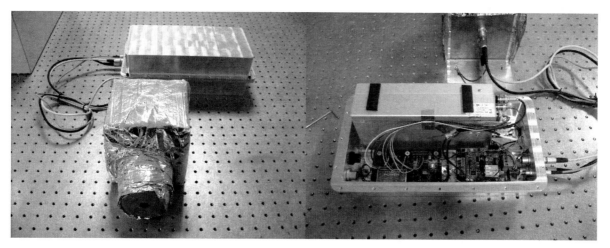

Fig. 8. Left: The NISBA photometer optical head and its control electronic box. Right: The control electronic

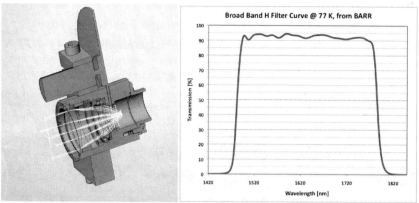

Fig. 9. Left: The opto-mechanical layout composed by the lens holder (left) and the chopper (right).
Right: Transmission Curve of the Broad Band H Filter from BARR.

The optical bandpass is selected by means of an interference broad band H filter from BARR, whose spectral response is plotted in Figure 9 *right*. To eliminate the largest part of the noise inside the detector electrical band-pass, the read-out is done with dual-phase lock-in amplifier that receives the signal from the detector output and the reference signal from the chopper driver. The photodiode is connected in direct current mode trough a trans-conductance preamplifier input provided by the lock-in electronics. A modulation frequency in the range 40-60 Hz (1000 rpm at the motor shaft) is selected and an analog high-pass filter is used with a turn-on of about 20 Hz to reject the worst components of the 1/f noise. A final low-pass 6dB/octave with a time constant of about 3s, is used to smooth the DC level of the reconstructed photocurrent. Such a device provides an effective detectivity of $D_{eff} \ll 10^{-11}$ W Hz^{-1}, well compatible with the optical signals to be detected. Power is provided by a battery package and a DC/DC converter. A PIC microcontroller with a four channel A/D converter digitizes both Lock In signals (in-phase and quadrature) together with the temperatures of the detector and of the NISBA's housing with a resolution of 12 bits. Data are finally transmitted via a RS232 line. This topology yields an easy interface the typical telemetry channels, used on balloon experiments, or to a local PC. To acquire data from the RS 232 both a control software and an acquisition system have been implemented in Lab-VIEW language that provides a Graphical User Interface to NISBA. Data are acquired and plotted in real time while are saved in a text file for a following analysis.

6.2 Ground tests and calibrations

Aiming at characterizing the H-band photometer of the NISBA (Near-Infrared Sky Background at Arctic pole), simultaneous observations of the same portion of sky have been obtained with both our photometer and the IR camera NICS installed at the Italian National Telescope (TNG) at La Palma (Canary Islands - Spain), in the period March 15-19 2010. Such modality has allowed us to obtain a quantitative relationship between the counts from the H-photometer and the calibrated sky emission in the same band.
Beside the calibration purpose, we also aiming at monitoring the sky emission, therefore NISBA observation procedure consists in pointing the same sky region for several hours during nigh time with a sampling time of 1.0 sec
NISBA data quality indicates that our device is well suitable for effectively monitoring the sky emission. In the best sky conditions (see Fig. 10), the monitored emission appears extraordinarily constant for many hours at a level of about Mag14.

This calibration when applied to the minimal detectable signal from NISBA electronics of 0.05nA shows a limiting magnitude of the sky background measurements of Mag H = 15.3 / arcsec^2. Such a level of emission is due to the well known OH line emission, which is the responsible for the sky background at these wavelengths. OH emission is also expected to vary both spatially and temporally by about 20% on time scales of hours [10]. The long term variation (of about 0.2 mag) detected by NISBA (Fig. 10) is due just to OH fluctuations. Moreover, we can rule out any stability problem related to our device that could affect substantially the measurements. To that scope, we have already presented the results of the output stability as a function of a drifting temperature from +20°C to - 30°C [9]. We concluded that the signal output (detector+lock-in) in that wide ($\Delta T = 50°$) temperature range, presents a small constant drift whose amplitude is less than 10% of the found OH fluctuation.

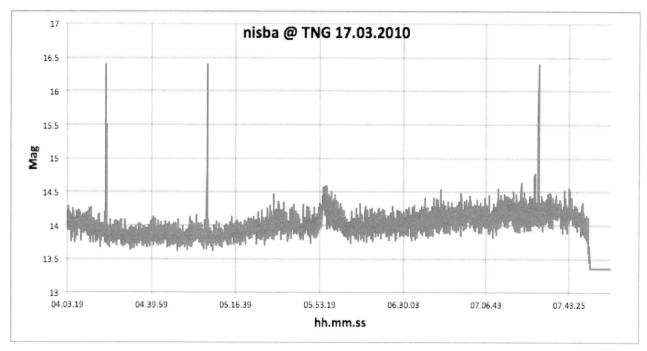

Fig. 10. NISBA acquisition at TNG on 17 March 2010. Spikes at about 16.5 mag are due to the commanded shuttering of the detector to verify the dark current level.

REFERENCES

[1] - Bahcall, J.N. & Soneira, R.M. 1980 ApJSS 44, 73
[2] - Beatty, J.J. et al. 2003 Proc. SPIE 4858, 248
[3] - Chatfield, C. 1995 The Analysis of Time Series - An Introduction, 5th Edition, Chapman & Hall/CRC
[4] - Danielson, R.E., Tomasko, M.G., & Savage, B.D. 1972 ApJ 178, 887
[5] - Fixsen, D.J. et al 1996 ApJ 470, 63
[6] - Kaneda, H. et al. 2007 Proc. SPIE 6666, 666607
[7] - Webb, K. 2007 Proc. SPIE 6666, 666606
[8] - Wieder, E.A. 1969 IEEE Trans. of Aerospace & Electr. Systems, vol.5, n.2, 330
[9] - Centrone, M., Giallongo, E., Lorenzetti, D., Pedichini, F., Pirrotta, S., & Vitali, F. 2010, OAR/10/IR1 Technical Report
[10] - Ramsay, S.K., Mountain, C.M., & Geballe, T.R. 1992 MNRAS, 259, 751
[11] - Ricci, M., Pedichini, F., & Lorenzetti, D. 2011, arXiv:1104.2181
[12] - "Numerical Recipes", pp. 559, 2002 Cambridge University Press.

SOFIA in Operation
Telescope performance during the Early Science flights

Hans J. Kärcher*[a], Jörg Wagner[b], Alfred Krabbe[b], Ulrich Lampater[c], Thomas Keilig[b], Jürgen Wolf[d]
[a]MT Mechatronics, Weberstrasse 21, 55130 Mainz, Germany,
[b]Deutsches SOFIA Institut, Universität Stuttgart, IRS, Pfaffenwaldring 29, 70569 Stuttgart, Germany,
[c]Deutsches SOFIA Institut, NASA Dryden Flight Research Center, Mail Stop DAOF S231, P.O. Box 273, Edwards CA 93523, USA,
[d]Deutsches SOFIA Institut, NASA Ames Research Center, MS211-3, Moffett Field, CA 94035, USA

ABSTRACT

The Stratospheric Observatory for Infrared Astronomy SOFIA started in December 2010 with the first series of science flights, and has successfully completed about 38 science missions until fall 2011. The science instruments flown included HIPO, FORCAST, GREAT and FLITECAM. Beside their scientific results (see related papers in these proceedings) the flights delivered an extensive data base which is now used for the telescope performance characterization and the operational optimization of the telescope in its unique environment. In this progress report we summarize recent achievements of the observatory as well as the status of the telescope and give an update of the SOFIA pointing system completed by intended future pointing optimization activities.

Keywords: Telescope, infrared, air-borne, astronomy, SOFIA, pointing

6. INTRODUCTION

SOFIA, the Stratospheric Observatory For Infrared Astronomy (Fig. 1) is a joint 80/20 project between NASA, USA and DLR (German Aerospace Center), Germany with the goal of opening up the entire infrared waveband and also adjacent wavebands for regular astronomical investigations. Since most of the infrared wavelengths (1 – 1000 microns) are not accessible from ground and since the life time of satellite based telescopes like that of HERSCHEL is limited, large territories in this spectral range still remain to be scientifically explored. Using an aircraft based telescope, SOFIA is therefore laid out as a full and long-time observatory. The telescope has been developed by a German consortium under contract with DLR, while the aircraft and all other related subsystems as well as the SOFIA Science Mission Operations (SMO) Center have been developed by the Universities Space Research Association (USRA) for NASA.

Past status reports on the SOFIA project and on its telescope as well as a detailed discussion of the requirements for such a observatory can be found in the proceedings of SPIE 3668, 4014, 4015, 4486, 4857, 5495, 7012, 7733 (Krabbe 2000, 2001, 2003, 2004), (Kärcher 2000, 2002, 2008), (Young et al 2010). Several examples of the first scientific results are provided within these proceedings as well as details on present activities for improving the performance of SOFIA.

Figure 1: SOFIA during its first open door flight Dec 18th, 2009

This paper presents a progress report and an overview describing recent achievements of SOFIA, the status of the telescope and its pointing system, as well as possible further pointing optimization activities. For this, SOFIA's flight program accomplished to date, which includes a first deployment to Germany, is summarized below, followed by a short, exemplary insight into the results of the first science flights as well as the telescope pointing status achieved. The article is completed by an outlook on further performance and handling optimization activities.

7. DEPLOYMENT TO GERMANY

A significant feature of SOFIA is its ability to not only operate from its home airport Palmdale, CA, but to also take off from other locations, e.g. for observing trips to the southern hemisphere. In addition, SOFIA also includes an important public outreach and education program. To demonstrate therefore SOFIA's deployment capabilities and to inform the German and US public about SOFIA, the observatory visited Cologne and Stuttgart in Germany as well as Washington, DC between September 16 and 23, 2011. During these days, about 7000 visitors became acquainted with the observatory and several high-ranking politicians, including German federal ministers, were given the opportunity to inform themselves about the present status of the project. The visit in Stuttgart (Fig. 2) was accompanied by an exhibition at the airport, which not only provided an insight into the astronomic purpose of SOFIA but also explained the program of including even schools into the science of the observatory and educating teachers in particular.

In addition, it was possible to demonstrate that the complex logistics of a deployment can be mastered by the observatory team. The flights from Palmdale to Cologne and from Stuttgart to Washington were also utilized for science observations.

Figure 2: SOFIA during its deployment in September 2011 at the Stuttgart airport

8. SOFIA FLIGHT PROGRAM ACCOMPLISHED TO DATE

Due to the substantial amount of modifications of SOFIA's airframe during the installation of the telescope, SOFIA required a new certification of airworthiness. Regaining the flight certification was the main purpose of the initial flight program. Meanwhile the focus of the flights has shifted significantly from technical issues to astronomic missions. However, engineering aspects still play a decisive role in the current flight program.

Compared, e.g., to the certification of commercial aircrafts, the technical flights or test flights of SOFIA had and still have three extra key aspects. The first one is to get experience with the handling of the (new) telescope, to characterize its performance, and to improve its operation. The second aspect is gaining experience with the astronomical instrument commissioning. The third aspect was the execution of the Early Science program, which included Short Science, Basic Science, and German Demonstration Science. Its purpose was to demonstrate to the American and German science community SOFIA's capability of obtaining excellent scientific astronomical results. Together with these first astronomic missions the flight program to date completed 89 flights. The flights of particular importance are summarized here:

04/26/2007:	First flight after installing the telescope.
May – Nov. 2007	7 closed door envelope expansion flights: - proofing the strength of the airframe, - flutter tests, - stepwise increasing the admissible aircraft speed and altitude.
12/19/2007	First flight with activated telescope.
Jan. 2008 – Aug. 2009	*Downtime for aircraft and telescope modifications (e.g. installing the door drive).*
Dec. 2009 – Nov. 2010	39 open door envelope expansion flights: - stepwise opening of the door, - 12/18/2009 first flight with completely open door, - stepwise increasing speed (up to Ma = 0.92) and altitude with open door, - stepwise increasing the telescope elevation, - proof of safe aircraft handling with open door (including open door landings), - flutter tests, - testing the effects of misconfigurations (e.g. wrong synchronisation between door and telescope, flights with engine failures etc.).
05/26/2010	First light flight (instrument for observation: FORCAST).
Dec. 2010 April 2011	3 Short Science flights (instrument: FORCAST), 3 Short Science flights (instrument: GREAT).
May 2011 – June 2011	10 Basic Science flights (instrument: FORCAST) incl. German Demonstration Flight hours.
06/22/2011	Pluto occultation flight (instruments: HIPO and Fast Diagnostic Camera (FDC))
July 2011	7 Basic Science flights (instrument: GREAT) including German Demonstration Flight hours.
Sept. 2011	Deployment to Cologne, Stuttgart (Germany) and Washington, DC.
Sept. – Nov. 2011 Oct. 2011	4 Basic Science flights (instrument: GREAT) including German Demonstration Flight hours, 4 characterization and integration flights (instrument: HIPO and FLITECAM).
Dec. 2011	3 characterization and integration flights (instruments: HIPO and FDC).
Jan. 2012 – July 2012	*Downtime for aircraft and telescope modifications (e.g. installation of new avionics in the cockpit).*

The four observation instruments FORCAST, GREAT, HIPO, and FLITECAM being mentioned in this summary are explained in more detail in the next section. Finally, it should be noted that the Full Operational Capability (FOC) milestone of the observatory currently is expected for about Summer 2013. In September 2014 the Readiness for Sustained Science Observations (RSSO) status is supposed to follow.

9. SOFIAS FIRST GENERATION OF SCIENCE INSTRUMENTS

One of SOFIA's strength is its spectroscopic sensitivity, which competes well with other infrared space observatories, in particular with HERSCHEL's high-resolution heterodyne spectrometer HIFI. Due to its more modern design and its multiplex advantage, SOFIA's Far-Infrared-Field-Imaging-Line Spectrometer (FIFI-LS) is only about a factor of four less sensitive compared to PACS onboard HERSCHEL if used in the line-mapping mode. SOFIA has also its own edge in particular below 55 µm, which is spectroscopically not accessible by HERSCHEL or any other current mission. Compared with other infrared observatories (Figure 3), SOFIA also provides excellent angular resolution, which will excel between 10 and 80µm.

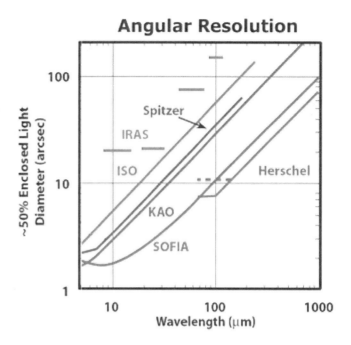

Figure 3: SOFIA in comparison with other infrared observatories

The spectral coverage and spectral resolution of the first generation of SOFIA instruments is shown in Fig. 4. Four camera systems HIPO, FLITECAM, FORCAST, and HAWC fully cover the wavelength region between the visible and the sub-mm wavebands employing very different techniques and detectors. Since these camera systems directly benefit from a diffraction limited imaging performance of the telescope, HIPO, FLITECAM, as well as FORCAST will deliver even higher angular resolution data as the telescope pointing jitter is minimized further (see Section 5).

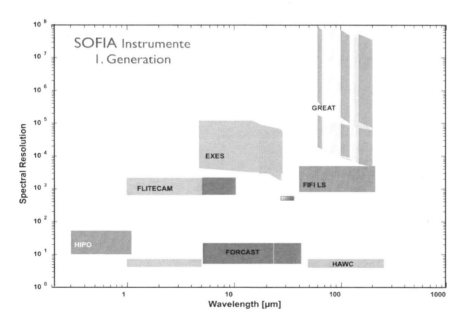

Figure 4: The first generation of SOFIA instruments

HIPO (Primary Investigator (PI) E. Dunham, Lowell Obs.) and FLITECAM (PI I. McLean, UCLA) can be mounted together on the telescope instrument flange (Figure 5) shows the situation on the instrument simulator at NASA Dryden). This allows for simultaneous observations in the infrared between 1.25 µm and 3.6 µm (FLITECAM) and in the optical between 0.3 µm and 1.0 µm (HIPO). Since simultaneous observations guarantee obtaining data under the identical and coherent conditions, such data is excellently suited for telescope alignment and characterization under flight conditions.

Figure 5: HIPO and FLITECAM on the telescope simulator at NASA Dryden

FORECAST (PI T. Herter, Cornell) is a wide-field camera for continuum and narrowband imaging in the infrared from 5 – 40µm. FORECAST has very successfully flown during 2011 (Fig. 6) and first results have meanwhile been published in a special edition of ApJ (749, L17 to 24, 2012) (Young et al 2012).

Figure 6: FORECAST installed on the instrument flange of the telescope

GREAT (PI R. Güsten, MPIfR) is a modular heterodyne instrument, providing a structure that can hold two receivers at a time. Both receivers can be independently operated giving the instrument a large amount of flexibility. On Fig. 7, only the rear of GREAT is visible showing the instrument racks while shadowing off the two receiver dewars. GREAT has very successfully flown during 2011 and the first results have meanwhile been published in a special edition of A&A (vol. 542, 2012) (Güsten et al 2012).

Figure 7: GREAT installed on the instrument flange of the telescope

10. STATUS OF POINTING

The SOFIA pointing system (Figure 8) is a complex inertial stabilization device. The gyros are the inertial equivalent of the main axes encoders of a ground-based telescope and deliver measurement signals of a high sampling rate. The imagers generate complimentary measurements using the absolute position of a tracking star in rather long intervals. The tracking data is used for telescope pointing and for compensating the gyroscope drift. The fine drive torque motors control the telescope's rigid body attitude with a bandwidth of approx. 4.5 Hz moving the 10 tons rotor structure. Disturbances that cannot be controlled by the fine drive are measured by the gyroscopes and forwarded to the much faster secondary mirror tilt-chop-mechanism for image motion compensation (M2 FF). Low frequency aircraft motion leads to a deformation of the telescope's Nasmyth tube, especially during turbulence. Accelerometer based image motion estimation is used to compensate for this effect (QS FBC). Wideband aerodynamic loads excite various natural frequencies of the telescope structure, and some of those mode frequencies have a substantial impact on image motion. Therefore, an active damping system was recently added to reduce the impact of flexible mode deformation on image motion (see Section 6).

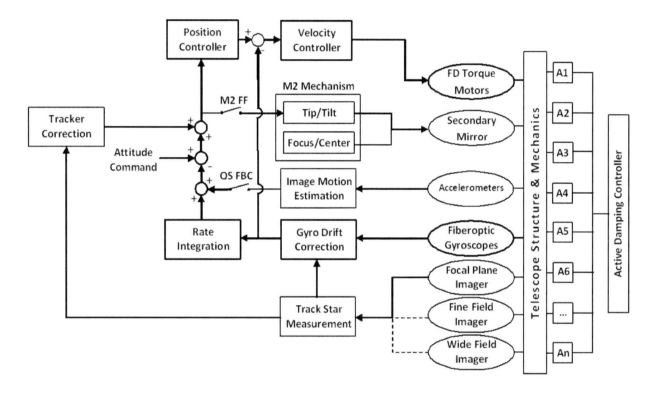

Figure 8: The SOFIA Pointing & Control System

11. JITTER REDUCTION BY ACTIVE MASS DAMPERS

Active mass damping is used to reduce image jitter that is created by dominant flexible modes of the telescope structure. The shape of such modes includes the secondary mirror and spider assembly motion around 90 Hz and the primary mirror motion around 70 Hz. Test flights showed that an active damping system tuned to those modes can reduce primary mirror modes very efficiently as can be seen in the cumulative RMS plot in Figure 9. In contrast to this, the best results for the secondary mirror modes around 90 Hz were achieved by removing the baffle plate, a non-load carrying structure that protects the Nasmyth tube from varying infrared scattered light levels (see Figure 10). It obviously acts as a strong disturbance source for the secondary mirror assembly, and the resulting motion cannot be compensated for by the active mass dampers mounted on the baffle plate. The necessity of the baffle plate is currently under discussion with the different science instrument groups.

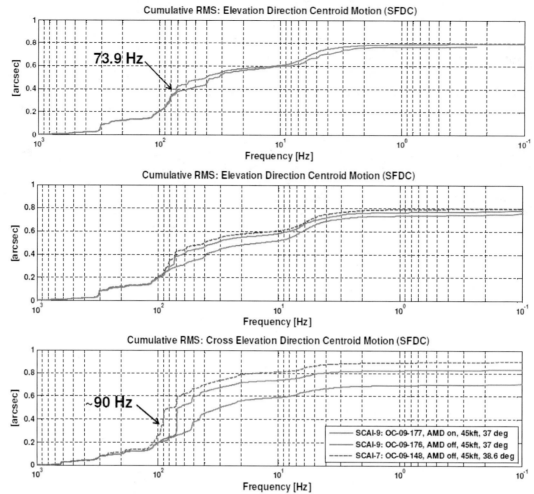

Figure 9: Cumulated pointing errors with (above) and without baffle plate;
OC-09-147 to 177 are test flight segments with height and telescope elevation

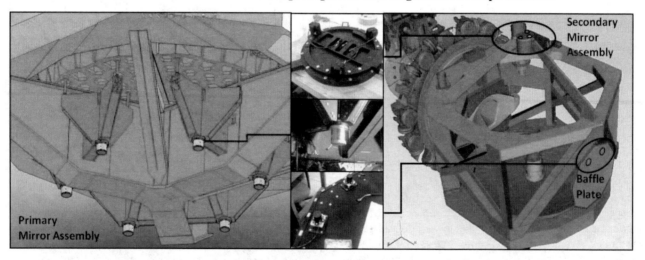

Figure 10: Reaction Mass Actuator Mounting Locations on the Telescope Assembly

The active mass dampers (AMD) and the removal of the baffle plate are two examples of improving the telescope performance of SOFIA. The AMD test system is currently being upgraded to make active damping available for regular science flights. The Baffle Plate will be modified to interrupt the disturbance path into the secondary mirror, and dampers will be added to the spider assembly to improve controllability of the mode shapes in the 90 Hz range.

12. OUTLOOK

The telescope activation and characterization flights to date as well as the first phase of science missions generated a lot of indispensable practical experience and performance data for the observatory. Based on this knowledge, the achievement of the top-level requirements for the pointing stability of the telescope is one of the central tasks of the next few years bringing SOFIA into its regular service. This requires a systemic approach on different levels. Beside the active mass dampers and the removal of the baffle plate for observation instruments that do not require this device, technical measures already known and proposed for the further reduction of pointing errors include but may not be limited to:

- Additional aeroacoustic investigations of the cavity in order to explore the potential of minimizing the turbulent air flow and the aeroacoustic noise. For this, modifying the geometry of the aperture and the cavity as well as using additional noise damping devices are possible options. Necessary aerodynamic simulations can now use a significant amount of new, real flight data.

- Use of an advanced chopper correcting higher vibration frequencies and therefore leading to sharper images.

- The open loop feed forward control for correction of structural eigenmodes will eventually be modified into a closed loop plus a smart control software allowing to detect eigenmodes and to take specific countermeasures.

- Additional improvement of the attitude control of the telescope. This includes also better imagers for tracking as well as other enhanced sensors.

The number of flights per week of SOFIA will be increasing from about two as of now to about four within the next one or two years. In order to achieve this elevated flight frequency, the telescope needs to be "ruggedized" in a way to make it most reliable and to minimize the failure rates. Several areas have been addressed where improvements will be significant in reducing the failure probability or failure effect. An example is an extended spare part stock including complete subsystems. Also, the increase in flight rate requires turn-around times for maintenance activities during the day as short as 8-12 hours. In order to achieve such short intervals, several subsystems of the telescope have been identified to be made maintainable within such short time interval or, alternatively, have to be completely replaced on short notice by a Line Replaceable Unit (LRU) for troubleshooting and repair.

Another issue is the continuous improvement of operation software and observation procedures during flights. The first astronomical missions generated a major quantity of unique experience, which is by now accordingly implemented into such tools. Therefore, the significant increase in SOFIA's effectivity and also efficiency over the last years can be expected to continue.

13. ACKNOWLEDGEMENTS

SOFIA, the "Stratospheric Observatory for Infrared Astronomy" is a joint project of the Deutsches Zentrum für Luft- und Raumfahrt e.V. (DLR; German Aerospace Centre, grant: 50OK0901) and the National Aeronautics and Space Administration (NASA). It is funded on behalf of DLR by the Federal Ministry of Economics and Technology based on legislation by the German Parliament, the state of Baden-Württemberg, and the University of Stuttgart. Scientific operation for Germany is coordinated by the German SOFIA-Institute (DSI) of the Universität Stuttgart, in the USA by the Universities Space Research Association (USRA). The development of the German Instruments is financed by the Max Planck Society (MPG) and the German Research Foundation (DFG).

REFERENCES

1. Krabbe A., "SOFIA telescope", SPIE 4014, 276 (2000)
2. Krabbe A., "SOFIA telescope", SPIE 4486, 71 (2001)
3. Kärcher, H. J. "Airborne environment – a challenge for telescope design", SPIE 4014, 282 (2000)
4. Kärcher, H. J., "The Evolution of the SOFIA Telescope Systems Design – Lessons Learned during Design and Fabrication", SPIE 4857, 257 (2002)
5. Krabbe A., "Becoming reality: the SOFIA telescope", SPIE 4857, 251 (2003)
6. Krabbe A. et al., " Preparing for first light: the SOFIA telescope", SPIE 5495, 251 (2004)
7. Kärcher, H.J. et al; "The SOFIA telescope: preparing for early science", SPIE 7012 (2008)
8. Young, E.T. et al, "SOFIA progress to initial science flights", SPIE 7733 (2010)
9. Güsten, R. et al, "GREAT; early science results", A&A, 542, L1 to L22 (2012)
10. Young, E.T. et al, "Early science with SOFIA, the Stratospheric Observatory for Infrared Astronomy", ApJ, 749, L17 to 24 (2012)

A new backup secondary mirror for SOFIA

Michael Lachenmann[a,b], Martin J. Burgdorf[a,b], Jürgen Wolf[a,b], and Rick Brewster[c]

[a]Deutsches SOFIA Institut, University of Stuttgart,
Pfaffenwaldring 29, 70569 Stuttgart, Germany;
[b]SOFIA Science Center, NASA Ames Research Center,
Mail Stop N211-1, Moffett Field, CA 94035, USA
[c]Orbital Sciences Corp., NASA Ames Research Center,
Mail Stop N232, Moffett Field, CA 94035, USA

ABSTRACT

The telescope of the Stratospheric Observatory for Infrared Astronomy (SOFIA) is a Cassegrain design with a convex, hyperbolic secondary mirror. It is 352 mm in diameter, was made from silicon carbide and weighs only 1.9 kg. As this material is brittle, and the secondary mirror is indispensable to observations with SOFIA, a backup with the same mass and moments of inertia was made of aluminium in 2004. This mirror, however, allows diffraction-limited observations only above 20 µm and it produces double peaked images. In this paper we discuss the requirements for a new backup secondary mirror that can be employed also at near-infrared and even visible wavelengths and describe the most important aspects of the manufacturing process. The starting point of our analysis was a high-precision measurement of the surface properties of the existing aluminium secondary mirror, using the NANOMEFOS technique, which was recently developed by TNO in Delft, the Netherlands. With the exact shape of the mirror as input for a Zemax model we could reproduce the results of actual measurements of its optical performance that had been carried out on SOFIA in 2004. Based on these findings we determined then the specifications to be fulfilled by a new backup secondary mirror in order to meet the requirements on improved optical performance. Finally, we discuss the dynamic deformation of the aluminium mirror during chopping motions.

Keywords: Stratospheric Observatory for Infrared Astronomy, SOFIA, light-weighted hyperbolic aluminium mirror, diamond turning, NANOMEFOS.

1. INTRODUCTION

SOFIA routinely uses a mirror made of Silicon carbide (SiC) that is mounted to the Secondary Mirror Mechanism (SMM) of the SOFIA telescope. This material is brittle, and the mirror sits at the entrance area of the open telescope cavity, putting it at risk from hits by objects, e.g., during an open-door landing. Given the fact that the secondary mirror represents a single-point failure and takes at least half a year to manufacture, an inexpensive backup H/W solution with reduced performance was made by NASA/USRA under the lead of Edwin Erickson in 2003/2004.[1] The emphasis at that time was rather to assure the continuation of flights with SOFIA than to make observations possible with all instruments. Therefore, the requirements for the backup concerned mainly mass and inertia being compatible with the chopper mechanism and an optical quality sufficient for diffraction-limited performance in the far infrared. The CAD model of the aluminium mirror with its mounting interface to the chopper mechanism is shown in Figure 1.

The actual requirement on the optical performance was to be diffraction-limited at wavelengths longer than 20 µm, which would result in a point spread function with 1.7 arcsec full width at half maximum (FWHM). The surface figure was checked with two profile measurements along 60% of the mirror diameter on a 3-axis measuring machine. With this measurement result, the mirror was expected to have a FWHM of 1.26 arcsec.

Further author information: (Send correspondence to M. Lachenmann)
M. Lachenmann: E-mail: lachenmann@dsi.uni-stuttgart.de, Telephone: +1 650 604 4536
Martin J. Burgdorf: E-mail: martin_burgdorf@hotmail.com
Jürgen Wolf: E-mail: jwolf@sofia.usra.edu, Telephone: +1 650 604 2126
Rick Brewster: E-mail: rick.brewster@nasa.gov, Telephone: +1 650 604 1872

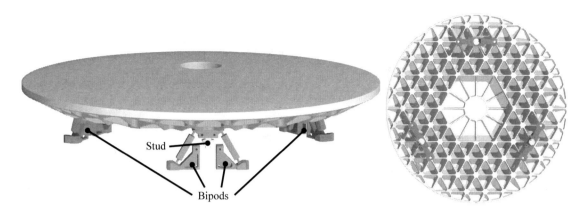

Figure 1. CAD model of the aluminium backup secondary mirror with its light-weighted backside to comply with the mass and moments of inertia of the SiC mirror. It is mounted to the SMM via three triangular shaped bipods (shown in red) which are fixed to the mirror with threaded studs (shown in yellow).

However, during ground tests in Waco in August 2004, it became apparent that the mirror produced double peaked images with peaks 5 arcsec apart. This was considerably worse than expected. It was speculated about several reasons for the limited performance, and since the later installed SiC mirror didn't show any significant aberrations, the reason had to be the aluminium mirror. One plausible explanation was that the surface figure was significantly worse at the outskirt area, where it could not be determined with the used measuring machine. Another was, that since only two profiles were measured, by chance they could have been "good" ones, while the rest of the mirror is severely deformed. Or it could be a structural problem due to gravity or temperature gradients which were not addressed during the development phase.

It seemed, therefore, desirable to carry out a more careful and complete characterisation of the mirror surface in order to find an explanation for the discrepancy between predicted and actual image quality before manufacturing a new one.

2. SURFACE FIGURE AND OPTICAL PERFORMANCE OF CURRENT ALUMINIUM MIRROR

Since it is a convex hyperbolical mirror (conic constant −1.2980, radius of curvature 955.0 mm, diameter 350 mm), the determination of its surface figure is usually either done with a tactile metrology device or in a contactless manner with a Hindle sphere, as it was used in the measurements of the SiC mirror. This, however, is a very elaborate task, because it needs a large and precise optical setup within a temperature-controlled room. As it turned out, there exists now a third possibility called NANOMEFOS (Nanometer Accuracy Non-contact Measurement of Freeform Optical Surfaces),[2] which was developed by TNO and uses a contactless optical probe to scan over the complete mirror surface with an accuracy of a few nanometers.

The NANOMEFOS measurements of the current aluminium mirror were performed in September 2011 at TNO in Delft (see Figure 2). Since one measurement run of the complete surface only takes approximately 10 minutes, ten measurement runs were conducted, using different mounting configurations. The one, that represents the way the mirror is mounted in SOFIA best was measurement #6—with one notable difference: in SOFIA, the mirror is placed upside down, while the NANOMEFOS measurement machine requires the mirror to be mounted face up.

During these surface measurements it was discovered that the bipods used to mount the mirror to the secondary mirror mechanism were partially deformed and pose a major influence on the surface figure error of the mirror. Especially one of the bipods seemed irreversible deformed leaving a clearly visible gap on one side, when it is not fixed to the mounting ring. This leads to a peak-to-valley surface error of more than 8.8 μm (see Figure 3a). Here, the optical surface has a saddle-like shape, which causes predominantly primary astigmatism in the image.

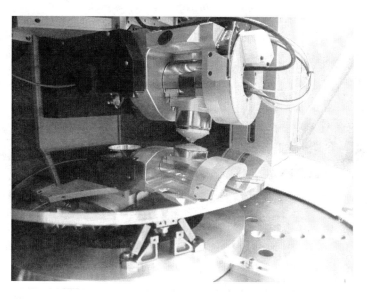

Figure 2. Current aluminium secondary mirror during the NANOMEFOS measurement #6 at TNO.

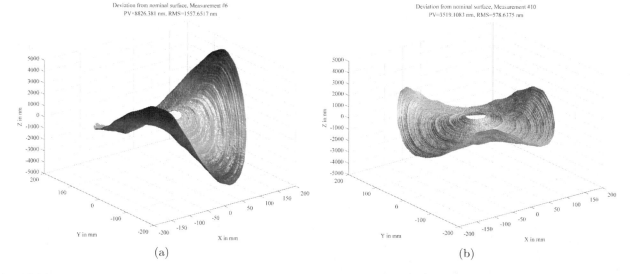

Figure 3. Surface error of measurement #6 ($k = -1.2980$, $r = 955.309$ mm) and measurement #10 ($k = -1.2980$, $r = 955.311$ mm) of the current aluminium mirror. The left one (measurements #6) is dominated by astigmatism, caused by the bipod mount, the right one (measurement #10) was obtained without the bipods, just resting on the studs.

The measurement configuration that showed the least surface error was measurement #10 (see Figure 3b). Here, the mirror rests plainly on its three studs to reduce deformations by the mounting. Since the studs are fixed permanently with the mirror, this measurement represents the inherent figure error contained in the mirror. It still shows a larger error than expected from the profile measurements, but much less than with the bipods. The discrepancy between the old measurements with a profilometer and the new ones can be explained by the limited number of profiles that were measured and the fact, that the deformations are highest at the outer edge of the mirror, where it was not possible to measure with the profilometer, due to size restrictions.

After the NANOMEFOS measurements, the question remained, if the measured surface error could explain the observed double peaked images. Therefore, the obtained secondary mirror surface was included into an optical model containing the SOFIA telescope and the HIPO (High-speed Imaging Photometer for Occultations) instrument. This was used to simulate the image of a point source and compare it to the images taken with HIPO during the ground test in 2004.

The best correlation of the simulations with the observations was achieved with measurement #6 data. This is not very surprising, since it represents the way the mirror is mounted to the telescope best. Figure 4 shows a comparison between these early observed images and the simulated images at different focus positions. To change and adapt to different focal plane positions, the SOFIA telescope can shift its secondary mirror. During the ground tests, the secondary mirror position was changed from -140 µm to $+160$ µm around the best focus position of the HIPO instrument. The simulations using the measurement #6 data show, like the HIPO data, some sort of double peak at best focus position, which becomes more and more separated at negative focus positions. However, the size of the double peak could not be fully reproduced with the simulations, which only show the peaks 3.3 arcsec apart, compared to the 4.9 arcsec of the observational data. A possible explanation could be, that the mirror was mounted with lower torques during the surface measurements. Higher torques could have led to a higher deformation of the mirror and thus a larger double peak distance in the image. Also, the simulations do not include blurring caused by seeing, which results in a sharper spot size compared to the observational data.

Using the measurement #10 surface as secondary mirror, a double peak could not be distinguished within the spot size at best focus, however, it's FWHM is with 2.3 arcsec still larger than specified. So, even with undeformed bipods and a stress-free mirror mount, the telescope would be diffraction-limited only at wavelengths >19 µm.

3. NEW MIRROR

A FWHM of 2.3 arcsec might be acceptable for observations in the visible or very near-infrared spectral range, where shear layer seeing is dominating the point-spread function (PSF),[3] and in the far-infrared, where the PSF is diffraction limited. It will, however, reduce the image quality for instruments in the near-infrared and, to a lesser extent, mid-infrared, e.g., FLITECAM (First Light Infrared Test Experiment Camera) with a range of 2.5 µm – 5 µm. This gap could be closed with a new aluminium backup mirror.

The existing aluminium mirror was diamond-turned in 2003. Since then, lathes have been continually improved and by combining them with surface measurements during the manufacturing process it is nowadays possible to manufacture the same aluminium mirror with a more accurate surface figure in the range of 0.5 µm peak-to-valley. A simulation with the above-mentioned optical model and a downscaled surface from measurement #10 with a surface error of 0.5 µm peak-to-valley showed that a new aluminium mirror would have the potential to be diffraction limited at wavelengths >3.3 µm, which is right in the middle of the FLITECAM spectral range.

Another modification that was made to the existing design was to replace the aluminium alloy from a conventional Al-6061 to RSA-6061. This new aluminium alloy has basically the same material properties in terms of density, Young's modulus, strength, thermal expansion, thermal conductivity, and hardness as the conventional Al-6061 alloy and could therefore replace the material without altering the design of the aluminium mirror. However, this alloy is produced with a special meltspinning process developed by RSP Technology, which reduces the grain size in the material. Thus, a surface roughness of less than 5 nm is feasible instead of the 20 nm present in the current aluminium mirror. This would reduce the total integrated scatter at 500 nm wavelength from 25% to 1.5%. To achieve the specified reflectance of 88% with this alloy, the front surface will be later coated with pure aluminium.

Since the mirror is 76% light-weighted at the back, the actual optical surface is less than 3 mm thick across most of its area. Because of the results of the profilometer measurement, it was assumed, that the diamond turning tool might leave a deeper imprint at the pocket positions of the light-weighted structure, resulting in a quilting effect. The small dimensions of the pockets of the structure would also lead to irregularities in thickness in the same scale. If that would be the case, either the wall thickness had to be increased, or the surface had to be reworked with a low force technique, for example ion-beam figuring, after the diamond turning. Using the results of the NANOMEFOS measurements, it was tried to find a correlation of the small scale or high frequency surface error with the positions of the pockets of the light-weighted structure. This was done by calculating the Zernike representation of the surface figure using the first 36 Zernike terms. They represent the large scale or low frequency surface error. Calculating the residuum between the measured error and the Zernike representation, thus shows only the high frequency part of the surface error. Figure 5 shows the superposition of this residuum

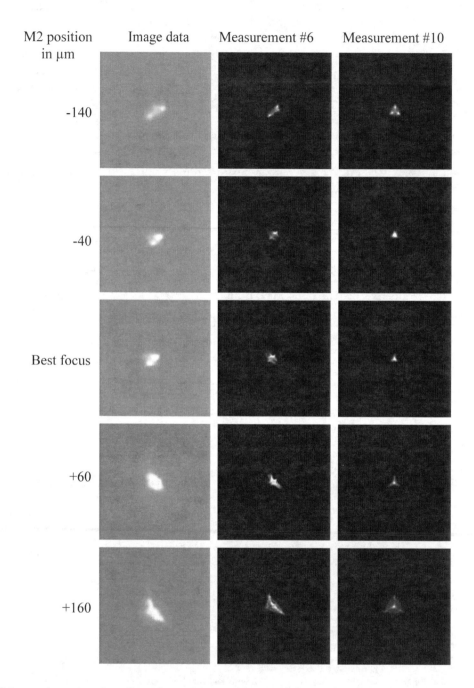

Figure 4. HIPO image data at various Secondary mirror positions and the results of the Zemax simulations. Each box is 200 pixels or 65 arcsec on a side. The left column shows HIPO images acquired during ground-tests in 2004. The middle and right column shows simulated images using the measured surface figure with different mirror mounts.

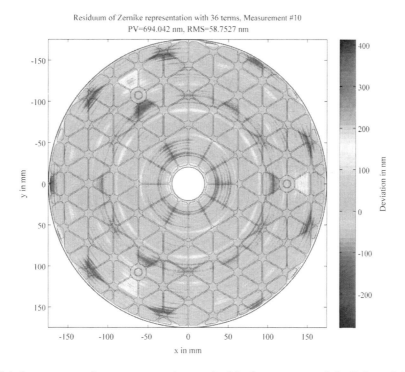

Figure 5. High frequency surface error superimposed with the structure of the light-weighted mirror.

with the structure of the light-weighting pockets on the back of the mirror. It can be seen, that there is only some correlation limited to the pocket positions around the mounting holes of the studs. This means, that the tool force was low enough and the wall thickness was high enough to do not cause quilting.

Therefore, the general construction of the mirror with the light-weighted structure on the backside was considered very good and was basically not altered. Only the mirror diameter was slightly increased by 2 mm to match exactly the as-built dimensions of the SiC mirror. The design matches pretty good the mass and the moments of inertia of the SiC mirror and were required to lay within a 5% margin to the CAD model. Since, a verification of the moments of inertia by measurement is elaborate, the dimensional tolerances were chosen close enough to assure the inertias. Table 1 lists further requirements.

The gravitational influence on the surface sag was analysed with different telescope orientations using a finite element model of the mirror. It was shown that its static surface deformation, due to its own mass, is in the range of 90 nm to 130 nm depending on the elevation angle of the telescope. This is much smaller than the manufacturing tolerance, but nevertheless, adds to the total surface error.

Finite element modeling and deformation analysis of the aluminium mirror was performed to evaluate the existing design. The analyses determined the potential deformation of the mirror during chopping motions for a chopping motion profile roughly comparable to present operations. The analyses indicate that the present design of the aluminium mirror is sufficiently stiff in and of itself not to excessively deform, given a reasonably well damped Tilt-Chop Mechanism (TCM) motion. However insufficiently damped oscillations of the TCM can drive deformation in the aluminium mirror beyond the presently considered mirror surface machining tolerance levels (0.3 – 1.4 µm).

The analysis for the mirror by itself, without inclusion of additional support components, was first performed for modal behavior. The fundamental bending mode of the mirror when pin constrained at its three mounting boss locations was determined to be 899 Hz (see Figure 6a).

Chop rotation motion was simulated on this model (see Figure 6b) and rigid body motion subtracted. For well damped rigid body oscillation, the present aluminium mirror shows acceptably low deformation. For a sweep of

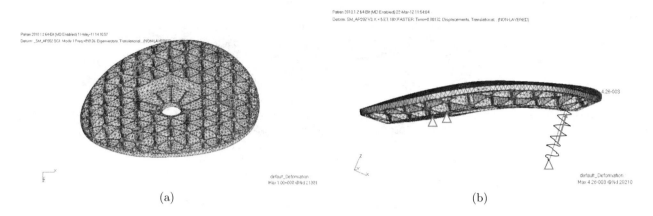

Figure 6. (a): Fundamental bending mode of the aluminium mirror ($f = 899$ Hz); (b): Deformation shape (exaggerated) of the aluminium mirror during rotation accelerations.

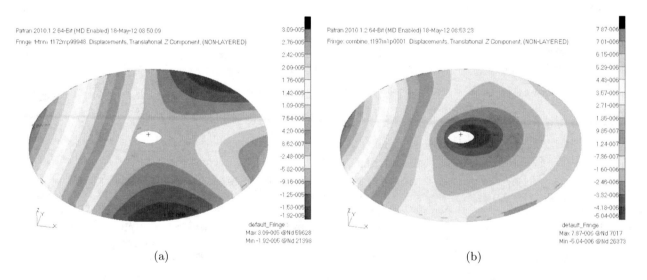

Figure 7. Z deformation of the aluminium secondary mirror undergoing a 33.6 arcmin sweep in 0.014 s; (a): $Z_{\max} = 0.79$ μm at $\Delta t = 0.014$ s; (b): $Z_{\max} = 0.2$ μm at $\Delta t = 0.017$ s.

Table 1. As-built properties of the current aluminium mirror and requirements of the new aluminium backup mirror. The mass and moments of inertias properties include the mounting hardware.

	As-built properties of current aluminium mirror	New aluminium mirror requirements
Diameter	350.2 mm	$352.6^{+0.1}_{-0.3}$ mm
Conic constant	-1.2980	-1.2980 ± 0.0005
Surface radius	955.311 mm	955 ± 1 mm, convex
Reflectance	not verified	more than 88% between 500 nm and 500 µm
Surface roughness	8 – 22 nm rms	less than 5 nm rms
Surface figure error	3.5 µm peak-to-valley 0.58 µm rms	less than 0.5 µm peak-to-valley less than 0.15 µm rms
Mass	1.968 kg	1.975 ± 0.1 kg
Moments of inertia	not verified	$I_{XX} = 0.0153$ kg m^2, $I_{YY} = 0.0153$ kg m^2, $I_{ZZ} = 0.0302$ kg m^2, 5% margin

33.6 arcmin in a time of 0.014 s, the mirror deformation is a maximum of 0.79 µm (see Figure 7a), which reduces to 0.2 µm at 0.017 s (see Figure 7b).

It is to be noted that for these analytical results the actual damping level within the mirror assembly is not determined, and the damping used herein was of value arbitrarily introduced to effect a reasonably well-damped chop motion. In the future, it is planned to measure the dynamic motion of the aluminium mirror during chopping using a mock-up of the chopper and fast CCD cameras.

4. CONCLUSION

The main lesson learned from this investigation is, that the mirror mounting has a major influence on the optical shape of an aluminium mirror. The large image aberrations of the current aluminium mirror are—to a large extent—due to a deformed mirror mounting, which distorted the mirror surface. Also the dynamic motion during chopping is to a fair amount due to the interface and requires further damping.

The design of the current mirror in general is very good and is still applicable to a new mirror with improved optical quality requirements. The gravitational deformation and the dynamic deformation during chopping were simulated and are less than the feasible surface figure error of state-of-the-art diamond turning machines. The diffraction limit of the new mirror would be reached at wavelengths greater than 3.3 µm, which would be within the wavelength range of FLITECAM.

5. ACKNOWLEDGEMENTS

The authors are grateful to Edwin Erickson for the discussions regarding the current aluminium mirror and his advice on manufacturing a new one. Also, we would like to thank Edward Dunham for providing the HIPO Zemax model.

SOFIA, the "Stratospheric Observatory for Infrared Astronomy" is a joint project of the Deutsches Zentrum für Luft- und Raumfahrt e.V. (DLR; German Aerospace Centre, grant: 50OK0901) and the National Aeronautics and Space Administration (NASA). It is funded on behalf of DLR by the Federal Ministry of Economics and Technology based on legislation by the German Parliament the state of Baden-Württemberg and the Universität Stuttgart. Scientific operation for Germany is coordinated by the German SOFIA-Institute (DSI) of the Universität Stuttgart, in the USA by the Universities Space Research Association (USRA).

REFERENCES

1. Erickson, E. F., Honaker, M. A., Brivkalns, C. A., Brown, T. M., Kunz, N., Preuss, W., and Haas, M. R., "Backup secondary mirror and mechanism for SOFIA," in [*Society of Photo-Optical Instrumentation Engineers (SPIE) Conference Series*], Oschmann, Jr., J. M., ed., *Society of Photo-Optical Instrumentation Engineers (SPIE) Conference Series* **5489**, 1012–1020 (Oct. 2004).
2. Henselmans, R., Cacace, L., Kramer, G., Rosielle, P., and Steinbuch, M., "The NANOMEFOS non-contact measurement machine for freeform optics," *Precision Engineering* **35**(4), 607 – 624 (2011).
3. Erickson, E. F. and Dunham, E. W., "Image stability requirement for the SOFIA telescope," in [*Society of Photo-Optical Instrumentation Engineers (SPIE) Conference Series*], Melugin, R. K. and Röser, H.-P., eds., *Society of Photo-Optical Instrumentation Engineers (SPIE) Conference Series* **4014**, 2–13 (June 2000).

Upgrade of the SOFIA target acquisition and tracking cameras

Manuel Wiedemann[*ab], Jürgen Wolf[ab] and Hans-Peter Röser[c]

[a] Deutsches SOFIA Institut, University of Stuttgart, Pfaffenwaldring 29, 70569 Stuttgart, Germany
[b] SOFIA Science Center, NASA Ames Research Center, Mail Stop N211-1, Moffett Field, CA 94035, USA
[c] Institute of Space Systems, University of Stuttgart, Pfaffenwaldring 29, 70569 Stuttgart, Germany

ABSTRACT

The Stratospheric Observatory for Infrared Astronomy (SOFIA) uses three CCD cameras with different optics for target acquisition and tracking. The Wide Field Imager (WFI with 68 mm optics) and the Fine Field Imager (FFI with 254 mm optics) are mounted on the telescope front ring and are therefore exposed to stratospheric conditions in flight. The Focal Plane Imager (FPI) receives the visible light from the 2.5 meter telescope and is mounted inside the pressurized aircraft cabin at ca. 20°C. It is planned to replace all three imagers' CCD sensors with commercial Andor iXon cameras to significantly increase the sensitivity allowing for tracking on fainter stars. Andor cameras were temporarily mounted on the FPI flange as stand-alone systems to optically measure the telescope's pointing stability and the performance of various telescope sub-systems during engineering flights. Three DU-888 cameras will now be integrated in SOFIA's telescope system, so their image data can be used for target acquisition and tracking. To replace the WFI and FFI, the cameras will also need to be tested under stratospheric conditions, to ensure that they can be operated safely and without degradation of performance. In this paper we will report about the results of the environmental tests with the cameras, the integration of the camera in the SOFIA tracker and the current status of the upgrade project.

Keywords: Stratospheric Observatory for Infrared Astronomy, SOFIA, CCD, Tracking Cameras, Environmental Testing

1. INTRODUCTION

SOFIA's three target acquisition and tracking cameras are identical camera heads that use different optics to achieve different fields-of-view and sensitivities. The Focal Plane Imager (FPI) receives the visible light from the SOFIA main telescope that passes through the dichroic tertiary mirror, while the Science Instrument (SI) receives the reflected IR light, as shown in figure 1. The FPI is not only the most sensitive of the three tracking cameras (due to the large optics), it also sees any image movements due to the chopping secondary mirror or due to telescope bending, thus it is closest to what the SI sees in terms of image motion. This makes the FPI the main tracking camera for SOFIA. Like the SI the FPI resides inside the aircraft cabin in a temperature and pressure controlled environment (T ≈ 20°C, p ≈ 0.8 bar in flight). The Fine Field Imager (FFI) and Wide Field Imager (WFI) have their dedicated optics and are mounted on the head ring of the SOFIA telescope. The cameras are therefore exposed to stratospheric conditions in flight (T ≈ -40°C, p ≈ 0.1 bar in flight).

1.1 Current CCD Sensors

The current imagers use *Thompson TH7888A* frame transfer CCD sensors. The sensors' geometry (1024 x 1024 pixels with a pixel size of 14 µm x 14 µm) and speed (readout rates are 2 MHz and 5 MHz) are well suited for the application as a tracking camera, but it can't compare to modern CCDs in terms of sensitivity. The TH7888A sensor is a front-illuminated sensor with a peak quantum efficiency (QE) of 19%. Furthermore, the cameras

---

[*] mwiedemann@sofia.usra.edu, phone: +1 650 604 2484, www.dsi.uni-stuttgart.de & www.sofia.usra.edu

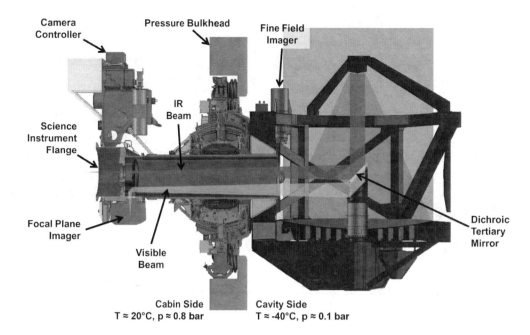

Figure 1. Sectional view of the SOFIA Telescope Assembly (TA). The incoming light beam is depicted in red, which is split by the dichroic tertiary, reflecting the IR light to the SI flange and transmitting the visible light. The visible light beam (depicted in green) is reflected by plane mirror through the Nasmyth tube underneath the IR beam to the FPI. The Fine Field Imager (FFI) is shown on the Telescope head ring, the Wide Field Imager (WFI) is mounted on the opposite side, not visible in this sectional view.

are not equipped with active cooling, which results in a high dark current rate for the FPI inside the warm aircraft cabin. Due to those shortcomings all three imagers but especially the FPI do not fulfill their sensitivity requirements. The current FPI's sensitivity limit is $m_v \approx 12$ mag and the current FFI's limit is $m_v \approx 11$, falling short of the required limit by 4 mag and 2 mag respectively (see table 1).

The CCD in the camera head is driven by the clock signals generated by the camera controller, which is mounted on the top of the TA balancer plate (see figure 1). The camera controller also contains the analog-to-digital converters (ADC) which means that the analog image signal is transferred over a significant distance from the camera head to the controller (ca. 6m for the FPI, 10m for the FFI and WFI). The digitized video signal is then transferred over a proprietary optical fiber interface to the Tracker Controller which contains a frame grabber and calculates the centroids. This system architecture allows that a large part of the sensitive electronics is contained in a temperature controlled environment. However, the great distance of analog image data transfer introduces additional noise before digitization, the transferred data is affected by EMI and the supply of DC power is difficult over long cables.

Table 1. Overview over the optical specifications of the three SOFIA target acquisition and tracking cameras.

	WFI	FFI	FPI
Field of View	6 deg	70 arcmin	8 arcmin*
Pixel FoV	21.4 arcsec	4.06 arcsec	0.55 arcsec
Focal Length	135 mm	710 mm	5230 mm
Aperture	67 mm	254 mm	2500 mm
F-Ratio	f/2	f/2.8	f/2.1
Optics	Petzval lens	Schmidt-Cassegrain	SOFIA telescope
Mag Limit Requirement	11 mag	13 mag	16 mag

* Defined by optics vignetting

1.2 Upgrade Plans

It was planned to upgrade the three cameras with *e2v CCD47-20 BI AIMO* sensors and thermo-electric coolers. This will enable the cameras to fulfill the requirement, increasing the limiting magnitude for the FPI to $m_v = 17$ mag and for the FFI to $m_v = 14$ mag. This CCD sensor is also a frame transfer chip with nearly the same geometry, having only somewhat smaller pixels with 13 μm x 13 μm. The back-illuminated version with mid-band A/R coating reaches a peak QE of 93% and the "inverted mode operation" and active cooling greatly reduce the dark current. Using a test camera with the e2v CCD47-20 sensor, DSI was able to verify these parameters and ensure the suitability of the sensor for the upgrade.[1]

The development of a prototype camera in the industry was initiated. The idea was to keep the same system architecture as the current imagers, with the clock voltage generation and the ADCs in the camera controller. This way the new cameras would use the same electrical and mechanical interfaces, allowing an easy integration, minimizing the changes to the existing systems. The development of the prototype proved to be more difficult than expected. The main problems was the fine tuning of the clock voltages to optimize frame transfer and to minimize read noise. These technical difficulties were never fully solved, as they are in part due to the system architecture where the analog image signals and clock voltages have to be transfered over long cables.

Simultaneously to the prototype development, DSI was also conducting the first flights with the "Fast Diagnostic Camera" (FDC), based on a commercial iXon 888 camera from Andor. The FDC uses a very similar e2v sensor with the same geometry and efficiency, but is equipped with an electron multiplying register, that allows on-chip signal amplification to increase low signal levels. It was successfully used on 5 engineering flights as a stand alone system to optically analyze the pointing stability of the SOFIA telescope.[2,3] On these flights the FDC was installed in lieu of the Focal Plane Imager and was commanded from a laptop, not interacting with any other SOFIA systems.

The good performance of the FDC, together with technical difficulties and delays in the development of the imager prototype using a CCD47-20 sensor, led to the idea of using an Andor iXon camera as a permanent tracking camera. A splitter box offered by Andor was used to make the image data stream available on a second output. A custom logic transformed this data stream into a format compatible with the SOFIA tracker. This setup was intended as a proof of concept only, as it didn't allow commanding the camera by the SOFIA mission control software. In two flights it was shown that tracking with the Andor camera was possible, which lead to the decision to upgrade all three imagers with Andor iXon cameras. Table 2 shows a comparison of the current imagers and the Andor iXon DU-888.

1.3 New Upgrade Approach

The new upgrade approach foresees industry standard parts and interfaces where possible. Aside from the camera interface that requires an Andor PCI or PCIe card, the interfaces are based on Ethernet or other industry standards. The camera controller is a ruggedized PC system running an open-source Linux. The camera control software is written in C/C++ and uses a software development kit provided by Andor. The change from mostly proprietary interfaces in the existing imager system to widely used standard interfaces is intended to ensure that replacements or upgrades of single components will be easily possible during the projected lifetime of SOFIA of 20 years.

Upgrading the FPI is straight forward and will be done first. Once the current FPI is replaced with the new camera, it will be fully integrated into the SOFIA systems. Additionally to the standard tracking application the camera will still have the option to be operated in a stand alone mode, like the FDC. This mode will allow high frame rate data acquisitions for engineering purposes at any time without the overhead of switching cameras.

The FFI and WFI will be upgraded in the next step. They will use the same camera controllers and software interfaces as the upgraded FPI. The camera heads however will need to be somewhat modified from the standard Andor iXon product. The cameras will need to be prepared for continuous operation under stratospheric conditions. This will include some modifications to the electronics of the camera, as well as a newly designed housing that can be attached to the existing mechanical interfaces and withstands all the necessary crash loads without disintegrating. These modifications will require some more time and will be done after the FPI upgrade.

Table 2. Comparison of the current SOFIA imagers and the upgraded imagers based on Andor iXon 888 cameras.

	Current Imagers	Upgraded Imagers
Pixel Format	1024 x 1024 pixels	1024 x 1024
Pixel Size	14 μm x 14 μm	13 μm x 13 μm
Peak QE	19% at 690 nm	93% at 570 nm
Min. Read Noise	40 e- rms	6 e- rms
Electron Multiplying Output	no	yes
Readout Rates	2 MHz (14 bit)	1 MHz (16 bit)
	5 MHz (8 bit)	3 MHz (14 bit)
		5 MHz (14 bit)
		10 MHz (14 bit)
T_{CCD} for FPI	ca. 25°C	≤ -30°C
Dark Current for FPI	ca. 10,000 e-/s	≤ 0.2 e-/s
FPI Mag Limit for Tracking*	12 mag	17 mag
T_{CCD} for FFI	ca. -35°C	ca. -35°C
Dark Current for FPI	20 e-/s	0.1 e-/s
FPI Mag Limit for Tracking*	11 mag	14 mag

* with $t_{exp} = 3$s

2. ENVIRONMENTAL TESTS

The main concern of using a commercial product like the Andor iXon cameras was the harsh operating conditions for the WFI and FFI cameras, mounted to the TA head ring. These two cameras experience typical operating temperatures of -30°C to -40°C in flight at around 40,000 ft pressure altitude. The Andor iXon cameras have a specified operating temperature range of 0°C to 30°C, so using these cameras on the TA head ring would be significantly out of their specified range. It was unclear if the standard cameras would operate under these conditions or even survive without taking permanent damage. This is why environmental tests were carried out, simulating these conditions, to see if the approach of using commercial cameras is feasible.

2.1 First Test Run

The first environmental tests were aimed at finding the minimum temperature at which the camera is still operational. The tested camera was an engineering grade iXon DU-885 camera on loan from Andor for this test. The test was conducted in the temperature/altitude chamber of the NASA Ames Engineering Evaluation Lab. During this test the camera was powered on throughout the cool down phase.

Five temperatures and the pressure altitude were recorded. Two sensors were mounted on the camera housing, one on the chamber's shroud and one recorded the air temperature. The fifth recorded temperature was the camera's built-in temperature sensor that reads out the CCD temperature. Figure 3 shows the temperature and altitude profiles on both test days. The locations of the two sensors on the camera housing are depicted in figure 2.

To avoid condensation the chamber was evacuated to a pressure altitude of 25,000 ft during the cool down

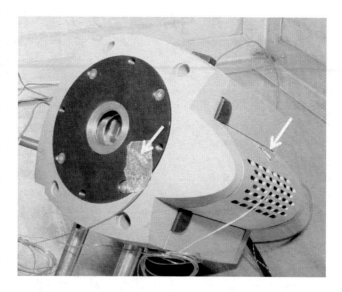

Figure 2. Andor iXon DU-885 inside a thermal vacuum chamber with thermocouples mounted on the camera housing.

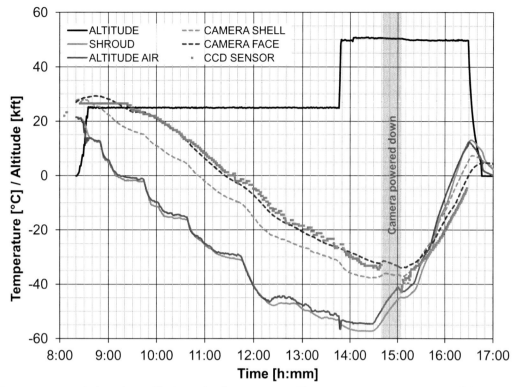

Figure 3. Temperature and altitude profile from the first test run. The temperature sensor on the CCD does not go over 26.5°C, which is why the CCD sensor's profile flattens out at this temperature.

to 0°C. After reaching an air temperature of 0°C, the camera was running idle (taking no images) for 15 minutes, allowing some time for all camera components to cool down. Then a series of images was taken:

- 5 dark frames with $t_{exp} = 30$ s
- 20 bias frames
- 5 flat field frames with $t_{exp} = 0.1$ s

The flat field frames were acquired using the diffuse light reflecting from the temperature/altitude chamber's back wall. The dark and bias frames were taken with the camera's internal shutter closed. This series was taken in the "13 MHz EM" output mode with a "Pre-Amp" setting of 3.8x and "Baseline Clamp" off. This procedure was repeated at -15°C, -30°C, -45°C and -55°C.

With the camera still performing well during the checkout at -55°C, the pressure altitude was increased to 50,000 ft and a long term test was started which included taking dark frames over 30 minutes. During this test, the compressors that are responsible for the cooling of the chamber failed (at 14:30) and the chamber started to warm up. This limited the test time at -55°C to roughly 45 minutes.

Since the camera was working without any noticeable issues to that point, it was then powered down for 20 minutes to see if it would start up again. At an air temperature of just below -40°C the camera was powered up and another checkout was conducted, before the tests were concluded for the day and the chamber was warmed up.

The image data taken during this test was later evaluated to see if the camera showed any degradation in performance, not obvious from visual inspection of the images. The image data was evaluated for bias level, read noise, gain and vertical charge transfer efficiency (CTE).

2.1.1 Bias Level

Before the first test run the mean bias level, averaged over all pixels in the frame, was measured at an air temperature of 20°C for different sensor temperatures in the 13 MHz mode (Pre-Amp: 3.8x, Baseline Clamp: off) for reference. During the test in the temperature/altitude chamber, the bias levels were measured again using the same camera settings. The CCD temperatures were reached due to the cold ambient temperatures and not stabilized with the thermoelectric cooler. Figure 4 shows the bias levels vs. sensor temperatures for both cases.

At a constant ambient temperature when only the sensor is cooled, the bias level drops nearly linearly going to lower temperatures. During the test in the temperature/altitude chamber, the bias level increased with lower temperatures. Even though the bias level increases, the noise in the frame (pixel-to-pixel variation) is nearly constant in all cases with a standard deviation of 12 - 13 ADU rms. So the only effect of the higher bias level is that it slightly reduces the dynamic range, but does not degrade the image quality in any way.

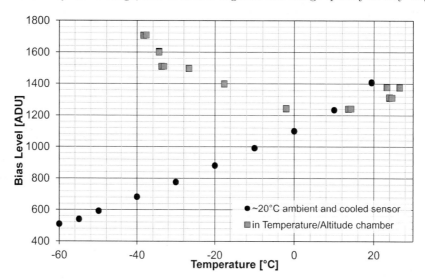

Figure 4. Bias level vs. sensor temperature, with the CCD temperature regulated by the thermoelectric cooler (dark circles) and with the CCD cooled due to the cold environment in the TV chamber (light squares).

2.1.2 Gain and Read Noise

The integrated shutter of the DU-885 allowed taking flat field as well as dark and bias frames, which in turn can be used to determine the gain and read noise at different temperature levels. The values for the gain and read noise were calculated from the frames taken in the test chamber and are summarized in table 3.

Both the gain and the read noise are slightly increasing going to colder temperatures. The increase of both figures is minor (around 10% over a drop of 70 Kelvin in air temperature) and acceptable for the application as a tracking camera.

2.1.3 Charge Transfer Efficiency

The Charge Transfer Efficiency (CTE) was measured using the change in variance of a flat-field method, as it is described by Christen et al.[4] The charge in the top rows of a CCD needs to be shifted more often before reaching the readout register than charge in the lower rows. A very small percentage of the charge is deferred in these shifting processes. This has a "smoothing" effect on the flat-field frame, which is more severe for pixels in the top rows, which means that the signal variance within a row decreases somewhat after several shifts. From the slope of the linear fit in a variance vs. row plot the vertical CTE can be derived.

The results for the CTE measurement at the different measurement points are also shown in table 3. The vertical CTE values show no clear tendency of decrease during the cool-down of the camera. It appears that CTE is not an issue for the cold camera electronics.

Table 3. Gain, read noise and vertical CTE values measured at different ambient temperatures and reference values measured by DSI and Andor before the environmental tests.

T_{CCD} [°C]	T_{air} [°C]	Gain [e-/ADU]	Read Noise [e- rms]	vert. CTE	Comments
23.3	23	1.07	13.2	99.9994 %	before chamber was sealed
> 26.5	14	1.07	13.8	99.9999 %	at 25 kft altitude
24.6	-1	1.07	13.1	99.9996 %	
13.6	-15	1.08	13.3	99.9995 %	
-2.0	-29	1.11	13.2	99.9991 %	
-17.6	-45	1.14	13.5	99.9990 %	
-26.0	-52	1.15	14.3	99.9997 %	
-33.8	-50	1.16	14.7	99.9994 %	at 50 kft altitude
-37.7	-42	1.16	14.3	99.9991 %	after cold start
-55.0	20	1.12	13.5	99.9999 %	DSI measurement before environmental test
-75.0	N/A	1.0	11.9	N/A	Andor measurement before shipment

2.2 Second Test Run

The second test run was intended to test the "cold start" behavior of an iXon camera. The camera was to be turned off for the duration of the cool down phase and when all components had reached a temperature below -55°C the camera should be powered up.

Since the first test run was successful and the engineering grade DU-885 showed no defects during or after the environmental test, it seemed to be an acceptable risk to start testing a science grade DU-888. So the second test run was conducted with a DSI owned DU-888 camera. For this run, a different environmental chamber was used, which is cooled with LN_2. It is possible to achieve lower air temperatures with this smaller chamber, but temperature and altitude need to be controlled manually by opening and closing valves to control the flow of LN_2 and to the vacuum pump.

With the camera turned off, the altitude was increased to 50,000 ft and the chamber was cooled down. In order to quickly cool all camera components down to -55°C, the air temperature was lowered significantly below -55°C during the cool down phase. It still took nearly 4 hours for all camera temperatures to drop below the desired temperature (see temperature / altitude profile in figure 6). To determine the temperatures of the camera components, 4 additional temperature sensors were mounted inside the camera. Two were mounted on the A/D converters and two more on a copper block and a cooling fin, which were the two parts assumed to have the largest heat capacity inside the camera. The location of these additional thermocouples is shown in figure 5.

Figure 5. Additional thermocouples inside the iXon DU-888 for the second test run.

When all temperatures had dropped below -55°C (at 12:45 in the temperature plot) it was attempted to power up the camera. Upon startup of the camera software, an error message came up that the camera could not be detected. After a few more attempts the temperature of the chamber was increased again to see if the camera would power up at warmer temperatures. After two more failed attempts, the camera was successfully initialized at around 13:30. At this point the warmest component inside the camera that was measured was the 1 MHz ADC, which had a temperature of -52°C. The copper block

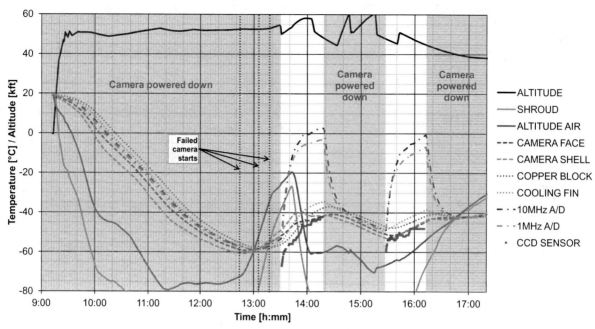

Figure 6. Temperature and altitude profile from the second test run. The camera was powered down for most of this test run to test the cold start behavior of the camera.

had a temperature of -55°C. Once the camera powered up, a series of bias and dark frames was taken in all readout modes and "Pre-Amp Gain" settings.

After about 50 minutes of operation the camera was turned off again and the air temperature was lowered for another cold soak. After about 65 minutes the cold start was successful on the first attempt. At this point the air temperature was at -67°C and the copper block was at -47°C. The camera was operated for 45 minutes before testing was concluded for the day and the chamber's temperature and pressure were brought back up.

2.2.1 Bias Level and Noise

A script was used to acquire image data in all possible combinations of readout rate (1 MHz, 3 MHz, 5 MHz and 10 MHz), output amplifier (Conventional or Electron Multiplying) and Pre-Amp Gain (1x, 2.5x and 5.2x). This allowed taking a series of bias frames in all readout modes very quickly.

Right after the camera software was started and successfully initialized the camera, the bias frames taken with the 5.2x Pre-Amp Gain setting had a very high signal level, close to or at saturation. With a Pre-Amp Gain setting of 1x or 2.5x the bias levels were just slightly higher than in normal operation (ambient temperature at ca. 20°C and CCD sensor cooled to -60°C). After some time the bias levels dropped in all readout modes, reaching values close to normal operation. The bias levels of all readout modes after ca. 5 and 45 minutes of operation and during normal operation are summarized in figure 7.

The noise in the bias frames (σ: standard deviation of the pixel to pixel variation) is also higher right after the camera is turned on. This is especially true for the modes with Pre-Amp gain of 5.2x. The standard deviations of all readout modes and after 5 min and 45 minutes of operation and during normal operations are summarized in figure 8.

The frames also showed patterns in some frames right after it was started up. After several minutes of operation the patterns disappeared again and the bias levels and noise levels dropped to the levels that are observed during normal operation. Even though the camera's performance wasn't acceptable in a few operating modes right after start-up, there were always modes that showed somewhat degraded performance, but in an acceptable range for the tracking application.

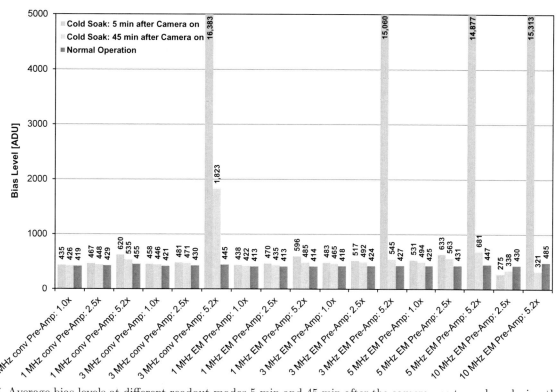

Figure 7. Average bias levels at different readout modes 5 min and 45 min after the camera was turned on during the cold soak test and during normal operation. The related temperatures can be found in figure 6.

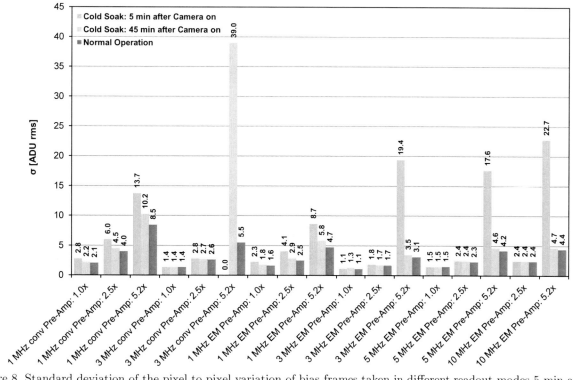

Figure 8. Standard deviation of the pixel to pixel variation of bias frames taken in different readout modes 5 min and 45 min after the camera was turned on during the cold soak test and during normal operation.

2.3 Conclusion from the environmental tests

This first series of environmental tests has shown that the standard Andor iXon cameras are able to operate at the cold temperatures and low pressures expected in flight, although these conditions are significantly out of the specified operational environment of the cameras. The second test run was designed to simulate harsher than normal flight conditions in the telescope cavity. In standard operations the cameras will be turned on before takeoff and be operational throughout the entire flight, so the electronics will likely never be as cold as during the second test run. The camera did have some problems during this run, as it didn't initialize right away and showed increased noise in the first minutes of operation. However the performance of the camera returned to normal and there were no permanent damages.

These first test results are very promising, but they don't prove that the camera is also suited for long-term operation with many temperature and pressure cycles. This is why DSI is working with Andor Technology in Belfast, North Ireland and Kayser-Threde in Munich, Germany to build and test hardened versions of the iXon 888 cameras, which will improve the cold-start capabilities and ensure the long-term reliability for operation under stratospheric conditions. This work is expected to include exchanging single electrical components for Military Specification (Mil-Spec) components and some modifications on the camera's FPGA programming on the electrical side and the design of a new housing and small changes to the cooling concept on the mechanical side. The custom camera housing will not only allow easy mounting to the existing mechanical interfaces, it will also fulfill all the safety aspects under crash loads.

3. PHOTOMETRIC SENSITIVITIES

The SOFIA requirements ask for tracking capabilities with the FPI on stars as faint as 16th magnitude.[5] Since both, the current FPI and the Andor camera mounted to the FPI flange have flown on SOFIA, the photometric sensitivities of both cameras in flight could be measured and compared to that requirement.

There was no dedicated test time to measure the FPI's sensitivity, so image data that was acquired during its routine tracking operation was used. The analyzed image data was acquired on flight 084 (SCAI #6). The targets were chosen from three star fields around HD62146, HD17156 and HD16906. The exposure time of all images was $t_{exp} = 1s$ and chopping was off.

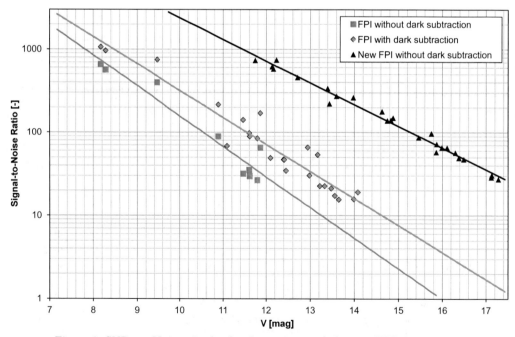

Figure 9. SNR vs. V magnitudes for the current and the new FPI at $t_{exp} = 1s$.

The photometry data for the new FPI (i.e. FDC) was acquired during flight 043 (OCF #2). On that flight the FDC had to be used with an OD = 1.2 neutral density (measured attenuation factor: 15.27) and additionally an R-filter (which reduced the incoming light by another factor of 3.2), to make simultaneous measurements with FORCAST possible, which required bright IR sources. The actual exposure time of the analyzed images was $t_{exp} = 16$s. The data was scaled to match an unfiltered observation with $t_{exp} = 1$s, so it is directly comparable to the FPI data.

Figure 9 shows the SNR as a function of V magnitudes for the current FPI (with and without dark correction) and the FDC, which will be the new FPI (without dark correction). The plot shows that the current FPI is not able to detect a $m_v = 16$ star with $t_{exp} = 1$s and even with a good dark field subtraction it can only achieve SNR ≈ 3. The new FPI achieves an SNR > 60 on a $m_v = 16$ star, so it will easily fulfill the SOFIA sensitivity requirement for tracking.

4. OUTLOOK

The designing phase of the flight worthy camera controller and software is well under way. The hardware to upgrade the FPI will be ready in late September 2012 and the integration and testing of the upgraded system on the airplane is planned shortly after. The work on the modifications to the Andor cameras in preparation for the upgrade of the FFI and WFI are expected to begin soon, so that these to cameras will be ready for implementation in 2013.

ACKNOWLEDGMENTS

We would like to thank our colleagues of USRA, NASA and DSI for valuable discussions. Special thanks to Lynn Hofland of the NASA Ames EEL for his support during the environmental tests. Thank you to Andor Technology for supplying an engineering camera and the productive collaboration. The authors also thank Dr. E. W. Dunham of Lowell Observatory, Flagstaff for in depth discussions and his continuous support of our FDC work and upgrade project.

M. Wiedemann thanks DLR, USRA and DSI for financial and administrative support for his PhD position.

SOFIA, the "Stratospheric Observatory for Infrared Astronomy" is a joint project of the Deutsches Zentrum für Luft- und Raumfahrt e.V. (DLR; German Aerospace Center, grant: 50OK0901) and the National Aeronautics and Space Administration (NASA). It is funded on behalf of DLR by the Federal Ministry of Economics and Technology based on legislation by the German Parliament the state of Baden-Württemberg and the University of Stuttgart. Scientific operation for Germany is coordinated by the Deutsches SOFIA Institut (DSI) of the University of Stuttgart, in the USA by the Universities Space Research Association (USRA).

REFERENCES

[1] Wiedemann, M., Wolf, J., and Röser, H.-P., "Testing the e2v CCD47-20 as the new sensor for the SOFIA target acquisition and tracking cameras," in [Ground-based and Airborne Telescopes III], SPIE Conference Series **7733** (July 2010).

[2] Pfüller, E., Wolf, J., Hall, H., and Röser, H.-P., "Optical Characterization of the SOFIA Telescope using fast EM-CCD Cameras," in [Ground-based and Airborne Telescopes IV], SPIE Conference Series **8444** (July 2012).

[3] Temi, P. et al., "SOFIA Observatory Performance and Characterization," in [Ground-based and Airborne Telescopes IV], SPIE Conference Series **8444** (July 2012).

[4] Christen, F., Kuijken, K., Baade, D., Cavadore, C., Deiries, S., and Iwert, O., "CCD Charge Transfer Efficiency (CTE) Derived from Signal Variance in Flat Field Images," in [Scientific Detectors for Astronomy 2005], Beletic, J. E., Beletic, J. W., and Amico, P., eds., 543 (Mar. 2006).

[5] NASA SPO, [Telescope Assembly (TA) Requirements], SOF-1011, Rev.8 (Aug. 2001).

The 3,6 m Indo-Belgian Devasthal Optical Telescope: Assembly, Integration and Tests at AMOS

Nathalie Ninane, Christian Bastin, Jonathan de Ville, Fabrice Michel, Maxime Piérard, Eric Gabriel, Carlo Flebus and Amitesh Omar*

Advanced Mechanical and Optical Systems (AMOS s.a.), LIEGE Science Park, B-4031 ANGLEUR (Liège), BELGIUM

* Aryabhatta Research Institute of Observational Sciences (ARIES), Nainital, India

ABSTRACT

AMOS SA has been awarded of the contract for the design, manufacturing, assembly, tests and on site installation (Devasthal, Nainital in central Himalayan region) of the 3.6 m Indo-Belgian Devasthal Optical Telescope (IDOT).
The telescope has a Ritchey-Chrétien optical configuration with a Cassegrain focus equipped with one axial port and two side ports. The meniscus primary mirror is active and is supported by pneumatic actuators. The mount is an Alt–Az type with for the azimuth axis a 5 m diameter hydrostatic track.
The telescope was completely assembled and tested in AMOS workshop. This step is completed and successful. The telescope is now ready for shipment to Nainital.
This paper describes the test campaign at sub-system and system level that has taken place to demonstrate that the telescope satisfies the main system requirements. Besides of the functionality of the telescope, the units interacting with the image quality or the tracking performance were plenty tested. Some selected tests directly connected to the performance of the telescope are also looked specifically in this paper.

Keywords: telescope testing, IDOT

1. INTRODUCTION

In 2007, AMOS SA signed a contract for the design and building of a 4M class telescope with ARIES (Aryabhatta Research Institute of Observational Sciences) India. Today the telescope is constructed and ready to be shipped in India. The telescope will be installed at Devasthal, in the Himalayan range, 2500 m above the sea; 350 km North of Delhi.

The Telescope design[1] was presented 4 years ago at Marseille; since then it has been fully assembled and tested at subsystem and system levels in AMOS assembly hall.
All along the integration and for each sub-system the functioning and the performances were verified as defined by the system engineering plan.
Dummy mirrors as well as dummy instruments were used to perform the mechanical tests, to characterize and adjust the drive of the axes, and to check the mirror integration procedures.
When ready the mirrors were integrated and aligned in the telescope. Some optical tests were performed in sighting stars through an aperture made in AMOS integration hall roof. First images with a FWHM equal to 2.1 arcsec were recorded despite of the bad seeing and thermal conditions. During these sky tests important functional tests were realised. The capability of the primary mirror correction with the active optic system was demonstrated. The mirror deformation modes were generated and measured. The wavefront error correction loop was closed. Its convergence was checked.
Tracking tests in open and close loops were also performed.

2. TELESCOPE OVERVIEW

The main characteristics of the telescope is summarised in Table 1. The optical combination is a Ritchey-Chrétien type with a Cassegrain focus where the light beam can be directed toward a main port designed for interfacing an instrument with a mass up to 2 tons or toward two side ports for smaller instruments. The mount type is an alt-azimuth. The telescope weights 150 tons. It rotates around the azimuth axis thanks to a hydraulic track[3]. The telescope is equipped with an active optic system[4] (AOS) that controls the primary mirror figure and the secondary mirror positioning to keep the telescope wavefront error in the specification for any operational conditions. The primary mirror is a meniscus 165 mm thick, 3700 mm diameter supported by 69 axial actuators and 24 lateral astatic levers. The set of forces applied by the actuators to the mirror are adjusted continuously to correct the telescope wavefront error. An Acquisition and Guiding Unit (AGU) that is aligned on a guide star at the edge of the telescope field of view measures the wavefront and tracking errors. The Telescope Control System[5] (TCS) computes the telescope trajectory taking into account of the weather conditions, the pointing model of the telescope and the tracking errors measured by the AGU. A more detailed description of the telescope is given in another SPIE paper; see reference (2).

Figure 1 shows pictures of the telescope in AMOS assembly hall. The telescope sizes are: height 13 m, width 7 m and total weight 150 tons.

Figure 1: IDOT in AMOS integration hall

Type:	Ritchey - Chrétien
Focal length:	32.4 m (telescope F#/9 with M1 F#/2)
Aperture stop:	3.6 m on M1
2 focal plane configurations:	Side Port and Axial Port
Field of View:	10 arcmin on side ports, 30 arcmin on axial port, (35 arcmin for the AGU)
Operational waveband:	350 nm to 5000 nm
M1 characteristics:	R = 14638.87 mm CC K = -1.03296 Optical Φ = 3600 mm, mechanical Φ = 3700 mm
M2 characteristics:	R = -4675.30 mm CX K = -2.79561 Optical Φ = 952 mm, mechanical Φ = 980 mm
Distance M1-M2:	5.51 m
Back focal length:	2.5 m
Focal plane Radius of curvature:	1935.78 mm
Scale plate:	157 μm / arcsec (0.006 arcsec/μm)

Table 1: Summary of the telescope characteristics

3. TELESCOPE REQUIREMENTS AND TEST PLAN

Table 2 summarizes the main performance requirements specified for 3.6 m IDOT.

Optical main requirements	
Optical image quality	- Encircled Energy 50% < 0.3 arcsec, - Encircled Energy 80% < 0.45 arcsec, - Encircled Energy 90% < 0.6 arcsec, For the waveband 350 nm to 1500 nm; without corrector for 10 arcmin FOV.
Mechanical main requirements	
Sky coverage (elevation range)	15 to 87.5°
Pointing accuracy	< 2 arcsec RMS
Tracking accuracy	< 0.1 arcsec RMS for 1 minute in open loop, < 0.1 arcsec RMS for 1 hour in close loop, < 0.5 arcsec Peak for 15 minutes in open loop.

Table 2: Main telescope requirements

For each system performance requirement, an error budget was established in order to: i) identify the lower-level contributors ii) analyse and discuss the error generation mechanism as well as the error combination modes iii) set up the computation logic of the instrument performances. In parallel, a test plan is developed to verify all along the telescope assembly the subsystem contributions to the error budget. The error budget is then consolidated with actual values and corrective actions are undertaken if necessary.

Both image quality and tracking performance requirements are tight and moreover impossible to verify at system level in the lab. For measuring the image quality of a 3.6 m aperture telescope a stable 3.6 m diameter collimated beam is necessary. This is practically not feasible. For measuring the tracking error, a collimated beam with a controlled displacement representative of the star movements with accuracy better than the requirement is needed. Once again this is not feasible.
The only reasonable way to verify the telescope at system level is to test it on the sky; but with the limitations given by the seeing and the thermal homogeneity of the telescope and the building.

The telescope was first integrated with dummy mirrors and instruments. The use of dummy mirrors presents some major advantages during AIV phase. First the mirrors are fragile and cost effective items; it is safer to check the integration procedures and perform a maximum of tests without any risk for them. Moreover mirrors are long lead items. The integration and tests can then start without waiting for the mirror delivery. And at last, dummies can be instrumented easily which simplifies the test preparations.

After the complete assembly of the telescope a electromechanical test and tuning campaign was undertaken. It consisted mainly in:
- The characterisations and the tuning of the axes (azimuth, elevation, adapter, rotator) that can be made only telescope fully assembled and balanced;
- Drive tests for actual axis trajectory cases, all axes working together. The result for one trajectory case is given and explained in chapter 4.
- Functional tests of the telescope (except the tests that need a star light beam) ;
- Test of the integration procedures. The pieces to handle are big and heavy, and in the case of the mirrors fragile. No hazard for people or for the mirrors is allowed.
- Tube deflexion tests. The relative movements of the mirrors and of the instrument main interface have been measured with a heterodyne interferometer for rotation of the tube between 15° and 90° and for temperature variations. Alignment correction laws of M2 in function of the elevation and temperature are introduced in the active optic system.
- Primary mirror jitter measurements that consists in recording and analysing the movements of the mirror in its cell up to 10 Hz for different tracking configurations.

The last AIV step in Belgium was the integration of the actual mirrors in the telescope to start then the test campaign "on the sky". An aperture in the integration hall roof was foreseen to have the opportunity to sight stars with the telescope before sending it in India (see Figure 2). The aperture gives a telescope field of regard with a complete in lighted pupil for telescope elevation between 68° and 75° and azimuth between 303° and 313°.

The aims of those tests were to make a start-up of the complete system, to verify some telescope tuning procedures, to verify that the telescope control loops (TCS, guider and AOS) are working properly and to train AMOS team for the installation and testing in India.

Because of the seeing and big thermal variations between daytime and night time it was not expected to be in measurement conditions for measuring the performance requirements.

Some Active Optic System and tracking test results are given in chapter 5.

Figure 2: IDOT pointing through the aperture in the roof in AMOS integration hall

4. AXIS DRIVE TESTS

The axis drive test consists in making a tracking simulation and recording the axis control following error. The coordinates of a star is entered in the telescope control system that computes the axis trajectories (angle set points) and sent them to the axis controller. The telescope tube moves accordingly, its position measured by the axis encoders is recorded. The tube following error is the difference between the TCS set points and the encoder measured angles. The resulting tracking error is then computed from the azimuth and elevation axis errors.

Figure 3 gives a measurement example that is the particular case of a star passing by the telescope blind spot close to the zenith. For that case the azimuth axis accelerates to reach its maximum tracking speed. The resulting error on the sky is given in Figure 4. This test was repeated for several cases of trajectory. The following errors found in any cases were compliant with the allocated values in the error budgets.

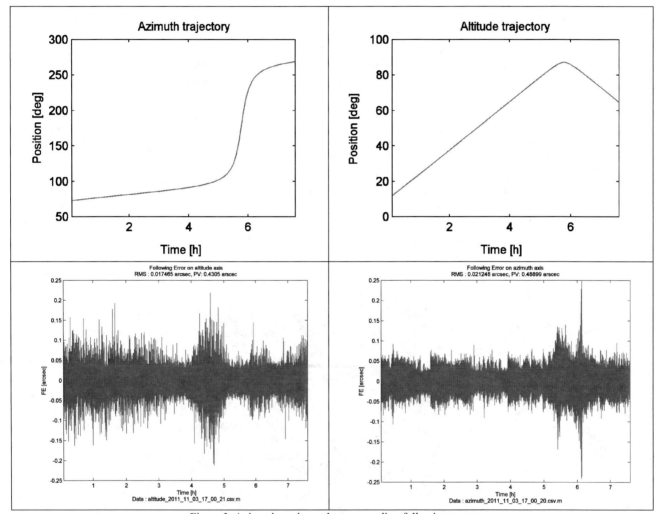

Figure 3: Axis trajectories and corresponding following errors

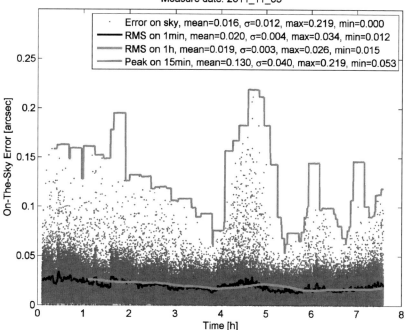

Figure 4: Tracking error resulting from the following errors given in Figure 3; in blue the amplitude of the error, in black the RMS value of the error computed on a running time interval of 1 minute, in red the RMS value of the error computed on a running interval of 1 hour, in green the peak error on a running interval of 15 minutes.

5. SKY TESTS

5.1 Thermal environment

The sky tests were performed in AMOS integration hall (Belgium) between mid February and mid May 2012. Depending of the night, the temperatures were at the beginning of the night between 19°C and 8°C; with a drop during the night of about 10°C. During daytime, the hall temperature is around 20°C. In the evening the roof aperture and the hall outdoor were open to establish an air flux and reduce as much as possible the in and out temperature differences. The thermal inertia of the primary mirror (M1) and the secondary (M2) are such that they stayed always few degrees above the ambient air. Table 3 gives typical temperature differences met during the sky tests.

Temperatures (°C)	Beginning of the night	End of the night
Outside temperature (weather station)	17	7
M1 edge +X	18.8	14
M1 centre front	18.4	14.8
M1 centre back	18.6	15.5
Air just above M1	18.2	11.8
M2 edge	19.4	14
M2 centre	19	15.2

Table 3: Typical night temperature during the sky tests

5.2 FWHM and Wavefront Error

Depending of the night, the Full Width at Half Maximum (FWHM) of the star images were between 2.1 and 6 arcseconds. Figure 5 gives an image recorded with the guider camera that gets a FWHM equal to 2.1 arcsec.

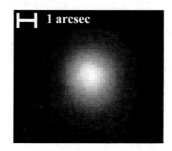

Figure 5: Image of HIP64532 (FWHM = 2.1 arcsec)

The wavefront error of the telescope was measured with the wavefront sensor of the AGU. It showed that the repeatability of the measurement was poor but coherent with the seeing; and that the spherical aberration drifted during the night; this is correlated with the radial thermal gradients in the primary mirror.
Figure 6 a. shows Focus and Spherical Aberration repeatability for an integration time of 10 s and a FWHM = 2.8 arcsec. The repeatability's for different exposure times and the coefficients measured by the WFS are in Figure 6 b.

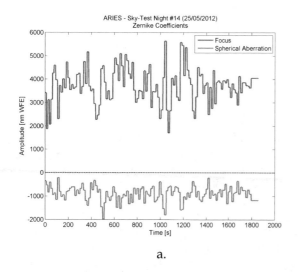
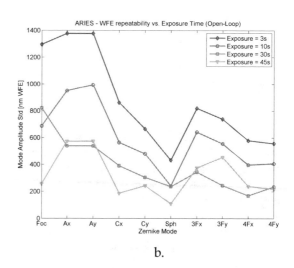

a.   b.

Figure 6: WFE repeatability (FWHM = 2.8 arcsec),); a) Focus and spherical aberration (t =10 s) , b) Focus (foc), Astigmatism (Ax & Ay), Coma (Cx & Cy), Spherical aberration (Sph), Trifold (3Fx & 3Fy), Quadrifold (4Fx & 4Fy) for different exposure times

The wavefront error repeatability is the test limitation for the active optic system.

5.3 Active optic tests

To check the deformation capability of the active optic system each deformation mode was generated and measured with the wavefront sensor of the AGU.
The AOS is used in open-loop mode. Each elastic mode is generated individually with sufficient amplitude to have a good signal-to-noise ratio. The WFS is acquired on a star of magnitude 4.40 (HIP30060) during an exposure time of 75 seconds.
Figure 7 and Figure 8 show the results for 5µm of respectively astigmatism and 3-fold.
The measured WFE is given on the left. The general shape of the phase map is according to the generated mode.
On the right, the amplitudes of Zernike coefficients are plotted in function of the time (at 1000 s an offset of 5 µm astigmatism X is applied and removed at 1500 s). The astigmatism measured in the WFE gets the same order of magnitude as expected (i.e. about 5µm). The difference is in the measurement noise.
For 3-fold, the measured amplitude is a little bit smaller than expected (i.e. about 3.5µm instead of 5µm).
The other conclusion of this test is about the cross-coupling. The curves show that when one mode is generated, the others do not vary out of the noise level.

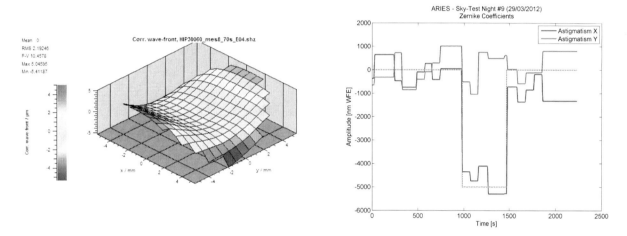

Figure 7: Verification of Active Optics dynamics (5µm WFE of Astigmatism X)

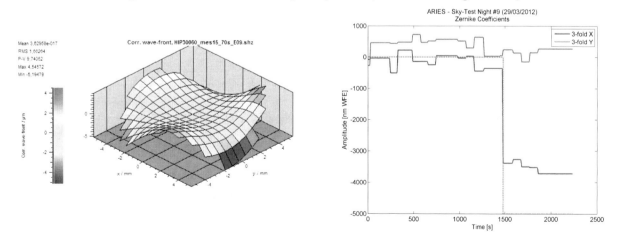

Figure 8: Verification of Active Optics dynamics (5µm of 3-fold X)

The last test performed at AMOS is the most representative of the telescope operational conditions. The complete loop is closed on the WFS, and the measured WFE is corrected thanks to the active M1 mirror and the M2 hexapod.
The system was let intentionally "misaligned" at the beginning of the measurement. AOS is configured to compensate for

astigmatism, 3-fold and spherical aberration with M1 active support. Focus and coma are corrected in moving M2 thanks to the hexapod.

The closed-loop test is run on Edasich (magnitude 3.25) with an exposure time of 10 seconds. In addition to the WFS integration, the Active Optics System is configured to compensate only 20% of the measured coefficients at each iteration, in order to filter the seeing (and avoid "correcting" it). The active optics correction has thus a time-constant of 50 seconds.

Figure 9 presents the results. The amplitude of focus and spherical aberration are given on the left. The first 800 seconds are in open-loop. In average, the WFE contains about 3500 nm WFE of focus and -800 nm WFE of spherical aberration. After the closed-loop is switched on, these modes clearly converge around zero.

The WFE RMS (without piston and tilts) is shown on the right. In open-loop, the optical quality was about 2200 nm RMS WFE. Closing the loop improves the WFE down to 750 nm RMS. It oscillates between 500 and 1000 nm RMS due to the seeing conditions. An analysis of the system shows that a WFE RMS = 750 nm corresponds to an EE50% ~1 arcsec at 500 nm.

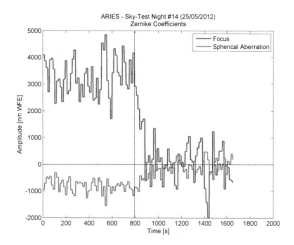 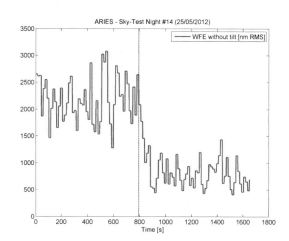

Figure 9: Active Optics in closed-loop (left: focus and spherical aberration – right: WFE rms)

5.4 Tracking tests

After elaboration of a pointing model through the field of regard, some tracking error tests were performed in open and close loop. The guider camera sights one star in the telescope field of view while a test camera records and analyses the images of a second star at the centre of the telescope field of view. The positions of the successive star centroids on the test camera give the tracking error. Figure 8 shows the measurement results. They give a tracking error of 0.25 arcsec RMS, the image integration time equal to 30 s and the FWHM was 2.5 arcsec. Other measurements showed that the results depend directly on the seeing and the integration time.

This test gives an upper limit of the tracking error; it cannot distinguish the effect of the seeing from the telescope itself.

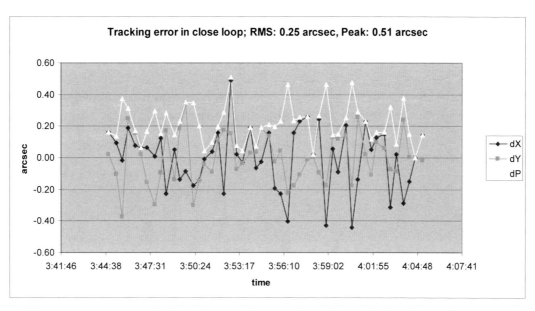

Figure 8: Tracking error measurement; in blue the centroid moves along X, in pink along Y, in yellow the amplitude of the error

6. CONCLUSIONS

IDOT was completely assembled in AMOS premises. During this AIV phase, tests at subsystem and system level were performed to guaranty the telescope specifications. The last test campaign was performed with actual mirrors, pointing stars through an aperture in the integration hall roof. Despite that AMOS hall is not a dome and Liège not an astronomical site, the sky tests were very useful. They have enabled to test all the functionalities of the telescope. All the subsystem interfaces were checked by this way. The guiding loop was tested and used. The active optic system functionalities were tried and adjusted.

Images with FWHM of 2.1 arcsec were a very good surprise given the test conditions.

Moreover the sky tests were a very good training exercise for AMOS AIV engineering team. The procedures were checked and corrected. These last points are useful for the preparation of the commissioning campaign.

Now the telescope is ready to be sent on site in India where it will be assembled by AMOS. Final tuning and performance tests will then be done in actual operating conditions.

7. ACKNOWLEDGEMENT

This work has been performed under ARIES contract reference 1985-14-02. AMOS is very grateful towards ARIES team for having put their confidence in AMOS team for the design and manufacturing of the 3.6 m telescope.

8. REFERENCES

(1) Flebus C., Gabriel E., Lambotte S., Pierard M., Rausin F., Schumacher J.M. and Ninane N., "Opto-mechanical design of the 3,6 m Optical Telescope for ARIES", Proc. SPIE 7012-09 (2008).
(2) N. Ninane & Co., "The 3.6 m Indo-Belgian Devasthal Optical Telescope: general description", Proc. SPIE 8444-67 (2012)
(3) Deville J., Bastin C. and Pierard M., "The 3.6 m Indo-Belgian Devasthal Optical Telescope: the hydrostatic azimuth bearing", Proc. SPIE 8444-150 (2012)
(4) Pierard M., Schumacher J.M., Flebus C. and Ninane N., "The 3.6 m Indo-Belgian Devasthal Optical Telescope: the active M1 mirror support", Proc. SPIE 8444-186 (2012)
(5) Gabriel E., Bastin C., Pierard M. - "The 3.6 m Indo-Belgian Devasthal Optical Telescope: the control system", Proc. SPIE 8451-82 (2012)

First Tests of the compact, low scattered-light 2m-Wendelstein Fraunhofer Telescope

Ulrich Hopp[a,b], Ralf Bender[a,b], Frank Grupp[a,b], Hans Thiele[c], Nancy Ageorges[c], Peter Aniol[d], Heinz Barwig[a], Claus Goessl[a], Florian Lang-Bardl[a], Wolfgang Mitsch[a], Michael Ruder[d]

[a]Universitäts-Sternwarte München, Scheinerstr. 1, D81579 München, Germany;
[b]Max Planck Institut fuer Extraterrestrische Physik, Giessenbachstrasse, D85748 Garching, Germany;
[c]Kayser-Threde GmbH, Wolfratshauser Str. 48, D81379 München, Germany;
[d]Astelco Systems GmbH, Fraunhoferstrasse 14, D82152 Martinsried, Germany

ABSTRACT

The integration of the 2m Fraunhofer telescope started in August 2011 at the Mt. Wendelstein observatory. The logistics of the project are a key problem of the integration as the observatory has no road access. All large or heavy components inlcuding the primary mirror were successfully delivered by helicopter. Meanwhile, they are integrated in the telescope. The special design features of this alt-az telescope are its compactness and the low-ghost wide field optics (0.7 deg. f.o.v. diameter).

We will briefly report on tests of the building and of the telescope system before the telescope moved to the mountain. The integration at the observatory and the first astronomical performances tests of the telescopes are discussed, and a brief update on the status of its instruments is presented. We comment on the cleaning and recoating strategy for the primary mirror based on sample tests.

Keywords: telescope, optics, robotic telescope, Wendelstein observatory

1. INTRODUCTION

The Mt. Wendelstein observatory is operated by the University Observatory of the Ludwig-Maximilian University Munich (USM) for an exclusive use by the astronomers of the USM. As such, it supports the student education at all levels and the science programs of the staff. These scientific projects include programs in combination with other facilities where USM is a partner like e.g. the 9 m Hobby-Eberly-Telescope (HET),[1] and participation in follow-up observations for large surveys where USM is involved (like e.g. PanStarrs1,[2] HETDEX,[3] or eROSITA[4]).

The telescope project was presented in detail during the last SPIE Astronomical Telescopes and Instrumentations conference.[5] The aim of the project was to equip the Mt. Wendelstein observatory[6] with a modern and competitive facility of the 2 m telescope class, together with a suite of instruments.

Here, we report the installation at the observatory which was preceded by an integration and several night of tests at the facility where steel cutting took place. We further briefly report the first performance tests at the observatory including those at the night sky.

2. SITE PREPARATION

The new building was finalized by installing its dome in May 2010. The dome was produced by *Baader Planetarium GmbH*, Mammendorf. The dome offers a slit of 2.8 m free aperture between zenith and horizon where the access near horizon requires to open an extra flap. Its motion is computer-controlled. Two 1 t cranes installed in the dome support maintenance and installation of telescope and instruments. A third crane can lift the telescope primary mirror in a dedicated position into its support cell.

Further author information: (Send correspondence to U.H.)
U.H.: E-mail: hopp@usm.lmu.de Telephone: +49 (0)89 21 80 59 97

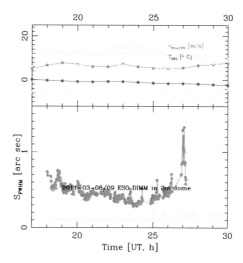

Figure 1. Seeing as measured during the night March 8 to 9, 2011. The seeing monitor was placed on the 2m telescope pier inside the dome. The DIMM was pointed towards various directions (south, north etc as indicated in the lower panel). The measurements of all nights showed little indications that the seeing differs for various directions. At the end of the night, clouds appeared and disturbed the so far excellent seeing. The upper panel show the log of several meteorological parameters as air temperature, wind speed, and position angle (partly recorded by two independent sensors). Units on the Y-axis in degree Celsius, m/s, and 10° degree steps, respectively.

For delivery of telescope parts, especially the primary mirror, a special platform was installed east of the dome building. Platform and building are connected through a 3m wide door and a rail system which supports a carriage system. On that carriage system, the primary mirror (or other heavy loads) can be moved in and out the dome. This operation is foreseen to take place a couple of times over the lifetime of the telescope[7] as there is no space at the observatory to install a mirror coating facility for the primary. The platform and the surrounding roofs of the observatory can serve as storage place for further items (see below).

Preparation of the telescope building further include the installation of an air conditioning system which regulates the day time temperature of the dome interior to the night level of the surrounding atmosphere. Installation, tests, and operational optimization turned out to be a surprisingly challenging task. The same air condition system supports a second circuit which lead off heat capacities from the instruments and provides air conditioning in several technical rooms of the observatory. First experiences with the interactions of telescope structure, telescope optics and air conditioning are discussed by Thiele et al.[7]

Since the installation of the dome was finished more than a year before the arrival of the new telescope, there was ample time to check dome and building on local seeing effects. The *European Southern Observatory* (ESO) kindly made one of its seeing monitors (DIMM) available. The DIMM observed for more than two months in the late winter season of 2011 (February through beginning of April). Most of the nights, the DIMM was installed on the pier of the 2m telescope and was thus operated inside the dome. The dome including its louvers and its daily preparation with the air conditioning system were operated already under the operational model for the 2m telescope.

DIMM measurements were obtained in 17 nights and a total of 132 hours (inside dome). This is certainly not good enough to derive a statistic of typical seeing conditions. The 4967 individual measurements result formally into a median seeing over these 17 nights of 1.0 arc sec FWHM*. But the measurements also proofed that outstanding good seeing of 0.5 arc sec FWHM are available over several hours (Fig. 1). Values around the 1998 median value of ~ 0.7 arc sec dominated in additional four clear nights. Parallel logging of meteorological

---

*A total of 678 seeing measurements in seven nights were obtained with the DIMM placed outside the dome. Those measurement result in a median value of 0.9 arc sec FWHM.

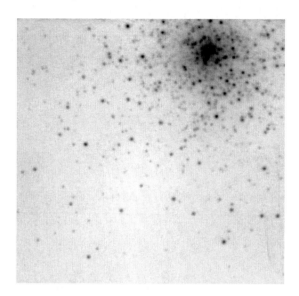

Figure 2. White light images of the globular cluster M15 taken with the new 2m telescope at Hallstadt on a parking lot. Apogee 2k CCD camera with Fairchild CCD. Stack of nine individual exposures of 3 seconds each. Seeing of the exposures about 2.6 arc sec.

data indicated that instability of the seeing or sudden raises of the seeing values could be always attributed to overall atmospheric changes like upcoming clouds. Therefore, we concluded that the dome and dome building does not harm the good seeing reported for the site.[6]

To support secure remote observing, sensitive web cameras for night time use were mounted. One inside the dome allowing to watch telescope and dome motion (Fig. 5), two very similar devices east and west of the dome to follow its motion from outside, and finally two *SBIG* all sky cameras to watch out for clouds.[6]

3. FIRST TELESCOPE INTEGRATION

The three mirrors of the telescope (primary, secondary, and flat third rotating mirror which guides the light into one of the two Nasmyth instrument ports) were all produced by *Lytkarino Optical Glass Factory*, Moscow; LZOS). All three mirrors were fabricated to specs according to interferometric tests down at the company.

The mirrors were securely packed and delivered to Germany by plane. They arrived early in 2011 in Hallstadt (near Bamberg) were the production of the mounting by *Leicht Maschinenbau* was under way following the design drawings of the main telescope contractor *Kayser-Threde GmbH*, Munich, and its prime sub-contractor *Astelco Systems GmbH*, Martinsried. More details on the production of the telescope can be found in the report of Thiele et al.[7] in this proceeding. The system was integrated in late spring 2011, including mirrors, and a first optical adjustment took place.

3.1 First tests

As the transport to the observatory is risky and expensive, a test of the complete system was scheduled in Hallstadt. After integration and mechanical and laser adjustment in the integration hall, the telescope was moved outside that hall and placed on an adjacent parking lot. This enabled three first observing nights for first tests, but naturally under rather poor seeing conditions.

A preliminary pointing model and a preliminary alignment was established. No severe optical aberrations or other failures either from the support systems, the mounting, or the drives could be identified at the level of the available seeing (2 to 3 arc sec). A technical first light images was obtained (see fig. 2) which also indicated a rather good quality. These tests took place in early July. Packing of the telescope parts was scheduled for three weeks and installation of the telescope including optics was scheduled for August and September to not compromise this challenging effort with a possible early winter season start. Thus, testing in Hallstadt was

Figure 3. Left: The dome surrounded by the outdoor crane which lifted heavy parts of the mounting through the dome slit into the dome for immediate integration. The two boxes contain the forks of the mounting. Right: Arrival of the primary mirror in its transport box by helicopter.

limited to three nights only, and we decided to ship the telescope immediately to the observatory. This left a major part of the test program, especially on optical quality, for the observatory.

4. INSTALLATION AT THE OBSERVATORY

From the very beginning, the system design had to take care of an installation concept which allowed to transport the telescope pieces including its optics by helicopter to the observatory.[5,7] Individual pieces (including shipping boxes and tools) should not exceed ~ 4 t as the nearby available heavy-lift helicopters cannot transport heavier pieces to the mountain top.

The delivery of all telescope parts but the mirrors was done on August 9, 2011 after an outdoor crane was installed at the observatory (Fig. 3). The helicopter placed most of the pieces on one of the roofs and the ground plate (Azimuth stator plate) was directly delivered onto one of the mirror carriages. Either carriage or outside crane lifted the parts into the dome where they were directly integrated into the system. The outdoor crane lifted its pieces through the dome slit which required good weather conditions. Major pieces were all in place within a week.[7]

The three mirrors of the telescope optics followed the beginning of September (Fig. 3), again with the same heavy-lift helicopter of *HeliSwiss*. The delivery of the mirrors, which were transported in massive protective housings which included spring mechanism to avoid strong accelerations of the mirrors, was smooth, without problems, and precisely to the thread points of the mirror carriage. September and October saw the integration of the optics and its first alignment[7] (without star), the integration and test of the electronics, and several verification tests of the telescope control software.

5. OPERATIONAL ASPECT

As said, the space at the observatory is too limited for a coating facility which can renew the coating of the primary. The basic strategy to handle this aspect is a protected AL-coating of all three major mirrors which allows to apply frequently cleaning strategies (e.g. washing, CO_2 cleaning). This should lengthen the duty cycle for recoating to a affordable interval (expected 5 yr or more). To enable the recoating outside the observatory, the eastern platform is permanently installed and prepared with a rail system, and the mirror carriages as well as the mirror transport boxes are on store. Thus, the procedure used to install the mirrors[7] (especially the primary) can be also used to disassemble them and transport those heavy optical pieces between Mt.Wendelstein and coating facilities in Europe.

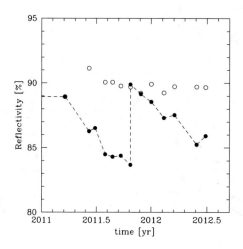

Figure 4. Reflectivity of the Sample 'A' with time. The sample was exposed to typical observing conditions at Mt. Wendelstein by placing it into the dome of the 40 cm telescope which was in nightly use throughout the test. Reflectivity was measured with an IRIS 908 reflectometer under an angle of incidence of 45° roughly every second month. Measurements at 650 nm are shown, open circles show data for the sample kept under clean conditions, filled dots for the one installed in the dome. The jump at ~ 2011.8 for the dome probe demonstrates the effect of washing.

To test the reliability of this scheme, *LZOS* has supplied us with mirror samples coated as applied to the three mirrors of the 2m telescope. These sets of samples were delivered early in the project. One set was kept under clean conditions in a laboratory while another was placed in the dome of the Wendelstein 40 cm telescope, which was full-time operational during the test time. Thus, the second test samples were exposed in a realistic manner to observing conditions. Both sets of samples were measured for their reflectivity every other month with an IRIS 908 reflectometer of Segrif, Montegnee, Belgium.

The test samples placed in the dome were washed about twice a year. Washing was done first with purified water, than with a mix of purified water and Ph-neutral soap, and again with purified water. The samples were immediately dried with clean pressured air.

The measurements show a decay rate of the reflectivity of about ~ 0.08 yr^{-1} (1.00 = 100% reflectivity, 0.00 = 0%), rather independent of wavelength (4 bands between 470 and 880 nm are tested). Further, the sample protected with SiO and the one protected with SiO_2 behaved identical. Washing was applied when the samples approach a reflectivity of $\sim 85\%$, after washing, the a reflectivity of $\sim 90\%$ was recovered (as measured with an angle of incidence of 45°). Over the $\sim$ one years of testing, no overall degrading was so far detected (Fig. 4). A similar test with a different probe was done over 2.5 yr. In this case, the AL coating including a *Wacker-PLASIL* protective coating was kindly supplied by the staff of the *Calar Alto observatory*, Spain. The probe exposed to the observing conditions was meanwhile cleaned four times and the reflectivity always returned near to the starting value of $\sim 90\%$. In other words, no degrading of the protected aluminum could be detected so far. This indicates that the above strategy is realistic.

The primary mirror is under reflectivity control since roughly a year. Its reflectivity decay is about ~ 0.05 yr^{-1} so far (slightly smaller at 880 nm), but it has been much less exposed to observing conditions than the samples.

6. PERFORMANCE TEST AT THE OBSERVATORY

Having the telescope installed at the observatory (Fig. 5) first on sky tests with simple CCD camera systems (a *FLI* system on the one Nasmyth port and an *Apogee* system on the other one) for imaging started in Dec. 2011

Figure 5. Screen shot of the dome internal web camera which allows a remote observer to track telescope and dome behavior. Taken during one of the telescope test nights with minor illumination inside the dome. The very sensitive webcam easily show stars in the open dome slit.

under much better seeing conditions than available in Hallstadt. Some of the images indicate a seeing around 0.5 arc sec FWHM. First light was performed in the night Dec. 19/20, 2011 under less good conditions.

The first aim was to establish a simple pointing model independently for both ports. Observing 20 and 16 Tycho stars, respectively, a formal rms pointing accuracy of 1.9 arc sec was archived. The stars were distributed over the full visible sky.

Tracking and pointing are directly related as the software forces the telescope on a trajectory taking the pointing model and refraction correction continuously into account. Thus, testing the tracking accuracy is a different way of verifying the pointing model. Tests were done on either port and with two position angles separated by 90°. The tests resulted in a similar accuracy for both ports, about 0.35 arc sec per 600 sec, a value which we expect to improve to the requested specification by establishing a denser sampled pointing model.

7. INSTRUMENTS

The first set of instruments consists of two imagers and two spectrographs. A wide field CCD imager (WWFI[8,9]) is almost ready to be installed at the telescope with all its individual components tested.

The field spectrograph (VIRUSW[10,11]) is even already operational and on loan at the McDonald 2.7 m telescope. It is producing scientific data on a regular basis.

A high resolution echelle spectrograph (FOCES[12]) undergoes an upgrade mostly to achieve an improved wavelength stability.[13,14] Right now, it is operated in Munich under laboratory conditions and housed into an additional envelope for pressure and temperature stability tests.[14,15]

An optical-near infrared camera (3kk[16]) which takes simultaneous images in three filters is still under construction. While the CCD and NIR detector systems including their optics, shutters, and filter wheels are already assembled, the containment including the beam splitter housing is still to be done. The relative delay of 3kk is due to limited person-power.

An update of some of these instruments is provided in the Ground-based Instrument Conference of this Amsterdam SPIE meeting.[9,11,15]

Besides guider units on both Nasmyth ports, a Shack-Hartmann system of *OPTOCRAFT GmbH*, Erlangen, will be permanently mounted one of the ports.[17] Grupp et al.,[18] this SPIE meeting, report about the first successful use of the system at the Wendelstein 40cm telescope. The guider unit for the wide field imager is an integral part of the camera and also almost finished. The guider for the other port has just started construction with the detector system already available for laboratory testing.

8. SUMMARY AND OUTLOOK

The rather compact[5] 2m-Fraunhofer Wendelstein telescope was successfully installed at the observatory despite the challenging logistics. The helicopter delivery of mounting and especially optics revealed to be much smoother than expected. Overall, the schedule was severely impacted by the weather which is responsible for some of the delay of the project. Detailed optical testing has still to be performed while writing this contribution. Subsequent alignment might take some further time and on sky testing of the instruments has to be taken into account. In total, we expect first science operation to start in late summer with the wide field camera while the installation of the spectrograph and their testing is foreseen for 2013.

ACKNOWLEDGMENTS

Financial support for the building comes from the *Freistaat Bayern*, the telescope is financed together by the *Freistaat Bayern* and the *Federal Government of Germany (BMBF)*. Support for the instruments comes from the *Excellence Cluster Origin and Structure of the Universe*, and the *German Science foundation (DFG)*. It is a pleasure to thank the employees of the *Staatliche Bauamt München 2* and the *Bauamt Rosenheim* and their project partners for a smooth and successful collaboration. We acknowledge the collaboration with Klaus Hodapp and colleagues in Hilo and Klaus Reif and colleagues in Bonn. Dr. M. Sarazin of the *European Southern Observatory* kindly made available on of the ESO DIMM's for further on-site testing. We thank Dr. Sarazin for his extensive introduction into the system. The project has benefitted from intensive discussions with Drs. L. Noethe, S. Guisard, J. Pirad (all at ESO), U. Thiele (Calar Alto Obs.), S. Barnes, G. Hill, P. McQueen (McDonald Obs.), and R. Hessman (Göttingen).

REFERENCES

[1] Ramsey, L., Adams, M., Barnes, T., Booth, J., Cornell, M., Fowler, J., Gaffney, N., Glaspey, J., Good, J., Hill, G., Kelton, P., Krabbendam, V., Long, L., MacQueen, P., Ray, F., Ricklefs, R., Sage, J., Sebring, T., Spiesman, W., and Steiner, M., "The Hobby-Eberly Telescope," in [*Society of Photo-Optical Instrumentation Engineers (SPIE) Conference Series*], *Presented at the Society of Photo-Optical Instrumentation Engineers (SPIE) Conference* **3352**, 34–46 (1988).

[2] Burgett, W. and Kaiser, N., "The Pan-STARRS Project: The Next Generation of Survey Astronomy Has Arrived," in [*Proceedings of the Advanced Maui Optical and Space Surveillance Technologies Conference, held in Wailea, Maui, Hawaii, September 1-4, 2009, Ed.: S. Ryan, The Maui Economic Development Board., p.E39*], (2009).

[3] Hill, G. J., Gebhardt, K., Komatsu, E., Drory, N., MacQueen, P. J., Adams, J., Blanc, G. A., Koehler, R., Rafal, M., Roth, M. M., Kelz, A., Gronwall, C., Ciardullo, R., and Schneider, D. P., "The Hobby-Eberly Telescope Dark Energy Experiment (HETDEX): Description and Early Pilot Survey Results," in [*Astronomical Society of the Pacific Conference Series*], T. Kodama, T. Yamada, & K. Aoki, ed., *Astronomical Society of the Pacific Conference Series* **399**, 115–+ (Oct. 2008).

[4] Predehl, P., Andritschke, R., Becker, W., Böhringer, H., Bornemann, W., Bräuninger, H., Brunner, H., Brusa, M., Burkert, W., Burwitz, V., Churazov, E., Dennerl, K., Eder, J., Elbs, J., Freyberg, M., Friedrich, P., Fürmetz, M., Gaida, R., Guglielmetti, F., Hälker, O., Hartner, G., Haberl, F., Herrmann, S., Huber, H., Kendziorra, E., von Kienlin, A., Kink, W., Kreykenbohm, I., Lamer, G., Lapchov, I., Meidinger, N., Merloni, A., Mican, B., Mohr, J., Mühlegger, M., Müller, S., Nandra, K., Pavlinsky, M., Pfeffermann, E., Reiprich, T., Robrade, J., Rohé, C., Santangelo, A., Sasaki, M., Schächner, G., Schmid, C., Schmitt, J., Schreib, R., Schrey, F., Schwope, A., Steinmetz, M., Strüder, L., Sunyaev, R., Tenzer, C., Tiedemann, L., Vongehr, M., and Wilms, J., "eROSITA," in [*Society of Photo-Optical Instrumentation Engineers (SPIE) Conference Series*], *Society of Photo-Optical Instrumentation Engineers (SPIE) Conference Series* **8145** (Sept. 2011).

[5] Hopp, U., Bender, R., Grupp, F., Barwig, H., Gössl, C., Lang-Bardl, F., Mitsch, W., Thiele, H., Aniol, P., Schmidt, M., Hartl, M., Kampf, D., and Schöggl, R., "The compact, low scattered-light 2m Wendelstein Fraunhofer Telescope," in [*Society of Photo-Optical Instrumentation Engineers (SPIE) Conference Series*], *Society of Photo-Optical Instrumentation Engineers (SPIE) Conference Series* **7733** (July 2010).

[6] Hopp, U., Bender, R., Goessl, C., Mitsch, W., Barwig, H., Riffeser, A., Lang, F., Wilke, S., Ries, C., Grupp, F., and Relke, H., "Improving the Wendelstein Observatory for a 2m-class telescope," in [*Society of Photo-Optical Instrumentation Engineers (SPIE) Conference Series*], *Presented at the Society of Photo-Optical Instrumentation Engineers (SPIE) Conference* **7016** (July 2008).

[7] Thiele, H., Ageorges, N., Kampf, D., Hartl, M., Egner, S., Aniol, P., Ruder, M., Hopp, U., Bender, R., Grupp, F., Barwig, H., Gössl, C., Lang-Bardl, F., and Mitsch, "New Fraunhofer Telescope Wendelstein: Assembly, Installation, and current Status," in [*Society of Photo-Optical Instrumentation Engineers (SPIE) Conference Series*], *Society of Photo-Optical Instrumentation Engineers (SPIE) Conference Series* **8444** (2012).

[8] Gössl, C., Bender, R., Grupp, F., Hopp, U., Lang-Bardl, F., Mitsch, W., Altmann, W., Ayres, A., Clark, S., Hartl, M., Kampf, D., Sims, G., Thiele, H., and Toerne, K., "A 64 Mpixel camera for the Wendelstein Fraunhofer Telescope Nasmyth wide-field port: WWFI," in [*Society of Photo-Optical Instrumentation Engineers (SPIE) Conference Series*], *Society of Photo-Optical Instrumentation Engineers (SPIE) Conference Series* **7735** (July 2010).

[9] Gössl, C., Bender, R., Lang-Bardl, F. Hartl, M., Kampf, D., Sims, G., "Commissioning of the WWFI for the Wendelstein Fraunhofer Telescope," in [*Society of Photo-Optical Instrumentation Engineers (SPIE) Conference Series*], *Society of Photo-Optical Instrumentation Engineers (SPIE) Conference Series* **8446** (2012).

[10] Fabricius, M. H., Barnes, S., Bender, R., Drory, N., Grupp, F., Hill, G. J., Hopp, U., and MacQueen, P. J., "VIRUS-W: an integral field unit spectrograph dedicated to the study of spiral galaxy bulges," in [*Society of Photo-Optical Instrumentation Engineers (SPIE) Conference Series*], *Presented at the Society of Photo-Optical Instrumentation Engineers (SPIE) Conference* **7014** (Aug. 2008).

[11] Fabricius, M. H., Grupp, F., Drory, N., Bender, R., Hopp, U., Arns, J., Barnes, S., Gössl, C., Hill, G. J., and Lang-Bardl, F., "VIRUS-W: commissioning and first-year results of a new integral field unit spectrograph dedicated to the study of spiral galaxy bulges," in [*Society of Photo-Optical Instrumentation Engineers (SPIE) Conference Series*], *Presented at the Society of Photo-Optical Instrumentation Engineers (SPIE) Conference* **8446** (2012).

[12] Pfeiffer, M. J., Frank, C., Baumueller, D., Fuhrmann, K., and Gehren, T., "FOCES - a fibre optics Cassegrain Echelle spectrograph," *Astronomy & Astrophysics Supplement Series* **130**, 381–393 (June 1998).

[13] Grupp, F., Udem, T., Holzwarth, R., Lang-Bardl, F., Hopp, U., Hu, S.-M., Brucalassi, A., Liang, W., and Bender, R., "Pressure and temperature stabilization of an existing Echelle spectrograph," in [*Society of Photo-Optical Instrumentation Engineers (SPIE) Conference Series*], *Society of Photo-Optical Instrumentation Engineers (SPIE) Conference Series* **7735** (July 2010).

[14] Grupp, F., Brucalassi, A., Lang, F., Hu, S. M., Holzwarth, R., Udem, T., Hopp, U., and Bender, R., "Pressure and temperature stabilization of an existing chelle spectrograph II," in [*Society of Photo-Optical Instrumentation Engineers (SPIE) Conference Series*], *Society of Photo-Optical Instrumentation Engineers (SPIE) Conference Series* **8151** (Sept. 2011).

[15] Grupp, F., Brucalassi, A., Feger, T., Holzwarth, R., Udem, T., Lang, F., and Bender, R., "Stability achieved for environmentally stabilized FOCES echelle spectrogarph (FOCES stability IV," in [*Society of Photo-Optical Instrumentation Engineers (SPIE) Conference Series*], *Society of Photo-Optical Instrumentation Engineers (SPIE) Conference Series* **8446** (2012).

[16] Lang-Bardl, F., Hodapp, K., Jacobson, S., Bender, R., Gössl, C., Fabricius, M., Grupp, F., Hopp, U., and Mitsch, W., "3kk: the Optical-NIR Multi-Channel Nasmyth Imager for the Wendelstein Fraunhofer Telescope," in [*Society of Photo-Optical Instrumentation Engineers (SPIE) Conference Series*], *Society of Photo-Optical Instrumentation Engineers (SPIE) Conference Series* **7735** (July 2010).

[17] Grupp, F., Lang, F., Bender, R., Gössl, C., and Hopp, U., "A multi-instrument focal station for a 2m-class robotic telescope," in [*Society of Photo-Optical Instrumentation Engineers (SPIE) Conference Series*], *Presented at the Society of Photo-Optical Instrumentation Engineers (SPIE) Conference* **7014** (Aug. 2008).

[18] Grupp, F., Hu, S., Lang, F., Bogner, S., Becker, M., Bode, A., Lambrecht, J., Hopp, U., and Bender, R., "Test system for a Shack-Hartmann sensor based telescope alignment demonstrated at the 40cm Wendelstein Telescope," in [*Society of Photo-Optical Instrumentation Engineers (SPIE) Conference Series*], *Presented at the Society of Photo-Optical Instrumentation Engineers (SPIE) Conference* **8444** (2012).

SALT's Transition to Science Operations

David A. H. Buckley[1,a], J. C. Coetzee[b], Steven M. Crawford[b], Kenneth H. Nordsieck[c], Darragh O'Donoghue[b] & Theodore B. Williams[d]

[a]SALT Foundation, c/o SAAO, Observatory 7935, Cape Town, South Africa
[b]South African Astronomical Observatory, Cape Town, South Africa
[c]Space Astronomy Lab, University of Wisconsin, Madison, Wisconsin, USA
[d]Department of Physics and Astronomy, Rutgers University, New Jersey, USA

ABSTRACT

The Southern African Large Telescope (SALT) began its re-commissioning phase in April 2011 following the completion of remedial engineering work on the telescope and the major science instrument, the Robert Stobie Spectrograph (RSS). The engineering work required modifications to the spherical aberration corrector, in order to improve the telescope's image quality, and RSS, to improve its throughput. Positive test results included delivery of sub-arcsecond images, essentially meeting the original telescope image quality specifications and exhibiting none of the previous field-dependent aberrations, while the RSS has shown greatly improved efficiency performance. SALT has since transitioned to science operations, as from 1 September 2011, following the first open call for charged science proposals from the SALT partners. This paper discusses the current performance of SALT and it First Generation instruments, some initial science results, the proposal process and the operational model for the telescope.

Keywords: large telescopes (SALT); telescope and observatory operations, telescope control systems, telescope efficiency; telescope software systems and tools; automated astronomical data reductions

1. BACKGROUND AND HISTORY

The Southern African Large Telescope (SALT) is now one of five 10-m class segmented mirror telescopes, the only one in the southern hemisphere (Buckley, Meiring & Swart 2006), and is closely modeled on the pioneering design of the fixed-altitude Hobby Eberly Telescope (HET) at McDonald Observatory, Texas (Ramsey et al. 1998). HET and SALT, being analogues of the Arecibo radio telescope (although *not* zenith pointing), comprise of a 10-m × 11-m hexagonal primary mirror array of 91 identical 1.2-m ×1.0-m hexagonal mirrors with spherical surfaces. A 4-mirror spherical aberration corrector (SAC) is mounted on a moving tracker at the prime focus, which also carries all of the science instruments. The telescope can only access an annular region of the sky, 12° wide, centred at the zenith, which represents ~12% of the sky (at an airmass of < 2.5). The geometry implies that only objects in the declination range +10° to −75° are observable, and the total uninterrupted observing times typically range from 1 – 2 hours, although they can be extended to 4 – 5 hours at the extreme declinations if the telescope is re-pointed in azimuth. This design also means that the effective aperture varies during an observation, at worst being equivalent to an un-obscured 7-m diameter mirror, and at best equivalent to a ~8.5-m diameter mirror. The changing entrance pupil geometry poses additional complexities when it comes to calibrations.

The SALT Foundation Pty. Ltd. is a private non-profit company, set up to fund SALT's construction (including instruments) and operation. The endeavour is a collaboration between 13 universities and institutes from South Africa, Germany, India, New Zealand, Poland, the United Kingdom and the United States of America. South Africa is the largest partner, with ~1/3 of the current observing shares, with the USA comprising another ~1/3 share and the remaining partners taking up the remaining ~1/3. Participation in SALT was attractive to its partners for a number of reasons, including:

- Access to an affordable 10-m class telescope with versatile observational capabilities
- Access to the southern hemisphere
- Affordable ownership (~$20M for telescope; ~$8M for first-light instruments; ~$3M per annum for operations)
- A good observatory site (50% photometric; 75% spectroscopic; 0.9 arcsec median seeing; dark; dry)

- Synergies with other facilities accessible by SALT partners (e.g. HET, WIYN, SOAR)
- A relatively inexpensive and flexible queue-scheduled service operation
- Assisting in the development of science & technology and educational opportunities in South Africa

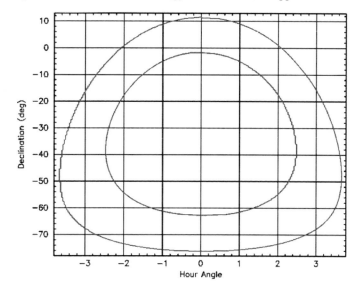

Figure 1: The SALT visibility annulus. Only objects inside the annulus are visible to SALT.

2. COMPLETION AND COMMISSIONING

SALT was inaugurated in Nov 2005, when the construction phase was essentially completed (Buckley et al. 2006). Two of the First Generation instruments, were mounted on the telescope and a third was still in its planning stages. Following this, SALT entered an intensive period of commissioning and performance verification, which included the initial science observations. ``First Science'' was announced in 2006, following the beginning of commissioning in mid-2005 of the UV-Visible imaging camera, SALTICAM (O'Donoghue et al. 2003), which culminated in the first SALT science paper (O'Donoghue et al. 2006). In parallel, the commissioning of the Robert Stobie Spectrograph (RSS; formerly known as PFIS; Burgh et al. 2003), began in October 2005 and continued through to Nov 2006 (Buckley et al. 2008). RSS is a versatile, but highly complex, imaging spectrograph, supporting multiple modes, namely:

- Longlsit and Multi-Object Spectroscopy from 320 – 900nm using a suite of fully tunable VPH diffraction gratings
- Fabry-Perot imaging spectroscopy, including dual etalon and tunable filter modes
- Polarimetry, both imaging and spectro polarimetry, in linear, circular or all-Stokes modes
- High time resolution observations, by virtue of frame transfer CCDs

It was during this commissioning period that problems with the telescope's image quality and the throughput of RSS were uncovered, which led to the eventual removal (in Nov 2006) of RSS for optical repairs (Buckley *et al.* 2008; Nordsieck, Nosan & Schier 2010).

Despite the problems with image quality, SALT was scheduled regularly (~50% of nights) for science observations from 2006 – 2009, albeit with a restricted capabilities, also in part due to the continuing engineering work that needed to be carried out by the operations team. Much of this was related to various subsystems not fully completed before the Project Team left in late 2005/early 2006. Three major performance issues were uncovered during commissioning: the aforementioned problems with telescope image quality and low spectrograph throughput, and the failure of the capacitive mirror edge sensor system to meet its performance requirements. These are discussed in some more detail in the next section. The image quality was finally fixed by mid-2010, which was followed by the reinstallation of the repaired RSS and the refurbished SALTICAM imager. Science commissioning programs were then initiated to characterise the performance of the telescope and instruments (Buckley 2012). This phase transitioned into the first science semester in September 2011, when 75% of the telescope time was set aside for charged science time for the SALT partners. The remaining 25% of the time was retained to complete the commissioning of outstanding instrument modes.

3. INITIAL PERFORMANCE PROBLEMS

Science commissioning began in 2006, although there was still significant continuing engineering work being carried out by the operations team, much related to various subsystems not completed before the Project Team left in late 2005/early 2006. Examples of these included the CO_2 mirror cleaning system, the atmospheric dispersion compensator (ADC) and exit pupil baffles, the SALTICAM auto-guiding system, the Observation Control System (OCS) and the mirror edge sensing system (SAMS). Most of these were completed by the time SALT started to be re-commissioned in early-2011, the exceptions being the full-up OCS and the SAMS. The former is an on-going software project to fully integrate the instrument control and configuration with the TCS in order to improve observing ease and efficiency. A first stand-alone build has already been implemented and the integrated system is expected to be fully complete during 2012.

Three major performance issues were uncovered during initial commissioning in 2006: 1.) the failure of the SAMS edge sensing system to meet its design requirements, 2.) the aforementioned problems with telescope image quality, and 3.) lower than expected spectrograph throughput, particularly in the blue (<400nm). The nature of these problems and their cure are now discussed.

Figure 2: A sub-arcsecond (FWHM ~ 0.85") quality SALT image following the image quality fix.

3.1 SAMS: the SALT mirror edge sensor system

SALT, like all segmented mirror telescopes, requires an active optics system to maintain the alignment of the mirror segments against deformation caused by flexure or temperature changes. This is achieved by a closed loop system consisting of mirror actuators (3 per segment), which tip, tilt and piston the individual segments, and a mirror position-measuring system, using edge sensors (SAMS). The initial system on SALT (Swiegers & Gajjar 2004) was based on a capacitive-sensing system, as used on the Keck telescopes. However, following protracted tests, the capacitive sensors on SALT were found to be inadequate to measure relative mirror motion, primarily owing to their intrinsic sensitivity to relative humidity (R.H.) variations, which is worse at Sutherland compared to Hawaii (diurnal change in R.H. 50% for ~40% of the time). The most likely reasons for the capacitive sensors not meeting the SALT specifications are micro-condensation on the surface of the sensor, leading to moisture absorption by the sensor components, exacerbated by dust. In early 2008 SALT declared the SAMS capacitive system a failure and we have since been pursuing a program to identify alternative technologies for mirror edge-sensing (e.g. Buckley *et al.* 2010; Menzies *et al.* 2010). This work has paralleled investigations by other groups building segmented mirror telescopes (e.g. HET, LAMOST, TMT and E-ELT). As part of this program we have been involved in the development and testing of inductive edge sensors (which are immune to R.H. effects), from three manufacturers, and an optical interferential edge sensor (Buckley *et al.* 2010). The prototype testing phase is now complete and a contract has just been awarded for a new SAMS system, based on inductive sensing.

3.2 Fixing SALT's image quality

The image quality problems that were apparent during the early commissioning phase (2005/2006) were mostly manifested as field-dependent aberrations, usually becoming noticeable for field angles in excess of ~2 arcmin. An extensive campaign to diagnose the cause of the image quality problem (O'Donoghue *et al.* 2008), lasting the better part of a year, indicated that the alignment of the four mirrors inside the Spherical Aberration Corrector (SAC) was the cause. In particular the problem was due to deficiencies in the interface of the SAC to the telescope, which allowed for thermally and mechanically generated stresses to transfer through to the opto-mechanics of the SAC mirrors, causing de-centre and tip/tilting of the mirrors which then resulted in the field dependent aberrations (O'Donoghue et al. 2010). The design, fabrication, testing and installation of a new interface, employing a kinematic mount, then followed from April 2009 to July 2010. Thereafter the telescope's optical performance was thoroughly tested, and results show that SALT is now delivering acceptable image quality (see Fig. 2).

3.3 Fixing RSS Throughput

Not long after the installation of RSS and the commencement of commissioning observations in 2006, it was discovered that the overall throughput performance was significantly less than expected, as low as 20 – 40% of predictions, and much worse at shorter wavelengths (see Fig. 4). In-situ testing of the RSS optics using two different wavelength lasers (at 365 and 670 nm) and calibrated diode detectors indicated that the throughput losses occurred in both the collimator and camera optics, which necessitated their removal and return to the US optomechanical vendor for further diagnosis and repair. The throughput losses were found to be due to two reasons: (1) a poor multi-layer anti-reflection coating on the camera's field-flattening lens, and (2) absorption losses in all of the lens multiplets. The former mostly affected the mid-wavelengths (with a dip at 550nm), and was subsequently corrected by recoating that element. The multiplet losses were mostly in the UV-blue region (< 400nm) and were diagnosed to be a result of material incompatibility between index-matching lens-coupling fluid, the bladder and tubing containing it, and O-ring seals. This was despite the vendor's spec sheets indicating that the materials were compatible. It was found that a chemical reaction between these materials produced a UV-absorbing polymer, which was responsible for the attenuation (Buckley et al 2008; Nordsieck et al. 2010). Following this discovery an investigation began to find alternative materials, which were then tested successfully before replacing the original materials in the multiplets. Unfortunately, during this period of repair, a CaF_2 element was broken and a NaCl lens damaged, requiring re-fabrication of the former and re-polishing of the latter. The repaired and reassembled optics were then returned to SALT in July 2009 and reinstalled into RSS.

4. SALT RECOMMISSIONING

During the repair period for RSS and the telescope engineering stand-down to affect the modifications to the SAC, the SALT Operations and Instrument teams undertook a number of tasks to improve the performance, reliability and operational efficiency of both the telescope subsystems and the instruments. For the former these focussed on both hardware and software improvements, particular in regard to the Prime Focus Payload. The SALTICAM autoguider was also completed during this time, as well as some modifications to the control system of the instrument. These were completed in early 2011 when the refurbished instrument was reinstalled on the telescope. Improvements to the RSS mechanics were also affected during the stand-down, and thorough testing of all the mechanisms and controls were undertaken. This culminated in early April 2011 with the reinstallation of RSS onto the telescope and the start of its recommissioning.

A call for commissioning proposals in 2010 resulted in ~90 science programs from the SALT community being accepted on a best-efforts basis, and which were subsequently initiated once the instruments were reinstalled. These programs were mostly completed by the beginning of the first semester of "normal" science operations, which began in September 2011. One of the first observations with RSS was narrow-band imaging of the globular cluster ω Centauri, to test both the image quality and to determine the astrometric parameters of the telescope and instrument. Narrow-band images in several filters were taken for this purpose and a composite image is shown in Figure 2.

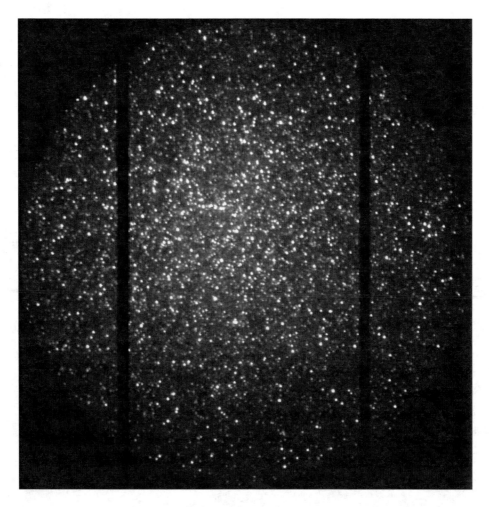

Figure 3: A composite three-filter image of the globular cluster ω Cen taken with RSS during commissioning in April 2011.

During the recommissioning period, a fortuitous and rare (last one in 1966) outburst of the recurrent nova, T Pyx, occurred, which became a Target of Opportunity (ToO) program for SALT. This ably demonstrated SALT's ability to react quickly to such transient events, which is one of its strengths and a prominent science driver. A variety of observations were undertaken, and UV spectroscopy with RSS has confirmed the instrument's improved response following the optics repair, with a clear detection of spectral features down to the atmospheric cut-off, with detection of HeII 320.4 nm and the Bowen fluorescence line at 313.3 nm; see Fig. 4).

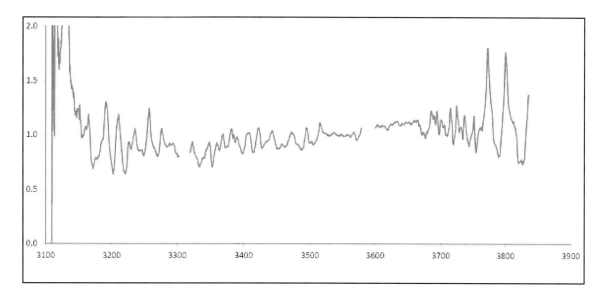

Figure 4: A UV-blue spectrum of the recurrent nova T Pyx taken with the RSS during its outburst in 2011.

RSS spectropolarimetry of T Pyx was also conducted on 6 nights over ~1 month (11 May – 14 June). A 900l/mm VPH grating was used, with wavelength coverage 308 – 621 nm and at a spectral resolution of R ~ 900 with a 1.5 arcsec slit. The H I, He I, N II P Cyg lines all show depolarization, indicative of intrinsic polarization at a level of ~0.6%. The data exhibit P Cyg line de-polarization and an unusually low value for the interstellar polarization (0.5 – 0.7%), given its purported distance of ~3 kpc. An example of T Pyx spectropolarimetry is shown in Fig. 5.

The polarimetric features are likely due to electron scattering polarization in a distorted nova atmosphere, as seen in supernova polarization. Strong complex polarization features are seen in the Bowen fluorescence features, which to our knowledge has not been seen before. This suggested looking for other fluorescence lines in polarization, for which some are likely to be magnetically sensitive. Polarized flux features_remain after the interstellar polarization is removed and P Cyg features remain in absorption, implying that the line absorption is not patchy, which could cause intrinsic polarization enhancements, as in SNe.

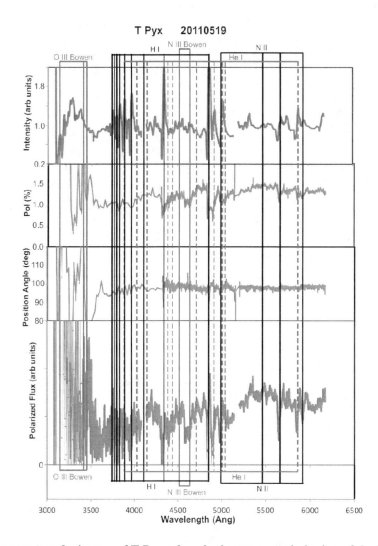

Figure 5: Linear spectropolarimetry of T Pyx taken during re-commissioning of the RSS in May 2011.

Regrettably, further polarimetry observations with RSS was halted in Nov 2011 when the polarizing beamsplitter mosaic (a 3 × 3 array of Wollaston prisms) sprang a leak of lens coupling fluid, necessitating its removal for repair. This is currently being modified to avoid the use of lens fluid altogether in favour of an optical coupling grease or pliant glue.

5. SALT OPERATIONS

In designing SALT, the basic overall operational criteria aimed at for the total system (i.e. telescope) can be condensed into several key efficiency parameters, as presented in Table 1. The relatively large fraction for "Engineering & Calibrations" reflects the additional regular loss of observing time needed for mirror alignment, which at present can amount to 10-20% of a night, depending upon conditions.

A new mirror segment edge sensing system has been designed, based on inductive sensor technology, and will be installed over the next ~2 years. This will reduce the need for optical alignments (using the Shack-Hartmann wavefront camera at the mirror array centre of curvature) of the mirror array from several times a night, to once every ~5 days. In principle the alignment can also be started during twilight and even in overcast conditions, which will help in lessening the impact on science operations. SALT's design goals with an active mirror control system using edge sensors and actuators, reduced this loss – by design – to 0%, but the current lack of this on SALT reflects this larger fraction as at May 2011. In addition, the active mirror control will improve overall image quality by delivering a more optimal PSF, thus boosting throughput. At present observations are done with no control, which mean that image quality is inevitably sub-optimal, particularly as conditions change.

Table 1: SALT efficiency

Telescope	Bad weather	Faults	Engineering and Calibrations	Science time (incl. overheads)
SALT (original spec)	25%	3%	7%	65%
SALT (May 2012)	35%	8%	9%	47%

Procedures for conducting calibration observations are still being optimally defined, and are done both during the night (e.g. spectral calibrations with arc lamps) and during the day (e.g. flat fields simulating the pupil motions of completed observation). SALT's instruments, particularly the main work-horse instrument, the Robert Stobie Spectrograph (RSS), has various observing modes, and potentially many such modes will be used in a single night, all requiring relevant calibration observations.

In all aspects of both telescope and instrument design, the operational philosophies adopted have been purposely developed for a queue scheduled operation where getting to the target and observing it in whatever available instrument mode happens in the fastest and most efficient manner. In addition, instrument mode changes (i.e. from SALTICAM to RSS) can take place by simply inserting a fold mirror (< 60s).

Unlike all of the existing SAAO telescopes, the approved observational programs undertaken with SALT are not conducted by their respective proposers, but rather by SALT astronomy operations staff, namely the SALT Operator and SALT Astronomer. The manner in which a typical observation with SALT progresses, from the proposal phases to the relevant Time Allocation Committees (TACs), through to the eventual observation and data dissemination is summarized in the Table 2. SALT users can apply for SALT observations twice a year, with observing semesters running from 1 May – 31 October and 1 November – 30 April.

Table 2: The full SALT proposal & observation sequence

Step #	Activity	Description	Responsibility
1	Complete Phase I web-based proposal form using a basic template	Write scientific justification and use web based tools to ascertain viability	Proposal Principal Investigator (PI): - completion of Phase 1 form SALT Astronomers Operations: - provision of web tools, forms
2	Submit Phase I proposal	Send to SALT Astronomy Operations	PI
3	Phase I technically reviewed and sent to respective partner TACs	Phase I proposals loaded into database. SALT Astronomy Operations perform tech feasibility & determine statistics	SALT Astronomy Operations
4	Phase I proposal reviewed and time allocated	PI alerted of allocation	TAC and SALT Astronomy Operations
5	Complete Phase II web-based proposal form	More detailed time-line and telescope and instrument configurations defined; SALT Astronomers (SAs) check viability, etc.	PI to complete and submit to SALT; SAs provide Phase II tools & assistance
6	Plan weekly observation priorities	Broad look at viable programs for coming week; identify high priority programs; coordinate engineering or repair activities; define required filters/etalons for the week	SALT Astronomy Operations; SALT Technical Operations for engineering
7	Schedule observations, including calibrations	Review target availability (continuously during the night) and observing conditions; schedule suitable observing block(s) for viable accepted Phase II programs target(s)	SA on duty at the telescope, using the Observation Planning Tool (OPT) & program display GUIs
8	Prepare to observe	Select specific observing block/target; send info to Observation Control System (OCS)	SA, using the OCS
9	Acquire target; choose & configure instrument; focus; acquire and setup	Point telescope and guide probes	SALT Operator (SO) and SA; using the telescope control GUI (SOMMI) and instrument GUIs.

	guide/focus star		
10	Conduct observation	Take any requested calibration observations (before, during, after) if requested (e.g. arcs); begin on-sky observations	SA using instrument GUI; SO using calibration system GUI
11	Monitor observation	Check telescope guidance & focus; Run quick-look data reduction tool SALTFIRST; assess data quality (e.g. S/N)	SO using guidance/focus image display GUI SA using SALTFIRST and instrument detector GUIs
12	Complete observation; store and transfer data; complete logs	Data stored on local instrument computer & copied to quick-look & quality control computer (QCPC); complete on-line logging as required	Automatic processes plus manual log entries by SO and SA
13	Primary mirror alignment	Assess delivered telescope image quality; look at MASS-DIMM measurement; optically realign mirror with Shack-Hartmann wavefront sensor at centre of curvature	SA (assessment) SO (alignment; 20-30 mins)
14	If observing conditions acceptable, return to step 7	Assess weather, seeing, transparency	SO & SA using Environmental Display System (EDS; includes meteorological data; all-sky camera; MASS-DIMM seeing)
15	End of night procedures	Do twilight calibrations (observe standards) if appropriate; park telescope & instruments and close down	SO & SA
16	Plan and schedule day calibrations	Mainly for accurate flat fielding using moving pupil baffle	SA & SO
17	Data delivery and problems	Check data was sent to Cape Town HQ & pipeline completed OK; look at nightlogs for any issues; contact PIs if needed; liaise with Tech Ops staff	Cape Town-based SALT Astronomy Operations staff

The observational efficiency of SALT is mostly determined by the time it takes various telescope and instrument subsystems to perform their respective tasks and the timelines, specified in the SALT Systems Specification, is shown in Table 3. The times for completion of these tasks are given as upper limits, where applicable. Where no times are given, the process is considered to be instantaneous (e.g. issuing of a stop/start command via a mouse click). The telescope slewing will involve the parallel execution of several TCS commands (e.g. dome rotate, structure rotate, tracker position, payload configuration). Most of these commands are transparent to the SALT Operator, who will simply issues a command to 'go to next target' via a single mouse click on a GUI, rather than issuing a sequence of subsystem commands.

Likewise, the SA and SO actions involving the observation itself occur in parallel, like the assessment of data quality (e.g. S/N ratio, spectral resolution, flux) and adjustment of the observing schedule as conditions or circumstances change. An observation block (the smallest scheduled unit of a program) will more often than not consist of a set of repeat exposures, during which the data quality of the penultimate observation can be assessed in real time as the current observation is still in progress using the quick-look analysis tool (SALTFIRST). Adjustment of subsequent exposure times, as necessary, is then possible, provided the observation block time is not exceeded. Many observations have observing blocks closely matching the maximum allowable track time (as determined by the target position and time), inclusive of any acquisition and setup overheads (typically 300 – 600s depending on complexity).

Table 3: The SALT observation timeline

System	Step 1	Step 2	Step 3	Step 4	Step 5	Step 6	Step 7
TCS (SA/SO actions)	Confirm target selection from list. Point telescope				Set onto target, guide star and adjust field orientation, including slitmask adjustments (<360 s)	Observation of science target. Assess data Check schedule (< 2 h)	Stop track. Select next targets

Structure		Lift up (<20 s)	Rotate azimuth (<130 s)	Lower (<30 s)			
Tracker		Slew to target (x, y, z, θ, φ, ρ) (<120 sec)			Open loop tracking of target and command of guidance probe offsets (<60 s)	Start closed loop guiding	
Dome		Rotate dome to required azimuth (<180 s)					
Active Mirrors*		Suspend primary mirror alignment control			Continue alignment control of the primary mirrors (<15 sec)		
Calibration System					Arc/flat calibration (<60s)		Arc/flat (<120s)
OCS Configure Telescope/ Instrument	Select program. Send mode & config to instrument computer	Instrument configuration done (slit, grating, grating angle, articulation angle, filters, etalons/filters (<180s)					

SALT observations are scheduled in real time by running the Observation Planning Tool (OPT), a LabVIEW program which interrogates the MySQL Science Database (SDB) which contains all observing program information. A search can be made of all active observing blocks currently accessible to SALT, and the resulting list can then be filtered to select appropriate programs (e.g. for the current Moon, seeing and transparency conditions). The start/end track times are displayed, as well as other information, and the SALT Astronomer (SA) can then select appropriate observing blocks and add them to a schedule.

The schedule can then be sent to the SALT Operators control GUI from which the telescope can be pointed to a particular object with a single "point to target" command, which command the Telescope Control System. This sets in motion all of the various telescope subsystems to move the telescope to the target. In parallel, the SA can load the defined instrument configuration for the observing block into the Observation Control System, which configures the instrument (e.g. selects and inserts appropriate grating, filter, etalon, slit and rotates to the requested grating and camera angles, and position angle). The OCS is a sophisticated tool which allows for telescope – instrument communications, for example setting up dithering patterns, control nod & shuffle modes, etc.

A graphical display of the observing blocks is also now provided by a GUI, based on the SALT visibility plot (Fig. 1), which displays the instantaneous positions (at any given time) of all active targets, colour coded for project priority. This allows, at a glance, the SA to see which alternative programs are available, or will be available soon, or will be lost from the visibility annulus. Blocks can also be filtered on current observing conditions.

6. SALT PROPOSAL SUBMISSION TOOL

Observing with SALT involves three separate phases. In phase one the Principal Investigator (PI) uses the Principal Investigator Proposal Tool (PIPT) to submit a proposal containing a scientific justification and basic information about the planned observations. Submitted proposals are stored in the Science Database (SDB). The Time Allocation Committees of the partners included in the proposal use the Web Manager to view the proposal and to allocate observing time to it. In phase two the PI downloads the accepted proposal (together with the information how much time has been allocated to it) into the PIPT, adds the detailed information required by SALT Astronomy Operations for performing the observations and resubmits the proposal. In the third and final phase the SALT Astronomers carry out the requested observations, and the observational data, and ancillary information (e.g. environmental and guide star data) are made available to the PI through a data download FTP site and eventually through a so-called Web Manager interface, accessible through any internet browser.

Various tools are provided to aid the PI in planning the observations: instrument setups can be configured with the SALTICAM simulator and the RSS simulator. The availability of targets can be checked with the Visibility Tool. Slitmasks for multiple object spectroscopy (MOS) with RSS can be created with the RSS Slit Mask Tool. These tools are explained in detail in the following sections. With the exception of the Web Manager, all the tools are Java applications and can be downloaded as a Windows executable, a Mac OS X application or a platform-independent jar file from (http://www.salt.ac.za/wm).

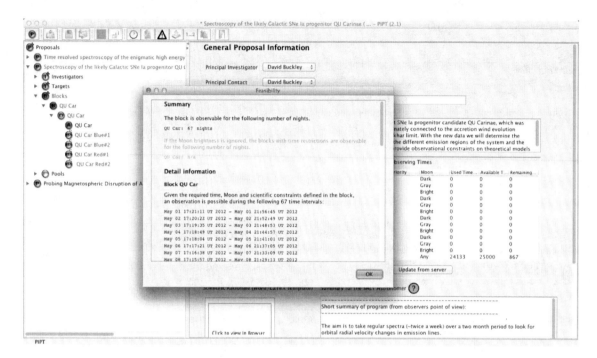

Figure 6: Block visibility information display in the SALT PIPT

For Phase 1 proposals, only a table of target names and positions need to be completed in the Phase 1 form, including the Moon conditions required for each. Other conditions, like maximum seeing and degree of atmospheric transparency are also stipulated for the proposal as a whole. Both mandatory and optional targets are allowed, where the latter are a list of up to 50 targets (with potentially different priorities), from which a subset is expected to be observed, the choice of which is entirely up to the SALT Astronomers, for flexibility. The last information supplied in the Phase 1 proposal is a list of instrument modes and configurations.

Once a proposal is accepted, PIs can then complete their Phase 2 proposal, starting by converting their existing Phase 1 proposal, thus avoiding the need to duplicate information already entered. The major additional information added in the Phase 2 proposal is: 1.) finding charts for all targets or fields, for which a tool is provided to generate these in a standard format (see Fig. 7), 2.) guide star information (optional), 3.) observation block definitions, including instrument and observation configurations, and 4.) calibration requirements (e.g. arcs, flat fields), whether done immediately prior to or after the observations (which are charged), or during twilight or daytime (which are free).

Observation blocks are the smallest schedulable unit, and contain one acquisition observation and a list of science observations, potentially with different instruments/modes and repeat exposure iterations or cycles of repeated configurations (e.g. filter cycles). Each observation in turn contains a list of instrument configurations. The software checks total observing time for a block doesn't exceed the maximum track length available for the observed target.

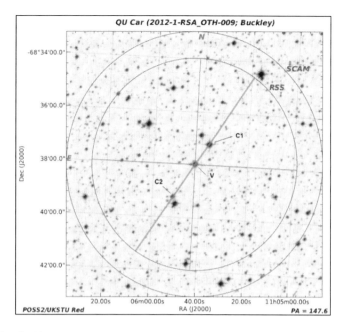

Figure 7: An example of a finding chart generated by the SALT finding chart tool, showing the field of view of the two current instruments (SALTICAM and the RSS), with the slit position for the latter indicated.

7. SALT VISIBILITY TOOL

As mentioned in section 1, due to the fixed elevation angle of SALT's primary mirror (37°), target visibility is constrained to an annulus in the sky. The available "track length", i.e., the amount of time SALT is able to follow continuously any given target, changes according to date, time and target position in the sky. In order to check the visibility and available track length the PI may use the SALT Visibility Tool, a Java application which can be launched from the PIPT or run as a standalone tool. After the observation date and the target location has been entered, this tool displays a plot of the target visibility during the specified night. If the PI clicks in this plot, the track length available from the clicked time onwards is calculated (see Figure 8).

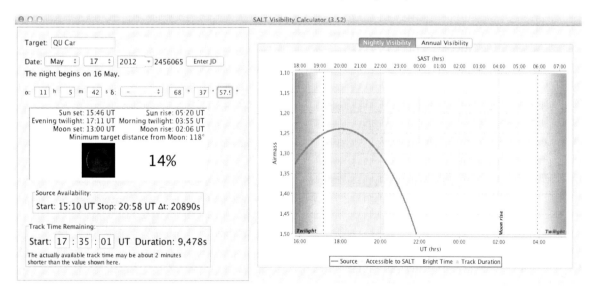

Figure 8: The SALT Visibility Tool, used to see when objects are visible to SALT on a given night.

The inverted parabolic curve shows the location of the target as a function of time. The target is visible if it is located in the horizontal shaded area, defined in the airmass range 1.16 (elevation = 59°) to 1.37 (elevation = 47°). The available track length is highlighted by a horizontal pink shaded area and is also given at the bottom left of the window. Other quantities shown in tool for the given date are the sun rise and sun set times, the start and end times of the evening and morning twilight, the moon rise and moon set times, the lunar phase and the minimum angular distance between the target and Moon. Furthermore, the Visibility Tool contains a plot displaying the visibility over the course of the whole year, including when new & full Moon occur.

8. INSTRUMENT SIMULATION TOOLS

In order to plan an observational setup with any of the SALT instrument, PIs can use the respective Java simulation tools. Currently they exist for the two First Light instrument, SALTICAM and RSS. These tools can be run as a standalone application, but it may also be launched from the PIPT. The simulation is a two step process, the first step being to define a target spectrum. The simulators offers the user a choice from a list of asteroid, blackbody, galaxy, stellar atmosphere (Kurucz models) and power law spectra, some of which may be further customized. In addition, spectra may be imported from an ASCII file. The user may superpose an arbitrary number of these spectra to model a more complex composite target spectrum. A redshift can also be applied to the target spectrum.

Along with the spectrum, the location of the Sun and Moon relative to the target, the lunar phase and the year of observation have to be specified, as these are required for the background calculation. The sky background used for the SNR calculation is made up of three components. First, the airglow, based on a high resolution moonless sky UVES spectrum (Hanuschik 2003) which has been corrected to the dark sky UBVRI at Sutherland. Second, the zodiacal light contribution is obtained by normalizing the solar spectrum at 550nm, correcting for atmospheric extinction, and multiplying by a fit of zodiacal light photometry by Levasseur-Regourd & Dumont (1980). Thirdly, if the Moon is above the horizon, the simulated moonlight is based on a model for the V-band brightness of moonlight (Krisciunas & Schaefer 1991), extrapolated to other wavelengths.

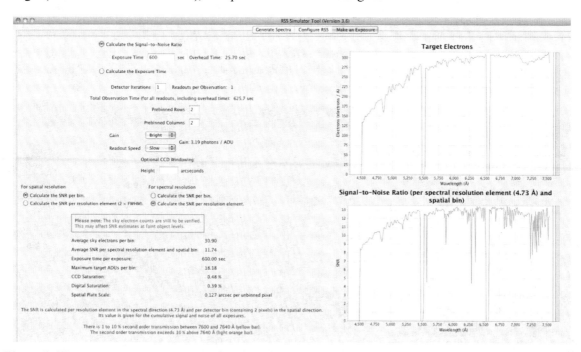

Figure 9: The exposure and detector setup page for the RSS simulator tool, showing predicted counts (in electons) and SNR spectral plots.

The second step is to provide the instrument setup details, for either SALTICAM or RSS. If the simulators are launched from the PIPT, updates made in the simulators are automatically reflected in the PIPT, and vice versa.

After supplying all the information, the PI may calculate the expected photon count rates as well as the signal-to-noise ratios ("SNRs") for the given exposure times. The next step is to select the observation mode (e.g. various imaging modes for SALTICAM and imaging, spectroscopy or Fabry-Perot in the case of RSS) and provide the required setup information.

The third and final step is to specify the exposure time and detector setup, including the pixel binning parameter, gain setting and readout speed. After all information has been supplied, the PI may calculate the expected photon count rate as well as the S/N ratio for a given exposure time. In case of a spectroscopic observation, the detection rate and the S/N are plotted as a function of the wavelength. Wavelengths for which more than 5% of the second order light is let through the order blocking filter are clearly marked in these plots. Alternatively, the PI may calculate the required exposure time for a requested SNR.

9. THE SALT SLITMASK GENERATION TOOL

When requesting an RSS observation with multiple object spectroscopy (MOS), the PI has to design the slitmask to use. This is used to generate the required g-code used to laser-cut slitlets, typically up to ~50, in a thin (200μm) carbon fibre mask. This can be created with the RSS Slit Mask Tool (RSMT), a standalone Java application based on JSky (http://jsky.sourceforge.net/index.html).

The Slitmask Tool lets the user add slits, boxes for reference and alignment stars, or arbitrary shaped slitlets, on top of a FITS image. Annotations in form of text and simple graphics may be added. A created slitmask can be validated to ensure that all slits, slitlets and reference stars boxes are situated in the field of view, that all spectra fall on the CCD and, most importantly, that there are no overlapping spectra.

In order to support survey mode observations with many targets in a field, a recent Python-based version includes an optimizer. This assigns weights to the slitmask elements and varies the position and angle on sky of the slitmask to find the combination for which the sum of weights of the valid slitmask elements reaches its maximum. It can also be used to create a set of several slitmasks for the same field.

10. THE SALT PROPOSAL WEB MANAGER

The Web Manager (WM) is a web application for accessing observing proposals, submitted to the SALT Database by the PIPT, which is written in PHP and JavaScript and uses the template engine Smarty (http://www.smarty.net/.) for generating its web pages. The purpose of the WM is to access information about observing programs and their status (e.g. completed observing blocks), both for users and SALT Astronomers. The only requirements for using it are internet access and a web browser.

After logging in, the Web Manager shows a page that is freely configurable by the user. Possible information on this page is weather data, proposal statistics, some news about the SALT software itself and one or more lists of proposals. The lists show the results of user-defined queries of the SDB. The default configuration is to show the list of proposals in which the currently logged-in user is included as an investigator. The information shown includes all the data that has been entered into the PIPT prior to the submission, as well as some additional information such as the proposal status.

The WM consists of different sections which display the following information, all derived from the Science Database:
- Title, abstract, total time allocated in different priority and Moon classes
- Flags for ToO or time critical observations, name of Liaison SALT Astronomer, program status
- Status summary (dates and observing times of completed or attempted observing blocks)
- Instructions and critical aspects of programs (written by the PIs for the SALT Astronomers carrying out the observations)
- Feedback comments to the PIs from the SALT Astronomers attempting the observations
- Details on the investigators (names, institutions, contact details) and contact person
- Observations Block details (total observation time, number of visitis, frequency of visits)

- Observation Configuration details
- Target information (coordinates, type, brightness, proper motion, finding charts)
- Instrument Configurations (filters, gratings, angles, detector modes, individual exposure times, number of iterations, number of repeat cycles)

An additional task for which the Web Manager is used is the approval of proposals. When a proposal is first submitted, each investigator receives an email containing a link with a unique identifier. This link leads to a page where the investigator may accept (or reject) the proposal. As the user is identified by the identifier in the link, there is no need for the user to have a Web Manager account for accepting a proposal, although an account is required for viewing it. Finally, after observations are done, an email is sent to the investigators to inform them that there exists new observational data. The investigators can then use the Web Manager to request the download of their data.

11. THE SALT DATA REDUCTION PIPELINE: PYSALT

In order to add maximum value to the data which SALT users receive, SALT Astronomy Operations have been developing a SALT data reduction package based on PyRAF ("Pythonized IRAF"), which has been customized specifically for SALT instruments and has been named PySALT. This is a challenging task, given the complexity of instrument modes. However, since the start of the first semester of charged science (1 Sep 2011) all PIs have routinely been receiving both raw FITS data and pipeline reduced data, albeit only covering basic reductions at present. The same data reduction procedures that have been developed for the quick-look data analysis tool used at the telescope, SALTFIRST, is being expanded into the full PySALT versions.

The primary goal for the development of SALT science software is to increase the scientific productivity of the telescope. By providing fast access to high quality data, we can minimize the effort required to publish scientific results. Furthermore, by having a central resource provide the tools necessary for data reduction, the amount of redundant work done by the diverse SALT partnership is minimized. To achieve this goal, we have identified three critical areas of software development:

1. Provide science quality reductions for the major operational modes of SALT.
2. Create analysis tools for the unique modes of SALT.
3. Create a framework for the archiving and rapid distribution of SALT data.

Obviously providing science quality data reductions to users will have an immediate return by minimizing the amount of work required for the principle investigators to publish their results. As it is a large consortium with many members having similar data reduction needs, the existence of a common framework for data reduction minimizes the amount of redundant work that needs to be done.

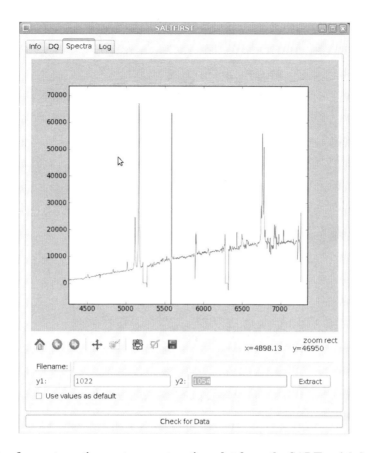

Figure 10: An example of an automatic spectrum extraction plot from the SALT quick-look data reduction tool, SALTFIRST, which runs at the telescope continuously during observations.

An important aspect of this development is the open-source nature of these tools. In the South African and many other African research communities, software with restrictive or expensive licenses are prohibited due to the lack of resources. Along with contributing to the cost-effective nature of SALT, open source software is far more accessible to these communities. As such, we have also adopted an open source license for all SALT science software and have avoided requiring any closed source or cost prohibitive software.

Basic CCD data reductions are implemented as part of the SALTRED package in PySALT. For imaging, it will provide tools for image reduction including gain and cross-talk correction, bias subtraction, flat-fielding and fringe correction, distortion correction, and astrometric solutions. For other modes, it will provide the same reductions through distortion correction, at which point further reductions will be handled by packages specific to that mode. Further descriptions of each task are provided on the SALT website (http://pysalt.salt.ac.za/).

All of the tasks have been implemented with both speed and flexibility in mind. When available, the first implementation of each task utilized an IRAF task (or other existing software) and took advantage of existing tools for these corrections. To increase the speed and performance of the tasks, we have replaced the IRAF task with a Python implementation for certain modes of operation. In the end, we have leveraged the extensive history of selected IRAF tools to provide the user with reliable existing tools while enhancing the performance with new, more powerful code.

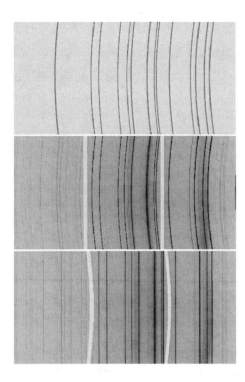

Figure 11: An Example of an RSS long slit observation of an Xe arc lamp using the PG3000 grating. The strong spatial curvature of the spectral lines in the upper two panels is a consequence of the nature of the off-axis behaviour of Volume Phase Holographic (VPH) diffraction gratings.

To further enhance the performance of the data reductions, the files are only opened once and all of the reductions tasks are performed sequentially on the file. Only three corrections are applied currently as part of the pipeline process, which include gain, cross-talk, and overscan subtraction. Future development of the code is focused on increasing performance and usability of the functions. An important addition will be to implement variance and bad pixel frames. Furthermore, we hope to replace some of the IRAF tasks with faster algorithms for data reduction to further improve the performance of the pipeline.

Spectroscopic reductions are part of the SALTSPEC package and include all the steps necessary to produce wavelength calibrated 1-D spectra from long slit spectra. The tasks that form part of this package are listed in Table 5. The package was designed to provide automated reduction of SALT images aimed at producing high quality scientific reductions and are built for general use. They should be applicable to different observing modes and targets. However, different reductions tools may provide better results, but these should be applicable to most tasks.

The main task of the package is *specidentify*, which determines the wavelength solution for the arc lamps. Although the task includes an interactive mode, it is primarily designed to work without any human intervention and calculate the wavelength solution over the full frame. The *specidentify* task is built around the PySpectrograph package that produces a model of the spectrograph. The PySpectrograph model uses information about the system to model the spectrograph and is built upon the grating equation. Because the configuration of RSS is known very well, the model of the spectrograph provides a solution that is within a few percent of the actual solution. An example of this process is given in Figure 11, which in the top frame provides an example of the spectra created using the PySpectrograph model for RSS. In the next frame, we provide an example of a Xe arc lamp image from SALT. Finally, we show the same spectra after wavelength calibration in the bottom frame, showing the removal of the distortion of the arc lines. The wavelength solution was generated automatically. Future development of the SALT spectroscopy software package will include providing support for multi-object and high speed modes. An important part of the spectrograph package will be integrating it into the data reduction pipeline to automatically produce wavelength calibrated data. The quick-look tool, SALTFIRST, currently does this using the PySpectrograph model.

12. PROGRESS AND CONCLUDING REMARKS

SALT is now a fully operational telescope and has been routinely obtaining science observations for the SALT partners, delivering ~650GB of data in the first semester (30 April 2012). Regrettably not all of the accepted proposals were completed, for a number of reasons, primarily due to a combination of the following:

1. Incomplete commissioning (particularly MOS, polarimetry dual etalon and High Resolution Fabry-Perot spectroscopy)
2. Too difficult programs where targets were simply too faint or required too good a seeing conditions.
3. Complex proposals that required significant time to fully set up and complete
4. Lack of an adequate means to comprehensively plan observing options in the queue ahead of time
5. Some difficulties with as-submitted Phase 2 proposals (e.g. poor finding charts, inaccurate astrometry)
6. Lack of responses from PIs (e.g. for Phase 2 submissions)
7. Technical issues with the telescope (e.g. SALTICAM fold mirror)
8. Technical issues with the instruments (e.g. SALTICAM filter problems; RSS mechanism issues
9. Poorer than expected weather conditions (average of 35% loss compared to the expected 25%).

This first semester saw the transition from commissioning to science observations and as such was still somewhat of a learning experience. Many of the issue above were addressed (e.g. MOS now commissioned) and other are continuing (e.g. better scheduling visualization tools and more automated scheduling). Notwithstanding the above, 93 out of a total of 144 accepted proposals had observations taken. One positive result was that the average overhead time for setting up on a target (including slewing, guide star setup and instrument configuration) decreased significantly, with the result that the canonical overhead time used in the proposal tools has been decreased from 900s to 600s. Further improvements on this are expected as the OCS (Observation Control System) is developed further.

Although SALT, by design, imposes restrictions on how observations can be conducted, the overall design of the telescope – from hardware to software – has been done with efficiency and ease of operation in mind. A variety of tools have been produced for SALT users to enable the writing and submitting of observing proposals, modifications of their observing programs, monitoring of progress, collections of data and pipeline data reductions.

SALT has some significant competitive advantages, mostly associated with the queued service observing mode, coupled with the flexible range of instrument capabilities, which can be switched quickly (within minutes) from one instrument or mode to the next. Data is routinely available to users within 24 hours of the observation blocks having been completed, allowing for a quick response by PIs to the results as they occur and allowing for reconfigurations of observations or instruments, if appropriate. This has a particular benefit for Target of Opportunity type observations. Likewise the queue-scheduled *modus operandii* is also well suited to time critical observations and synoptic monitoring on a range of timescales (days, weeks, months, years).

At the time of writing the first science semester with SALT is just behind us and we are embarking on the second. While there is still some commissioning to complete for some instrument modes (e.g. RSS polarimetry), these are expected to be mostly completed within the semester. Late in 2012 a new instrument, SALT HRS, a stable high resolution ($R \sim 70,000$) fibre-fed vacuum spectrograph will be installed on SALT, giving additional capabilities, and as with the exiting two instruments, able to be swopped between with minutes.

SALT users have keenly anticipated the potential for exciting new science afforded by SALT, and this was reflected in the first semesters over-subscription numbers for many partners (2 – 4 times). While current research papers highlighting SALT science is only just beginning to flow (only 35 refereed publications to date), this is anticipated to vastly accelerate over the next year or two. In addition, when a new near IR mode (currently under construction) is added to the RSS, in ~2 years time, the range of capabilities will be greatly enhanced again.

ACKNOWLEDGEMENTS

The authors acknowledge the enormous effort by many individuals within the SALT Technical and Astronomy Operations divisions and others at SAAO, plus the instrument teams, in helping to see SALT to completion and achieving full science operations. A team effort has resulted in this success and special mention goes to the following people for the development of the many SALT observational software support tools that are described in this paper: Christian Hettlage, Encarni Romero Colmenero, Steve Crawford, Tim Oliver Husser, Janus Brink, Anthony Koeslag, Deney Maartens, Luis Balona, Tim Pickering, Keith Smith and Brent Miszalski.

REFERENCES

1. Buckley, D.A.H., 2001, "The Southern African Large Telescope: an alternative paradigm for an 8-m class telescope", *New Ast. Rev.*, **45**, 13
2. Buckley, D.A.H., Charles, P.A., Nordsieck, K.H., O'Donoghue, D., 2006. The Southern African Large Telescope project, in *Scientific Requirements for Extremely Large Telescopes*, eds P. Whitelock, M. Dennefeld, B. Leibundgut. , *IAU Symp. 232,* 1 (Cambridge University Press, Cambridge)
3. Buckley, D.A.H., Swart, G. P. and Meiring, J.G., 2006, *Proc. SPIE* **6267**, 62670Z
4. Buckley, D.A.H, et al., 2008, *Proc. SPIE* **7014**, 701407-1
5. Buckley, D.A.H., et al., 2010, *Proc. SPIE* **7739**, p. 773912-1
6. Buckley, D.A.H., 2012, in Proc. of the Middle East and African Regional IUA Meeting (MEARIM) #2 (April 2011), *African Skies* **16,** 89
7. Burgh, E. et al., 2003, *Proc. SPIE* **4841**, 1463
8. Hanuschik, R. W., 2003, *Astron.Astrophys.* **407**, 1157
9. Krisciunas, K., Schaefer, B. E., 1991, *PASP* **103**, 1033
10. Levasseur-Regourd, A. C. and Dumont, R., 1980, *Astron. Astrophys.* **84**, 277
11. Menzies, J., Gajjar, H., Buous, S., Buckley, D., Gillingham, P., 2010, *Proc. SPIE,* **7739**, 77390X-1
12. Nordsieck, K.H., Nosan, F., Schier, J.A., 2010, *Proc. SPIE* **7735**, 773582-1
13. O'Donoghue, D., et al., 2003, *Proc. SPIE* **4841**, 465
14. O'Donoghue, D., et al., 2006, *MNRAS* **372**, 151
15. O'Donoghue, D., et al., 2010. *Proc. SPIE* **7739**, 77390Q-1
16. Ramsey, L.W. et al., 1998, *Proc. SPIE* **3352**, 34

The QUIJOTE-CMB Experiment: studying the polarisation of the Galactic and Cosmological microwave emissions

J.A. Rubiño-Martín[a,b], R. Rebolo[a,b,h], M. Aguiar[a], R. Génova-Santos[a,b], F. Gómez-Reñasco[a], J.M. Herreros[a], R.J. Hoyland[a], C. López-Caraballo[a,b], A.E. Pelaez Santos[a,b], V. Sanchez de la Rosa[a], A. Vega-Moreno[a], T. Viera-Curbelo[a], E. Martínez-Gonzalez[c], R.B. Barreiro[c], F.J. Casas[c], J.M. Diego[c], R. Fernández-Cobos[c], D. Herranz[c], M. López-Caniego[c], D. Ortiz[c], P. Vielva[c], E. Artal[d], B. Aja[d], J. Cagigas[d], J.L. Cano[d], L. de la Fuente[d], A. Mediavilla[d], J.V. Terán[d], E. Villa[d], L. Piccirillo[e], R. Battye[e], E. Blackhurst[e], M. Brown[e], R.D. Davies[e], R.J. Davis[e], C. Dickinson[e], S. Harper[e], B. Maffei[e], M. McCulloch[e], S. Melhuish[e], G. Pisano[e], R.A. Watson[e], M. Hobson[f], K. Grainge[f], A. Lasenby[f,g], R. Saunders[f], and P. Scott[f]

[a]Instituto de Astrofisica de Canarias, C/Via Lactea s/n, E-38200 La Laguna, Tenerife, Spain;
[b]Departamento de Astrofísica, Universidad de La Laguna, E-38206 La Laguna, Tenerife, Spain;
[c]Instituto de Fisica de Cantabria (IFCA), CSIC-Univ. de Cantabria, Avda. los Castros, s/n, E-39005 Santander, Spain;
[d]Departamento de Ingenieria de COMunicaciones (DICOM), Laboratorios de I+D de Telecomunicaciones, Plaza de la Ciencia s/n, E-39005 Santander, Spain;
[e]Jodrell Bank Centre for Astrophysics, School of Physics and Astronomy, University of Manchester, Oxford Road, Manchester M13 9PL, UK;
[f]Astrophysics Group, Cavendish Laboratory, University of Cambridge, Madingley Road, Cambridge CB3 0HE, UK;
[g]Kavli Institute for Cosmology, Univ. of Cambridge, Madingley Road, Cambridge CB3 0HA;
[h]Consejo Superior de Investigaciones Cientificas, Spain

ABSTRACT

The *QUIJOTE* (Q-U-I JOint Tenerife) *CMB* Experiment will operate at the Teide Observatory with the aim of characterizing the polarisation of the CMB and other processes of Galactic and extragalactic emission in the frequency range of 10–40 GHz and at large and medium angular scales. The first of the two QUIJOTE telescopes and the first multi-frequency (10–30 GHz) instrument are already built and have been tested in the laboratory. *QUIJOTE-CMB* will be a valuable complement at low frequencies for the *Planck* mission, and will have the required sensitivity to detect a primordial gravitational-wave component if the tensor-to-scalar ratio is larger than $r = 0.05$.

Keywords: cosmic microwave background, polarisation, cosmological parameters, early Universe, telescope, instrumentation

1. INTRODUCTION

The study of the Cosmic Microwave Background (CMB) anisotropies is one of the most powerful tools in modern cosmology, and it has played a crucial role in our understanding of the Universe. With the latest results from WMAP satellite,[1] and with the information provided by ground-based experiments such as VSA,[2] ACBAR,[3] CBI,[4] SPT[5] or ACT,[6] it has been possible to determine cosmological parameters with accuracies better than five per cent.[7] *Planck* satellite, launched in May 2009, is expected to improve the accuracy on the determination of the cosmological parameters, reaching precisions of less than a percent.[8]

Corresponding author: J.A. Rubiño-Martín. Email: jalberto@iac.es, Telephone: +34 922 605 276

Until now, the majority of the CMB constraints are obtained from intensity measurements. However, the CMB contains a wealth of information encoded in its polarisation signal. Since the first detection of polarisation by the DASI experiment,[9] other experiments have provided measurements of the angular power spectrum of the polarisation.[10–17] Despite their relatively poor signal-to-noise ratio, they still show excellent agreement with the predictions of the standard ΛCDM model.

The standard theory predicts that the CMB is linearly polarized, the physical mechanism responsible for its polarisation being Thomson scattering during the recombination or reionization epochs. Thus, the polarisation state on any direction $\hat{n}$ on sky can be well described by the two Stokes parameters Q and U. Full-sky maps of these two parameters can be decomposed into complex spin-2 harmonics

$$Q(\hat{n}) \pm iU(\hat{n}) = \sum_{\ell m} a_{\pm 2,\ell m} \, _{\pm 2}Y_{\ell m}(\hat{n}), \qquad (1)$$

However, in practice these coefficients ($a_{\pm 2,\ell m}$) are not used in CMB studies to describe full-sky polarisation maps. Instead, these polarisation maps are decomposed in terms of two scalar components usually called a E-field (gradient) and a B-field (rotational),[18,19] and which are given by the coefficients

$$a_{E,\ell m} = -\frac{a_{2,\ell m} + a_{-2,\ell m}}{2}, \qquad a_{B,\ell m} = -\frac{a_{2,\ell m} - a_{-2,\ell m}}{2i} \qquad (2)$$

From here, the angular power spectra can be written as

$$C_\ell^{XY} = \frac{1}{2\ell+1} \sum_{m=-\ell}^{m=+\ell} a_{X,\ell m}^* a_{Y,\ell m} \qquad (3)$$

where X and Y can take the values T, E, or B. Thus, in addition to the temperature power spectrum TT (C_ℓ^{TT}), we have three parity-independent angular power spectra to describe the polarisation field: the cross-correlation of temperature T and E mode, TE (C_ℓ^{TE}); and the auto-correlation of the E and B modes, EE (C_ℓ^{EE}) and BB (C_ℓ^{BB}), respectively. All the other combinations (TB and EB) are expected to be zero for the CMB field.

The importance of this decomposition is connected with the physics of generation of the CMB anisotropies. If the fluctuations in CMB intensity are seeded by scalar perturbations (i.e fluctuations in the density alone), one would only expect primordial E modes in the CMB polarisation. However, vector and tensor perturbations, like those due to gravitational waves (GW) in the primordial Universe,[20] are mechanisms that could generate primordial B-modes in the polarisation on large angular scales. Therefore, if we can measure these modes, we may have a unique way to carry out a detailed study of the inflationary epoch. In particular, the energy scale V at which inflation occurred can be expressed in terms of r, the ratio of tensor to scalar contributions to the power spectrum, as[21]

$$r = 0.001 \left(\frac{V}{10^{16} \text{ GeV}} \right)^4 \qquad (4)$$

Based on BB upper limits alone, the best current constraint on the inflationary GW background is[15] $r \leq 0.72$ (95% C.L.). When combining this information with the measurements of the other three CMB power spectra (TT, TE and EE), the WMAP data[16] alone gives $r \leq 0.36$ (95% C.L.). Finally, when BAO and SNIa constraints are included,[7] we have $r \leq 0.2$ (95% C.L.). These numbers translate into a constraint of $\lesssim 4 \times 10^{16}$ GeV.

Because of the importance of detecting primordial gravitational waves,[22,23] there is a huge interest to develop experiments to measure (or constrain) the amplitude of B-modes power spectrum of the CMB polarisation. Here we present one of these efforts.

The *QUIJOTE* (Q-U-I JOint TEnerife) *CMB Experiment*[24] is a scientific collaboration between the Instituto de Astrofísica de Canarias, the Instituto de Física de Cantabria, the IDOM company, and the universities of Cantabria, Manchester and Cambridge, with the aim of characterizing the polarisation of the CMB, and other galactic and extragalactic physical processes in the frequency range 10–40 GHz and at angular scales larger than 1 degree. Updated information can be found on the project website.[25]

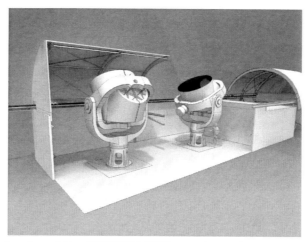

Figure 1. A 3D drawing of the *QUIJOTE-CMB* experiment dome and the two telescopes.

2. PROJECT BASELINE

The *QUIJOTE-CMB* experiment consists of two telescopes and three instruments (see Fig. 1), which will observe in the frequency range 10–40 GHz with an angular resolution of ∼ 1 degree, from the Teide Observatory (2400 m) in Tenerife (Spain). Experience over more than 27 years[26] with several CMB experiments (Tenerife Experiment, JBO-IAC Interferometer, COSMOSOMAS, Very Small Array) shows that this is an excellent place for CMB observations. The project has two phases already funded:

- *Phase I.* Construction of the first *QUIJOTE-CMB* telescope (QT1) and two instruments which can be exchanged in the QT1 focal plane. The first instrument (MFI) is a multichannel instrument providing the frequency coverage between 10 and 20 GHz, and it will start commissioning during the summer of 2012. The second instrument (TGI) will consist of 31 polarimeters working at 30 GHz, and it is expected to start operations at the end of 2013. This phase also includes a two-element interferometer operating at 30 GHz, which will be used as a "source-subtractor" facility to monitor and correct the contribution of polarized radio-sources in the final *QUIJOTE-CMB* maps.

- *Phase II.* Construction of the second *QUIJOTE-CMB* telescope (QT2), and a third instrument (FGI) with 40 polarimeters working at 40 GHz.

There are also plans for a future *Phase III* of the project, which considers the construction of a new instrument with at least 100 receivers at W-band. However, this third phase is not funded yet.

Table 1 summarizes the basic (nominal) characteristic of these three instruments in phases I and II. The noise equivalent power (NEP) for one stabilized polarimeter channel is defined here as

$$\text{NEP} = \sqrt{2} \frac{T_{\text{sys}}}{\sqrt{\Delta \nu \, N_{chan}}}, \qquad (5)$$

where T_{sys} stands for the total system temperature, $\Delta \nu$ is the bandwidth and N_{chan} is the number of channels (computed here as the number of horns times the number of output channels per horn). From here, the noise sensitivity is obtained as $\text{NEP}/\sqrt{t}$, being t the integration time. We note that the system temperature (T_{sys}) values appearing in Table 1 have several contributions: the receiver contribution; the estimated contribution of the opto-mechanics; the spillover contribution (i.e., the background contribution when the instrument is placed in the focal plane of the telescope); the atmospheric contribution at the considered frequency; and the CMB contribution (2.7 K).

Table 1. Nominal characteristics of the three *QUIJOTE-CMB* instruments: MFI, TGI and FGI. Sensitivities are referred to Stokes Q and U parameters. See text for details.

	MFI					TGI	FGI
Nominal Frequency [GHz]	11	13	17	19	30	30	40
Bandwidth [GHz]	2	2	2	2	8	8	10
Number of horns	2	2	2	2	1	31	40
Channels per horn	4	4	4	4	2	4	4
Beam FWHM [°]	0.92	0.92	0.60	0.60	0.37	0.37	0.28
$T_{\rm sys}$ [K]	25	25	25	25	35	35	45
NEP [$\mu K\,s^{1/2}$]	280	280	280	280	390	50	50
Sensitivity [$Jy\,s^{1/2}$]	0.30	0.42	0.31	0.38	0.50	0.06	0.06

Figure 2. Left: *QUIJOTE-CMB* enclosure at the Teide Observatory. Right: Inside the *QUIJOTE-CMB* dome, before the installation of QT1.

3. EXPERIMENT DESCRIPTION

3.1 Telescopes and Enclosure

The *QUIJOTE-CMB* experiment consists of two telescopes (hereafter QT1 and QT2) that will be installed inside a single enclosure at the Teide Observatory. The enclosure and the building hosting the control room were finished in June 2009 (see Fig. 2).

The layout of both QT1 and QT2 telescopes is based on an altazimuth mount supporting a primary (parabolic) and a secondary (hyperbolic) mirror disposed in an offset Gregorian Dracon scheme, which provides optimal cross-polarisation properties (designed to be ≤ -35 dB) and symmetric beams. Each primary mirror has a 2.25 m projected aperture, while the secondary has 1.89 m. The system is under-illuminated to minimize side-lobes and ground spillover. Each telescope is mounted on its own platform that can rotate around the vertical axis at a maximum frequency of 6 rpm (i.e., $36\,{\rm deg\,s^{-1}}$).

The telescope control software for QT1 was implemented during 2009, and the different observing modes (raster, scanning, tracking, etc.) have been tested.[27] The construction scheme, as well as the fabrication techniques for QT1 have been already presented.[28] We note that the QT1 mirrors have been designed to operate up to 90 GHz (i.e., rms $\leq 20\,\mu$m and maximum deviation of $d = 100\,\mu$m). However, QT2 has been specified to have a better surface accuracy, so the telescope could in principle operate up to 200 GHz.

The installation of QT1 at the Teide Observatory took place during May 2012 (see Fig. 3).

3.2 Instruments

3.2.1 Multi-frequency Instrument (MFI)

This is a multi-channel instrument with five independent sky pixels: two operate at 10–14 GHz; the other two at 16–20 GHz, and finally a central polarimeter at 30 GHz that is being used as a demonstrator of the second

Figure 3. Left: QT1 at the IAC workshop (June 2009). Right: Installation of QT1 at the Teide Observatory (May 3rd, 2012).

instrument during the laboratory tests and commissioning phase. The main science driver for the MFI is the characterization of the Galactic emission. A complete description of the MFI and the details on the software are presented elsewhere.[27, 29] Here, we only provide the basic aspects of the instrument (see Fig. 4).

The optical arrangement includes five conical corrugated feedhorns (designed by the University of Manchester). Each horn feeds a novel cryogenic on-axis rotating polar modulator which can rotate at speeds of up to 1 Hz. We consider two possible operational modes: either continuous rotation of the polarimeters, or discrete changes of the positions of the motors in steps of 22.5° (note that the polar modulation occurs at four times the rotation angle). The orthogonal linear polar signals are separated through a wide-band cryogenic Ortho-Mode-Transducer (OMT) before being amplified through two similar LNAs (a Faraday-type module in the case of 30 GHz). These two orthogonal signals are fed into a room-temperature Back-End module (BEM) where they are further amplified and spectrally filtered before being detected by square-law detectors. All the polarimeters except the 30 GHz receiver have simultaneous "Q" and "U" detection i.e. the 2 orthogonal linear polar signals are also correlated through a 180° hybrid and passed through two additional detectors. The band passes of these lower frequency polarimeters have also been split into an upper and lower band which gives a total of 8 channels per polarimeter (see Table 1).

The FEM for the low frequency channels was built by the IAC. The receivers for these channels use MMIC 6-20 GHz LNAs (designed by S. Weinreb and built in Caltech). The gain for these amplifiers is approximately 30 dB, and the noise temperature is less than 9 K across the band. The 30 GHz FEM was built at the University of Manchester, and the design used an existing Faraday module (same as the one used for OCRA-F*). The BEM for the 30 GHz instrument was built by DICOM, with collaboration of IFCA at the simulation level. The cryogenics and the mechanical systems were provided by the IAC.

3.2.2 Thirty-GHz Instrument (TGI)

This instrument will be mainly devoted to primordial B-mode science. TGI will be fitted with 31 polarimeters working in the range of 26-36 GHz. After the laboratory tests with the 30 GHz polarimeter of the MFI, we found

---

*OCRA-F: http://www.jodrellbank.manchester.ac.uk/research/ocra/ocraf.html.

Figure 4. Left: Close view of the MFI, during the integration phase (May 2011). Center: Integration of the MFI in the QT1 focal plane (December 2011). Right: MFI already installed at the QT1 focal plane (January 2012). The electronic boxes controlling the telescope and the instrument are also installed.

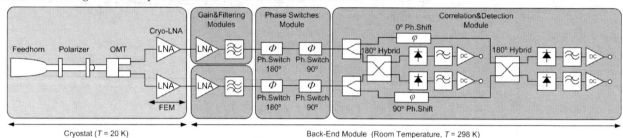

Figure 5. Configuration of each of the 31 receivers of the TGI, in the *QUIJOTE-CMB* experiment.

that the MFI design, based on the spinning polar modulators in a cryogenic environment, is not appropriate for the long-term operations required for the TGI. Thus, we have modified the receiver configuration by replacing the rotating polar modulator with a fixed polarizer. The current new design is presented in Fig. 5. It includes a fixed polarizer and 90° and 180° phase switches to generate four polarisation states to minimize the different systematics in the receiver. A detailed description of the system has been already presented.[30]

3.2.3 Forty-GHz Instrument (FGI)

Also devoted to primordial B-mode science, the FGI will be fitted with 40 polarimeters working at 40 GHz. The conceptual design of a polarimeter chain for the FGI is identical to the one used for the TGI (see Fig. 5).

3.3 Source subtractor facility

An upgraded version of the VSA source subtractor (VSA-SS) facility,[31] which is being carried out by the Cavendish Laboratory and the University of Manchester, will be used to monitor the contribution of polarized radio-sources in the *QUIJOTE-CMB* maps. The VSA-SS is a two element interferometer, operating at 30 GHz, with 3.7 m dishes and a separation of 9 m (see Fig. 6). The VSA-SS system only measured one linear polarisation of the incoming radiation, so it is being upgraded to include a half-wave plate (HWP) in front of each of the antennas in order to allow for successive measurements of Stokes Q and U. Here, we use a dielectrically embedded mesh-HWP designed and produced at the University of Manchester (see right panel of Fig. 6).

Using the method described by Tucci et al. (2004)[32] to simulate the polarisation properties of radio sources at the *QUIJOTE-CMB* frequencies, we have estimated that in order for the residual source contribution to our measurements be equal to or smaller than the expected B-mode signal for the case of $r = 0.1$ at 30 GHz, we must remove the effects of all sources whose Stokes I intensity is higher than 300 mJy (see Fig. 7). Our strategy is therefore to measure the 30 GHz Stokes I intensity of known radio sources (e.g., from the GB6 catalogue) and then measure the polarisation of those that we find have Stokes I greater than 300 mJy. The total number of sources to be monitored in the whole *QUIJOTE-CMB* surveyed area will be around 500. The expected polarised flux sensitivity per source of the VSA-SS is 2–3 mJy.

An upgrade of this VSA-SS facility to operate at 40 GHz during the Phase II of the project is currently under discussion.

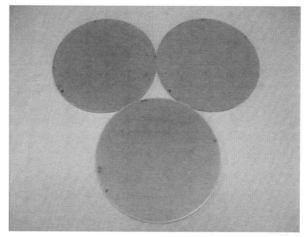

Figure 6. Left: One of the two antennas of the VSA source subtractor. This facility will be re-used to measure the polarisation of radio sources to correct the *QUIJOTE-CMB* 30 GHz maps. Right: Picture of the HWPs used for the VSA-SS.

4. SCIENCE GOALS AND SCIENCE CASES

4.1 Core science

The *QUIJOTE-CMB* experiment has two primary scientific goals:

- to detect the imprint of gravitational B-modes if they have an amplitude $r \geq 0.05$;

- to provide essential information of the polarisation of the synchrotron and the anomalous microwave emissions from our Galaxy at low frequencies (10–40 GHz).

For these scientific objectives, *QUIJOTE-CMB* will conduct two large surveys in polarisation (i.e., Stokes Q and U maps):

i) a shallow "Galactic" survey. It will cover around $10\,000\,\text{deg}^2$ of sky. It is expected to be finished after 2–3 months of effective observing time with each instrument, reaching sensitivities of ~ 10–$15\,\mu\text{K}$ per one degree beam in the Stokes Q and U maps with the MFI (11–19 GHz), and $\lesssim 3\,\mu\text{K}$ per beam with the TGI and FGI at 30 and 40 GHz.

ii) a deep "Cosmological" survey. It will cover around $3\,000\,\text{deg}^2$. Here, we shall reach sensitivities of ~ 3–$4\,\mu\text{K}$ per one degree beam after one year of effective observing time with the MFI (11–19 GHz), and $\lesssim 1\,\mu\text{K}$ per beam with TGI and FGI at 30 and 40 GHz.

According to these nominal sensitivities, *QUIJOTE-CMB* will provide one of the most sensitive 11–19 GHz measurements of the polarisation of the synchrotron and anomalous emissions on degree angular scales. This information is extremely important given that B-modes are known to be sub-dominant in amplitude as compared to the Galactic emission,[33] as illustrated in Fig. 7. The *QUIJOTE-CMB* maps will also constitute an unique complement of the *Planck* satellite[†], helping in the characterization of the Galactic emission. In particular, the combination of *Planck* and *QUIJOTE-CMB* will allow us: (a) to determine synchrotron spectral indices with high accuracy, and to fit for curvature of the synchrotron spectrum to constrain CR electron physics;[34] (b) to study the large-scale properties of the Galactic magnetic field;[35] or (c) to assess the level of a possible contribution of polarized anomalous microwave emission.[36,37]

Using the MFI maps from the deep survey, we plan to correct the high frequency *QUIJOTE-CMB* channels (30 and 40 GHz) to search for primordial B-modes. As an illustration, Fig. 8 presents two cases. The left panel shows the scientific goal for the angular power spectrum of the E and B modes after 1-year of effective observing time, assuming a sky coverage of 3 000 square degrees, with the TGI only. In this particular case, the final noise

---

[†]*Planck*: http://www.rssd.esa.int/index.php?project=Planck

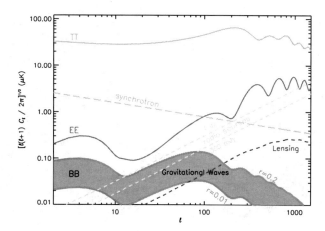

Figure 7. Expected foreground contamination in the 30 GHz *QUIJOTE-CMB* frequency band. It is shown the contribution of polarized synchrotron emission and radio-sources for the case of subtracting sources down to 1 Jy in total intensity (upper dashed line for radio-sources) and 300 mJy (lower dashed-line). The physical models for these emissions are described in.[33] The shaded red area shows the expected level of primordial gravitational waves for r in the range 0.01–0.2.

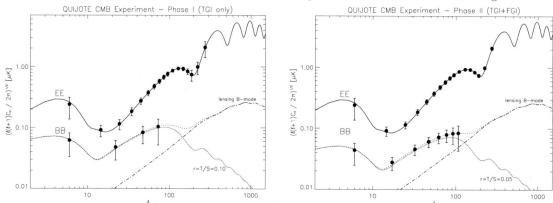

Figure 8. Left: Example of the *QUIJOTE-CMB* scientific goal after the Phase I of the project, for the angular power spectrum of the CMB E and B mode signals. It is shown the case for 1 year (effective) observing time, and a sky coverage of $\sim 3,000\,\text{deg}^2$. The red line corresponds to the primordial B-mode contribution in the case of $r = 0.1$. Dots with error bars correspond to averaged measurements over a certain multipole band. Right: Same computation but now for the *QUIJOTE-CMB* Phase II. Here we consider 3 years of effective operations with the TGI, and that during the last 2 years, the FGI will be also operative. The red line now corresponds to $r = 0.05$.

level for the 30 GHz map is $\sim 0.5\,\mu\text{K/beam}$. The right panel shows the scientific goal for the *QUIJOTE-CMB* Phase II. Here, we consider 3 years of effective observing time with the TGI, and 2 years with the FGI. Note that, once the two instruments (FGI and TGI) are available, they can be operated simultaneously, as we will have two telescopes. Finally, we stress that the computations presented in Fig. 8 correspond to the optimal situation in which the foreground removal leaves a negligible impact on the power spectrum. More realistic estimates will be published in a future paper.

4.2 Non-core science

Apart from the scientific goals described in the previous section, we have identified a number of secondary science projects. The characteristics of *QUIJOTE-CMB* make it a suitable experiment for performing (relatively-short) observations in specific regions that would allow us to tackle scientific objectives different to those for which it was conceived. Some of these possible projects are:

i) Study of the polarisation of Galactic regions and extragalactic sources. One of the main science drivers of *QUIJOTE-CMB* is to characterize the polarisation of the large-scale synchrotron emission from our Galaxy.

However, it is also interesting to study this polarisation in specific Galactic regions, and also in extragalactic regions like M31, or in some of the 22 polarized sources detected in WMAP data.[38] This could be done either with the MFI or with the polarized source subtractor, depending on the angular resolution.

ii) Study of the North Polar Spur. This a huge feature, visible mainly in radio wavelengths, which covers about a quarter of the sky and extends to high Galactic latitudes. Two main hypotheses have been proposed for its origin, namely a superbubble inflated by stellar winds and supernovae activity from the Scorpius-Centaurus OB association, on one hand, and an interaction between the loop I superbubble and the local superbubble.[39] *QUIJOTE-CMB* data in this region may help to disentangle these two hypotheses.

iii) Study of the polarisation of the anomalous microwave emission (AME) in the Perseus molecular cloud and in other bright Galactic clouds. Apart from the synchrotron, it is also mandatory to have a good characterization of the AME polarisation in order to assess what level of contamination current and future B-modes experiments will suffer. At present, only upper limits of the polarisation percentage have been obtained;[40] these stand at $\sim 1\%$ at the 95% C.L. We estimate that 35 h of observations with the 30 GHz channels of the MFI on Perseus could allow us to obtain a $\sim 1\%$ upper limit at the 99% C.L. Other possible targets include the ρ-Ophiuchi molecular cloud,[41] the dark nebula LDN1622 or the Pleiades reflection nebula.[42] These *QUIJOTE-CMB* measurements will provide a unique tool to understand the physical mechanism responsible for the AME, helping to distinguish between the electric dipole and the magnetic dipole radiation models.[43,44]

iv) Study of the WMAP haze in polarisation. This is an excess of microwave emission towards the centre of the Galaxy that was found at 23 GHz in WMAP data, with a significantly flatter spectrum than synchrotron, and which has recently been shown to have a Gamma-ray counterpart in Fermi data.[45] This is a burning subject at the moment, mainly owing to one of the proposed hypotheses for its origin, which is based on hard synchrotron radiation driven by relativistic electrons and positrons produced in the annihilations of one (or more) species of dark matter particles.[46] *QUIJOTE-CMB* data could have an important contribution here, as it could allow us to measure, or to constrain the expected level of polarisation of this synchrotron emission.

v) Study of the polarisation of the WMAP cold spot. This is a non-Gaussian feature in the CMB, in the form of an extremely extended and cold region, which was found in WMAP data.[47] After several considerations, it was proposed as a possible scenario for its origin the presence of a texture, a kind of topological defect which is predicted to occur in the primordial Universe. If this hypothesis were correct, a lack of polarisation would be expected in this region as compared with typical values of the primordial CMB. Therefore, the *QUIJOTE-CMB* data could help to disentangle the Gaussian and the texture hypotheses. In particular, it has been estimated[48] that these data would be able to reject the Gaussian hypothesis with a significance of $\sim 1\%$.

5. PROJECT STATUS AND TIMELINE

QT1 is already installed at the Teide Observatory, and is now in the commissioning process. Immediately after this, the MFI will be commissioned, probably during this summer (2012). In parallel, the commissioning phase of the source-subtractor facility will take place.

We expect to install the QT2 at the end of 2013. Concerning the other two instruments, the TGI will be installed in the focal plane of the QT2, and will be commissioned by the end of 2013. The TGI will be available by the end of 2014. Note that once the TGI is operative, the FGI will be permanently installed in QT1, while the TGI will be placed in QT2.

ACKNOWLEDGMENTS

The *QUIJOTE-CMB* experiment is being developed by the Instituto de Astrofisica de Canarias (IAC), the Instituto de Fisica de Cantabria (IFCA), and the Universities of Cantabria, Manchester and Cambridge. Partial financial support is provided by the Spanish Ministry of Economy and Competitiveness (MINECO) under the projects AYA2010-21766-C03 (01, 02 and 03), and also by the Consolider-Ingenio project CSD2010-00064 (EPI: Exploring the Physics of Inflation[49]).

REFERENCES

[1] Jarosik, N. et al., "Seven-year Wilkinson Microwave Anisotropy Probe (WMAP) Observations: Sky Maps, Systematic Errors, and Basic Results," *ApJS* **192**, 14–+ (Feb. 2011).

[2] Dickinson, C. et al., "High-sensitivity measurements of the cosmic microwave background power spectrum with the extended Very Small Array," *MNRAS* **353**, 732–746 (Sept. 2004).

[3] Reichardt, C. L. et al., "High-Resolution CMB Power Spectrum from the Complete ACBAR Data Set," *ApJ* **694**, 1200–1219 (Apr. 2009).

[4] Readhead, A. C. S. et al., "Extended Mosaic Observations with the Cosmic Background Imager," *ApJ* **609**, 498–512 (July 2004).

[5] Keisler, R. et al., "A Measurement of the Damping Tail of the Cosmic Microwave Background Power Spectrum with the South Pole Telescope," *ApJ* **743**, 28 (Dec. 2011).

[6] Hlozek, R. et al., "The Atacama Cosmology Telescope: A Measurement of the Primordial Power Spectrum," *ApJ* **749**, 90 (Apr. 2012).

[7] Komatsu, E. et al., "Seven-year Wilkinson Microwave Anisotropy Probe (WMAP) Observations: Cosmological Interpretation," *ApJS* **192**, 18–+ (Feb. 2011).

[8] Planck Collaboration I, "Planck early results. I. The Planck mission," *A&A* **536**, A1 (2011).

[9] Kovac, J. M., Leitch, E. M., Pryke, C., Carlstrom, J. E., Halverson, N. W., and Holzapfel, W. L., "Detection of polarization in the cosmic microwave background using DASI," *Nature* **420**, 772–787 (Dec. 2002).

[10] Montroy, T. E. et al., "A Measurement of the CMB <EE> Spectrum from the 2003 Flight of BOOMERANG," *ApJ* **647**, 813–822 (Aug. 2006).

[11] Sievers, J. L. et al., "Implications of the Cosmic Background Imager Polarization Data," *ApJ* **660**, 976–987 (May 2007).

[12] Wu, J. H. P. et al., "MAXIPOL: Data Analysis and Results," *ApJ* **665**, 55–66 (Aug. 2007).

[13] Bischoff, C. et al., "New Measurements of Fine-Scale CMB Polarization Power Spectra from CAPMAP at Both 40 and 90 GHz," *ApJ* **684**, 771–789 (Sept. 2008).

[14] Brown, M. L. et al., "Improved Measurements of the Temperature and Polarization of the Cosmic Microwave Background from QUaD," *ApJ* **705**, 978–999 (Nov. 2009).

[15] Chiang, H. C. et al., "Measurement of Cosmic Microwave Background Polarization Power Spectra from Two Years of BICEP Data," *ApJ* **711**, 1123–1140 (Mar. 2010).

[16] Larson, D. et al., "Seven-year Wilkinson Microwave Anisotropy Probe (WMAP) Observations: Power Spectra and WMAP-derived Parameters," *ApJS* **192**, 16 (Feb. 2011).

[17] QUIET Collaboration, Bischoff, C., et al., "First Season QUIET Observations: Measurements of Cosmic Microwave Background Polarization Power Spectra at 43 GHz in the Multipole Range $25 < \ell < 475$," *ApJ* **741**, 111 (Nov. 2011).

[18] Zaldarriaga, M. and Seljak, U., "All-sky analysis of polarization in the microwave background," *Phys. Rev. D* **55**, 1830–1840 (Feb. 1997).

[19] Kamionkowski, M., Kosowsky, A., and Stebbins, A., "Statistics of cosmic microwave background polarization," *Phys. Rev. D* **55**, 7368–7388 (June 1997).

[20] Polnarev, A. G., "Polarization and Anisotropy Induced in the Microwave Background by Cosmological Gravitational Waves," *Soviet Astronomy* **29**, 607 (Dec. 1985).

[21] Partridge, B., "CMB observations and Cosmological Constraints," in [*The Cosmic Microwave Background: from quantum fluctuations to the present Universe*], Edited by J.A. Rubiño-Martin, R. Rebolo and E. Mediavilla **CUP 2009**, pp. 1–52 (2009).

[22] Bock, J. et al., "Task Force on Cosmic Microwave Background Research," *ArXiv Astrophysics e-prints* (Apr. 2006).

[23] Peacock, J. A. et al., "ESA-ESO Working Group on "Fundamental Cosmology"," tech. rep. (Oct. 2006).

[24] Rubiño-Martín, J. A. et al., "The QUIJOTE CMB Experiment," in [*Highlights of Spanish Astrophysics V*], Diego, J. M., Goicoechea, L. J., González-Serrano, J. I., and Gorgas, J., eds., 127 (2010).

[25] "QUIJOTE-CMB webpage." http://www.iac.es/project/cmb/quijote.

[26] "CMB webpage at the IAC." http://www.iac.es/project/cmb.

[27] Gómez-Reñasco, F. et al., "Control system architecture of QUIJOTE multi-frequency instrument," *Society of Photo-Optical Instrumentation Engineers (SPIE) Conference Series* (2012).

[28] Gómez, A. et al., "QUIJOTE telescope design and fabrication," in [*Society of Photo-Optical Instrumentation Engineers (SPIE) Conference Series*], *Society of Photo-Optical Instrumentation Engineers (SPIE) Conference Series* **7733** (July 2010).

[29] Hoyland, R. et al., "The status of the QUIJOTE I Multi-Frequency Instrument," *Society of Photo-Optical Instrumentation Engineers (SPIE) Conference Series* (2012).

[30] Cano, J. et al., "Multi-Pixel Ka-Band Radiometer for the QUIJOTE Experiment (Phase II)," *Proc. of the 42nd European Microwave Conference (EuMC 2012), Amsterdam, The Netherlands* (2012).

[31] Watson, R. A. et al., "First results from the Very Small Array - I. Observational methods," *MNRAS* **341**, 1057–1065 (June 2003).

[32] Tucci, M., Martínez-González, E., Toffolatti, L., González-Nuevo, J., and De Zotti, G., "Predictions on the high-frequency polarization properties of extragalactic radio sources and implications for polarization measurements of the cosmic microwave background," *MNRAS* **349**, 1267–1277 (Apr. 2004).

[33] Tucci, M., Martínez-González, E., Vielva, P., and Delabrouille, J., "Limits on the detectability of the CMB B-mode polarization imposed by foregrounds," *MNRAS* **360**, 935–949 (July 2005).

[34] Kogut, A., "Synchrotron Spectral Curvature from 22 MHz to 23 GHz," *ArXiv e-prints* (May 2012).

[35] Ruiz-Granados, B., Rubiño-Martín, J. A., and Battaner, E., "Constraining the regular Galactic magnetic field with the 5-year WMAP polarization measurements at 22 GHz," *A&A* **522**, A73 (Nov. 2010).

[36] Watson, R. A., Rebolo, R., Rubiño-Martín, J. A., Hildebrandt, S., Gutiérrez, C. M., Fernández-Cerezo, S., Hoyland, R. J., and Battistelli, E. S., "Detection of Anomalous Microwave Emission in the Perseus Molecular Cloud with the COSMOSOMAS Experiment," *ApJL* **624**, L89–L92 (May 2005).

[37] Battistelli, E. S., Rebolo, R., Rubiño-Martín, J. A., Hildebrandt, S. R., Watson, R. A., Gutiérrez, C., and Hoyland, R. J., "Polarization Observations of the Anomalous Microwave Emission in the Perseus Molecular Complex with the COSMOSOMAS Experiment," *ApJL* **645**, L141–L144 (July 2006).

[38] López-Caniego, M., Massardi, M., González-Nuevo, J., Lanz, L., Herranz, D., De Zotti, G., Sanz, J. L., and Argüeso, F., "Polarization of the WMAP Point Sources," *ApJ* **705**, 868–876 (Nov. 2009).

[39] Wolleben, M., Landecker, T. L., Reich, W., and Wielebinski, R., "An absolutely calibrated survey of polarized emission from the northern sky at 1.4 GHz. Observations and data reduction," *A&A* **448**, 411–424 (Mar. 2006).

[40] López-Caraballo, C. H., Rubiño-Martín, J. A., Rebolo, R., and Génova-Santos, R., "Constraints on the Polarization of the Anomalous Microwave Emission in the Perseus Molecular Complex from Seven-year WMAP Data," *ApJ* **729**, 25 (Mar. 2011).

[41] Dickinson, C., Peel, M., and Vidal, M., "New constraints on the polarization of anomalous microwave emission in nearby molecular clouds," *MNRAS* **418**, L35–L39 (Nov. 2011).

[42] Génova-Santos, R., Rebolo, R., Rubiño-Martín, J. A., López-Caraballo, C. H., and Hildebrandt, S. R., "Detection of Anomalous Microwave Emission in the Pleiades Reflection Nebula with Wilkinson Microwave Anisotropy Probe and the COSMOSOMAS Experiment," *ApJ* **743**, 67 (Dec. 2011).

[43] Draine, B. T. and Hensley, B., "Magnetic Nanoparticles in the Interstella Medium: Emission Spectrum and Polarization," *ArXiv e-prints* (May 2012).

[44] Draine, B. T. and Hensley, B., "The Submm and mm Excess of the SMC: Magnetic Dipole Emission from Magnetic Nanoparticles?," *ArXiv e-prints* (May 2012).

[45] Dobler, G., "A Last Look at the Microwave Haze/Bubbles with WMAP," *ApJ* **750**, 17 (May 2012).

[46] Hooper, D., Finkbeiner, D. P., and Dobler, G., "Possible evidence for dark matter annihilations from the excess microwave emission around the center of the Galaxy seen by the Wilkinson Microwave Anisotropy Probe," *Phys. Rev. D* **76**, 083012 (Oct. 2007).

[47] Vielva, P., Martínez-González, E., Barreiro, R. B., Sanz, J. L., and Cayón, L., "Detection of Non-Gaussianity in the Wilkinson Microwave Anisotropy Probe First-Year Data Using Spherical Wavelets," *ApJ* **609**, 22–34 (July 2004).

[48] Vielva, P., Martínez-González, E., Cruz, M., Barreiro, R. B., and Tucci, M., "Cosmic microwave background polarization as a probe of the anomalous nature of the cold spot," *MNRAS* **410**, 33–38 (Jan. 2011).

[49] "Consolider-Ingenio 2010 project: Exploring the Physics of Inflation." http://www.epi-consolider.es/.

The Next Generation of the Canada-France-Hawaii Telescope: Science requirements and survey strategies

Alan McConnachie[a], Patrick Côté[a], David Crampton[a], Daniel Devost[b], Doug Simons[b], Kei Szeto[a] and the NGCFHT Concept Study Team

[a]NRC Herzberg Institute of Astrophysics, 5071 West Saanich Road, Victoria, British Columbia, Canada, V9E2E7
[b]CFHT Corporation , 65-1238 Mamalahoa Hwy, Kamuela, Hawaii 96743 USA

ABSTRACT

A concept study is underway to upgrade the existing 3.6 meter Canada-France-Hawaii Telescope (CFHT) to a 10 meter class, wide-field, dedicated, spectroscopic facility, which will be the sole astronomical resource capable of obtaining deep, spectroscopic follow-up data to the wealth of photometric and astrometric surveys planned for the next decade, and which is designed to tackle driving science questions on the formation of the Milky Way galaxy and the characterization and nature of dark energy. This unique facility will operate at low ($R \sim 2000$), intermediate ($R \sim 6000$) and high ($R \sim 20000$) resolutions over the wavelength range $370 \leq \lambda \leq 1300$nm, and will obtain up to 3200 simultaneous spectra per pointing over a 1.5 square degree field. Unlike all other proposed or planned wide field spectroscopic facilities, this "Next Generation CFHT" will combine the power of a 10m aperture with exquisite observing conditions and a mandate for *dedicated* spectroscopic studies to enable transformative science programs in fields as diverse as exoplanetary host characterization, the interstellar medium, stars and stellar astrophysics, the Milky Way galaxy, the Local Group, nearby galaxies and clusters, galaxy evolution, the inter-galactic medium, dark energy and cosmology. A new collaboration must be formed to make this necessary facility into a reality, and currently nearly 60 scientists from 11 different communities - Australia, Brazil, Canada, China, France, Hawaii, India, Japan, South Korea, Taiwan, USA - are involved in defining the science requirements and survey strategies. Here, we discuss the origins of this project, its motivations, the key science and its flow-down requirements. An accompanying article describes the technical studies completed to date. The final concept study will be submitted to the CFHT Board and Science Advisory Committee in Fall 2012, with first light for the facility aiming to be in the early 2020s.

Keywords: Next Generation Canada-France-Hawaii Telescope; spectroscopy; new facilities; surveys

1. INTRODUCTION

The scientific capabilities of a dedicated wide-field, multi-object spectrograph on a 6.5-10m telescope are well documented, having been discussed almost continuously for more than a decade now.[1-4] But while the potential impact of such a facility is universally recognized by the international astronomical community, there remains, to date, no approved and/or funded facility of this sort on the horizon. This situation is remarkable given that the scientific need for such a facility has only sharpened with time, particularly in light of the many wide-field imaging and astrometric surveys scheduled for the coming decade (e.g., Pan-STARRS, VISTA, VST, Hyper-SuprimeCam, GAIA, the Dark Energy Survey, Skymapper, LSST, Euclid and WFIRST).

The Next Generation CFHT (NGCFHT) is a proposal to redevelop CFHT by replacing the existing 3.6m telescope with a 10m, segmented-mirror telescope equipped with a dedicated wide-field, highly multiplexed fibre spectrograph capable of carrying out transformational spectroscopic surveys of the faint universe. Achieving first light in the early 2020s, the proposed NGCFHT would fill what will arguably be the single most important missing capability in astrophysics at the end of this decade.

Further author information: (Send correspondence to Alan McConnachie and Pat Côté)
Alan McConnachie: E-mail: alan.mcconnachie@nrc-cnrc.gc.ca
Pat Côté: E-mail: patrick.cote@nrc-cnrc.gc.ca

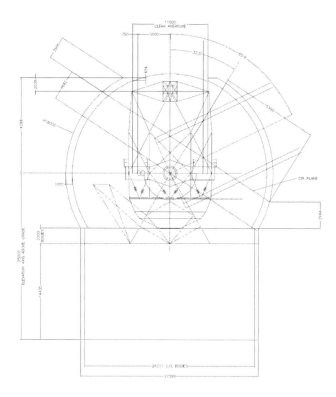

Figure 1. Conceptual layout of a 10m telescope and enclosure mounted on the CFHT pier, as baselined for the NGCFHT. Figure provided by Empire Dynamic Structures Ltd. (DSL). See companion paper by Szeto et al. (2012)

NGCFHT began as a grassroots movement within the CFHT user communities in 2010[5]). Beginning in early 2011, a concept study of the NGCFHT was initiated to explore both the facility's scientific capabilities and requirements, and its technical feasibility and readiness*. In this article, we discuss the scientific development of this facility and the resulting flow down requirements. A companion paper discusses the technical progress to date.[6]

2. CFHT: REDEVELOPMENT AND NEAR-TERM FUTURE

The 3.6m CFHT is operated by Canada (through the National Research Council), France (through the Centre National de la Recherche Scientifique) and the University of Hawaii. The telescope began operations in 1979, situated on the north end of the Mauna Kea summit ridge at an altitude of 4204 m. This location is one of the best astronomical sites in the world, enjoying outstanding atmospheric stability (a median seeing of the free atmosphere of 0.4 arcsecs with a 10th percentile quality of 0.25 arcsecs), low precipitable water vapor (median ~ 0.9 mm) and a high percentage of usable nights ($\sim 70\%$ annually)†. In addition to the three main partners, Taiwan (through the Academia Sinica Institute of Astronomy and Astrophysics, ASIAA), Brazil (through the Brazilian Ministry of Science, Technology and Innovation, MCTI), China (through the National Astronomical

---

*More information, including links to relevant documents, is available at http://orca.phys.uvic.ca/~pcote/ngcfht/The_Next_Generation_CFHT.html

†http://www.cfht.hawaii.edu/Instruments/Imaging/MegaPrime/observingstats.html

Observatory of China, NAOC) and South Korea (through the Korean Astronomy and Space Science Institute, KASI) currently have access to the available observing time through arrangements with CFHT.

Prior to the completion of the Keck telescopes and the refurbished Hubble Space Telescope in the early 1990s, it could be argued that CFHT was the best optical telescope in world, and it remains one of the "best-in-class" to the present day. Early instruments on CFHT included various CCD and IR imagers, HRCam, HIFI, Herzberg, coudé and UV Prime spectrographs, CIGALE, among others. From the early-to-mid 1990's, CFHT focused on the development of powerful instruments that gave it unique capabilities for a telescope of intermediate size. Notable instruments from this period include Gecko, MOS-SIS, OASIS and the AOB, as well as a succession of ever-larger CCD cameras: e.g., FOCAM, MOCAM and CFH12K. In the mid-2000s, CFHT moved towards a leaner, more specialized instrumentation suite, including MegaCam, ESPaDOnS, and WIRCam, that aimed for a high level of impact through a combination of unique capabilities and a larger number of nights dedicated to ambitious surveys. The CFHT Legacy Survey[‡], in particular, demonstrated this approach to be an highly effective one. In the near future, new instruments are being developed for the telescope, and include SITELLE (a wide-field Fourier transform spectrometer[12]), SPIRou (a near-IR, high-resolution spectrograph and spectropolarimeter[13]) and possibly 'IMAKA (a wide-field, GLAO-corrected mosaic camera equipped with orthogonal transfer CCDs[14]).

The idea of replacing the 3.6m CFHT with a much larger telescope is not a new one,[7–10] and - uniquely - redevelopment of the Observatory can potentially be allowed by the Mauna Kea Master Plan. Many concepts have been considered but perhaps the most comprehensive report is that of Grundmann et al.[11] who carried out a study commissioned by the CFHT Science Advisory Committee to identify the largest telescope which could reasonably be installed making use of the existing pier. In answer to the question, Grundmann concluded that a $12-15$m segmented mirror alt-az telescope similar to the Keck 10m design would be able to fit on the pier inside the CFHT dome. At the time, a major motivation for this and other studies relating to "upgrading" CFHT was to identify options for building large aperture facilities that would surpass the power of the new 8-10m class telescopes. This desire ultimately led the partners to engage in the forthcoming Extremely Large Telescopes (ELTs), and this took the focus off CFHT redevelopment for over a decade.

The ongoing technical studies that are being conducted as part of the NGCFHT concept study are described in the companion article. However, it is worth noting here some of the findings that confirm the earlier expectations of Grundmann and others. Namely, despite the huge increase in collecting area, the weight of the Keck mirror and cell is only $\sim 10\%$ larger than that of CFHT, and the telescope weight is similar (~ 270 tonnes versus 266 tonnes). Because the CFHT 3.6m primary has a slow ($f/3.8$) focal ratio, a much larger modern telescope can be accommodated within a dome of similar size to that of CFHT (modern telescopes have primaries with f/ratios ~ 1). Of course, the current dome, which has a shutter width of only 6.5m, will need to be replaced with one having a larger aperture.

In the time since the early studies on CFHT redevelopment, the worldwide suite of 6.5-10m class telescopes has grown and matured to enable a vast range of astrophysical studies (there are now ~ 15 international facilities in this aperture class). The power of these telescopes will ultimately be superseded by the 30-40m class telescopes, and it is possible that the 8m telescopes will take on a supporting (or "work-horse") role, currently more associated with the present 4m-class telescopes. The science-case for another *general-purpose* large telescope is therefore not as well defined as it was in the 1990s, and it is here that the concept for the NGCFHT differs from its forebears and from any existing or proposed large telescope. Uniquely, NGCFHT will be a *10m class, dedicated, spectroscopic facility*.

3. POINTS OF DEPARTURE

3.1 Photometry, astrometry, and the need for a spectroscopic facility

The most productive astronomical survey of all time, the Sloan Digital Sky Survey (SDSS) demonstrated the scientific potential of very large sky surveys, in particular the combination of large area photometry and highly multiplexed spectroscopy. There are now a large number of either ongoing or planned photometric surveys which together will map the entire sky at optical and/or IR wavelengths. These include (but are not limited to) the

---

[‡]http://www.cfht.hawaii.edu/Science/CFHTLS

DES, surveys with the VST (ATLAS, KIDDS) and VISTA (VIKING, VHS), PanSTARRS PS1 (and plans for PS2), Skymapper and the Subaru HyperSuprimeCam. A key priority in the USA is the LSST, and in Europe is the forthcoming EUCLID space mission (with perhaps WFIRST happening in the USA). In 2013, ESA will launch GAIA, which will image the entire sky to create an ultra-precise astrometric map, and obtain photometry for all objects down to $V \sim 20$.

While wide-field imaging is burgeoning, wide-field spectroscopy has not kept pace. There are multiple drivers for ever wider and deeper photometry, and prominent among them is cosmology and dark energy. Here, sampling the structure of a cosmologically significant volume to high redshift allows the expansion history of the Universe to be examined, and so place constraints on fundamental parameters. While no specific recommendations for highly multiplexed spectroscopy were made, the "Astro2010 Decadal Review" in the USA noted that

> "The properties of dark energy would be inferred from the measurement of both its effects on the expansion rate and its effects on the growth of structure (the pattern of galaxies and galaxy clusters in the universe). In doing so it should be possible to measure deviations from a cosmological constant larger than about a percent. Massively multiplexed spectrographs in intermediate-class and large-aperture ground-based telescopes would also play an important role."

For this science, low-to-intermediate resolutions ($R \sim 2000 - 5000$) are generally sufficient.

Another key driver of wide field studies is in relation to the structure and content of the Milky Way galaxy, and here GAIA will be transformative. In contrast to many of the extra-galactic/cosmology themes that require highly multiplexed spectroscopy, Galactic science (including "near-field cosmology" and resolved stellar populations) generally demands higher resolutions ($R > 20000$) to resolve features with small equivalent widths and determine high-order abundance information in addition to velocities and various stellar parameters. The "Report by the ESA-ESO Working Group on Galactic populations, chemistry and dynamics"[§] makes several recommendations regarding improving wide-field spectroscopic facilities, including the consideration of dedicated 4m and 8m highly multiplexed spectroscopic telescopes. The final paragraph of this report makes the following observation:

> "Our terms of reference were to propose a set of recommendations to ESA and ESO for optimizing the exploitation of their current and planned missions. However, the Galaxy is an all-sky object; in fact, from the ground, the outer parts of the Galaxy are best observed from the Northern hemisphere, as the extinction is on-average lower there. In parallel with Recommendations 2(a) to 2(d), there is a real need for dedicated highly multiplexed spectrographs in the Northern hemisphere."

In response to this emerging void in ground based astronomy, several projects that address specific science questions are underway or proposed, and will help deal with the current shortcomings. These studies include (but are not limited to):

- AAT/HERMES ($N = 400, R28K$[15]); the 3.9m AAT will shortly undertake the GALAH survey with this instrument, observing $\sim 10^6$ stars and measuring the abundances of > 15 elements per star to around $V \sim 14$;

- The ESO-GAIA survey;[16] some 300 co-investigators are involved in a dedicated effort to use the VLT high resolution multiplexed spectrographs to observe $\sim 10^5$ stars with FLAMES/GIRAFFE ($V \lesssim 20$) and 10^4 stars with FLAMES/UVES ($V \lesssim 15$). This project saw first light in December 2011 and has been allocated 300 nights over 5 years;

- ESO/MOONS,[17] ESO/4MOST;[18] In 2010, ESO made a Call for Letters of Intent for "Wide-Field Spectroscopic Survey Facilities". Two of these - MOONS and 4MOST - were selected for further study. MOONS is a "GIRAFFE-like", highly multiplexed (N=hundreds) near IR spectrograph working to around 1800nm. 4MOST is a 4-m class dedicated spectroscopic facility that would be hosted at either the NTT or VISTA. It would work in the optical and would operate in both low and high resolution modes;

[§]http://sci.esa.int/science-e/www/object/index.cfm?fobjectid=43454#

- KPNO4m/BigBOSS ($N = 4000; R5K^{19}$); BigBOSS is a proposed DOE-NSF dark energy experiment using the Mayall telescope. BigBOSS would perform a 500 night survey in collaboration with NOAO, and would address a large array of community science in addition to the primary BAO science case for which the instrument is designed;

- Subaru/PFS ($N = 2400, \sim R4K^{20}$); The Prime Focus Spectrograph for the 8-m Subaru telescope recently passed the Conceptual Design Stage and is now under construction with first light around 2017. Upon completion, it will be the most powerful multiplexed spectrograph in the world at an exquisite site, and plans are underway for a Subaru Strategic Program to use this instrument over $\sim$ 300 nights. PFS emerged following the cancellation of the Wide Field Multi-Object Spectrograph, and is in some ways a spectroscopic complement to the Subaru/HyperSuprimeCam, which itself will likely become the premier imaging instrument when it shortly sees first light. The reader is referred to Ellis et al. (2012) for a full description of the excellent science enabled by PFS.

As the above list makes clear, there is currently considerable effort by the international astronomical community to address the pressing need for highly multiplexed wide field spectroscopy. Unique amongst these efforts is the NGCFHT, which combines the power of a 10m aperture with an exquisite observing site and a mandate to operate as a dedicated spectroscopic facility, thus fulfilling a unique role in the coming era of GAIA, LSST, Euclid, etc.. This proposal was first put forward in a White Paper submitted to the Canadian Long Range Plan 2010 (LRP2010,[5] and subsequently presented at the CFHT Users Meeting in November of that year. The Canadian LRP2010 recognises that "the science case for an ngCFHT is unassailable" and recommended that the proposal is developed further to obtain a better understanding of its scope, cost, schedule and technical readiness.

3.2 Baseline specifications and formation of the Science Working Groups

As a point of departure for the NGCFHT Concept study, a baseline design was introduced in the White Paper submitted to the Canadian LRP2010 that was adopted for the first stage of the science study. This baselined to a 10m, segmented telescope similar in design to the Keck telescopes, with a wide-field corrector providing a 1.5 square degree field of view. The spectrograph capabilities were based heavily on the Gemini/WFMOS design from Ellis et al. (2009).[3] Key parameters are summarised in Table 1, where we show the adopted specifications for the telescope and instrument, as well as the basic parameters for two "strawman" surveys, in which the bulk (80%) of the observing time is equally divided between bright- and dark-time surveys. In this model, high-spectral resolution observations of compact and/or bright objects (stars, quasars, nearby galaxies) are collected during times of high moon illumination, while dark time is reserved primarily for lower-resolution spectroscopy of the faintest objects (distant galaxies, quasars, etc). In this model, the remaining observing time (20%) is reserved for PI-led programs and/or "Key Projects" that will be executed contemporaneously with these dedicated, wide-field (10,000 deg^2) surveys. Note that a key distinction between NGCFHT and previous efforts to develop this type of capability is the opportunity for these smaller, but high-impact, science programs, as described in the next section.

To examine these baseline specifications and the observing strategies described above, ten science working groups (SWGs) were assembled during 2011. The SWGs were chosen to cover the broadest possible range of science topics: exoplanets; the interstellar medium; stellar astrophysics; the Milky Way; the Local Group; nearby galaxies and clusters; galaxy evolution; QSOs and AGNs; the intergalactic medium; cosmology and dark energy. Five of these SWGs are led from Canada, and five are led from France. Their composition is given in Table 2, although membership continues to grow steadily with time. As of May 2012, the SWGs include nearly 60 scientists from Canada, France and Hawaii, as well as Australia, Brazil, China, India, Japan, South Korea, Taiwan, and the USA.

4. SCIENCE DRIVERS AND FLOWDOWN REQUIREMENTS

In the Fall of 2011, the SWGs submitted interim reports, based on preliminary investigation of the science enabled by wide field multiplexed spectroscopy, as applied to their specialism. In particular, the SWGs were tasked with

Table 1. Baseline Design for NGCFHT

Parameter	Original baseline	New specs, if different
Primary Mirror	10m (segmented)	
Field of View	1.5 deg² (hexagonal)	
Wavelength Range	370 - 970 nm	370 - 1300 nm
Number of Fibers:	3200 (lower resolutions)	
	800 (high resolution)	
Spectral Resolution ($\mathcal{R}$)	1500 (370-650 nm)	2000 (370-1300 nm)
	3500 (630-970 nm)	6000
	5000 (480 - 550 nm / 815-885 nm)	20000 (480-680 nm)
	20000 (480-680 nm)	
Bright Survey (40%)		
Areal Coverage	$\Omega_{\rm tot} = 10{,}000$ deg²	
Spectral Resolution	$\mathcal{R} = 5000, 20000$	
Sky Brightness	$\mu(g) < 21.1$ mag arcsec^{-2}	
Depth	$g \approx 19.7$ mag (1hr at R =20000; S/N = 20 per Å)	
Dark Survey (40%)		
Areal Coverage	$\Omega_{\rm tot} = 10{,}000$ deg²	
Spectral Resolution	$\mathcal{R} = 1500, 3500$	
Sky Brightness	$\mu(g) > 21.1$ mag arcsec^{-2}	
Depth	$g \approx 23.1$ mag (1hr at R =3500; S/N = 5 per Å)	

determining what transformational science programs could be conducted using NGCFHT as defined by the "point of departure" requirements in Table 1, and what changes to the baseline configurations would improve the impact of the facility. These reports amply demonstrate that NGCFHT has the potential to perform cutting-edge research across a wide range of disciplines, and resulted in modification of the baseline specifications, as detailed in Table 1. Section 4.1 and 4.2 describe two of the main science areas that drive the requirements for NGCFHT and the main survey strategies, and Section 4.3 lists examples of other transformational science programs that NGCFHT will perform, to give a flavor of the versatility of this type of facility. Section 4.4 summarised the main changes to the baseline requirements that have emerged as a result of the SWG studies.

4.1 Key science I: Decoding the DNA of the Galaxy

A detailed study of the kinematical and chemical structure of the Milky Way is absolutely vital for progress in near field cosmology. To complement the GAIA data, spectroscopic observations from ground based facilities are essential. NGCFHT can become *the* major contributor to this subject. Due to space constraints, here we summarise a few studies that involve *only the halo* of the Galaxy. While this component has the largest volume (with a radius exceeding 100kpc), it contains only a few percent of the stellar mass. Some of these are the oldest stellar populations of the Galaxy, that have been formed a few Myr after the Big Bang, and the dynamics and the chemical composition of its member star provide us with strong constraints on the formation and evolution of our Galaxy. We note that NGCFHT will be similarly transformative for studies of the disk (thin and thick), the bulge, and star clusters; ultimately, the longer term culmination of projects such as NGCFHT and GAIA must be in the construction of a holistic understanding of the formation and interconnectedness of each of these supposedly distinct "components" of our Milky Way galaxy.

4.1.1 Chemical tagging

Detailed abundances, providing abundance ratios of elements formed through different nucleosynthetic channels, are a key factor to unravel the formation and history of stellar Galactic halo components. To separate the components that have undergone a different chemical evolution by their abundance pattern (so-called chemical tagging[21]) a precision in the abundances of the order of 0.1 dex is required. For example, we are becoming aware of the existence in the halo of stars with different [α/Fe] ratios, at the same [Fe/H].[22]

The NGCFHT will enable to measure detailed abundances of a sample of about 500 000 distant halo stars at magnitudes $V = 18 - 19$. In this magnitude range, the stellar density would be 600 stars/square degree at the Galactic poles. The radial velocity should be matched to the accuracy of the GAIA transverse velocities, therefore of a few km/s at the faint end. The dwarf/Turn-Off stars will be within the distance range $2 - 10$kpc and the giants in the range $10 - 100$kpc. About two-thirds of the stars will be dwarfs/TO. Up to now the study

Table 2. NGCFHT Science Working Group Membership

Member	Institute	Science Working Group
Nobou Arimoto	NAOJ, Japan	Milky Way Structure and Stellar Populations
Michael Balogh	University of Waterloo, Canada	Galaxy Evolution (chair)
Patrick Boiss	IAP, France	The Interstellar Medium
James Bolton	University of Melbourne, Australia	The Intergalactic Medium
Piercarlo Bonifacio	GEPI, Université Paris Diderot, France	Milky Way Structure and Stellar Populations (chair)
Francois Bouchy	IAP, France	Exoplanets
Andrew Cole	University of Tasmania, Australia	The Local Group
Len Cowie	Institute for Astronomy, Hawaii, USA	QSOs and AGNs
Scott Croom	University of Sydney, Australia	QSOs and AGNs
Katia Cuhna	NOAO, USA	Stars and Stellar Astrophysics
Richard de Grijs	Kavli Institute, Beijing, China	Nearby Galaxies and Clusters
Magali Deleuil	LAM, France	Exoplanets (chair)
Ernst de Mooij	University of Toronto, Canada	Exoplanets
Simon Driver	ICRAR, Australia	Nearby Galaxies and Clusters
Patrick Dufour	Université de Montréal, Canada	Stars and Stellar Astrophysics
Sara Ellison	University of Victoria, Canada	The Intergalactic Medium
Sebastien Foucaud	NTNU, Taiwan	Galaxy Evolution
Ken Freeman	ANU, Australia	Milky Way Structure and Stellar Populations
Karl Glazebrook	University of Swinburne, Australia	Galaxy Evolution
Pat Hall	York University, Canada	QSOs and AGNs (chair)
Zhanwen Han	Yunnan Observatory, China	Stars and Stellar Astrophysics
Michael Hudson	University of Waterloo, Canada	Nearby Galaxies and Clusters (chair)
John Hutchings	HIA, Canada	QSOs and AGNs
Rodrigo Ibata	Université de Strasbourg, France	The Local Group
Pascale Jablonka	Observatoire de Paris, France	The Local Group
Jean-Paul Kneib	LAM, France	Cosmology and Dark Energy (chair)
Chiaki Kobayashi	Australian National University, Australia	Stars and Stellar Astrophysics
Rolf-Peter Kudritzki	Institute for Astronomy, Hawaii, USA	Stars and Stellar Astrophysics
Rosine Lallement	LATMOS-IPSL, France	The Interstellar Medium (chair)
Damien Le Borgne	IAP, France	Galaxy Evolution
Yang-Shyang Li	KIAA, China	The Local Group
Lihwai Lin	ASIAA, Taiwan	Galaxy Evolution
Yen-Ting Lin	IPMU, University of Tokyo, Japan	Nearby Galaxies and Clusters
Martin Makler	Centro Brasilerio de Pesquisas Fisicas, Brasil	Cosmology and Dark Energy
Nicolas Martin	Université de Strasbourg, France	The Local Group
Alan McConnachie	HIA, Canada	The Local Group (chair)
Norio Narita	NAOJ, Japan	Exoplanets
Changbom Park	KIAS, South Korea	Galaxy Evolution
Eric Peng	Peking University, China	Nearby Galaxies and Clusters
Patrick Petitjean	AIP, France	QSOs and AGNs
Cline Peroux	LAM, France	The Intergalactic Medium (chair)
Ryan Ransom	Okanagan College/DRAO, Canadaa	The Interstellar Medium
Swara Ravindranath	IUCAA, India	Galaxy Evolution
Bacham Eswar Reddy	IIA, India	Milky Way Structure and Stellar Populations
Marcin Sawicki	St. Mary's University, Canada	Galaxy Evolution
Carlo Schimd	LAM, France	Cosmology and Dark Energy
Luc Simard	HIA, Canada	Galaxy Evolution
Raghunathan Srianand	IUCAA, India	The Intergalactic Medium
Else Starkenburg	University of Victoria, Canada	Stars and Stellar Astrophysics
Thaisa Storchi-Bergmann	UFRGS, Brasil	QSOs and AGNs
Charling Tao	CPPM, France and Tsinghua, China	Cosmology and Dark Energy
Sivarani Thirupathi	IIA, India	Milky Way Structure and Stellar Populations
Keiichi Umetsu	ASIAA, Taiwan	Cosmology and Dark Energy
Kim Venn	University of Victoria, Canada	Stars and Stellar Astrophysics (chair)
Chris Willott	HIA, Canada	QSOs and AGNs
Ting-Gui Wang	USTC, China	QSOs and AGNs
Jong-Hak Woo	Seoul University, South Korea	QSOs and AGNs
Xue-Bing Wu	Peking University, China	QSOs and AGNs

of the chemical properties of the halo has relied, except for a few exceptions, on local samples of halo stars that pass near enough to the Sun to be observed at high resolution. A great improvement in the chemical and dynamical properties of the halo will come with in situ studies that will sample the halo throughout its depth and scan for gradients and variations in chemical properties.

4.1.2 The metal-weak tail of the halo Metallicity Distribution Function

With the proposed NGCFHT surveys, we will obtain a clear picture of the halo Metallicity Distribution Function (MDF), including its metal-poor tail. This has a direct bearing on the formation of the first stars, and on the dark baryonic content of galaxies. The first stars to be formed after the Big Bang were formed with the primordial chemical composition, i.e. hydrogen and helium and traces of lithium. Such a gas may have difficulty in providing cooling mechanisms efficient enough to allow the formation of low-mass stars. Several theories on star formation postulate the existence of a critical metallicity, below which only extremely massive stars can be formed ([23] and references therein), while other theories allow fragmentation to produce low-mass stars at any metallicity ([24] and references therein). The implication for the baryonic content of galaxies is obvious: if the first generation of massive stars that reionised the Universe formed along with low-mass stars, a large fraction of these is currently present as old, cool white dwarfs. On the other hand a small fraction of these (essentially those of mass less than $0.8\,M_\odot$) is still shining today and can be observed. If there is a critical metallicity the metal-weak tail of the MDF ought to show a sharp drop at this value. A sharp drop does indeed appear around [Fe/H]= -3.5.[25, 26] Remarkably many stars have already been discovered below this metallicity, the most metal-poor star to date being SDSS J102915+172927[27] with $Z \leq 6.9 \times 10^7$.[28] Below [Fe/H]= -3.5, the metallicity has to be determined from high resolution spectra, since the metal lines become too weak to measure at $R \sim 2000$ (that has been used extensively for this kind of work). The Hamburg-ESO MDF is based on only 1638 stars. NGCFHT will provide the definitive answer to this issue, with metallicities for a sample of 500 000 stars at resolution $R = 20000$.

4.1.3 The inner and outer halos

The Galactic stellar halo is not a homogeneous component and some sub-components are now clearly detailed.[29] To understand and describe its nature is a major question of the galactic formation. The inner and outer halo - with a transition usually defined at about 10 kpc from the Sun - can be distinguished considering the time of mixing of stars in the 3D-3V phase space, the inner halo being more mixed and smooth within the 3D configuration space than the outer halo (where the mixing time scale is a fraction of the age of the Galaxy). The identification of these structures will be a key to understand the mechanism of galactic accretion and formation.

A major fraction of the halo must result from the accretion and dissolution of about two hundreds infalling small galaxies.[30] With adapted dynamical techniques and tools, a radial velocity survey combined with photometric and/or GAIA distances will allow the identification of the remnants of disrupted galaxies. Combined with GAIA proper motions, identification will be more straightforward and less model dependent. The three integrals of motion of individual stars will be measurable, and will allow us to separate and kinematically classify the streams within the outer halo.

In the inner halo, where accreted dwarf galaxies may still be major constituents, the situation is more tricky. Here, the integrals of motion do not evolve as "adiabatically" for stars in streams crossing the inner part of the Galaxy, and this makes their kinematic identification more difficult. Tens of thousands of radial velocities must be measured to identify streams above the "background" in phase space that this introduces. Nevertheless, the vast database of high resolution spectra to $V \sim 19 - 20$ that NGCFHT will obtain means that both the inner and outer halo can be studies in situ.

4.1.4 The thick disk – halo transition

NGCFHT observations of the Galactic halo at $V \sim 18 - 19$ will detect a majority ($\sim 60\%$) of thick disk stars at $1 - 4\,\text{kpc}$ above the Galactic plane. These observations will probe in detail the kinematics of the intermediate Galactic stellar populations, and they will be a critical complement to the science case of the chemical labeling of the Galactic discs, unraveling the kinematics of stellar populations at much lager distances, and giving a more complete kinematic view of the thick disk on a wide scale. The current observations[31] seem to indicate that the thick disk stars have similar abundance pattern to the halo high α stars. However this conclusion is based on a local sample and ultimately must be explored using the NGCFHT in-situ sample.

4.2 Key science II: Expanding on Universal expansion

The power of a 10m aperture combined with the ability to conduct efficient and wide spectroscopic surveys means that a massive spectroscopic dataset will be accumulated that will be uniquely powerful for exploring the expansion history of the Universe through a range of cosmological probes. In what follows, we discuss some of the principle studies in isolation, but note that for most of the science discussed, it is the combination of the cosmological probes (each with different systematics) that allows for the highest precisions and most valuable insights.

4.2.1 Baryonic Acoustic Oscillations

Numerous studies (including those discussed in earlier sections) plan to use BAO to provide precise characterization of dark energy; NGCFHT will extend current and future BOSS BAO measurements by focusing on the high redshift window $1 < z < 6.5$. This will be achieved using both Extremely Luminous Galaxies (ELGs) with $1 < z < 2$, and possibly Lyα emitters to extend measurements in the $3 < z < 6.5$ redshift window.

4.2.2 Redshift Space Distortions

The peculiar velocities of galaxies are produced by the gravitational pull of the matter content of the universe. On large scales these peculiar motions become coherent bulk flows towards over-dense regions and away from under-dense regions. The velocity bulk flows depend on the amplitude of the matter fluctuations and therefore are a measure of the growth rate of structure.[32] These coherent flows imprint a distinct anisotropic feature in the distribution of galaxies in redshift space, generally referred to as Redshift Space Distortions (RSDs). For example, the two point correlation function gets squashed in the line-of-sight direction.

As a tracer of the growth rate of structure, RSDs are in principle capable of distinguishing dark energy from modified gravity models.[33,34] RSD detection by the standard method - measurement of the quadrupole anisotropy in the redshift-space two-point correlation functions - is limited by the sample variance of the power spectrum in the surveyed volume V. This method yields a measure of fG, where G is the linear growth factor, and $f = \frac{d\ln G}{d\ln a}$. The sample-variance limit for standard RSD analyses is $\sigma_{\ln fG} \geq 35/\sqrt{k_{max}^3 V}$ if accurate peculiar-velocity theory is available for $k < k_{max}$. With its large aperture, survey mode and long lifetime, NGCFHT will provide the tightest cosmological constraints ever achieved using RSDs by measuring a high density of sources over an extremely wide field of view.

4.2.3 Galaxy evolution, large scale structure, weak lensing through calibration of photometric redshifts

There are many synergies between photometric and spectroscopic surveys. One of the most important is photometric redshift calibration: in this way, a limited amount of spectroscopy can be leveraged into redshift estimates for considerably more objects. Photometric redshifts can be calibrated directly by obtaining spectroscopic redshifts for representative subsets of a photometric sample. However, many authors have pointed out that the completeness of such samples need to be very high, whereas actual redshift surveys to depths somewhat brighter than those probed by DES or future surveys like LSST have failed to obtain secure ($> 95\%$ confidence) redshifts for $30 - 60\%$ of galaxies targeted. Newman[35] suggested an alternate calibration method, based upon measuring the cross-correlation on the sky between galaxies in a spectroscopic sample with those in a photometric sample, as a function of spectroscopic z. Unlike direct calibration, this method requires only that photometric and spectroscopic samples cover the same redshift range and overlap on the sky, but they need not match in other properties; e.g. a survey of only the brightest galaxies at a given redshift would suffice, so long as it contains a large number of objects. This would complement and enhance any results from direct calibration of photo-zs. Similar methods have already been successfully applied to data,[36] and may supersede direct calibration of photo-zs for faint objects.

NGCFHT will target both LRGs at the low redshift end and ELGs and QSOs at higher redshifts, and will be the preferred dataset for calibrating future photometric surveys to high accuracy. At the bright end, this calibration will be direct; but most photometric objects will be fainter than spectroscopic surveys will probe, making its use for cross-correlation even more valuable. Accurate photometric redshifts impact a vast spectrum of research, such as galaxy evolution, large-scale structure statistics, and weak lensing. For the latter, and in addition to understanding the inter-relation of halos and galaxies, the measurement of weak lensing shear on a large sample of galaxies in the background of the NGCFHT spectroscopic survey sample will enable unique tests of General Relativity, such as whether the two scalar gravitational potentials in the Newtonian gauge are equal.[37,38]

4.2.4 Neutrino Masses

The sheer size of the NGCFHT extra-galactic dataset, together with the high density of the sample and the high bias of the quasars, will lead to an accurate measurement of the power spectrum on large scales. This will in turn provide accurate constraints on several cosmological parameters. As an example, neutrinos free stream through the universe and if they have mass they suppress the growth of cold dark matter on scales below the free streaming scale, while no suppression is expected on large scales. This leads to a broad feature in the matter power spectrum on scales of tens to hundred Mpc. The amplitude of the feature depends on neutrino mass. One needs a very large survey volume to detect this effect. The quantity of quasar redshifts in the NGCFHT dataset will be the largest volume redshift survey available and ideally suited to search for this effect.

4.2.5 Non-Gaussianity

Non-Gaussianity in initial conditions of density perturbations has emerged as one of the most important discriminators among the inflationary models of the earliest universe. A new method has recently been proposed based on theoretical models that suggest bias of galaxies becomes scale dependent on very large scales due to non-Gaussianity.[39] This allows one to use galaxy and quasar clustering to learn about non-Gaussianity of the initial conditions in the universe. The method relies on tracers being highly biased, since the effect scales as bias and vanishes for unbiased tracers. High redshift quasars are particularly suitable for this purpose; the first application of this method to SDSS photometric data[40] gave results comparable to the latest CMB constraints from WMAP, while having completely different systematic errors. The NGCFHT extragalactic survey will allow one to go from a 2D to a 3D analysis, greatly increasing the number of available modes. The errors could be further reduced by bispectrum analysis, which contains information complementary to the power spectrum. The end result is likely to be a constraint on non-Gaussianity comparable and independent to the expected Planck constraint of 5-10 on f_{nl}.

4.2.6 Strong Lensing

In the SDSS-I, the Sloan Lens ACS Survey (SLACS[41,42]) has discovered over 100 strong gravitational lens systems through the spectroscopic signature of two redshifts along the same line of sight. This has enabled a unique experimental approach to the study of the structure and dynamics of massive galaxies at relatively low redshift[43] through follow-up imaging of the spectroscopic lens candidates. The BOSS Emission-Line Lens Survey is successfully extending this technique to significant cosmological look-back time at redshift $z \sim 0.4-0.6$, thereby directly probing the structural and dynamical evolution of LRGs. The NGCFHT spectroscopic surveys will extend this technique to $z = 1$ and beyond using ELG lenses, and so yield a direct measurement of massive galaxy structure over an unprecedented baseline in cosmic time.

4.3 Key Science III: A versatile facility for astrophysical research

In addition to science cases relating to the Milky Way and Cosmology (previous sections), other examples of transformational science programs identified by the SWGs that are uniquely enabled by NGCFHT, and which demonstrate the broad impact of this type of facility, include:

1. **[The Interstellar Medium]** The definitive three-dimensional map of the density and kinematics of the Galactic ISM through spectroscopy of molecular, atomic and ionized gas along many thousands of sight lines.

2. **[Stars and stellar astrophysics]** The measurement of fundamental stellar parameters spectroscopic masses, distances, metallicities, rotation rates for an unprecedented sample of high-mass stars in the Milky Way, which will revolutionize our understanding of stellar evolution and feedback in star-forming galaxies locally and at high redshift.

3. **[The Local Group]** The first chemo-dynamical deconstruction of an $L\star$ galaxy (M31), and a systematic survey of the structure of dark matter halos in both low-mass (dwarf) galaxies and high-mass galaxies in nearby groups and clusters.

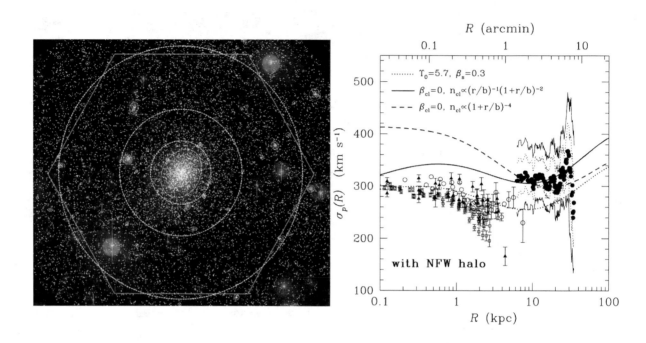

Figure 2. (Left Panel) The central region of the Virgo B subcluster based on deep, wide-field CFHT imaging from the Next Generation Virgo Cluster Survey. The 1.5 deg^2 field of NGCFHT is shown by the hexagon. White symbols show confirmed or probable baryonic substructures (e.g., giant galaxies, dwarf galaxies, UCDs, and star clusters) within the Virgo cluster, selected to have $g_{AB} \leq 23.5$. The yellow circles have radii of 50, 100 and 200 kpc. (Right Panel) Stellar velocity dispersion profile for the central galaxy (M49) in the preceding panel, as well as for 263 globular star clusters based on multi-object spectroscopy from the Keck 10m telescope. The curves show models for different baryonic tracer populations embedded in the cluster dark matter potential, for a variety of velocity anisotropy parameters. A 210-hr ngCFHT survey of the Virgo cluster could be supplemented with data from the Dark-Wide survey to derive radial velocities for more than 600,000 sources located inside the clusters virial radius. Spectroscopically confirmed members of Virgo would then be used as dynamical test particles to measure the distribution of dark matter in the cluster and the orbital properties of the embedded baryonic tracers.

4. [**Nearby galaxies and clusters**] A spectroscopic determination of the faint end of the galaxy luminosity function at $z < 0.01$ down to $M_r \sim 13$, and over a cosmological volume containing thousands of $L\star$ galaxies.

5. [**Galaxy evolution**] A detailed study of the connection between halo mass and stellar mass for $0 < z < 2$, and the evolution of galaxies with stellar masses as low as $10^9 M_\odot$ from $z \sim 1$ to the present day.

6. [**Galaxy evolution**] An unprecedented measurement of the evolution of massive galaxies and their star formation rates at high redshifts, possibly reaching to $z = 7$.

7. **QSOs and AGN** The determination of the redshift evolution of dark energy through baryon acoustic oscillations in the Ly$-\alpha$ forest, and an AGN Hubble diagram calibrated through reverberation mapping.

8. **QSOs and AGN** The most thorough study of AGN feedback ever conducted through high-S/N, high-resolution spectroscopy and time-domain low-resolution spectroscopy.

9. [**The Intergalactic Medium**] The identification of unprecedented samples of quasar spectra to be used in tracing the large-scale structure of the intergalactic medium as well as its ionization and thermal history, and 2000 damped Lyman- systems that would dramatically improve upon our knowledge of early nucleosynthesis and the evolution of metals out to redshifts of $z \sim 4$.

4.4 Changes to the baseline requirements

As the above discussion demonstrates, the baseline NGCFHT facility presents exciting opportunities for transformational research across a remarkably broad range of fields. At the same time, the SWGs have identified a number of possible design modifications that have the potential to enhance the already dramatic scientific impact expected for the facility. Examples of such modifications include:

1. The implementation of a simplified, three-resolution design: $R = 2000, 6000 - 7500$ and 20000.

2. The adoption of a wedding cake dark-time survey strategy, with approximate survey areas and depths of: (1) 4300 deg^2 to $I_{AB} = 23.5$; (2) 100 deg^2 to $I_{AB} = 24.25$; and (3) 1.5 deg^2 to $I_{AB} = 26$.

3. The consideration of a nod and shuffle capability for improved sky subtraction beyond $\simeq 700$ nm.

4. The possible implementation of a duel fibre system, to aid in sky subtraction and/or an extension of wavelength coverage to ~ 350 nm and ~ 1.3 m, depending on spectrograph observing mode. These recommendations are now being reviewed by the technical team, which is adjusting the scope and schedule of the technical study accordingly (see companion paper by Szeto et al. 2012).

5. NEXT STEPS

The SWGs are in the process of compiling final reports on the key science programs that NGCFHT will enable. These reports will form the basis of the science component of the Concept Study that will be submitted to the CFHT Board and SAC in the Fall of 2012. The science drivers described above are currently leading the development of the top level Science Requirements for the NGCFHT facility, and will continue to do so throughout the lifetime of the facility.

Key technical studies that are underway and will be reported on in the Concept Study include:

1. Load capacity of telescope pier;

2. Load capacity of telescope enclosure;

3. Telescope enclosure configuration;

4. Aero-thermal study;

5. Telescope optical design;

6. Low and high resolution spectrograph conceptual designs;

7. Telescope down time.

Progress to date is summarised in the companion article by Szeto et al. (2012). The final Concept Study will additionally include cost estimates, revised schedule, and a Business Plan. ROM costing based on work completed to date suggest the total cost for NGCFHT will be in the vicinity of USD170-200M, or ∼USD30M per partner based on six equal partners.

We fully expect the completed Concept Study to demonstrate that upgrading the existing telescope to a dedicated, 10-m class spectroscopic facility is an exciting, feasible, and affordable endeavor, that will place the new facility in the top echelon of astronomical facilities throughout the next decade and beyond. Further, it will recycle much of the existing infrastructure and will keep within the existing three-dimensional envelope of the enclosure.

A key component to ensuring the realisation of NGCFHT is the formation of a new partnership. Crucially, the new partnership will not contain only Canada, France and Hawaii: here, the SWGs demonstrate the broad appeal of this proposal to a large fraction of the international community. Through the SWGs, these communities are presently involved in defining the basic science requirements, which will set the direction of all future developments. The synergies enabled by NGCFHT are extensive; a wide field spectroscopic facility on Mauna Kea ideally complements the existing and proposed facilities that are co-located on the same site. The same applies even for facilities and surveys based at some southern sites due to the significant overlap in the observable sky. In order to meet schedule and have the NGCFHT operating by the early 2020s, the year following the submission of the Feasibility Study will build upon efforts begun in the past 18 months, and be dedicated to forming the new collaboration to turn this exceptional scientific promise into a reality.

ACKNOWLEDGMENTS

More information, including links to relevant documents, is available at
http://orca.phys.uvic.ca/∼pcote/ngcfht/The_Next_Generation_CFHT.html

REFERENCES

[1] Brown, M., and Dey, A. "Next Generation Wide-Field Multi-Object Spectroscopy," in *ASP Conf. Series* **280** (2001)
[2] Dey, A., et al., "KAOS: The Kilo-Aperture Optical Spectrograph" (2003)
[3] Ellis, R., et al., "Wide-Field Fiber-Fed Optical Multi-Object Spectrometer WFMOS Study Summary" (2009)
[4] Barden, S., et al., "WFMOS: A Feasibility Study for Gemini" (2009)
[5] Côté, P., et al., "The Next Generation CFHT" (2010) in [*Society of Photo-Optical Instrumentation Engineers (SPIE) Conference Series*], Society of Photo-Optical Instrumentation Engineers (SPIE) Conference Series (July 2012).
[6] Szeto, K., et al., "The Next Generation CFHT"
[7] Richer, H., et al., "Illuminating the Dark Universe: Report of the Next Generation CFHT Committee" (1998)
[8] Carlberg, R., et al. "A Wide-Field 8 Meter Telescope for Canadians" (1999)
[9] Geyl, R., et al., "The New CFHT" (2000)
[10] Burgarella, D., et al., "The Next Generation Canada-France- Hawaii Telescope" (2000)
[11] Grundmann, W., "A CFH 12-16m Telescope Study" (1997)
[12] Drissen, L., Bernier, A.-P., Rousseau-Nepton, L., Alarie, A., Robert, C., Joncas, G., Thibault, S., and Grandmont, F., "SITELLE: a wide-field imaging Fourier transform spectrometer for the Canada-France-Hawaii Telescope," in [*Society of Photo-Optical Instrumentation Engineers (SPIE) Conference Series*], Society of Photo-Optical Instrumentation Engineers (SPIE) Conference Series **7735** (July 2010).

[13] Artigau, É., Donati, J.-F., and Delfosse, X., "Planet Detection, Magnetic Field of Protostars and Brown Dwarfs Meteorology with SPIRou," in [*Astronomical Society of the Pacific Conference Series*], Johns-Krull, C., Browning, M. K., and West, A. A., eds., *Astronomical Society of the Pacific Conference Series* **448**, 771 (Dec. 2011).

[14] Evans, C., Lin, H., McColgan, A., Rowlands, N., and Salmon, D., "IMAKA: imaging from MAuna KeA optical design," in [*Society of Photo-Optical Instrumentation Engineers (SPIE) Conference Series*], *Society of Photo-Optical Instrumentation Engineers (SPIE) Conference Series* **7735** (July 2010).

[15] Barden, S. C., Jones, D. J., Barnes, S. I., Heijmans, J., Heng, A., Knight, G., Orr, D. R., Smith, G. A., Churilov, V., Brzeski, J., Waller, L. G., Shortridge, K., Horton, A. J., Mayfield, D., Haynes, R., Haynes, D. M., Whittard, D., Goodwin, M., Smedley, S., Saunders, I., Gillingham, P. R., Penny, E., Farrell, T. J., Vuong, M., Heald, R., Lee, S., Muller, R., Freeman, K., Bland-Hawthorn, J., Zucker, D. F., and de Silva, G., "HERMES: revisions in the design for a high-resolution multi-element spectrograph for the AAT," in [*Society of Photo-Optical Instrumentation Engineers (SPIE) Conference Series*], *Society of Photo-Optical Instrumentation Engineers (SPIE) Conference Series* **7735** (July 2010).

[16] Gilmore, G., Randich, S., Asplund, M., Binney, J., Bonifacio, P., Drew, J., Feltzing, S., Ferguson, A., Jeffries, R., Micela, G., Negueruela, I., Prusti, T., Rix, H.-W., Vallenari, A., Alfaro, E., Allende-Prieto, C., Babusiaux, C., Bensby, T., Blomme, R., Bragaglia, A., Flaccomio, E., Francois, P., Irwin, M., Koposov, S., Korn, A., Lanzafame, A., Pancino, E., Paunzen, E., Recio-Blanco, A., Sacco, G., Smiljanic, R., van Eck, S., and Walton, N., "The Gaia-ESO public spectroscopic survey.," *The Messenger* **147**, 25–31 (2012).

[17] Cirasuolo, M., Afonso, J., Bender, R., Bonifacio, P., Evans, C., Kaper, L., Oliva, E., and Vanzi, L., "MOONS: The Multi-Object Optical and Near-infrared Spectrograph," *The Messenger* **145**, 11–13 (Sept. 2011).

[18] de Jong, R., "4MOST: 4-metre Multi-Object Spectroscopic Telescope," *The Messenger* **145**, 14–16 (Sept. 2011).

[19] Schlegel, D. J., Bebek, C., Heetderks, H., Ho, S., Lampton, M., Levi, M., Mostek, N., Padmanabhan, N., Perlmutter, S., Roe, N., Sholl, M., Smoot, G., White, M., Dey, A., Abraham, T., Jannuzi, B., Joyce, D., Liang, M., Merrill, M., Olsen, K., and Salim, S., "BigBOSS: The Ground-Based Stage IV Dark Energy Experiment," *ArXiv e-prints* (Apr. 2009).

[20] Ellis, R., Takada, M., Aihara, H., Arimoto, N., Bundy, K., Chiba, M., Cohen, J., Dore, O., Greene, J. E., Gunn, J., Heckman, T., Hirata, C., Ho, P., Kneib, J.-P., Le Fevre, O., Murayama, H., Nagao, T., Ouchi, M., Seiffert, M., Silverman, J., Sodre, Jr, L., Spergel, D., Strauss, M. A., Sugai, H., Suto, Y., Takami, H., Wyse, R., and the PFS Team, "Extragalactic Science and Cosmology with the Subaru Prime Focus Spectrograph (PFS)," *ArXiv e-prints* (June 2012).

[21] Freeman, K. and Bland-Hawthorn, J., "The New Galaxy: Signatures of Its Formation," ARA&A **40**, 487–537 (2002).

[22] Bonifacio, P., Sbordone, L., Caffau, E., Ludwig, H.-G., Spite, M., González Hernández, J. I., and Behara, N. T., "Chemical abundances of distant extremely metal-poor unevolved stars," A&A **542**, A87 (June 2012).

[23] Schneider, R., Omukai, K., Bianchi, S., and Valiante, R., "The first low-mass stars: critical metallicity or dust-to-gas ratio?," MNRAS **419**, 1566–1575 (Jan. 2012).

[24] Greif, T. H., Springel, V., White, S. D. M., Glover, S. C. O., Clark, P. C., Smith, R. J., Klessen, R. S., and Bromm, V., "Simulations on a Moving Mesh: The Clustered Formation of Population III Protostars," ApJ **737**, 75 (Aug. 2011).

[25] Salvadori, S., Schneider, R., and Ferrara, A., "Cosmic stellar relics in the Galactic halo," MNRAS **381**, 647–662 (Oct. 2007).

[26] Schörck, T., Christlieb, N., Cohen, J. G., Beers, T. C., Shectman, S., Thompson, I., McWilliam, A., Bessell, M. S., Norris, J. E., Meléndez, J., Ramírez, S., Haynes, D., Cass, P., Hartley, M., Russell, K., Watson, F., Zickgraf, F.-J., Behnke, B., Fechner, C., Fuhrmeister, B., Barklem, P. S., Edvardsson, B., Frebel, A., Wisotzki, L., and Reimers, D., "The stellar content of the Hamburg/ESO survey. V. The metallicity distribution function of the Galactic halo," A&A **507**, 817–832 (Nov. 2009).

[27] Caffau, E., Bonifacio, P., François, P., Sbordone, L., Monaco, L., Spite, M., Spite, F., Ludwig, H.-G., Cayrel, R., Zaggia, S., Hammer, F., Randich, S., Molaro, P., and Hill, V., "An extremely primitive star in the Galactic halo," Nature **477**, 67–69 (Sept. 2011).

[28] Caffau, E., Bonifacio, P., François, P., Spite, M., Spite, F., Zaggia, S., Ludwig, H.-G., Monaco, L., Sbordone, L., Cayrel, R., Hammer, F., Randich, S., Hill, V., and Molaro, P., "X-Shooter GTO: chemical analysis of a sample of EMP candidates," A&A **534**, A4 (Oct. 2011).

[29] Carollo, D., Beers, T. C., Lee, Y. S., Chiba, M., Norris, J. E., Wilhelm, R., Sivarani, T., Marsteller, B., Munn, J. A., Bailer-Jones, C. A. L., Fiorentin, P. R., and York, D. G., "Two stellar components in the halo of the Milky Way," Nature **450**, 1020–1025 (Dec. 2007).

[30] Helmi, A., "The stellar halo of the Galaxy," A&A Rev. **15**, 145–188 (June 2008).

[31] Nissen, P. E. and Schuster, W. J., "Two distinct halo populations in the solar neighborhood. II. Evidence from stellar abundances of Mn, Cu, Zn, Y, and Ba," A&A **530**, A15 (June 2011).

[32] Kaiser, N., "Clustering in real space and in redshift space," MNRAS **227**, 1–21 (July 1987).

[33] Guzzo, L., Pierleoni, M., Meneux, B., Branchini, E., Le Fèvre, O., Marinoni, C., Garilli, B., Blaizot, J., De Lucia, G., Pollo, A., McCracken, H. J., Bottini, D., Le Brun, V., Maccagni, D., Picat, J. P., Scaramella, R., Scodeggio, M., Tresse, L., Vettolani, G., Zanichelli, A., Adami, C., Arnouts, S., Bardelli, S., Bolzonella, M., Bongiorno, A., Cappi, A., Charlot, S., Ciliegi, P., Contini, T., Cucciati, O., de la Torre, S., Dolag, K., Foucaud, S., Franzetti, P., Gavignaud, I., Ilbert, O., Iovino, A., Lamareille, F., Marano, B., Mazure, A., Memeo, P., Merighi, R., Moscardini, L., Paltani, S., Pellò, R., Perez-Montero, E., Pozzetti, L., Radovich, M., Vergani, D., Zamorani, G., and Zucca, E., "A test of the nature of cosmic acceleration using galaxy redshift distortions," Nature **451**, 541–544 (Jan. 2008).

[34] Percival, W. J. and White, M., "Testing cosmological structure formation using redshift-space distortions," MNRAS **393**, 297–308 (Feb. 2009).

[35] Newman, J. A., "Calibrating Redshift Distributions beyond Spectroscopic Limits with Cross-Correlations," ApJ **684**, 88–101 (Sept. 2008).

[36] Ho, S., Lin, Y.-T., Spergel, D., and Hirata, C. M., "Luminous Red Galaxy Population in Clusters at 0.2 z 0.6," ApJ **697**, 1358–1368 (June 2009).

[37] Zhang, P., Liguori, M., Bean, R., and Dodelson, S., "Probing Gravity at Cosmological Scales by Measurements which Test the Relationship between Gravitational Lensing and Matter Overdensity," *Physical Review Letters* **99**, 141302 (Oct. 2007).

[38] Reyes, R., Mandelbaum, R., Seljak, U., Baldauf, T., Gunn, J. E., Lombriser, L., and Smith, R. E., "Confirmation of general relativity on large scales from weak lensing and galaxy velocities," Nature **464**, 256–258 (Mar. 2010).

[39] Dalal, N., White, M., Bond, J. R., and Shirokov, A., "Halo Assembly Bias in Hierarchical Structure Formation," ApJ **687**, 12–21 (Nov. 2008).

[40] Slosar, A., Hirata, C., Seljak, U., Ho, S., and Padmanabhan, N., "Constraints on local primordial non-Gaussianity from large scale structure," JCAP **8**, 31 (Aug. 2008).

[41] Bolton, A. S., Burles, S., Koopmans, L. V. E., Treu, T., and Moustakas, L. A., "The Sloan Lens ACS Survey. I. A Large Spectroscopically Selected Sample of Massive Early-Type Lens Galaxies," ApJ **638**, 703–724 (Feb. 2006).

[42] Bolton, A. S., Burles, S., Koopmans, L. V. E., Treu, T., Gavazzi, R., Moustakas, L. A., Wayth, R., and Schlegel, D. J., "The Sloan Lens ACS Survey. V. The Full ACS Strong-Lens Sample," ApJ **682**, 964–984 (Aug. 2008).

[43] Koopmans, L. V. E., Treu, T., Bolton, A. S., Burles, S., and Moustakas, L. A., "The Sloan Lens ACS Survey. III. The Structure and Formation of Early-Type Galaxies and Their Evolution since z ~ 1," ApJ **649**, 599–615 (Oct. 2006).

The Optics and Detector-Simulation of the Air Fluorescence Telescope FAMOUS for the Detection of Cosmic Rays

Tim Niggemann[*a], Thomas Hebbeker[a], Markus Lauscher[a], Christine Meurer[a], Lukas Middendorf[a], Johannes Schumacher[a] & Maurice Stephan[a]

[a]III. Physikalisches Institut A, RWTH Aachen University, Physikzentrum, 52056 Aachen, Germany

ABSTRACT

A sophisticated method for the observation of ultra-high-energy cosmic rays (UHECRs) is the fluorescence detection technique of extensive air showers (EAS).

FAMOUS will be a small fluorescence telescope, instrumented with silicon photomultipliers (SiPMs) as highly-sensitive light detectors. In comparison to photomultiplier tubes, SiPMs promise to have a higher photon-detection-efficiency. An increase in sensitivity allows to detect more distant and lower energy showers which will contribute to an enrichment of the current understanding of the development of EAS and the chemical composition of UHECRs.

Keywords: Air fluorescence, astroparticle-physics, extensive air shower, Fresnel lens, ray-tracing, silicon photomultiplier, ultra-high-energy cosmic rays, Winston cone

1. ULTRA-HIGH-ENERGY COSMIC RAYS

Cosmic rays are ionised atoms such as hydrogen, helium, etc. and elementary particles, e.g. electrons, which originate from within our solar system, our galaxy and yet undetermined extragalactic sources. The relative abundances of the constituents of the cosmic radiation, normalised to oxygen, for a primary particle energy of $10.6\,\text{GeV}$, are displayed on the left-hand side in figure 1. Cosmic rays can be measured over 30 orders of magnitude in kinetic energy[1] by Earth-bound, airborne and space experiments. The measured all particle spectrum is displayed on the right-hand side in figure 1.

1.1 The Energy Spectrum

The all-particle energy spectrum follows a steep power law

$$\frac{\mathrm{d}^4 n}{\mathrm{d}E\,\mathrm{d}t\,\mathrm{d}\Omega\,\mathrm{d}A} \propto E^{-\gamma} \qquad (1)$$

with the energy E, time t, the solid angle Ω, the area A and most importantly, the spectral index $\gamma \approx 2.7$ for $E \lesssim 4 \cdot 10^{15}\,\text{eV}$. Several peculiar features have been revealed:

Knee: At an energy of $E \approx 4 \cdot 10^{15}\,\text{eV}$, the spectrum gets even steeper with a change in the spectral index from $\gamma = 2.7$ to $\gamma = 3.1$[1]. Direct measurements of the cosmic rays revealed a large suppression in the flux of light elements of the cosmic radiation in this energy region. Currently, many possible explanations are discussed in the literature[1]. The two most popular correspond to an upper energy limit in the acceleration process of galactic supernovae and a leakage of cosmic rays from the galaxy due to a very weak galactic magnetic field ($B \approx 0.3\,\text{nT}$) which is not able to bend these particles to a contained track.

2nd Knee: Above $E \approx 4 \cdot 10^{17}\,\text{eV}$, the spectrum gets slightly steeper. This feature coincides with a predicted occurrence of heavy nuclei[4] (atomic number $Z \geq 28$).

---

* Corresponding author: e-mail: niggemann@physik.rwth-aachen.de, telephone: +49 (0)241 8027338.

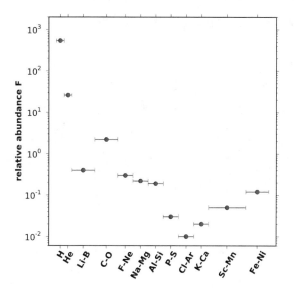 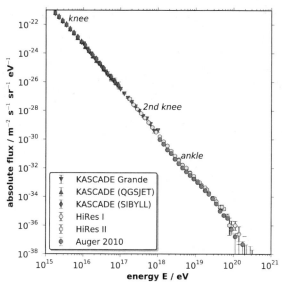

Figure 1: **Left:** Relative abundances F of the elements of cosmic ray particles at $E = 10.6\,\text{GeV}$ normalised to oxygen. Adapted from[2]. **Right:** Differential flux of cosmic rays from various extensive air shower measurements as function of the energy of the primary particle. Adapted from[3].

Ankle: For unknown reasons, the spectrum flattens again at $E \approx 4 \cdot 10^{18}\,\text{eV}$ to $\gamma = 2.6$ until it shows a strong depression at $E > 10^{20}\,\text{eV}$ with at least 5σ significance[1]. In 1966, K. Greisen, G. Zatsepin and V. Kuzmin proposed a cut-off (GZK-cut-off): the primary particles of the cosmic radiation are likely to interact with the photons of the cosmic microwave background[5] on their way through space and lose kinetic energy. In consequence, cosmic rays with energies above $10^{20}\,\text{eV}$ must originate from within $100\,\text{Mpc}^*$.

Due to the lack of knowledge about the nature of cosmic rays in the highest energy regime above $E = 10^{15}\,\text{eV}$, today's astroparticle physics is also eager to reveal the composition of the elements and the sources. The extremely low particle flux at ground of less than 1 particle per m^2 and year requires earthbound experiments to investigate a large part of our atmosphere. A sophisticated way to achieve this goal is the fluorescence detection technique.

2. THE FLUORESCENCE DETECTION TECHNIQUE

As the primary particle penetrates the atmosphere of the Earth, it hits a nucleus of an air molecule. Starting with this first hadronic interaction, numerous secondary charged and uncharged particles are created along a trajectory, called shower axis, down to Earth[1].

2.1 Extensive Air Showers

An important measure of extensive air shower physics is the slant depth X in units of $[X] = \text{g cm}^{-2}$ which equals the amount of traversed matter. It can be expressed as a function of the altitude h and the zenith angle θ:

$$X(h, \theta) \approx \frac{1}{\cos\theta} \int_h^\infty \rho(h')\,\text{d}h' = \frac{1}{\cos\theta} \int_h^\infty \rho_0 e^{-h'/h_0}\,\text{d}h' \approx 1000\,\text{g cm}^{-2} \frac{e^{-h/h_0}}{\cos\theta} \quad (2)$$

with $h_0 \approx 7.25\,\text{km}$ and the density $\rho_0 = 1.35\,\text{kg m}^{-3}$.

---

*The parsec (pc) is an astronomical unit of length whereas $1\,\text{pc} = 3.26\,\text{ly}$.

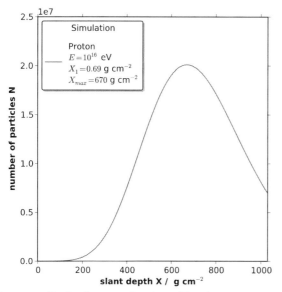 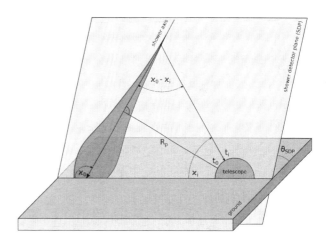

Figure 2: **Left:** Longitudinal profile of an extensive air shower. The plot has been produced using the simulation program CONEX[6]. **Right:** Schematic of the shower-detector-plane used for the event reconstruction. At a time t_i the signal arrives at the telescope at an angle χ_i with respect to the ground. The distance of the closest point of the shower axis to the telescope is defined as the shower-telescope-distance R_p and detected at t_0. Adapted from[7].

An empiric parametrisation of the longitudinal profile of the electro-magnetic shower component has been obtained by T. K. Gaisser and A. M. Hillas in 1977 and describes the number of electrons and positrons N_e as a function of the slant depth X as follows

$$N_\mathrm{e}(x) = N_\mathrm{max} \left(\frac{X - X_1}{X_\mathrm{max} - X_1} \right)^{\frac{X_\mathrm{max} - X_1}{\lambda_\mathrm{int}}} \exp\left(\frac{X_\mathrm{max} - X_1}{\lambda_\mathrm{int}} \right) \qquad (3)$$

with the depth of the first interaction X_1, the depth of the shower maximum X_max, the maximum number of particles N_max (see left-hand side of figure 2). The shape parameter λ_int is the mean interaction length for the primary particle. Values for protons (p) and iron nuclei (Fe) are:

$$\lambda_\mathrm{int,p} \approx 80\,\mathrm{g\ cm^{-2}}, \quad \lambda_\mathrm{int,Fe} \approx 20\,\mathrm{g\ cm^{-2}} \qquad . \qquad (4)$$

2.2 Fluorescence Yield

The most prominent secondary particles in an extensive air shower are electrons (and positrons) which themselves can hit nitrogen molecules of the air. A fraction of the collision energy gets absorbed by the nitrogen molecules and they get into an excited state. This energy deposit follows the Bethe-Bloch formula[9]. As the molecules de-excite, ultraviolet light is emitted isotropically (see figure 3).

The number of emitted fluorescence photons N_γ^0 is proportional to the energy deposit $E_\mathrm{dep}^\mathrm{tot}$ [9]:

$$\frac{\mathrm{d}^2 N_\gamma^0}{\mathrm{d}X \mathrm{d}\lambda} = Y_\lambda(\lambda, P, T, u) \cdot \frac{\mathrm{d}E_\mathrm{dep}^\mathrm{tot}}{\mathrm{d}X} \qquad . \qquad (5)$$

The proportionality factor is called fluorescence yield Y_λ. It depends on the wavelength λ, the atmospheric pressure P, the temperature T and the humidity u. In figure 3, right-hand side, the fluorescence light flux is presented as a function of the energy of an extensive air shower.

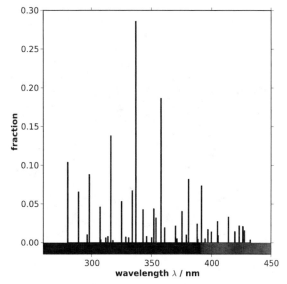
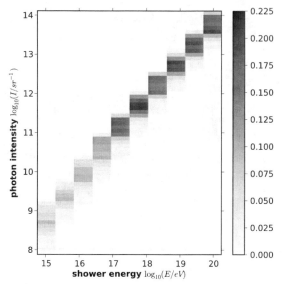

Figure 3: **Left:** Fluorescence light spectrum as a function of the photon wavelength λ. The data have been obtained by a fluorescence light simulation by the software package Offline[8] by the Pierre Auger Collaboration. **Right:** Number of fluorescence photons per solid angle arriving at a toy fluorescence telescope with an aperture diameter of $D = 510$ mm and a field of view of $30° \times 30°$. The data have been obtained by Offline with a shower simulation in logarithmic energy bins in steps of $10^{0.5}$ eV from 10^{15} eV to 10^{20} eV. Each energy bin contains 100 extensive air showers per shower-to-telescope distance $R_p = 1, 2, 3, ...12$ km simulated with CONEX. The integral in each energy bin is scaled to 1.

A parametrisation of the fluorescence yield has been compiled by M. Nagano et al.[10] and expressed by means of photons per energy by F. Arqueros et al.[9] as

$$Y_\lambda = \frac{1}{\mathrm{d}E_{\mathrm{dep}}^{\mathrm{tot}}/\mathrm{d}X} \cdot \frac{\rho A_\lambda}{1 + \rho B_\lambda \sqrt{T}} \tag{6}$$

with the absolute temperature T and the density of the gas ρ. The constants A_λ and B_λ have been measured by M. Nagano et al. for 10 different wavelengths between 300 nm and 400 nm[10]. The values for the reference wavelength $\lambda = 337$ nm are

$$A_{337\,\mathrm{nm}} = (45.6 \pm 1.2)\,\mathrm{m}^2\mathrm{kg}^{-1} \tag{7}$$
$$B_{337\,\mathrm{nm}} = (2.56 \pm 0.10)\,\mathrm{m}^3\mathrm{kg}^{-1}\mathrm{K}^{-1/2} \quad . \tag{8}$$

The results of the parametrisation are in good agreement with exact theoretical calculations at percent level[9].

2.3 Reconstruction of Extensive Air Showers

A fluorescence light detector usually has a camera consisting of single pixels. As the shower passes through the field of view of the detector, the pixel i receives a signal at the time t_i. As shown in figure 2, right-hand side, the shower axis and the pointing directions to the camera span the shower-detector-plane (SDP)[7]. The timing information t_i can be used to fit the shower axis

$$t(\chi) = t_0 + \frac{R_p}{c}\tan\left[(\chi_0 - \chi)/2\right] \tag{9}$$

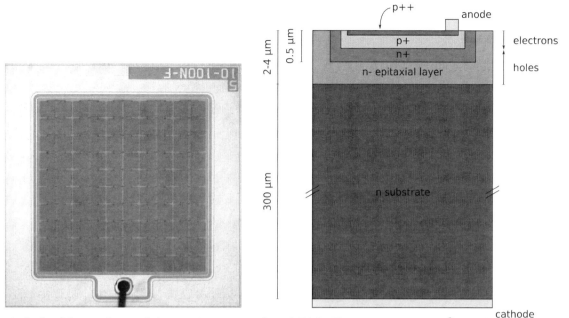
Figure 4: **Left:** Macro-photo of the sensitive area of an SiPM. The area is $1 \times 1\,\text{mm}^2$ and features a matrix of 10×10 individual, square Geiger-mode avalanche photodiodes. The space in between the grey squares is not sensitive. Adapted from[11]. **Right:** Schematic of an avalanche photodiode. Adapted from[12].

whereas c is the speed of light, χ the observation angle and χ_0 the angle between the shower-axis and the ground. The shower-telescope-distance R_p is defined as the distance to the closest point of the shower axis detected at t_0.

Once the geometry of the shower has been determined, the observation angle χ can be translated to the slant depth X and plotted against the number of electrons N_e. The number of electrons can be derived with the aid of the fluorescence yield Y_λ from the amount of collected light in a given pixel. Additionally, the attenuation and variable density of the atmosphere have to be considered.

The fit of the Gaisser-Hillas-function (see equation 3) to the data reconstructs the shower maximum X_max and, by using N_max, the energy E_0 of the primary particle[1]. This gives hints on the yet unknown chemical composition and the origin of the features of the all particle spectrum of the cosmic rays.

Up to now, fluorescence detectors use a mature technology, photomultiplier tubes as light detector[1], which rely on the photo-electric effect[7]. Silicon photomultipliers promise to have a higher sensitivity which could enrich the amount of physical information on the air shower obtained.

3. SILICON PHOTOMULTIPLIERS

A silicon photomultiplier (SiPM) is a silicon based chip capable of detecting single photons. These devices have several advantages over the established photomultiplier tubes (PMTs).

A typical PMT has an entrance window of ca. 1 cm diameter, is several centimeters long and has to be driven by a high voltage of ca. 1000 V. The photon detection efficiency (PDE) of current PMTs can reach up to 40 %[13]. In contrast, the SiPM has a sensitive area of several square millimeters, is a few millimeters thick and can be operated with several 10 V. The PDE of commercially available SiPMs is already at 40 %[14] and future devices promise to reach at least 60 %[15]. The intrinsic gain of the signal of both sensors is comparable and at a level of 10^6.

However, the operation parameters (PDE, gain, etc.) of an SiPM depend heavily on the ambient temperature. Noise phenomena are more challenging to control. If these effects can be understood well, the SiPM may replace the PMT. In addition, SiPMs have the potential of mass production.

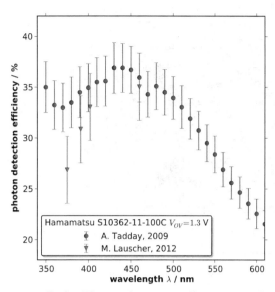
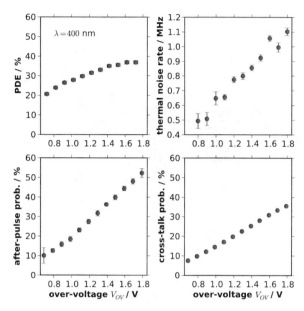

Figure 5: **Left:** Photon detection efficiency as a function of the photon wavelength λ for a Hamamatsu S10362-11-100C $1\times 1\,\mathrm{mm}^2$ SiPM with $100\,\mu\mathrm{m}$ cell pitch. The blue circles denote data from[16], the red triangles from[17]. Not shown are the systematic errors. **Right:** Dependence of the SiPM operation parameters $PDE(\lambda)$, thermal noise rate, crosstalk- and afterpulse-probability as a function of the over-voltage V_{ov} for a Hamamatsu S10362-11-100C SiPM[17]. The ambient temperature is $T = (25.5 \pm 1.0)\,^\circ\mathrm{C}$.

3.1 Geiger-mode Avalanche Photodiode

An SiPM consists of a matrix of cells, whereas each cell is a Geiger-mode avalanche photodiode (GAPD). Each GAPD delivers a standard signal[†], also referred to as "1 photon equivalent" (1 p.e.), if it gets hit by one ore more photons within its recovery time τ_{rec}. The SiPM signal is the sum of all GAPD signals. Since up to several 1000 single GAPDs are situated on an SiPM, many coincidental photons can be observed. The number of cells limits the dynamic range of the SiPM.

A cross-section through a GAPD is shown in figure 4. The "p++" and "p+" layers consist of silicon that has been enriched with positive charge carriers (acceptors). Correspondingly, the "n+", "n-" and "n" layers are enriched by negative charge carriers (donors). By applying a reverse voltage V_{bias} which exceeds the breakdown voltage V_{b} by an overvoltage V_{ov}

$$V_{\mathrm{bias}} = V_{\mathrm{b}} + V_{\mathrm{ov}} \quad , \qquad (10)$$

a depletion zone with an intense electric field, in which no charge carriers are present, forms at the p-on-n-junction where the "p+" and the "n+" layers adjoin. An electrical current is suppressed.

As a photon hits the GAPD, it might get absorbed in the depletion zone and an electron-hole-pair is created. The charge carriers accelerate in the electric field of the depletion zone and induce further electron-hole-pairs on their way through the lattice: an avalanche is created. The avalanche weakens the depletion zone, a current flow is measurable. The quenching resistor on top of the GAPD stops the avalanche and the depletion zone can be replenished. After the recovery time τ_{rec}, the GAPD is again able to trigger at full 1 p.e. level.

3.2 Photon detection efficiency

The photon detection efficiency (PDE) is one of the most important properties of a photon counting device like an SiPM. It is defined by

$$PDE(\lambda) = \epsilon_{\mathrm{geom}} \cdot \epsilon_{\mathrm{avalanche}} \cdot QE(\lambda) \qquad (11)$$

[†]Therefore, it is called "Geiger-mode".

with the geometrical fill factor ϵ_{geom}, the trigger probability $\epsilon_{\text{avalanche}}$ and the quantum efficiency $QE(\lambda)$. The geometrical fill factor can be increased by minimising the dead space between the single GAPDs. The quantum efficiency $QE(\lambda)$ is the probability of an incoming photon to create the first electron-hole-pair. The trigger probability $\epsilon_{\text{avalanche}}$ is the probability for this first electron-hole-pair to create an avalanche. The PDE varies as a function of the photon wavelength λ and the overvoltage V_{ov} of the SiPM as presented in figure 5.

3.3 Noise Phenomena

As any sensor or detector, an SiPM also suffers from different kinds of noise phenomena.

Thermal Noise: The atoms in the silicon lattice can be thermally excited and accidentally create an electron-hole-pair. This triggers an avalanche in an SiPM cell. Along with the trigger probability $\epsilon_{\text{avalanche}}$ of the SiPM cell, the thermal noise rate increases with the applied overvoltage V_{ov} (see figure 5). Thermal noise can be reduced by a factor of 2 every 8 °C cooling down. Additionally, two types of correlated noise occur: optical crosstalk and afterpulses.

Optical Crosstalk: During an avalanche, an electron can recombine with a hole. A photon gets emitted. This photon can be transmitted to a neighbouring cell and trigger an avalanche there. Figure 5 illustrates the probability for this process.

Afterpulses: During an avalanche, the possibility of electrons to fill defects of the electron-configuration of the silicon lattice is enhanced. Few 10 ns later, the trapped electron might be released and trigger another avalanche. The time lag between the original trigger and the afterpulse follows a superposition of two exponential decays with the time constants

$$\tau_{\text{short}} = (45.4 \pm 2.6) \text{ ns} \qquad (12)$$
$$\tau_{\text{long}} = (123.2 \pm 14.0) \text{ ns} \qquad (13)$$

for a Hamamatsu $3 \times 3 \text{ mm}^2$ SiPM with 100 μm pitch operated at an over-voltage $V_{\text{ov}} = (1.31 \pm 0.01) \text{ V}$[17]. The fraction of the number of short-time-constant pulses to the long-time-constant pulses is

$$R_{\text{short-long}} = 0.46 \qquad . \qquad (14)$$

Figure 5 shows the dependence of all described noise phenomena on the over-voltage.

4. THE OPTICS OF FAMOUS

The telescope FAMOUS[‡] is a fluorescence light telescope, currently in construction, for the observation of extensive air showers which utilises SiPMs for the instrumentation of its camera. The design goal is the construction of a 64 pixel camera with a field of view of $1.5° \times 1.5°$ per pixel. To minimise dead space between pixels, they are arranged in a hexagonal grid. This results in a wide total field of view of approximately $12° \times 12°$. Due to the low intensity of the fluorescence light, a large telescope aperture is needed. Therefore, the focal ratio has been set to

$$\frac{f}{D} \approx 1 \qquad (15)$$

with the focal length f and the aperture diameter D. The aperture diameter has been set to $D = 510 \text{ mm}$.

Since FAMOUS is also a development prototype, the camera has to be placed outside the path of light rays to ensure easy access and to keep requirements on the size of the read-out-electronics moderate. In combination with the wide field of view, this design goal disqualifies most of the standard mirror-systems. Thus, a simple refractive design with a Fresnel lens is used as imaging optical element (see figure 8).

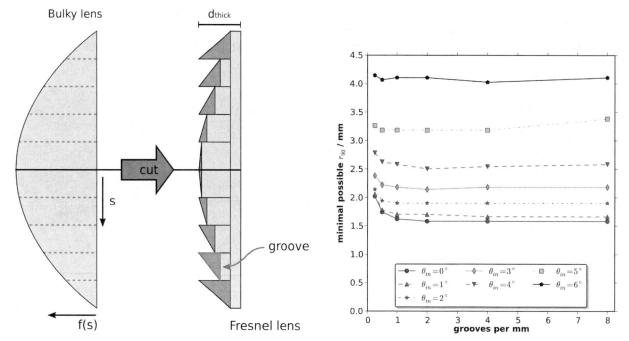

Figure 6: **Left:** Illustration of the Fresnel lens construction principle. A bulky lens is divided into annular concentric sections, grooves, whose thicknesses are reduced to a minimum. **Right:** The minimal aberration radius R_{90} possible of the Fresnel lens of FAMOUS with $D = f = 510$ mm as a function of the number of grooves per millimeter situated on the lens. The different curves denote different incident angles of parallel light. The data have been obtained by a custom ray-tracing simulation[18] with the toolkit Geant4[19].

4.1 Fresnel Lens

Starting with a bulky lens of the desired refraction power and diameter, the surface of a Fesnel lens is divided into annular concentric sections named "grooves". The thickness of each groove gets reduced to the minimum possible. This procedure is displayed in figure 6. Since the slope of each groove can be chosen separately, aspheric lens surfaces can be realised easily which helps to overcome optical aberrations such as spherical aberration or coma[20]. Furthermore, the more grooves per millimeter are situated on the lens, the better the imaging quality.

The lens for FAMOUS is manufactured of ultraviolet transparent acrylic[21] with 10 grooves per millimeter and has a transmission efficiency of approximately 70 %[18].

4.1.1 Image Size

The simplification of the lens surface has an impact on imaging quality. Although FAMOUS does not need a good spatial resolution like an optical telescope, the image from within a certain part of the field of view has to be fully covered by the corresponding pixel. The size of the image in the focal plane can be measured by the aberration radius R_{90} which encircles 90 % of the energy contained by the image. The theoretical optimum R_{90} can be calculated by the Airy pattern[20] to $R_{90} = 0.67\,\mu\text{m}$[18]. In figure 6, the R_{90} of the Fresnel lens of FAMOUS is displayed as a function of the number of grooves and the incident angle of a parallel light beam. Although R_{90} is in this case of the order of millimeters, the image still fits into a camera pixel of FAMOUS (radius is $r = 6.7$ mm). The geometry of a camera pixel of FAMOUS will be presented in the next section.

4.2 Camera Pixel

A camera pixel of FAMOUS is instrumented by a Hamamatsu-S10985-100C SiPM, i.e. an array of four $3 \times 3\,\text{mm}^2$ SiPMs yielding in a total area of $d_{\text{sipm}}^2 = 6 \times 6\,\text{mm}^2$, and a cell pitch of $100\,\mu$m. Larger extents are currently not commercially available.

[‡] "<u>F</u>irst <u>A</u>achen <u>M</u>ulti pixel photon counter camera for the <u>O</u>bservation of <u>U</u>ltra-high-energy-cosmic-ray air <u>S</u>howers"

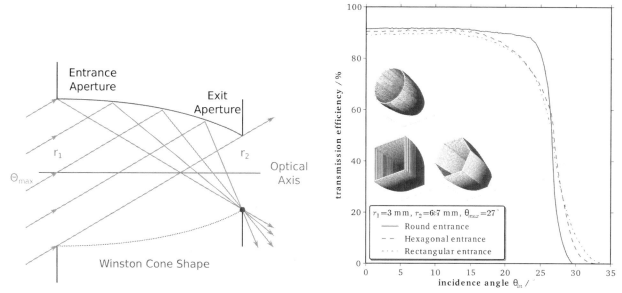

Figure 7: **Left:** Schematic of a Winston-cone. The blue straight line signifies the reflective border of the cone which follows a parabola tilted by the angle $\theta_{\max}$. This shape gets rotated around the optical axis. The yellow rays symbolise photons entering under in incidence of $\theta_{\max}$. **Right:** Winston cone transmission efficiency for different shapes depending on the incident angle θ_{in}.

Since each camera pixel should have a field of view of $1.5° \times 1.5°$, the pixel must be large enough to capture the whole light out of its corresponding solid angle fraction of the sky. By every $1.5°$ change in incident angle of a parallel light beam on the Fresnel lens, the image walks approximately

$$d_{\mathrm{img}} \approx D \cdot \tan(1.5°) = 13.4\,\mathrm{mm} \tag{16}$$

on the image plane. Therefore, the light has to be concentrated by a factor of

$$c = \frac{d_{\mathrm{img}}}{d_{\mathrm{sipm}}} = \frac{13.4\,\mathrm{mm}}{6.0\,\mathrm{mm}} = 2.2 \quad . \tag{17}$$

This requires the use of a light funnel.

4.2.1 Winston Cone

The Winston cone is a compound parabolic concentrator[22] which reaches the theoretically maximum possible concentration of

$$c = \frac{r_1}{r_2} = \frac{1}{\sin(\theta_{\max})} \tag{18}$$

whereas r_1 is the radius of the entrance of the cone, r_2 the radius of the exit ($r_2 < r_1$) and $\theta_{\max}$ the maximum allowed incident angle. The length is given by

$$l = \frac{r_1 + r_2}{\tan \theta_{\max}} \quad . \tag{19}$$

A schematic of the Winston cone construction is presented on the left-hand side in figure 7. The shape follows a parabola whose symmetry axis has been tilted by $\theta_{\max}$ with respect to the optical axis of the cone. This shape gets rotated around the optical axis. The focal point of the parabola coincides with the opposite side of the Winston cone exit. In consequence, parallel light, incident under $\theta_{\max}$, can be transmitted, light with an incident angle $\theta > \theta_{\max}$ can not.

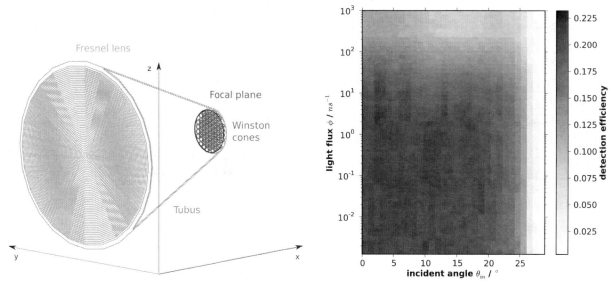

Figure 8: **Left:** Illustration of the Geant4 ray-tracing simulation of FAMOUS. **Right:** Dynamic range of a pixel of FAMOUS composed by a Winston cone with $r_1 = 6.7\,\text{mm}$ entrance radius and $6 \times 6\,\text{mm}^2$ Hamamatsu SiPM at room temperature and $V_{\text{ov}} = 1.3\,\text{V}$. The data have been obtained by a custom Geant4 simulation.

The geometry of the Winston cone used for FAMOUS is $r_1 = 6.7\,\text{mm}$, $r_2 = 3\,\text{mm}$ and $l = 19.6\,\text{mm}$. The cone is made of polished aluminium with a circular entrance and exit. The simulated transmission efficiency for this geometry as a function of the incident angle of the light is displayed in figure 7, right-hand side. With over 90 % transmission efficiency, the Winston cone is a good light funnel for the SiPM.

5. DETECTOR SIMULATION OF FAMOUS

The ray-tracing and detector response simulation of FAMOUS has been carried out with the software framework Geant4 [§]. This framework simulates the passage of particles[¶] through matter while considering a reasonable range of physics processes[19]. Many optical effects such as reflection, absorption, polarisation etc. are also included. It is commonly used in high energy physics to expose complex detector geometries like the big particle detector CMS[23] at the Large Hadron Collider, CERN, to study their response. Written in the programming language C++, the object oriented design allows a fast adaptation. A screenshot of the simulation of FAMOUS is presented left-hand side of figure 8.

With the aid of this simulation, the behaviour of FAMOUS, when illuminated by the night-sky and the faint fluorescence light of extensive air showers (EAS), can be studied. Most important is the determination of the lower and the upper limit of its sensitivity.

5.1 Dynamic Range of a Pixel of FAMOUS

Within the dynamic range of a pixel of FAMOUS, the signal should be proportional to the number of incoming photons. Near the lower bound of the dynamic range, it is not possible to distinguish the signal from the noise. Near the upper bound of the dynamic range, the signal begins to saturate. The linear relation between the photon flux and the signal is no longer fulfilled.

As presented in figure 8, right-hand-side, the detection efficiency of a pixel is 22.5 % per photon up to incident angles of light of $\theta_{\text{in}} = 23°$ and light fluxes of $\phi \approx 30\,\text{ns}^{-1}$. The detection efficiency is expected to rise significantly with future generations of SiPMs.

[§]<u>G</u>eometry <u>an</u>d <u>t</u>racking.
[¶]And also photons which are quantised portions of light.

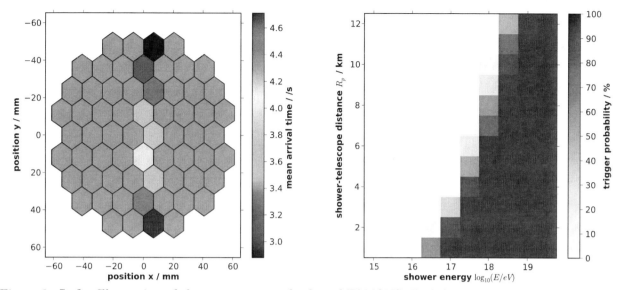

Figure 9: **Left:** Illustration of the camera event display of FAMOUS. Each hexagon represents a pixel. The color-code denotes the mean arrival time of the fluorescence light signal of a 10^{18} eV vertical EAS in $R_p = 4$ km distance. The data have been obtained by the Geant4 simulation of FAMOUS. **Right:** Trigger probability of FAMOUS, operated in Aachen, Germany, as a function of the shower energy E and the shower-to-telescope-distance R_p.

5.2 Fluorescence Light Simulation

The extensive air shower simulation program CONEX[24] has been used to create a library of vertical EAS with logarithmic energy bins of $\log_{10}(\Delta E/\text{eV}) = 0.25$ in the regime $E = [10^{15}, 10^{20}]$ eV. Each energy bin contains 100 EAS.

The Offline framework of the Pierre Auger Collaboration[8] is capable of simulating the fluorescence light emitted by EAS. For each energy bin, the 100 EAS have been placed each in $R_p = 1, 2, 3...12$ km distance to the telescope. The information on the arrival direction and position of the photons on the aperture of the telescope has then been exported to the custom Geant4 ray-tracing simulation of FAMOUS.

5.3 Trigger Probability of FAMOUS

The probability of FAMOUS to trigger a measurement for a certain EAS is a function of the telescope-to-shower-distance R_p and the shower energy E. For the calculation, a simple signal-over-threshold trigger has been defined. The threshold is set by the noise detected by the camera.

5.3.1 Night-Sky Background

Besides the noise created by the SiPM itself, the intrinsic brightness of the night-sky creates a more or less constant noise level in the telescope. In urbanised environments, light pollution is the dominant contribution. The upper limit for the night-sky radiance near Aachen in Eynatten, Belgium[||] in the fluorescence-light-regime $\lambda \in [300, 400]$ nm has been measured to[25]:

$$L_{\text{NSB,UV}}^{(\text{max})} = 1.9 \cdot 10^{12} \, \text{m}^{-2} \text{s}^{-1} \text{sr}^{-1} \quad . \tag{20}$$

As determined by the ray-tracing simulation, the night-sky creates a constant background rate of approximately $\phi = 0.3 \, \text{ns}^{-1}$ in a pixel of FAMOUS which contributes over 90 % to the overall noise.

[||] June 1st, 2011, 01:14 h for a duration of 120 s.

5.3.2 Results

For each pixel, the simulation returns a signal trace: the voltage drop $V = V_\text{s} + V_\text{th}$ over the SiPM as a function of the time. Pixels not illuminated by the EAS do not exceed a certain threshold V_th. Everything above the threshold V_th can be considered as EAS signal V_s. The signal-to-noise ratio is defined as

$$SNR = 10 \log_{10}\left(\frac{\text{Signal}}{\text{Noise}}\right) \text{dB} = 10 \log_{10}\left(\frac{V_\text{s}}{V_\text{th}}\right) \text{dB} \quad . \tag{21}$$

For each simulated EAS, the analysis determines the maximum signal-to-noise-ratio SNR_max of the camera. If $SNR_\text{max} > 0$, the signal is as high as the background created by the SiPM noise and the NSB and is considered to be detected. On the left-hand side of figure 9, the image of an EAS produced in the camera of FAMOUS is illustrated. The grey pixels did not satisfy $SNR > 0$, the color of all the rest denotes the mean arrival time determined by the maximum SNR within the pixel signal trace.

The ratio of detected EAS versus not detected EAS in a shower energy E and distance R_p bin defines the trigger probability. Figure 9, right-hand side, illustrates the trigger probability of FAMOUS as simulated for an urbanised observation site like Aachen. Finally, the simulations yield almost 100 % trigger probability for EAS with energies $E \geq 10^{17}\,\text{eV}$ and $R_\text{p} = 1\,\text{km}$.

6. SUMMARY & OUTLOOK

The refractive optic of the air fluorescence telescope prototype FAMOUS consists of a big Fresnel lens with a diameter of $D = 510\,\text{mm}$ equal to the focal length f. To match a field of view of $1.5° \times 1.5°$ per pixel, each pixel is composed by a Winston cone, i.e. a light concentrator, and a Hamamatsu S10985-100C array of four $3 \times 3\,\text{mm}^2$ SiPM. The Winston cone has an entrance radius of $r_1 = 6.7\,\text{mm}$. The focal plane of FAMOUS features 64 hexagonally arranged pixels in total.

Detailed detector-response-simulations demonstrated that the current design of FAMOUS might be capable of detecting extensive air showers even in an urbanised, bright environment. Considering the small aperture diameter of FAMOUS and the moderate sensitivity of currently available SiPMs, the energy and spatial range, which can be covered with FAMOUS, is quite promising. FAMOUS is currently in construction and is expected measure the first light this year.

ACKNOWLEDGMENTS

The authors would like to thank their colleagues of the CMS and Auger working groups of the III. Physikalisches Institut A, RWTH Aachen University, as well as Barthel Philipps, head of the mechanical workshop of the institute, for the construction of the telescope components. They are also obliged to Franz Beissel, III. Physikalisches Institut B, RWTH Aachen for the development of an SiPM amplifier and to colleagues from the universities of Granada and Lisbon for the vivid exchange of experience and experimental results. Furthermore, the authors would like to acknowledge the tremendous amount of computing time supplied by the RWTH RZ computer cluster. This work is funded by the German Federal Ministry of Education and Research BMBF and the European astroparticle physics network ASPERA.

REFERENCES

[1] Blümer, J., Engel, R., and Hörandel, J., "Cosmic rays from the knee to the highest energies," *Progress in Particle and Nuclear Physics* **63**(2), 293–338 (2009). arXiv:0904.0725v1.

[2] Particle Data Group, [*Review of particle physics*], Journal of physics, Institute of Physics Publishing (2011).

[3] Meurer, C. and Scharf, N., "HEAT - a low energy enhancement of the Pierre Auger Observatory," *Astrophysics and Space Sciences Transactions* **7**, 183–168 (2011). arXiv:1106.1329v1.

[4] Hörandel, J., "On the knee in the energy spectrum of cosmic rays," *Astroparticle Physics* **19**(2), 193–220 (2003).

[5] Greisen, K., "End to the Cosmic-Ray Spectrum?," *Physical Review Letters* **16**, 748–750 (Apr 1966).

[6] Pierog, T., Alekseeva, M., Bergmann, T., Chernatkin, V., Engel, R., Heck, D., Kalmykov, N., Moyon, J., Ostapchenko, S., Thouw, T., et al., "First results of fast one-dimensional hybrid simulation of EAS using CONEX," *Nuclear Physics B-Proceedings Supplements* **151**(1), 159–162 (2006). arXiv:astro-ph/0411260v1.

[7] Abraham, J., Abreu, P., Aglietta, M., Aguirre, C., Ahn, E., Allard, D., Allekotte, I., Allen, J., Allison, P., Alvarez-Muñiz, J., et al., "The fluorescence detector of the Pierre Auger Observatory," *Nuclear Instruments and Methods in Physics Research Section A: Accelerators, Spectrometers, Detectors and Associated Equipment* **620**(2-3), 227–251 (2010).

[8] Argirǫ, S., Barroso, S., Gonzalez, J., Nellen, L., Paul, T., Porter, T., Prado Jr, L., Roth, M., Ulrich, R., and Veberic, D., "The offline software framework of the Pierre Auger Observatory," *Nuclear Instruments and Methods in Physics Research Section A: Accelerators, Spectrometers, Detectors and Associated Equipment* **580**(3), 1485–1496 (2007). arXiv:0707.1652v1.

[9] Arqueros, F., Hörandel, J., and Keilhauer, B., "Air fluorescence relevant for cosmic-ray detection - Summary of the 5th fluorescence workshop, El Escorial 2007," *Nuclear Instruments and Methods in Physics Research Section A: Accelerators, Spectrometers, Detectors and Associated Equipment* **597**(1), 1–22 (2008). arXiv:0807.3760v1.

[10] Nagano, M., Kobayakawa, K., Sakaki, N., and Ando, K., "New measurement on photon yields from air and the application to the energy estimation of primary cosmic rays," *Astroparticle Physics* **22**(3), 235–248 (2004). arXiv:astro-ph/0406474v2.

[11] Rennefeld, J., "Studien zur Eignung von Silizium Photomultipliern für den Einsatz im erweiterten CMS Detektor am SLHC," (2010).

[12] Renker, D. and Lorenz, E., "Advances in solid state photon detectors," *Journal of Instrumentation* **4**, P04004 (2009).

[13] Querchfeld, S., "Test neuer Photomultiplier für die Entwicklung einer Auger-Nord Fluoreszenz Kamera," (2010).

[14] Hamamatsu, "http://jp.hamamatsu.com/resources/products/ssd/pdf/s10362-11_series_kapd1022e05.pdf," (Nov. 2009).

[15] Assis, P., Brogueira, P., O., C., M., F., Hebbeker, T., Lauscher, M., Lorenz, E., Mendes, L., Meurer, C., Mirzoyan, R., Niggemann, T., Pimenta, M., Rodrigues, P., Schweizer, T., Stephan, M., and Teshima, M., "R&d for future sipm cameras for fluorescence and cherenkov telescopes," ICRC (2011).

[16] Eckert, P., Schultz-Coulon, H., Shen, W., Stamen, R., and Tadday, A., "Characterisation studies of silicon photomultipliers," *Nuclear Instruments and Methods in Physics Research Section A: Accelerators, Spectrometers, Detectors and Associated Equipment* **620**(2-3), 217–226 (2010).

[17] Lauscher, M., "Characterization Studies of Silicon Photomultipliers for the Detection of Fluorescence Light from Extensive Showers," (2011).

[18] Niggemann, T., "New Telescope Design with Silicon Photomultipliers for Fluorescence Light Detection of Extensive Air Showers," (2012).

[19] Agostinelli, S., Allison, J., Amako, K., Apostolakis, J., Araujo, H., Arce, P., Asai, M., Axen, D., Banerjee, S., Barrand, G., et al., "Geant4 - a Simulation Toolkit," *Nuclear Instruments and Methods* **506**(3), 250–303 (2003).

[20] Gross, H., Singer, W., and Totzeck, M., [*Handbook of Optical Systems: Physical image formation*], Wiley-VCH (2005).

[21] Fresnel Optics GmbH, "Private communications," (May 2011).

[22] Winston, R., Miñano, J., Welford, W., and Benítez, P., [*Nonimaging optics*], Academic Press (2005).

[23] CMS Collaboration, "CMSSW software cross-reference," (May 2012). http://cmslxr.fnal.gov/lxr/.

[24] Bergmann, T., Engel, R., Heck, D., Kalmykov, N., Ostapchenko, S., Pierog, T., Thouw, T., and Werner, K., "One-dimensional hybrid approach to extensive air shower simulation," *Astroparticle Physics* **26**(6), 420–432 (2007).

[25] Stephan, M., Hebbeker, T., Lauscher, M., Meurer, C., Niggemann, T., and Schumacher, J., "Future use of silicon photomultipliers for the fluorescence detection of ultra-high-energy cosmic rays," *SPIE* **8155**, 81551B (2011).

Experimental characterization of the turbulence inside the dome and in the surface layer

Aziz Ziad[a], Dali-Ali Wassila[a], Julien Borgnino[a], and Marc Sarazin[b]

[a] UMR 7293 Lagrange, Université de Nice-Sophia Antipolis,
CNRS, OCA, Parc Valrose F-06108 Nice Cedex 2, France;
[b] European Southern Observatory, Karl-Schwarzschild-Strasse 2, D-85748 Garching, Germany

ABSTRACT

We present the concept of a new instrument dedicated to modeling turbulence inside the dome and in the surface layer. It consists of using parallel laser beams separated by non redundant baselines between 0.1 and 2-3m and measuring Angle-of-Arrival (AA) fluctuations from spots displacements on a CCD. We use weighted least-square method to fit the measured AA longitudinal and transverse covariances with theoretical forms deduced from the usual models of turbulence. Then, the whole parameters characterizing this turbulence are provided from a complete spatio-temporal analysis of AA fluctuations. Thus, the surface layer turbulence energy in terms of C_N^2 constant is provided from the AA structure function as in the DIMM instrument.

Keywords: Site-testing, Atmospheric turbulence, Adaptive Optics, Long Baseline Interferometry

1. INTRODUCTION

Within the framework of site qualification for the E-ELT and its dome design, local turbulence due to the immediate environment of the telescope must be taken into account. A better knowledge of the phenomenon and its modeling becomes crucial if one wants to optimize the performances of the telescope. However, in contrast with the rest of the atmosphere, the surface layer turbulence is affected by ground and environment characteristics (convective flow, roughness, etc.). Some models were proposed to link the image quality to the heat dissipation in the surface layer.[1] In this work we propose an optical experiment to characterize the surface layer and extensively the turbulence inside the telescope enclosure. It consists of the wavefront analysis by means of parallel laser beams on a multi-baseline configuration. The fluctuations of the angle of arrival (AA) are measured with CCD camera and their longitudinal and transverse covariances at different baselines are compared to the theoretical ones. The choice of the configuration of the parallel beams is important to obtain an optimal sampling of these AA covariances.

The principle of the use of laser beams in the study of optical turbulence was introduced by various authors particularly for horizontal propagation for the measurement of inertial range limits corresponding to the inner and outer scales. Indeed, the scintillation of laser beams with different apertures or wavelengths has been used for the estimation of the inner scale .[2-4] On the other hand, AA fluctuations have been also used for characterization of local turbulence by [5,6] particularly for the outer scale measurement.

2. THEORETICAL BACKGROUND

2.1 Statistics of Angle-of-Arrival fluctuations

The proposed experiment for local turbulence modeling is based on AA covariance measurement at different baselines. Indeed, we consider the normalized AA longitudinal covariances as used in the GSM instrument [7]

$$\Gamma_\alpha = \frac{C_\alpha(B, D, L_0)}{\sigma_\alpha^2(D, L_0)} \quad (1)$$

Further author information: (Send correspondence to Aziz Ziad)
Aziz Ziad: E-mail: ziad@unice.fr, Telephone: 0033492076338

where $\sigma_\alpha^2(D, \mathcal{L}_0)$ is the variance of AA fluctuations. This expression is Fried's parameter r_0 and wavelength λ independent.

The AA covariance expression for two telescopes of diameter D separated by a baseline B is given by :[7,8]

$$C_\alpha(B, D, L_0) = \pi\lambda^2 \int f^3 W_\phi(f, L_0)[J_0(2\pi fB) - \cos(2\gamma)J_2(2\pi fB)][\frac{2J_1(\pi fD)}{\pi fD}]df \quad (2)$$

where J_n represents the nth order Bessel functions of the first kind and γ is the baseline angle with the x-direction. Useful formulae for this AA covariance were given by .[9] Introducing W_ϕ expression given by the model of von Karman $0.0229 r_0^{-5/3}(f^2 + \frac{1}{\mathcal{L}_0^2})^{-11/6}$, it appears the factor of $\lambda^2 r_0^{5/3}$ which is not depending on the wavelength as well as the AA covariance. Another possibility, is the use of cross-covariances of AA measured in the longitudinal and transverse directions.

The AA variance $\sigma_\alpha^2(D, \mathcal{L}_0)$ is obtained from Eq. 2 at the origin $B = 0$. In the case of von Karman model and $L_0 >> D$, an approximation of this AA variance is given by :[10]

$$\sigma_\alpha^2(D, \mathcal{L}_0) \simeq 0.179\lambda^2 r_0^{-5/3}[D^{-1/3} - 1.525\mathcal{L}_0^{-1/3}] \quad (3)$$

The Fried parameter r_0 is the size of a spatial coherence area of the perturbed wavefront and it depends physically on the structure constant of refractive index fluctuations $C_N^2(h)$ integrated along the propagation path as:

$$r_0 = \left[0.423\left(\frac{2\pi}{\lambda}\right)^2 \int C_N^2(h)dh\right]^{-3/5}. \quad (4)$$

The AA structure function σ_d^2 is also an interesting quantity for turbulence characterization particularly for seeing estimation by means of DIMM instrument .[11] This quantity is differential allowing to avoid vibration effects and less sensitive to outer scale. Since $\sigma_d^2 = 2[\sigma_\alpha^2(D, \mathcal{L}_0) - C_\alpha(B, D, \mathcal{L}_0)]$, the expression of the AA structure function is completely deduced from Eq. 2. For the transverse and longitudinal cases, the expression of σ_d^2 is given by [9,12] when $B > D$:

$$\sigma_d^2 = K_{l/t}\lambda^2 r_0^{-5/3} D^{-1/3} \quad (5)$$

where $K_{l/t}$ is constant depending on the ratio B/D as:

$$K_l = 0.34(1 - 0.570(\frac{B}{D})^{-1/3} - 0.040(\frac{B}{D})^{-7/3}) \quad (6)$$

$$K_t = 0.34(1 - 0.855(\frac{B}{D})^{-1/3} + 0.030(\frac{B}{D})^{-7/3}) \quad (7)$$

Thus, using the DIMM method one can have estimation of Fried's parameter r_0 or seeing in both longitudinal and transverse direction. These estimations should lead to the same results if exposure time is short enough. On the other, variance and covariance are appropriate for estimation of outer scale $\mathcal{L}_0$ if r_0 is provided by DIMM method.

2.2 Modeling the dome & surface layer turbulence

The proposed experiment uses several separated laser beams propagating simultaneously in a turbulent media and leading to several different baselines. The corresponding longitudinal ($\gamma = 0^o$) and transverse ($\gamma = 90^o$) covariances are then compared to the theoretical ones to test the validity of the four commonly used models which are the Kolmogorov, von Karman, Greenwood-Tarazano and Exponential model. This method has been successfully used in the GSM instrument .[13] The existing atmospheric turbulence models are characterized by their phase power spectra :[14]

For the von Karman (vK) model, one has:

$$W_\phi^{vK}(\vec{f}, \mathcal{L}_0) = 0.0229 r_0^{-5/3} \left[f^2 + \frac{1}{\mathcal{L}_0^2} \right]^{-11/6} \quad (8)$$

and in the case of the Greenwood-Tarazano (GT) model ,

$$W_\phi^{GT}(\vec{f}, \mathcal{L}_0) = 0.0229 r_0^{-5/3} \left[f^2 + \frac{f}{\mathcal{L}_0} \right]^{-11/6} \quad (9)$$

or in the case of the Exponential (Ex) model :

$$W_\phi^{Ex}(\vec{f}, \mathcal{L}_0) = 0.0229 r_0^{-5/3} f^{-11/3} (1 - exp(-f^2 \mathcal{L}_0^2)). \quad (10)$$

All these models tend to the Kolomogorov one when the outer scale is considered infinite. In the case where these models do not adjust the measurements, we can construct other models by changing the powers and constants of the existing ones .[14]

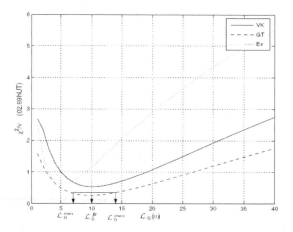

Figure 1. Example of χ^2 obtained with normalized covariances measured by means of the GSM instrument .[13]

The instrument provides for each acquisition two sets of AA covariances in longitudinal and transverse directions. Then, we use weighted least-square method on data obtained to test usual models of turbulence. This is built on the hypothesis that the optimum description of a set of data is one which minimizes the weighted sum of squares of deviations of the experimental normalized covariances Γ_α^i from the theoretical fitting function $\Gamma_\alpha(B_i, \mathcal{L}_0)$. It can be written as

$$\chi^2 = \sum_{i=1}^n \frac{[\Gamma_\alpha^i - \Gamma_\alpha(B_i, \mathcal{L}_0)]^2}{\sigma_i^2}. \quad (11)$$

where n is the number of baselines and σ_i^2 is the variance of Γ_α^i that acts as a weighting factor for each spatial covariance. Minimization of χ^2 is obtained by varying the non-fixed parameter $\mathcal{L}_0$ of the model with the help of pre-calculated grids from Eq. 1 and Eq. 2. Solely $\mathcal{L}_0$ needs to be adjusted to minimize the value of χ^2, so the number of degrees of freedom is $\nu = n - 1$. The variance σ_i^2 is a suitable choice of weights and can be estimated from the data. Advantage of this method is that it gives more weight to accurate covariances and deduced outer scale estimations take into account errors of covariance measurements on which normally distributed errors are assumed. Errors on the normalized covariances are described in section 4. Typical examples of χ^2 obtained for each model on a particular normalized covariance measured are represented in Figure 1. The $\mathcal{L}_0$ value is deduced from the χ^2 minimum. $\mathcal{L}_0^{min}$ and $\mathcal{L}_0^{max}$ correspond to a χ^2 difference $\Delta\chi^2 = 0.1$ as a measure of the interval with significant variations of the χ^2 quantity.

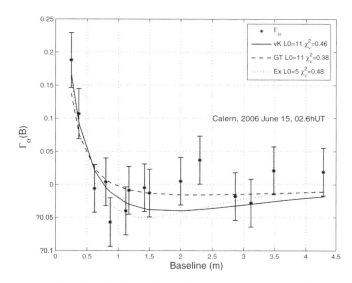

Figure 2. Example of normalized AA longitudinal covariances (star plot) measured with the GSM instrument at Calern on 2006 June 15 (2.6 UT). Theoretical covariances are fitted to data in order to minimize the χ^2.[13]

Figure 2 represents an example of experimental normalized covariances obtained at Calern on 2006 June 15 with 6 aligned modules of the GSM instrument leading to 15 non-redundant baselines. Best fitting obtained for usual turbulence models (Eqs 8,9 and 10) are also represented. Some outlier points appear but their effects on the χ^2 minimization are diminished by the weighting factor of $1/\sigma_i^2$ which decreases with large values of Γ_α^i.

If the normalized Γ_α are well adapted to check the validity of the surface layer model, we will use the AA structure function and variance in Eqs 5 and 3 respectively for the estimation of the energy C_N^2 and consequently the equivalent seeing.

2.3 Temporal spectrum of AA fluctuations

Different authors have demonstrated the interest of AA spectral analysis for the characterization of the atmospheric turbulence. Indeed, the PSD of AA fluctuations presents different regimes of temporal frequency ν [7,15] : a $\nu^{4/3}$ for low frequency domain, a $\nu^{-2/3}$ for intermediate frequencies and for high frequency a $\nu^{-11/3}$ or a $\nu^{-8/3}$ depending on spatial filtering. The cutoff frequencies separating these different regimes are proportional to $v/\mathcal{L}_0$ and $D/\mathcal{L}_0$ where v is the wind speed and D the telescope diameter. Thus, as shown in Figure 3, the estimation of the high cutt-off frequency lead to the wind speed v and then combined with the low cutt-off frequency estimation provide the outer scale $\mathcal{L}_0$ value. On the other hand, estimation of the wind speed could be used to estimate the coherence time which is given by Roddier formulae $\tau_0 = 0.31 r_0/v$.[16,17]

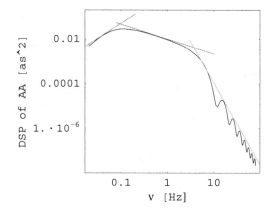

Figure 3. Example of AA temporal spectrum for D=2cm, $v = 0.2 m/s$, $\mathcal{L}_0 = 3m$ and $r_0 = 20cm$. The asymptotic behaviors indicate in red the $\nu^{4/3}$, in blue $\nu^{-2/3}$ and in green $\nu^{-11/3}$.

3. DOME & SURFACE LAYER TURBULENCE MONITOR

3.1 Instrument description

The instrument consists of a laser multibeam which is initially provided by one laser diode. Optical fibers and fiber couplers are used to split in several parallel beams (4 or 5). A filter of density (ND) is also used to reduce flux intensity and to avoid, therefore, the detector saturation. Then a collimator with achromatic lenses is used to obtain parallel beams of 2cm diameter (Figure 4). This beam diameter has been chosen in respect of inner scale value (between few mm to cm) which represents the smallest detail in the perturbed wavefront. The different beams propagate inside a local turbulence which we want to characterize. The separation between the different beams is very important to obtain an optimal sampling of the AA covariance curves. A specific study has been dedicated to this issue based on numerical simulation (see next paragraph). Then, using mirrors and beam splitters, the parallel beams are oriented on a small telescope (D=12-14cm and f/10) (Figure 4) which will make them converge in the CCD located in its focal plane. A Barlow lens X2 is associated to the telescope to elongate its focal distance to increase the sensitivity to AA fluctuations. The different beams are separated by adjusting mirrors and beam splitters orientations. Thus, each laser parallel beam leads to a light spot on the CCD allowing the reconstruction of the longitudinal and transverse covariance curves of AA fluctuations to be compared to the theoretical models. As explained above, the choice of beam size is of order of inner scale of turbulence but beam should be larger enough to reduce the spot spread on the CCD. The size of each beam is, therefore, fixed to $2cm$, this corresponds to the pupil diameter of our optical system. With a focal distance $f \sim 3m$, the detector sampling is 0.5"/pix for a CCD Prosilica GC650 (659x493 pix^2 and $pixel = 7.4 microns$). It is possible to use Barlow's lens X3 to elongate the focal distance to increase the instrument sensitivity which could reach 0.2"/pix with a Barlow X3. This method is already used and works well in the DIMMs running at Dome C in Antarctica. The estimation of the AA fluctuations is based on the centroid method as used in the DIMM instrument.[18]

3.2 The experimental configuration

The AA fluctuation statistics are well adapted for probing atmospheric turbulence as it has been demonstrated by means of DIMMs, GSM, MOSP and more generally of Shack Hartmann wavefront sensors. For local turbulence detection, the sensitivity should be increased since the fluctuations are smaller than in the atmospheric turbulence measured by these instruments. If we consider a local turbulence intensity of $C_N^2 = 10^{-14} m^{-2/3}$ which is typical for most known sites. The path of light over a distance of 30 meters of such turbulence leads to an equivalent seeing of 0.61" at $\lambda = 0.5 \mu m$. Our DIMM at Dome C measures, using the same method, seeing 3 times lower

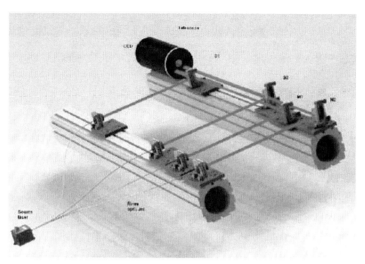

Figure 4. The schematic optical device of the Dome & Surface Layer turbulence instrument.

with a sampling of $0.75''/pix$. As said above in section 3.1, the sampling of the proposed instrument could reach $0.2''/pix$ giving enough sensitivity for the detection of the surface layer AA fluctuations.

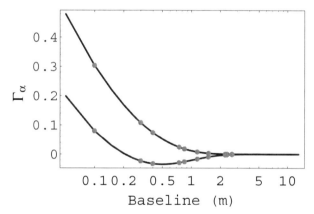

Figure 5. Example of AA longitudinal and transverse covariances versus baseline for $r_0 = 16cm$, $\mathcal{L}_0 = 3m$ and $D = 2cm$. An example of possible configuration consists of 5 laser sources at 0, 150, 225, 235 and 265mm leading to baselines of 10, 30, 40, 75, 85, 115, 150, 225, 235 and 265mm. The corresponding AA normalized covariances are shown for transverse and longitudinal cases.

There is a strong sensitivity of the AA covariance to the outer scale for metric and submetric values of the baseline.[8] The choice of baseline values is a trade-off between instrumental constraints and the need to sample regularly longitudinal and transverse AA covariances in the 0-3 m range. The upper limit of this range was chosen in regards to the outer scale in the surface layer which should be smaller than in the whole atmosphere. In addition, using Eq. 3, the rms AA fluctuations for surface layer typical conditions with seeing of 0.61" at $0.5\mu m$ and $L_0 = 3m$ is $\sigma_\alpha = 0.1"$. It is easy to measure such quantity even without using a Barlow lens. Different tests at Laboratory have been made with a prototype of this instrument with laser beams and the results are very interesting even in weak and strong turbulence. Each observation campaign will be preceded by a study based on numerical simulation to decide the instrument configuration. This simulation will serve also for the choose of the best sampling for the reconstruction of the longitudinal and transverse covariances. Using Eq. 2, the normalized AA covariances for longitudinal and transverse configurations are shown in Figure 5 considering typical dome & surface layer turbulence conditions given above. The longest baseline could be limited to the outer scale value since the curves tend to saturation. The trend of the covariances at small baselines could be described by few baselines. Then, we will concentrate the sampling for intermediate baselines. An example of possible configuration is presented in Figure 5 deduced from the GSM one [13] by changing the scale factor. As said above, this issue is solved by numerical simulation that why we decided to keep the instrument configuration changeable to be adjusted to the turbulence conditions.

4. NOISE SOURCES

The error sources affecting the proposed instrument are similar to DIMM's ones. Typical errors in this kind of experiment are essentially due to the centroid determination, the statistical error and finally the exposure time. This last is not very important since we use a laser beam as a source with stabilization to ensure the flux constancy for a large time. The source is fixed and then the instrument is not affected by vibrations contrary to the seeing monitors where a tracking system is used. We could plan to use wind-screen to protect the instrument from the wind effect. As described by [12] the centroid determination error comes from the CCD readout noise (RON) and the signal Poisson noise (the images are considered corrected with the flat field). The RON error is now well modeled (Eq. 9 of [12]) and estimations are possible which are expected to be reduced since the source flux is not limited.

The photon noise σ_p^2 has less impact on the measurement than the RON error since the CCD is directly illuminated. The variance σ_p^2 could be estimated using models (Eq. 11 of [12]) but some factors are determined by means of numerical simulation and laboratory tests.

The statistical error will be minimized by making a long series of images acquisition and in the other hand, we will use very short exposure time ($\sim 1ms$) for each frame to avoid the exposure time debiasing.

As shown in Figure 4, the beams have the same length but we have different residual paths in the detector block. For small baselines the energy due to these residual propagation should be weak and would not affect the results. On the other hand, if the AA variances for the different spots are not equal, one have to conclude that is due to these residual propagations. This could be estimated by placing the source bench close to the detector bench and then we have access only to the residual propagation variances that we have to subtract from the measurements.

5. FIRST PROTOTYPE AND FIRST RESULTS

In the framework of the E-ELT project, a prototype of this instrument has been developed (Figure 6) and tested in optical laboratory in different conditions of turbulence.[19] This instrument is now installed in a dome at La Silla Observatory and first results will be available soon. This prototype consists of four parallel laser beams leading to six different baselines in addition to the variance measurement and then these seven measurements are used to fit with the theoretical models. The tests in the laboratory revealed the presence of vibrations due to external sources even if the instrument is isolated. The vibration spectrum has been characterized and then the prevailing frequencies are removed by soft filtering. The sensitivity has been checked and the instrument is able to measure weak turbulence of order of $10^{-3} as^2$ at $\lambda = 635nm$. The first results obtained in laboratory are very interesting [19] leading particularly as expected to small values of outer scale 2-3m (Figure 7). These results and their analysis will be presented in a near future in another paper.

5.1 Acknowledgments

This work is supported by ESO in the framework of the E-ELT project.

Figure 6. Photo of the monitor of dome & surface layer turbulence installed in a laboratory at the Observatoire de la Côte d'Azur. Top and bottom panels show respectively the source and the reception benches. In the middle panel are presented reception bench components (reflecting mirror, beam splitter, telescope and CCD).

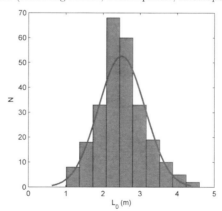

Figure 7. Example of histogram of outer scale measured in laboratory for data series obtained successively.

REFERENCES

[1] Racine, R., S. D. C. D. and Sovka, J. *PASP* **103**, 1020–1032 (1991).
[2] Livingston, P. *Appl. Opt.* **11**, 648 (1972).
[3] Hill, R. and Ochs, G. *Appl. Opt.* **17**, 2430 (1978).
[4] Ochs, G. and Hill, R. *Appl. Opt.* **15**, 3068 (1985).
[5] Consortini, A., R. L. and Stefanutti, L. *Appl. Opt.* **11**, 2543 (1970).

[6] Consortini, A., I. C. and Paoli, G. *Optics Communications* **214**, 9 (2002).
[7] Avila, R., Z. A. B. J. M. F. A. A. and Tokovinin, A. *J. Opt. Soc. Am. A* **14**, 3070–3082 (1997).
[8] Borgnino, J., M. F. Z. A. *Optics Communications* **91**, 267–279 (1992).
[9] Conan, R., B. J. Z. A. and Martin, F. *J. Opt. Soc. Am. A* **17**, 0807 (2000).
[10] Ziad, A., B. J. M. F. and Agabi, A. *A&A* **282**, 1021 (1994).
[11] Sarazin, M. and Roddier, F. *A&A* **227**, 294 (1990).
[12] Tokovinin, A. *PASP* **114**, 1156 (2002).
[13] Maire, J., Z. A. B. J. M. F. *MNRAS* **386**, 1064 (2008).
[14] Voitsekhovich, V. V. *J. Opt. Soc. Am. A* **12**, 1346 (1995).
[15] Conan, J. *J. Opt. Soc. Am. A* **12**, 1559–1570 (1995).
[16] Roddier, F., G. J. and Lund, G. *J. Opt. (Paris)* **13**, 263 (1982).
[17] Ziad, A., B. J. D.-A. W. B. M. M. J. and Martin, F. *J. Opt.: Pure & Applied Optics* **14**, 045705 (2012).
[18] Aristidi, E., A. A. F.-E. A. M. M. F. S. T. T. T. V. J. and Ziad, A. *A&A* **444**, 651 (2005).
[19] Dali-Ali, W., PhD thesis, Univ. Nice Sophia-Antipolis (2011).

Seeing trends from deployable Shack-Hartmann wavefront sensors, MMT Observatory, Arizona, USA

J. Duane Gibson[a], G. Grant Williams, Thomas Trebisky,
MMT Observatory[b], University of Arizona, Tucson, Arizona, USA 85721-0065

ABSTRACT

Deployable Shack-Hartmann wavefront sensors (WFS) for the f/5 and f/9 secondary configurations have been used at the 6.5-meter MMT Observatory (MMTO) since 2003. Probe mirrors for these WFS's are moved into the optical path of the telescope between scientific observations multiple times each night. Results from the wavefront measurements are then used to bend the primary mirror (M1) and to reposition the secondary mirror (M2) to correct for wavefront errors. In addition to measuring the optical wavefront error, the Shack-Hartmann data are used to determine the delivered seeing using the measured spot sizes. This study attempts to analyze the more than 75,000 WFS measurements and associated seeing values obtained at the MMTO since 2003. The overall WFS data reduction and analysis procedure is discussed. This data analysis includes: 1) finding the spots in each image, 2) centroiding the spots, 3) measuring a point-spread function, 4) determining an average spot width and a derived seeing value, and 5) computing the best-fit Zernike polynomial coefficients. Wavefront slopes are calculated from spot displacements and wavefront aberrations are fit with a 19-term Zernike polynomial. As part of this study, the WFS-derived seeing values are correlated with other observing parameters, such as mirror-air temperature contrasts. Finally, seasonal climate and local weather (*e.g.*, prevailing wind direction) effects on astronomical seeing are evaluated.

Keywords: Astronomical seeing, wavefront sensors, telescope observing, image analysis, MMT Observatory

1. INTRODUCTION

Improvement of local seeing at the 6.5-meter MMT Observatory (MMTO) is a primary objective of the Observatory. Seeing results from optical aberrations produced by changing density inhomogeneities in the atmosphere along the optical path. On-axis wavefront sensors (WFS's) are used to determine the wavefront error and a calculated seeing for the f/5 and f/9 secondary mirrors. Data from the MMTO adaptive f/15 secondary mirror is not discussed here. Although the overall data reduction pipeline for the deployable MMTO WFS systems is presented here (Figure 1), the emphasis of this study is to evaluate the overall data trends in the derived seeing data and to correlate these seeing data sets with thermal and environmental data. As will be shown, these observing factors can strongly influence and, at times, limit seeing at the site.

2. WFS OVERVIEW

2.1 WFS data acquisition

Details of the f/5 and f/9 WFS hardware, software, and data acquisition and reduction at the MMTO are presented elsewhere[1,2,3]. The MMTO WFS software was initially developed for the f/9 secondary mirror, followed by independent development of the f/5 WFS software. These two software packages were later combined, although there are still unique aspects for each software mode because of different hardware. The MMTO WFS software also includes an option for the MMT and Magellan Infrared Spectrograph (MMIRS), a cryogenic multi-slit spectrograph that operates from 0.9 to 2.4 microns with the f/5 secondary mirror. Delivered seeing data from these three configurations ("f/5", "f/9", and "MMIRS") are presented here.

---

[a] Contact email: jdgibson@mmto.org
[b] The MMT Observatory is a joint facility of the Smithsonian Institution and the University of Arizona.

WFS operation begins with the collection of one or more images (Figure 1, Step 1). Separate image acquisition systems are used for the f/5 and f/9 WFS systems. The f/5 WFS system uses a 14x14 square geometry lenslet array. Spots for the f/5 WFS are captured by a Santa Barbara Instrument Group (SBIG) ST9XE charge-coupled device (CCD) camera. The 512x512 pixel images consist of 20x20 micron pixels. Exposures from 0.12 to 3600 seconds are possible with 10-millisecond resolution. WFS images are acquired on a Windows-based computer. An ASCII-based, client-server protocol, the "msg" protocol that was developed by the Smithsonian Astrophysical Observatory (SAO), is used to transfer images from the f/5 WFS computer to the WFS data reduction software.

In contrast to the f/5 WFS, the f/9 WFS system uses a 13x13 hexagonal geometry lenslet array and an Apogee Instruments KS-260 CCD camera for spot acquisition. As with the f/5 system, the camera acquires 512x512 images with 20-micron pixels. Images are captured on a Linux-based computer through a remote shell login.

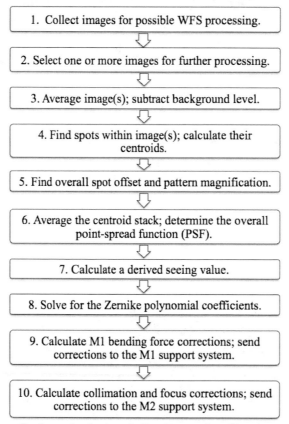

Figure 1. The MMTO WFS data collection and reduction pipeline. Intermediate files are stored for each step of the data reduction. See text for details.

During WFS image acquisition, the telescope operator (TO) moves the target WFS star onto the optical center of rotation axis for the telescope, using zero-coma moves of the secondary mirror. With this mode of secondary mirror movement, the M2 mirror is tilted and translated so that no coma is introduced into the image during secondary mirror movement. After the WFS star is centered, one or more WFS image files are saved in FITS format for further image processing and data analysis. Imaging parameters, such as exposure time, may be varied when acquiring these images.

2.2 WFS data reduction and analysis

After WFS images have been acquired, the TO selects one or more images for further WFS data analysis (Figure 1, Step 2). The selected WFS images are averaged and passed through IRAF's imarith() function (via the Python "pyraf" interface) to do background subtraction (Figure 1, Step 3). Spots are identified and their centroids found (Figure 1, Step

4) via the Scientific Python (SciPy) package scipy.ndimage. The overall spot offset and pattern magnification (Figure 1, Step 5) within the WFS image(s) are found, also via the scipy.ndimage package. These values will be applied later to the centroid locations for registration purposes. After the centroids are averaged, the results are passed to the IRAF psfmeasure() library via pyraf, which determines a position-dependent model of the point-spread function (PSF) for each image (Figure 1, Step 6).

Astronomical seeing is determined (Figure 1, Step 7) as the diameter, *i.e.*, the full width at half maximum (FWHM), of the seeing disc (the PSF for imaging through the atmosphere). The PSF diameter is a reference to the best possible angular resolution that can be achieved by the telescope during long exposures. The size of the seeing disc is set by the atmospheric conditions at the time of the observation.

Roddier[4] reviewed the standard model for astronomical seeing. Using this standard model, observed seeing can be calculated using the relationship:

$$\text{seeing} = 206265 * 0.98 * \lambda/r_0. \quad (1)$$

where: "r_0" is the Fried parameter, the optical quality of the atmosphere, and "λ" is the wavelength of observation. WFS seeing is determined by deconvolving a reference FWHM to obtain an effective r_0, similar to techniques described elsewhere[5,6]. Seeing is then corrected to zenith using the following equation:

$$\text{seeing(zenith)} = \text{seeing}(z) * \sec(z)^{0.6}. \quad (2)$$

where: "z" is the zenith angle (90 degrees minus the telescope elevation).

Next, the spots are scaled based upon the overall spot magnification. A reference grid center offset and grid magnification are determined. Instrument-specific corrections are subtracted. These results are used to: 1) re-centroid all of the spots based upon the optimized grid offset, 2) calculate the pupil coordinates, and 3) find the wavefront gradients and phase differences. The Zernike polynomial coefficients are then found (Figure 1, Step 8) from the wavefront gradients and, optionally, from the phase differences.

Results from each WFS measurement are displayed to the TO. The TO then has the choice of whether to use all or a portion of the Zernike polynomials to determine the primary mirror actuator forces to be sent to the M1 control system (Figure 1, Step 9). Corrections for collimation and focus can also be sent to the M2 control system (Figure 1, Step 10).

The remainder of this study will focus on the WFS-determined seeing values for the past nine years at the MMTO and correlations between these seeing values and observing parameters, such as glass-air temperature contrasts and seasonal weather trends.

3. SEEING DATA ANALYSIS

Figure 2 presents histograms for MMTO WFS seeing data from March, 2003, through March, 2012. A total of 77956 WFS seeing values are included in the data of which 55229 samples are from the f/5 WFS, 19913 from the f/9 WFS, and 2814 from the MMIRS WFS. Bin size for the histograms in Figure 2 is 0.1 arcsec. The f/9 data shows the overall best seeing values with a median value of 0.79 arcsec and a mean value of 0.85 arcsec. The f/5 median seeing is 0.84 arcsec while the mean seeing is 0.93 arcsec. The MMIRS WFS data set is relatively small with only 2814 samples and represents only a few nights of data. The median MMIRS WFS seeing is 0.94 arcsec while the mean seeing is 1.03 arcsec. The overall median seeing for the combined WFS configurations is 0.84 arcsec while the mean seeing is 0.92 arcsec. For this study, the combined data sets from the three WFS configurations (*i.e.*, f/5, f/9, and MMIRS) are referred to as the "combined" WFS data.

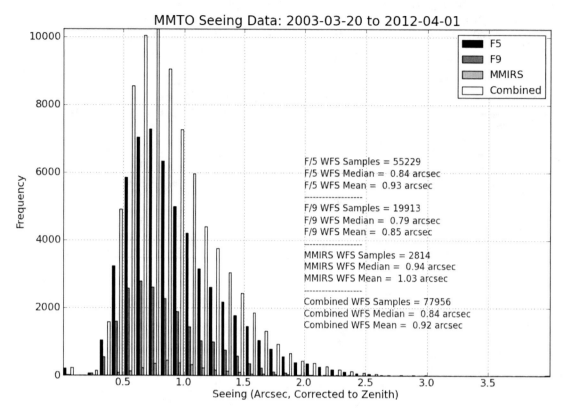

Figure 2. Histograms of WFS seeing values from March, 2003 through March, 2012. Individual histograms are shown for the f/5, f/9, and MMIRS WFS data. A histogram for the combined data set from these three WFS systems is also shown. Bin sizes for the histograms are 0.1 arcsec.

3.1 Effects of WFS image exposure time on seeing

Figure 3 shows the variation of the median seeing value with different WFS image exposure times. The most common exposure time used for WFS determinations is five seconds, representing approximately 40% of all WFS-related images. As seen in this figure, there is an overall increase in measured seeing values with longer exposures. The median observed seeing values range from 0.7 arcsec for two-second exposures to over 1.0 arcsec for exposures of ten or more seconds. This trend is believed to result mainly from the operational procedures used during WFS measurements. The exposure time for WFS images is increased to produce a set signal/noise ratio to ensure a good WFS measurement. Therefore, as seeing degrades, exposure time is increased, introducing a bias of larger seeing values for longer exposures. Telescope tracking and secondary mirror oscillation are believed to be minor effects compared to this operational influence on the data set.

3.2 Effects of primary mirror-to-air temperature contrast ("mirror seeing")

A contributor to the overall astronomical seeing is mirror seeing. Mirror seeing results from localized variations in the refractive index of air at the M1 mirror-air interface. Previous studies have modeled the related temperature-induced M1 surface distortions at the MMTO[7]. Convection cells that are driven by mirror-air temperature contrasts cause these variations in the local refractive index. Several sets of temperature measuring devices are available at the MMTO. For this study, only the E-series thermocouples related to the primary mirror cell are used to determine the overall relationships between temperature contrasts and calculated seeing.

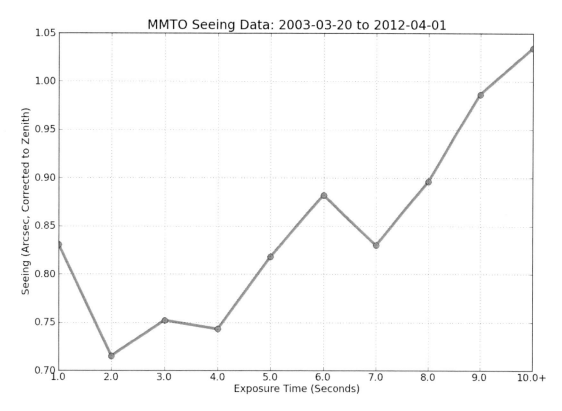

Figure 3. Variation of median determined seeing values with exposure time for the combined WFS data sets. An overall increase in determined seeing is observed with longer exposure times. This increase is believed to result mainly from a strong bias between seeing values and exposure times in the WFS operational procedures as described in the text. The most common exposure time for WFS images is five seconds, representing approximately 40% of all samples.

The MMTO primary mirror has an array of E-series thermocouples attached within the honeycomb hexagonal core structure of the mirror. These thermocouples are used to measure glass temperatures within the primary mirror as well as air temperatures behind the mirror in the lower plenum, in the telescope chamber, and outside of the telescope enclosure. Sixteen E-series thermocouples are attached to glass immediately behind the frontplate of the mirror, the portion of the mirror closest to the reflecting surface. The average of these sixteen thermocouples represents the mirror glass temperature in Figures 4 through 9 of this study. The chamber air temperature in these figures is obtained from an E-series thermocouple that is located adjacent to the Nasmyth platform on the telescope mount. The outside air temperature values are from an E-series thermocouple located outside of the telescope enclosure. All of these thermocouples are connected to the same isothermal junction box to provide accurate relative temperatures between different thermocouples. Using this one set of temperature sensors eliminates possible temperature calibration issues between different types of sensors.

Figure 4 shows the impact of the M1 mirror-air temperature contrast on the overall determined seeing for the combined WFS data sets. As seen in this figure, the lowest seeing values (*i.e.*, less than 0.8 arcsec) are found when the glass temperature is between 0.0 °C and -1.0 °C colder than the adjacent chamber air temperature. This range of mirror-air temperature contrasts, where the primary mirror is at or slightly colder than the adjacent air temperature, is a major operational goal for optimal local seeing conditions. Seeing deteriorates rapidly when the M1 glass temperature is warmer than the chamber air temperature. Seeing is also degraded when the glass temperature is more than -1.0 °C colder than the chamber air temperature, although to a lesser extent than when the mirror is warmer than the adjacent air. There results are similar to findings at other sites[8].

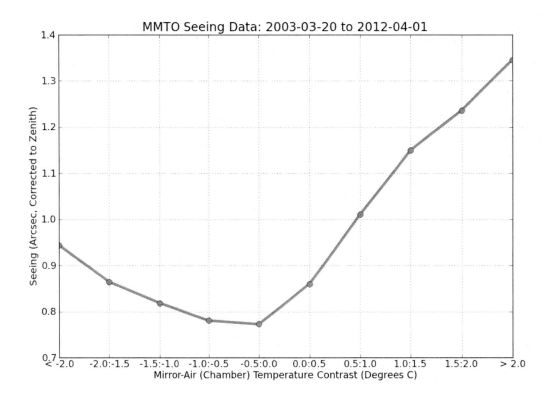

Figure 4. Median seeing values as determined by combined WFS measurements with varying mirror-air temperature contrasts. The lowest seeing values are seen when the mirror glass temperature is from 0.0 °C to -1.0 °C colder than the adjacent chamber air temperature. Seeing is degraded when the glass temperature is warmer than the adjacent chamber air temperature or when the glass is more than -1.0 °C colder than the adjacent air temperature.

The MMTO primary mirror has a forced air ventilation system that attempts to equilibrate glass temperatures within the primary mirror as well as to cool the primary mirror to a temperature at or below the chamber air temperature. The recent changes in the control software for this ventilation system are presented elsewhere[9]. This control software attempts to maintain the M1 glass temperature at approximately 0.5 °C below the adjacent air temperature. Future work is planned to evaluate the performance of this control software.

Figure 5 shows the distribution of seeing values with the primary mirror glass colder and warmer than the chamber air temperature. As seen in the figure, the median seeing is 0.15 arcsec lower (0.79 versus 0.94 arcsec) when the glass temperature is colder than the chamber air temperature. Although the primary mirror is actively cooled, approximately a third of the seeing measurements included in Figure 5 were collected under thermal conditions when the primary mirror is warmer than the chamber air temperature. These conditions commonly occur when temperatures fall rapidly at sunset and the ventilation system is unable to cool the primary mirror as fast as the falling air temperatures. Daytime activities in the telescope chamber can substantially warm the chamber relative to the outside air temperature. In addition, WFS measurements are more commonly taken during the beginning of the night, soon after opening the telescope enclosure. The WFS measurements are therefore biased towards these more undesirable conditions that occur soon after sunset and at the beginning of the observing night. Another undesirable situation is when the chamber is substantially warmer than the outside air temperature at the time of opening the telescope chamber. The telescope enclosure itself has a large thermal inertia that must be overcome under these conditions. The temperature of the primary mirror is somewhat decoupled from this thermal inertia since the mirror is actively cooled.

Figure 6 presents histograms of the entire combined WFS seeing data set and a subset of that data where the mirror is at the same temperature to slightly colder (0.0 °C to -1.0 °C) than the adjacent air temperature. An overall reduction of 0.07 arcsec (median values of 0.77 compared to 0.84 arcsec) is found under these optimal thermal conditions.

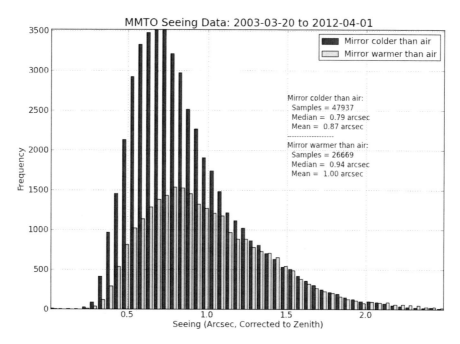

Figure 5. Distribution of seeing values when: 1) the primary mirror temperature is colder than the chamber air temperature, and 2) the primary mirror glass temperature is warmer than the chamber air temperature. Substantially lower seeing values are seen when the glass temperature is colder than the adjacent air temperature.

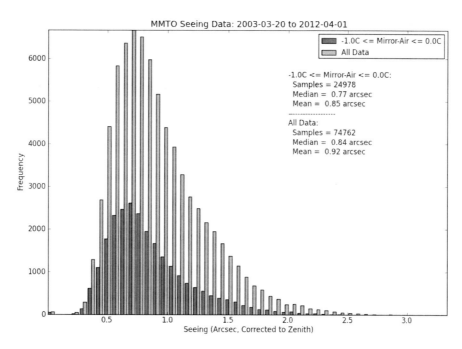

Figure 6. Distribution of combined WFS seeing values with low mirror-glass temperature contrast compared to the entire combined WFS seeing values. The low mirror-glass temperature contrast data represents observing conditions where the glass-chamber air contrast is between 0.0 °C and -1.0 °C.

Figure 7 presents a kernel density estimation (KDE) diagram of the seeing (Y-axis) versus the temperature contrast between the primary mirror glass and the chamber air temperature (X-axis). These KDE diagrams are similar to histograms, but use kernel distributions (*e.g.*, Gaussian surfaces of unit area or volume) rather than the bins used with histograms. Smoothing of data is influenced by the bandwidth of the kernel that is applied to the data set. This figure illustrates the desirable thermal conditions where the primary mirror is at or slightly below the chamber air temperature for optimal seeing conditions. It also shows that as the mirror glass temperature becomes warmer than the adjacent air temperature, seeing deteriorates rapidly, as was also seen in previous figures. Having a glass temperature 1.0 °C or more colder than the adjacent air temperature is preferred to having the glass warmer than the adjacent air. This criterion is included in the control algorithms for the ventilation system for the primary mirror.

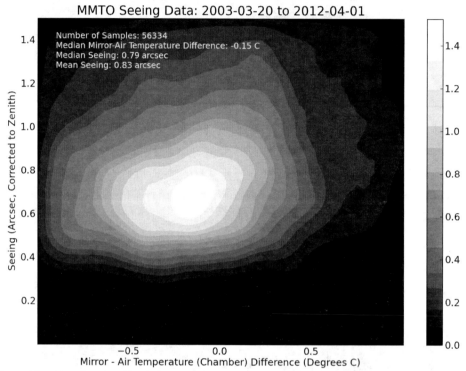

Figure 7. A kernel density estimation (KDE) diagram of seeing versus mirror-air temperature contrast for the past nine years at the MMTO. The default SciPy Scott's covariance factor for the Gaussian kernel is used in this figure. As in all KDE diagrams in this study, the shade of gray represents the data density resulting from the KDE smoothing. Mirror–air contrasts are restricted to a range from -1.0 °C to 1.0 °C in the figure.

3.3 Effects of telescope enclosure thermal conditions ("dome seeing")

Dome seeing results from convection between the telescope enclosure and the outside air and is typically treated separately from seeing contributed by the primary mirror ("mirror seeing"). If ventilation of the enclosure is sufficient, dome seeing can be largely eliminated by wind flushing the air within the telescope chamber. Figure 8 illustrates the effects of temperature contrast between air within the telescope chamber and outside of the telescope enclosure during WFS measurements. The figure suggests that the measured seeing is largely independent of the chamber-outside temperature contrast, implying that the MMTO has little thermally-induced dome seeing. As seen in Figure 8, the overall seeing is approximately 0.82 to 0.86 arcsec under differing chamber-outside thermal conditions. The abnormally high value of over 1.0 arcsec for the "-2.0:-1.5" °C bin results from a small data set (only 52 samples). The MMTO is well ventilated and large negative air contrasts between the chamber and outside are relatively uncommon. All other bins in the figure have hundreds to thousands of samples within each bin.

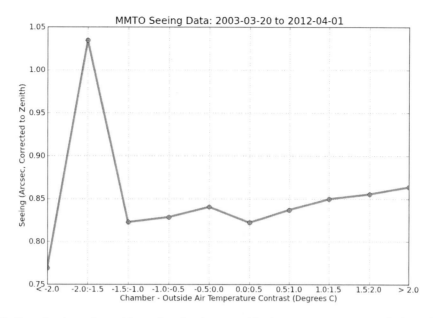

Figure 8. Distribution of seeing values with varying chamber – outside air temperature contrasts. Seeing values remain at approximately 0.85 arcsec regardless of the chamber – outside air temperature contrast. This suggests that dome seeing effects are small on the overall seeing values determined from WFS measurements. The high values at "-2.0:-1.5" °C are associated with a relatively small data set (*i.e.*, 52 samples).

Figure 9 shows the frequency of WFS-determined seeing values for different mirror – air temperature contrasts. The M1 mirror frontplate temperature is contrasted with two different air temperatures: the chamber temperature and the outside temperature. The median mirror – chamber air temperature is -0.19 °C while the median mirror – outside air temperature is 0.61 °C, indicating that the chamber is typically warmer than the outside air temperature during WFS measurements. As explained elsewhere, this partially results from a bias of WFS measurements during the early portion of the night, with rapidly falling temperatures soon after sunset, and from thermal inertia of the telescope enclosure. It should be noted that the primary mirror is pre-cooled before night operations.

3.4 Effects of seasonal climate patterns on seeing

The MMTO site is characterized by monsoonal weather during the summer months, large diurnal temperature changes during the spring and fall, and weather systems originating to the west over the Pacific Ocean during the winter. Figure 10 presents seasonal variations in the combined WFS seeing values from 2003 to the present. Seasons are defined as: January-March (winter), April-June (spring), July-September (summer), and October-December (fall). This figure includes seeing determinations from March, 2003, through March, 2012.

Figure 10 shows little overall variation in seeing values observed with the different seasons. Values are comparable to the overall median seeing value (*i.e.*, 0.84 arcsec) presented earlier. The total number of seeing measurements from the summer months (July through September) are lower because this is traditionally the period of "summer shutdown" when telescope maintenance and, occasionally, *in-situ* re-aluminization of the primary mirror occurs. This is also the monsoon season in southern Arizona, characterized by intense afternoon thunderstorms.

3.5 Effects of local wind patterns on seeing

Wind data are currently available from three sensors at the MMT. Two sensors are located on the west side of the building (*i.e.*, "Vaisala4" and "Young") at approximately 20 feet above the ground level. The third unit ("Vaisala3") is located on the east side of the enclosure on a flagpole at around 20 feet above the ground level. The Vaisala3 and

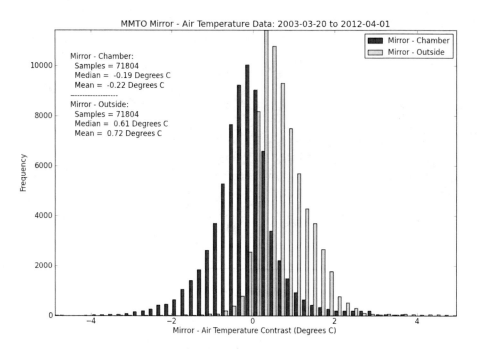

Figure 9. Distributions of the mirror-glass temperature contrast with the chamber and outside air temperatures during WFS determinations from 2003 to 2012. The glass-chamber air temperature contrast values are determined by the E-series thermocouples attached to the backside of the frontplate of the primary mirror and an E-series thermocouple located in the chamber, near the Nasmyth platform. The glass-outside air temperature contrast values are derived from the same frontplate E-series thermocouples and a single E-series thermocouple located outside of the telescope enclosure

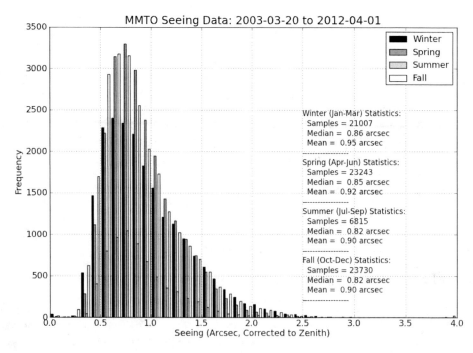

Figure 10. Seasonal seeing values (in arcsec, corrected to zenith) for the combined WFS data sets from 2003 to 2012. Relatively little difference in overall median and mean measured seeing values are seen with the different seasons.

Vaisala4 sensors are Vaisala model WXT520 remote weather stations, which measure air temperature, relative humidity, wind speed and direction, and barometric pressure data. These two Vaisala devices obtain wind data from ultrasonic sensors on top of each unit. The Young wind sensor is a rotating 3-cup anemometer. Although the Vaisala and Young units have different types of wind sensors, their data is comparable. Wind speed and directions can vary greatly between the east and west side of the MMTO enclosure, depending on the prevailing wind direction. This effect is most notable for winds from the east or west where the building enclosure provides a wind barrier for some sensors. The Vaisala units also record a maximum (*i.e.*, gust), average, and minimum wind speed and direction over a five-second sampling interval and a 20-second averaging interval. Data have been logged since 2005-04-14 for the Vaisala3, since 2007-12-10 for the Vaisala4, and since 2009-02-24 for the Young unit. The wind data used for the following figures is the highest wind speed (and corresponding wind direction) from any of the three sensors (or of the available wind sensors). It is assumed that this highest value indicates the prevailing wind direction, *e.g.*, a sensor on the west side of the enclosure would be used when winds are from the west. Since these weather stations are exposed to the elements throughout the year, repair and servicing of the units is commonly required. When only one or two wind sensors are available, data from those sensors is used in this study.

Figure 11 illustrates the complex wind relationship seen at the summit of Mt Hopkins at 2617 meters (8585 feet), the site of the MMTO. Only wind data concurrent with WFS measurements are included in the figure. This complex relationship results from several superimposed wind patterns that occur throughout the year. Winter (January through March) winds are dominantly from the south and average around 6.5 meters/second (m/s) or 14 miles/hour (mph). Spring winds (April through June) are highly variable, but typically from the south to west. Median wind speeds are relatively high at 7 m/s (15.5 mph) or more, driven by the large diurnal temperature variations. Summer (July through September) weather patterns are dominated by afternoon monsoons and winds from the east to southeast, averaging around 5.5 m/s (12 mph). Finally, fall (October through December) sees the lowest wind speed of around 4.5 m/s (10 mph), typically from the southwest to northwest. Wind patterns and atmospheric turbulence are also complicated by the presence of Mount Wrightson at 2,881 meters (9,452 feet) to the east and slightly north of the MMTO site. As will be shown later, poor seeing conditions are found associated with prevailing wind from the east.

Figure 12 presents the median seeing values at different wind speeds. The lowest seeing values are found when the wind speed is between 3.0 and 4.0 m/s (7 – 9 mph). Seeing progressively degrades at higher wind speeds with seeing values exceeding 1.0 arcsec at wind speeds above 11.0 m/s (24 mph). This may indicate non-laminar airflow in high wind conditions.

Figure 13 presents histograms of calculated seeing under "low", "medium", and "high" wind conditions. For this study, "low" wind conditions are defined as calm wind speeds of less than or equal to 2.0 m/s (4.4 mph), "medium" wind ranges from 2.0 m/s to 10 m/s (4.4 to 22.0 mph), and "high" wind is above 10 m/s (22.0 mph). These histograms suggest that seeing is similar under "low" to "medium" wind conditions with medium seeing values of around 0.8 arcsec. This contrasts with a medium seeing of over 1.0 arcsec at wind speeds of greater than 10 m/s.

Seeing is also dependent on wind direction. Figure 14 presents the median seeing values for winds from differing directions. The best seeing is found when the prevailing wind direction is from the south or west. The worst seeing occurs when the wind is from the east. As mentioned earlier, Mount Wrightson lies just north of east from Mount Hopkins and the MMTO site. This mountain has a major impact on seeing when winds are from an easterly direction. In addition, almost all of the heat removed from the MMTO site through air conditioning is dumped on the northeast side of Mt. Hopkins, opposite the prevailing wind.

Finally, Figure 15 investigates the contribution of variability of wind speed to overall astronomical seeing at the MMTO site. This variability of wind speed could be an indication of non-laminar airflow surrounding the telescope enclosure. The two Vaisala remote weather stations log both an average wind and a maximum (*i.e.*, "gust") wind speed. The devices calculate these values internally with configurable parameters for sampling and averaging intervals. The units are set to the default values of sampling wind values every five seconds and averaging the values over 20-second intervals.

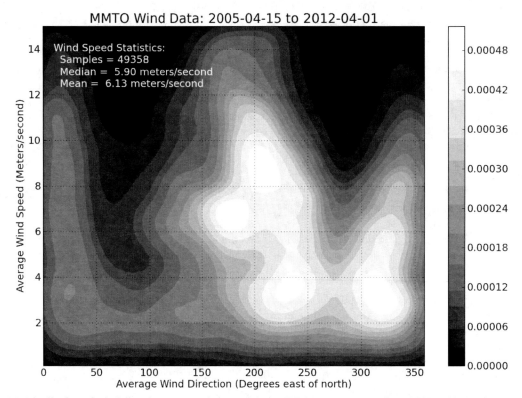

Figure 11. Distribution of wind direction versus wind speed during WFS measurements from 2005 to 2012. The complex pattern results from superimposed seasonal climate patterns. Fewer data are shown in this figure compared to other figures in this study since wind data have not been continuously available over the past nine years.

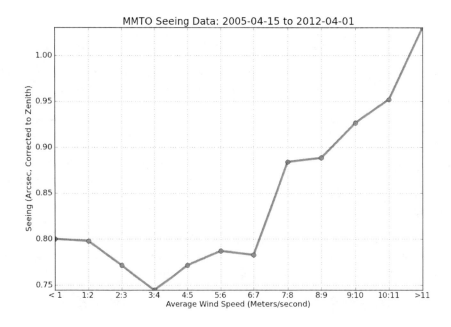

Figure 12. Distribution of median seeing values at different wind speeds during WFS measurements from 2005 to 2012. The median seeing is at or below 0.8 arcsec for wind speeds up to 7.0 m/s (15 mph). The best seeing (< 0.75 arcsec) is at wind speeds in the 3.0 to 4.0 m/s (7 - 9 mph) interval.

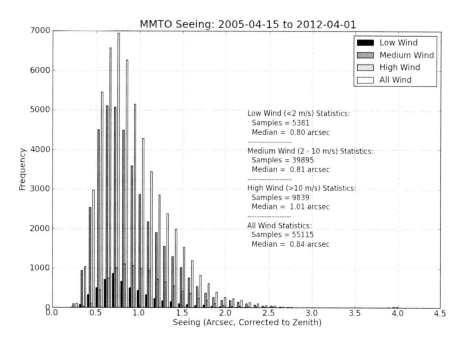

Figure 13. Distribution of median seeing values under different wind regimes during WFS measurements from 2005 to 2012. "Low wind" conditions are defined as an average wind speed of 2.0 m/s or less. "Medium wind" conditions range from 2.0 m/s to 10.0 m/s. "High wind" conditions exist when winds are 10.0 m/s and above. Substantially poorer seeing is found under the high wind conditions.

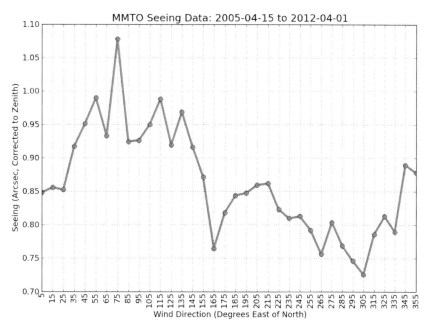

Figure 14. Distribution of median seeing values at different wind directions during WFS measurements from 2005 to 2012. The best seeing conditions are when the prevailing wind is from the west while the worst seeing conditions are with winds from the east.

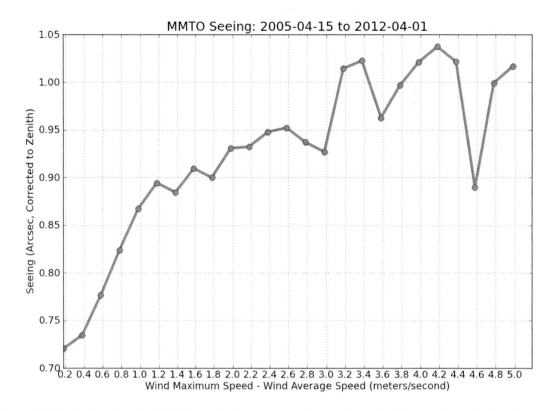

Figure 15. Distribution of median WFS-determined seeing versus the maximum wind (*i.e.*, gust) speed minus the average wind speed for the two Vaisala Model WXT520 weather stations at the MMTO during WFS measurements from 2005 to 2012. Progressively higher calculated seeing values are seen with increased maximum – average wind speed contrast.

4. CONCLUSIONS

An overview of the data acquisition and analysis for the deployable WFS systems at the MMTO has been presented. These systems have been in routine use for the past nine years and have yielded a rich set of data, including measured astronomical seeing values under a wide range of observing conditions.

Analysis of these seeing data and the associated observing parameters at the time of their measurement results in several conclusions:

1. Observing parameters, such as thermal conditions and weather patterns, can have a significant impact on the local seeing at the MMTO.
2. The observed correlation between increased seeing and longer exposure times is believed to result mainly from operational procedures.
3. Optimal seeing conditions are when the primary mirror is 0.0 °C to -1.0 °C below the chamber air temperature. This minimizes air turbulence and refraction of light at the M1 glass-air interface.
4. The localized temperature contrast at the M1 glass-air interface ("mirror seeing") contributes more to determined seeing than temperatures farther from this interface, for example, at the secondary mirror ("tube seeing", data not presented here) or outside of the telescope enclosure ("dome seeing").
5. Although the MMTO site has significant seasonal weather patterns, such as the summer monsoons, these patterns do not have an apparent significant impact on long-term seeing trends. These weather patterns do affect time lost to weather.

6. Seeing at the MMTO site is best when the prevailing wind direction is from the west and worst when winds are from the east. This results, in part, from air turbulence over Mt Wrightson to the east of the MMTO site.
7. Seeing degrades when the average wind speed exceeds 10 meters/second (22 miles per hour). It also degrades with increased variability of wind speeds at a given time.
8. When seeing is very bad, fewer WFS measurements are taken since the WFS systems are commonly not able to perform the necessary image analysis.

Efforts are underway at the MMTO to address several of these findings. The control ventilation system for the primary mirror was automated approximately three years ago, which has reduced the mirror-air contrast and associated mirror seeing. The mirror seeing results presented here combine data from before and after this ventilation system upgrade. Further work could help refine the control of this ventilation system. Finally, years of experience have taught the TO's many of the trends found here. For example, the typically bad seeing conditions at the MMTO site when the wind is from the east.

Some of the current operational practices at the MMTO include: 1) controlling the thermal impact of daytime activities within the telescope chamber, 2) starting the ventilation system for the primary mirror well in advance of night-time observations, 3) opening the rear shutters to improve air flow through the telescope enclosure, even when observing is not possible, 4) selecting science targets, in part, based upon current wind conditions, and 5) the scheduling of extended telescope maintenance periods, such as "summer shutdown", during the summer monsoon season. Through these and other efforts, the MMTO will continue to strive for improved local seeing during telescope operation.

REFERENCES

[1] West, S. C., "Interferometric Hartmann wave-front sensing for active optics at the 6.5-m conversion of the Multiple Mirror Telescope," Applied Optics 41, 3781-3789 (2002).
[2] West, S. C., S. P. Callahan, R. James, P. Spencer, H. Olson, B. Kindred, R. Ortiz, and T. Pickering, "F/9 top-box Shack-Hartmann: practical design and implementation," MMTO Technical Memo #03-6 (2003)
[3] Pickering, T. E., S. C. West, and D. G. Fabricant, "Active optics and wavefront sensing at the upgraded 6.5-meter MMT," Proc. SPIE 5489, 1041-1051 (2004).
[4] Roddier, F., "The effects of atmospheric turbulence in optical astronomy." Progress in optics 19, 281-376 (1981)
[5] Tokovinin, A, "From Differential Image Motion to Seeing", Pub. Astronomical Society of the Pacific, 114, 800, 1156-1166 (2002).
[6] Martinez, P, J. Kolb, M. Sarazin, and J. Navarrete, "Active optics Shack-Hartmann sensor: using spot sizes to measure the seeing at the focal plane of a telescope", Mon. Not. R. Astron. Soc, 421, 3019-3026 (2012).
[7] Pickering, T. E., "Modeling temperature-induced surface distortions in the MMTO 6.5 meter primary mirror." Proc. SPIE 7017, 701718-701718-11 (2008).
[8] Salmon, D., J.-C. Cuillandre, G. Barrick, J. Thomas, K. Ho, G. Matsushige, T. Benedict, and R. Racine, "CFHT image quality and the observing environment," Pub. Astronomical Society of the Pacific ,121, 882, 905-921 (2009).
[9] Gibson, J. D., G. G. Williams, S. Callahan, B. Comisso, R. Ortiz, and J. T. Williams, "Advances in thermal control and performance of the MMT M1 mirror," Proc. SPIE 7733, 77333-77333-12 (2010).

An Updated T-series Thermocouple Measurement System for High-accuracy Temperature Measurements of the MMT Primary Mirror

D. Clark*, J.D. Gibson

MMT Observatory, University of Arizona, 1540 E. Second St., Tucson, AZ USA

ABSTRACT

Starting in 2009, MMTO began design and installation of a new set of electronics to measure a set of radially-distributed type T thermocouples installed after the primary mirror polishing was completed. These thermocouples are arranged in both single measurement points and as thermopiles for differential temperature sensing. Since the goal of the primary mirror temperature control system is to minimize mirror seeing and mirror figure errors induced by temperature variation across the primary mirror, it depends on excellent accuracy from the temperature sensing system. The new electronics encompass on-board cold-junction compensation, real-time ITS-90 curve fitting, and Ethernet connectivity to the data servers running in the MMTO software infrastructure. We describe the hardware design, system wiring, and software used in this system.

Keywords: Thermocouples, cold-junction compensation, data acquisition, real-time software, databases

1. PRIMARY MIRROR THERMOCOUPLES

The 6.5m MMT primary mirror has 4 radially-distributed type T thermocouple arrangements that distribute several measuring points across the mirror organized partly in thermopiles to measure differential temperatures between the front, middle, and back plates of the mirror[1]. These measurements can be used to drive setpoints in the MMT mirror ventilation system to minimize mirror figure errors induced by temperature gradients in the glass[2]. These thermocouples are in addition to the type E units installed after polishing by the Steward Observatory Mirror Lab (SOML), which we currently use for thermal setpoint information.

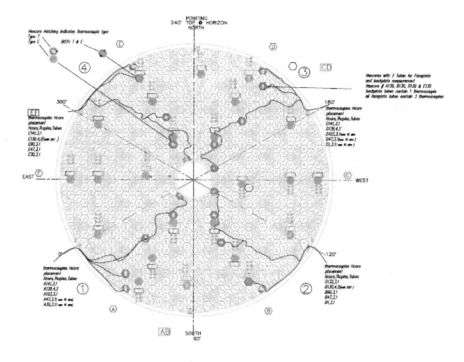

Figure 1. Thermocouple locations in the MMT primary mirror.

*dclark@mmto.org; phone 1 520 621-1558; fax 1 520 621-4144; www.mmto.org

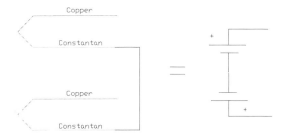

Figure 2. Type T thermocouples arranged in a thermopile to measure temperature differences.

The thermocouples are bonded with RTV to the mirror; a PVC tube surrounds the attachment point to keep circulating air from disturbing the measurement. Most of the thermocouples are bonded to the interior side of the mirror front plate. A total of sixteen thermocouples exit each of the 4 radii portholes in the mirror cell, arranged as in figure 3, below:

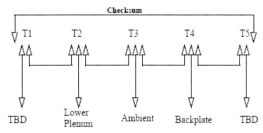

Figure 3. Thermopile connections on the mirror. TBD are free connections "to be determined".

The motivation for replacing the original HP DAU (data acquisition unit) was that we experienced several performance and operation issues: poor cold-junction compensation, the relays in the DAU board wore out, and the wiring was difficult to deal with when being removed for in-situ coating of the primary mirror.

2. NEW MEASUREMENT HARDWARE

In the new system, we wished to provide connectors for easy removal in preparation for mirror coating, better cold-junction compensation, and comparable noise and error performance. Since the thermocouple arrangement within the mirror presents two different measurement possibilities – absolute temperature and differential temperature, the design concept was to separate the measurement types into two different boards, called slices.

2.1 Absolute measurement board

To measure absolute temperatures, we require only one thermocouple per measurement channel, plus cold-junction compensation. In our system, we directly digitize six thermocouple voltages using an LTC2412 24-bit 8-channel analog-to-digital converter (ADC).

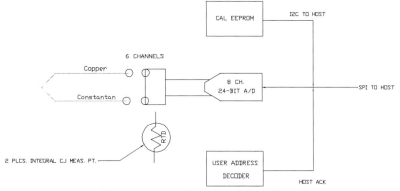

Figure 4. Absolute measurement board block diagram. The yellow links indicate thermal bonding.

Two platinum RTDs are mounted on the board with thermally-conductive epoxy[3] in between the thermocouple connectors to bring the connector copper-constantan joints and the RTDs into thermal equilibrium. The last two channels on the ADC are used to measure the RTD value and averaged to find the cold-junction temperature for use in ITS-90 thermocouple measurement calculations discussed below.

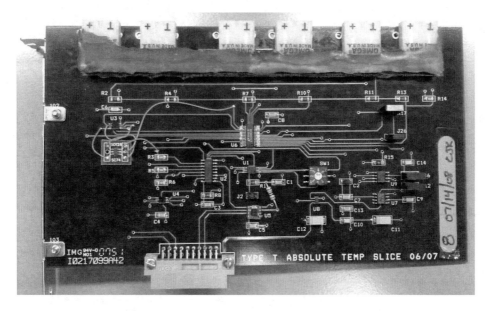

Figure 5. Absolute temperature type T thermocouple input board.

2.2 Thermopile measurement board

The thermopile-input board uses the same ADC circuit, but has the thermocouple inputs arranged in 4 banks of 2 pairs. Stacking the thermopiles in pairs increases the signal levels and allows differential points across more than a single pair of points to be measured if desired. Single-pair thermopile measurements can still be done by jumpering out the unused thermocouple connections.

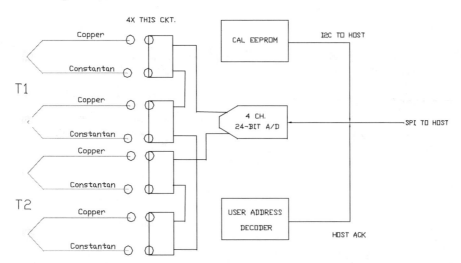

Figure 6. Thermopile circuit block diagram. The 4 connections are thermally bonded together.

The thermocouple connectors in each bank are thermally bonded together with thermally-conductive epoxy, as in the absolute thermocouple board. The connectors are wired such that the constantan-copper junctions at the board connectors all cancel, so long as each joint is in thermal equilibrium with the others, so no cold-junction compensation is needed.

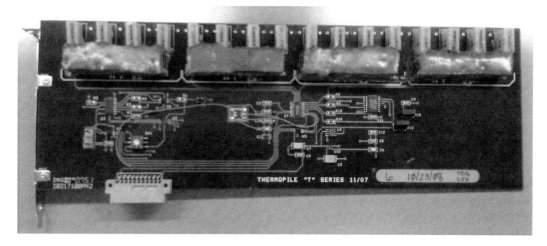

Figure 7. Thermopile differential thermocouple input board.

2.3 Host CPU for data collection

The two thermocouple input boards are designed to plug into a small motherboard that brings the SPI and I2C serial lines from the boards to a bus connected to the data-collection CPU. The CPU is a Digi International (formerly Rabbit Semiconductor) RCM3305. This small Z-80 based microprocessor module communicates with the thermocouple boards via its internal serial-port hardware and has on-board Ethernet to connect with the MMTO database mini-server that collects data for inclusion in the MMT MySQL database.

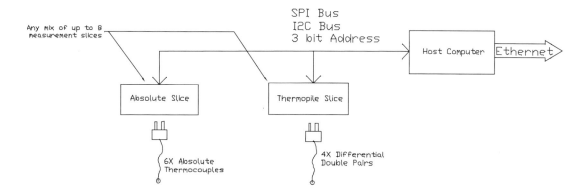

Figure 8. Host CPU system block diagram.

The CPU has 3 GPIO pins assigned to address bits used to access a given slice card for a total of up to 8 boards in the system. The boards can be inserted in any order and any mix of types, so long as their individual board addresses are unique. Address conflicts are detected at startup in the host software during the discovery phase when the input data structures are being initialized.

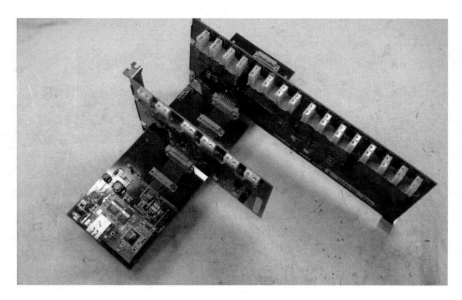

Figure 9. Sample assembly with host RCM3305 and motherboard.

3. MEASUREMENT SOFTWARE

The RCM3305 application software was written in C using the Rabbit Semiconductor version 9.52 IDE along with the extra-cost μC-OSII[4] and networking libraries. The data-collection software passes the measurement data over Ethernet to the MMTO MySQL database mini-server assigned to it. Some details of the implementation follow.

3.1 RCM3305 application software

The data-collection software uses μC-OSII tasks to collect data from the measurement slices, convert them to temperatures with ITS-90 corrections, and pass them over the network. At startup, the software first sets up the local i/o hardware on the RCM3305 module itself for communications with the temperature slice bus, then searches for boards that respond to each of the 8 available addresses on the bus. When a live board is detected, the software reads the calibration EEPROM on the board to discover what kind of board it is and its calibration data. Depending on the board type, the appropriate data structures are initialized and filled in with the board data.

During this board-discovery process, an error in setting the board addresses will result in more than one calibration EEPROM responding to the I2C bus transaction; the resultant garbling of the bus data is flagged as an error and that address is marked invalid, so only unique and readable addresses end up with allocated data structures. Once all data structures are initialized, the data-collection and processing tasks are spawned and the host continuously runs them.

The host tasks are broken down into a) reading the ADCs on each board, b) conversion of the ADC values to voltages, c) conversion of voltages to temperatures, e) support of UDP communications with the MySQL database mini-server, and f) support of TCP communications via a telnet session for engineering access. The ADC readout is simply continuous scanning of the channels available on each live board via the SPI serial port, and the voltage conversion is done in a separate task since the floating-point calculations involved were found to slow the temperature-calculation task down; separating out the floating-point operations helped increase the total sample throughput.

The voltage-to-temperature conversion task looks at the list of installed boards from the discovery phase and depending on their type either calculates temperatures for a single thermocouple (absolute slice), or a thermopile. When the calculations are done for an absolute board channel, a reverse quadratic fit to a subset of the ITS-90 table for type T thermocouples is applied, along with the appropriate cold-junction temperature adjustment. For a thermopile, we simply use the mean type T thermocouple sensitivity of 38.6μV/°C to find the differential temperature for a given pair, as we normally assume the thermopile connections are isothermal thanks to the thermally-conductive epoxy.

In the UDP communications task, we listen for a connection to be established on a given socket/port combination, and if the appropriate string is received (e.g. 'get all\n') the current set of measurements are read from the slice data structures, converted to ASCII strings, and written back over the socket. A mutex protects access to the slice data structures to prevent overwriting the data by the other tasks running on the CPU. The connection socket is closed when the client closes or aborts the connection and the task reverts to the listen state. Otherwise, samples are sent over the socket whenever a supported sample string is received. For simplicity, we only allow a couple of commands over the UDP connection. Incorrect and garbled commands get ignored and the socket closed immediately.

For more advanced access, the TCP communications task accepts a telnet connection with a modest number of allowed commands. Most of these have to do with filling in the data in the calibration EEPROM and getting an inventory of the installed boards. The host can write to the EEPROM only if a jumper is set on the slice board, and this is normally done during post-assembly checkout of each card. The data to be filled in include the board type, serial number, measured RTD excitation current (for the cold-junction measurement on the absolute boards), and gain/offset values for each of the available ADC channels on that board. The calibration data is collected for each board using an insulated copper block with thermocouples bonded into it with thermally-conductive epoxy; a pair of platinum RTDs is also bonded into the block and their resistance is measured with an HP 34401A 6½-digit multimeter to find the actual temperature of the block. The copper block's temperature can be varied with an electronically-controlled Peltier cooler[5].

3.2 MySQL mini-server

To minimize the computing overhead, we use UDP to transport the measurements over the network to the database server. In the MMTO software infrastructure, data sources like these are connected as clients to database middleware we call "mini-servers".

The MMTO mini-servers share a Perl library, MMTServer.pm, that uses a multiplexing socket server architecture to handle multiple clients and network responses from hardware devices. The specific parameters for this mini-server are contained in a MySQL table, miniservers:parameters, which has 135 parameters for a T-series mini-server to cover all 8 possible board locations and total channels. Configuration parameters for the T-series communications mini-server are in a separate MySQL table: miniservers:miniservers, which contains the hardware device name, update interval, MySQL background log names, and other configuration parameters for that mini-server. The hostname and port number for the hardware device associated with the mini-server is defined by a DNS Service (SRV) record, using the hardware device name obtained from miniservers:miniservers. These configuration parameters are all read from the MySQL tables when the mini-server is initialized. At the moment, we are logging all 135 parameters for the available T-series data at a 5-second interval. Data are logged on the MMT site server into the mmtlogs:ne_cell_background_log MySQL table, which can be accessed from any common programming language.

4. MEASUREMENT PERFORMANCE

The T-series thermocouples are located in adjacent hex cores in the mirror to some of the original SOML type E units. In the figure below, we show data from an observing night from 1700 to 0500 hours with the type E temperatures and the telescope chamber temperatures measured by a Vaisala HMP-247 temperature/humidity sensor located at the telescope's Nasmyth platform.

No averaging is at applied to the T-series measurements. The RMS variation of the absolute temperature measurements are about 0.07°C, with a peak-to-peak variation of 0.4°C. For differential measurements, we see typically 0.2°C RMS and 1.3°C peak-to-peak. With filtering, these could certainly be improved. As can be seen, the absolute data agrees well with commercial Vaisala unit's output, and the mirror thermal control is tracking the ambient temperature as intended. The type E system's cold-junction compensation offset was collected at SOML after "cold-soaking" the mirror for many hours and could still have a small offset that explains the difference with the T-series measurement.

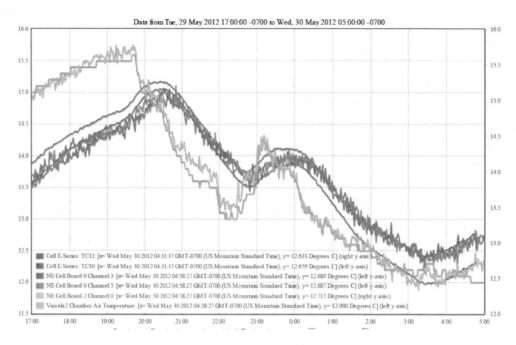

Figure 10. Absolute temperature measurement during telescope operation.

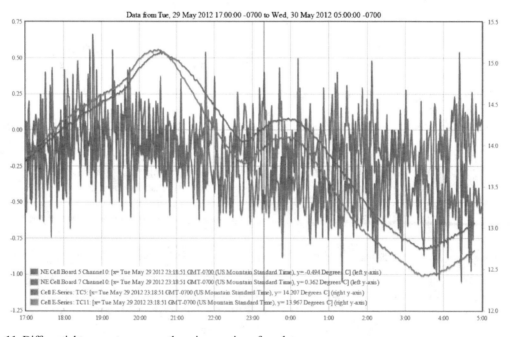

Figure 11. Differential temperature across the primary mirror faceplate.

The thermal control system should keep the mirror isothermal during operation. We can see this above, where the differential temperature from the mirror faceplate about 0.5m from the center hole to about 0.25m in from the outer edged is measured with a differential slice board. The y-axis on the left is for the differential data; the right y-axis is for the type E thermocouples' absolute temperature at adjacent hex cores. The mirror gradient is slightly increasing during the night in the differential measurement; this conforms with the type E data measured separately.

5. CONCLUSION

Personnel and scheduling constraints during and after the 2010 primary mirror re-coating kept us from installing the last two T-series radii enclosures. In the ensuing time, the RCM3305 units have become obsolete (though newer versions are available), and less-expensive, more powerful Linux-based solutions for the host CPU module have become available. MMTO would like to install the last two units and turn to increasing the data quality from calibration and getting faster, filtered data samples from the thermocouple system. That said, the µC-OSII software has proven very reliable, running continuously for months at a time without issue, and the 24-bit ADC units with their continuous internal calibration produce acceptable measurements.

We would like to acknowledge the efforts of the MMTO electronics staff in building and deploying the hardware for this system, particularly Brian Comisso, Tom Gerl, and Cory Knop for their efforts.

REFERENCES

[1] West, S.C., Spencer, P., and Callahan, S., "A high-precision differential-thermocouple temperature system for the 6.5m MMT primary mirror", MMTO Conversion Technical Memo #99-2, (1999)
[2] J. D. Gibson, G. G. Williams, S. Callahan, B. Comisso, R. Ortiz and J. T. Williams, "Advances in thermal control and performance of the MMT M1 mirror", Proc. SPIE 7733, 77333Y (2010)
[3] http://www.epoxies.com/tech/50-3100R.pdf
[4] Labrosse, J., [MicroC/OS-II the Real-time Kernel], CMP Books, Kansas, (2002)
[5] http://www.thermoelectric.com/2005/pr/cp/ahp-301cp.htm

A spectropolarimetric focal station for the ESO E-ELT

Klaus G. Strassmeier*[a], Igor Di Varano[a], Ilya Ilyin[a], Manfred Woche[a], Uwe Laux[b]

[a]Leibniz Institute for Astrophysics Potsdam (AIP), An der Sternwarte 16, D-14482 Potsdam, Germany; [b]Landessternwarte Thüringen, Sternwarte 5, D-07778 Tautenburg, Germany

ABSTRACT

We present a conceptual design for a spectropolarimetric focal station for ESO's *European Extremely Large Telescope* (E-ELT). It uses the intermediate f/4.4 focus, the only symmetric focus of the telescope. A dual channel, full Stokes-vector polarimeter provides on-axis light for the wavelength range 380-1600nm to up to two spectrographs simultaneously via two pairs of fibers. With such spectropolarimetric capability and a proper spectrograph for the optical and the near infrared wavelengths, the E-ELT would be able to provide the full parameter space of an incoming wavefront. Because of the on-axis entrance location of the polarimeter collimator and an entrance aperture of just 1.3 arcsec, the expected poor image quality of the intermediate telescope focus is not directly relevant.

Keywords: Telescopes, E-ELT, focal stations, polarimetry, spectropolarimetry, fibers

1. INTRODUCTION

Spectropolarimetry often offers results highly complementary to the ones of other observing techniques. It can provide clues not otherwise obtainable, is a strictly differential and therefore very accurate method, can exploit non-photometric nights and is photon starved even for some solar applications and thus requires, and optimally exploits, extremely large telescopes (ELTs). The strength of spectropolarimetry is the direct anchorage in physics of astronomical observations, making sure that the E-ELT can be credited also for the physical explanation of its discoveries.

Since most polarizing processes rarely exceed a few percent in the degree of polarization, many applications will typically aim for an accuracy of ≈0.1%. This level of accuracy is not obviously in conflict with the state of the art achievable with current astronomical instrumentation (Keller 2002, Baade 2006). Collecting enough photons for spectropolarimetry at this level will not be problematic for the E-ELT, but the requirements on the dynamic range of the detectors and intrinsic stability of the polarimeter and its attached spectrograph remain demanding.

Per wavelength resolution element, polarization detectable at the 0.1% level translates into a few million photons. For many classes of objects, even 8-m telescopes quickly run out of power when confronted with such a task. Therefore, as soon as the scientific objectives under consideration reach beyond the mere detection of an object, there is no stronger justification for increased telescope diameter than for spectropolarimetry. In fact, since modern detectors will reach negligible read noise levels in the near future, equipment that can measure polarization to better than 0.1% can also detect very dim sources.

In this paper, we present an update of our ongoing work (Strassmeier & Ilyin 2009, Strassmeier 2011) towards a Smart Focal Plane Polarimeter (SFPP) for the E-ELT's intermediate f/4.4 focus. Its heart is a spectropolarimeter of dual-beam type with a Schwarzschild collimator. The operating wavelength range is 380-1600nm with either an ultraviolet option towards the atmospheric cut off or an IR option towards 2.5μm. Our concept allows at least two different spectrographs to receive light from the spectropolarimeter via two pairs of fibers simultaneously. Its aim would be to detect a differential polarimetric precision of 10^{-5} in the optical through minimizing the instrumental impact on spectral-line polarization as well as continuum polarization by as much as possible. The focal station could be extended to also provide calibration light for other polarimeters in instruments mounted at the Nasmyth focii and thus could become a future smart focal plane. The entire focal station is retractable to a parking position within the telescope's Adaptive Optics (AO) tower.

*kstrassmeier@aip.de

2. TELESCOPE RELATED INSTRUMENT CONSTRAINTS

2.1 Focus availability

The intermediate f/4.4 focus is originally not intended for scientific use due to its poor image quality (mostly limited by coma) and hidden location within the central hole of the M4 mirror (Fig. 1). However, a pick-up of light just on the optical axis via fibers would be only mildly affected by poor image quality and could be exploited by a dedicated front end for one or two fiber-fed, single-object spectrographs located somewhere else. At the current stage, our focal station is designed as a pure polarimetric light feed and currently not intended to be used as a non-polarimetric light feed, albeit principally possible. To physically reach the focus, a robotic lever arm is required to position and park the polarimeter.

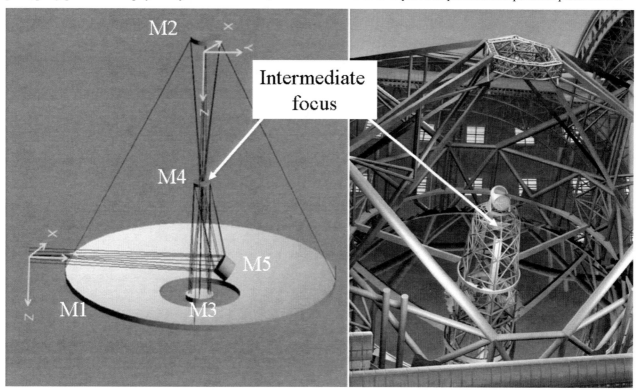

Figure 1. **Left**. Optical design of the 39m E-ELT. M1: irregularly shaped 39.3m outer diameter with 37m entrance pupil. M2: secondary with diameter of 4090mm allowing a „free focus position". M3: circular mirror with diameter of 3756mm. M4: adaptive mirror, elliptic with average diameter 2380mm. M5: tip-tilt mirror, elliptic with 2600mm. **Right**. CAD rendering of the AO tower. The intermediate focus is located within the top end of the tower coinciding with the M4 surface as indicated. Both images courtesy of ESO.

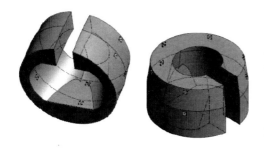

Figure 2. 3D rendering of the parking volume.

2.2 Parking volume within the AO tower

The available parking volume within the AO tower is determined by the optical beam diameter and the mechanical trusses that hold the various mirrors. ESO's E-ELT project office kindly provided a document for the free space (E-DWG-ESO-405-0987). Fig. 2 shows its complex shape. It consists of a cylinder of 5200mm diameter with a height of 1600mm like a truncated cone with a cut-out of a coaxial inner cone determined by the elliptic M4 (2440 mm × 2400 mm). The light path to M5 requires the "unavailability" of an 1100 mm wide section on the side. The volume is again shown in Fig. 3 within the AO tower.

2.3 Maximum dimensions

The stringent space constraints due to the requirement for off-axis wave-front sensing through the Nasmyth foci (while observing on axis in the intermediate focus) ask for a slim polarimeter shape in the intermediate focus. The second requirement is the safety distance of 300 mm from the vertex of M4. The total available diffraction-free and vignetting-free maximum space is then shown in Fig. 4. It is again of complex shape but roughly a cone with diameter 200 mm and length of approximately 800 mm.

Figure 3. An example robotic arm within the telescope's AO tower moves and holds the polarimeter in the focus. Shown is the tilted M4 mirror with its central hole (top), the positioning arm (left) with the cone-shaped polarimeter (middle). The parking volume when the instrument is not in use and out of the focus is shown as the shaded green volume (in the left half of the picture). See also Fig. 5.

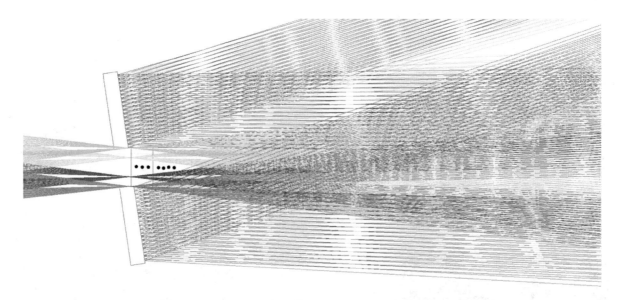

Figure 4. Optical detail around the intermediate f/4.4 focus. M4 is indicated as the tilted bar to the left. Note that the diameter of the central hole in M4 is 500mm (beam diameter of 481 mm) and that the primary-ray focus is located on the theoretical vertex of the surface of M4. Rays are shown for a FOV of 5 and 10 arcmin around the optical axis, respectively. The central ray-free region in the figure is completely unvignetted. The volume with three dots and within a distance of 300 mm from the surface of M4 is the technical „no-go zone" while the cone-like volume with the four dots is available for the polarimeter.

2.4 Summary of requirements for the positioning mechanics

The main performance requirements for the mechanical structure of the focal station are:

(a) The relative position of the collimator optical axis with respect to the telescope optical axis must be good to within ± 2 mm (proportional to approximately ±2.5 arcsec in the focal plane). Fine acquisition (centering) must be provided through two CCD cameras in the two polarimetric sub beams with a 5-arcsec FOV centered at the fiber entrance.
(b) The displacement of the center of the diaphragm in front of the fiber input with respect to the telescope axis must be controlled to an accuracy of better than 1% of the fiber core diameter, i.e. with a 4 mm "microlens" coupled to a 600-µm fiber, to currently 94 µm (tbc).
(c) The temperature variation of all optical components should be less than ±1°C within a nightly observation and ±2-3°C between night observation and day storage.

3. MECHANICAL DESIGN OF A POSITIONING SYSTEM

3.1 Introduction

Two rows with five equally distributed interface flanges (each 200×200×50 mm) are available within the AO tower truss. The positioning mechanism is foreseen to be mounted on some of these interfaces, not necessarily at all, in agreement with the geometric tolerances specified by ISO2768cK. Three main configurations were studied for the positioning mechanism in this paper.

Firstly, we divided the total parking volume into three sections with different sizes, and the minimal volume was considered our first choice. This volume is shaded light green in Fig. 5. The remaining two sub-volumes are currently occupied by subassemblies planned for other projects. This constraint led to the development of a foldable crane-like mechanism. Next we considered a swing arm analogous to that employed for the prime Gregorian foci of the Large Binocular Telescope (LBT). This would be a more stiff solution and with comparable good performance of the existing structures on the LBT. Unfortunately, a simple rescaled version does not completely fit inside the red and green area in Fig. 5. A third option is based on a lightweight structure with just two motions. It contains fewer parts, utilizes the available overall height of the available volume of 2800 mm and needs just two actuators to reach the focus: a screw jack and a telescopic bar to lift the polarimeter up towards M4. For this configuration, we performed a static, a kinematic and a dynamic analysis. A design decision can only be made once the available parking volume has been finalized.

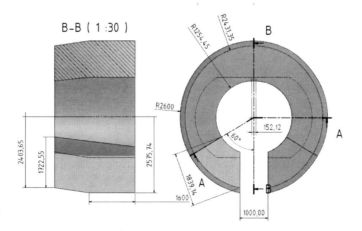

Figure 5. Subdivision of the available parking volume from Fig. 2 into three sections: a minimal one for which no other reservations or claims are foreseen from any other proposed project (green or "brightest" section); a blue one symmetric to the green one but on the other side of the telescope light path and also free of claims; and a red volume extending for 240° currently reserved for other facilities, most notably the AO calibration unit.

3.2 A crane-like arm

The crane-like mechanism (Fig. 6) includes ten movable bodies and permits an unfolding motion plus a turning around the (vertical) z axis, using up two vertical mechanical interfaces in the AO tower. During the unfolding procedure the instrument is lifted up via two consecutive prismatic joints designed as telescopic bars. The external constraints consist of a lower universal joint with two degrees of freedom (DOF) around z and x, and a revolute joint on the top. These joints are attached to the external frame. A motor would be installed in a fixed position on the external frame, on-axis with the revolute joint, and it drives (actuates) the turning bar and moves the instrument around z. In order to determine the DOFs, the system can be reduced to a 2D mechanism with totally ten bodies, including the fixed frame.

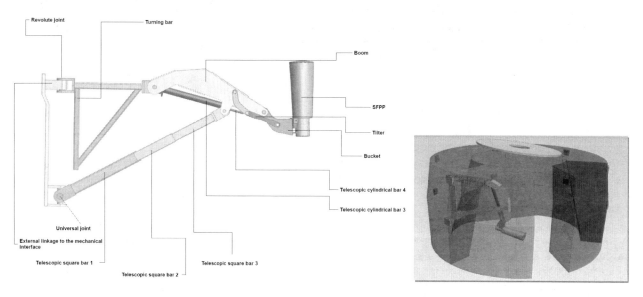

Figure 6. **Left**: Layout of the crane mechanism. The ten constitutive bodies can be distinguished with a total of three DOFs (two lifting-up motions via telescopic bar 2 and 3, and one tilting motion via the telescopic cylindrical bar). **Right**: The arm and the polarimeter (SFPP) in parking position. It fits within the small "light green" volume currently not occupied by any other request. The tilted M4 mirror is indicated on the top.

3.3 A swing arm

For the swing arm design (Fig. 7), we took one of the existing swing arms mounted on the LBT as a reference. Such a system needs seven revolute joints as external constraints mounted on plates on the tower truss. For these, it is necessary to implement an additional supporting structure (tbd). Figure 7 shows the main elements of the mechanism. It is possible to determine the DOF's and simplifying it to a 2D case. The articulation H together with the joint plate AB welded on the instrument holding frame can be considered a single body. Adding up all the linkages on the upper plane, i.e. the intermediate arm, an articulation brace, the support D, and the connecting joints, there are in total seven bodies. Of these, five are revolute joints of multiplicity 1, two of multiplicity 2 as they are connecting three links and two lower pairs on the guiding rod and the screw jack joint, meaning that it provides rotation and sliding at the same time.

The seven revolute joints are located on two parallel planes orthogonal to the optical axis at a distance of 770 mm and distributed on two circles of radii 4860 and 4600 mm extending over an angle of 72° (occupying almost half of the red zone and partially the green one during parking). Such a mechanism must ensure the rotation of the holding braces around z, which size would be L×W×H=1750×80×160 mm. This is accomplished via a jack screw installed on the fork beam and coupled to the motor box with pivot around the (vertical) z axis. In parking mode the swing arm is inside the allowed zone, while the long screw and the articulation braces are leaning into the service subvolume (see Fig. 8). In order to rotate the instrument frame by the required 43°.5, it is necessary to have a linear displacement of the screw jack of 820 mm. However, in operations mode one guiding rod is leaning out of the parking volume by 96 mm (tbc).

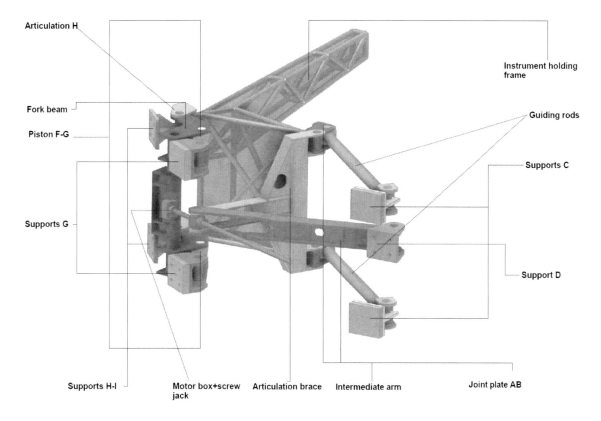

Figure 7. 3D model of the swing-arm solution with supports to the tower structure at points C, D, G, and H-I as indicated. The jack screw actuator is coupled to a motor box and would enables rotation around the vertical z axis.

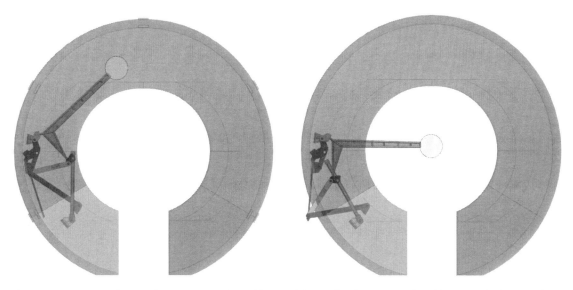

Figure 8. **Left**: Top view of the swing arm in parking position. All substructures fit within the principle parking space. **Right**: Top view of the swing arm when the polarimeter is in operating position.

3.4 A screw jack actuator with telescopic bar

Such a mechanism is simpler than the previous ones and can be summarized as a combination of two dyads connected by a hinge (Fig. 9). It comprises six bodies with in total two degrees of freedom. The f/4.4 focus position is approached by 1) a rotation around the revolute joints A located on the bottom plane of the available space at an angular distance of 33.6° and a linear distance of 1400 mm; and by 2) a translation by means of a telescopic arm contained in the hollow bar of the main frame (for details see Fig. 9). Two guiding rods at both sides, as for the swing-arm case, contribute to a more uniform motion.

Interpreting the system as a planar mechanism, it can be noted that there are effectively four bodies. As the guiding rods are only auxiliary they can be omitted. There are three revolute joints, two external and one articulated joint between the screw jack and the frame; and two sliders (screw and telescopic bar). Therefore, the total DOF is two.

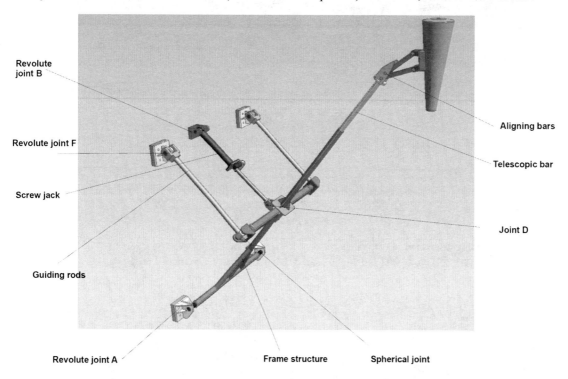

Figure 9. CAD model of the screw-jack mechanism. The mechanism is mounted via five interfaces (joints A, B and F) to a supporting substructure (not shown) and has only two degrees of freedom.

4. A STOKES-VECTOR SPECTROPOLARIMETER

4.1 Constraints due to acquisition & guiding and wave-front sensing for M1

The polarimeter is not required to be operated with laser guide stars, only with natural guide stars. Therefore, the f/4.4 focus position remains as for an infinitely distant target, as shown in Fig. 4. The deformable M4 mirror is only needed in a stiff position and only to reflect off-axis light >2.5 arcmin away from the optical axis towards the wave-front sensors. The polarimeter concept is then constrained by the requirement that the initial target acquisition (and possibly guiding) is done through the guide probes in the ⌀5-10 arcmin ring FOV in one of the two Nasmyth foci (which one is tbd). For a complete unvignetted operation of all three wave-front sensors simultaneously, the enclosure dimension of the polarimeter has to be at most the linear size of that ring volume at the location of the intermediate focus (see Sect. 2.3).

This volume is shown in Fig. 4 and indicated there with four dots. Note that no additional image derotation is foreseen for the polarimeter because all polarimetric targets are observed on the telescope optical axis. We proceed with a science FOV for the polarimeter of 1.3 arcsec and try to fit the instrument diameter, its length, and the distance to the telescope focus into the free inner FOV. This comes down to designing a collimator system that matches these dimensions.

4.2 A "Schwarzschild" collimator system

The collimator converts the divergent f/4.4 telescope beam into a parallel 15-mm-diameter beam with an intermediate pupil in a location where the polarimetric optical components can be easily positioned. Larger beam sizes are optically more forgiving but the more problematic will be the production of homogeneous calcite with birefringent behavior. We set the collimated-beam diameter proportional to the expected maximum size of a calcite crystal likely produce able in the near future (see Sect. 4.3). This diameter then sets the maximum ray angles for the collimator.

The overall design is an inverted Schwarzschild system solely based on two reflections (Fig. 10). It has the big advantage over a refractive system that the cross talk due to variable birefringency introduced by residual thermal and mechanical stresses during transmission is minimized. We refer to DiVarano et al. (2011) for the design of the refractory collimator system of PEPSI, the polarimetric focal station of the Large Binocular Telescope. This will lead to a minimized net instrumental polarization. Another advantage is that it is free of chromatic errors and thus useable for optical and NIR spectrographs alike. The disadvantage is that the incidence angles become very large, up to 8 degrees for the f/4.4 case, and the available FOV very small (1.3 arcsec in our case). The latter is an advantage from the polarimetric point of view because it minimizes the contribution from the sky polarization.

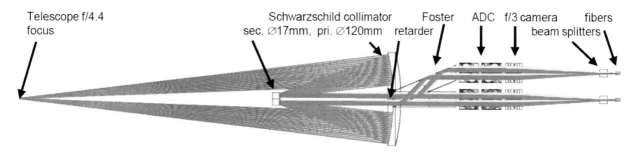

Figure 10. Optical design of the dual-beam high-precision polarimeter proposed for the E-ELT. The telescope's f/4.4 focus is indicated to the very left. The fiber entrance for one ordinary and one extra-ordinary beam for one spectrograph is shown at the very right. The collimated beam diameter is 15mm. The reflective collimator system has a primary mirror of diameter 120mm (middle). The system's total length from the telescope focus to the fiber entrance is 800mm, which fits the requirements in Fig. 4.

4.3 The linear-polarizer "Foster" unit

This unit is a scaled version of the same component employed for the PEPSI polarimeter for the Large Binocular Telescope (Strassmeier et al. 2008, Ilyin et al. 2011) and similar to the one employed for the HARPS polarimeter at ESO's 3.6m telescope (Snik et al. 2008, 2011).

Figure 11, top panel, shows the conceptual optical layout of the respective unit for the E-ELT. Two calcite crystals of dimensions 17×17×50mm split the incoming (retarded) beam into an ordinary and an extra-ordinary beam orthogonally linearly polarized with respect to each other. A third prism, made of e.g. standard Schott Bk-7, redirects the ordinary beam such that it exits the optical assembly parallel to the extra-ordinary beam. Their physical separation is 34mm. This small separation is required to minimize the overall dimensions of the polarimeter due to the stringent space constraints.

The interface surfaces between the individual prisms must be immersed and their separation optimized so that any residual fringing itself is minimized. The entrance surfaces would be without AR coating while the exit surfaces would carry an AR coating. Ilyin (2012) presented a detailed error propagation of the configuration in Fig. 11 and concluded on an expected cross-talk between the two polarized beams of about 0.02% (i.e. 1/5000) at 900 nm, and less than that for the NIR wavelengths, but with a maximum of 0.2% (=1/500) in the blue at 400nm.

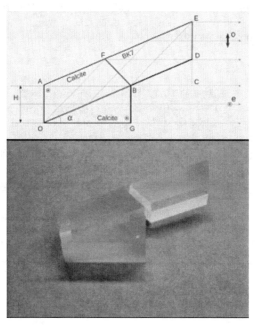

Figure 11. The linear-polarizer "Foster" beam-splitter unit. **Top**: Schematic design. The beam diameter H is set to 15 mm. **Bottom**: Prototype with two 14×14 mm calcite prisms and one BK7 transfer prism. For a successful realization of this concept, a calcite block of up to 17×17 mm is needed, possibly 20mm. Such a crystal dimension is currently not available.

4.4 The retardation unit

The retardation unit consists of a rotatable polymethylmethacrylate (PMMA) super-achromatic quarter-waveplate retarder (Samoylov et al. 2004). These waveplates are commercially available for various wavelength ranges. In the present concept, we require a single elliptical retarder for the entire wavelength range covered by a (future) spectrograph or a combination of two spectrographs, i.e. from 380 to 1600 nm, possibly to 2500 nm. A single retarder is desired in order to minimize the number of moving parts and due to the space constraints within the collimator system. The retardation accuracy is standard, $\pm \lambda/100$, the wave-front deformation should be < 0.5 wave/cm at 633 nm and the beam deviation shall be < 5 arcsec.

Note that instead of using classic half-wave retarders for linear polarization, we would rotate the entire polarimeter package (including the collimator) in steps of 45°, i.e. obtain linear polarization without the retarding element in the optical path (and thus no direct circular-linear cross talk). This requires that the retarder element can be moved out of the optical path, which is a challenging task for the current design and no solution is suggested at the moment.

4.5 Atmospheric dispersion correction

The telescope's own atmospheric dispersion corrector (ADC) is located on top of the AO tower optically before the M3 mirror and physically on the backside of the M4 mirror support. It is retractable and must be retracted for polarimetric use of the telescope because its counter-rotating prisms would introduce instrumental polarization larger by a factor 1000 or so compared to the signals to be measured. Its crosstalk can not be calibrated accurately enough in practice when it comes down to the $\delta P/P=10^{-5}$ level (P, degree of polarization).

Therefore, our current design adds an ADC behind the linear polarizer but still within the collimated beam, as shown in Fig. 10. Due to the extreme pupil ratio (39000/15=2600) the atmospheric dispersion will be magnified proportionally. The extreme angles in the ADC can be partly compensated for by choosing high refraction-index glasses but beyond a zenith distance of approximately 45°, we expect an increasing vignette due to the size of the Foster unit. We note that enlarging the entrance size of the Foster to 20mm (instead of 17mm) does not relief the situation significantly (tbc).

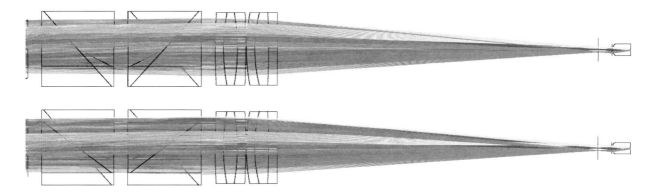

Figure 12. Design of the ADC and fiber coupling. The collimated beam diameter exiting the Foster unit (very left; not shown) is 15mm. The top beam is the ordinary beam, the bottom beam is the extraordinary beam. A Risley-combination of two counter rotating prisms per beam make up the ADC (the two left box-like units). The camera optics are a six-lens system in two groups (middle) with an f-ratio of f/5 and an effective focal length of $F=125$ mm. An aluminized and tilted aperture stop in front of the microlens (right) removes as much sky straylight as possible and redirects light within a FOV of 5 arcsec to a viewing camera (not shown). The "microlens" ($\varnothing$ 4 mm) is laid out for a fiber injection at f/3 into the fiber core of $\varnothing$ 600µm.

4.6 Camera and fiber injection

The camera optics and the fiber microlens injection are shown in Fig. 12 for both polarimetric beams. The magnification factor of the camera optics with $F1 = 125$mm and the 4 mm fiber core "microlens" with $F2 = 10.7$mm is $F1/F2 = 11.7$. The allowed lateral displacement of the pupil image from the Foster prism onto the fiber core is to be at most 1% of the fiber core, i.e. a mere 8 µm. It corresponds to a maximal lateral shift of the parallel beam from the calcite-prism axis of 94 µm. This defines the internal mechanical stability of the Foster prism components and their mounts versus temperature and gravitational stresses.

4.7 Preliminary sensitivity analysis

In order to determine the required mechanical positioning accuracy of the polarimeter, we first employ a tolerance analysis in Zemax for a wavelength of 550 nm and the telescope in zenith pointing. Preset displacement values were Δx, Δy, Δz and tilt angles $\Delta\alpha$, $\Delta\beta$, where z is the direction of the telescope optical axis and α and β the angles along the perpendicular x and the y axis. The summary of this preliminary analysis is the following:

- Defocusing (Δz) is not critical. The anticipated range for initial positioning is: -3 mm $\leq \Delta z \leq +3$ mm.
- Decentering the instrument optical axis from the telescope optical axis in both directions with $\Delta x = \Delta y = 0.1$ mm is already significant, and if it is 0.3 mm the light beam would already fall out of the fiber core. Such displacements may occur only on a single axis, e.g. $\Delta y = 0.1$ mm and $\Delta x = 0$, but still lead to a displacement radius at the fiber entrance of 380 µm. Obviously, these numbers require an internal target acquisition (and possibly guiding) system based on two fiber-viewing CCD cameras for each of the polarimetric subbeams.
- Imminent tilt angles of the instrument optical axis ($\Delta\alpha$, $\Delta\beta$) of 1 arcmin (0.017°) are insignificant.

We considered eight test configurations relative to ADC rotations and with respect to zenith angles of 0-15-25-35-45-55-65°. Relative tilt is defined over the full length of the instrument of 800 mm, meaning that 1 arcmin of tilt is equivalent to 232 µm at the end of the polarimeter.

The polarimeter structure in itself has rather stringent mechanical tolerances: the first one comprises the alignment of the collimator M1 and M2 with defocusing/de-center errors no larger than ±2-5µm; the second one comprises the Foster prisms plus the ADC doublets with allowed displacements of ±0.1 mm; a third tolerance includes the λ/4 plates and the microlenses with an allowable mechanical tolerance, mounting plus preload, of the order of ±10 µm.

Such tolerance can be achieved using a tube structure with a top ring, a tube, and a mirror cell made of the same material, either Super Invar or Kovar, or carbon-fiber-reinforced polymer (CFRP) with layers orientation optimized to have a CTE very close to that of Zerodur. This would keep the collimator M2 permanently aligned with the collimator M1. A tube of outer diameter 144 mm, 2 mm thickness, and 192 mm length with four molded spiders to support the M2 mirror could be one solution. On the M2 support structure, three thin flexure arms (0.4 mm) are mounted via three tiny M2.5 screws. The M2 mirror would be properly prepared and glued to the inner faces of the flexures arms (see detail in Fig. 13, top left). The primary mirror would be foreseen to be radially supported via three braces of inverted V-shape on a cell plate that is dismountable by three screws.

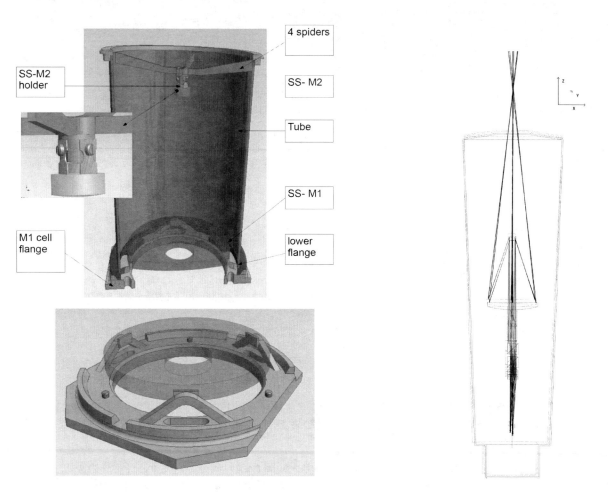

Figure 13. **Top left:** Cut view of the tube structure with a zoomed-in detail on the three flexures belonging to the secondary mirror holder. The entire structure would be either made of Invar/Kovar or CFRP. **Bottom left**: primary mirror cell with V-shaped flexures, with four tapered profiles to match the lower flange interface, mounted via 3 M5×8 screws. **Right**: Sketch of the light beam and the mechanical frame rotational translation when the instrument undergoes a tilt in y of 0.6 arcmin and a defocus $\Delta z = +3$ mm.

REFERENCES

[1] Keller, C. U., "Instrumentation for astrophysical spectropolarimetry", in: J. Trujillo-Bueno, F. Moreno-Insertis, F. Sanchez (eds.), Proc. of Astrophysical Spectropolarimetry, Cambridge Univ. Press, 303 (2002).

[2] Baade, D., Wang, L., Hubrig, S., Patat, F., "The Scientific Requirements for Extremely Large Telescopes", IAU Symp. 232, Cambridge Univ. Press, 248 (2006).

[3] Strassmeier, K. G., Ilyin, I. V., "The E-ELT: A Chance to Measure Cosmic Magnetic Fields", Astrophysics and Space Science Proceedings, Volume ISBN 978-1-4020-9189-6, Springer Netherlands, 255 (2009).

[4] Strassmeier, K. G., "Observational MHD: Advances in Spectropolarimetry and the Prospects for the E-ELT", Proc. IAU Symp. 274, Bonanno, A. et al. (eds.), Cambridge Univ. Press, 274 (2011).

[5] DiVarano, I., Strassmeier, K. G., Ilyin, I., Woche, M., Kärcher, H. J., "Integration of a Thermo-Structural Analysis with an Optical Model for the PEPSI Polarimeter", SPIE 8336, 31 (2010).

[6] Ilyin, I., "Second-order error propagation in the Mueller matrix of a spectropolarimeter", AN 333, 213 (2012)

[7] Strassmeier, K. G., Woche, M., Ilyin, I., Popow, E., Bauer S.-M., Dionies, F., Fechner, T., Weber, M., Hofmann, A., Storm, J., Materne, R., Bittner, W., Bartus, J., Granzer, T., Denker, C., Carroll, T., Kopf, M., DiVarano, I., Beckert, E., Lesser, M., "PEPSI: The Potsdam Echelle Polarimetric and Spectroscopic Instrument for the Large Binocular Telescope", in Proc. of SPIE 7014, 21 (2008).

[8] Ilyin, I., Strassmeier, K. G., Woche, M., Dionies, F., Di Varano, I., "On the design of the PEPSI spectropolarimeter for the LBT", AN 332, 753 (2011).

[9] Snik, F., Kochukhov, O., Piskunov, N., Rodenhuis, M., Jeffers, S., Keller, C., Dolgopolov, A., Stempels, E., Makaganiuk, V., Valenti, J., Johns-Krull, C., "The HARPS Polarimeter", in Proc. on Solar Polarization 6, J. R. Kuhn et al. (eds.), San Francisco: ASP, 237 (2011).

[10] Snik, F., Jeffers, S., Keller, C., Piskunov, N., Kochukhov, O., Valenti, J., Johns-Krull, C., "The upgrade of HARPS to a full-Stokes high-resolution spectropolarimeter", SPIE 7014, 22 (2008).

[11] Samoylov, A.V., Samoylov, V.S, Vidmachenko, A.P., Perekhod, A.V., "Achromatic and super-achromatic zero-order waveplates", Journal of Quantitative Spectroscopy & Radiative Transfer 88, 319 (2004).

Performance of industrial scale production of ZERODUR® mirrors with diameter of 1.5 m proves readiness for the ELT M1 segments.

Thomas Westerhoff, Peter Hartmann, Ralf Jedamzik, Alexander Werz

SCHOTT AG, Hattenbergstrasse 10, D-55122 Mainz, Germany

ABSTRACT

The two Extremely Large Telescopes under discussion, the Thirty Meter Telescope (TMT) and the European Extremely Large Telescope (E-ELT), will use a multitude of hexagonal shaped mirror segments to achieve the large aperture of 30 m and 39 m, respectively. The proper functionality of both telescopes will relay on the reproducibility of specified material properties material of the individual segments.
SCHOTT has a well proven experience in production of mirror substrates for segmented telescopes. Today five of the world's six segmented telescopes are using ZERODUR® as mirror substrate material. SCHOTT had delivered roughly 250 substrates between 1 m and 1.9 m for those segmented telescopes. Since 2003 SCHOTT delivered more than 260 mirrors of 1.5 m in diameter for an industrial application not related to astronomy. In this paper some achievements for the astronomical telescope segments are reported together with those on the serial production for the industrial application. This presentation includes data on the fulfillment of the CTE specification and stress birefringence and achieved tolerances on surface figure and flatness. The data to be presented will demonstrate the excellent reproducibility of ZERODUR®'s material properties and its manufacturing process. The production capabilities at SCHOTT for the successful delivery in time of the multitude of ZERODUR® segments are presented and discussed. They will demonstrate that ZERODUR® is well prepared for the demands of industrial scale production for the two large segmented ELT's in quality, quantity, and in the requested time period.

Keywords: ZERODUR®, Extremely Large Telescopes, coefficient of thermal expansion, coefficient of thermal expansion homogeneity, 1.5 m mirror substrates, serial production, ELT M1 segments.

1. INTRODUCTION

For more than 40 years ZERODUR® of SCHOTT has been successfully used in many challenging application in astronomy [1]. The material was chosen due to its extremely low coefficient of thermal expansion. The most prominent achievement was the delivery of the world largest monolithic mirror substrates of 8.2 m in diameter for the Very Large Telescope (VLT) of the European Southern Observatory (ESO) on Paranal in Chile [2]. The technology of segmented telescopes will outpace this performance by far.

Project	Mirror Diameter	Location	Year	Number of Segments	Dimension [mm]
KECK I	10 m	Hawaii / USA	1990	43	Ø 1900 x 77
KECK II	10 m	Hawaii / USA	1993	42	Ø 1900 x 76
Hobby Eberle	9,5 m	Texas / USA	1995	96	Ø 1170 x 56
GTC	10,4 m	Canarian Island / ESP	2003	42	Ø 1860 x 84
LAMOST	6 m	China	2003	40	Ø 1100 x 55

Table 1: SCHOTT ZERODUR® deliveries for segmented telescope

The two segmented Extremely Large Telescopes (ELTs) under discussion intend to use several hundred of hexagonal shaped mirrors to build the extreme large primary mirror apertures of larger than 30 m. The TMT is planning to utilize 492 segments to yield a primary mirror diameter of 30 m. A number of 82 spare segments are also requested in order to achieve a continuous recoating of the segments without impact on the planned observation up time. The primary mirror of the E-ELT comprises of 798 segments and 133 spares. The crucial requirement for the ELT segments is the very low coefficient of thermal expansion and the homogeneity of material properties throughout the segment volume. Additionally an exceptional low amount of bubble and inclusions needs to be achieved in the critical volume of the mirror substrates. This excellent quality level needs to be reproduced in every segment. The polishing of hundreds of segments with a diameter of 1.4 m in 4 - 5 years requires a very effective and robust polishing process. Segment material properties need to be identical from segment to segment in order to avoid polishing process parameter adjustments causing time and budget constraints endangering the success of the telescope. With its history of successful deliveries of more than 250 segments for 5 of the world's 6 segmented telescopes SCHOTT is the only company proven the capability of supplying mirror blanks with excellent reproducibility of material properties for a longer period of time.

ZERODUR® is not only the material of choice for astronomical telescopes mirror. It is also widely used in a many challenging industrial application in which high precision is required. One example is the usage of ZERODUR® as mirror substrate material in mask alignment systems of the LCD Lithography industry in flat screen production. SCHOTT has delivered more than 260 mirrors of 1.5 m in diameter for his application and gained many valuable experiences in realizing this serial production over several years.

This paper is presenting some of the achievements and improvements in material properties for the segmented telescope substrate deliveries in chapter 2. In chapter 3 the serial production for the industrial application is outlined and results are presented. The subsequent chapter will outline the intended production sequence for ELT mirror segment production at SCHOTT followed by a summary and outlook.

2. REPRODUCIBILITY AND IMPROVEMENT OF MATERIAL PROPERTIES ACHIEVED FOR SEGMENTED TELESCOPE SUBSTRATES

With the introduction of telescopes using primary mirrors consisting of segments manufacturing requirements changed significantly. Single mirror production allowed paying special attention to the individual properties and also possibly existing imperfections of given raw material blanks. Such process is not possible for mirrors to be built from segments. This holds especially when numbers needed are as high as those for the extremely large telescopes being planned now. Adjustment of polishing processes to the very quality of a given blank or even correction methods compensating problematic properties were acceptable when all effort concentrated on to a single precious large mirror blank when the main value of this blank was being existent in one large piece at all. Now large mirrors of leading edge quality will be possible only, when a large number of medium sized segments can be processed to achieve highest surface figure and roughness quality. The large number requires polishing processes capable of delivering in spec segments in short time without the need of individual adjustments to varying blank quality. Varying quality would require painstaking inspection of each blank, documentation of the results and feeding them into the process for compensation. Such procedure will easily become a threat for the total telescope project since it would lead to time and cost increase not just by per cent but by factors. Thus, the need is not only for high quality of single blanks. It is also inevitable to realize a very reliable high reproducibility of this quality.

SCHOTT has enabled technical optics and optical design with all its inestimable consequence for general technology not only by introducing new optical glasses in the 1880s but also by making their properties and quality highly reproducible thus removing random variations as main influence on the quality of optical systems. ZERODUR® from its very beginning was subject to efforts to master it highly reproducibly due to the company's philosophy. Since its production is a variant of the optical glass production it also benefitted from a rich heritage of melting, casting and annealing experience.

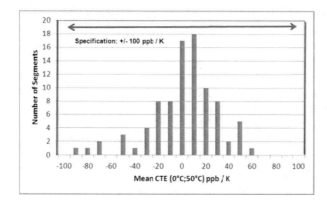 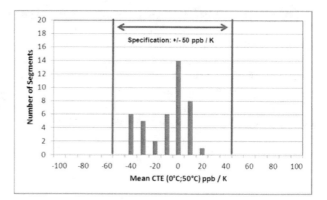

Figure 1: Absolute CTE values achieved for 1.9 m segments for KECK and for GTC ten years later.

The outstanding reproducibility was proven for the first time with a large data basis with the segmented telescopes KECK I and II, which asked for this high reproducibility in large castings of 2 m diameter within a set of more than 80 blanks in total. Also the x-ray telescope CHANDRA challenged reproducibility utilizing 8 large cylinders for Wolter mirrors in orbit since 13 years with another 16 spares still on the ground. Later almost 100 1 m segments for the Hobby-Eberley-telescope followed as well as 42 2 m for the 10 m GRANTECAN telescope, which is similar to the KECK telescopes and 40 1 m blanks for the LAMOST telescope.

Figure 1 displays the absolute CTE values of segments with an edge to edge dimension of 1.9 m achieved for the 84 blanks of KECK and 10 years later for 36 blanks for the Gran Telescopio Canarias (GTC). The data represent the excellent quality of ZERODUR® and clearly reflect improvements in full filling the twice as tight tolerance for the CTE value of the GTC.

Even more impressive is the improvement achieved for the CTE homogeneity in ZERODUR® blanks for segmented telescopes. Figure 2 shows the peak to valley homogeneity measured on the KECK and GTC segments together with the results of investigations of 36 segments with edge to edge dimension of 1.5 m performed in the last 5 years.

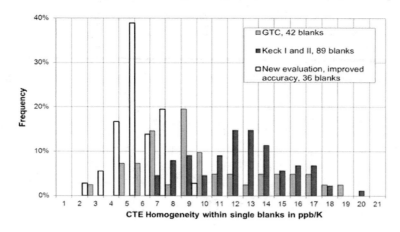

Figure 2: Peak to valley CTE homogeneity of 1.9 m segments for KECK and GTC and 36 1.5 m segments.

The CTE homogeneity is excellent full filling the tight tolerance of 20 ppb / K for the KECK and the GTC. The centre of the value distribution has shifted to smaller values already for the GTC segments in comparison to the KECK segments. Improvements are much more obvious for the data of the latest investigation full filling the very tight tolerance of 10 ppb / K of the E-ELT. The latest investigation was performed utilizing the improved dilatometer accuracy with a reproducibility of ± 1.2 ppb / K whereas the measurements on the KECK and GTC segments were realized on the older dilatometer with an reproducibility of only ± 5 ppb / K. There is good chance that also the segments for KECK and GTC had similar small values for the CTE homogeneity which could not be detected due to the noise of the measurements.

More details on CTE homogeneity and variations of the blanks for KECK and GTC and have already been published [3, 4, 5, 6]. In 2009 SCHOTT presented an intensive investigation on CTE homogeneity gradients in ZERODUR® blanks for segmented telescopes. The results did not show any systematic axial or radial gradients of the excellent level of CTE homogeneity in every investigated blank [7].

Next to the CTE the bulk stress of the blanks measured via birefringence is of interest for the challenging polishing of that many segments. Computer controlled polishing and stress mirror polishing are two options[8, 9, 10]. In both options the blanks will first be polished in disk shape and then cut into hexagonal shape. A low and reproducible level of bulk stress of the blanks is essential for his procedure to avoid irregular distortion of the blanks under cutting. Too high stress levels will change the optical figure away from the entrance specification for the final polishing after fixing the blanks on the support mounts.

Figure 3 demonstrates the progress in bulk stress mastering. The already very low values of the KECK segments have been even undercut by the results for the GTC segments. The maximum bulk stress value of the latter set being produced about a decade later is 3 nm/cm. This is equivalent to 0.1 MPa mechanical stress, which is a very low value. The bulk stress of all segments is fairly below the specification limit of 10 nm/cm, which was valid for both projects.

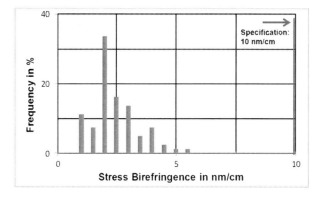
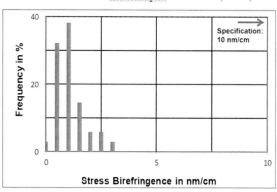

Figure 3: ZERODUR® stress birefringence variation in a set of more than 80 disks of 1.9 m diameter as used for the KECK telescopes (left) and in a set of 42 disks of 1.9 m diameter for the GTC (right).

Data presented in figure 4 provide some more evidence for the high reproducibility of ZERODUR® properties and quality. Those data have been measured for every segment delivered to the KECK I and KECK II telescope produced in many different production runs during a time period of almost 5 years. They reflect the reproducibility of material properties for many individual batches.

Figure 4 left shows the statistical variation of the density. One standard deviation amounts to 0.05% of the mean value. The width of the x – axis in diagrams of figure reflects an interval of ± 2 % of the mean value. Figure 4 centre and right show the statistical variations of Young's modulus and Poisson's ratio observed for the KECK segments. Their higher standard deviation compared with that of density comes from the lower measurement accuracy for these quantities. The single outstanding bar at 90.0 GPa in figure 4 centre is due to the fact that for KECK I measurement data were given only in 0.5 GPa intervals.

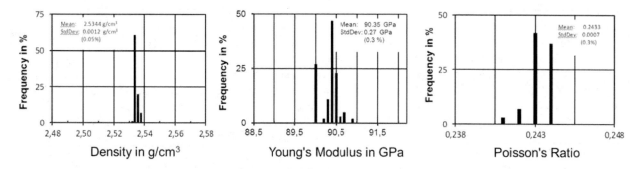

Figure 4: Density, Youngs's Modulus and Poisson's Ratio as determined for the KECK segments.

Such high reproducibility must not be taken as granted. It is the result of many years of systematic development and the wide experience gained thereof. For economic reasons segments for ELTs must be cast in multiple thicknesses. With producing large glass and glass ceramic pieces difficulties grow with thickness square or an even higher exponent. Up to now only SCHOTT has proven to be capable of delivering highly homogeneous zero expansion glass ceramic with highest reproducibility for many hundred blanks in the range between 1 and 2 m diameter.

3. SERIAL PRODUCTION OF ZERODUR® MIRRORS 1.5 M IN DIAMETER

ZERODUR® is not only the material of choice for astronomical telescopes mirror. It is also widely used in a many challenging industrial application in which high precision is required. One example is the flat screen production in which ZERODUR® is used as mirror substrate material in mask aligners in the LCD Lithography process. One of the optical designs for LCD Lithography is using a ZERODUR® mirror of a diameter of 1.5 m and a thickness of 340 mm. The raw material casting for this mirror is of 350 mm in height, which would yield six segments for the ELT's after appropriate cutting. Since 2003 SCHOTT delivered more than 250 of those large mirrors with an equivalent of more than 1500 ELT segments (refer to figure 5).

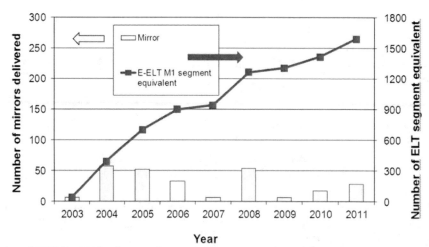

Figure 5: Deliveries of ZERODUR® mirrors of 1.5 m in diameter and resulting ELT segment equivalents.

The yearly quantity of delivered mirrors depends on the investments cycles of the LCD industry. The large amount of mirrors delivered in 2004 to 2006 is reflecting the steep ramp up of flat screen production after market introduction. To increase productivity larger glass sheet sizes have been introduced into volume production requiring lithography tools equipped with larger optic. Today flat screens have completely pushed aside the Cathode Ray Tube screens. The future investments of the LCD industry will have to be adjusted to the average market growth for television and computer screens which was reported in the single digit percentage region as long time average. For the future it is expected that deliveries for LCD Lithography tools will come back in low quantities not interfering with potential deliveries for the ELT segments.

As an example for the reproducibility and robustness of the production process achievements in specified geometrical parameters for those mirrors are commented in the following. Four 5 – axis CNC machines with a rotary table of 2 m in diameter have been used for grinding the mirrors. A 3D coordinate measurement machine (CMM) with a free workspace of 1.8 m has been used to measure the dimensions achieved through grinding. The mirrors of 1.5 m in diameter have a specified flatness on the rear surface of 200 µm. The concave curvature of the reflecting front surface is specified to 100 µm deviation from the theoretical figure with a radius of curvature of 2500 mm. Figure 6 displays the distribution of values of flatness (right) and curvature (left) achieved during the production period of in the meantime 8 consecutive years. The curvature tolerance is evaluated as shape fit versus the specified radius and as best fit through the data points measured with a 3D CMM.

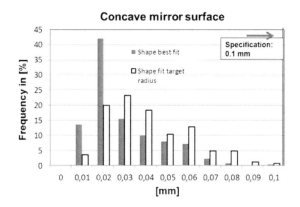
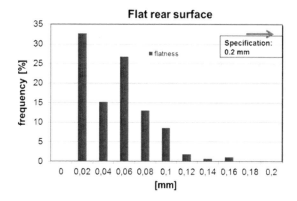

Figure 6: Achievements for concave surface tolerance and flatness tolerance for mirrors 1.5 m in diameter

The mean value of the distribution for both features is well below the specified value and reflects an excellent performance in machining of such large mirrors over such a long time. A state of the art CAD/CAM process chain is available and secures specified quality from construction, CNC programming, machining and testing.

Taking the 250 mirrors delivered for LCD Lithography and their potential cutting into 6 ELT segments each into account SCHOTT has already delivered the material for the E-ELT and the Thirty Meter Telescope (TMT) in the last years. With the data presented SCHOTT proved the availability of a robust industrial serial production for such mirrors. The installed ZERODUR® melting capacity at SCHOTT is sufficient even for a parallel delivery of segments to E-ELT and TMT.

4. ELT MIRROR SUBSTRATE PRODUCTION

Based on the successful deliveries for the segmented telescope and the LCD Lithography SCHOTT has gained sustainable experiences in handling mirrors of 1.5 m in diameter. Nevertheless for the production of on segment per day the logistics have to be carefully planned in order to full fill the delivery requirements of the E-ELT. The current specification for the E-ELT segments foresees Hydro Fluoride (HF) acid etching of the rear and the circumferential surface. SCHOTT already outlined a preferred production sequence for the E-ELT segment production to incorporate this requirement as shown in figure 7. The melting capacity of the ZERODUR® tank is large enough to be of any concern. More than 100 annealing furnaces are available for coarse annealing and ceramization. As reported above 4 CNC machines with sufficient workspace and capacity are also in place. The CTE measurement capacity can easily be adapted since it is not utilized to its full extend. For blank cutting a state of the art diamond wire saw has been installed in 2011. Next to its utilization of confectioning larger boules for customer order manufacturing it can be used to effectively cut the raw boules into the thinner segments. A recently installed HF acid etching facility for 4 m blank etching will be adapted to the segment etching and has enough capacity to handle the ELT quantities. Several digital control devices are available for intermediate control of achieved dimensions during CNC grinding. The final inspection is foreseen to take place on the 3D CMM described above. Capacity can be easily increased via shift models if necessary. SCHOTT has a large amount of experience for successfully package larger ZERODUR® mirrors and is able to design transport containers which can be used shuttling between SCHOTT and the polishing contractors.

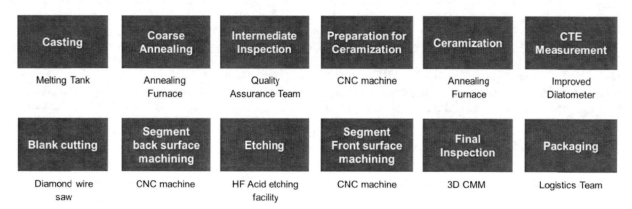

Figure 7: SCHOTT proposed production sequence of E-ELT segments based on current specification.

With all the experience already accumulated SCHOTT is prepared for the ELT segment production [4]. The standard inventory situation will allow early production runs on a short term notice.

5. SUMMARY AND OUTLOOK

ZERODUR® has the material of choice for 5 of the world's 6 segmented telescopes and is granting excellent quality levels for absolute CTE, homogeneity of CTE, bulk stress and internal quality inevitable for ELT segment production. The reproducibility of those properties has been proven with > 250 mirror blanks for segmented telescopes and with additional investigations in the last 5 years. Material properties, as density, Young's Modulus, and Poisson number have been proven to be independent over many different ZERODUR® batches. Geometrical parameters achieved with CNC grinding on mirrors of 1.5 m in diameter for LCD Lithography tools are much better than the specification values. Achieved deliveries of > 250 of such mirrors are equivalent to > 1500 ELT segments; in other words SCHOTT has already delivered the material quantity required by the E-ELT and TMT in total. The equipment covering the full process chain for the ELT segment production is already in place and successfully utilized at SCHOTT. SCHOTT ZERODUR® is ready to start ELT segment substrate material production immediately.

6. REFERENCES

[1] Döhring, T., Jedamzik, R., Thomas, A., Hartmann, P., "Forty Years of ZERODUR® mirror substrates for astronomy; review and outlook" Proc. SPIE Vol. 7018, 70180P (2008)

[2] Morian, H., Mackh, R., Müller, R., Höness, H., "Performance of the four 8.2 m ZERODUR® mirror blanks for the ESO/VLT" Proc. SPIE 2871, 405 (1997)

[3] Jedamzik, R., Müller, R., Hartmann, P., "Homogeneity of the linear thermal expansion coefficient of ZERODUR® measured with improved accuracy" Proc. SPIE Vol. 6273 (2006)

[4] Morian, H., Hartmann, P., Jedamzik, R., Hoeness, H., "ZERODUR® for Large Segmented Telescopes" Proc. of SPIE Vol. 4837, 805 (2003)

[5] Jedamzik, R., Döhring, T., Müller, R., Hartmann, P., "Homogeneity of the coefficient of linear thermal expansion of ZERODUR®" Proc. of SPIE Vol. 5868, 58680S-1 (2005)

[6] Döhring, T., Jedamzik, R., Hartmann, P., Thomas, A., Lentes, F. T., "Properties of ZERODUR® mirror blanks for extremely large telescopes" Proc. SPIE Vol. 6148, 61480G (2006)

[7] Jedamzik, R., Döhring, T., Johansson, T., Hartmann, P., Westerhoff, T., "CTE characterization of ZERODUR® for the ELT century" Proc. SPIE Vol. 7425, (2009)

[8] Sporer, S., "TMT: stressed mirror polishing fixture study" Proc. SPIE Vol. 6267, (2006).

[9] Rodolfo, J., Chouarche, L., Chaussant, G., Harny, A., Carel, J., Pernet, B., Billet, J., Lepan, H., Ruch, E., "Prototypes segment polishing and testing for ELT M1" Proc. SPIE Vol. 8450, 8450-84 (2012)

[10] Hugo, E., Floriot, J., Rousselet, N., Laslandes, M., Ferrari, M., Lemaitre, G., "Stress polishing of segments for future extremely large telescopes: results obtained on a full scale demonstrator" Proc. SPIE Vol. 8450, 8450-84 (2012)

E-ELT project: Geotechnical investigation at Cerro Armazones.

Paolo Ghiretti [a], Volker Heinz [a], Daniela Pollak [b], Jose Lagos [b]

[a] ESO, Karl Schwarzschild Strasse 2, D85748, Garching bei München, Germany
[b] ARCADIS Chile S.A., Antonio Vargas 621, Provencia, CP 7500966, Santiago, Chile

ABSTRACT

The design and construction of large telescopes include significant geotechnical challenges. In order to guarantee reliable and stable operations, a giant telescope like the European – Extremely Large Telescope (E-ELT) requires a foundation performance according to the level of accuracy of the other telescope's components. This paper describes the main geological and geotechnical activities conducted on site along with the studies completed in specialized geotechnical laboratories with the objective to achieve a thorough characterization of the ground conditions. This study shows that, the properties of the foundation materials are appropriate to guarantee a good performance of the E-ELT.

1. INTRODUCTION

The geotechnical and geological ground characterization is an imperative task to ensure good design and performance of any building. This is particularly true for highly sensitive structures such as telescopes, as their high-precision components cannot be affected by foundation settlement and/or angular deformations. This is even more critical in highly seismic environments such as in Chile. This paper presents the description of the investigation conducted for the evaluation of the foundation conditions of the European – Extremely Large Telescope (E-ELT), with the objective to obtain the geotechnical and geological properties for design. The E-ELT will be built over a platform on the top of Cerro Armazones, at an elevation of 3.046 m.a.s.l. located in the Second Region of Antofagasta, Chile, at the Coastal Cordillera (*Cordillera de la Costa*).

2. FIELD WORK AND LABORATORY TEST

A thorough field work program was completed for the geological and geotechnical characterization of the E-ELT foundation, including:

- Eight boreholes (vertical and inclined) with lengths between 19 m and 75 m, with continuous sample recovery and core logging and mapping. The selection of the lengths and inclination of boreholes was made considering the elevation of the foundation platform, dip direction of the structures observed in the field and the placement of the foundation of the telescope (see Figure 2-1). The drilling was made almost entirely in rock (soils were found on the surface with less than 2 m of thickness). No water level was encountered during the drilling of any borehole.
- Geophysical survey was completed during and after the boreholes drilling, to supplement the information provided by these. The geophysical tests conducted were:

- o Refraction Seismic and Multichannel Analysis of Surface Waves (MASW) profiles were completed obtaining the compressional (Vp) and shear (Vs) wave velocities,
- o Optical Televiewer technique was also used to obtain orientation of geological structures,
- o Down Hole wave velocity measurements were performed within all the boreholes, obtaining compressive and shear velocity values.

Figure 2-1: Boreholes location on E-ELT Platform

- Laboratory testing were conducted on rock samples taken from drill cores and on soils samples obtained from the ground surface. Rock test as Density, Porosity, Vs-Vp Velocity, UCS, Triaxial and direct shear were developed. Soil laboratory testing were only chemical test (Ph, Organic matter, Total soluble salts, chloride, sulphats and coal-lignite) to establish possible aggressiveness of the ground and its adequacy to be used as rock material for the access road or concrete aggregates. All laboratory testing were performed according to European standards or equivalent.

3. GEOLOGY

3.1 REGIONAL GEOLOGIC AND GEOMORPHOLOGIC INFORMATION

The area under study is located in the Second Region of Antofagasta, Chile, at the Coastal Cordillera (*Cordillera de la Costa*) ant it is constituted by a basement of Mesozoic intrusive rocks dissected by North-South faults and barely covered by Cenozoic sediments and volcanic rocks.

The Cerro Armazones is a residual relief remnant originated by a Jurassic and Cretaceous granitoid basement. It has a relatively circular shape with moderately steep side slopes. The base elevation is approximately 2.850 m.a.s.l. and it's slopes show significant erosion channels that transport the colluvial material from the summit towards the surrounding plains, where they come to rest and form a continuous piedmont around the Cerro Armazones, very typical in the area.

The Cerro Armazones is located inside the Atacama Fault Zone formed by series of very long faults trending N-S, close to the East limit defined by the Quebrada Grande Fault located 3 km to the East of the area being studied. Even though, out of the study area, diverse authors have indicated the existence of very recent fault movements in the Atacama Fault area, these evidences are only observed to the west of the Paposo Fault, while to the east between no evidences have reported (see Figure 3-1).

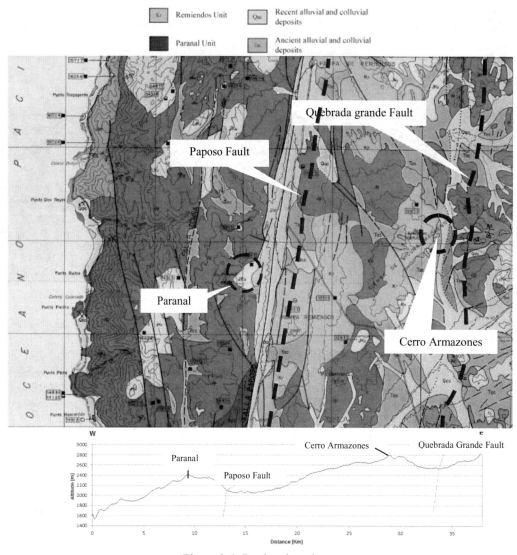

Figure 3-1: Regional geology

3.2 DISTRICT GEOLOGY

The rocks in Cerro Armazones were classified into three units according to the process which originated them: an intrusive plutonic unit and two dykes of a different composition.

- <u>Granitoids:</u> Found across the entire area under study making up the largest part of Cerro Armazones. A great part of the hill surface is highly fragmented and weathered. (See Picture 3-2)
- <u>Andesitic dykes:</u> Intrusions formed from volcanic rocks of an intermediate chemical composition and 1 m to 5 m width. The unit corresponds to a high strength rock, with medium weathering affecting both the rock and the discontinuities surfaces. These dykes cut across the unit formed by the plutonic rocks described above. (See Picture 3-2)

- **Aplitic dykes:** This unit is formed from volcanic rocks of an acid composition. The rock color varies from pink to white. The unit corresponds to a strong to very strong rock. It is fractured with low-medium weathering principally affecting the discontinuities' surfaces. The width of these intrusions does not exceed 6 to 7 m.
- **Soil units:** The soil layer found at the Cerro Armazones is less than 10 cm thick and corresponds to a mixture of sand and gravel with limited fines content. This layer results from the weathering of the underlying rock units caused by physical and chemical processes, with little or no transport, generating an oxidized cover in situ which is, very often, difficult to distinguish from the rock (See Picture 3-4) Colluvial deposits are found near the hillside borders close to the valley floor level, with a maximum thickness of 5 m.

Picture 3-2 Photograph of a granitoid outcrop, showing slightly weathered discontinuity surfaces.

Picture 3-3: On the left side, surface trace of an andesitic dyke displaced by faulting. The right side picture shows another andesitic dyke outcrop

Picture 3-4: Example of the reduced soil cover on the rocks found at the hillside. The white line shows the contact between both materials.

4. GEOLOGICAL-GEOTECHNICAL MODEL

As a part of the geological-geotechnical characterization of Cerro Armazones for E-ELT placement, a tridimensional geologic model was developed representing the main lithologies in order to associate them with the geomechanical classification (See Figure 4-1).

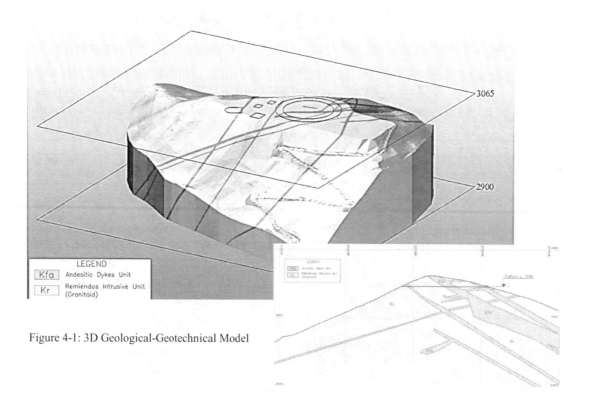

Figure 4-1: 3D Geological-Geotechnical Model

5. GEOMECHANICAL CLASSIFICATION

From the 3 lithologies described above, those that define the geomechanical quality of the rock mass at Armazones are granitoids and dykes with an andesitic composition, which represent approximately 80% of the materials found in the area. Geomechanical classification systems were used to characterize the rock mass. Figure 5-1 presents the distribution of RQD and RMR, compression (Vp) and shear (Vs) velocities for each lithology with depth on two of the boreholes excavated.

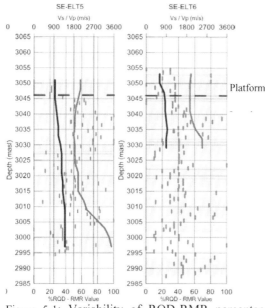

Figure 5-1: Variability of RQD-RMR percentages and Vs-Vp with depth on Boreholes

Rock Quality Designation (RQD): Parameter based on the recovery of rock borings, as the core percentage recovered in whole fragments with a length equal or greater than 100 mm, from the entire drill length.

Rock Mass Rating (RMR): System used to characterize the rock mass corresponds to Bieniawski classification (1976) from the following parameters:
- Uniaxial compressive strength of rock material
- Rock quality designation (RQD)
- Spacing of discontinuities
- Condition of discontinuities
- Groundwater conditions
- Orientation of discontinuities

The analysis of Televiewer, REMI and Downhole tests, as RQD and RMR results, indicate that there is not an increase of geotechnical quality of the rock mass under the platform level.

6. GEOMECHANICAL PROPERTIES AND FOUNDATION DESIGN BASES

In order to obtain the rock mass strength parameters, the intact rock properties obtained from laboratory tests on rock where scaled to rock mass level, which was made through the application of a failure criteria. For highly fractured hard rock, as is the case of the rock mass where the telescope will be founded, the bearing capacity of the rock used for foundation design was estimated using the Hoek and Brown failure criterion.

Rock mass strength, Elastic and Dynamic Modulus, were obtained for each lithology, and Winkler modulus and settlements estimation where obtained also as an equivalent value for the telescope foundations, according to the 3D Model. (See Figure 6-1)

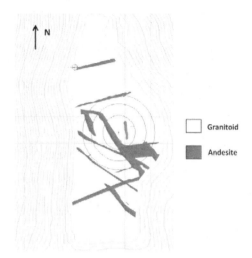

Figure 6-1: Granitoid and intrusive partial areas Rock Formations at Foundation Level

7. RIPPABILITY

The construction of the platform for the foundation of the telescope will required cutting about 190.000 m^3 of the hill, between 3.046 m.a.s.l and 3.064 m.a.s.l.

The propagation velocity of longitudinal waves (Vp), correlated with the rock type is the most widely used parameter to determine the depths to which the material it is excavated, rippable or requires blasting. In the case of study it is estimated that for igneous rocks the rock can be considered rippable with compression speeds until the order of 2.000 m/s. According to the result of the geophysical survey developed, the compression velocities above the elevation of the platform at 3.046 m.a.s.l. are less than 2.000 m/s, so most of the material should be excavable, and only specific areas will require the use of explosives.

As a reference, Figure 7-1 have the results of one of the Profiles developed, including Down hole test, with the platform level.

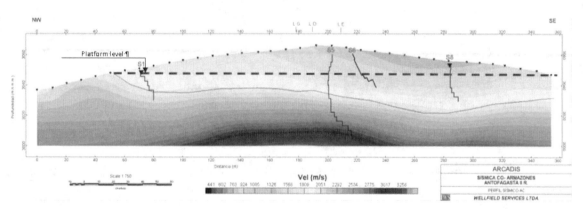

Figure 7-1: Compression velocities obtained from Seismic Refraction Profile and Down Hole Tests results

8. CONCLUSIONS

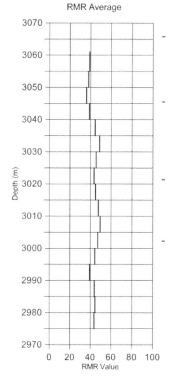

- The E-ELT will be founded on a rock mass, consist in three lithologies; Granitoids, Andesitic dykes and aplitic dykes.

- Geological and geotechnical information generated in this stage of the study was considered sufficient to determine the geological and geotechnical model of the E-ELT, and the estimation of parameters for the design of foundations.

- The rock mechanical properties vary with depth; however, at greater depths not necessarily better conditions have been observed. Due to alternating occurrences of the lithologies with depth the rock mechanical properties also vary.

- Figure 9-1 presents the RMR_{76} average considering all the boreholes as function of depth, in which it is observed that the variation of the elevation of the platform will not involve a substantial improvement in the geotechnical characteristics of the massif.

Figure 9-1: RMR Average of Boreholes as depth function

9. REFERENCES

The available data reviewed for this report are:

[1] ARCADIS, "Ground Investigation Report", Report N°E-TRE-ARC-145-0023 developed for ESO, February 2012.
[2] ARCADIS, "Geotechnical Report", Report N°E-TRE-ARC-145-0024 developed for ESO, February 2012.
3 Bieniawski, Z.T. (1976)."Rock mass classification in rock engineering". Exploration For Rock Engineering (ed. Z.T. Bieniawski), Vol.1, pp 97-106. Balkema: Cape Town.
[4] EN 1998-5: 2004. Eurocode 8: Design of structures for earthquake resistance. Part 5.Foundations, retaining structures and geotechnical aspects.
[5] EN 1998-1: 2004. Eurocode 8: Design of structures for earthquake resistance. Part 1. General rules, seismic actions and rules for buildings

Technological developments toward the Small Size Telescopes of the Cherenkov Telescope Array

R. Canestrari[*a], T. Greenshaw[b], G. Pareschi[a] and R. White[c] for the CTA Consortium

[a] INAF-Osservatorio Astronomico di Brera - Via Bianchi, 46 23807 Merate (Lc) Italy
[b] University of Liverpool, Liverpool, UK
[c] University of Leicester, Leicester, UK

ABSTRACT

In the last two decades a new window for ground-based high energy astrophysics has been opened. This explores the energy band from about 100 GeV to 10 TeV by making use of Imaging Atmospheric Cherenkov Telescopes (IACTs). Research in Very High Energy (VHE) gamma-ray astronomy is progressing rapidly and, thanks to the newest facilities such as MAGIC, HESS and VERITAS, astronomers and particle physicists are obtaining data with far-reaching implications for theoretical models.

The Cherenkov Telescope Array (CTA) is the ambitious international next-generation facility for gamma-ray astronomy and astrophysics that aims to provide a sensitivity of a factor of 10 higher than current instruments, extend the energy band coverage from below 50 GeV to above 100 TeV, and improve significantly the energy and angular resolution to allow precise imaging, photometry and spectroscopy of sources. To achieve this, an extended array composed of nearly 100 telescopes of large, medium and small dimensions is under development. Those telescopes will be optimized to cover the low, intermediate and high energy regimes, respectively.

In this paper, we focus our attention on the Small Size Telescopes (SSTs): these will be installed on the CTA southern hemisphere site and will cover an area of up to 10 km^2. The energy range over which the SSTs will be sensitive is from around 1 TeV to several hundreds of TeV. The status of the optical and mechanical designs of these telescopes is presented and discussed. Comments are also made on the focal surface instruments under development for the SSTs.

Keywords: CTA, Imaging Atmospheric Cherenkov Telescope, gamma-rays, segmented optics

1. INTRODUCTION

With the advent of ground-based IACTs in late 1980's, observations of VHE gamma-rays became possible and, since the discovery of the TeV emission from the Crab Nebula by Whipple in 1989 [1], this astronomy has achieved exceptional results. More then 130 TeV sources, both galactic and extragalactic, have been detected to date (see Figure 1).

The international communities working in TeV astronomy in Europe, America and Japan are now involved in the study of a huge array of Cherenkov telescopes called the Cherenkov Telescope Array [2] observatory. CTA aims to (a) increase sensitivity by another order of magnitude for deep observations, (b) boost significantly the detection area and hence the detection rates, particularly important for transient phenomena and at the highest energies, (c) improve the angular resolution and hence the ability to resolve the morphology of extended sources, (d) provide wide and uniform energy coverage from some 10 GeV to beyond 100 TeV in the energy of the photons, and (e) enhance the all sky survey capability, the monitoring capability and the flexibility of operation.

---

[*] rodolfo.canestrari@brera.inaf.it

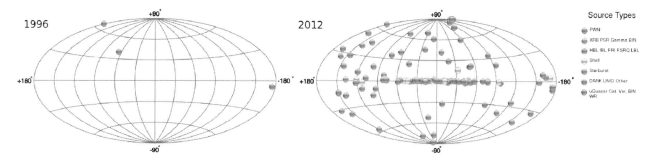

Figure 1. Sky maps showing the improvement in number of sources detected in the VHE astrophysics from 1996 (left panel) to 2012 (right panel). (Images created using http://tevcat.uchicago.edu/)

CTA will implement a large array with a modular design composed of three different telescope types, namely the Small Size Telescopes (SST), the Medium Size Telescopes (MST) and Large Size Telescopes (LST), of 4-7m, 12m, and 24m diameter respectively. An artist's view of CTA is represented in Figure 2.

The SSTs are the component of CTA that will allow observation and measurement of photons in the energy range from about 1 TeV to 300 TeV. The lower limit of this range is determined by the threshold at which the SSTs trigger and is matched to the sensitivity of the MST section of CTA, while the upper limit is given by the area of the array on the ground, which must be of order 10 km^2 to achieve the high energy differential sensitivity required by CTA. The performance of this section of the array in terms of angular and energy resolution is 0.04° and 10%, respectively. These are achievable by the deployment of an array populated by many tens of telescopes (40-60 units) exploiting large fields of view of the order of 8°-10° allowing the imaging of showers whose axis is far from a given telescope.

CTA was listed in the emerging proposals of ESFRI 2006 and it is now ranked in ASPERA and ASTRONET. Moreover, in December 2009, the European Commission within the 7th Framework Program (FP7) provided funding to the CTA Consortium, comprising 129 institutes from 22 countries, for the Preparatory Phase of the Cherenkov Telescope Array (CTA-PP, FP7-INFRASTRUCTURES-2010-1).

Figure 2. Artist's view of a possible CTA configuration, with three different telescopes types covering overlapping energy ranges, and area coverage which increases with increasing gamma-ray energy.

This new, complex and very large facility needs, amongst other things, a substantial technological development of many telescope subsystems. From the installation, operation and maintenance of each single telescope to the efficient coordination of the complete array in a structure organized as an observatory, from the science data center to the huge power supply.

Within the framework presented above, this paper has a particular focus on the development of the SST section of the CTA observatory. We describe the design of the telescope optics and structure and discuss the mirror and the camera systems for the different telescope options considered.

2. SINGLE-REFLECTION TELESCOPES: THE DAVIES-COTTON DESIGN

The SST array can be constructed in several alternative ways. The first approach, which is that traditionally used for IACTs, uses Davies-Cotton optics (SST-DC). The attraction of this approach is that it is a proven technology and allows the use of a camera based on conventional photomultipliers (PMTs) which is very similar to that needed for the MST. It is, however, possible that this is not the best way for CTA to proceed. As the camera dominates the cost of telescopes of this size, large numbers of SSTs-DC could not be easily affordable.

2.1 Optical design performance

Davies-Cotton telescopes are equipped with a single reflector which is constructed of many identical mirror facets. The facets are typically hexagonal in shape and are sections of a spherical surface with focal length F. These are arranged on a mirror cell (the dish) having a spherical profile with radius F. This design has been adapted from solar concentrators [3]. For F/D values around 1.5, this leads to optical performance that is somewhat better than that of an equivalent parabolic reflector for off-axis incoming rays and, for telescopes of the size of the SST, is acceptable for field angles of up to a few degrees. This is illustrated in Figure 3, which shows the 80% encircled energy diameter as a function of the field angle.

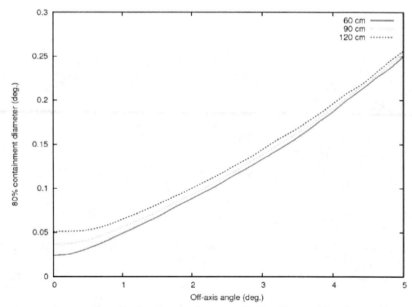

Figure 3. The 80% containment diameter for a Davies-Cotton telescope with F = 11 m and D = 7 m and hexagonal mirror segments of sizes (flat-to-flat) 60, 90 and 120 cm.

2.2 Mechanical structure designs

The Institute of Nuclear Physics (IFJ) of the Polish Academy of Sciences (PAN) is designing a telescope of the above type, consisting of a mast, a dish and dish support structure, as well as the columns which support the telescope. The mast positions the camera with respect to the mirrors, which are mounted on the dish, which in turn is attached to the dish support structure (see Figure 4). The rigidity of the mast is increased by the use of pre-tensioned steel rods. The dish structure is mounted on the telescope support columns. All steel profiles and tubes can be obtained as off-the-shelf products from industry.

The dish is built of square steel profiles with increased thickness in areas where the dish is connected to the dish support. It is a welded structure of about 6 m in diameter composed of two identical, symmetric parts which are bolted together. It accommodates 36 hexagonal mirror tiles of size 1.05 m (flat-to-flat). In a similar way, its support structure is welded out of square steel profiles with the wall thickness increased in parts of the structure in which the toothed-rack of the elevation drive system, the main axis, and the mast are installed. It can be split into two symmetric parts for easy transportation. The number of profiles was minimized to simplify the design and to reduce the cost. The back structure supports a 4 ton counter-weight. Note that the mast is attached directly to the dish-support structure, not the dish, so that no stress is transferred directly to the dish. The design of the mast guarantees the proper location of the camera with respect to the reflecting surface. The mast is built of circular steel tubes and steel rods pre-stressed with a force of 3500 N by means of turnbuckles.

Two vertical columns support the elevation axis that allows rotation from the parking position at -13° up to 95°. The columns are built of steel square profiles and bolted to the rotating platform. A crown roller bearing with diameter of 2 m is bolted to the base of the rotating platform and to the concrete foundation and allows for rotation of the telescope in azimuth (±270°).

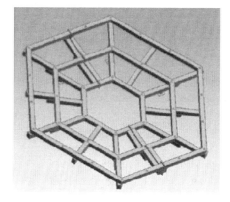
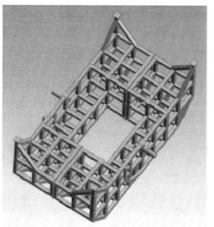

Figure 4. (Left panel) Illustration of the SST-DC telescope structure. The camera, mast, dish, dish-support structure, supporting columns and platform are shown. (Top-right panel) Detail of the dish structure. (Bottom-right panel) Detail of the disc-support structure.

Finite Element Analysis (FEA) has been used to guide the design and to perform structural checks. The main loads applied on the structure include its dead weight (including camera), the wind (for operative and survival conditions), ice and snow. Studies so far have confirmed that the structure meets the requirements for the observations and is able to withstand the maximum survival wind speed of 200 km/h. The eigenfrequencies at which the structure oscillates have also been calculated at various angles of elevation. At a telescope elevation of 60°, the lowest value obtained is 2.2 Hz; Figure 5 shows the three lowest oscillation modes.

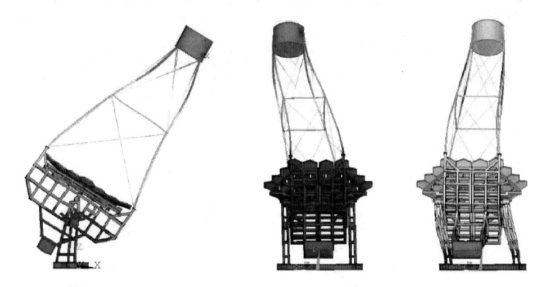

Figure 5. The three lowest frequency eigenmodes, (i), (ii) and (iii) from left to right, of the oscillations of the SST-DC structure.

The elevation drive system is composed of two 1.5 m radius-of-curvature toothed-racks (off-the-shelf products), each driven by an electric motor of power about 8 kW. The motors are fixed to the telescope support columns and provide backlash free steering of the telescope's elevation. This drive system will be equipped with 4 encoders (2 in the motors and 2 independent devices on the axis). The steering program will check the consistency of readouts from all encoders. The system will also be equipped with end-switches on each side. There will be one brake per side. The azimuth drive system is composed of a roller crown bearing driven by two motors of about 7 kW power. The system will operate in two modes. In the first mode both motors work in the same direction. This mode will be used for quick movements of the telescope during the positioning/parking phase. In the second mode, the motors work in opposition in order to avoid backlash. This mode will be used for tracking. The azimuth drive system will be equipped with end-switches equivalent to those used for the elevation drive. There will be one brake fixed to the crown bearing.

2.3 Mirror design

The radius of curvature of the hexagonal mirror segments required for the SST-DC is approximately 23 m and their size is 1.05 m flat-to-flat. These are being developed by IFJ PAN using glass cold slumping technology, developed at the INAF-Brera Astronomical Observatory [4][5], and an open structure composite mirror design. FEA studies have demonstrated that it may be difficult to achieve the required radius of curvature using the cold slumping technique: the stresses induced in the glass sheet as it is moulded are potentially too high and cause the mirror to have a short lifetime. Alternatives are therefore under consideration.

2.4 Camera designs

The attraction of the SST-DC option is not only the wealth of experience with Davies-Cotton telescopes that exists within past and present IACTs experiments but also the fact that, for appropriately chosen telescope parameters, the cameras of the SST-DC and the MST could be similar, leading to some economies of scale. In particular, we choose to use the same 1.5'' PMTs for the MST and SST, giving pixels of physical diameter $d = 50$ mm. The required angular pixel pitch for the SST is $0.25°$. Assuming a telescope field-of-view of $10°$, the camera should host about 1459 pixels within a diameter of about 2.05 m and have a mass of about 1600 kg. More details on the developments on going for the MST camera can be found in [6][7].

3. DUAL-MIRROR TELESCOPES: THE SCHWARZSCHILD-COUDER DESIGN

We are also investigating the use of dual mirror SSTs (SST-DM). As has previously been demonstrated [8], dual mirror Schwarzschild-Couder telescopes allow better correction of aberrations at large field angles and hence the construction of telescopes with a smaller focal ratio. This implies that, for a given primary mirror and angular pixel size, the physical pixels are smaller. The cameras for these telescopes can thus be based on multi-anode PMTs (MAPMTs) or Silicon PMs (SiPMs) and can be considerably cheaper than those envisioned for the SST-DC. The SST group within CTA has designed dual-mirror telescopes which have the potential to provide the required optical performance and allow exploitation of these technologies. For similar reasons, groups working on the development of the MST are also pursuing this option [9].

The Italian ASTRI (Astrofisica con Specchi a Tecnologia Replicante Italiana) project run by INAF (Istituto Nazionale di AstroFisica) is producing a complete design of a SST-DM [10]. Rapid progress is being made by this group towards the construction of an end-to-end prototype telescope composed of the telescope structure itself, the mirrors, the camera based upon SiPMs and the control software.

A second telescope is being prototyped in France as part of the GATE (GAmma-ray Telescope Elements) project [11]. It is based on the optical design produced in the UK in Durham, as was the preliminary mechanical design. UK, US and Japanese groups within CTA are designing a further camera for the dual mirror SST, the Compact High-Energy Camera (CHEC), based on MAPMTs.

In the following, the optics and mechanical structure of the ASTRI and GATE telescopes are described. A further section focuses on the two cameras. Only the telescope and camera that will give the best performances versus cost will then be adopted for the final implementation of the CTA-SST array.

3.1 Optical design performance

Matching the physical size of the pixels offered by MAPMTs or SiPMs sensors (a few millimetres) to the required angular pixel size of the SST implies that the focal length of the telescope $F \sim 2$ m. Ensuring sufficient collection area to obtain efficient triggering in the SST energy range, that is, a primary mirror of diameter about 4 m, then requires that the telescope's focal ratio be about 0.5.

The proposed designs have the Schwarzschild-Couder configuration, optimized using the commercial software ZEMAX, ensuring a light concentration higher than 80% within the dimension of the pixels over a wide field. The mirrors profiles are aspheric with substantial deviations from the main spherical component (see Figure 6 bottom panels). The ASTRI and UK designs somewhat differ a bit one to each other; hereinafter we describe the ASTRI design as a reference. This design has been optimized taking into account also a realistic way to implement the telescope, such as the segmentation of the primary mirror M1 and the arrangement of detection units into the detector. The optical system is shown in the top left panel of Figure 6. It has a plate scale of 37.5 mm/°, a pixel size of approximately 0.17° and an equivalent focal length of 2150 mm. This setup delivers a corrected field of view up to 9.6° in diameter as shown by the enclosed energy curves plotted in Figure 6 (top right panel). Concerning the throughput, a mean value of the effective area of about 6.5 m^2 is achieved, taking into account: the segmentation of the primary mirror, the obscuration of the secondary mirror, the obscuration of the detector, the reflectivity of the optical surfaces as a function of the energy and incident angle, the losses due to the detector's protection window and finally the efficiency of the detector as a function of the incident angles (ranging from 25° to 72°). In Table 1 we summarize the geometry of the optical system: the resulting telescope is compact, having an M1 diameter of 4 m and a primary-to-secondary distance of 3 m.

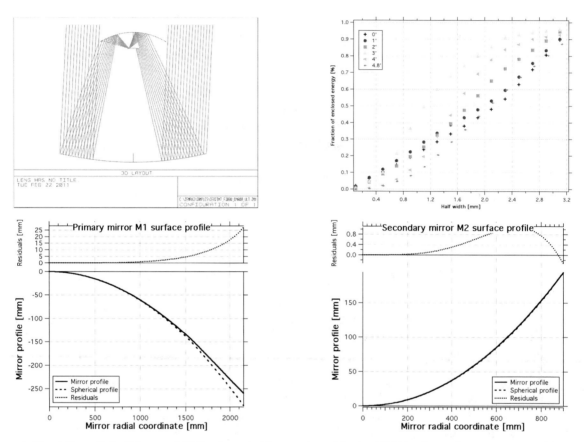

Figure 6. (Top-left panel) the Schwarzschild-Couder optical layout adopted for the ASTRI end-to-end prototype. (Top-right panel) the fraction of the energy from a point source contained within a square as a function of the half-width of the square for various field angles. (Bottom panels) radial profiles of the primary M1 (left) and secondary M2 (right) mirrors, and deviations from sphere.

Table 1. Main geometrical dimensions of the ASTRI optical system.

ELEMENT NAME	DIAMETER [mm]	RADIUS OF CURVATURE [mm]	SHAPE	DISTANCE TO... [mm]
M1	4306	-8223	Even asphere	M2: 3108.4
M2	1800	2180	Even asphere	DET: 519.6
DET	(side) 360	1000	--	--

3.2 Mechanical structures designs

Both the ASTRI and GATE telescopes adopt an alt-azimuthal design in which the azimuth axis will permit a rotation range of ±270°. The mirror cell is mounted on the azimuth fork which allows rotation around the elevation axis from -5° to +95°. Fixed on the mirror cell is the mast structure that supports the secondary mirror and the camera. The mast has a quadrupod layout, but with different solutions between the two projects. In order to balance the torque due to the overhang of the optical tube assembly with respect to the horizontal rotation axis, two counterweights are also supported from the mirror cell. Figure 7 shows the overall assemblies of ASTRI and GATE.

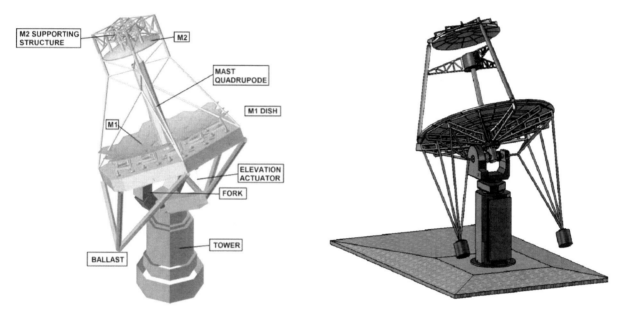

Figure 7. (Left panel) The ASTRI telescope structure showing the tower and fork, the counterweights, the M1 and M2 mirror support structures, the quadrupod and its central tube and the elevation actuator. (Right panel) The GATE telescope structure, shown here is the version with a shallow fork.

The ASTRI design

In the ASTRI design, the dish is a ribbed steel plate of about 40 cm thickness constructed in two halves to allow easy transportation. To this are attached the supports for the mirror segments, each of which includes a single and a double axis actuator and a bearing, which allows steering of the segments for alignment purposes. The quadrupod legs, with their radial bracing, counteract the lateral deformations of the mast structure due to gravitational and wind loads, while the central tube increases the torsional stiffness of the structure. Attached to the foundations is the tower, which ensures that the remaining telescope structure is at the correct height above the ground. At the top, it is interfaced to the azimuth fork. From the structural point of view the tower transmits loads (vertical and lateral) from the azimuth fork to the foundation. It is envisaged that some mechanical and electrical elements of the telescope equipment will be located in the space in the tower, so openings to access this equipment are foreseen. The tower will consist mainly of a steel tube, with section variations along its height. At the top of the tower is attached the azimuth fork, the interface with the tower being the azimuth bearing. The fork supports the elevation assembly (dish, counterweight and quadrupod) through the linear actuators and the bearings at its upper ends. The fork transmits loads coming from the elevation assembly to the tower and is composed of welded steel box sections. All steel structures are protected against corrosion by paint. The dish structure is attached to the azimuth fork using two preloaded tapered roller bearings, one on each of the fork's arms. At the upper end of the mast is the structure for supporting the monolithic secondary mirror. This consists of three actuators attached to the mirror via whiffletrees to ensure the load is spread over a sufficiently large area. In addition, three lateral arms support the transverse components of the secondary mirror's weight as they vary with the orientation of the telescope.

The azimuth drive is located at the base of the tower and is composed of two pinions, driven by electric motors, that couple with a rim gear. Axially pre-loaded ball bearings complete the azimuth assembly. The linear actuator, which drives the elevation, attaches to the tower at its lower and the dish structure at its upper end. The actuator consists of a preloaded ball screw driven through a gearbox by an electric motor. The orientation of the telescope is determined using absolute encoders located on each of the azimuth and altitude axes. Safety is ensured during movement by a system of software and electro-mechanical switches.

Also in this case, FEA has been used to evaluate the performance of the system. The lowest frequency eigenmode of the oscillations of the dish and mast is 8.5 Hz. FEA has also been carried out to determine the effects on the telescope of temperature gradients. This indicates that temperature differences of ±1°C from the base of the mast and its central tube to the top of the mast are inconsequential. However, similar temperature shifts across the dish support structure can lead to significant contributions to the misalignment of the primary mirror segments. Temperature variations in the tower structure can also lead to noticeable effects on the telescope pointing. Studies to understand the likely magnitude of these effects in operation are underway, as are strategies for dealing with any residual problems in these areas.

The GATE design

The main elements of the GATE structure are, again, the foundation on which the tower/fork structure is mounted via the azimuth drive system. The version shown has a shallow fork structure for which two counter weights are needed, mounted exterior to the fork. Also under study is a design with a shorter tower and a deeper fork, allowing a single counterweight to be mounted centrally. In both cases, the fork is fastened via the azimuth bearings and drive to the primary dish structure, which carries the secondary mirror support and the camera. Aluminum and carbon fiber have also been considered, but finally not selected to optimize costs and easiness of manufacturing and mounting.

The drive systems under study consist either of crown gears which are driven by pinions connected to electric motors via a gearbox, or worm gears, again driven by electric motors.

FEA has determined that the lowest frequency eigenmodes of the oscillations of the telescope involve transverse motion of the secondary with respect to the primary and are about 5 Hz. Rotational eigenmodes of oscillation have lowest frequencies of about 12 Hz. These values refer to a preliminary design not yet optimized.

3.3 Mirror designs

For the ASTRI telescope it is proposed to construct the primary mirror as a set of 18 hexagonal-shaped panels having 850 mm face-to-face dimension. Three different types of mirror profiles are necessary to reproduce the M1 profile. GATE will exploit petal-shaped segments. The number and dimension of these is still under investigation. It is hoped that only two types of segment will be required. For both telescopes, the aim is to build a monolithic secondary mirror. More details about the technology under development can be found in [12].

3.4 Camera designs

The ASTRI camera based on SiPM sensors

The ASTRI camera design uses SiPMs as photosensors. The pixels contain 3600 cells, each of which is an avalanche photodiode operated in quenched Geiger mode. These cells are of dimensions 50×50 μm^2, giving a fill factor of 70%. The particular device chosen is the Hamamatsu S11828-334 monolithic multi-pixel SiPM consisting of 4x4 pixels of roughly 3×3 mm^2. Four of these are grouped together to form one pixel of physical size 6.2×6.2 mm^2, matching the required angular size. As is shown in the top left panel of Figure 8, four of the Hamamatsu devices are put together to form a unit. Four such units then form a module called a Photon Detection Module (PDM – shown in the top right panel of the same figure). This module is composed of 16 Hamamatsu devices and has dimensions of 56x56 mm^2. The PDMs are constructed by plugging the Hamamatsu devices into connectors attached to a printed circuit board (PCB). Figure 8 also shows one unit attached to a PCB on which the connectors for the remaining 15 units are visible (bottom left panel). Under each unit on the PCB is a small temperature sensor, allowing the temperature of the SiPMs to be monitored, providing a route through which the temperature dependent SiPM gain can be stabilised. A schematic layout for the Front End Electronics (FEE) is aslo shown in the remaining panel of Figure 8. The FEE boards supply the power for the SiPMs, perform the readout and form the first trigger signals. Several Application Specific Integrated Circuits (ASICs) are available for the readout of the camera. Under consideration are in particular the TARGET series of chips and the EASIROC (Extended Analogue SiPM Readout Chip). A range of simulation tools has been produced to aid the design of the readout and trigger. Moreover, as visible from the same figure, there are small gaps between the Hamamatsu devices when they are mounted to form units and PDMs. In order to recover the light that would be lost in these gaps, light guides are placed over them. These are of thickness 2.5 mm and made of high refractive index glass; these are designed to utilize the total reflection occurring at the light guide's side walls.

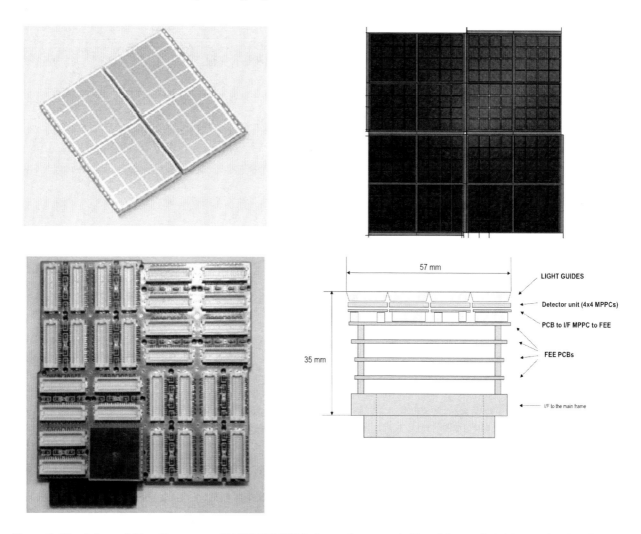

Figure 8. (Top-left panel) Four Hamamatsu S11828-334 SiPMs form a detector unit (Top-right panel) and four such units a PDM. (Bottom-left) The PCB on which the connectors for the Hamamatsu units are soldered; one Hamamatsu device is plugged into its socket in the bottom left of the picture. (Bottom-right panel) Support structure for the FEE of one PDM unit, on top of each SiPM the light guides are visible.

In addition to the FEE mentioned above, design of the back end electronics is underway. This will use a Field Programmable Gate Array (FPGA) and local memory to provide interfaces to the CTA data acquisition, to the camera controller and to the CTA clock. There will also be circuitry to provide the various DC voltages needed to power the elements of the camera.

A schematic diagram showing the location of the module and the various electronics boards is shown in Figure 9 (left panel). This also gives the dimensions of the camera. The total height of the camera is about 30 cm. A preliminary design of the chassis of the complete camera is also shown (right panel). The camera lid can be seen in this picture, this can be closed to protect the sensors from the elements. As the focal plane of the dual mirror telescopes is convex, with a radius of curvature of 1 m, the PDMs must be attached to a precisely machined curved plate. Below the sensor plane is the support structure to which further electronics boards and the cooling system can be attached.

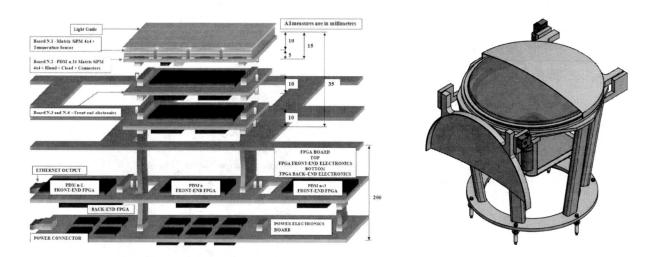

Figure 9. (Left panel) Schematic diagram of the camera electronics for the ASTRI telescope. (Right panel) The preliminary layout of the camera chassis.

The CHEC camera based on MAPMTs sensors

The default photosensor for the Compact High-Energy Camera is the Hamamatsu H10966 MAPMT. This consists of 64 pixels of size 6x6 mm^2, in a unit of dimensions 52x52 mm^2. Thirty-two MAPMTs can be used to cover the focal plane, providing a field of view of about 9°.

Each FEE module provides the high voltage supply needed by a MAPMT, samples the signals produced by its 64 channels at a frequency of about 500 MHz, forms a first level trigger by applying thresholds to sums of four pixels and outputs a digitized waveform for each of the MAPMT's channels. CHEC will make use of modules designed at SLAC for the Advanced Gamma-ray Imaging System (AGIS) based on the TARGET ASIC (see Figure 10 top panel) [13]. Minor modifications to this are needed and these will be carried out in collaboration with SLAC. Unfortunately, the TARGET modules, as they are, cannot be connected directly to the MAPMTs in the CHEC as the curvature of the focal plane would then require large gaps between the MAPMTs. The MAPMT is instead attached to a preamplifier board and then a structure containing a twisted length of ribbon cable, allowing bending in two planes, which carries the signals from the MAPMT to the electronics. The preamplifier board behind the MAPMT allows the PMTs to be operated at low gain, important given the high counting rate they will experience due to background photons. The preamplifier also allows shaping of the MAPMT signal and hence optimization of the performance of the FEE. Further, if it becomes clear that SiPMs will offer better performance per unit cost than MAPMTs, the MAPMT photosensor plane with its preamplifiers can be replaced with a SiPM-based system with new preamplifiers ensuring the correct signal shape enters the TARGET module. A mechanical frame that provides the required rigid support has been designed. This system is illustrated in Figure 10 (bottom panel). Prototype mechanical structures have shown that this system functions as hoped, allowing the MAPMTs to be placed on a curved surface.

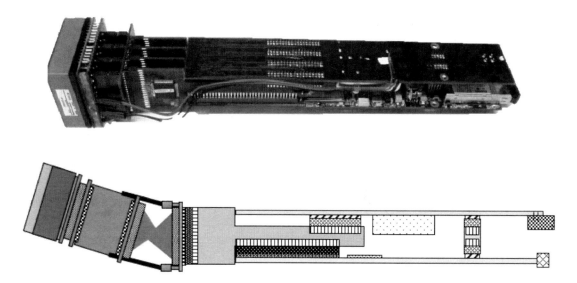

Figure 10. (Top panel) A TARGET module attached to a Hamamatsu MAPMT; note that the direct attachment shown here is replaced by a flexible system in the CHEC, as is described in the text. (Bottom panel) Illustration of the mounting of a MAPMT on a TARGET module, showing the bending allowed by the structure.

The trigger signals provided by the FEE modules must be combined and examined to select candidate Cherenkov events against the night-sky background. This camera trigger system forms part of the Back-End Electronics (BEE), which is also responsible for processing data from the FEE modules and distributing clock signals with the required level of precision to the FEE modules. Recent developments mean that it is feasible to process trigger signals with nanosecond accuracy and sub-nanosecond delay correction from all FEE modules in one or two FPGAs. The entire camera trigger can be performed on a high-density, active backplane, housing the FPGAs and connected directly to the FEE modules. The backplane also routes the serial data from the FEE modules to a single location for processing. The read-out electronics may be placed on a mezzanine card, and use a dedicated FPGA to compress the data for transmission to a central array location, or on an embedded PC located at the rear of the camera and connected to the BEE trigger backplane via Ethernet. An active backplane is under development by CTA collaborators at Washington University; the BEE design will be developed with this team.

The mechanical aspects of the camera include a support matrix for the MAPMTs, internal elements for support of the electronics, a cooling system and an external structure, which includes a lid and the interface to the telescope. The MAPMT support matrix must ensure precise focal-plane positioning. The internal structure must allow adequate cooling of the electronics and remain stable on decade timescales to repeated camera movements. Thermal modelling of the camera will be done during the camera design to assess the cooling and control requirements, which will be implemented during the mechanical prototyping. The external structure must be weather-proof and minimise dust ingress, and provide minimal additional shadowing of the primary mirror. The most complex element of the external structure is the lid, which must be remotely operated, highly reliable and provides a screen for the imaging of stars for alignment purposes.

Each aspect must be designed with assembly and access procedures in mind. Figure 11 shows a schematic diagram of the CHEC.

Figure 11 Conceptual design of the CHEC showing: (A) MAPMTs, (B) preamplifier boards, (C) interface to FEE modules in which the curvature of the focal plane is removed, (D) SLAC TARGET-based FEE, (E) back-plane, which will include BEE; the MAPMT position matrix, external mechanics, power-supplies, cooling system and control computer are not shown.

4. CONCLUSIONS

The SST array is the section of CTA which dominates the sensitivity of the instrument in the high photon energy regime. The area covered by the SST array therefore determines the highest energies to which CTA is sensitive and the quality of the measurement of the highest energy photons. This is determined by the quality of the individual telescope images, but also crucially by the number of telescopes observing the showers caused by those photons. Both the area the SST array can cover and the multiplicity of the measurements it makes increase if more telescopes can be built, making it important to reduce the cost of the telescopes to the lowest possible level. Hence, although the SST array can be built using the Davies-Cotton telescopes with PMT-based cameras that have traditionally been used for IACTs and will be used for the LST and the MST, this may not represent the best option for CTA. In parallel with the design of a SST-DC telescope, the SST group is therefore investigating alternative designs. The dual mirror telescope allows the reduction of the plate scale to levels that make it possible to use SiPM arrays or MAPMTs in a compact camera. Compared to the conventional SST-DC design, this reduces the camera cost enormously, at the expense of introducing a more complex optical system, with its attendant costs. If the group is successful in producing functioning dual mirror telescopes, the indications are that these will be cheaper and will represent a better option.

The potential benefits of the alternatives to the SST-DC make it important that these are investigated fully: the result may be a significantly improved CTA instrument. The risks and uncertainties associated with these novel designs mean that a conventional telescope of the required size must also be constructed. This implies that the SST group must continue to pursue all the above options in parallel, focusing its efforts on the telescopes offering the best performance per unit cost as and when it becomes clear these function as hoped and the costs are as expected.

REFERENCES

[1] Weekes, T. C., et al., "Observation of TeV gamma rays from the Crab nebula using the atmospheric Cerenkov imaging technique", The Astrophysical Journal, 342:379{395}, July 1989.

[2] Actis, M. for the CTA consortium, "Design concepts for the Cherenkov Telescope Array CTA: an advanced facility for ground-based high-energy gamma-ray astronomy", Experimental Astronomy, volume 32, pages 193-316, (2011)

[3] Davies, J. M., Cotton, E. S., "Design of the Quartermaster Solar Furnace", Solar Energy Sci. Eng. 1 (1957) 16–22.

[4] Pareschi, G., et al., "Glass mirrors by cold slumping to cover 100 m^2 of the MAGIC II Cerenkov telescope reflecting surface". Proc. SPIE 7018, (2008).

[5] Vernani, D., et al., "Development of cold-slumping glass mirrors for imaging Cerenkov telescopes". Proc. SPIE 7018 (2008).

[6] Kubo H. for the CTA consortium, "Development of the Readout System for CTA Using the DRS4 Waveform Digitizing Chip", 32nd ICRC (2011).

[7] Puehlhofer G. for the CTA consortium, "FlashCam: A camera concept and design for the Cherenkov Telescope Array", 32nd ICRC (2011).

[8] Vassiliev, V. V., Fegan, S. J., Brousseau, P. F., "Wide field aplanatic two-mirror telescopes for ground-based gamma-ray astronomy", Astropart. Phys, 2007, vol. 28, issue 1, p. 10-27.

[9] Cameron, R. A., et al., "Development of a mid-sized Schwarzschild-Couder Telescope for the Cherenkov Telescope Array". Proc. SPIE, 8444-43 (this conference).

[10] Canestrari, R. for the ASTRI collaboration, "The Italian ASTRI program: an end-to-end dual-mirror telescope prototype for Cherenkov light imaging above few TeV", 32nd ICRC (2011).

[11] Laporte, P., et al., "An Innovative Telescope for the Very High Energy Astronomy". Proc. SPIE, 8444-254 (this conference).

[12] Canestrari, R., et al., "Techniques for the manufacturing of stiff and lightweight optical mirror panels based on slumping of glass sheets: concepts and results". Proc. SPIE 7437 (2009).

[13] Vandenbroucke J. for the CTA consortium, "Development of an ASIC for Dual Mirror Telescopes of the Cherenkov Telescope Array", 32nd ICRC (2011).

SST-GATE: An Innovative Telescope for the Very High Energy Astronomy

Philippe Laporte[a], Jean-Laurent Dournaux[a], Hélène Sol[b], Simon Blake[c], Catherine Boisson[b], Paula Chadwik, Delphine Dumas[a], Gilles Fasola[a], Fatima de Frondat[a], Tim Greenshaw[d], Olivier Hervet[b], James Hinton[e], David Horville[a], Jean-Michel Huet[a], Isabelle Jégouzo[a], Jürgen Schmoll[c], Richard White[e], Andreas Zech[b]

[a]GEPI – Observatoire de Paris, CNRS, Univ. Paris Diderot
5, Place Jules Janssen, 92190 Meudon, France

[b]LUTH – Observatoire de Paris, CNRS, Univ. Paris Diderot
11, Av. Marcellin Berthelot, 92190 Meudon, France

[c]Durham University
CfAI (Centre for Advanced Instrumentation), NetPark Research Institute
Joseph Swan Road, Netpark, Sedgefield, TS21 3FB, United Kingdom

[d]University of Liverpool
Liverpool L69 7ZE, United Kingdom

[e]University of Leicester, University Road
Leicester LE1 7RH, United Kingdom

ABSTRACT

The Cherenkov Telescope Array (CTA) is an international collaboration that aims to create the world's largest (ever) Very High Energy gamma-ray telescope array, consisting of more than 100 telescopes covering an area of several square kilometers to observe the electromagnetic showers generated by incoming cosmic gamma-rays with very high energies (from a few tens of GeV up to over 100 TeV). Observing such sources requires – amongst many other things - a large FoV (Field of View). In the framework of CTA, SST-GATE (Small Size Telescope – GAmma-ray Telescope Elements) aims to investigate and to build one of the two first CTA prototypes based on the Schwarzschild-Couder (SC) optical design that delivers a FoV close to 10 degrees in diameter. To achieve the required performance per unit cost, many improvements in mirror manufacturing and in other technologies are required. We present in this paper the current status of our project. After a brief introduction of the very high energy context, we present the opto-mechanical design, discuss the technological trade-offs and explain the electronics philosophy that will ensure the telescopes cost is minimised without limiting its capabilities. We then describe the software nedeed to operate the telescope and conclude by presenting the expected telescope performance and some management considerations.

Keywords: very high energy, Cherenkov, gamma-ray astronomy, CTA, Small Size Telescope, Telescope prototyping, Schwarzschild-Couder, Dual-mirror, GATE.

1. INTRODUCTION

The current generation of ground-based Imaging Air Cherenkov Telescopes (IACTs) has opened a new window in the area of very high energy (VHE) astronomy and has demonstrated the large diversity of sources in this energy range. Current instruments cover energies from about 100 GeV up to a few 10 TeV, providing thus access to the most energetic, non-thermal emission regions in the sky. The spectacular astrophysics results from the current IACTs [1,2] has led the international community to engage the CTA project to investigate the cosmos more deeply in the VHE domain with an instrument that provides about an order of magnitude increase in sensitivity, significant improvements in the spectral coverage, the angular, spectral and temporal resolution, as well as an improved flexibility in the observational modes.

With its enhanced performance, CTA is expected to lead to major discoveries in the fields of astronomy, astrophysics, cosmology, particle physics and fundamental physics. The CTA consortium expects to detect of the order of a thousand sources of different types (e.g. supernovae remnants, pulsar wind nebulae, binary systems, star clusters, the Galactic Centre, active galactic nuclei, starburst galaxies and possibly new classes of VHE emitters, including dark matter). Three key-questions will be explored with CTA:
- The origin of cosmic rays and their role in the universe,
- The nature of particle acceleration in systems containing black holes,
- The ultimate nature of matter and physics beyond the Standard Model.

The first topic refers to the physics of galactic cosmic accelerators such as pulsars and , supernovae,and gamma-ray loud binary systems. The interaction of accelerated particles with their environment and their cumulative effects at different scales, from star forming regions to starburst galaxies, will be investigated. A better understanding of the VHE emission from active galactic nuclei (AGN) or possibly from gamma-ray bursts (GRB) will help to track down the origin of ultra-high energy cosmic rays. This question is also related to the second topic, the acceleration mechanisms of charged particles close to (super-)massive black holes, and the nature of these particles. GRB, as well as galaxy clusters, are promising candidates for VHE emission, which should be accessible with the increased performances of CTA. Probing the infrared extragalactic background light (EBL) with VHE gamma-rays from distant sources and the search for signals from dark matter will provide important input into our understanding of cosmology, while the search for Lorentz invariance violation using VHE flares at high redshifts will test the fundaments of physics.

CTA presents a logical evolution of the currently operating experiments H.E.S.S., MAGIC and VERITAS, experiments with up to five telescopes, which have launched the cosmic exploration at VHE after the pionnering works by Whipple, HEGRA, CAT and other experiments. CTA will consist of two arrays of several tens of telescopes to provide an unprecedented sensitivity and sky coverage by the use of a new generation of telescopes. CTA will be the first VHE facility operated as an open, proposal-driven observatory, analogous to optical ones, that shall be available for the whole scientific community.

In the next section we argue for the need of a new kind of telescope to cope with the aforementioned scientific goals. We then provide details on the performance of an ideal dual mirror telescope in the framework of CTA (third section) and describe the telescope we are designing to fulfil these performances (fourth section). We conclude by exposing the future work and the schedule of our project.

2. WHY A SCHWARZSCHILD-COUDER TELESCOPE?

VHE gamma-rays are detected indirectly through the mostly ultra-violet Cherenkov light emitted from cascades of charged particles (air showers) they trigger upon interaction with the atmosphere. The geometry and energy of the air shower are reconstructed from its image, taken with several telescopes to improve the stereoscopic view, and are used to derive the arrival direction and energy of the primary gamma-ray. At energies of a few 10 to a few 100 GeV, the Cherenkov signals from air showers are weak and telescopes with large mirror surfaces are required to capture them. At energies of a few 10 to a few 100 TeV, the Cherenkov light emission is strong, but the detection is limited by the low statistics of the usually steep energy spectra of the astrophysical sources. In this domain, telescopes may have small mirror areas, but they need to cover a large area on the ground. For this reason, CTA will consist of different types of telescopes of different sizes. A few large-size telescopes (LSTs), with parabolic dishes, will cover the lowest energies; a few tens of medium-size telescopes (MSTs), of Davies-Cotton (D-C) design, are optimised for the TeV energy range; and many tens of small-size telescopes (SSTs), of D-C or S-C design, will cover the highest energies up to 100 TeV.

Cherenkov telescopes operate generally in a photon starved regime. To avoid contamination of the air shower image with the night sky background, the exposure must match the duration of the Cherenkov light pulse, of the order of a few nsec. Therefore, unlike in conventional optical astronomy, the image on the camera, corresponding to the indirect signal from a single gamma-ray, cannot be improved through increased exposure. This motivates the development of optical systems with very large primary mirrors, having diameters in the range of 10–30m. In addition, the optical systems of Cherenkov telescope must be composed of the minimal number of optical elements to circumvent light loss [11, 12, 13, 14]. For this reason, the first Cherenkov telescopes were designed according to the Davies-Cotton (D-C) optical formula, which is based on a single spherical reflector segmented into individual mirrors. Initially imagined to concentrate solar energy for power production, the optical quality of the D-C allowed the discovery of the first sources in the VHE domain. All the current IACT arrays use either D-C or parabolic designs. The latter minimize time dispersion of the Cherenkov signal and are thus preferred for the largest telescopes, such as the LST component of CTA. The D-C design is the baseline for the MST component.

In these designs, the rapidly increasing effect of optical aberrations with the off-axis angle presents a major disadvantage, if one aims at a wide FoV. Especially for the SST component of CTA, a wide FoV is of great interest, since it permits to capture images of air showers at large distances from the telescopes, since air showers at the highest energies provide relatively strong signals. This allows a larger spacing of the SST part of the telescope array, and thus an increased effective area for the same number of telescopes. Moreover, in order to accurately estimate the background that is subtracted from the gamma-ray signal from an astrophysical source, a wide FoV is generally required to have within the same field both the putative source and a few equivalent regions of empty sky. This can be a constraint in particular for extended galactic sources, which will be the main target at the highest energies.

Moreover, in the D-C and parabolic designs, the camera must be situated in the focal plane of the single reflector and when the size of the reflector is increased, to allow a larger field of view, the size and weight of the camera and its distance from the dish increases. This implies strong constraints on the mechanical structure of the telescope.

Finally, an optical system that has both large aperture and high f-ratio (ie field of view) leads to large plate scale, typically 50 mm per arcmin. This implies large and expensive cameras, consisting usually of at least several hundred photo-multiplier tubes, with the drawback of an enlarged effect of vignetting.

An ideal Cherenkov telescope would provide a small plate scale, a large diameter and a large FoV. As explained in [7], increasing the aperture diameter to improve the light-gathering power unavoidably results in a corresponding decrease in the f–ratio, in turn amplifying all primary aberrations, such as spherical aberration $\propto 1/f^3$, coma $\propto \delta/f^2$, as well as astigmatism and field curvature $\propto \delta^2/f$, with δ being the field angle. For a given plate scale, an aplanatic design radically outperforms the single reflector designs in terms of the effective light gathering power, the ability to accommodate a wide FoV, and the amount of time dispersion (in the case of the D-C)

To achieve the enhancement in the performances it is necessary to develop telescopes based on an aplanatic optical formula that minimises the aberration over a large FoV. This precisely was the aim of Schwarzschild's studies at the end of the nineteenth century when he developed a dual mirror optical formula. Despite the very promising performance of this aplanatic formula, without any coma and spherical aberrations, and an optimisation proposed by Couder at the beginning of the twentieth century, it has never been built. The Observatoire de Paris, represented by the LUTh and the GEPI, and the University of Durham, Liverpool and Leicester) are proposing to build such a telescope to demonstrate that it is technically possible to achieve the theoretically predicted performances. The objective of this work is to propose to the CTA consortium a new kind of telescope able to enhance all key performances at the same time instead of only the light collecting power to the detriment of the uniformity of field of view, as in the case of the single reflector telescopes. The S-C prototype also aims to demonstrate that a significant reduction in the cost of the SST telescopes can be achieved for this design (smaller camera, lighter structure…) compared to the baseline D-C option.

3. THE CHALLENGES OF BUILDING AN S-C TELESCOPE

As we intend to develop a prototype of S-C telescope in the frame of CTA, we have derived the high level scientific requirements of CTA in the frame of a dual mirror telescope. They are gathered in Table 1.

Optical		Mechanical	
Designation	**Value**	**Designation**	**Value**
Field of view	9°	Pointing precision	< 7 arcsec
PSF	0.1° @ 80% [a]	Tracking precision	< 5 arcmin [d]
Mirror diameter	4 m	Source localisation	< 5 arcsec
Pixel size	6x6 mm²	Slew speed	> 90°/min
Plate scale	0.025°/mm	Reliability of operation	97 % of the observational hours (e)
Angular resolution	0.02° [b]	Total lifetime	30 years
Throughput	> 60% [c]	Night lost for maintenance	< 3 observational nights/yr
Effective mirror area	> 5 m²	Cost running	< 312 person.hours/yr
		Unit cost	< 250,000 euros
		Power consumption	< 10 kW

Table 1: Scientific requirements for the S-C telescope under construction at the Meudon's site of the Observatoire de Paris. [a] PSF size is determined by the area in which 80% of the energy is spread. [b] The angular resolution depends on the energy; we give here the most constraining. [c] The throughput includes the vignetting. [d] This precision includes the systematic and the statistic errors. [e] About 1500 observational hours per year are foreseen for CTA.

These optical and mechanical specifications have been derived in three levels which are "essential" (the minimum we can afford), "optimal" and "goal". They give to the system engineering team the range in which each parameter must be situated in order to allow for an optimisation of the various budgets without back-looping permanently with the scientific team. This important scientific work is essential to get a reactive technical team to the unavoidable changes in the specifications in a project that has the size of CTA.

Table 1 presents the most constraining requirements. The optical requirements have been derived in optical surfaces by Zemax simulation whereas the mechanicals ones have been transformed in high level technical requirements for the engineering team.

However beyond these technical specifications, CTA requires to build two arrays of telescopes which, indeed, imply to take into account the environment, the array control and the consistency through all the whole CTA projects namely:
- Each telescope must be ordered independently to optimise the scientific targets and their ability to point and track any cosmic source whatever the target of the other telescopes is.
- The telescope must be operated from a remote control room.
- The telescope must be compliant with the existing camera (volume, weight…) and provide their positioning in their focal plate with the proper accuracy.
- The telescope must have a parking position that minimise the wind effective area to resist to a wind speed up to 200 km/h.
- The telescope must cope with the power supply that will be available on site (10 kW maximum per telescope).
- The maintainability must be made with less than 6 person.hours per week and must not let the telescope unavailable more than 3 observational nights per year.
- The telescope life time shall be at least 30 years without any protection against the environmental conditions.

- Each part of the telescope must be subcontracted.
- The choice of materials and software must be compliant with others CTA telescopes.

Moreover, the environmental conditions also constrain the telescope design. CTA will consist in two sites, in the north and the south hemispheres, each one having several dozen of SST telescopes. In order to be compliant with the future CTA requirements related to the site location, we designed the SST-GATE telescope to be compliant with the most constraining parameters over the 4 sites still in competition. The specific cost as well as maintenance specificities will be recorded in order to give to the Project Committee the real cost of our prototype once the site will be chosen. The environmental conditions are gathered in TableTableau 2.

Parameter	Observing conditions	Critical conditions	Emergency conditions	Survival conditions
Temperature range (°C)	-10 to +40	-15 to +45	-20 to +50	<-20 to > 50
Wind speed range (km/h)	0 to 50	50 to 65	65 to 100	>100
Humidity range (%)	5 to 95	5 to 95	5 to 95	0 to 100

Tableau 2: Climate conditions assumed to design the SC prototype.

In the critical situation, the telescope can observe but with degraded performance. Velocities and accelerations of movements are reduced to 70% of their nominal capacities. In the emergency scenario, velocities and accelerations of the movements are reduced to 10% of their maximum capacities and the telescope must return to its parking position. Under survival conditions, the telescope is parked and cannot be moved.

All these additional constraints have guided the opto-mechanical design described in the next section. For instance, the telescope will remain without any protection so each system shall be sealed or placed in a hermetic box to prevent any water damage. A long duration life-time requires choosing long-life equipment with a mature technology or it means to design the telescope in order to make the maintenance rapid and easy. Moreover, the electronic architecture must use long-life components that shall still exist in 30 years.

The prototype we are building will be situated at the Meudon's site of the Observatoire de Paris and will be devoted to measure the performances an SC telescope can achieve. Once this operation will be done (phase A), it will be used for testing scientific cameras such as the CHEC one – developed by a UK-US-Japan consortium – as well as for student training and outreaches (phase B). Some equipment are specific to each phase. We split the cost and the procurement catalogue for each phase, the first being dedicated to telescope building and to make it available for observation (including the maintenance). The second one is devoted to the specificity of our prototype and includes all the equipment and parts related to the outreaches. For instance, considering the use of our telescope for our own research program in Meudon and for educational trainings, and in order to decrease the maintainability cost, we will get a shelter.

4. THE UK-FRENCH TELESCOPE DESIGN
4.1. High level conceptual design
The high level technical requirements derived from the scientific specifications led us to the functional diagram presented in Figure 1. One can see that the telescope can be described in 6 different elements:
- The FSS (Foundation and Slab Structure) will support the telescope weight and will provide the torque resistance.
- The AAS (Alt-Azimuthal Structure) provides the ability of the telescope to point in any direction in the sky and to track any scientific source with the required accuracy in the limit of the environmental conditions (wind speed up to 50 km/h). It also has the function to support the optical part of the telescope.
- The elevation structure (including several functions, see the PBS in Figure 2) which gathers all the movable parts (mirrors M1 and M2 plus their supporting structure and the camera),

- The TCA (Telescope Control and Alignment) which is in charge of controlling the whole telescope adequately to

perform the scientific operations.
- The SDS (Software Data System) will operate the telescope, the data storage and all the operations to perform the alignment.
- The PSS (Protective Shelter Structure) which will protect the telescope during its outreach life.

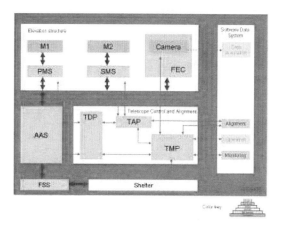

Figure 1 : Functional diagram of the UK-French design. See text and Figure 2 for the meaning of the acronyms.

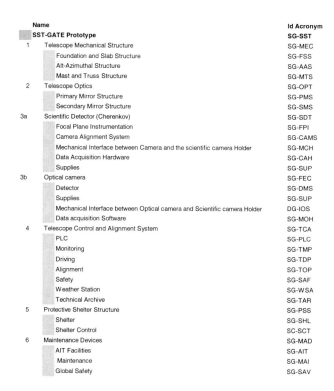

Figure 2: Two first levels of the PBS developed for the UK-French prototype.

For cost reasons we have preferred an alt-azimuthal structure instead of an equatorial one because the latter requires an additional pillar. Moreover, it constraint the orientation of the parking position to the East or to the West which are not necessary the less windy directions on the future CTA sites.

The functional analysis helped us to split the telescope in as independent as possible parts in order to simplify the procurement, the tests, the feasibly and the mounting. That explains why the PBS (Product Breakdown Structure) is composed of 6 main functions, as described in our High Technical Level Requirements. The first and the second level are showed in Figure 2.

4.2. Trade-offs

When the cost, the maintainability and the life-time are the drivers to build a telescope, without sacrificing the performances, a certain number of trade-offs must be solved to determine the best design. We pointed out:
- The kind of material (aluminium, steel or carbon fibre).
- The different parts that require a safety system redundancy.
- The fastening system of the camera.
- The camera removing.
- The risk analysis of the mirror procurement.
- The size of the alt-azimuthal fork.
- The use of a single or double motor in the AAS.

The choice of the material is essential because it largely impacts the cost. Indeed, a metallic structure is heavier than a carbon fibre one and will require a more expensive foundation. As the CTA arrays will be situated in deserts, this item may increase rapidly. In the other hand, designing and mounting carbon fibre structure is complex due to (1) the difficulty to find companies that provides several profiles in this material and (2) the strong dependence of the characteristics with respect to the dimension (along the fibre or not). The steel won this trade-off with a small advantage compared to the aluminium option, thanks to the low price of the material considering the product market and due to the easiness of manufacturing and mounting steel pieces.

In terms of safety, we identified that the elevation axis would need a safety system in addition to a passive return mechanism. Actually, a failure of this axis may generate an undesirable movement that could be dangerous for telescope integrity and for human beings during the maintenance activities. If a redundant elevation axis could be an interesting idea, it is not relevant in the frame of CTA, because this solution does not solve the problem of a failing motor leading to a blocking axis or a failure of power supply (a battery is foreseen for each telescope, but no requirement is available). This explains why we choose a passive mechanical system plus an electric engaging mechanism. In case of an electrical failure, the elevation motor is disengaged and the passive system makes the telescope able to go to its parking position without help. This solution works for any of the previous hazard situations.

Having the detector between M1 and M2 is an advantage as it provides a lighter camera compared to the DC telescopes. Nevertheless it imposes to be able to remove the detector for maintenance. We found an innovative solution that avoids any human intervention in the telescope and thus minimise the optics hazards. It consists in two connections "kneecap" and one "punctual" – all fastened to the M2 mirror – that allows the detector to rotate while the telescope is in its parking position (see Figure 3). This method won the trade-off as it cumulates several advantages: the camera is situated at a height of about one meter so that it facilitates the operations by limiting the amount of staff needed, the equipment required and the human, M1, M2 and camera hazards. It is also lighter than the three other solutions considered.

The mirror procurement trade-off was the most difficult to solve because there is no suppliers capable of making mirror cumulating several meters in diameter, a relatively complex shape, a small radius of curvature and a low cost. Fortunately, some laboratories are working on this topic for CTA. The IRFU has succeeded in making mirrors with the appropriate radius of curvature (about two meters) with a variant of the cold-slumping method. The M2 mirror is under procurement for our prototype.

The diameter of the M1 mirror makes it difficult or impossible to be produce in one part if we want to remain in the cost envelope so a tessellated option is considered in all prototypes. The LAM laboratory (Marseille, France) works on the hydro-forming method which consists in applying a pressure on a pre-rectified metallic sheet to get the desired shape and form. Petals mirrors could be made by this method. However this technology will not be available soon enough for our prototype. Electro-forming is another solution but the mirror transportation from US represents a too high cost for one telescope. We are so investigating the feasibility of M1 with the CEA technology which has the advantages of being feasible and very light.

Figure 3: Picture simulated of the camera removing after full rotation while the telescope is in its parking position

Two others trade off have been made on the alt-azimuthal system. The first one deals with the size of the fork which could be either long with a small tower or the reverse. The trade off significantly conclude for the small fork which allows lower axial stresses comparing to a long fork configuration. In addition, small fork is lighter and less expensive at the first order, so this configuration is compliant with CTA requirement.

The last trade off consists in the use of single or double motors for the alt-azimuthal system. We decided the use of a single motor in order to simplify the control of the elevation and azimuthal systems and because the MTBF (Mean time Between Failure) of the foreseen motors is larger than 10,000 hours (8 years of observation). Indeed, two motors imply a higher accuracy in the control and the synchronisation and would not offer an easier solution in the case of a power breakdown.

4.3. The telescope design
4.3.1. The Alt-Azimuthal Structure

The AAS has to support several constraints which must not cause any irreversible damage or deformation. The first load is the compression due to the mass of the MTS. If this load exceeds the Euler one, a buckling may occur in the AAS and will imply to unmount the whole telescope to repair the permanent deformation. The second load is the torsion due to a brutal stop of the azimuth or the elevation drive. The strength must remain in the steel elastic domain. The third constraint is the bending load caused by the wind. We considered a maximal wind speed of 200 km/h for this effect. Unfortunately, period and strength of the wind are not constant so we have to take the fatigue damage into account that implies to add a security coefficient in the calculation of the wind load. We also assumed the worst case of a laminar profile for the wind which is also included in the security coefficient.

Several kind of AAS have been analysed in our preliminary study in which the elements have been modelled by circular steel tubes. It showed that the internal stresses follow a linear function with the mass of the MTS and that it increases with the square of wind speed, as theoretically expected. It also allows us performing a rough analysis of several possibilities in order to choose the best kind of structures among: conic and cylindrical tower, fork structure made by steel tubes or reinforced and long or short fork. In addition to the mechanical performance, the criteria for choosing the

AAS design were the mechanical performance, the cost, the maintenance cost and the safety hazards.

Despites a conic tower offers more stability than a cylindrical one, we chose the latter for cost reasons, and we compensate the lower performance with a larger thickness. Concerning the fork mount, an assembly of tubes can be sufficient but it is expensive to integrate because of the welding manpower and the maintenance frequency. We chose a closed fork mount in order to (1) reinforce the fork structure, (2) simplify the welding operations and (3) close the fork so that the maintenance is minimal inside.

A fine analysis has been performed by Finite Element Method (MD.Patran, MD.Nastran) to choose the best solution. The corresponding model is shown in Figure 4. The fork mount is made of an assembly of steel tubes, stiffened by 15 mm thickness steel sheets. The tower has a diameter of 530 mm and is 15 mm thick.

In this model, the bottom of the tower is clamped while its top is submitted to the combined effects of the MTS mass and the wind. The elevation subsystem is modelled by a single point located at its centre of gravity with the corresponding mass of this assembly. This point is linked to the top of the AAS with 24 points by a MPC (Multi Point Constraint) element (RBE3) in order to define a relationship between the degrees of freedom of the 24 previous points and those in the centre of gravity of the elevation assembly.

We focused theses analyses on the static behaviour, the eigenmodes and the buckling load. The static analysis shows that the internal stresses remain lower than the steel yield stress times the security factor. It also shows that the deformation is located in the fork part and not in the tower. The long fork shows a lower stiffness while it is heavier (1.9 tons instead of 1 ton). This last result means that a short fork is stiffer than a long fork and it confirms our trade-off choice of a short fork.

Figure 4: Von Mises stresses for equivalent long fork and short fork

The modal analysis also shows that the first two modes are the bending of the tower and occur at about 75 Hz. At this frequency, the structure is not excited by the wind. The following mode is a torsion mode and occurs at 188 Hz. Regarding the buckling analysis, the Euler load (53 500 kN) is greater than the compression load (35 kN) and no buckling should occurs.

4.3.2. The optical structure

The telescope is based on a Schwarzschild-Couder optical formula as discussed in Section 2. It is composed of a primary mirror (M1) with a diameter of 4 meters and a secondary mirror (M2) with a diameter of 2 meters. The camera is located between the two mirrors which is significantly different from the classical DC telescopes. The detecting surface will be a disc of 362 mm in diameter. A distance of 510.7 mm separates M2 and the detector.

The telescope is composed of the PMS (Primary Mirror Structure) which holds the mirror tiles, their actuators, the supporting structure and the counterweight. The SMS (Secondary Mirror Structure) is related to M2 and consists of the actuators and the frame structure. The PMS and the SMS are maintained at the proper distance by three tubes of 100 mm in diameter (named MTS for Mast and Truss Structure). We chose kneecaps to fasten the SMS to the MTS and "pivot" at the interface between M1 and the MTS in order to ease the mounting procedure.

The PMS is fastened to the elevation axis and is located at 2.5 meters above the ground. The design has been made to minimise it so that the torque due to the wind is lowered. It also eases the maintenance because most of the elements are reachable at human height. We remind that our telescope will have an elevation range from -5° to +185° in order to (1) proceed to the alignment (theodolite method) and (2) access to any part of the mirrors at human height. Finally, it allows removing easily the camera (see further).

The azimuth and elevation driving system have been studied via trade-offs to solve the cost-maintenance-hazard equation. The baseline consists in using ball bearings to define the rotation axis. This solution is less efficient for slow movements than the fluid film but fluids are forbidden by the CTA requirements. Each driving system will be an assembly of a crown moved by a worm gear and an electric motor. A servo-motor wired with Ethernet to the SST-GATE backbone (see section 4.4) will complete the assembly. As requested by the CTA requirements, the elevation movement will range from -5° to +90° to address the CTA requirements and +90° to +185° to perform a simpler maintenance of the telescope. The azimuth will range from -90° to +450° with respect to the parking position.

As explained previously, the safety system is foreseen in this assembly for the two axes. It consists in a passive system that mechanically works against the movement generated by the motor. Thus, even in case of power supply or motor failure, the telescope returns to its parking position without any help. An electrical disengaging system is foreseen to free the axis in such a condition. This passive system has also the advantage of annihilating the play in the movement chain and will enhance the tracking accuracy.

4.3.3. Telescope behaviour

An optical simulation has been performed with Zemax (1) to optimise the mirror's shape, the distance between the optical surfaces to provide the best PSF over the whole field of view and (2) to define the sensitivity of the optical elements to a misalignment.

To perform the optimal PSF over the large field of view (9°), no simple shape fits the needs so a polynomial formula has been developed to express the height variation of the mirror's shape with respect to a sphere along the radius. The flatness error, which means the discrepancy that is tolerable between the real surface and the theoretical one, is about 10 mm for the M1 and M2 mirrors along the normal surface [8]. This is explained by the weak dependency of the PSF to this kind of movement. The sensitivity is a little bit more constraining in the other directions (2 mm) but remains easily feasible.

For cost reasons, the mirrors will be made by an assembly of tiles that must fit the mirror shape. In [8], the PSF degradation has been studied versus decentring, tilt and focus. The most constraining movement is the rotation along the axial and tangential axes of the tiles (0.1 mrad). This can occur only when the structure is locally deformed by a thermal gradient or a mechanical constraint. We will study in detail within the next months these two possibilities in order to implement a passive (preferred) or active corrective system to prevent local deformations larger than 0.1 mm at the scale of the tile (0.1 mrad times the typical tile size) and 0.2 mm at the telescope scale.

The consequence is very important in terms of mirror manufacturing because they can be made of several tiles to be assembled (M2 mirror) or not (M1 mirror) with no consequence on the optical performances. Having this kind of mosaic does not strongly constraint the movable mechanical structure of the telescope as it tolerates deformations and rotation at the scale of the mm which is usually manageable with passive mechanical solutions. This is the same conclusion for the distance because our prototype will be able to translate and tilt independently M1 and M2.

Finally, the optical quality is not sensitive to a variation of the distance between the optical elements. A displacement of +/- 5 mm does not change the PSF's width by no more than 10%. We are thus confident for the telescope construction and the performance to achieve.

4.3.4. The shelter

The location of the telescope on the site of the Meudon's Observatory implies the use of a shelter in order to work on the telescope despite the environmental conditions. Moreover, after the optical performance test in 2014 the prototype will be dedicated to outreaches for which the shelter is required to minimise the maintenance during this phase. Finally, the Meudon site is classified, which requires integrating our telescope SST-GATE in the landscape

The shelter must fit with the following requirements, which come from the telescope, the location site and the lifetime:
- Due to the telescope, the shelter must have a minimal length of 6 meters, a minimal width of 5 meters, and a

minimal height of 5 meters. The power supply must be lower than 5 kW.
- Due to the characteristics of the Meudon site, the maximal wind speed is 150 km/h, the maximum snow weight is 110 kg/m^2, it has to avoid the propagation of any tear and an M2 certification against fire hazard has to be considered for the telescope security but also, in our case, for the outreaches.
- Due to the safety and the lifetime (30 years), the opening and the closing of the shelter must be done only when the telescope is in parking position. It must be possible to operate manually the shelter in case of a power failure. A guarantee of a minimal lifetime of 10 years is required.

Four different concepts are in competition.
- Mobile garage: it is a rigid structure in galvanized steel with a coverage made with PVC fabric sliding to discover the telescope on two rails.
- Retractable garage: it is an assembly of four rigid parts (an aluminium structure protected with polycarbonate plates) sliding and stored one into each other.
- PVC dome: It is a dome with an oval footprint composed of a structure in aluminium and tubes covered with a PVC fabric; the opening solution is made by a full 180° rotation of the structure.
- Resin dome: It is a moulded resin parts which rotate and can be stored one into each other, each part is a portion of a sphere.

Currently, a call for tender is opened; decision will be made during the summer for an assembly and mounting within the 6 following months. The shelter option, first not considered by the CTA consortium, may be adopted for the SST telescopes.

4.4. Telescope Control

CTA will be composed of many distributed systems and sub-systems. The UK-French telescope control system has to be designed in such a way that its implementation in the CTA framework shall require as little developments as possible. With this objective in mind, the choice of standard industrial solutions can't be avoided; these have to provide reliable, scalable, manufacturer-independent, and long-term support systems.

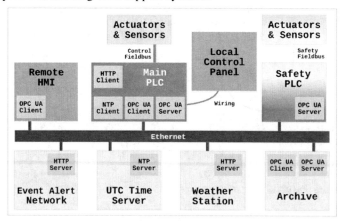

Figure 5: Communication architecture

For the implementation of the communication architecture, Ethernet is selected as the low-level layer and OPC UA as the software layer (see Figure 5). Other client-server protocols have to be added for instance the synchronisation. If a control software is used, this architecture may be integrated by using an interfacing layer such as DevIO in a CORBA environment to be compliant with the ACS environment.

For scalability and easy interfacing, the software will be based on Object-Oriented Programming (OOP); depending on the functions and the targets, LabVIEW, C++ and Java are preferred languages.

The environment of telescopes is always demanding in terms of temperature and altitude. In consequence the hardware has to be rugged: the considered main PLC is a NI CompactRIO, selected among different solutions.

5. EXPECTED PERFORMANCES
5.1. Alignment
For the SST-GATE prototype, the performances are measured by the PSF size and the ability of the telescope to track a scientific target on the sky within the 7 arcsec required by the scientific specifications.

From the Zemax simulation, the expected PSF over the field of view is presented in Figure 6. This performance is related to the mechanics (relative movement of the optical elements) and to the alignment (absolute position of the optical elements). We discussed in section 4 that the sensitivity to a mechanical deformation will be weak. A thermal-mechanical simulation will help us to identify the weaknesses of the design and to modify it to ensure the proper behaviour. Constraint gauges will also be installed in the telescope to make the telescope control able to compensate for the deformation.

For the alignment point of view, we intend to use to following procedure:
- To mount the PMS, the MTS and a sight in place of M2. With the M1's sight, we mechanically define the optical axis of the telescope. It has the advantage to link the optical axis position to the mechanical rotation axes.
- To mount the tiles on the PMS and align them by looking the image of a distant point source – aligned with the sights – in the image plane of M1.
- To mount the MTS then M2 with three reference lengths in order to place the M2 references parallel to the M1 ones.
- Same for the M2-detector distance.
- To look a star and refine the alignment according to the recorded PSF (to be minimised).

The use of reference length is possible for the SST telescope because the accuracy needed on the optical distances is larger than the millimetre. We remind that the M2's tiles will be assembled during the manufacture to alleviate the alignment procedure. We chose to identify the optical axis with the mechanical one because the error budget allocates a discrepancy between them of a fraction of millimetre which is reachable with classical mechanics.

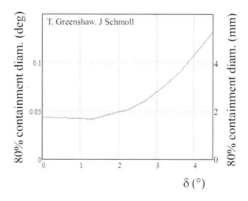

Figure 6: Expected SST-GATE PSF over the field of view. The pixel size of the scientific camera will be 0.2 degree.

We foresee to develop an elevation axis that ranges from - 5° to + 185° to ease the alignment of the Alt-azimuthal mechanical axes to the optical axis. Following the theodolite procedure to define the horizontal plane, we need to reverse the azimuth and the elevation angle (e.g. azimuth goes from 0° to 180° thus elevation goes from 0° to 180°) to look at the same direction in the sky but by changing the mechanical configuration. This capability enlarges the specifications with respect to the CTA requirements but it will ease the alignment process without any additional equipment.

As discussed in section 4, the mirrors must be moved in translation and tip-tilt in order to be tuned with respect to the optical axis (alignment procedure). Due to the tessellation of M1, we shrewdly assembled 4 tiles (two adjacent tiles per crown) to get 6 "big" tiles to be moved instead of 24. This reduced the number of actuators and, because single tiles are assembled four by four, the specifications for the M1 actuators become similar to the M2 ones, with strength of 1000 N for a stroke of 30 mm. We will thus provide a single kind of actuators for our prototype in order to optimise the cost equation.

5.2. Tracking performance
All the telescope assembly is involved in the tracking performance. The main identified parameters are presented in Table 3. We divided the error into systematic and statistic errors because the systematic errors can be compensated (if measured) whereas the statistical ones cannot be.

Systematic error	Statistic error
Horizontality of the azimuth rotation axis	Azimuth wobble
Perpendicularity between the azimuth and the elevation axes	Azimuth rotating smoothness
MTS deformation	Elevation wobble
Misorientation of the telescope optical axis with respect to the centre of rotation of the fork mount	Elevation rotating smoothness
	MTS vibration

Table 3: List of the main parameters that are involved in the source tracking performance.

A first errors budget has been established on the strength of the datasheet and on the method we foresee to position and to orient the different parts of the telescope. The estimate of the total statistic error equals 11 arcsec (no margin) so a discrepancy of 4 arcsec with respect to the requirements, if we consider that the all systematic errors can be corrected. The conclusion is that the orientation of the different parts will have to be refined by using reference stars. We are working on this aspect to prepare the AIT phase.

6. CURRENT STATUS OF THE PROJECT
As presented before, the SST-GATE telescope is an ambitious project using a new generation of optical design. The full design had to be created and optimized to fit the CTA requirement. Presently, the status of the project is well going considering the CTA schedule.

The main difficulties of the mechanical structure have been highlighted and solved. Doing so, we are now able to move to the building steps and to think about the control and monitor aspects, and to make a stiffer link with the science requirement considering the design.

6.1. The foundation
The project has been already brought to reality. Indeed, foundations of the telescope have been realized at the Observatoire de Paris in Meudon. The location has been chosen within the Meudon's site in order to be compliant with

the observation constraints (other telescopes are in use), the environmental positioning and for an easiness of access and control.

Following the mechanical simulation, foundations have been made. It is composed of a large rectangular shape of about 5x6 m² and a stiffer square has been built in order to support the telescope itself. Power supply, Ethernet and fibre cables could be easily provided to the telescope thanks to the already built gallery which links the telescope to the dedicated control room.

6.2. Future work

As presented in the previous paragraphs, the main aspects of the telescope structure have been frozen; the validations have been made by successful mechanical analysis, In order to frozen definitively the mechanical design, further simulations will be made concerning the alt-azimuthal structure and the dishes of mirror M1 and M2. The telescope would be then slightly optimized. The risk analysis which is currently done would help us to complete the mechanical design and the definition of the software control. Last optimisation of the structure will be done by the parallel analysis in terms of mechanical and thermal simulations scheduled for the end of the year and the beginning of 2013. The goal will be to decrease as possible the weight and amount of steel in the structure in order to decrease the cost and assembling effort.

Moreover, the telescope design will be analysed through the scientific point of view. Currently, the scientific simulations we are performing show the various dependencies of the PSF but further work remains to take into account the entire structure of the telescope. For instance, the tubes connecting the camera to the mirror M2 create an obscuration on the detector that lead to a lower PSF quality. Thus, telescope design has to be related to the science quality that we want to obtain.

In the same way, alignment process which guarantees the required optical quality of our telescope has to be analysed through both the mechanical and science simulations. The error budget of the mechanical parts would have an influence on the PSF and must be taken into account as soon as possible.

Based on these studies, we would be able to build the telescope progressively by the beginning of 2013, starting with the alt-azimuthal system which is almost fixed now.

Moreover, the procurement of the shelter will be an important milestone for us because it will be the very first step of the construction allowing a continuous work on the site during the winter; power would be provided to test the well working of the shelter by the end of this year, and later on to test the functionality of the telescope drive.

Following the alt-azimuthal structure achievement, the whole telescope would be progressively built, for instance the mirror M2 should be received and mounted by the end of the summer 2013.

7. PERSPECTIVES AND CONCLUSION

CTA is presently a really challenging project in both technical and scientific aspects. The Observatoire de Paris is involved to demonstrate that a new generation of telescope based on Scharzschild Couder design is a really good solution to achieve the resolution needed by CTA requirement.

From a technical point of view, the telescope is almost fixed in its mechanical design, advances in control, monitoring and alignment systems as software, hardware and components are under studies or procurement and will benefit from the definitive frozen structure due at the end of the year. Analyses are currently done to finalise and optimise simulations and the telescope will be built in current 2013.

In addition to prove the S-C design concept, the SST-GATE team has proposed since 2011 to use our prototype as a camera test bench. We are currently taking into account the camera designs – especially the CHEC camera one. This particularity devotes our design to be fully compliant with others CTA teams. For instance, the camera holder has to fit with all SST cameras from both a mechanical and an electrical point of view.

Moreover, the telescope is devoted to a long life of use in Meudon thanks to the research teams interested by this innovative telescope but also by the educational outreaches which positively drive the manufacturing requirement of our telescope like the need of low cost of maintenance and the procurement of a shelter that represents an interest directly usable for CTA.

8. ACKNOWLEDGMENTS

This work is done under the Convention n° 10022639 between the Région d'Ile-de-France and the Observatoire de Paris. We gratefully acknowledge the Ile-de-France, the CNRS (INSU and IN2P3), the CEA and the Observatoire de Paris for fundings and for support

REFERENCES

[1] J.A. Hinton & W. Hofmann, "Teraelectronvolt Astronomy", Ann. Rev. Astron. Astrophys., 47:523 (2010).
[2] CTA Consortium, "Design Concepts for the Cherenkov Telescope Array", Experimental Astronomy, 32, p 193, 2011.
[3]. Bernl¨ohr, K., et al., "Theoptical system of the H.E.S.S. imaging atmospheric Cherenkov telescopes. PartI: layout and components of the system", Astropart.Phys. 20,111–128(2003).
[4]. Fernandez, J., et al.,Optics of the MAGIC telescope, AppendixA of J. Barrio, et al., "The MAGIC Telescope", MPI-PhE/98-5, Max-Planck-Institut für Physik (1998).
[5]. Weekes, T.C. et al., "VERITAS: the Very Energetic Radiation Imaging Telescope Array System", Astropart. Phys. 17 ,221–243 (2002).
[6]. Kawachi, A., et al., "The optical reflector system for the CANGAROO-II imaging atmospheric Cherenkov telescope", Astropart. Phys., 14, 261–269 (2001).
[7] Vassiliev V., Fegan, S., Brousseau, P., "Wide field aplanatic two-mirror telescopes for ground-based gamma-ray astronomy", Aph, 28, 10 (2007).
[8] J. Rousselle, "Schwarzschild-Couder Telescope: Study of non-linear optical system", General CTA meeting, Amsterdam, 14-18 May, 2012.

A new era for the 2-4 metres class observatories: an innovative integrated system telescope-dome

G. Marchiori*[a], A. Busatta[a], S. De Lorenzi[a], F. Rampini[a], C. Perna[b], P. Vettolani[b].
[a]EIE Group, via Torino 151a, Venezia, Italy 30172
[b]INAF, Via del Parco Mellini, X Roma, Italy 45030.

ABSTRACT

The experience and the lessons learned gained in two decades of activity in astronomical industry in projects like NTT, VLT, LBT, VST VISTA and finally E-ELT brought to study a flexible fully integrated system which could address every astronomic institute to approach astronomy with a complete self standing facility including a dome, a telescope with 2 to 4 meters class optics and relative instruments with which it is possible to match the desired science cases and objectives. This paper describes the aspects of the flexibility which is so important to adapt the design to the specifications in order to fulfil the institutes science goals in the least time possible through the latest design tools such as CAD CAE FEM etc and the best and more cost effective technology experienced along the projects mentioned before.

Keywords: telescope, dome, observatory.

1. INTRODUCTION

The project aims at developing a detailed analysis of scientific/industrial importance, as well as of the feasibility of the creation of an integrated system of astronomical services for the emerging countries and/or for those countries interested in developing an educational course in the astronomical field.

The main goal of the project is to maximize the scientific and industrial knowledge acquired in more than 20 years of active participation in the main international astronomical projects.

The goals of the project are aimed at the production and supply of radio telescopes or astronomical observatories complete with building, telescope and basic instrumentation, as well as the supply of a technical and scientific educational course: from the functional and managerial knowledge of the equipment, to the training of the scientific personnel, and to the constitution of specific observatory plans in radio astronomy or optical/infrared astronomy.

Worldwide, there are nations that are interested on starting and developing a national astronomy program despite little experience in this field. In this context the issues to be faced mainly regard:

- the growth of a class of astronomers able to study science with telescopes of at least 2m class;
- the construction of an observatory with which develop specific science cases;
- the development of a class of technicians able to use and maintain an entire observatory.

While the first of these points is linked directly to the astronomical science, the second and third derive from it but they are based on engineering and technology.

Considering that the design, procurement, construction, installation, commissioning, use and maintenance of an observatory is complicated to be managed by a newborn astronomical community, it is much easier to purchase a complete "quasi-standard" product to enhance the interoperability of all subsystem and at the same time reduce drastically the risks for the customer.

The experience developed by European Industrial Engineering in more than 25 years of activity in astronomical field, being involved in the design and construction of several of the most powerful observatories across the world (NTT dome, VLT telescopes and domes, LBT telescope, ALMA prototype and series, VST telescope and dome, VISTA dome,

E-ELT telescope and dome), allows to "summarize" the most powerful, reliable and cost effective technology applied to astronomy.

Moreover the customer shall be guided and involved directly on the entire development process in order to favour the confidence on the whole system and to guarantee the entire knowledge of its possibilities and limits.

2. OBSERVATORY SUB-SYSTEMS

2.1 Dome

In an observatory, the dome is the essential part which necessarily protects the telescope from the external agents like rain, dust, hailstorm, wind, etc. Moreover, the dome is provided with all the sub-systems which are necessary to allows the telescope to work as per specifications. Finally it hosts all the telescope sub-systems: from the electrical cabinets till the coolant supplies and even the hydrostatic plant required for the bearings.
Since the dome is tailored on telescope requirements, its design characteristics depends directly to it: from the concrete pillar till the cooling plant which guarantees to avoid thermal shocks.

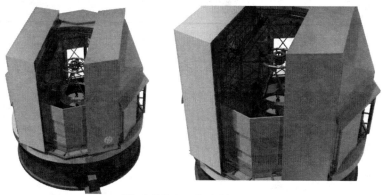

Fig 1. EIE Standard Observatory

In particular the CAD and CAE software allowed to build a quickly adaptive design which allows to modify parameter like height, circumference, observing doors and ventilation doors aperture. The design is based on parts that throughout the years showed their reliability in several observatories (e.g. NTT, VLT, VST, VISTA) matching the performance requested without degradation. That is possible also following strictly the preventive maintenance foreseen during observatory lifetime.

This kind of dome has a standard series of sub-systems studied in order to achieve performance at the lowest cost possible; these sub-systems are:

- Rotation Mechanisms, based on a rail/trolleys system;
- Observing Doors, based on two doors sliding by means of trolleys upon dedicated rails;
- Ventilation Doors, which allows the telescope to be ventilated by fresh air and at the same time protect from lateral moonlight rays;
- Wind Screen and Moon Screen, based on a systems of sliding panels to protect the telescope from wind and moon rays;
- Calibration Screen to start up observations with the telescope;
- Dome Crane (10tons), to allow maintenance around telescope areas;
- Electrical systems: power supply, lightning protection, fire alarm,
- Cooling system, based on air treatment units which guarantee no thermal shocks to the telescope and prevent at the same time the dust entrance.

All these sub-systems have been studied and analyzed deeply, also by means of Finite Element Models which kept into account the survival and operational loads due to gravity, wind, temperature and earthquake; since the observatories mentioned before are installed in harsh environment the loads imposed by design are severe: winds greater than 200km/h, ice, intense earthquake (Maximum Likely Earthquake) etc.

Moreover, all components were studied in a FMECA analysis in order to obtain the correct maintenance intervals to achieve the lifetime foreseen, with preventive actions limited only to daytime so to prevent down time.

Finally all the facilities necessary to complete the observatory such as auxiliary building to host observatory technical areas like, mirror coating areas and plants, control rooms, instruments rooms, electrical power supply rooms (MV/LV transformer, etc.) can be easily added in order to shape the observatory as per customers requests.

2.2 Telescope

The telescope is the main instrument with which it is possible to perform the required studies. For this reason its characteristics and sub-systems depend directly from the science to be observed; thus, the performance under the point of view of stiffness and control stability are strictly dependant from the hardware installed on it. Since the hardware can increase costs very quickly, it is necessary to keep well in mind the real performance objectives in order to use the best set of solution with the aim of holding down the costs. For this reason it is necessary experience in the use of several different hardware solution, and the knowledge of each single contributions to the performance.

To be able to respond to the widest requests possible European Industrial Engineering developed a flexible design in order to reduce as much as possible the design time between using an hardware configuration instead of another. This flexibility interests especially the optical and instruments hosted, driving systems and bearing systems for azimuth and altitude axis, as they are essential parts involved in telescope stiffness and in its axis motion precision. In particular the different hardware solutions bring to structure concepts which can diverge significantly one from the others. For example, the use of an hydrostatic bearings in the azimuth axis improves smoothness of motion, stiffness and precision (thus, overall performance) compared to a slewing bearing but it is more complicated and requires power to work so at the end it is more expensive to be implemented and to be used. The azimuth structure, in this case must transmit loads in the best way possible to have a correct behaviour during observation.

The EIE telescope design has been studied in order to be modular and guarantees the best behaviour taking in account the hardware installed. It is very important to highlight that the design flexibility does not mean that the telescope structure is studied as a compromise between different hardware, but rather, the telescope design concept "tailors" itself to the hardware installed in order to obtain the maximum performance possible; in this way the telescope is not only "hosting" the hardware but it contributes to gather the best behaviour with an eye to the overall costs.

Here below it is possible to see different telescope concepts obtained working with the EIE telescope flexible design: a nasmyth and a cassegrain solution

Fig 2. EIE Standard 2 to 4m telescope adapted for different focal length and optical configuration (nasmyth and cassegrain)

It is necessary to mention that, the telescope design studies were carried out for telescope structures that host mirrors from 2 to 8m. The widest flexibility in terms of motion systems is obtainable with the range within 2 to 4m while for bigger telescopes some choices are more strict to obtain sufficient level of performance.

The flexibility includes also the technical analysis of every key part by means of:

- Finite Element Modelling;
- servomechanism control response;
- Failure Modes, Effects and Criticality.

The instruments generally are designed by customer astronomers and opticians in order to comply to the science cases requested. If required the choices in this particular delicate piece of equipment can be driven by astronomers and technicians coming from national and international scientific institutions.

3. CONCLUSIONS

The success of this project will allow to propose, thug a process of scientific cooperation, innovative dynamics of growth for the developing countries, as well as new and/or renewed commercial, scientific and industrial exchanges also in similar fields. But it will, above all, allow to maximize the efforts of scientific and industrial growth at all levels.

AKNOWLEDGEMENTS

European Industrial Engineering S.r.l. would like to express its special thanks to INAF (Italian Institute of Astrophysics).

Low frequency - high sensitivity horizontal monolithic Folded-Pendulum as sensor in the automatic control of ground-based and space telescopes

F. Acernese[a,b], R. De Rosa[b,c], G. Giordano[a], R. Romano[a,b], F. Barone[a,b]

[a] Dip. di Scienze Farmaceutiche e Biomediche, Universitá degli Studi di Salerno, Italia
[b] Istituto Nazionale di Fisica Nucleare (INFN), Sezione di Napoli, Italia
[c] Dip. di Scienze Fisiche, Universitá degli Studi di Napoli "Federico II", Italia

ABSTRACT

We describe a new mechanical implementation of a monolithic inertial sensor and its application to the control of the top stage of a multi-stage mechanical suspension for seismic attenuation. In particular, we discuss the sensor theoretical model and the experimental results in connection with a new control strategy that can be applied to the control of mechanical suspensions (seismic attenuators), inertial platforms and low frequency monitoring and control of the mechanics of earth-based and space telescopes. The monolithic inertial sensor is a compact, light, fully scalable, tunable in frequency ($< 100\,mHz$), with large measurement band ($10^{-6}\,Hz \div 10\,Hz$) and high quality factor ($Q > 1500$ in air) instrument, with immunity to environmental noises guaranteed by an integrated laser optical readout. The measured sensitivity curve is in very good agreement with the theoretical one ($10^{-12}\,m/\sqrt{Hz}$) in the band ($0.1 \div 10\,Hz$). Although its natural application is in the fields of earthquake engineering and geophysics, its performances make it suitable also for applications as sensor in the control of mechanical suspensions and inertial platforms of interferometric detectors of gravitational waves, where a residual horizontal motion better than $10^{-15}\,m/\sqrt{Hz}$ in the band $0.01 \div 100\,Hz$ is a requirement, and in the control of the mechanics of ground-based or space telescopes.

Keywords: Folded Pendulum, Monolithic Sensor, Seismometer, Inertial Platform, Mechanical Suspension, Control System.

1. INTRODUCTION

Inertial control is a well-known technique, today used also for large band control of multi-stage suspensions (seismic attenuators) and inertial platforms for a variety of scientific and technical applications. This control technique requires suitably positioned acceleration sensors to provide the error signals and position sensors to provide the displacement signals. In particular, this technique is successfully used for the mechanical suspensions controls of the present interferometric detectors of gravitational waves (GEO600,[1] LIGO,[2] TAMA[3] and VIRGO[4]) and already scheduled for their upgrades (*Advanced* LIGO,[5] *Advanced* VIRGO,[6] LCGT[7]) and for a new generation of interferometric detectors, like the Einstein Telescope (ET),[8–10] whose design is more oriented to the observational aspects than to the detection capability.

The reduction of the fundamental and technical noises limiting the present and next generation of interferometric detectors assumes great relevance in the low frequency band ($0.01\,Hz \div 10\,Hz$), where a large population of detectable gravitational waves sources is present. In particular, seismic and newtonian noises must be largely reduced. The reduction of seismic noise requires the installation of the detectors in carefully chosen underground sites and the improvement, at the same time, of the mechanical suspensions performances. The reduction of Newtonian noise is, instead, still an open theoretical and experimental problem,[11] that requires suitable detector geometry and an optimized mechanical design of the suspensions in connection with the characteristics of the hosting underground site.

These requirements explain the general scientific interest in the development of new seismometers and accelerometers, with architectures that both guarantee high sensitivities ($< 10^{-10}\,m/\sqrt{Hz}$), expecially in the low

---

Send correspondence to Prof. Fabrizio Barone - E-mail: fbarone@unisa.it

frequency band ($1\,mHz \div 10\,Hz$), coupled to large immunity to environmental noises and suitable dimensions for their integration on suspension chains, on inertial platforms or in supporting mechanical structures that have to satisfy stringent requirements in the low frequency bands. In fact, sensors with performances that could make them suitable for these specific applications (e.g. STS-2,[12] Trillium-240[13]) have both unsuitable dimensions and weights and not adequate characteristics for working in ultra-high vacuum or cryogeny, having they been developed for geophysics applications.

Among the possible and available sensors, developed in the past and available in literature and/or currently used in experiments of physics or in commercial instruments, the Folded Pendulum (hereafter FP) architecture seems to be very promising. Folded Pendulum,[14] better known as *Watt-linkage*, has been applied as ultra-low frequency seismic attenuator for vibration isolation in interferometric detectors of gravitational waves.[15] More recently, single-axis monolithic accelerometers have been developed as acceleration sensors for the control system of advanced seismic mechanical attenuators and as tiltmeters in geophysics and environmental monitoring.[16–18]

The great advantage of an acceleration sensor is its force-feedback control architecture, which largely improves linearity and dynamic range of the device. The working principle is very simple: the inertial force generated by the vibrational signal on the test mass is compensated with a feed-back force, generated by a suitable control system and applied to the test mass, using an electromagnetic transducer. Being the feed-back force proportional to the ground acceleration, then the current used to drive the transducer is directly proportional to the mass acceleration. Although technology has largely improved in these years, nevertheless it is always very hard to design sensors that satisfy both requirements of high sensitivity and large band, expecially in connection with requirements of large dynamic range, long term stability and low sensitivity to environmental noises.

We opened a dedicated research line aimed to the development of low frequency large-band high-sensitivity sensors for geophysical applications (exploration of the low frequency band of the seismic spectrum), as a stand-alone sensors or as part of large and geographically distributed seismic networks, specializing some of them also for applications of interest in the field of gravitational waves detection.[19–21, 25–29] This research led us to build broadband single-axis monolithic FP sensors of reasonable size with very low natural resonance frequencies (down to $\approx 66\,mHz$), very large measurement bands and large dynamics. It is worth underlining that the great advantage of an inertial FP sensor is the removal of the limitations in the measurement band and sensitivity due to the internal feed-back electronics (like in the force feed-back accelerometers) and, consequently, less stringent requirements in the design and the implementation of the control system of the mechanical multi-stage suspensions (seismic attenuators), inertial platforms and, in general, supporting mechanical structures. Moreover, their full scalability makes it possible the design of monolithic sensors of suitable dimensions for positioning them also on the lower stages of multi-stage suspensions or on small dimension mechanical benches. Finally, the innovative application of laser optics techniques for the implementation of the monolithic FP sensor readouts (optical levers and interferometers) has largely improved their sensitivity, especially in the low frequency band, and immunity to environmental noises.[19–24]

In the following sections we will describe the new architecture of the horizontal monolithic FP sensor developed at the University of Salerno (UNISA Seismometer[30]), that can be configured both as seismometer and as accelerometer, in both cases with excellent performances. We will compare the theoretical performances, as predicted by the theoretical/numerical models, the experimental measurements and sensitivities in connection with different optical configurations of the integrated readout and experimental results obtained with its application at the INFN Gran Sasso National Laboratory, where some prototypes are operational since December 2010.[24, 26, 28] We, then, will describe in detail the preliminary results of their application as sensors to the top stage of a multistage mechanical suspension for seismic attenuation, discussing the model developed for their inclusion within the control system and a possible control strategy for a general non inertial control of mechanical suspensions and inertial platforms, applications that is of direct interest especially for high sensitive low frequency monitoring and control of the mechanics of earth-based and space telescopes.[29]

2. THEORETICAL MODELS

2.1 UNISA Monolithic Sensor Theoretical Model

A FP can be modeled according to the mechanical scheme shown in Figure 1, with two vertical beams of equal length, l, a pendulum of mass m_{p_1} and an inverted pendulum of mass m_{p_2}, concentrated in their centers of mass in P_1 and P_2, respectively, at $l_b = l/2$. The central mass, m_C, is modeled, instead, with two equivalent masses, m_{c_1} and m_{c_2} ($m_C = m_{c_1} + m_{c_2}$), concentrated in the pivot points C_1 and C_2, respectively at the same distance, l_p, measured from the pivot points of the pendulum and of the inverted pendulum arms.[23] The distance between the pivot points C_1 and C_2 is fixed and equal to l_d. All these hypotheses, are well satisfied in all our mechanical implementations of monolithic FP sensors.

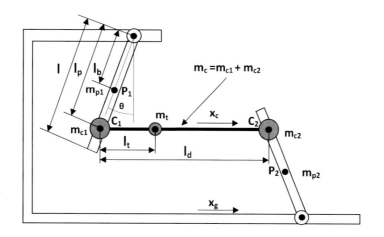

Figure 1. Folded Pendulum Mechanical Scheme.

Then, applying the simplified Lagrangian model developed by Liu, et al.[15] for small deflection angles, θ, and integrating it with a dissipation term, for a better description of the FP performances,[21] we were able to describe the basic FP dynamics, main characteristics and expected performances. It is, nevertheless important to underline here that design and implementation of optimized high performance FP sensors require a more detailed analysis of the FP dynamics, that can be obtained only with simulations based on very accurate numerical models, like the one we developed for this specific task.[19] According to this model, for small deflection angles, θ, the FP resonance frequency, f_o, is

$$f_o = \frac{\omega_o}{2\pi} = \frac{1}{2\pi}\sqrt{\frac{K_{g_{eq}} + K_{e_{eq}}}{M_{eq}}} = \sqrt{\frac{K_{eq}}{M_{eq}}} \qquad (1)$$

where $K_{g_{eq}}$, the equivalent gravitational linear stiffness constant, and $K_{e_{eq}}$, the equivalent elastic constant, are defined, respectively, as

$$K_{g_{eq}} = (m_{p_1} - m_{p_2})\frac{gl}{l_p^2} + (m_{c_1} - m_{c_2})\frac{g}{l_p} \qquad K_{e_{eq}} = \frac{k_\theta}{l_p^2} \qquad (2)$$

while M_{eq}, the equivalent mass, is defined as,

$$M_{eq} = (m_{p_1} + m_{p_2})\frac{l^2}{3l_p^2} + (m_{c_1} + m_{c_2})\frac{g}{l_p} \qquad (3)$$

Equation 1 is the classic expression of the resonance frequency of a spring-mass oscillator with an equivalent elastic constant, K_{eq}, and mass, M_{eq}. Note that with a suitable mechanical design of the FP mechanical components, the equivalent gravitational linear stiffness constant, $K_{g_{eq}}$, can assume negative values, partially compensating the equivalent elastic constant, $K_{e_{eq}}$, reducing the FP resonance. This means that the low frequency sensitivity of a FP seismometer (open loop configuration) can be largely improved with a careful design of the FP mechanics, and, in particular, with a careful choice of its resonance frequency: the lower the resonance frequency, the higher the low frequency sensitivity.

The FP sensitivity can be also changed on already implemented FPs, optimizing them for specific applications and requirements. This can be obtained using a specially developed tuning procedure, based on the addition of a suitable tuning mass, m_t. The addition of a tuning mass, m_t, positioned at a distance l_t from the pendulum-central-mass pivot point, C_1, as shown in Figure 1, changes the values of the equivalent masses m_{c_1} and m_{c_2}, that are increased by fractions of the tuning mass, as function of its position, l_t, according to the relations

$$\Delta m_{c_1} = m_t \left(1 - \frac{l_t}{l_d}\right) \qquad \Delta m_{c_2} = m_t \left(\frac{l_t}{l_d}\right) \qquad (4)$$

The equivalent gravitational linear stiffness constant, $K_{g_{eq}}$, and the equivalent mass, M_{eq}, change accordingly,

$$\Delta K_{g_{eq}} = (\Delta m_{c_1} - \Delta m_{c_2})\frac{g}{l_p} = m_t \left(1 - \frac{2l_t}{l_d}\right) \qquad \Delta M_{eq} = m_t \qquad (5)$$

According to the position of the tuning mass, this term assumes positive or negative values. Hence, the FP resonance frequency, f_o, increases or decreases, respectively. In Figure 2, a picture of f_o vs. value (m_t) and position (l_t) of the tuning mass is shown for a typical monolithic sensor.

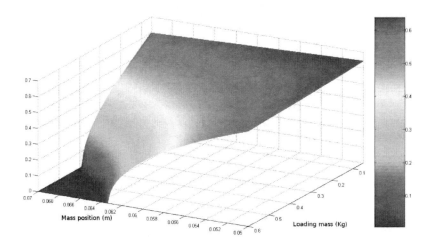

Figure 2. Resonance frequency, f_o, vs. tuning mass value, m_t, and its position, l_t, for a typical monolithic FP.

It is here worth underlining the importance of the FP tuning sensitivity for a comfortable and stable tuning, obtained deriving Equation 1 with respect to the position of the tuning mass, that is[21,25]

$$S_{f_o} = \frac{df_o}{dl_t} = \frac{g}{2\pi l_d l_p} \frac{m_t}{\sqrt{M_{eq}(m_t)K_{eq}}} \qquad (6)$$

As expected, Equation 6 demonstrates that the FP sensitivity is function of the value of the tuning mass, m_t, and that to obtain the same resonance frequency change, Δf_o, the heavier is the tuning mass, the smaller is its necessary displacement.

Defining, then, the coordinate of the FP frame (fixed to the ground) as x_g and the coordinate of the FP central mass (m_c) as x_c, then the mass displacement transfer function with respect to the ground displacement in the Laplace domain is[23]

$$H(s) = \frac{(x_c(s) - x_g(s))}{x_g(s)} = \frac{-(1-A_c)s^2}{s^2 + \frac{\omega_o}{Q(\omega_o)}s + \omega_o^2} \qquad (7)$$

where $Q(\omega_o)$ is the global Quality Factor and

$$A_c = \frac{\left(\frac{l_p}{3l} - \frac{1}{2}\right)(m_{p_1} - m_{p_2})}{M_{eq}} \qquad (8)$$

is the parameter related to the center of percussion effects.[15] Equation 7 can then be rewritten in the Fourier domain as

$$H(\omega) = \frac{x_c(\omega) - x_g(\omega)}{x_g(\omega)} = \frac{(1 - A_c)\omega^2}{-\omega^2 + \frac{\omega_o}{Q(\omega_o)}j\omega + \omega_o^2} \quad (9)$$

The dependence of $Q(\omega_o)$ on the FP resonance angular frequency, ω_o, has been experimentally demonstrated and will be discussed in the following section.

It is important to underline that the FP configuration couples horizontal forces and frame tilts. This problem is intrinsic to all horizontal inertial sensors, and cannot, in principle, be solved only without the introduction of a tiltmeter of comparable sensitivity, but that in practice, being the tilt and horizontal signals well separate in band for many applications, often it does not represent and unsolvable problem.[17,18]

2.2 Platform Control Theoretical Model

To test the open loop FP monolithic sensors, in particular, for the mechanical suspension control, we used a simplified model of one of the most efficient Seismic Attenuation System (SAS):[18] a chain of pendulums, suspended from a very low-frequency stage called Inverted Pendulum, that is actually, a large circular platform, mounted on three equi-spaced legs supporting the weight of all the other stages (Figure 3).

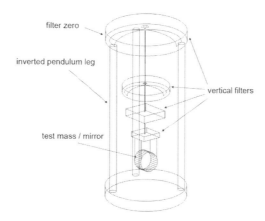

Figure 3. Model of SAS in the Napoli VIRGO laboratory.

Schematically, the mechanical suspension consists of four stages: the pre-isolation stage (Inverted Pendulum mounting at the top a circular rigid table that can is, in practice, a platform supporting instrumentation), the vertical filters, the chain of pendulums, the mirror for the measurement of the seismic attenuation at the lower stage. The vertical isolation is obtained with Monolithic Geometric Anti-Spring filters (MGAS).[31] The MGAS, based on linear anti-spring effect, is a set of radially arranged cantilever springs, mounted from a common retainer ring structure and opposing to each other via a central disk. The inverted pendulum (IP) is the element on which the horizontal seismic attenuation system is based. This system consists of three flexible joints, each supporting a leg. At the top, the three legs are connected to a rigid table by means of small flexures. Such a rigid table, the so called Filter Zero (F0), is a vertical filter which uses blades to suspend the chain of pendulums. The IP has the function of pre-filtering low frequency seismic noise, providing attenuation at frequencies of microseismic peaks, and providing a quasi-inertial stage to actively damp the motion of the suspended chain, avoiding actuation noise re-injection. The IP is a three degrees of freedom system: it has two translational modes and one torsional mode. It is clear that to implement an isolated lower stage in SAS, it is necessary to control the IP.

To control the position of Filter Zero, a set of sensors, for error signals measurement, and actuators, for acting on the platform, are necessary. In general, three position sensors are sufficient for DC correction of Filter Zero (LVDT) and three FP sensors for damping. To understand the relationship between the FP signal, the LVDT signal and the displacement of Filter Zero, we built a model described by the input-output block scheme of the whole system (sensor-system)(Figure 4).

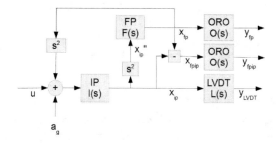

Figure 4. SAS Control Scheme.

Defining the control signal, $u(t)$, the position of filter zero respect to the tower, fixed to the ground, $x_{ip}(t)$, the position of the test mass of the FP respect to the filter zero, $x_{fp}(t)$, and the position of test mass of the FP respect to the tower, fixed to the ground, $x_{fpip}(t)$, then the transfer function of the inverted pendulum, $I(s)$, is

$$I(s) = \frac{x_{ip}(s)}{a_g(s)} = \frac{G_I}{s^2 + \frac{\omega_0}{Q}s + \omega_0^2}, \qquad (10)$$

the transfer function of the FP, $F(s)$, is

$$F(s) = \frac{x_{fp}}{a_{ip}} = \frac{(A_c - 1)}{s^2 + \frac{\omega_0}{Q}s + \omega_0^2}, \qquad (11)$$

the transfer function of LVDT, $L(s)$, is

$$L(s) = G_L x_{ip} \qquad (12)$$

where the transfer function of the optical readout, $O(s)$, is

$$O(s) = G_O x_{fp} \qquad (13)$$

where G_I, G_L and G_0 are the inverted pendulum, the LVDT and the Optical readout gain respectively. The relationship among the control signals and the readout signals can be, therefore, written as

$$\begin{aligned}
\frac{y_{fp}}{(u + a_g)} &= \frac{F(s)I(s)}{O(s)(1 - s^2 F(s)I(s))} \\
\frac{y_{LVDT}}{(u + a_g)} &= \frac{L(s)I(s)}{(1 - s^2 F(s)I(s))} \\
\frac{y_{fpip}}{(u + a_g)} &= \frac{s^2 L(s)I(s)F(s)}{O(s)(1 - s^2 F(s)I(s))} = \frac{(F(s) - L(s))I(s)}{O(s)(1 - s^2 F(s)I(s))}
\end{aligned} \qquad (14)$$

Substituting an explicit expression for the transfer function in the previous equations it is possible to evaluate the relationship among ground acceleration, Filter Zero displacement noise, sensor outputs and control signals. These expressions, necessary for the control filter design and error signal extraction, are in general obtained experimentally.

3. EXPERIMENTAL RESULTS

3.1 The UNISA Monolithic Sensor Experimental Results

The UNISA Horizontal Seismometer[30] is a new FP implementation, that allows very compact and robust FP implementations, with large gaps among the arms, the central mass and the frame, increasing the sensor dynamics to $\approx \pm 0.8 cm$, connected with elliptically shaped flexures. Furthermore the innovative application of laser optics techniques for the implementation of the FP monolithic readout (laser optical levers and laser interferometers), has improved its sensitivity, expecially in the low frequency band, increasing, at the same time, its immunity to environmental noises.[19] It is important to underline here that the sensitivity of the interferometric optical readout depends on many parameters, like the optical configuration, the quality of the laser, the open loop error signal extraction techniques used, etc.. Many configurations and techniques exist in literature suitable for this purpose, and applied to seismic sensors since long time, so that any improvement of the optical readout may be simply a matter of suitable choice and cost, and does not require special dedicated studies.

The UNISA Horizontal Seismometer[30] is characterized by a symmetric FP configuration, that allows the optimization of the moments of inertia of both the arms according to the specific application, with the two joints of the inverted pendulum working in compression. All the sensors were implemented in Aluminium Alloy 7075-T6, with the same dimensions (134 mm 134 mm 40 mm), the same ellipticity (16/5) and thickness (100 μm) of the hinges. In Figure 5 the UNISA Horizontal Seismometer is shown. The application of the sensor with no

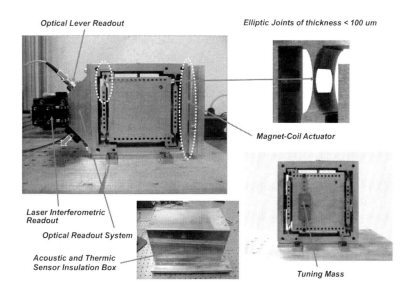

Figure 5. the UNISA Horizontal Seismometer.

feed-back control system has the great advantage that no limitations to the band and sensitivity are introduced by the control electronics, so that the quality of the instrument depends mainly on a careful and optimized mechanical design. The real limitations to the performances of a mechanical monolithic sensor become then the thermal noise, the sensitivity to the external temperature and acoustic noise (in air) and the readout sensitivity. Being the latter actually a laser optical readout (optical lever and laser interferometer), its quality is a problem of cost and portability of the sensor.

The performances of the UNISA Horizontal Seismometer were checked through a series of tests aimed both to demonstrate that the prototype follows the predictions of the theoretical/numerical models developed for simulation and design. Their reliability was proved by comparing the experimental monolithic FP transfer function at its natural design resonance frequency,[19] made using a standard measurement procedure used in control theory to obtain the transfer function of a linear system injecting white noise, and checking the effective quality of the tuning procedure developed (a resonance frequency of $66\,mHz$ has been obtained). This result is still more relevant if the small dimensions of the monolithic FP are taken into account. It is anyway important to underline that tuning the FP at its lowest possible natural resonance frequency improves the sensor measurement band at low frequencies, but at the same time reduces the restoring force of the pendulum to external perturbations, increasing the probability for the test mass to touch the frame, saturating the sensor output. Although this may be again only a problem of dynamics for the UNISA Horizontal Seismometer (that can be partially solved enlarging the gaps among the central mass-arms and arms-frame), it is not at all a problem if it is configured as accelerometer, being the central mass always forced in its rest position by the force feed-back control.

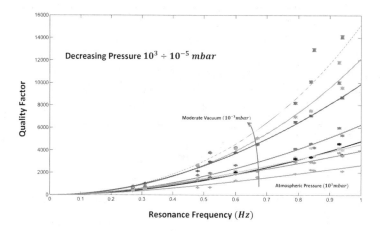

Figure 6. Quality Factor vs. Resonance Frequency for the UNISA Horizontal Seismometer.

The real problem is that the FP quality factor, Q, like every mechanical oscillator, decreases together with its natural frequency, so that the FP performances decrease moving its resonance frequency towards the low frequencies region. This effect is fully taken into account in Equation 9, where the theoretical prediction and/or the experimental measurements of the function $Q = Q(\omega)$ of the mechanical system become relevant. We remind that this function depends also on the value of the tuning mass, m_t. This dependence is function of the ratio m_t/m_c: the larger is this ratio, the larger is the increase of Q. In order to validate the UNISA Horizontal Seismometer we performed a series of tests to experimentally evaluate the function $Q = Q(\omega)$. We made all the tests positioning the sensor in a vacuum chamber and measuring the function $Q = Q(\omega)$ for different values of the pressure in the chamber with a $m_t = 240\,g$ tuning mass. We repeated the measures for different values of the resonance frequency, using the calibration procedure. The results are reported in Figure 6. As expected, the Quality Factor increases at the increase of the resonance frequency and shows the expected parabolic dependence, $Q(f_o) = a \cdot f_o^2$ (the value of a depending on the air pressure). In particular, Figure 6 shows that already at the atmospheric pressure and at the prototype design natural resonance frequency ($f_o \approx 0.721\,Hz$) the value of the quality factor is $Q > 1800$, thus demonstrating that the UNISA sensor perfectly fits for applications in air. Moreover, a quality Factor, $Q > 6000$ was measured for the prototype at the design natural resonance frequency with a moderate vacuum ($p = 10^{-5}\,mbar$), reaching values up to $Q > 14000$ for a natural resonance frequency of $f_o = 0.94\,Hz$. In synthesis, the sensor works very well both in vacuum and in air (the main goal of the FP monolithic new design). It is again relevant to underline that for this monolithic Aluminium (7075-T6) prototype values of $Q > 100$ are measured also for resonance frequencies below $100\,mHz$.

Finally, in Figure 7 the best theoretical and experimental sensitivities curves are shown at ($T = 300\,K$, assuming the FP tuned at a resonance frequency ($f_o = 100\,mHz$). The measurements were made with the FP central mass clamped to the frame, in air and with no thermal stabilization both for the optical lever (with PSD photodiodes) and for the interferometric readouts. The sensitivity of the STS-2 by Streckeisen[12] and the Trillium-240 by Nanometrics,[13] representing the state-of-art of the low frequency seismic sensors are reported for comparison in this figure, together with the Peterson New Low Noise Model (NLNM)[32] and the McManara and Bouland Noise Model,[33] representing the minimum measured Earth noise evaluated from a collection of seismic data from several sites located around the world.

Experimental applications of the FP have started since 2009. After preliminary tests performed in 2009 at DUSEL, in the Homestake Mine,[34] UNISA FP prototypes are in continuous acquisition since the end of 2010 in a seismic station located in the INFN Gran Sasso National Laboratory, for low frequency characterization both of the site and of the sensors. Figure 8 shows the Power Spectral Density of the ground acceleration in the band $10^{-6} \div 1\,Hz$. The microseismic peaks are still not well separated because the signal was not cleaned from the many earthquakes measured by the sensor during the 6000 hours of data. It worth underlining the large measurement

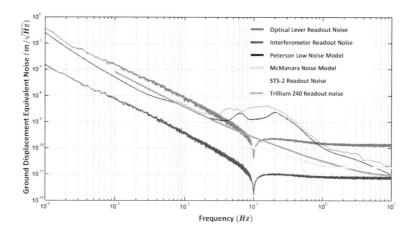

Figure 7. Measured readout sensitivity curves of the UNISA Horizontal Seismometer, compared with the Peterson Low Noise Model, with the McManara Noise Model and with two commercial sensors: STS-2 by Streckeisen and Trillium-240 by Nanometrics.

band ($10^{-6} \div 10\,Hz$) of the UNISA Horizontal Seismometer,[30] coupled with a good sensitivity, although largely limited by our technical choice of using a PSD optical lever as readout and the safe resonance frequency chosen for this first test in the LNGS ($200\,mHz$). Tuning the seismometer at a lower resonance frequency would simply translate the sensitivity curve towards the low frequency region, enlarging the low frequency band, while the change of the optical readout would largely increase the sensitivity through the whole band. Nevertheless, even with this not optimal configuration, the seismometers are already measuring in the bands $10^{-5} \div 5 \cdot 10^{-5}\,Hz$ and $50\,mHz \div 500\,mHz$.

The evaluation of the power spectral density of the acceleration using more data will allow the exploration of lower frequency regions. In fact, the UNISA Horizontal Seismometer[30] is an open loop sensor, so that there are no limitations coming from the feed-back control, as it happens in the majority of the commercial instruments used for seismic noise acquisition. The main limitations of this seismometer are the readout system electronic noise, the mechanical joints thermal noise and the air damping for seismometers not operating in vacuum.

In conclusion, the measurement band of the present implemented version of the UNISA Horizontal Seismometer with a laser optical lever readout is at least seven decades, $10^{-6}\,Hz \div 10\,Hz$. The quality of the data as sensor and its sensitivity let us think that it can be very effectively applied also as sensor in the mechanical

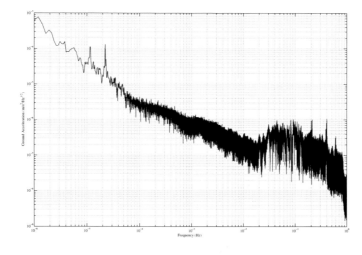

Figure 8. Power Spectral Density of the ground acceleration in the band $10^{-6} \div 1\,Hz$.

control systems, like, for example, in the control system of mechanical suspensions (seismic attenuators).

3.2 Platform Control Experimental Results

We used a triplet of linear variable differential transformers (LVDT), that are high-precision position sensors, mounting them in a triangular configuration at the edge of Filter Zero plate. They are low-power, ultra-high-vacuum compatible, non-contacting position sensor with nanometer resolution and centimeter dynamic range. LVDTs are used to measure the relative position of the IP with respect to the ground. Three horizontal open loop monolithic FP sensors with optical lever readouts, calibrated at resonant frequency of about $1\,Hz$ and with sensitivity of $4 \cdot 10^{-9}\,m/s^2/Hz^{1/2}$, are mounted at same angle of LVDTs for measuring the IP acceleration with respect to the ground. An example of Filter Zero position noise, as measured with open loop monolithic FP sensors is shown in Figure 9. This Figure shows that the sensor sensitivity is limited by ADC noise, for frequency above few Hertz, due to mechanical low pass filter of inverted pendulum.

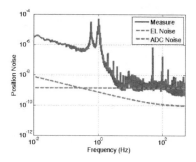

Figure 9. Power spectral density of filter zero as measured by open loop monolithic FP sensors, compared with theoretical noise.

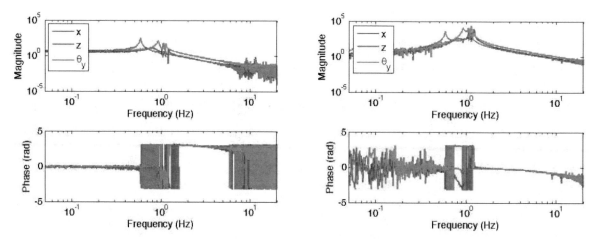

Figure 10. Measured diagonalized transfer functions of filter zero position measured by LVDT (on the left) and acceleration measured by open loop monolithic FP sensors (on the right).

The horizontal actuation system is obtained with a triplet of magnet-coil actuators, each made of a couple of coils and a central magnet. The magnet is orthogonal to the coil axis and a current passing through the coil generates a proportional force. Actuators are mounted in a triangular configuration for the LVDT and for the open loop monolithic FP sensors but, for practical reasons, not in the same position. The sensors and actuators signals, sampled at $4\,kHz$, are processed with a standard digital system consisting of a 16 bit analog-to-digital converter (ADC), a central processing unit (CPU) and a 16 bit digital-to-analog converter (DAC).

The control strategy is based on a two step procedure: in the first step a diagonalization of error signals is made to take advantage of controlling a Single Input Single Output (SISO) system instead of a Multiple Input

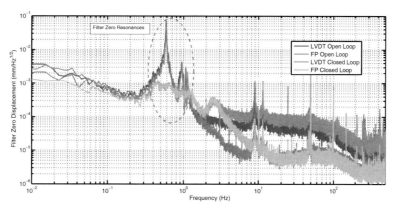

Figure 11. Preliminary results about Filter Zero control performance.

Multiple Output (MIMO) system, as described in.[35] In the second step three PID control filters, one for each degree of freedom, are implemented to control the IP dynamics. The transfer functions 14, after diagonalization procedure, shown in Figures 10, have been modeled for PID control filter design.

Preliminary results of the expected control performances are shown in Figure 11, where the Power Spectral Density of the Filter Zero displacement is shown in open loop and in closed loop configuration (using the LVDTs error signals). This figure shows that the FP sensors signals have a better signal-to-noise ratio with respect to the LVDT signal, so that they can be effectively used to perform a more stable control of Filter Zero. Furthermore, the same figure shows that the error signal provided by the FP sensors is sufficient to perform the inertial damping of Filter Zero resonances, resulting in a reduction of about $25\,dB$.

4. CONCLUSIONS

In this paper, after the description of the new typology of the sensors used, we have described the preliminary results of the control of the top stage (Filter Zero) of a mechanical multi-stage suspension (seismic attenuator). We have demonstrated that three open loop monolithic FP sensors, equipped with a simple optical lever readout, provide accurate error signals for the inertial damping of a multi-stage mechanical suspension or for an inertial platform, with a readout sensitivity of about $3 \cdot 10^{-9}\,m/Hz^{1/2}$, performances that are interesting also for the control of the mechanics of ground-based or space telescopes. Of course, the obtained results are only preliminary, because neither the physical limits of the sensors have been reached nor the new control strategy has been optimized. In fact, we expect large improvements by the next planned steps, that are the replacement of the FP sensor optical lever readout with already tested interferometric one, that will improve the sensors sensitivity (at least to $10^{-12}\,m/Hz^{1/2}$), and the design and optimization of the damping control with open loop monolithic FP sensors for tests in high vacuum.

REFERENCES

[1] H. Grote, et al., the LIGO Scientific Collaboration *Class. Quantum Grav.* **25**, 114043, doi: 10.1088/0264-9381/25/11/114043 (2008).

[2] B.P. Abbott, et al., the LIGO Scientific Coll., *Reports on Progress in Physics* **72** 076901, doi: 10.1088/0034-4885/72/7/076901 (2009).

[3] K. Arai, et al., the TAMA Collaboration, *Journal of Physics: Conference Series* **120** 032010, doi: 10.1088/1742-6596/120/3/032010 (2008).

[4] T. Accadia, et al., the VIRGO Collaboration, *Class. Quantum Grav.* **28** 114002, doi: 10.1088/0264-9381/28/11/114002 (2011).

[5] The Advanced LIGO Team, *Advanced LIGO Reference Design*, LIGO-M060056 (2011).

[6] The VIRGO Collaboration, *Advanced Virgo Technical Design Report VIR-0128A-12* (2012).

[7] K. Kuroda, the LCGT Coll., *Class. Quantum Grav* **27**, 084004, doi: 10.1088/0264-9381/27/8/084004 (2010).

[8] M. Punturo, et al., ET Science Team, *Class. Quantum Grav.* **27**, 194002 (p.12), doi: 10.1088/0264-9381/27/19/194002 (2010).

[9] M. Abernathy, et al., ET Science Team, *The Einstein Gravitational Wave Telescope Conceptual Design Study*, ET-0106C-10, issue 4, June 28 (2011).

[10] B. Sathyaprakash, et al., ET Science Team, *Class. and Quant. Grav.* **29** 124013 (p.16), doi:10.1088/0264-9381/29/12/124013 (2012).

[11] G. Cella, E. Cuoco, et al., *Class. and Quant. Grav.* **15** p.3339, doi: (1998).

[12] Nakayama Y, et al., *Performances test of STS-2 seismometers with various data loggers*, Proc. IWAA2004, CERN, Geneva, 4-7 October, (2004).

[13] http://www.nanometrics.ca/products/trillium-240.

[14] E. S. Fergusson, *US Nat. Museum Bull.*, **228**, 185 (1962).

[15] J. Liu, L. Ju, D.G. Blair, *Phys. Lett. A*, **228**, 243-249, doi: 10.1016/S0375-9601(97)00105-9 (1997).

[16] A. Bertolini, R. DeSalvo, F. Fidecaro, M. Francesconi, S. Marka, V. Sannibale, D. Simonetti, A. Takamori, H. Tariq, *Nucl. Instr. and Meth. A*, **556**, 616-623, doi:10.1016/j.nima.2005.10.117 (2006).

[17] A. Bertolini, R. DeSalvo, F. Fidecaro, A. Takamori, *IEEE Trans. on Geosci. And Rem. Sens.*, **44**, 273-276 doi: 10.1109/TGRS.2005.861006 (2006).

[18] A. Takamori, A. Bertolini, R. DeSalvo, A. Araya, T. Kanazawa, M. Shinohara, *Meas. Sci. Technol.*, **22**, 115901, doi: 10.1088/0957-0233/22/11/115901 (2011).

[19] F. Acernese, R. De Rosa, G. Giordano, R. Romano, F. Barone, *Rev. Sci. Instrum.*, **79**, 074501, doi:10.1063/1.2943415 (2008).

[20] F. Acernese G. Giordano, R. Romano, R. De Rosa, F. Barone, *Nucl Instrum. and Meth. A*, **617**, pp.457-458, ISSN: 0168-9002, doi: 10.1016/j.nima.2009.10.112 (2010).

[21] F. Acernese, R. De Rosa, F. Garufi, G. Giordano, R. Romano, F. Barone, *Journ of Phys. Conf. Series*, **228**, 012035, doi: 10.1088/1742-6596/228/1/01203 (2010).

[22] F. Acernese, R. De Rosa, G. Giordano, R. Romano, S. Vilasi, F. Barone, *Proc. SPIE* Vol. 7981, SPIE, Bellingham, 79814J (p.11), ISBN: 9780819485434, doi: 10.1117/12.879415 (2011).

[23] F. Acernese, R. De Rosa, G. Giordano, R. Romano, F. Barone, *Proc. SPIE* Vol. 8345, 83453F (p.9), doi: 10.1117/12.913389 (2012).

[24] F. Acernese, R. De Rosa, G. Giordano, R. Romano, S. Vilasi, F. Barone, *Journ of Phys. Conf. Series*, **339**, 012001 (p.10), doi: 10.1088/1742-6596/363/1/012001 (2012).

[25] F. Acernese, R. De Rosa, R. DeSalvo, F. Garufi, G. Giordano, J. Harms, V. Mandic, A. Sajeva, T. Trancynger, F. Barone, *Journ. Phys. Conf. Series*, **228**, 012036 (p.6), doi:10.1088/1742-6596/228/1/012036 (2010).

[26] F. Acernese, R. De Rosa, G. Giordano, R. Romano, S. Vilasi, F. Barone, *Proc. SPIE* Vol. 7981, 79814R (p.7), doi: 10.1117/12.879423 (2011).

[27] F. Acernese, R. De Rosa, G. Giordano, R. Romano, S. Vilasi, F. Barone, *Proc. SPIE* Vol. 7981, 798156 (p.10), doi: 10.1117/12.879420 (2011).

[28] F. Acernese, R. De Rosa, G. Giordano, R. Romano, F. Barone, *Proc. SPIE* Vol. 8345, 83453D (p.7), doi: 10.1117/12.913336 (2012).

[29] F. Acernese, R. De Rosa, G. Giordano, R. Romano, F. Barone, *Proc. SPIE* Vol. 8345, 83453C (p.6), doi: 10.1117/12.913334 (2012).

[30] F. Barone, G. Giordano, *Low frequency folded pendulum with high mechanical quality factor, and seismic sensor utilizing such a folded pendulum International application published under the patent cooperation treaty (PCT)* WO 2011/004413 A3 (2011).

[31] A. Bertolini, G. Cella, R. DeSalvo, V. Sannibale, *Nucl. Instr. Meth. A*, **435**, 475-483, doi: 10.1016/S0168-9002(99)00554-9 (1999).

[32] J. Berger, P. Davis, *2005 IRIS 5-Year Proposal*, 38 (2005).

[33] D.E. McManara and R.P. Buland, *Bull. Seism. Soc. Am.*, **94**, 1517-1527 (2004).

[34] F. Acernese, R. De Rosa, R. Desalvo, F. Garufi, G. Giordano, J. Harms, V. Mandic, A. Sajeva, T. Trancynger, F. Barone, *Journ. of Phys. Conf. Series* **228** p.012036 (2010).

[35] G. Persichetti, A. Chiummo, F. Acernese, F. Barone, R. De Rosa, F. Garufi, S. Mosca, *IEEE Trans. Nucl. Sci.*, Vol. 58, 1588, doi: 10.1109/TNS.2011.2159846 (2011).

Herzberg Institute of Astrophysics' vibration measurement capabilities with applications to astronomical instrumentation

P.W.G. Byrnes[a]
[a] National Research Council Canada, Herzberg Institute of Astrophysics,
5071 West Saanich Road, Victoria, B.C., Canada V9E 2E7

ABSTRACT

The Herzberg Institute of Astrophysics, Astronomy Technology Research Group's vibration measurement capabilities include modal test via impulse hammer or electrodynamic exciter, structural response monitoring via piezoelectric accelerometers, and data acquisition via LabVIEW virtual instruments. This paper will review our existing capabilities, and give examples of past and future applications relevant to astronomical instrumentation.

Keywords: vibration test, modal test

1. INTRODUCTION

Astronomical instrumentation is known to be highly sensitive to vibration. Although astronomical observatories generally go to great efforts to avoid unnecessary vibration sources and to mitigate the effects of those which are unavoidable, the transmission of vibration energy remains a concern for instrument designers and users.

Modal testing, in which the resonant vibration modes of a structure are deliberately excited with a low amplitude mechanical input, is a broad class of vibration testing that is popular in industry. The resonant frequencies, frequency response amplitude, damping and other properties of a structure or mechanism can be obtained through such a test. Often these results are used to refine finite element models and simulations, which usually depend on simplifications and assumptions.

Operational vibration testing, in which the vibration sources are items of operating equipment, is often used in industry to detect machine deterioration, to anticipate component failures and to schedule preventative maintenance. Ambient vibration monitoring is a related category of vibration testing, in which operational vibrations propagate from their sources through buildings and mechanical structures to points of interest where they are measured.

Knowledge of the ambient vibration environment provides an opportunity to make astronomical instrument designs more robust and tolerant of environmental vibrations. In addition, the vibration sources and transmission characteristics of future facilities can be better understood and environmental vibration concerns minimized through careful design choices.

The Herzberg Institute of Astrophysics (HIA) has assembled a set of multi-purpose vibration test equipment (Table 1), and has devised software tools to facilitate vibration testing of components, sub-systems and entire instruments. This paper summarizes the HIA's existing capabilities and describes some recent and future applications.

[a] Peter.byrnes@nrc-cnrc.gc.ca

Table 1 HIA vibration test equipment manifest

Item	Manufacturer	Model	Specifications	Qty.
Dynamic signal acquisition module	National Instruments	PXI-4461	24-bit; 4-channel input	1
Dynamic signal acquisition module	National Instruments	PXI-4462	24-bit; 2-ch. input; 2-ch. output	1
Dynamic signal acquisition module	National Instruments	PXI-4498	24-bit; 16-channel input	1
PXI Chassis	National Instruments	PXI-1036	6-slots	1
PXI Starfabric interface	National Instruments	PXI-8310	PCMCIA Type II	1
Impedance head	PCB Piezotronics	288D01	100 mV/ g; 22.4 mV/ N, IEPE	1
Triaxial accelerometer	Bruel & Kjaer	4524B	100 mV/ g, IEPE	1
Triaxial accelerometer	PCB Piezotronics	356B18	1 V/g, IEPE	3
Uniaxial accelerometer	Bruel & Kjaer	4507B004	100 mV/ g, IEPE	3
Seismic accelerometer	Bruel & Kjaer	8340	10 V/ g, IEPE	1
Impulse hammer	Bruel & Kjaer	8206	220 N; 22.3 mV/ N, IEPE	1
Electrodynamic exciter	TMS	2100E11	440 N, 2 Hz-3 kHz	1
Power amplifier	TMS	2100E18		1
Notebook computer	Dell	Precision M70		1
Control and data acquisition	National Instruments	LabVIEW	2011 Full development suite	1

2. MODAL TESTING

2.1 Common methods

Modal testing of a device involves intentionally stimulating its mode(s) of vibration while monitoring its response(s), usually via accelerometer(s) attached to the device under test. The stimulus can be either an impulse or a continuous waveform applied at one or more point(s) on the device under test.

In an impulse test, the device under test is struck with an instrumented hammer that includes a force transducer, which provides an output of the time-dependant force profile (closely approximating a half-sine waveform) during the brief interval of contact with the device under test. Impulse hammer testing is ideal for testing of small, low mass devices, although impulse hammers are also available in large sizes for testing of civil structures. In the frequency domain, a narrow impulse has a broad frequency band and will excite any resonant modes found within that band.

Another means of stimulating the device under test is with an electrodynamic exciter, which includes an electromagnetic voice coil that is driven with a time varying waveform. For modal testing, a low-amplitude swept sine waveform is often generated, amplified, and supplied to the exciter. Depending on the relative sizes of the device under test and of the electrodynamic exciter, the device under test may be mounted directly onto the armature of the electrodynamic exciter. Often, however, the electrodynamic exciter is connected to the device under test using a thin rod referred to as a stinger, which transfers only axial forces and eliminates lateral force transfer between the exciter and device under test. A force transducer is usually included in series with the stinger at the point of attachment with the device under test. When this transducer also includes an accelerometer it is sometimes called an impedance head.

Some characteristics of these two approaches to modal testing are compared in Table 2.

Table 2 Comparison of modal test approaches

Criterion	Impulse hammer	Electrodynamic exciter
Set up	Simple	Complex
Test duration	Short	Long
Test article size	Small to medium	Small to large
Repeatability	Low	High
Mass loading	No	Yes

2.2 Frequency response functions

The standard data product provided by a modal test is the frequency response function, which expresses the response of the device under test to a force input vs. frequency. When the measured response is acceleration, the frequency response function is called an accelerance plot, having magnitude units of acceleration per unit force or, in SI units kg^{-1}. Other forms of the frequency response are given in Table 3.

Table 3 Forms of the frequency response function (FRF) given a force input F [1]

Response parameter R	FRF (R/F)	Inverse FRF (F/R)
Displacement	Receptance	Dynamic stiffness
Velocity	Mobility	Mechanical impedance
Acceleration	Accelerance	Apparent mass

One method of evaluating the frequency response function is given in equation (1), where $G_{FR}(f)$ is the cross-correlation of the hammer (input) signal with the response (output) signal and $G_{FF}(f)$ is the hammer (input) signal auto-correlation. This method, known as the H_1 method, is preferable for impulse hammer tests where noise may exist in the output signal [3]. An alternative formulation of the frequency response function (H_2) has the auto-correlation of the response (output) signal in the numerator and the cross-correlation term in the denominator, and is preferred when noise may exist in the input signal. The H_3 method is an average of the H_1 and H_2 methods.

$$H_1(f) \equiv \frac{G_{FR}(f)}{G_{FF}(f)} \tag{1}$$

The frequency response is complex valued function. Often, it is presented as separate magnitude and phase plots versus frequency. If frequency responses are measured at multiple well-chosen points on the device under test, a matrix of FRFs with a spatial dimension can be compiled during a test. This can allow the mode shapes of the device under test to be reconstructed [2], however mode shape reconstruction is beyond the scope of this paper.

Measured frequency response functions can be used to verify finite element analysis predictions. Often, differences between experimental and analytical results can arise when experimental boundary conditions are not identical to the ideal boundary conditions used in the analysis. Other differences can arise due to inexact material properties or dimensional errors in the as-built device. Hence, modal testing can be a powerful quality assurance tool. Modal testing is often performed before and after an environmental test or an accelerated life test to detect and quantify any mechanical degradation of the device caused by exposure to the transportation, handling, or operational environments that have been simulated.

2.3 Impulse test virtual instrument

To facilitate modal testing by the impulse method, the HIA has developed a modal test virtual instrument using LabVIEW. The front panel of this virtual instrument is shown in Figure 1. All necessary data acquisition settings are graphically controlled via this interface. Raw voltage time-series data from the accelerometers and impulse hammer are displayed for each impulse struck. A cumulative frequency response function and coherence function are also displayed in real-time during an experiment. The immediate availability and display of both raw and processed data allows initial interpretation and subjective affirmation of the result quality during the test work flow. Both the raw time-series data and the frequency response function can be saved for further analysis and review.

The impulse hammer signal is shown on the front panel with an expanded timescale to allow close inspection. It is important that the chosen sampling rate resolves the impulse well. The acuity of the impulse waveform is a function of the compliance of both the device under test and the impulse hammer tip. For this reason, several interchangeable tips of various stiffnesses are available. Stiffer tips produce narrower impulses in the time domain resulting in greater bandwidth in the frequency domain, but the input energy is then distributed over the larger frequency range. The most compliant tip that can excite the resonant modes of interest should usually be chosen. A graphical indicator located in the top right corner of the front panel shows the auto-spectrum of the hammer impulse, allowing the usable bandwidth to be estimated.

Data acquisition is triggered by the rising edge of the hammer impulse signal. Buffered pre-trigger samples are acquired to resolve the initial points of the hammer impulse waveform. Provided that the sample length is sufficiently long to ensure that structural responses are well damped before being truncated, windowing is not usually required for an impulse test.

A signal flow diagram for this virtual instrument is shown in Figure 2.

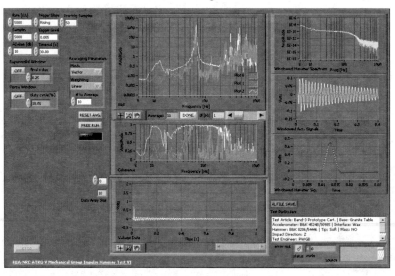

Figure 1 Impulse test virtual instrument front panel, showing raw data (bottom, centre) and frequency response functions [$\times$ 2.23 kg^{-1}] (top, centre). Other indicators show the hammer spectrum (top, right), and accelerometer signals at expanded scale (centre and bottom, right)

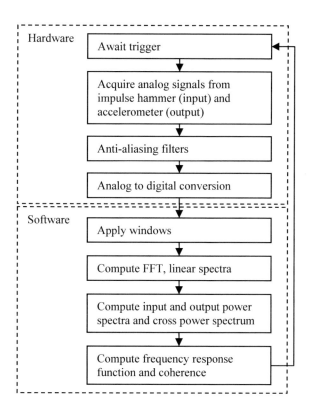

Figure 2 Impulse test signal flow diagram [4]

The accelerometer and impulse hammer signals are routed via shielded cables to one of three 24-bit dynamic signal acquisition module housed in a six-slot PXI- chassis. The dynamic signal acquisition modules include both analog and digital anti-aliasing filters. The digitized data is communicated from the PXI- chassis to a PCMCIA card installed in a notebook computer and connected by a pair of CAT5-e cables. The complete data acquisition set-up is shown in Figure 3.

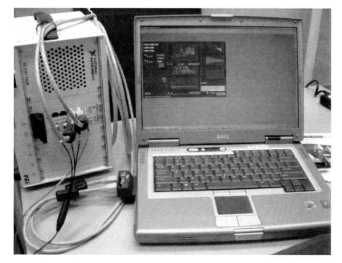

Figure 3 Modal test apparatus configuration showing PXI- chassis (left) containing the PXI-4461 dynamic signal acquisition module and the impulse test virtual instrument running on a notebook computer (right)

2.4 Case study: ALMA Band-3 receiver cartridge modal test

The HIA has been responsible for the development and production of the band-3 receiver cartridges for the Atacama Large Millimeter Array (ALMA). Modal testing of the prototype ALMA band-3 receiver cartridge was performed in 2007 using the impulse hammer approach [5].

During this test, the cartridge was secured on a handling fixture, which was clamped to a laboratory optical bench. A miniature tri-axial IEPE accelerometer with 100 mV/g sensitivity was affixed with beeswax to the 4 K-stage of the cartridge, just below the cold optics and mixer assembly. The structural modes of interest were cantilever bending modes of the cartridge body, in particular the first mode, which is due to bending about cut-outs in the thermal support tube nearest the base interface. These structural modes were excited by impacting the edge of the 4 K-stage with a lightweight impulse hammer in a horizontal direction, as shown in Figure 4.

Figure 4 . Impulse hammer test of the ALMA band-3 cartridge showing the location of the tri-axial accelerometer and the direction of impact.

The impulse test virtual instrument described in section 2.3 was used to display and analyze the signals and to display the frequency response function results in real time.

Time series of 1 second length were acquired at a rate of 5000 samples/s giving 1 Hz resolution. The sample length was chosen to ensure that structural responses were well damped before being truncated. As a result, no windowing was applied to the signals. Accelerance frequency response functions in three axes were computed using the H_1 method. Several frequency response functions were averaged to remove random noise effects.

The measured structural responses were primarily cantilever bending modes of the cartridge body with frequencies of 60 Hz and 69 Hz. These results differ from finite element analysis predictions (an example of which is shown Figure 5) due to the additional compliance of the assembly and test fixture, present during the test but excluded from the FEA, which assumes a rigid support at the base.

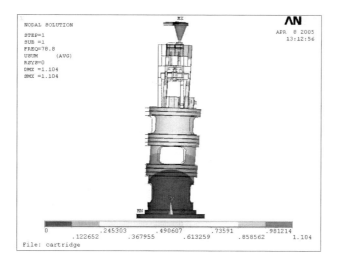

Figure 5 Finite element analysis prediction of first mode of vibration of the ALMA band-3 receiver cartridge

2.5 Structural damping

Although not computed automatically during the experiment, structural damping can be estimated by fitting an exponential curve to the peak value of each oscillation in the time series waveform as shown in Figure 6. If the structural mode of interest is well separated in frequency from others that may have been excited, a suitable band-pass filter can sometimes be used to help isolate the mode of interest. For the ALMA band-3 receiver cartridge a structural damping ratio of between 0.04 - 0.09 was measured. A future enhancement to the impulse testing virtual instrument will be to implement an automatic real-time damping calculation.

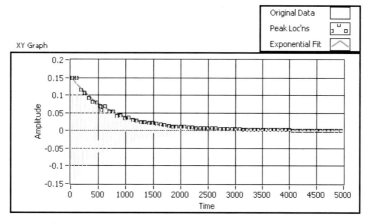

Figure 6 Structural resonance mode acceleration signal with exponential curve fit to estimate damping

3. VIBRATION MONITORING

Ambient vibration levels in many laboratory areas and in astronomical observatories is increasingly a concern. Vibrations originating from various sources propagate through structural paths to points of interest, such as optics that are highly sensitive to alignment, within instruments. Development of valid operational vibration requirements for future instruments will benefit from experimental data measured at existing facilities.

3.1 Vibration monitor virtual instrument

To facilitate vibration monitoring experiments, the HIA has developed a vibration monitor virtual instrument using LabVIEW. The front panel of this virtual instrument is shown in Figure 7. All necessary data acquisition settings are graphically controlled via this interface. Raw voltage time-series data from multiple accelerometers are displayed. In addition, processed data is displayed graphically during an experiment. The processed data include acceleration spectral density, displacement spectral density, integrated acceleration spectral density, rms acceleration, and rms displacement.. The virtual instrument can be programmed to operate intermittently on a pre-set duty cycle.

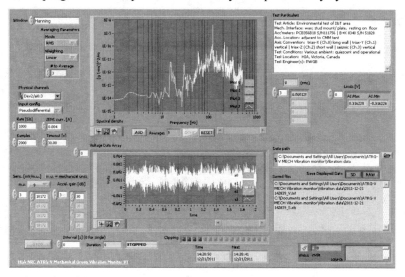

Figure 7 Vibration monitor virtual instrument front panel showing raw data (bottom, centre) and acceleration spectral density (top, centre). Alternative displays showing displacement spectral density, integrated acceleration spectral density and rms acceleration can be selected.

3.2 Case study: ambient vibration levels in the HIA integration and test area

The Herzberg Institute for Astrophysics maintains a well-equipped integration and test area in Victoria, B.C., Canada. The integration and test area occupies space on the ground floor of a modern office wing. Ambient vibration levels were monitored in this area during the winter of 2012. The vibration monitor virtual instrument described in paragraph 3.1 was configured to operate on a 50-minute duty cycle acquiring three two-second blocks of data from a single-axis high-sensitivity low-noise seismic accelerometer that was oriented vertically on the smooth concrete slab floor.

Typical acceleration spectral density curves from this experiment are shown in Figure 8. Values below approximately 10 Hz may not be reliable due to very low amplitude with respect to accelerometer noise, indicated by the dashed purple line. Figure 9 presents root mean squared acceleration amplitude for a period of several days and reveals very low vibration levels overnight with higher levels apparent during the business day. The generic vibration criterion applicable during these intervals of increased floor vibration is approximately VC-C [6]. Routine activities in the building and vehicular traffic outside are potential sources of these small but detectable floor vibrations.

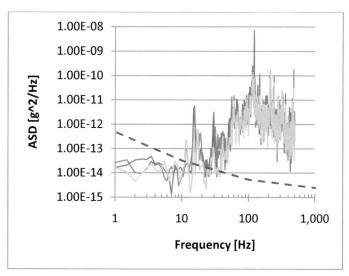

Figure 8 Typical vertical floor acceleration spectral densities in the HIA integration and test area; the dashed purple line is the transducer noise floor

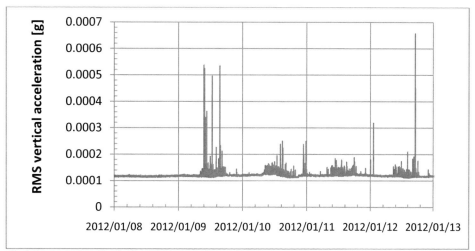

Figure 9 RMS vertical floor acceleration in the HIA integration and test area

3.3 Case study: PEARL facility observing platform

Comparative measurements were performed in the summer of 2011 to characterize and ascertain whether a potential temporary location on the observing platform of the Polar Environment Atmospheric Research Laboratory (PEARL) near Eureka, Nunavut (latitude 80°N) would be suitable for a 0.5 m robotic telescope campaign. The observing platform consists of a steel deck grating on a steel framework, which is supported by pillars that are structurally isolated from the building beneath.

During June, 2011, the 0.5 m robotic telescope was temporarily installed at the HIA Dominion Astrophysical Observatory's (DAO) Plaskett telescope. The robotic telescope's pedestal was placed on the observatory floor, which is steel plate supported on a steel framework structure, and which approximates the structural characteristics of PEARL observing platform. The floor near the telescope pedestal was instrumented with a single-axis high-sensitivity low-noise seismic accelerometer. Acceleration data were gathered while optical jitter observations of stars were undertaken. In addition, vibrations were deliberately induced during some observations by persons moving nearby within the observatory.

The vibration monitoring equipment was then deployed at the PEARL station in August 2011. The accelerometer was installed on a structural member supporting the observing platform floor near the proposed location for the telescope, as shown in Figure 10. Acceleration data was acquired during various wind conditions over several days. The resulting displacement spectra, shown in Figure 11, were compared with those obtained previously at the DAO.

The PEARL station observing platform was determined to be unsuitable for the robotic telescope use owing to intolerable levels of vibration. As a result of this study, a concrete pad providing a more rigid foundation for the robotic telescope will be constructed adjacent to the PEARL facility during the summer 2012 building season.

Figure 10 Seismic accelerometer installed under the observing platform of the PEARL atmospheric research

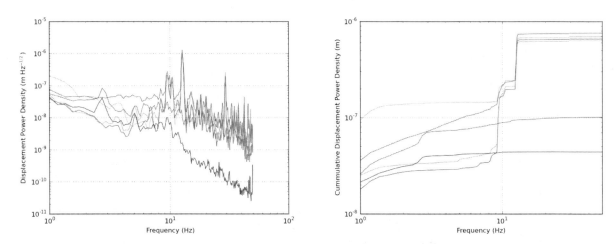

Figure 11 Displacement spectral density (left) and cumulative displacement spectral density (right) plots comparing vibrations at the PEARL observing platform and DAO Plaskett observatory floor for various wind speeds [7]. Yellow: PEARL 0.3m/s, Pink: PEARL 2m/s, Cyan: PEARL 4 m/s, Red: PEARL 8 m/s, Blue: DAO quiet, Green: DAO noisy.

4. FUTURE WORK

4.1 NFIRAOS structural FEM validation

The Herzberg Institute for Astrophysics is responsible for NFIRAOS, the Narrow Field Infra-Red Adaptive Optics System for the Thirty Meter Telescope (TMT) [8]. Equipment and systems, including NFIRAOS, mounted to the TMT structure are bound by earthquake survival requirements. Verification of compliance with these requirements by analysis is planned for NFIRAOS. It is also planned to validate the finite element models of the main mechanical structures of NFIRAOS by modal tests during the test phase of the project. These modal tests will provide as-built modal frequency results as well as structural damping ratios.

Critical sub-systems and assemblies, such as large mounted optics, will also be tested to affirm that their resonant characteristics are as predicted, ensuring that accelerations that may be realized during a seismic event will not exceed design values.

4.2 Accelerated life test of adhesive bonds

NFIRAOS will operate at 30 °C within a thermally controlled enclosure, and includes some large optics whose mounts are designed with bonded flexures. Thermal cycling and accelerated life testing of representative adhesive bonds is planned to qualify the adhesive joint design. Flaw detection during similar life testing has been demonstrated by monitoring the evolution of the frequency response function [9]. Changes in the FRF can provide early indication of adhesive joint dis-bonding. These tests are expected to use an electro-magnetic exciter to impart a low-level sine sweep or random vibration input to the test article, which will be instrumented with accelerometers.

5. CONCLUSION

The Herzberg Institute of Astrophysics has assembled a comprehensive set of vibration test equipment, has devised graphical software tools to facilitate both modal testing and ambient vibration monitoring, and has developed the methods and skills required to perform these tests. The productive use of this equipment and software has been demonstrated in several astronomical instrumentation applications to date. Further development of software capabilities is continuing, and there are numerous future applications planned or being considered.

REFERENCES

[1] Ewins, D.J., Modal testing, 2nd ed., Research Studies Press, Herts., UK, 241-248 (1997)
[2] Allemang, R.J.; Brown, D.L., Experimental modal analysis, Harris' Shock and vibration handbook, 5th ed., McGraw-Hill (2002)
[3] Dossing, O., Structural testing, part I: Mechanical mobility measurements, Bruel & Kjaer (1998)
[4] Avitabile, P., Experimental techniques, Society of experimental mechanics, 26-4 (2002)
[5] ALMA Band-3 Cartridge impact modal test report, FEND-40.02.03.00-156-A-REP (2007)
[6] Gordon, C.G., Generic criteria for vibration-sensitive equipment, Proc SPIE, 1619 (1991)
[7] Murowinski, R.; Sivanandam, S., private communication (2011)
[8] Herriot, G., TMT NFIRAOS: adaptive optics system for the Thirty Metre Telescope, SPIE 8447-58 (2012)
[9] Cote, P.; Desnoyer, N., Thermal stress failure criteria for a structural epoxy, Proc SPIE, 8125 (2011)

ALMA Array Element Astronomical Verification

S. Asayama[*a,b], L.B.G. Knee[a,c], P. G. Calisse[a,d], P. C. Cortés[a,e], R. Jager[a,d],
B. López[a], C. López[a], T. Nakos[a,d], N. Phillips[a,d], M. Radiszcz[a],
R. Simon[a,e], I. Toledo[a], N. Whyborn[a,d], H. Yatagai[a,b], J. P. McMullin[f], P. Planesas[g]

[a]Joint ALMA Observatory (Chile); [b]National Astronomical Observatory of Japan (Japan);
[c]NRC-Herzberg Institute of Astrophysics (Canada); [d]European Southern Observatory(Chile);
[e]National Radio Astronomy Observatory (U.S.); [f]National Solar Observatory(U.S.);
[g]Observatorio Astronómico Nacional (Spain)

ABSTRACT

The Atacama Large Millimeter/submillimeter Array (ALMA) will consist of at least 54 twelve-meter antennas and 12 seven-meter antennas operating as an aperture synthesis array in the (sub)millimeter wavelength range. The ALMA System Integration Science Team (SIST) is a group of scientists and data analysts whose primary task is to verify and characterize the astronomical performance of array elements as single dish and interferometric systems. The full set of tasks is required for the initial construction phase verification of every array element, and these can be divided roughly into fundamental antenna performance tests (verification of antenna surface accuracy, basic tracking, switching, and on-the-fly rastering) and astronomical radio verification tasks (radio pointing, focus, basic interferometry, and end-to-end spectroscopic verification). These activities occur both at the Operations Support Facility (just below 3000 m elevation) and at the Array Operations Site at 5000 m.

Keywords: ALMA, millimeter-wave, sub-millimeter, antenna

1. INTRODUCTION

The Atacama Large Millimeter/submillimeter Array (ALMA) is a joint project between astronomical organizations in Europe, North America, and East Asia, in collaboration with the Republic of Chile. ALMA will cover all the available atmospheric frequency windows between 30 GHz and 950 GHz in 10 bands. State-of-the-art microwave, digital, photonic, and software systems will capture the signals, transfer them to the central building, and correlate them, while maintaining accurate synchronization [1]. The array consists of a 12-m array (50 12-m antennas) and of the Atacama Compact Array (ACA: 4 12-m antennas and 12 7-m antennas). The ACA 12-m antennas are used to make a single-dish image to provide the "zero-spacing" information, and the ACA7-m antennas are used to obtain the short baseline data [2]. The ALMA antennas are operated at the array operation site (AOS) at an elevation of 5000 m above sea level and need to be transferred and relocated on antenna stations at AOS to change antenna configurations by dedicated antenna transporters. East Asia (NAOJ) will deliver all 12-m and 7-m ACA antennas, Europe (ESO) will deliver 25 12-m antennas, and North-America (AUI) will deliver 25 12-m antennas.

The ALMA System Integration Science Team (SIST) is a group of scientists and data analysts whose primary task is to verify and characterize the astronomical performance of array elements (individual, fully equipped antennas) as single dish and interferometric systems. This work is done for antennas being processed for the first time as part of ALMA construction, and for operational array elements which require re-verification following maintenance, repair, or upgrade activities. They also support System Verification [3], Commissioning and Science Verification [4], and Array Systems activities as required. The verification tasks performed vary according to the requirement of the array element in question. The full set of tasks is required for the initial construction phase verification of every array element, and these can be divided roughly into fundamental antenna performance tests (verification of antenna surface accuracy, basic tracking, switching, and on-the-fly rastering) and astronomical radio verification tasks (radio pointing, focus, basic interferometry, and end-to-end spectroscopic verification). These activities occur both at the Operations Support Facility (OSF: just below 3000 meters elevation) and at the AOS at 5000 meters.

2. ASSEMBLY INTEGRATION AND VERIFICATION

AIV (Assembly Integration Verification) is a project led by the System Integration Group (SIG) with the collaboration of the wider ALMA Department of Engineering (ADE). The AIV project assembles and integrates the ALMA equipment into the antennas and verifies their technical and astronomical performance at the OSF [5]. In the construction phase, SIST expends most of its effort in AIV. Antennas which pass the suite of verification tests are delivered to the Commissioning and Science Verification group to be integrated into the operational array at the AOS. AIV is organized into a 4-station processing model (see Figure 1) as follows:

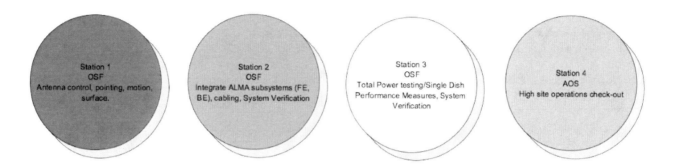

Figure 1. AIV processing station model

2.1 Station 1

Station 1 of AIV focuses on verifying antenna control, antenna motion, and measures and sets the antenna reflector surface. In this station, tests are run to confirm the interfaces to the antenna motion, reliable control of the antenna, and software safety mechanisms. Fast switching tests are performed to further confirm the antenna motion and settling times. Tower holography measurements are made to measure and set the antenna reflector surface to its specification.

2.2 Station 2

Station 2 of AIV focuses on the installation of component electrical/cryogenic systems and incrementally establishing the full connectivity of the antenna as an observing platform. In this station, the major electrical/cryogenic components are installed and their interfaces and control tested. Connectivity is established as well as the ability to monitor and control the established interfaces. Then tests of many other aspects of the system, including Front End amplitude stability, receiver noise performance and LO tuning are done. Once the system is fully operational, it is a functioning single dish telescope.

2.3 Station 3

Station 3 focuses on the total power checkout of the complete antenna system as an operational single dish telescope. Radio pointing measurements are obtained for the overall pointing characterization of the antenna as well as the beam squint (pointing offset between orthogonal polarizations). Focus curves are determined, and On-the-fly (OTF) mapping (a primary observing mode) and data acquisition are tested. Station 3 also performs a 2-antenna interferometric checkout to confirm phase stable cross-correlation to ensure proper array operation. The final radio pointing model is also obtained using interferometric pointing measurements, which are less sensitive to atmospheric fluctuations and variations in the system. Finally, a full end-to-end spectroscopic system checkout is done to test bandpass stability, baseline quality, to check for internally generated RFI, and the overall tuning performance.

2.4 Station 4

Station 4 of AIV takes place at the AOS and has the goal of confirming the operation of the characterized antenna at the high site and to do on-sky Station 3 tests in frequency bands not usable at the OSF due to atmospheric opacity. In this station, a quick assessment of the previous stations is done (e.g., receiver tempertures, radio pointing, focus, OTF) to confirm the antenna is still operating as expected. Additional high-frequency spectroscopic check out is done. The antenna is also checked to confirm cross-correlations with other antennas to insure proper array operation.

3. SPECIFICATIONS AND REQUIREMENTS

Table 1 summarizes a subset of the ALMA requirements and specifications most applicable to the AIV astronomical verification. Note that the antenna main reflector surface accuracy set target is chosen to ensure that the surface remains within specification over the full range of operational conditions – temperatures from -20 to +20ºC, day-time/night-time, elevation angles from 0 to 90 degrees, winds up to 10 m/s, etc. Due to better atmospheric characteristics, higher frequency bands above Band 8 (> 385 GHz) are largely characterized at the AOS as part of Station 4.

Table 1. Summary of specifications and requirements.

Task	Requirement (12-m antennas)	Requirement (7-m antennas)	Notes
Surface Verification	Antenna surface accuracy < 25 μm RMS for EL $\geq$ 20 deg.	Antenna surface accuracy < 20 μm RMS for EL $\geq$ 20 deg.	Main reflector surface set point for 12-m antennas $\leq$ 12 μm 7-m antennas $\leq$ 6 μm.
Fast Switching Verification	Time to settle after 1.5 deg step motion in any direction: $\leq$ 3.0" after 1.8 sec and $\leq$ 0.6" after 2.3 sec (EL $\leq$ 60 deg)	Same as 12-m antennas	
Radio Pointing Verification	All sky pointing accuracy < 2" RMS Polarization beam squint < 10% of primary beam FWHM.	Same as 12-m antennas	Higher frequency bands done at AOS.
Focus Curves Determination	Focus curves in Band 3 through Band 7	Same as 12-m antennas	Higher frequency bands done at AOS.
On-The-Fly Mapping Verification	Raster tracking: RMS < 1" for Scan speed v $\leq$ 0.05 deg/sec; RMS < 2" for Scan speed 0.05 < v $\leq$ 0.5 deg/sec	Same as 12-m antennas	There is also a "turn-around" RMS specification when moving from one leg of the raster to the next.
Interferometric Checkout	Establish stable single-baseline interferometer at OSF for Station 3 tasks.	Same as 12-m antennas	OSF uses prototypes of the LO system; line-length correctors and WVR phase correction not used.
Spectral Checkout	Frequency locking, bandpass, interference, polarization, signal paths	Same as 12-m antennas	Higher frequency bands done at AOS.

4. EXAMPLE OF INTEGRATION AND VERIFICATION

The ACA 7-m antenna designated "CM10" was the 39th antenna processed by AIV and delivered to the AOS on 24 May 2012. ACA 7-m fundamental antenna performances can be found in [7]. Here we present the astronomical radio verification results by SIST.

4.1 Surface measurements

The surface accuracy is measured using the near field radio holography technique using a nearby (~ 300 m distant) transmitter operating at ~ 100 GHz mounted on a tower. It is not technically feasible to manufacture a large diameter precision antenna as a single surface, therefore optimization of the surface is done by dividing the structure into smaller regions (i.e., panels) that can be individually adjusted in position to a desired shape. Thermal and gravitational collections are set into to the surface using models provided by the vendors/antenna Integrated Product Teams. Figure 2 shows that the final CM10 surface has a mean RMS = 5.7 microns including thermal and gravitational corrections after 4 surface adjustments. CM10 met the SIST surface set point goal of < 6 microns.

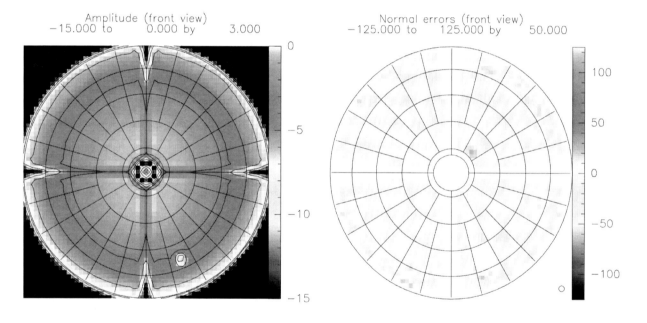

Figure 2. Final surface map of CM10 obtained with tower holography measurements. Left: Amplitude map in decibels showing shadowing due to the sub-reflector support legs, hexapod, and the hole for the optical pointing telescope. Right: Phase error map expressed in terms of surface error in microns. The scale in microns is denoted by a color bar, and the black lines outline the panels.

4.2 Radio pointing measurements

Since cross-correlation data is less affected by atmospheric and receiver fluctuations, the most sensitive way of performing pointing observations is using an interferometer. Verification can be done by pointing scans (5-points or cross scans) with monitoring the amplitude variation of the correlated signal and then performing a gaussian fit to the measurements. Figure 3 shows the interferometric pointing verification results used for model fitting. It contains 286 individual pointing determinations distributed over the sky. The model fit rms = 1.57" (population standard deviation = 1.62"), which meets the specifications.

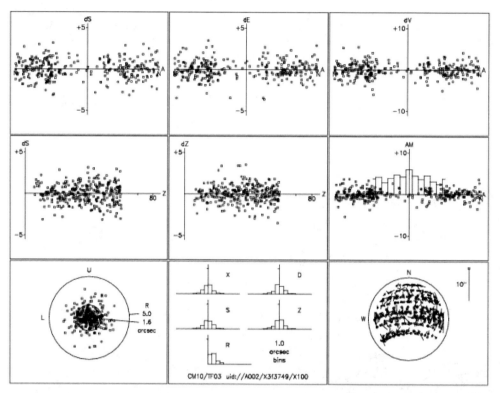

Figure 3. Interferometric pointing data plot, showing sky coverage (lower right panel), and bull's-eye plot (lower left panel), plus other panels showing residuals as a function of various angular quantities. Pointing is 1.57" rms.

4.3 Focus curve measurements

The gravitational and thermal deformation of the main reflector and the position change of the subreflector cause changes of the optimal focus position and each receiver band has its own focus offset. Focus models are determined to permit optimization of focus for these effects using simple parameterizations. Focus curve measurements are made using single-dish amplitude scans in X, Y, Z using planets, interleaved with pointing measurements. A Gaussian function is fitted to the amplitude as a function of the change in the subreflector location and the position of the peak amplitude with associated error is determined. This data along with the elevation and temperature data are used in a joint fit to produce the models. The focus plots (data and best-fit model) for Band 3 (100 GHz Band), Band 6 (230 GHz Band), and Band 7 (345 GHz Band) for T = +10°C are shown in Figure 4.

BAND 7	X [mm]	X0b7
	Y [mm]	Y0b7 + YC x cos (El)
	Z [mm]	Z0b7 + ZS x sin (El) +ZT x (T-10)
BAND 6	X [mm]	X0b7_6 + X0b7
	Y [mm]	Y0b7_6 + Y0b7 + YC x cos (El)
	Z [mm]	Z0b7_6 + Z0b7 + ZS x sin (El) +ZT x (T-10)
BAND 3	X [mm]	X0b7_3 + X0b7
	Y [mm]	Y0b7_3 + Y0b7 + YC x cos (El)
	Z [mm]	Z0b7_3 + Z0b7 + ZS x sin (El) +ZT x (T-10)

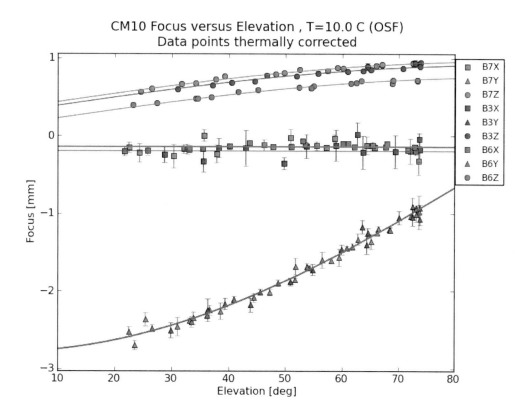

Figure 4. Top: Focus curve models for Band 3, 6, and 7. Bottom: CM10 Focus curves as a function of elevation corrected to T = +10°C. Squares: X axis. Triangles: Y axis. Circles: Z axis. Red: Band 7. Green: Band 6. Blue: Band 3.

4.4 Spectral checkout

The primary aim of the spectral checkout task is to test the functionality and performance of the frontend and IF of new antennas coming through AIV as a final product (spectra), and it is designed to look at general spectral quality by performing astronomical observations. We check for passband flatness, ripples from standing waves, frequency tuning reliability, and for internally generated interference, and a full-up end-to-end test of the spectroscopic signal path from the Front End through the Back End systems, including IF, digitization, and data transmission system. The test is performed through position switching (ON and OFF) observations on sources with known spectral line emission, as well as planets and empty patches of the sky (such as the celestial south pole) and also by using loads with known temperatures. The test also obtains antenna temperature-calibrated data by means of hot-ambient load-sky calibration. Figure 5 shows examples of spectra obtained for CM10. The strong $^{12}CO(2-1)$ 230.548 GHz and $^{12}CO(3-2)$ 345.796 GHz lines were observed, as shown in Figure 5. This observation also confirmed the correct physical connection of the USB and LSB channels.

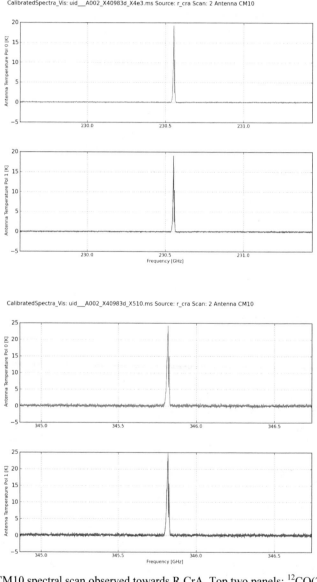

Figure 5. Examples of the CM10 spectral scan observed towards R CrA. Top two panels: $^{12}CO(2-1)$ spectrum in Band 6. Lower two panels: $^{12}CO(3-2)$ spectrum in Band 7. Red: Polarization 0. Blue: Polarization 1.

5. CONCLUSION

AIV has developed and demonstrated the capability of the "assembly line" verification of array elements. During the initial construction phase, the full set of tasks is required for every array element. Additional tests not described here are only performed on the first one or two examples of a new antenna design as a "design verification" activity. This allowed a speedup in processing in order to meet the schedule requirements. By automating SIST tests and the analysis procedures, AIV has ramped up to the current production phase processing rate which permits delivery of one fully-verified antenna every two weeks.

ACKNOWLEDGEMENTS

The Atacama Large Millimeter/submillimeter Array (ALMA), an international astronomy facility, is a partnership among Europe, Japan and North America, in cooperation with the Republic of Chile. ALMA is funded in Europe by the European Organization for Astronomical Research in the Southern Hemisphere (ESO), in Japan by the National Institutes of Natural Sciences (NINS) in cooperation with the Academia Sinica in Taiwan, and in North America by the U.S. National Science Foundation (NSF) in cooperation with the National Research Council of Canada (NRC) and the National Science Council of Taiwan (NSC). ALMA construction and operations are led on behalf of Europe by ESO, on behalf of Japan by the National Astronomical Observatory of Japan (NAOJ) and on behalf of North America by the National Radio Astronomy Observatory (NRAO), which is managed by Associated Universities, Inc. (AUI).

The authors dedicate this paper to the memory of our ALMA colleague Professor Koh-Ichiro Morita.

REFERENCES

[1] Wootten, A., and Thompson, A. R., "The Atacama Large Millimeter/Submillimeter Array, " Proceedings of the IEEE vol 97, Issue 8, pp. 1463-1471 (2009)

[2] Iguchi, S., et al., "The Atacama Compact Array (ACA) ", Publications of the Astronomical Society of Japan, vol.61, No.1, pp.1-12, (2009)

[3] Hills, R. E., and Peck, A. B., "ALMA commissioning and science verification", Proc. SPIE 8444, Paper 90 (2012)

[4] Bhatia, R., et al., "System engineering of the Atacama Large Millimeter/submillimeter Array", Proc. SPIE 8449, Paper 6 (2012)

[5] Lopez, B., McMullin J. P., Whyborn, N. D., and Duvall, E., "Engineering within the Assembly, Verification and Integration (AIV) Process in ALMA", Proc. SPIE 7733, 77335B-77335B-6 (2010)

[6] Lopez, B., Jager, R., Whyborn, N. D., and Knee, L. B. G., "Assembly, integration, and verification (AIV) in ALMA: series processing of array elements," Proc. SPIE 8449, Paper 24 (2012)

[7] Saito, M., Inatani, J., Nakanishi, K., Saito, H., and Iguchi, S., "Atacama compact array antennas," Proc. SPIE 8444, Paper 128 (2012)

Trajectory generation for parametric rotating scan patterns at the LMT

David R. Smith*
MERLAB, P.C., 357 S. Candler St., Decatur, GA 30030, USA

Kamal Souccar
Large Millimeter Telescope, Astronomy Department,
University of Massachusetts, Amherst, MA 01003, USA

ABSTRACT

As main axis controllers for large, high precision telescopes have become more sophisticated, astronomers are developing new approaches to observing. Rather than merely pointing at a particular set of coordinates and tracking to account for the rotation of the Earth or for the ephemeris of the source, these new observing techniques often call for a telescope to scan in a pattern around the nominal source coordinates. These motions are conducted at higher rates and accelerations than traditional single point or raster maps, which introduces additional complexity into the control. Additionally, these motions must work in conjunction with an existing trajectory generator.

This paper describes how a general parametric rotating scan pattern has been implemented at the Large Millimeter Telescope (LMT). The desired motion is combined with the existing trajectory generator and allows the addition of a parametric scan pattern, defined for each axis. The pattern can also be continuously rotated at a constant rate as the system tracks the source. The development allows for any parametric scan pattern, and the particular case of a continuously rotating Lissajous pattern is addressed in detail.

Keywords: Large telescopes, Tracking, Main axis control, LMT, Parametric scan patterns

1. INTRODUCTION

The Large Millimeter Telescope/Gran Telescopio Milimétrico (LMT) is a 50m diameter radio telescope designed for operation at 1–3 mm wavelengths[1]. It is located atop Sierra Negra in central Mexico. The LMT main axis servo loops were first closed about seven years ago, and since that time, there have been continuing improvements to its accuracy and capability.

In an early paper, Souccar and Smith presented the controller architecture along with some initial results[2]. Errors due to azimuth track joints and elevation gear rim misalignments were presented, as well as typical tracking performance. Later, the servo team had the opportunity to perform direct testing of the frequency response of the telescope structure using the control system. This allowed measurement of experimental frequency response functions in both azimuth and elevation with the telescope in different configurations. With this information, the team tuned the tracking and slewing performance of the LMT, characterized some of the known errors and reduced some of them[3]. These improvements included non-linear compensation of the system at very slow rates to reduce the stick-slip limit cycle of the telescope[4].

As a result of this earlier work, the LMT was able to track a given astronomical source, and even to perform raster scan patterns for on-the-fly maps. However, the fully-digital architecture of the main axis control system, together with the application of a trajectory generator[5] is capable of much more sophisticated observing strategies.

The scientific and instrument teams suggested that other telescopes employed more complicated scan patterns with good results[6]. Such scan patterns are added to a nominal tracking position in order to map the area

---

* David_Smith@merlab.com; Phone 404-378-2138

around the source. The usual approach for doing this is to employ a parametrically-defined Lissajous pattern, but it would be desirable to allow an arbitrary parametric scan pattern. Additionally, in order to obtain more uniform sky coverage, it is useful to rotate the scan pattern continuously on the sky. A simple scan pattern (a 2:3 Lissajous figure) is shown in Figure 1. Each of the points in the graph represents an even time spacing. When this pattern is put in sky coordinates at a 60° elevation angle and rotated continuously, it results in the scan pattern shown in Figure 2.

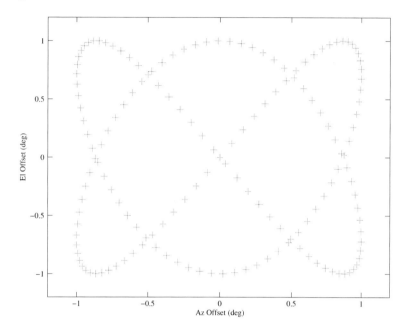

Figure 1: Example Scan Pattern (2:3 Lissajous Figure)

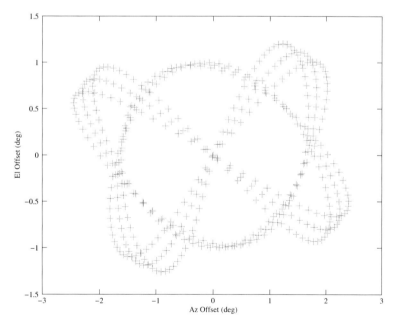

Figure 2: Example Rotating Scan Pattern (2:3 Lissajous)

This paper describes how the LMT can handle this type of observing while still retaining the benefits of the trajectory generator. The equations presented here allow for any parametric scan. The particular case of a Lissajous pattern is addressed in detail as an example.

2. THE LMT TRAJECTORY GENERATOR

The LMT trajectory generator[5] guarantees that the path commanded to the control system always remains within specified velocity, acceleration, and jerk limits. Further, it provides feedforward information, which enables improved controller tracking performance. The trajectory generator is called separately in each axis with a function call of the form:

$$\text{coeffs} = f(s_0, v_0, a_0, s_1, v_1, a_1, v_{\lim}, guess_v, guess_a, dt). \tag{1}$$

The initial conditions of the commanded position, velocity, and acceleration of the path (s_0, v_0, a_0) are always known from the previous call to the function and the state of the telescope. In the simplest case, the controller provides only the final position s_1 and sets the flags $guess_v$ and $guess_a$ to tell the trajectory generator to assume that either the system is attempting to track a source with slowly varying axis velocity or to assume that the new command is part of a larger motion for a position switch.

To accomplish this, the trajectory generator calculates, based on the current motion of the telescope, where it would be at the end of the next interval dt, which for the LMT is 40 ms (25 Hz update rate). This predicted value is compared to the user-supplied value of s_1. If the values are 'close', as determined by a tolerance, then the potential error is assumed to be due to a slight adjustment to the tracking rate, so the system assumes a final velocity and acceleration of $v_1 = (s_1 - s_0)/dt$ and $a_1 = 0$.

If the error is larger than some other tolerance, the trajectory generator assumes that the system is making a step but that it will continue tracking at the current velocity. This is appropriate for a position switching observation. Thus, it assumes that $v_1 = v_0$ and $a_1 = 0$. For cases in between the two tolerances, the trajectory generator linearly transitions between the two behaviors. Since it saves the previously commanded v_0 and s_0, the trajectory generator generally guesses well for tracking operations.

Unfortunately, when including a function such as a rotating Lissajous pattern, this type of guessing is inappropriate. Figure 3 shows the elevation position for a constant-rate tracking motion plus the example 2:3 Lissajous figure shown previously. It is clear that the basic assumption at each time step that the final acceleration will be near zero is fundamentally incorrect, and the velocity is constantly changing. As a result, the $guess_v$ and $guess_a$ flags are turned off and the user must provide not only the target position s_1, but also the target velocity and acceleration (v_1 and a_1) for every time step in the trajectory.

3. PARAMETRIC SCAN PATTERNS

It is helpful to consider the system as the sum of the target position, velocity, and acceleration and the rotating pattern. The vectors become

$$\begin{aligned}
\mathbf{s} &= \mathbf{s}_{\text{track}} + \mathbf{s}_{\text{rotatingpattern}} = \mathbf{s}_t + \mathbf{s}_r \\
\mathbf{v} &= \mathbf{v}_{\text{track}} + \mathbf{v}_{\text{rotatingpattern}} = \mathbf{v}_t + \mathbf{v}_r \\
\mathbf{a} &= \mathbf{a}_{\text{track}} + \mathbf{a}_{\text{rotatingpattern}} = \mathbf{a}_t + \mathbf{a}_r.
\end{aligned} \tag{2}$$

Each of these vectors have an azimuth and elevation component. For this paper, x is the azimuth angle and y is the elevation angle. The track in azimuth and elevation (i.e., x and y), as well as all of its derivatives are assumed to be known from the astronomical calculations or interpolated from the positions. This leaves the task

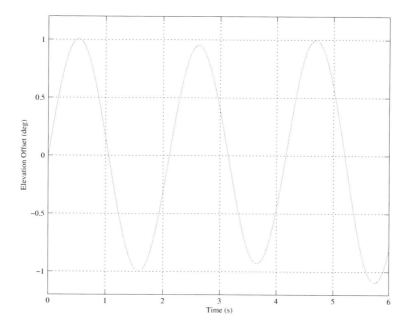

Figure 3: Example Elevation Rotating Scan Trajectory

of determining the effect of the pattern on the target position, velocity, and acceleration in azimuth $(x, \dot{x}, \ddot{x})$ and elevation $(y, \dot{y}, \ddot{y})$.

3.1 Calculating the scan pattern

First, it is necessary to calculate the desired scan pattern. The development below will work for any parametrically defined scan pattern, but a Lissajous pattern is presented here as an illustrative example. For now, these are left in joint-space coordinates. That is, they are driving the azimuth and elevation angle of the telescope rather than trying to produce the pattern on the sky. The required correction in azimuth due to the elevation angle must be applied after rotating the pattern.

In order to retain the benefits of the trajectory generator, not only the scan pattern position must be determined, but also its velocity and acceleration. The azimuth and elevation positions are defined parametrically as $x_l(t)$ and $y_l(t)$, respectively. For the Lissajous figure example, this becomes

$$\begin{aligned} x_l &= A\sin(at + \phi) \\ y_l &= B\sin(bt). \end{aligned} \tag{3}$$

where A, B, a, b, and ϕ are constants. The azimuth and elevation scan velocities are simply the time derivatives of the position. For the Lissajous figure example, they are

$$\begin{aligned} \dot{x}_l &= Aa\cos(at + \phi) \\ \dot{y}_l &= Bb\cos(bt). \end{aligned} \tag{4}$$

Finally, the azimuth and elevation scan accelerations are simply the time derivatives of the velocity. For the Lissajous figure example, they are

$$\begin{aligned} \ddot{x}_l &= -Aa^2\sin(at + \phi) \\ \ddot{y}_l &= -Bb^2\sin(bt). \end{aligned} \tag{5}$$

3.2 Calculating the rotated scan pattern

Since some observing techniques require continuously rotating a given pattern on the sky at a constant rate, it is convenient to separate this part of the calculation from the parametric scan pattern defined above. We define this constant angular rotation rate as Ω. If the pattern does not rotate, this section must still be applied, but set $\Omega = 0$. Again, it is necessary to calculate the position, velocity, and acceleration of the pattern.

The equations for rotating the pattern in telescope joint space are simply

$$\begin{aligned} x_r &= x_l \cos\Omega t - y_l \sin\Omega t \\ y_r &= x_l \sin\Omega t + y_l \cos\Omega t. \end{aligned} \quad (6)$$

However, since the goal is to follow the rotating pattern on the sky (a sphere), the azimuth angle must be corrected by $1/\cos\theta_{\rm el}$. Thus, to generate a rotating pattern on the sky, the equations become

$$\begin{aligned} x_r &= (x_l \cos\Omega t - y_l \sin\Omega t)/\cos\theta_{\rm el} \\ y_r &= x_l \sin\Omega t + y_l \cos\Omega t. \end{aligned} \quad (7)$$

Where $\theta_{\rm el} = y_t + y_r$. It is essential to include this effect, as it doubles the azimuth range even at the modest elevation angle of 60°, and increases sharply with increasing elevation.

The calculation of the velocity is, of course, merely the derivative of the position functions. For astronomical observations, the tracking portion of $\theta_{\rm el}$, which is y_t, changes slowly. However, since the elevation scan rate $\dot{y}_r$ may be significant, the velocity calculation must include time variability of $\theta_{\rm el}$. For the constant pattern rotation rate Ω, the derivatives become:

$$\begin{aligned} \dot{x}_r &= (\dot{x}_l\cos\Omega t - x_l\Omega\sin\Omega t - \dot{y}_l\sin\Omega t - y_l\Omega\cos\Omega t)/\cos\theta_{\rm el} \\ &\quad + (x_l\cos\Omega t - y_l\sin\Omega t)\dot{\theta}_{\rm el}\sin\theta_{\rm el}/\cos^2\theta_{\rm el} \\ \dot{y}_r &= \dot{x}_l\sin\Omega t + x_l\Omega\cos\Omega t + \dot{y}_l\cos\Omega t - y_l\Omega\sin\Omega t. \end{aligned} \quad (8)$$

These equations reduce to

$$\begin{aligned} \dot{x}_r &= [(\dot{x}_l - y_l\Omega + x_l\dot{\theta}_{\rm el}\tan\theta_{\rm el})\cos\Omega t - (\dot{y}_l + x_l\Omega + y_l\dot{\theta}_{\rm el}\tan\theta_{\rm el})\sin\Omega t]/\cos\theta_{\rm el} \\ \dot{y}_r &= (\dot{x}_l - y_l\Omega)\sin\Omega t + (\dot{y}_l + x_l\Omega)\cos\Omega t. \end{aligned} \quad (9)$$

Where $\dot{\theta}_{\rm el} = \dot{y}_t + \dot{y}_r$.

Obtaining the accelerations requires one more time derivative. Again, both $\theta_{\rm el}$ and $\dot{\theta}_{\rm el}$ are functions of time and must be treated as such in the derivative calculation. This leads to the equations

$$\begin{aligned} \ddot{x}_r &= [(\ddot{x}_l - \dot{y}_l\Omega + (\dot{x}_l\dot{\theta}_{\rm el} + x_l\ddot{\theta}_{\rm el})\tan\theta_{\rm el} + x_l\dot{\theta}_{\rm el}^2(1+\tan^2\theta_{\rm el}))\cos\Omega t \\ &\quad - (\dot{x}_l\Omega - y_l\Omega^2 + x_l\dot{\theta}_{\rm el}\Omega\tan\theta_{\rm el})\sin\Omega t - (\ddot{y}_l + \dot{x}_l\Omega + (\dot{y}_l\dot{\theta}_{\rm el} + y_l\ddot{\theta}_{\rm el})\tan\theta_{\rm el} \\ &\quad + y_l\dot{\theta}_{\rm el}^2(1+\tan^2\theta_{\rm el}))\sin\Omega t - (\dot{y}_l\Omega + x_l\Omega^2 + y_l\dot{\theta}_{\rm el}\Omega\tan\theta_{\rm el})\cos\Omega t]/\cos\theta_{\rm el} \\ &\quad + \dot{\theta}_{\rm el}[(\dot{x}_l - y_l\Omega + x_l\dot{\theta}_{\rm el}\tan\theta_{\rm el})\cos\Omega t - (\dot{y}_l + x_l\Omega + y_l\dot{\theta}_{\rm el}\tan\theta_{\rm el})\sin\Omega t]\sin\theta_{\rm el}/\cos^2\theta_{\rm el} \\ \ddot{y}_r &= (\ddot{x}_l - \dot{y}_l\Omega)\sin\Omega t + (\dot{x}_l\Omega - y_l\Omega^2)\cos\Omega t \\ &\quad + (\ddot{y}_l + \dot{x}_l\Omega)\cos\Omega t - (\dot{y}_l\Omega + x_l\Omega^2)\sin\Omega t. \end{aligned} \quad (10)$$

This expression reduces to the form

$$\ddot{x}_r = [(\ddot{x}_l - 2\dot{y}_l\Omega - x_l\Omega^2 + x_l\ddot{\theta}_{el}\tan\theta_{el} + x_l\dot{\theta}_{el}^2(1 + 2\tan^2\theta_{el}) + 2(\dot{x}_l - y_l\Omega)\dot{\theta}_{el}\tan\theta_{el})\cos\Omega t$$
$$- (\ddot{y}_l + 2\dot{x}_l\Omega - y_l\Omega^2 + y_l\ddot{\theta}_{el}\tan\theta_{el} + y_l\dot{\theta}_{el}^2(1 + 2\tan^2\theta_{el}) + 2(\dot{y}_l + x_l\Omega)\dot{\theta}_{el}\tan\theta_{el})\sin\Omega t]/\cos\theta_{el} \quad (11)$$
$$\ddot{y}_r = (\ddot{x}_l - 2\dot{y}_l\Omega - x_l\Omega^2)\sin\Omega t + (\ddot{y}_l + 2\dot{x}_l\Omega - y_l\Omega^2)\cos\Omega t.$$

In all cases, $\theta_{el} = y_t + y_r$, $\dot{\theta}_{el} = \dot{y}_t + \dot{y}_r$, and $\ddot{\theta}_{el} = \ddot{y}_t + \ddot{y}_r = \ddot{y}_r$.

This formulation applies for any parametric definition of x_l and y_l. It is important to note that for significant scan velocities or pattern rotation angular velocities, especially at higher elevation angles, the acceleration has significant Coriolis terms.

3.3 Pointing corrections

In a typical pointing model, there are no velocity or acceleration dependent terms. Further, the corrections vary slowly across the sky. As a result, the pointing offsets can be added to the final calculated position of the scan pattern and the velocity and acceleration contributions are zero.

4. SOFTWARE IMPLEMENTATION

To implement this in the software requires calculating the tracking profile, calculating the pattern, rotating the pattern, and including the pointing corrections. Each of these is discussed in detail below.

4.1 Calculate the target tracking path

The first part of the calculation is straightforward. It is merely the determination of the tracking path of the target on the sky. The steps required to obtain all of the required information are as follows:

1. The system already has (s_0, v_0, a_0) of the desired track from the previous calculation. Define the vector s_0 to be the ordered pair (x_0, y_0).

2. Calculate the desired position at the next dt, and save it as (x_1, y_1). Set $x_t = x_1$ and $y_t = y_1$.

3. If the derivatives of the tracking path are known, calculate them directly. However, if the tracking path is given only as a function of position, calculate the desired position $2dt$ in the future. Call this az/el position (x_2, y_2) and use standard difference equations to construct the derivatives. If the tracking system is structured such that it can only provide the next value (at dt), the derivatives can be generated by saving previous values and using the appropriate backward difference formulas. In such a case, the tracking path acceleration can typically just be set to zero for astronomical sources.

4.2 Calculate the scan pattern

In this case, we consider a Lissajous figure. Any parametric tracking path can be substituted here, but the calculation shortcuts are specific to the Lissajous pattern. Throughout this section, the time variable t is the current time plus dt so that the calculations are providing the next input for the trajectory generator.

It is generally convenient to save the values of the parametric position and velocity in each direction. For the Lissajous example, save the values of $xlsin = A\sin(at + \phi)$, $xlcos = A\cos(at + \phi)$, $ylsin = B\sin bt$, and $ylcos = B\cos bt$. In terms of these values, calculate the position, velocity, and acceleration of the parametric shape. The formulas are as follows:

$$\begin{aligned}
x_l &= xlsin \\
y_l &= ylsin \\
\dot{x}_l &= a * xlcos \\
\dot{y}_l &= b * ylcos \\
\ddot{x}_l &= -a * a * xlsin \\
\ddot{y}_l &= -b * b * ylsin.
\end{aligned} \qquad (12)$$

4.3 Calculate the rotated scan pattern

Calculating the rotated scan pattern requires the most computation, so it is worthwhile to look for values that appear repeatedly. Note that once the scan pattern (x_l, y_l) and its derivatives have been calculated, the formulas here are the same, whether the scan pattern is a Lissajous figure or any other parametrically-defined shape. The calculation steps are as follows:

1. Save the value of $somt = \sin(omt)$ and $comt = \cos(omt)$.
2. Calculate $y_r = x_l * somt + y_l * comt$.
3. Calculate $\theta_{el} = y_t + y_r$.
4. Calculate and save the value of $sthel = \sin\theta_{el}$, $cthel = \cos\theta_{el}$, and $tthel = \tan\theta_{el}$.
5. Calculate $x_r = (x_l * comt - y_l * somt)/cthel$.
6. Calculate and save $tmp1 = \dot{x}_l - y_l\Omega$ and $tmp2 = \dot{y}_l + x_l\Omega$.
7. Calculate $\dot{y}_r = tmp1 * somt + tmp2 * comt$.
8. Calculate and save $\dot{\theta}_{el} = \dot{y}_t + \dot{y}_r$.
9. Calculate and save $theldottan = \dot{\theta}_{el} * tthel$.
10. Calculate $\dot{x}_r = ((tmp1 + x_l * theldottan) * comt - (tmp2 + y_l * theldottan) * somt)/cthel$.
11. Calculate $tmp3 = \ddot{x}_l - 2\dot{y}_l\Omega - x_l\Omega^2$ and $tmp4 = \ddot{y}_l + 2\dot{x}_l\Omega - y_l\Omega^2$.
12. Calculate $\ddot{y}_r = tmp3 * somt + tmp4 * comt$.
13. Calculate and save $\ddot{\theta}_{el} = \ddot{y}_t + \ddot{y}_r$.
14. Calculate and save $thelddottan = \ddot{\theta}_{el} * tthel$, $theldotsq = \dot{\theta}_{el}^2$, $tanval = 1 + 2 * tthel * tthel$, and $tmp5 = thelddottan + theldotsq * tanval$.
15. Calculate $\ddot{x}_r = [(tmp3 + x_l * tmp5 + 2 * tmp1 * theldottan) * comt - (tmp4 + y_l * tmp5 + 2 * tmp2 * theldottan) * somt]/cthel$.

4.4 Calculate the pointing corrections

Now that the actual axis angle commands are known to track the desired sky path, we can apply the appropriate pointing corrections. Again, we assume that the pointing corrections are slowly varying so that their derivatives are insignificant. The calculation steps are simple.

1. x_p = Azimuth Pointing correction for an Az/El of $(x_t + x_r, y_t + y_r)$.
2. y_p = Elevation Pointing correction for an Az/El of $(x_t + x_r, y_t + y_r)$.

5. FINAL TRAJECTORY GENERATOR COMMANDS

With these values all calculated, the final position, velocity, and acceleration are as follows:

$$\begin{aligned} s_1 &= (x_t + x_r + x_p, y_t + y_r + y_p) \\ v_1 &= (\dot{x}_t + \dot{x}_r, \dot{y}_t + \dot{y}_r) \\ a_1 &= (\ddot{x}_t + \ddot{x}_r, \ddot{y}_t + \ddot{y}_r). \end{aligned} \qquad (13)$$

Finally, call the trajectory generator in each axis using (s_1, v_1, a_1).

6. CONCLUSIONS

We have developed the equations necessary to add a rotating parametric scan pattern to the basic tracking functions of the LMT without sacrificing the advantages of the trajectory generator. The basic scan pattern is defined by time-parametric functions in azimuth and elevation then optionally rotated continuously on the sky at a constant rate. As part of the rotation calculation, the $1/\cos\theta_{\text{el}}$ azimuth corrections are included so that the desired pattern appears correctly on the sky. With this approach, the feedforward velocities and accelerations of the scan pattern can help to mitigate the effects of high scan rates, high pattern rotation rates, large patterns, or high elevation angles. The approach is general for any scan pattern that is parametrically defined in time with differentiable functions.

ACKNOWLEDGMENTS

The authors wish to express appreciation to the LMT site crew for their technical and logistical support in carrying out this work. This material is based in part upon work supported by National Science Foundation Grant Number AST-0838222 and/or Contract Number NNX09AR37G from the National Aeronautics and Space Administration.

REFERENCES

[1] Hughes, D.H., *et al.*, "The Large Millimeter Telescope (LMT): current status and preparations for early science observations," *Proceedings of the SPIE Conference on Astronomical Telescopes and Instrumentation: Ground-Based Telescopes*, Amsterdam, The Netherlands, 2012.
[2] Souccar, K., and Smith, D.R., "The architecture and initial results of the Large Millimeter Telescope tracking system," *Proc. SPIE*, **7012**, 701209 (2008).
[3] Smith, D.R., and Souccar, K., "Main axis control of the Large Millimeter Telescope," *Proc. SPIE*, **7733**, 77332M (2010).
[4] Smith, D.R., and Souccar, K., "Friction compensation strategies in large telescopes," *Proc. SPIE*, **7733**, 77332N (2010).
[5] Smith, D.R., and Souccar, K., "A Polynomial-based trajectory generator for improved telescope control," *Proc. SPIE*, **7019**, 701909 (2008).
[6] Kovács, A., "Scanning strategies for imaging arrays," *Proc. SPIE*, **7020**, 5 (2008).

Atacama Compact Array Antennas

M. Saito[*a,b], J. Inatani[a], K. Nakanishi[a,b], H. Saito[a], S. Iguchi[a]

[a] National Astronomical Observatory of Japan 2-21-1 Mitaka, Tokyo 188-8588, Japan
[b] Joint ALMA Observatory, Av. Alonso de Cordova 3107, Vitacura, Santiago, Chile

ABSTRACT

The ACA (Atacama Compact Array) system is an important element of ALMA and consists of four ACA 12-m antennas and twelve ACA 7-m antennas. The ACA system aims to acquire the total power data with four 12-m antennas and the short baseline interferometer data with 7-m antennas. The ACA system also increases reliability of the interferometer maps of astronomical sources larger than the field view of the 12-m antenna. The science performance of these antennas has been extensively verified at OSF (operation support facility) at an elevation of 2900 m in Atacama desert in northern Chile since 2007. The pointing performance has been verified with a dedicated optical pointing telescope, the servo performance is tested with angle encoders, and the surface accuracy has been measured with a radio holography method. Both ACA 12-m antennas and 7-m antennas have been successfully demonstrated to meet the very stringent ALMA specifications.

Keywords: ALMA, radio, antenna, pointing, surface, holography, servo

1. INTRODUCTION

ALMA (Atacama Large Millimeter/submillimeter Array) is a radio interferometer array under construction in the Atacama Desert at an altitude of approximately 5000 meters in northern Chile in the frequency range from 31.3 to 950 GHz (10 – 0.35 millimeter in wavelength) [1-6]. ALMA consists of fifty 12-m antennas and "Atacama Compact Array (ACA)". The ACA system is composed of four 12-m antennas and twelve 7-m antennas both of which are delivered by NAOJ (National Astronomical Observatory of Japan). The science operation of ALMA will be performed for 24 hrs. Since the ACA antennas are exposed, the antenna should meet the specification with solar irradiation load and wind load.

Figure 1. The ACA 12-m antenna (left) and the ACA 7-m antenna (right)

---

* masao.saito@nao.ac.jp phone 81 422 34-3633; fax 81 422 34-3764;

1.1 Operating Conditions

ALMA defines primary operating conditions under which science observation can be performed having full performance of the antenna. The concrete parameters are summarized in Table 1.

Table 1. Operating Conditions.

Primary Operating Conditions			
Daytime		**Night**	
Ambient Temperature	-20 < T < 20 C	Ambient Temperature	-20 < T < 20 C
Wind*	6 m/s + gust	Wind*	9 m/s + gust
Solar irradiation	1290 W/m^2	Solar irradiation	no
Temperature Gradient	0.6 C/10min	Temperature Gradient	no
*at AOS (Array Operation Site) of 5050 m above sea level			

1.2 Specification of ACA 12-m and 7-m antennas

Table 2 shows the major specification summary of ACA 12-m and 7-m antennas.

Table 2. Summary of ACA 12-m and 7-m technical specifications

Major Specifications of ACA antennas*			
ACA 12-m antenna		**ACA 7-m Antenna**	
Absolute Pointing	≤ 2" rss	Absolute Pointing	≤ 2" rss
Offset Pointing	≤ 0".6 rss	Offset Pointing	≤ 0".6 rss
Surface Accuracy	< 25 μm	Surface Accuracy	< 20 μm
Fast Switching	< 1.8 s(settle to 3")	Fast Switching	< 1.9 s(settle to 3")
	< 0".6 rss (2.3-4.3 sec)		< 0".6 rss (2.4-4.4 sec)
*Under primary conditions:			

2. ACA ANTENNA DESIGN

2.1 Common Design for 12-m Antenna and 7-m antenna

The ACA antennas are designed, manufactured, and integrated by Mitsubishi Electric Corporation (MELCO). The antenna has a steel mount structure and is equipped with a receiver cabin made of steel. We have employed direct drive mechanism both to achieve a fast position switching capability and smooth drive at low to high velocities. The angle encoders are multi-pole resolvers with a 25-bit resolution that have demonstrated their high performance on the ASTE 10-m antenna [7]. The ACA antennas are equipped with a metrology system to correct pointing errors in real-time. The metrology system is composed of reference frame structures made of materials with low thermal expansion coefficient in the yoke and base structure.

2.2 ACA 12-m Antenna Main Reflector

The surface consists of seven rings of machined aluminum panels and the total number of panels is 205. The reflector surface has a suitable surface treatment to enable direct solar observing. Each reflector panel is mounted by four or five adjusters onto a reflector backup structure (BUS). Its backup structure (BUS) consists of low thermal expansion coefficient material; invar central hub, CFRP tube, invar joints, and CFRP board in order to guarantee the thermal stability demanded by the antenna. Further, to keep BUS temperature as uniform as possible, 16 blowers inside the BUS are installed. The subreflector support structure is also made of CFRP to ensure the performance under thermal load.

2.3 ACA 7-m Antenna Main Reflector

The primary reflector surface consists of machined aluminum panels. The reflector surface has a suitable surface treatment to enable direct solar observing. The reflector surface is mounted by three adjusters per panel onto a reflector backup structure (BUS). The BUS has a truss structure constructed of steel and has fans to stabilize the temperature distribution in order to minimize thermal deformation of the main reflector. The subreflector support structure is also made of steel and is thermally stable by circulating the air by a fan located at the foot of the structure.

3. PERFORMANCE MEASUREMENTS AND RESULTS

3.1 Absolute Pointing

All sky pointing measurements have been conducted with ACA antennas to evaluate the absolute pointing performance [8-9]. An optical pointing telescope onboard the BUS [10] is used for pointing measurements. Number of observed stars is typically around 120 which are well distributed on the whole sky and it takes about 30 minutes to complete a single set of observing run. The data were recorded as both processed stellar images and centroid positions of stars. We analyzed those data sets each of which contains at least more than 70 stellar images. We have obtained twenty or more all-sky pointing data sets for each antenna during two to four weeks of a pointing performance verification campaign.

The following procedures are applied as a standard analysis procedure. We exclude outliersand data taken at El lower than 20 degree. The pre-determined pointing model is applied for all the data sets. Fitting Az/El offset to the data and subtracting them from the data using the TPOINT telescope pointing analysis software [9], and one example of TPOINT output is given in Figure 2. The residual errors are evaluated by taking "population standard deviations".

In order to subtract the centroid fluctuation, a two minute-long continuous tracking of a bright star is performed and the stellar images with 1 second integration are collected. We take standard deviation of measured centroid fluctuation of the star within every 10 seconds, and we estimate seeing size as an average of them. The seeing measurement was performed before or after absolute pointing tests. The average optical seeing for 1 sec integration at the zenith was 0".2-0".4 dependent environment conditions. The value obtained from the nearest seeing measurement is adopted as the corresponded seeing value of each absolute pointing test.

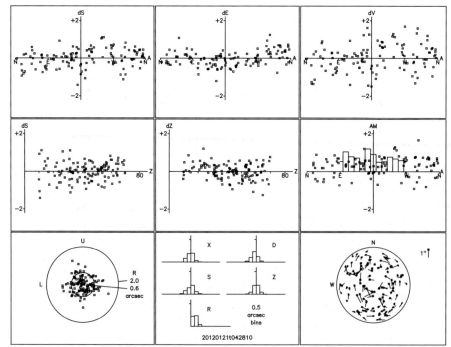

Figure 2. Example of result from an absolute pointing observing run of ACA 7-m antenna (CM09). The observation was carried out on January 21st, 2012. The plotted data are not original one, but have been removed offset in both Az and El axis. These plots were created by the TPOINT software

From the observed absolute error (APobs), we estimated that an absolute pointing error under windy condition (APw) at night is estimated as follows;

$$AP_w(night) = (AP_{obs}(night)^2 - d\theta_s(El=50)^2 - d\theta_{tw}(obs)^2 + d\theta_{tw}(9.5 \text{ m/s})^2 + d\theta_{mr}^2 + d\theta_{sr}^2)^{1/2} \quad (1)$$

The term AP_{obs} is apparent pointing error from nighttime measurements. The term $d\theta_s(El=50)$ is error of the centroid position within 1 second integration due to optical seeing at elevation being equal to 50 degree. Seeing induced error is subtracted individually from pointing error of each observation run using the estimated value. The term $d\theta_{tw}$ is wind induced tracking error for all sky at the AOS, which is estimated from a FEM model calculation. Wind induced tracking error during test observations [$d\theta_{tw}(obs)$] is estimated by considering the OSF and AOS pressure difference. The term $d\theta_{mr}$ is pointing error caused by the main reflector thermal metrology system. It consists of metrology estimation error and metrology sensor error. The term $d\theta_{sr}$ is pointing error caused by the subreflector position error which is not able to evaluate by the optical pointing telescope.

When we take root-square-sum of all the components and as a result the RMS value of $d\theta_{tw}(night)$) is evaluated compared against the absolute pointing specification of 2".0. So far all the ACA 12-m antennas ($AP_w(night)$=1".5 – 1".8) and ACA 7-m antennas ($AP_w(night)$= 0".9 – 1".3) meet the absolute pointing specification.

3.2 Offset Pointing

Offset pointing measurements were performed at nighttime with the optical pointing telescope. A group of three to five stars that are within 4 degree on the sky were selected and the telescope cycles between them over 15 minutes. Twenty four or more runs were carried out per antenna to achieve reasonable coverage of the entire sky.

The analysis of offset pointing observing runs proceeded as follows using the TPOINT software. We determined Az/El offsets using the first star of measurement. We subtract the offsets from the remaining data. After that a residual error shown in Figure 3 is evaluated. We adopted "population standard deviations" as offset pointing errors.

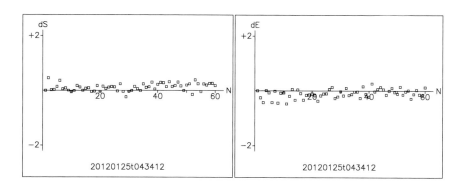

Figure 3. Example of result from offset pointing of three stars taken on January 25th, 2011 using ACA 7-m antenna. Residual pointing errors in Az (left panel) and El (right panel) are shown against observation sequence. These residuals that have been already removed offset in both Az and El axis that was determined by the firstly observed star. Apparent pointing error of this measurement is 0".27.

Similar to the absolute pointing, the offset pointing performance is also evaluated in the radio axis. Thus, seeing effect, wind effect, subreflector contribution (not measured by the optical pointing telescope), and metrology sensor errors are appropriately taken into account.

3.3 Surface Accuracy

The surface performance is evaluated mainly from holography measurements and FEM analysis under primary operating conditions.

Surface measurements were carried out using near-field holography at 104.2 GHz of ACA 12-m and 7-m antennas located at the OSF. The holography receiver was installed at the prime focus after removal of the subreflector. The transmitter is located on top of a ~50 m tower located at a linear distance to the antenna of 335.7 m. The holography maps were centered at an elevation of ~7 deg. Small variations from the optimum secondary position are corrected offline during the data reduction with the CLIC data reduction software [11]. Periodic pointing observations were also carried out during the observing runs. To reach a 20 cm spatial resolution on the main dish (with a SIGNAL feed taper factor of ~10 dB) in a reasonable time (i.e., ~1 hour), Over-sampled maps (HPBW~56" at 104.2 GHz) covering 1.24 degs were taken at scanning speeds of 300"/sec. Data samples come out of the DSP of the holography receiver at a rate of 48 ms/sample. More detailed description is given by [11].

The best performance of all the ACA 12-m antenna has achieved less than 8 μm after typically three rounds of panel setting on the basis of holography maps as an example shown in Figure 4. Measurement errors have been evaluated from the difference of two consecutive weighted apodized maps. The typical repeatability of these measurements at night ranges from 3-5 μm

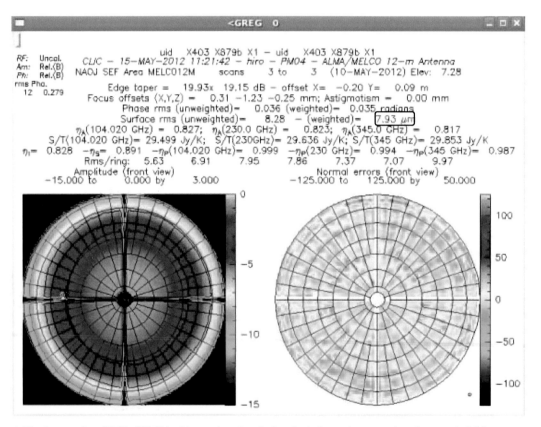

Figure 4. The best surface RMS of PM04 with panel setting during the holography campaign. Its error is 7.93 μm rms as shown as a yellow hatched square in the figure.

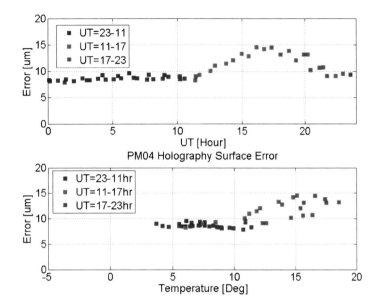

Figure 5. Surface RMS as a function of time. Blue crosses denote nighttime results. Daytime results are indicated with red and pink crosses. Bottom: Surface RMS as a function of Ambient Temperature. Symbols are the same as above.

Figure 5 presents the overview of holography maps of ACA 12-m antenna #4 (PM04); the total RMS as a function of UTC and ambient temperature and UTC. Even though day and night performance is different, a clear trend with temperature is not seen in the night data.

To evaluate the ambient temperature dependence quantitatively, the thermal behavior of the surface error of ACA antennas is modeled from night time holography maps. First, we apply Zernike polynomial expansion to fit the error pattern of night time holography maps after subtracting the reference (best) map. Second, each Zernike coefficient is linearly fit to ambient temperatures and then, a thermal model is constructed by summing all the Zernike components. Such a thermal model is used to evaluate at the extreme temperature under the primary operating conditions.

The worst surface error is usually observed during daytime as seen in Figure 5 and we found that such a degradation cannot be explained only by ambient temperature change or by wind load. Therefore, the temperature gradient of the structure caused by solar irradiation degrades the surface accuracy. We evaluate the thermal gradient effects from holography maps over a wide range of daytime conditions. First we subtract the reference map and the thermal model in the surface that is predicted by the thermal model due to the difference in the air temperature between the time when the reference map was derived and the air temperature of each map. Using the average map of the three worst residual maps the solar irradiation effect is quantitatively estimated by correcting for measurement error.

Small scale errors which were not sampled by the holography measurement are derived from coordinate measuring machine (CMM) measurements of the panel errors with a resolution of 30 mm. The removal of the larger spatial scales are done by convolving the data with a box function of 240 mm in size and subtracting the convolved version from the CMM data. On average, the fine scale errors are 69 % of all the measured errors with Coordinate Measuring Machine.

Other errors are taken from the FEM results provided by MELCO. The large term in the FEM is gravity deformation. To assess the FEM analysis of gravitational deformation of the main reflector, we have evaluated the bus stiffness of the ACA 12 m antenna by analyzing the sub-reflector focus position variation due to gravitational deformation. The FEM results are compared with the best fit focus curves determined from radio observations. The predicted values as a function of elevation angles from FEM analysis agrees with measurements within 4 %.

All above points are properly incorporated in the calculation and consequently, the surface error of both ACA 12-m antennas (23.8 – 24.9 μm) and 7-m antenna (19.0 – 19.4 μm) meet the ALMA specification by setting the dish at 0 degree in temperature.

3.4 Servo Performance

Fast switching test measurements were made in the following way. The antenna commanded to switch between two positions (with sidereal tracking) separated by 1.5 degrees on the sky. Metrology was activated during measurements. Measurements are made over a matrix of Azimuth (Az) and Elevation (El) positions; for Az 0 to 360 degrees in steps of 60 degrees, and for El 30 to 60 degrees in steps of 15 degrees. Switch position angles are 270 degrees (Az only), 0 degree (El only), and 45 degrees (both Az and El). At each Az/El matrix position switches of three position angles were done cyclic every 10 seconds and repeated thirty times for each position angle. As a result, total 1620 fast switches have been performed. We obtained the antenna positions from the encoder readings and estimated the antenna tracking error by comparing them to the commanded positions. Tracking errors in both Az and El angles in the fast switching are given in Figure 6. Figure 6 also shows a histogram of tracking errors of ACA 12-m antenna at 1.8 seconds after starting switches. The antenna positions are settled within 1".8 for 100% of the measurements. Figure 6 also shows a histogram of RMSs of antenna tracking error in the period between 2.3 and 4.3 seconds after start switching motion. For all of the switching performances, tracking errors at that period are much less than the verification condition (0".6).

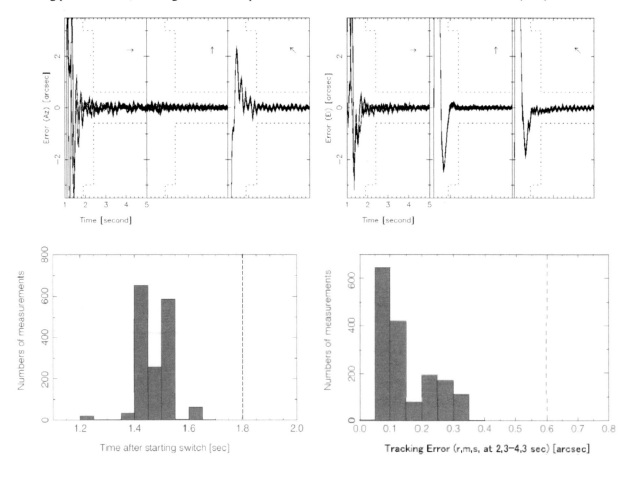

Figure 6. Plots of Az tracking errors (top left) and El tracking errors (top right) of ACA 12-m antenna with Az, El, and diagonal scans. Histogram of tracking errors at 1.8 seconds after starting switch (bottom left). Histogram of RMSs of antenna tracking error in the period between 2.3 and 4.3 seconds after start switching (bottom right)

4. SUMMARY

Table 3 summarizes the performance tests of both ACA 12-m and 7-m antennas. So far we have completed evaluation test all the 12-m antennas and nine out of 12 7-m antennas. All these values show that these antennas meet the major ALMA technical specifications.

Major Performance Test Results of ACA antennas*	
ACA 12-m antenna (4 antennas)	ACA 7-m Antenna (10 antennas)
Absolute Pointing 1.5 – 1".8 rss	Absolute Pointing 0.9 – 1".3 rss
Offset Pointing 0.58 – 0".59 rss	Offset Pointing 0.54 – 0".58 rss
Surface Accuracy 23.8 – 24.9 μm	Surface Accuracy 19.0 – 19.4 μm**
Fast Switching < 1.8 s(settle to 3")	Fast Switching < 1.9 s(settle to 3")
< 0".6 rss (2.3-4.3 sec)	< 0".6 rss (2.4-4.4 sec)
*Under primary conditions: ** from 3 antenna measurements	

ACKNOWLEDGMENTS

The authors wish to thank MELCO engineers and technicians for their state-of-art design, manufacturing, assembly and integration of ACA antennas and numerous individuals who have worked together in accomplishing the antenna construction and evaluation presented in this article. The authors also thank Joint ALMA Observatory staff for their kind support at the OSF.

REFERENCES

[1] Beasley, A. J., Murowinski, R., Tarenghi, M., "The Atacama Large Millimeter/submillimeter Array (ALMA)", Proc. SPIE 6267, 626702 (2006).

[2] Hills, R. E., Beasley, A. J., "The Atacama Large Millimeter/submillimeter Array", Proc. SPIE 7012, 70120N-70120N-8 (2008).

[3] Hills, R. E., Kurz, R. J., Peck, A. B., "ALMA: status report on construction and early results from commissioning", Proc. SPIE 7733, 773317-773317-10 (2010).

[4] de Graauw, M. W. M., Iguchi, S., McKinnon, M. M., Wild, W., Kurz, R. J., Hills, R. E., "Atacama large millimeter/submillimeter array (ALMA): construction and start of early science", Proc. SPIE 8444, (2012).

[5] Wootten, A., Thompson, A. R., "The Atacama Large Millimeter/Submillimeter Array," Proc. IEEE 97(8), 1463-1471 (2009).

[6] Iguchi. S. et al., "The Atacama Compact Array", PASJ, 61, 1-9 (2009).

[7] Ukita, N., Ezawa, H., Mimura, H., Suganuma A., Kitazawa, K., Masuda, T., Kawaguchi N., Sugiyama, R., Miyawaki, K., "A High-Precision Angle Encoder for a 10-m Submillimeter Antenna," Publ. of the Natl Astron. Obs. Japan 6, 59-64 (2001)

[8] Ukita, N., Saito, M., Ezawa, H., Ikenoue, B., Ishizaki, H., Iwashita, H., Yamaguchi, N., Hayakawa, T., "Design and performance of the ALMA-J prototype antenna", Proc. SPIE, 5489, 1085-1093 (2004)

[9] Mangum, J. G., Baars, J. W. M., Greve, A., Lucas, R., Snel, R. C., Wallace, P., Holdaway, M. "Evaluation of the ALMA Prototype Antennas", Publications of the Astronomical Society of the Pacific, 118, 1257-1301 (2006)

[10] Ukita, N., Ikenoue, B., Saito, M., "Optical Seeing Measurements with an Optical Telescope on a Radio Antenna", Publ. of the Natl Astron. Obs. Japan 11, 1-11 (2008)

[11] Baars, J. W. M., Lucas, R., Mangum, J. G., Lopez-Perez, J. A. "Near-Field Radio Holography of Large Reflector Antennas" IEEE Antennas and Propagation, Vol. 49, 24-41 (2007)

Very Large Millimeter/Submillimeter Array to Search for 2nd Earth

Satoru Iguchi*[a], Masao Saito[a,b]

[a]National Astronomical Observatory of Japan, 2-21-1 Osawa, Mitaka, Tokyo, Japan 181-8588;
[b]Joint ALMA Observatory, Av. Alonso de Cordova 3107, Vitacura, Santiago, Chile

ABSTRACT

ALMA (Atacama Large Millimeter/submillimeter Array) is a revolutionary radio telescope and its early scientific operation has just started. It is expected that ALMA will resolve several cosmic questions and will give us a new cosmic view. Our passion for astronomy naturally goes beyond ALMA because we believe that the 21st-century astronomy should pursue the new scientific frontier. In this conference, we propose a project of the future radio telescope to search for habitable planets and finally detect 2nd Earth as a migratable planet. Detection of 2nd Earth is one of the ultimate dreams not only for astronomers but also for every human being.

To directly detect 2nd Earth, we have to carefully design the sensitivity and angular resolution of the telescope by conducting trade-off analysis between the confusion limit and the minimum detectable temperature. The result of the sensitivity analysis is derived assuming an array that has sixty-four (64) 50-m antennas with 25-μm surface accuracy mainly located within the area of 300 km (up to 3000 km), dual-polarization SSB receivers with the best noise temperature performance achieved by ALMA or better, and IF bandwidth of 128 or 256 GHz.. We temporarily name this telescope "Very Large Millimeter/Submillimeter Array (VLMSA)". Since this sensitivity is extremely high, we can have a lot of chances to study the galaxy, star formation, cosmology and of course the new scientific frontier.

Keywords: Habitable Planet, Radio astronomy, millimeter and sub-millimeter, Radio interferometer

1. INTRODUCTION

Is the Earth the only planet where life exists in the vast universe? Are there any other planets where humans can live? Answers to these questions are yet to be discovered in spite of long-standing struggle of astronomers. Recently, vigorous efforts have been made to detect a "habitable planet" that has potential to sustain life. It is believed that habitable planets would be found in a "habitable zone" where water can exist constantly in liquid state to bring about the birth and evolution of life.

Is the birth of life a mere consequence of chemical reactions of terrestrial materials? Or, does outer space contain any material that could be the origin of life? There are many theories about the birth of life, but ALMA (Atacama Large Millimeter/submillimeter Array) aims to find evidence for it and unveil how life was formed and evolved [1-6].

Since the Big Bang, cosmic materials have evolved into various celestial objects and interstellar matter. From past observations at millimeter and infrared wavelengths, it has been found that interstellar matter takes many and varied forms and their evolution processes range widely. Although these interstellar matters are complex molecular aggregates, ALMA can identify their chemical composition using its unprecedented high resolution, sensitivity, and spectroscopic power. ALMA unveils the cosmic material evolution and is expected to find a clue to the origin of life.

Astronomy in the 20th century was the century of "astrophysics." Looking back in history, however, astronomy is a separate discipline that has developed independently of physics. Now in the 21st century, we should take a new approach for the development of modern astronomy without sticking to conventional ways. "Does the cosmic wall exist?" "What does exist outside the wall?" "Do aliens exist?" "Does 2nd Earth exist?" From these simple questions attracting everyone's attention, we had discussions and agreed to move forward with a plan to search for 2nd Earth which is exactly the same as the Earth we live on. Our ultimate goal is not to detect habitable planets that have potential to sustain life, but to discover Earth-like planets that have an environment suitable for the human race.

*s.iguchi@nao.ac.jp phone 81 422 34-3762; fax 81 422 34-3764; http://alma.mtk.nao.ac.jp/~iguchi/

2. WHAT IS VERY LARGE MILLIMETER/SUBMILLIMETER ARRAY (VLMSA)?

To directly detect 2nd Earth, we have to carefully design the sensitivity and angular resolution of the telescope by conducting trade-off analysis between the confusion limit and the minimum detectable temperature. We temporarily name this telescope "Very Large Millimeter/Submillimeter Array (VLMSA)". Since this sensitivity is extremely high, we have a lot of chances to study the galaxy, star formation, cosmology and of course the new scientific frontier. We assumed an array that has 64 50-m antennas with 25-μm surface accuracy mainly located within the area of 300 km (up to 3000 km), dual-polarization SSB receivers with the best noise temperature performance achieved by ALMA or better, and IF bandwidth of 128 or 256 GHz. Main parameters of VLMSA are listed in Table 1 and more details are presented in Section 2.2 to 2.4. Detailed sensitivity analysis was performed by referring to [6,7].

Table 1. Main parameters of VLMSA.

Item	Specification
Number of Antennas	64
Number of Antenna pairs	2016
Main Baseline Lengths	within 300 km
Expanded Baseline Lengths	up to 3000 km (> 3 antennas or 3)
Diameter of Antennas	50 m
Surface accuracy of Antennas	better than 25 μm
Pointing accuracy (offset, absolute)	better than 0.15 arcsec, 1.0 arcsec with metrology
Aperture Efficiency	Radiation efficiency (ohmic loss), Illumination efficiency, Spillover efficiency, Diffraction efficiency, Ruze efficiency and Blockage efficiency are utilized conveniently, which are the same as those of ALMA [6]
RF frequency coverage	64 GHz to 960 GHz (Band 1: 64-192 GHz, Band 2: 192-448 GHz, Band 3: 448-704 GHz, Band 4 704-960 GHz)
Single Sideband Receiving System	Image Rejection Ratio of better than -20 dB
Receiver Noise Temperatures	half values as compared with those of ALMA [1-6]
Polarization efficiency	> 99.5%
Beam squint	> 7 %
Intermediate Frequency Bandwidth	128 GHz around 100 GHz, 256 GHz at others
Quantization bit at digitizer	8 bit or more
Correlator input total bandwidth (max)	256 GHz at 8 bit
Frequency resolution at correlator	better than 4 kHz
Polarization products	1, 2 or 4
(Specific modes)	
Multi-beam Receiving System	at least 10x10 beams

2.1 Receiver Noise Temperature

The receiver noise temperatures of the SSB/2SB receiving system is defined as

$$T_{rx}(f) = \xi \times \frac{h \cdot f}{k_B} + B \, [\text{K}] \tag{1}$$

where h and k_B are the Planck constant and Boltzmann constant, and f was taken as the receiving frequency. The receiver noise parameters are defined in Table 2. These values are very challenging at this moment, but we hope we can achieve them in the future.

Table 2. Parameters of VLMSA receivers.

Frequency	Parameters
Band 1	$\xi = 2.5$, B= 10
Band 2	$\xi = 2.5$, B= 4
Band 3	$\xi = 4.0$, B= 4
Band 4	$\xi = 5.0$, B= 4

2.2 System noise temperature

We used the atmosphere transmission model for the ALMA site (see Figure 1), which was modeled by introducing two different continuum-like terms: collision-induced absorption of the dry atmosphere due to transient dipoles in symmetric molecules (N_2 and O_2), and continuum-like water vapor opacity. Moreover, the line absorptions of H_2O and O_2 and those isotope absorptions are included in this model.

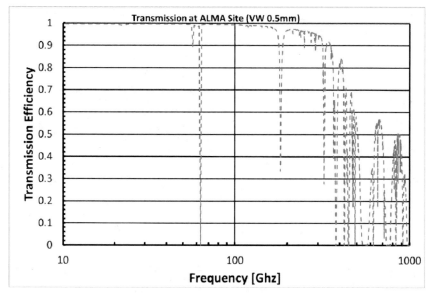

Figure 1. Assumed Atmospheric transparency at the ALMA site [Precipitable Water Vapor (PWV) = 0,5 mm] in 1-GHz steps.

The effective antenna noise temperature at the receiver plane is approximately defined by

$$T_{ant}(f) = \eta_{sp} T_{atm} \{1 - e^{-A\tau_0(f)}\} + (1 - \eta_{sp}) T_{amb} + \eta_{sp} T_{bg} e^{-A\tau_0(f)} \qquad (2)$$

where η_{sp} is the antenna spillover efficiency, $\tau_0(f)$ is the zenith opacity, A is the air mass, T_{atm} is the atmosphere temperature, T_{amb} is the ambient temperature, and T_{bg} is the Cosmic background temperature. Using the parameter values that the antenna spillover efficiency is 0.96, the air mass is 1.3 (at EL. 50 degrees), the atmosphere temperature is 257 K, the ambient temperature is 270 K, and the cosmic background temperature is 2.7 K, we can derive the effective antenna noise temperature at the receiver plane.

Finally, from Figure 1 and equation (2), we can estimate the antenna noise temperature of VLMSA. By adding the receiver noise temperature, we obtain the results as shown in Figure 2.

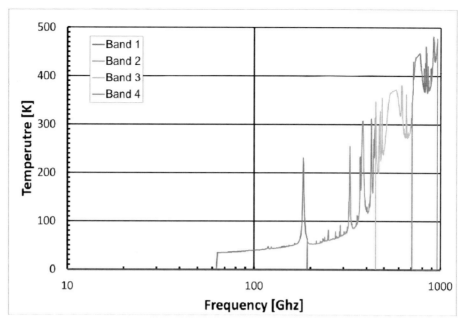

Figure 2. VLMSA receiver noise temperatures plus "assumed" antenna noise temperature at the receiver plane in 1-GHz steps.

2.3 Angular resolution

An image made from the *uv* plane track will have a synthesized beam with a slightly narrower minor axis of FWHM, which is written by

$$h_{uv} \times \frac{\lambda}{B_m} \qquad (3)$$

where λ is the observing wavelength [mm], B_m a maximum baseline length in array [mm], and h_{uv} a factor determined by the distribution of plots on the *uv* plane. The factor can be estimated with synthesized beam patterns calculated from the *uv* plot data. It was well known that there are no large discrepancies between our results and those at an h_{uv} of 0.7 [8]. At the baseline length of 300 km, we can achieve the angular resolution of 0.1 to 1 milli-arc-second (mas) (see Figure 3). When expanding to 3000 km, the angular resolution will be 0.01 to 0.1 mas.

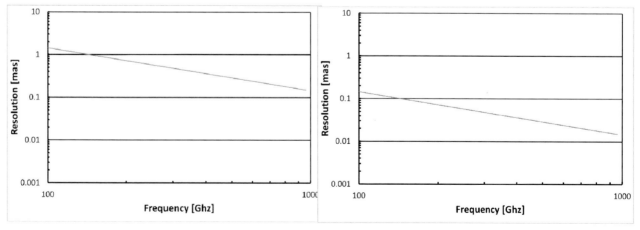

Figure 3. Angular resolution of VLMSA with the maximum baseline lengths of 300 km (left) and up to 3000 km (right).

2.4 Sensitivity Analysis: Sensitivity vs angular resolution

Figure 4 shows the sensitivity performances of VLMSA (3-σ detection) in 1-GHz steps of Band 1, 2, 3 and 4 and in all integrated bandwidth of each Band (see Table 1) under the 24-hour integration time. Also, this figure shows the relationship between the VLMSA detectable sensitivities and the radiation magnitude of planets like the Earth and Super Earth. Figure 5 shows the angular resolutions of VLMSA with baseline lengths of 300 km and up to 3000 km.

Note that we cannot use Band 4 to observe planets like Jupiter or the Earth under the baselines of up to 3000 km under the "24-hour" integration time, because the minimum detectable temperature at Band 4 in all integrated bandwidth (about 1500 K) exceeds the physical temperature of these planets.

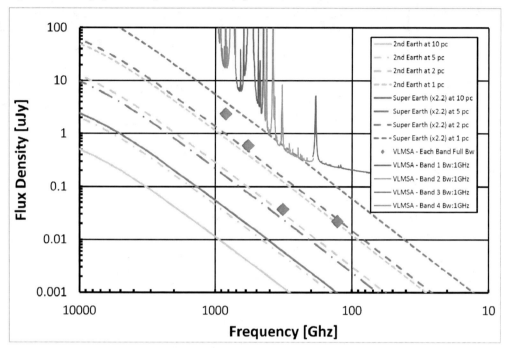

Figure 4. VLMSA detectable flux densities and the relationship between them and flux densities of 2nd Earth and Super Earth under the 24-hour integration time.

Figure 4 shows that VLMSA is successfully demonstrated to be able to detect "2nd Earth" at 2 pc with Band 2 (192-448 GHz: see Table 1) under the 24-hour integration time. The number of stars is small enough to be within 2 pc. The longer integration time, however, overcomes this limitation. Furthermore, in the current understanding of habitable planets, the liquid water can exist on planets larger than the Earth, "Super-Earths," thanks to green-house effects. In this case, such planets [9] can be detected up to 4 pc with Band 2 under the 24-hour integration time, and consequently about 30 stars can be a good target.

In addition, the impact of VLMSA is significant in planetary and stellar science. Giant planets with spectroscopy enable us to understand the planetary atmosphere in the evolution process of the solar system and also to investigate environments to harbor life. Images of red-giants shed light on the final stage of stars where our understanding is still rudimentary.

The powerful VLMSA opens a door to astrometry in the next level because high sensitivities allow us to detect stars with a distance of nearly 1 kpc with an angular resolution of 0.5 mas with the baseline lengths of 300 km (see Figure 3). The VLMSA therefore is a powerful instrument for astrometry. Such accurate and direct astrometry can cover a galactic structure in both plane and vertical, nearby star forming regions, and also stellar orbit anomaly with hidden planets. The advantage of VLMSA observations is being free from interstellar extinction, which can often be a severe limitation on optical and even near-infrared astrometry.

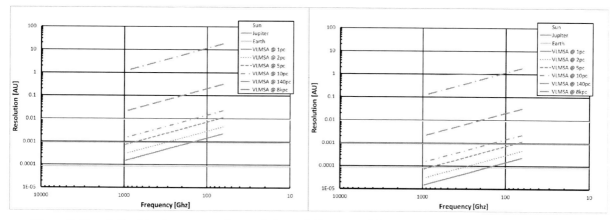

Figure 5. The relationships between angular resolution and diameters of the Sun, Jupiter and the Earth, (Left) in cases of baseline lengths of 300 km and (Right) up to 3000 km.

VLMSA can directly image Sun-like stars up to 10-20 pc at the maximum baseline lengths of 3000 km (see Figure 5 right). In addition, VLMSA will be able to image the planets like Jupiter within 5 pc by optimizing the brightness temperature sensitivity with two critical parameters of the baseline length and observing frequency. VLMSA will also be able to image the planets like Supper Earth and the Earth within 2 pc at the maximum baseline lengths of up to 3000 km (see Figure 5 right) due to more prolonged integration time. Thanks to direct detection of thermal emission, the radio approach is promising because the required dynamic range of 10^6 is smaller by 3 orders of magnitude than that at optical and near-infrared detecting reflection light.

3. PRIMARY SCIENCE REQUIREMENTS OF VLMSA

The primary science requirement for VLMSA is the flexibility to support the breadth of scientific investigation to be proposed by its creative scientist users over the decade's long lifetime of the instruments. However, three science requirements stand out in all the science planning for VLMSA. These three level-1 primary science requirements are: 1) The ability to directly detect thermal radiation from 2nd Earth in other planetary systems like our solar system at a distance of 2 pc, in less than 24 hours of observation; 2) The ability to clearly image a black hole with an accretion disk in the active central region of Sagittarius A and M87; and 3) The ability to identify the absolute position of stars like the Sun at a distance up to 1 kpc for the astrometry.

3.1 Search for 2nd Earth

VLMSA should have the ability to detect a planet like the Earth at a distance of 2 pc with 24-hour integrations time. Sensitivities to detect such a faint thermal continuum emission are mostly controlled by atmospheric transparency, aperture area, receiver temperatures, and correlator bandwidths. The aperture area is required to have more than 1.2×10^5 m^2 with receivers sensitive to both polarization having noise temperatures of only three times of quantum limit. At the same time, the antenna diameter is optimized to meet the pointing and surface accuracy requirements. In addition, the high dynamic range of 10^6 is required to detect 2nd Earth around a sun-like star.

If we make a direct image of a planet like the Earth, we will find a good solution to have the maximum baseline lengths of up to 3000 km.

3.2 Direct imaging of a black hole

To resolve a black hole with an accretion disk in the radio core, the angular resolution of 20 micro arc second may be necessary [9]. We will find a good solution to keep the extremely high sensitivity and to directly image a black hole with the angular resolution by locating a few antennas over the distance of up to 3000 km (see Figure 3). This sensitivity requirement can be easily realized with VLMSA, because these brightness temperatures exceed at least 100000 K.

3.3 Astrometry

VLMSA should have the capability to determine stellar positions up to 1 kpc with 24-hour integration time. Since the accuracy of astrometry is limited by the antenna location and phase stability, it is necessary to develop new metrology systems and phase correction methods to reduce coherence loss and antenna reference positions. This sensitivity requirement can be easily realized with VLMSA, because these stellar brightness temperatures like the Sun exceed at least 6000 K.

ACKNOWLEDGMENT

We thank Eiichiro Kokubo (NAOJ); Kotaro Kohno (Univ. of Tokyo); Munetake Momose (Ibaraki Univ.); Naoki Yoshida (Univ. of Tokyo, IPMU); Seiji Kameno (Kagoshima Univ.); Shogo Tachibana (Tokyo Univ., Frontier); Yasuhiro Murata (JAXA); Yuri Aikawa (Kobe Univ.); Yutaro Sekimoto (NAOJ, Advanced Technology Center); (in alphabetical order.) for their useful comments and suggestions in discussing big scientific goals to build the concept and technical requirements of the next-generation telescope.

REFERENCES

[1] Beasley, A. J., Murowinski, R., Tarenghi, M., "The Atacama Large Millimeter/submillimeter Array (ALMA)", Proc. SPIE 6267, 626702 (2006).
[2] Hills, R. E., Beasley, A. J., "The Atacama Large Millimeter/submillimeter Array", Proc. SPIE 7012, 70120N-70120N-8 (2008).
[3] Hills, R. E., Kurz, R. J., Peck, A. B., "ALMA: status report on construction and early results from commissioning", Proc. SPIE 7733, 773317-773317-10 (2010).
[4] de Graauw, M. W. M., Iguchi, S., McKinnon, M. M., Wild, W., Kurz, R. J., Hills, R. E., "Atacama large millimeter/submillimeter array (ALMA): construction and start of early science", Proc. SPIE 8444, 844489 (2012).
[5] Wootten, A., Thompson, A. R., "The Atacama Large Millimeter/Submillimeter Array," Proc. IEEE 97(8), 1463-1471 (2009).
[6] Iguchi, S., et al., "The Atacama Compact Array (ACA)," PASJ 61(1), 1-12 (2009).
[7] Iguchi, S., "Radio Interferometer Sensitivities for Three Types of Receiving Systems: DSB, SSB, and 2SB Systems," PASJ 57(4), 643-677 (2005).
[8] Thompson, A. R., Moran, J. M., & Swenson, G. W., Jr., "Interferometry and Synthesis in Radio Astronomy," 2nd ed. New York: John Wiley & Sons, (2001).
[9] Fressin, F. et al., "Two Earth-sized planets orbiting Kepler-20", Nature 482, 195-198 (2012).
[10] Doeleman, S. et al., "Event-horizon-scale structure in the supermassive black hole candidate at the Galactic Centre", Nature 455, 78-80 (2008).

ACA phase calibration scheme with the ALMA water vapor radiometers

Yoshiharu Asaki[a,b], Satoki Matsushita[c,d], Koh-Ichiro Morita[e,d] (deceased), and Bojan Nikolic[f]

[a]Institute of Space and Astronautical Science, 3-1-1 Yoshinodai, Chuou, Sagamihara, Kanagawa 252-5210, Japan;
[b]Department of Space and Astronautical Science, The Graduate University for Advanced Studies, 3-1-1 Yoshinodai, Chuou, Sagamihara, Kanagawa 252-5210, Japan;
[c]Academia Sinica Institute of Astronomy and Astrophysics, P.O. Box 23-141, Taipei 10617, Taiwan, R.O.C.;
[d]Joint ALMA Observatory, Alonso de Cordova 3107, Vitacura 763 0355, Santiago, Chile;
[e]National Astronomical Observatory of Japan, 2-21-1 Osawa, Mitaka, Tokyo 181-8588, Japan;
[f]Astrophysics Group, Cavendish Laboratory, University of Cambridge, J J Thomson Avenue, Cambridge CB3 0HE, United Kingdom

ABSTRACT

In Atacama Large Millimeter/submillimeter Array (ALMA) commissioning and science verification we have conducted a series of experiments of a novel phase calibration scheme for Atacama Compact Array (ACA). In this scheme water vapor radiometers (WVRs) devoted to measurements of tropospheric water vapor content are attached to ACA's four total-power array (TP Array) antennas surrounding the 7 m dish interferometer array (7 m Array). The excess path length (EPL) due to the water vapor variations aloft is fitted to a simple two-dimensional slope using WVR measurements. Interferometric phase fluctuations for each baseline of the 7 m Array are obtained from differences of EPL inferred from the two-dimensional slope and subtracted from the interferometric phases. In the experiments we used nine ALMA 12-m antennas. Eight of them were closely located in a 70-m square region, forming a compact array like ACA. We supposed the most four outsiders to be the TP Array while the inner 4 antennas were supposed to be the 7 m Array, so that this phase correction scheme (planar-fit) was tested and compared with the WVR phase correction. We estimated residual root-mean-square (RMS) phases for 17- to 41-m baselines after the planar-fit phase correction, and found that this scheme reduces the RMS phase to a 70 – 90 % level. The planar-fit phase correction was proved to be promising for ACA, and how high or low PWV this scheme effectively works in ACA is an important item to be clarified.

Keywords: Atacama Compact Array, Atacama Large Millimeter/submillimeter Array, Water Vapor Radiometer, Planar-fit Phase Correction

1. INTRODUCTION

Atacama Compact Array (ACA)[1] is designed to improve the short baseline coverage of Atacama Large Millimeter/submillimeter Array (ALMA)[2] especially for observations of extended radio structures. ACA consists of twelve 7-m antennas for interferometry (7 m Array) and four 12-m antennas for total power measurements of celestial sources (TP Array). In imaging simulation atmospheric phase fluctuations play a major role in synthesizing the visibilities with ALMA including ACA.[3] On the other hand effective phase calibration schemes for ACA should be carefully applied because the array configuration is so compact that the fast switching phase calibration will not work effectively.[4]

A phase calibration scheme for ACA in which a simple two-dimensional (2-D) slope of the excess path length (EPL) due to tropospheric water vapor variations aloft was proposed.[5] With this scheme phase correction

---

Further author information: (Send correspondence to Y.A.)
Y.A.: E-mail: asaki@vsop.isas.jaxa.jp, Telephone: +81 (50) 3362 2671

data is collected with water vapor radiometers (WVRs) attached to the TP Array antennas surrounding the 7 m Array. The WVR measurements are used for fitting the 2-D surface to obtain line-of-sight EPL values for each individual antenna of the 7 m Array. Interferometric phase fluctuations due to the water vapor for each interferometer baseline are estimated from the differences of the obtained EPL between the 7 m Array antennas. The interferometric phases are then corrected by subtracting the differences. At submillimeter wavelengths a similar phase correction scheme with the 2-D phase screen obtained from the interferometric phases has been tested using the Submillimeter Array, and this was proved to work well when the phase correction data of an arbitrary antenna is interpolated from the phase screen.[6]

Here we refer to the performed phase correction scheme in this study as a planar-fit phase correction. In this paper we report on the results of phase calibration experiments with the planar-fit scheme in ALMA commissioning and science verification (CSV). In section 2 we mention outlines of the experiments and the data analysis with the ALMA CSV data. The results are presented and discussed in section 3. We also make comparisons between the planar-fit phase correction and the baseline ALMA phase correction method, that is, WVR phase correction,[7–9] in section 3. We summarize this study in section 4. All the phase fluctuations mentioned in this paper are in path length, or interferometeric phase at the observing frequency.

2. EXPERIMENTS

2.1 Outlines

We have conducted four experiments in the ALMA CSV to investigate the feasibility of the planar-fit phase correction. We refer to each of the experiments as Exp-1 to Exp-4. Outlines of the experiments are summarized in Table 1. The first three experiments were conducted on December 19, 2010, at 86.3 GHz (Band 3), and Exp-4 was conducted on December 21, 2010, at 229.5 GHz (Band 6). The observation duration of all the experiments was about one hour. Precipitable water vapor (PWV) was 2.5 and 2.9 mm in the first three experiments and Exp-4, respectively, so that the tropospheric conditions at the telescope site were not good.

Antenna geometry of the experiments is shown in Figure 1. Antennas are all the ALMA 12-m full-spec instruments attached to the WVRs, so that the WVR measurements were simultaneously monitored during the experiments. We consider that the antenna configuration consists of two groups: one, referred to as the "interferometer array", has four supposed interferometer antennas to obtain interferometric complex data, and the other one, referred to as the "total-power array", has four antennas to which the WVRs are attached. In the first three experiments the interferometer array has six baselines ranging from 17 to 41 m length, surrounded by the four total-power array. In Exp-4 the interferometer array has a single 26-m baseline sandwiched by the two total-power array antennas. One WVR attached to antenna 3 (DV05) did not work properly through the experiments. In the discussion part we do not use the WVR corrected phases of the baselines of antenna 3 for the comparison of the WVR phase correction with the planar-fit phase correction.

2.2 Data analysis

In the present data analysis two data sets are available. One is phase-uncalibrated interferometer complex data, and the other is phase-calibrated complex data using the WVR phase correction. All the complex data was produced every 0.96 second. The WVR measurements for antennas A, B, C, and D in Figure 2 were retrieved from those two data sets for the baselines of antenna E (DV09). For example the WVR measurement $\phi_{\rm WVR}^{\rm A}$ for antenna A at time t is retrieved as follows:

$$\phi_{\rm WVR}^{\rm A}(t) = \phi_{\rm uncal}^{\rm A-E}(t) - \phi_{\rm WVRcal}^{\rm A-E}(t) - \frac{1}{T}\int_{t_0}^{t_0+T}\{\phi_{\rm uncal}^{\rm A-E}(t) - \phi_{\rm WVRcal}^{\rm A-E}(t)\}dt, \qquad (1)$$

where t_0 is the observation start time, T is the observation duration, $\phi_{\rm uncal}^{\rm A-E}$ is the uncalibrated interferometric phase, and $\phi_{\rm WVRcal}^{\rm A-E}$ is the calibrated interferometric phase with the WVR phase correction. This process was made using a personally developed python program under the Common Astronomy Software Applications (CASA) package environment.

The planar-fit phase correction has four steps. In the first step the EPL 2-D slope fitting is carried out every 0.96 second with the retrieved WVR measurements. The next step is to infer the line-of-sight EPLs of the

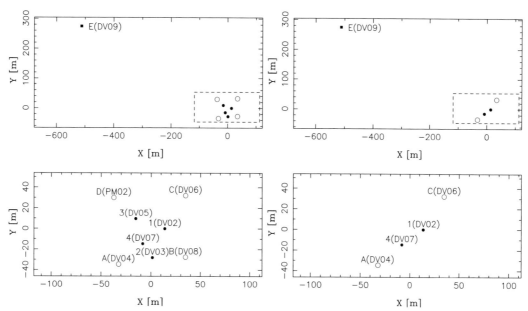

Figure 1. Antenna configuration of the experiments. The left and right plots show the configuration for Exp-1 to Exp-3 and Exp-4, respectively. The top plots show a wide overview, and the bottom ones show an enlarged view of the area surrounded by the dashed line in the tops. Large open circles and small filled circles represent the total-power array and the interferometer array, respectively. The filled square represents the reference antenna (E) to retrieve the WVR measurements of the total-power array.

Table 1. Outlines of the experiments.

Experiment name	Exp-1	Exp-2	Exp-3	Exp-4
Date	Dec 19, 2010	Dec 19, 2010	Dec 19, 2010	Dec 21, 2010
Time	04:02:31 – 05:07:17	05:17:21 – 06:22:12	06:30:52 – 07:35:44	09:41:22 – 10:45:55
Receiver	Band 3	Band 3	Band 3	Band 6
Frequency [GHz]	86.306	86.306	86.306	229.538
Source	0538–440	0538–440	0538–440	3C 279
Total-power array antenna (Fig. 2)	A, B, C, D	A, B, C, D	A, B, C, D	A, C
Interferometer array antenna (Fig. 2)	1, 2, 3[†], 4	1, 2, 3[†], 4	1, 2, 3[†], 4	1, 4
Baseline range [m]	16.6 – 41.2	16.6 – 41.2	16.6 – 41.2	26.4
PWV[‡] [mm]	2.5	2.5	2.5	2.9

[†] WVR attached to antenna 3 (DV05) did not work properly.

[‡] PWV: precipitable water vapor.[10]

interferometer array antennas from the normal unit vector of the slope and the EPL bias at the center of the total-power array. In the third step the interferometric phase fluctuations for each baseline are estimated from the differences of the inferred line-of-sight EPLs for each pair of antennas. In the last step the interferometric phases are then calibrated by subtracting the values in the previous step. Schematic examples of the planar-fit phase correction are shown in Figure 2. Since the 2-D slope fitting could not be performed for Exp-4, we conducted 1-D slope fitting using the two total-power array antennas (A and C). The planar-fit phase correction was made using a software developed for simulation studies of the planar-fit phase correction.[5] We estimated the root-mean-square (RMS) phase of the uncalibrated, the WVR corrected, and the planar-fit corrected time series after 10 point averaging, equivalent to 9.6 sec time averaging, for whole of the one-hour observing time.

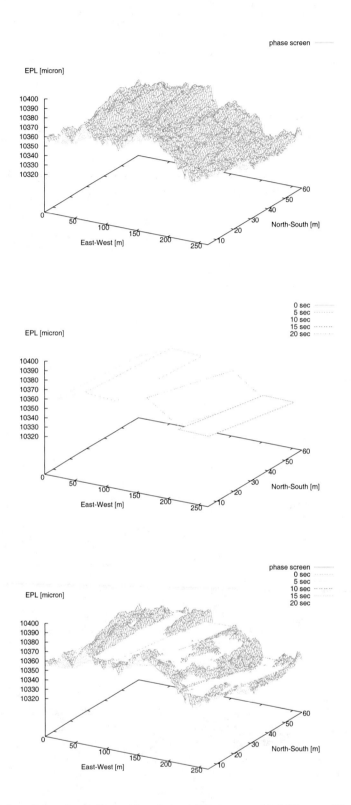

Figure 2. Schematic examples of the planar-fitting for the total-power array of 50 m × 50 m. The top plot shows a generated EPL phase screen (300 m × 50 m) moving with the wind velocity of 10 m s^{-1}, the middle one shows the fitting results with the four WVRs attached to the total-power array antennas, and the bottom one shows the superposition of the screen and the fitted planes.

3. RESULTS AND DISCUSSIONS

Figures 3 and 4 show the time series of the uncalibrated interferometric phase as well as the retrieved WVR measurements, the WVR corrected phase, and the planar-fit corrected phase. We calculated the RMS phases as listed in Table 2, and compiled them in Figure 5. The path length fluctuation of the interferometric phase should be smaller than 50 and 57 μm in RMS according to the ALMA specification in terms of the path length fluctuation (= $\{10\sqrt{2} \times (1 + \text{PWV}) + 0.02 \times \sigma_{ns}\}$ [μm], where σ_{ns} is RMS of the uncalibrated interferometric phase, in other words, natural seeing, in micron, and PWV is precipitable water vapor in millimeter) for the first three experiments and Exp-4, respectively. We confirmed from Table 2 that the natural seeing of the first three experiments meets the ALMA specification even under the rather high PWV condition. Moreover the RMS phases can be reduced when the planar-fit phase correction is applied to the interferometric phases. For Exp-4 the planar-fit phase correction reduces the RMS phase to meet the the ALMA specification while the natural seeing does not, so that the planar-fit phase correction is expected to be promising as the ACA phase calibration scheme under high PWV conditions in which the natural seeing cannot meet the ALMA specification.

We here define the "effectiveness of the phase correction", η_{wc} and η_{pf}, of the WVR phase correction and the planar-fit phase correction, respectively, as follows:

$$\eta_{wc} = \frac{\sigma_{ns} - \sigma_{wc}}{\sigma_{ns}}, \quad (2)$$

$$\eta_{pf} = \frac{\sigma_{ns} - \sigma_{pf}}{\sigma_{ns}}, \quad (3)$$

where σ_{wc} and σ_{pf} are the RMS phases using the WVR phase correction and the planar-fit phase correction, respectively. The effectiveness is listed in Table 3. Although the effectiveness of the WVR phase correction is superior to that of the planar-fit phase correction, it sometimes shows a comparable ability to the WVR phase correction. The RMS phase can be reduced typically by a factor of 10–30 % using the planar-fit phase correction.

Although the planar-fit phase correction was proved to be promising for ACA under rather high PWV conditions (2.5 – 2.9 mm), the ALMA specification may not be met using this scheme under much higher PWVs. On the other hand previous simulation studies to verify this scheme showed that it would not effectively work under low PWV conditions in which the natural seeing can meet the ALMA specification.[5] Since we have only a limited number of demonstrations, how high or low PWV the planar-fit phase correction effectively works in ACA is not clarified and will be one of the most important items to be investigated.

Table 2. RMS phases in path length (μm) of the interferometer baselines.

| Baseline | Exp-1 | | | Exp-2 | | | Exp-3 | | | Exp-4 | | |
(length)	[NS]	[WC]	[PF]	[NS]	[WC]	[PF]	[NS]	[WC]	[PF]	[NS]	[WC]	[PF]
2–4 (16.6 m)	21.5	11.7	18.8	25.2	16.3	22.0	24.2	16.2	18.2	–	–	–
3–4 (25.2 m)	31.3	–	25.9	32.8	–	26.2	28.7	–	21.9	–	–	–
1–4 (26.4 m)	45.8	37.9	43.2	38.0	24.9	32.3	35.0	29.4	29.4	77.9	26.6	53.7
1–2 (30.6 m)	47.5	38.2	45.5	39.5	23.1	35.2	39.0	25.6	31.9	–	–	–
1–3 (30.6 m)	45.4	–	39.1	42.1	–	32.7	35.3	–	29.1	–	–	–
2–3 (41.2 m)	40.8	–	31.2	44.0	–	32.0	41.4	–	24.3	–	–	–

[NS]: natural seeing   [WC]: WVR phase correction   [PF]: planar-fit phase correction

4. SUMMARY

We reported the phase calibration experiments conducted in the ALMA CSV to investigate the feasibility of the planar-fit phase correction for ACA. We estimated the residual RMS phases for the 17- to 41-m baselines after the planar-fit phase correction, and found that this scheme can typically reduce the RMS phase to a 70 – 90 % level. The planar-fit phase correction is promising for short baselines in ACA, and has an ability to achieve the ALMA specification in terms of the path length fluctuation especially under rather high PWV conditions.

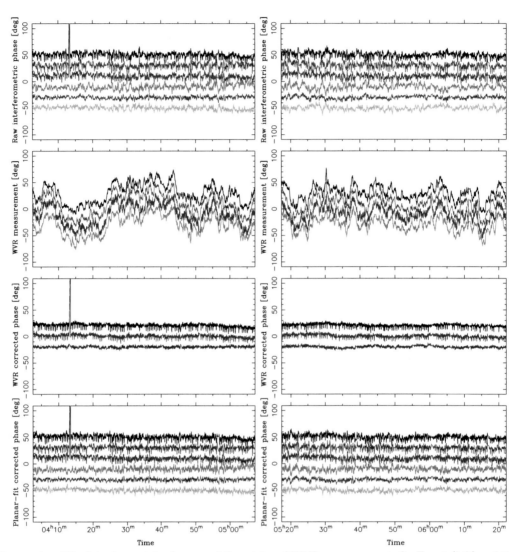

Figure 3. Time series of the interferometric phases and the retrieved WVR measurements for Exp-1 (left) and Exp-2 (right). The top plots show uncalibrated interferometric phase. The second row plots show the retrieved WVR measurements expressed in phase at the observing frequency. The third and bottom row plots show the WVR corrected and planar-fit corrected phase, respectively. Each of the time series has an artificial bias in order to be easily identified. For the interferometric phases black, red, blue, yellowish, pink, and orange represent baselines of 1–2, 1–3, 1–4, 2–3, 2–4, and 3–4, respectively. For the retrieved WVR measurements black, red, blue, and yellowish green represent antenna A, B, C, and D, respectively.

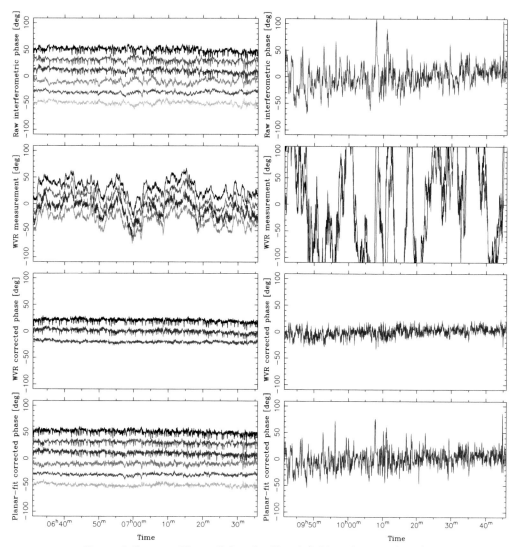

Figure 4. Same as Figure 3, but for Exp-3 (left) and Exp-4 (right).

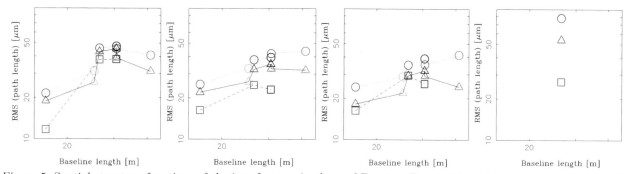

Figure 5. Spatial structure functions of the interferometric phase of Exp-1 to Exp-4. The abscissa is the baseline length in log scale, and the ordinate is the RMS phase in path length in log scale. Open circles connected with dotted lines are natural seeing, open squares connected with dashed lines are RMS phases of the WVR corrected phase, and triangles connected with solid lines are those of the planar-fitted corrected phase. Each color is identical with that used in Figures 3 and 4.

Table 3. Effectiveness of the phase correction of the interferometer baselines.

Baseline (length)	Exp-1 [WC]	Exp-1 [PF]	Exp-2 [WC]	Exp-2 [PF]	Exp-3 [WC]	Exp-3 [PF]	Exp-4 [WC]	Exp-4 [PF]
2–4 (16.6 m)	46 %	12 %	35 %	13 %	33 %	25 %	–	–
3–4 (25.2 m)	–	17 %	–	20 %	–	24 %	–	–
1–4 (26.4 m)	17 %	6 %	34 %	15 %	16 %	16 %	66 %	31 %
1–2 (30.6 m)	20 %	4 %	42 %	11 %	34 %	18 %	–	–
1–3 (30.6 m)	–	14 %	–	23 %	–	17 %	–	–
2–3 (41.2 m)	–	24 %	–	27 %	–	41 %	–	–

[WC]: WVR phase correction   [PF]: planar-fit phase correction

To clarify how high or low PWV the planar-fit phase correction effectively works in ACA is one of the most important items in the next verification phase.

ACKNOWLEDGMENTS

This paper makes use of the ALMA CSV data. ALMA is a partnership of ESO (representing its member states), NSF (USA) and NINS (Japan), together with NRC (Canada) and NSC and ASIAA (Taiwan), in cooperation with the Republic of Chile. The Joint ALMA Observatory is operated by ESO, AUI/NRAO and NAOJ. We thank all ALMA staff for their contribution to make these experiments possible. YA is supported by the NAOJ. SM is supported by the National Science Council (NSC) of Taiwan, NSC 100-2112-M-001-006-MY3. YA, SM, and BN express their heartfelt condolences to Professor Koh-Ichiro Morita who passed away on May 7, 2012. His contributions to this study have been very remarkable since the concept of ACA was considered.

REFERENCES

[1] Iguchi, S., Morita, K. I., Sugimoto, M., Vilaro, B. V., Saito, M., Hasegawa, T., Kawabe, R., Tatematsu, K., Sakamoto, S., Kiuchi, H., Okumura, S. K., Kosugi, G., Inatani, J., Takakuwa, S., Iono, D., Kamazaki, T., Ogasawara, R., and Ishiguro, M., "The Atacama Compact Array (ACA)," *Publ. Astron. Soc. Japan* **61**, 1–12 (2009).

[2] Hills, R. H., Kurz, R. J., and Peck, A. B., "ALMA: status report on construction and early results from commissioning," in [*SPIE conference: ground-based and airbone telescope III*], **7733**, 773317–773317–10 (2010).

[3] Tsutsumi, T., Morita, K.-I., Hasegawa, T., and Pety, J., "Wide-field imaging of ALMA with the Atacama Compact Array: imaging simulations," *ALMA memo* **488.1** (2004).

[4] Holdaway, M., "Does the ACA need phase compensation?," *ALMA memo* **491** (2004).

[5] Asaki, Y., Saito, M., Kawabe, R., Morita, K.-I., Tamura, Y., and Vila-Vilaro, B., "Simulation series of a phase calibration scheme with water vapor radiometers for the Atacama Compact Array," *ALMA memo* **535** (2005).

[6] Matsushita, S. and Chen, Y.-L., "Testing the Atacama Compact Array phase-correction scheme using the Submillimeter Array," *Publ. Astron. Soc. Japan* **62**, 1053–1062 (2010).

[7] Stirling, A., Hills, R., Richer, J., and Pardo, J., "183 GHz water vapour radiometers for ALMA: estimation of phase errors under varying atmospheric conditions," *ALMA memo* **496** (2004).

[8] Nikolic, B., "Inference of coefficients for use in phase correction I," *ALMA memo* **587** (2009).

[9] Nikolic, B., "Inference of coefficients for use in phase correction II: using the observed correlation between phase and sky brightness fluctuations," *ALMA memo* **588** (2009).

[10] Matsushita, S., Morita, K.-I., Barkats, D., Hills, R. E., Fomalont, E. B., and Nikolic, B., "ALMA temporal phase stability and the effectiveness of water vapor radiometer," in [*SPIE conference: millimeter and submillimeter wavelength telescopes II*], **Conference 8444**, 8444–125 (2012).

Functional safety for the Advanced Technology Solar Telescope

Scott Bulau*, Timothy R. Williams

National Solar Observatory, 950 N. Cherry Avenue, Tucson, AZ, USA 85726

ABSTRACT

Since inception, the Advanced Technology Solar Telescope (ATST) has planned to implement a facility-wide functional safety system to protect personnel from harm and prevent damage to the facility or environment. The ATST will deploy an integrated safety-related control system (SRCS) to achieve functional safety throughout the facility rather than relying on individual facility subsystems to provide safety functions on an *ad hoc* basis.

The Global Interlock System (GIS) is an independent, distributed, facility-wide, safety-related control system, comprised of commercial off-the-shelf (COTS) programmable controllers that monitor, evaluate, and control hazardous energy and conditions throughout the facility that arise during operation and maintenance.

The GIS has been designed to utilize recent advances in technology for functional safety plus revised national and international standards that allow for a distributed architecture using programmable controllers over a local area network instead of traditional hard-wired safety functions, while providing an equivalent or even greater level of safety.

Programmable controllers provide an ideal platform for controlling the often complex interrelationships between subsystems in a modern astronomical facility, such as the ATST. A large, complex hard-wired relay control system is no longer needed. This type of system also offers greater flexibility during development and integration in addition to providing for expanded capability into the future. The GIS features fault detection, self-diagnostics, and redundant communications that will lead to decreased maintenance time and increased availability of the facility.

Keywords: safety, functional safety, safety-related control system, interlocks, ATST

1. INTRODUCTION

Before we begin our discussion we need to define some of the key terms we will be talking about. Safety is freedom from unacceptable risk of injury to persons or damage to equipment or the environment. Functional safety is that part of safety that relies on active controls rather than simple safeguards such as barriers, signage and procedures.

In the past, safety systems have been hard-wired, typically using relay logic control circuits. Modern automation control systems no longer use large relay logic controls, but instead are developed using programmable logic controllers (PLCs). Until recently, functional safety systems were built using these same traditional hard-wired relay logic controls. The Advanced Technology Solar Telescope (ATST) is applying recent innovations in functional safety to the design of a large ground-based telescope.

Typically, functional safety design has been based on each subsystem providing safety functions for its own subsystem with little regard to integrating all the various subsystems in the final design. Each subsystem was added to the final design in an *ad hoc* basis. The ATST concept is to make a unified safety system that supervises all subsystems to ensure safety.

Rather than build a custom solution, the ATST has developed a Global Interlock System (GIS) based on a commercial off-the-shelf (COTS) system, using a safety-rated programmable electronic controller, which will decrease cost and development time.

2. BASIS OF DEVELOPMENT

2.1 Necessity of a safety-related control system

The GIS serves two main purposes. The primary purpose of the system is to eliminate the risk of injury to personnel and to prevent physical damage to the telescope, instruments, other infrastructure, or the environment. The secondary

---

*sbulau@nso.edu; phone 1 520 318-8389; fax 1 520 318-8500; atst.nso.edu

purpose of the GIS is to meet applicable governmental regulations concerned with the safety of persons and prevention of damage to the environment.

The GIS meets its goal by providing high integrity safety functions for connected subsystems and providing a complementary emergency stop function for the entire facility. This system also provides continuous status of the GIS to the operator and the Observatory Control System (OCS) for monitoring and troubleshooting.

Very early in the development of the ATST, a safety plan was implemented that required detailed hazard analysis of the ATST facility. Robert Hubbard, ATST lead systems engineer, has previously discussed the system safety and details of the hazard analysis for the ATST[1]. The project adopted MIL-STD-882D Standard Practice for System Safety, as the basis of hazard analysis. This analysis identified hazards at a preliminary stage in the design process that needed to be mitigated in the final design. The first step in designing the GIS was to utilize the existing ATST systems' hazards analysis as a baseline to develop the specification of the various safety functions.

Compliance with the Occupational Safety and Health Administration (OSHA) is a requirement for facilities built and operated in the United States. OSHA promotes the use of national consensus standards. There are numerous consensus standards which could apply in this case; none will be an exact fit for a complicated research facility like the ATST. ANSI/RIA R15.06 was chosen because it best fit the facility, as the standard defines a robot as an "automatically controlled, reprogrammable multipurpose manipulator programmable in three or more axes[2]."

ANSI/RIA R15.06 differs from MIL-STD-882 in the key aspect that after each hazard is evaluated it provides specific minimum requirements for mitigating the hazard, ranging from simple awareness means, such as training and signage, to complex redundant control circuits. The mitigation recommended is dependent on the severity, exposure frequency, and the likelihood of avoidance of the individual hazard.

2.2 Safety integrity levels

Until recently, safety was thought to be a product of the method rather than a measure of performance. The recently deprecated EN-954-1 standard (on which ANSI/RIA R15.06 is based) relied on different types of control architecture to determine the classification of a system's level of safety in which adding redundancy and cross-checking increases safety.

The newer IEC 61508 Functional safety of electrical/electronic/programmable electronic safety-related systems standard[3] instead relies on a mathematical definition of various levels of safety performance. Without regard to the method, the IEC standard bases measurement of safety on the probability that a system will experience a failure that will result in a hazardous condition.

This measure of safety under IEC 61508 is a series of safety integrity levels (SIL) that define either the probability of the system to fail to perform its design function on demand, or the probability of a dangerous failure per hour as shown in Table 1.

Table 1. Safety integrity levels as defined in IEC 61508

Safety integrity level	PFD Average probability of failure on demand	PFH_D Probability of dangerous failure per hour
4	$\geq 10^{-5}$ to $< 10^{-4}$	$\geq 10^{-9}$ to $< 10^{-8}$
3	$\geq 10^{-4}$ to $< 10^{-3}$	$\geq 10^{-8}$ to $< 10^{-7}$
2	$\geq 10^{-3}$ to $< 10^{-2}$	$\geq 10^{-7}$ to $< 10^{-6}$
1	$\geq 10^{-2}$ to $< 10^{-1}$	$\geq 10^{-6}$ to $< 10^{-5}$

2.3 Initial concept

An initial analysis served to identify those hazards that would be mitigated at least in part by an electrical/electronic/programmable electronic (E/E/PE) system. This preliminary hazard analysis showed that a safety-rated control system would need to be capable of SIL 3 to contend with hazards that might reasonably be foreseen at the facility. We identified several key properties that we wanted the final system to have: safety-rated, independent, distributed, reliable, maintainable, programmable, expandable and cost effective.

Because the function of the GIS is safety, any components of system needed to be designed for use in a safety system. This limits some possible solutions as the hardware is simply not suitable for use in a safety system. All components must be suitable and certified for use in safety systems and installed accordingly.

The system should be independent of the basic control system. We wanted to have a safety supervisor that was independent of the systems being monitored. Each subsystem is still designed using basic safety principles, the GIS is an additional high-integrity safety system that monitors the individual subsystems. This gives us diversity in our safety functions which would reduce the likelihood of a common cause failure.

The components would need to be distributed throughout the facility. The ATST is a large facility with a variety of control systems located throughout. It is not practical or desirable to centrally locate all components of the safety-related control system. I/O can be located where it is needed instead of located centrally. This reduces the amount of wiring required to be installed and maintained within cable wraps and throughout the facility.

Having a distributed system also aids during development as various major subsystems are being fabricated on different continents thousands of kilometers apart. The distributed design allows for division of safety functions to each subsystem. This enables the safety system to be initially installed and tested at the factory, rather than during integration at the site. This reduces engineering risks associated with the system.

The system needs to have as little downtime as possible. The GIS is a critical safety system, it will run 24 hours day every day for the life of the facility. If it is not running it can affect every subsystem in the facility. Therefore the system has to have a very high availability. Availability is affected by both the reliability and maintainability of the system.

One concern of having a large distributed system is how to locate faults when they too are distributed throughout the facility. The GIS utilizes a touch screen human machine interface (HMI) that displays the current status of all connected devices. For example, when an emergency stop device is activated, the HMI automatically displays a graphical representation of the device's location.

The system would need to be programmable and expandable as needed. The technology deployed at modern astronomical observatories is always progressing. It is very likely that during the expected lifetime of the facility new capabilities will be added. Having a modular and programmable system will allow for easier changes as specific requirements evolve. A modular approach also allows for improvements to the system itself as new products become available.

As with any system cost and development time are a consideration. By utilizing a commercial off-the-shelf solution, the project saves on the cost of developing and deploying a custom control system. Future staff will likely have knowledge and as this type of control system is used extensively in industrial automation.

3. IMPLEMENTATION OF SYSTEM

3.1 Understanding available technology

After the need for a safety-related control system was determined, a survey of the available technology was undertaken. At the time of the conceptual design, the Allen-Bradley GuardLogix controller was selected as the basis of the safety-related control system. The GuardLogix controller itself is based on the ControlLogix controller series which has been used extensively in industrial automation control since 1998.

The GuardLogix controller is essentially a pair of controllers that consists of a standard ControlLogix controller and a redundant safety partner. The standard processor can be used for normal automation duties and the addition of safety partner allows for its use in high integrity safety applications. The GuardLogix series of controllers and the associated I/O modules are TÜV-certified (TÜV Rheinland certifies components for use in functional safety systems) for use in up to SIL 3 applications. The original intent was that subsystem designers would choose to use the standard portion of the controller in the subsystem design.

The GuardLogix controller and I/O are modular and allow for quick replacement in the event of a failure. Past experience with ControlLogix processors show them to be robust and reliable. The GuardLogix processors have a mean-time-between-failure (MTBF) in the neighborhood of one million hours.

The GuardLogix is designed to interface with several types of distributed I/O modules over industrial networks. The Guard I/O modules are suitable for high integrity safety applications because the I/O modules have pulse test outputs that look for common types of failures, such as shorts to ground or broken wires. If these are detected the system is put into a safe state and must be repaired and tested prior to resuming operations.

The GuardLogix system is designed around a de-energize to safe principle. When the GuardLogix controller detects a fault within the system, power is removed from the affected outputs. Removing power causes all hazardous motion to stop.

3.2 Initial design

The initial design was to have a series of controllers, one for each subsystem, which would communicate through a master controller. The basic building block of this design is the Local Interlock Controller (LIC). Each LIC is responsible for ensuring the safety of its subsystem and relaying status information to the master controller (Figure 1). The master controller or Global Interlock Controller (GIC) is responsible for monitoring the various LICs and coordinating safety functions that involve multiple subsystems. Individual I/O modules would be located as needed throughout the facility. The system is tied together using an industrial Ethernet network that utilizes a safety protocol. This network is independent of all other observatory networks.

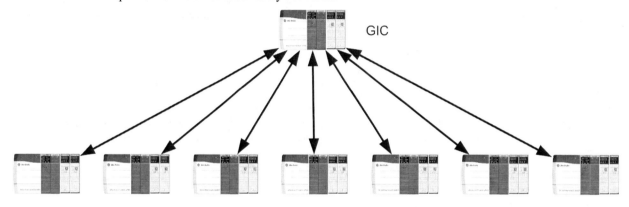

Figure 1. GIS Hierarchy

3.3 The independent safety network

The network is the backbone of the system relaying information to all the I/O in the system. As with many industrial controllers, notably those produced by Allen-Bradley, the network communicates via EtherNet/IP (Ethernet Industrial Protocol). The ability to use such a network in a safety system comes from the protocol that is used.

Communication across the Ethernet/IP network uses Common Industrial Protocol (CIP), a network independent protocol. The GIS communicates using safety CIP, an extension of standard CIP, which is TÜV-approved for use in SIL 3 applications according to the IEC 61508 standard.

The GIS uses CIP Safety over EtherNet/IP. This allows the use of common Ethernet technology, which provides a flexible network architecture and permits differing media types. The Independent Safety Network is standard IEEE 802.3 Ethernet at the physical and data link layers.

The network consists of a group of managed network switches. The switches themselves are connected via a redundant ring topology. Each LIC and its associated distributed I/O modules will be part of a separate virtual LAN (VLAN); this will reduce the amount traffic on each node (Figure 2).

Since the heart of high-integrity safety system is a network, security of the independent safety network is vital. Recent developments have shown that industrial programmable electronic control systems are vulnerable and have been targeted. ATST has adopted a defense-in-depth approach, which involves many layers to protect the system.

Network security will include a firewall between the one connection from the OCS and GIS. The independent safety network is physically isolated from all other facility networks. Access to the network and controllers is password protected and restricted. Procedures are implemented to track vulnerabilities and ensure all security updates are implemented. Procedures are also used to ensure that a compromised system is not connected to the independent safety network.

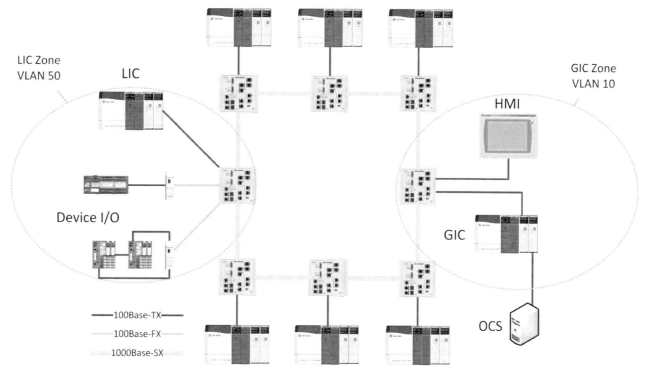

Figure 2. GIS network configuration

3.4 Controllers

The basic building block of the GIS is the Local Interlock Controller. Each subsystem's safety system is built around a GuardLogix controller and its associated distributed I/O. The LIC provides high integrity safety functions independently from the entire GIS. The LIC is responsible for monitoring the connected safety I/O, communicating status with the GIC, and applying interlocks based on the safety I/O and status received from the GIC.

Within the distributed GIS, there is only the one Global Interlock Controller (GIC). The GIC will be programmed as a centralized controller that monitors all the LICs status. It is through the GIC that inter-subsystem responses are recognized and distributed. It is the responsibility of the GIC to inform all LICs of global safety responses. The GIC will produce the necessary signals to corresponding LIC requiring their subsystem's action. It is also the responsibility of the GIC to send via a separate network, status information to the OCS at a 1 Hz rate.

The GIC provides for global safety functions, serves as an interface for safety functions that require the coordination of multiple LICs. The GIC also relays status of the entire GIS to the OCS and HMI.

3.5 Emergency-stop system

Integral to the GIS is a complementary emergency stop function that is capable of putting the entire facility in a safe state by activating any one of numerous emergency stop devices. Located throughout the facility are push buttons and pull cables that can activate an emergency stop. Having a single comprehensive emergency stop system will cause all hazardous motion in the facility to stop regardless of which system the emergency stop button is mounted on or near. The high-integrity design allows monitors the facility to ensure that the system has stopped prior to allowing access.

3.6 Application development

The hardware makes up the system, but obviously a programmable electronic control system will need software to run it. Part of what makes a programmable safety system possible is the use a limited variability language, in this case a limited set of commands for ladder logic. By limiting the number of commands available there are fewer possibilities of making a mistake. The commands available have all been certified for use in SIL 3 systems.

The configuration of the logic is essential to the safe operation of the facility. At every step of the development of various programs and routines the logic is checked to verify that code performs the safety function properly and that the specification of the safety function is valid.

The first step of the application development is to examine the hazard analysis for the subsystem that the LIC will control. Each hazard that is identified as having engineering safe guards is assessed to determine which safety functions the LIC will need to perform. Logic and timing diagrams are developed based on the hazard to mitigated. The LIC is then programmed using this logic.

The GuardLogix controller runs two tasks, the standard task and the safety task. Within the safety task are various programs and routines. All SIL 3 safety routines are run inside the safety task.

For each safety function a set of verification tests are designed to ensure that the application code performs the intended safety function correctly and can handle abnormal states, such as communication faults and component failure. After the functional verifications are completed the application program can be locked to prevent unauthorized or accidental modification of the program or manipulation of I/O data.

3.7 Maintainability

A major concern with many observatory safety systems is that a false alarm will cause a spurious shutdown of the system and lost observing time. While no system can be designed that is completely fault proof, we have strived to build a system that has very high reliability (which also increases safety) and also has a very high availability.

It is important to note that while the basic architecture of the safety system is to have redundant components this to only to ensure that the system can be put into a safe state in the event of failure. It is not to allow the continued operation of the facility in the event of a component failure. The one exception is the network. The network uses redundant Ethernet protocol that is fault tolerant. If one of the Ethernet channels fails the system recognizes the failure and reroutes network traffic and can continue to work uninterrupted. The user is notified via the HMI of the failure so that it can be repaired when the telescope is not in use.

This system's modular design allows for replacement of most failed components without tools and many components are hot-swappable so they can be replaced even while still powered up. The HMI aids in troubleshooting by quickly identifying failed components and even suggesting corrective actions that need to be taken.

The system will, after detecting a failure, prevent restarting of the affected safety function until a functional test of the replaced components is carried out to ensure that maintenance was properly completed.

3.8 Enhancement and development of design

The initial design was well-thought and planned, so little of the technical aspects have changed from the early concept stage to the final design. The redundant hard-wired emergency stop system was removed. The network was changed to a redundant configuration. The network was also divided into several virtual LANs, one for each subsystem, rather than one large network. New modular I/O was introduced by the vendor and was included in the final design.

We expect that during integration of the various systems and final installation at the facility there will be changes. The modular, programmable nature of the system will ease in making these changes. The design has been guided by the principle that we don't want to limit future capabilities.

The initial plan was for the project to build and configure the GIC and LICs and the vendors supply and connect the I/O modules. Currently several of the vendors will be supplying the LIC and associated I/O with the project providing oversight. A series of workshops discussing functional safety and GuardLogix controllers was held with the various vendors to ensure that overall system design was still unified.

The GIS schedule provided for a series of successive reviews, from the initial conceptual development through the final design over a period of several years. As a result of these reviews, elements were added or changed to better enhance the system as a whole. The Final Design review is scheduled for the later part of July of 2012.

4. METHOD OF ESTABLISHING ELEMENTS (I/O, INTERLOCKS AND LIMITS)

For the GIS, as with any safety system, it is necessary to establish the safety elements (interlocks, limits and I/Os), whether they are local, or distributed. It is necessary to understand the safety I/O the system will have and the required response expected of the system with respect to each of these I/O elements. In some cases a logical combination of triggered I/O may need a response. With safety systems most of the actions result in the removal of hazardous energy. This may occur by inhibiting a device from applying power or in removing power from a device in which power is currently applied.

The ATST developed the following method of establishing these I/O. An initial list of potential safety I/O for each subsystem was selected based on previous experience with large telescope systems. This established a baseline of interlocks and limits to be included in the GIS. This was then followed by a preliminary hazard analysis using the MIL-STD-882D. Once an initial mitigation was proposed a risk assessment using the ANSI/RIA R15.06 showed the minimal required type of mitigation necessary to reduce the hazard (risk) to an acceptable level. The resulting mitigation was a direct prescription of I/O circuitry required by the GIS to implement. Figure 3 shows an example entry from the telescope mount's hazard analysis.

Subsystem:						Preliminary Hazard Analysis			Prior to Safeguard Selection ANSI_RAI_R15.06-1999_(R2009)					Last Update: May-2012			
WBS,						Risk MIL-STD-882D								Risk MIL-STD-882D			
HWBD / Nr	System Item	Hazard	Causes	Effects	Project Phase(s)	IMRI	Value	Cat.	Severity	Exposure	Avoidance	Risk Category	Recommended Action	FMRI	Value	Cat.	Comments
Nr. 10	Optical Support Structure/TEOA access platform	Collision hazard Pinch hazard	Rotating Mount interfaces with deployable TEOA access platform	Personnel injury, damage to equipment	Maint.	2-C	6	Serious	S2	E1	A2	R2B	Allow deployment of platform ONLY in defined Mount Azimuth, Mount Altitude and Enclosure Azimuth position. Provide interlock for Mount and Enclosure when platform deployed	2-E	15	Medium	GIC, Mount LIC

Figure 3. Hazard Analysis Example

Once we had established the method for establishing the safety system's I/O, the procedure was incorporated into the requirements of the design. Later this method was passed to the contracted vendors as design requirements. Initially the project provided the preliminary list; the vendor was then responsible for following up with a hazard analysis, risk assessment and mitigation mapped to GIS I/O.

As each vendor produced the I/O that their subsystems LIC would need to respond, the interfaces with the GIS were modified. The project required the hazard analysis and risk assessment from each vendor on their subsystem. This establishes the database of mitigated hazards for documentation follow through of the project. The project additionally used this as the basis for an Intersystem Hazard Analysis from which many of the GIC functions were developed. An example of this is the interaction of the Enclosure aperture cover, the Telescope Mount M1 mirror cover, and the TEOA heat stop cover. All of these must close in a particular sequence in response to a safety function.

5. SAFETY MANAGEMENT PLAN

As a means of providing documented records for future support, the project has a procedure to establish safety functions for the GIS; however, it was also necessary to establish a safety management plan for the project to follow throughout its design, development, integration, and operations for years to come. Initially the plan provides for the systematic progression of developing the design of the GIS from the subcomponents (LICs and I/Os). Testing individual subsystems independently and then integrated into the entire system. The plan calls out for verification of the LICs with their I/Os and then a validation of the entire GIS. Once validated the plan specifies the significant operational training, ability to update the system then followed by verification and validation and once again training. This loop continues throughout the life of the facility (Figure 4).

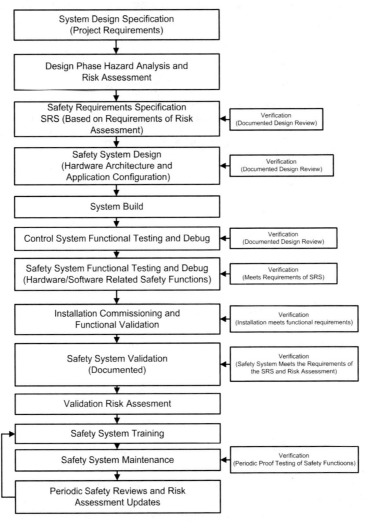

Figure 4. Safety lifecycle

6. DEVELOPMENT OF HARDWARE

6.1 Prototyping

In the spring of 2011, a prototype system consisting of two LICs and a GIC was assembled at the NSO facility in Tucson, Arizona. This prototype serves as a platform for various proof-of-concept tests. We verified key concepts of the design such as whether messages could be passed across VLANs from controller to controller. We also tested various input and output elements and developed example code for use by the vendors and subcontractors.

The prototype allowed us to test the reaction time of the system using actual hardware. For example, to activate an emergency stop which is connected to one subsystem and to have a safety contactor that is connected to another LIC de-energize, the system needs to monitor the input on one LIC, send a message to the GIC on the change of the input which in turn sends another message to the other LIC to respond by disabling the corresponding output. Our testing showed that this typically took 40-55 milliseconds.

The prototype system has run for over a year with only a single fault, which was corrected in less than five minutes. This demonstrates an availability of over 99.999%.

7. CONCLUSION

In the end we have a working prototype system that gives us great confidence that when the system is integrated and installed, it will meet our goals and provide the function safety necessary for the facility.

After considering the design and development thus far, one recommendation is to ensure the safety related control system requirements are clearly spelled out as part of each subsystem's requirements. It is also advised to meet with subcontractors and vendors and have workshops that explain the safety system and provide guidance at a very early stage and throughout development. Many of the practices of using E/E/PE control systems in functional safety are now becoming more common in industry, but many are still unfamiliar with the techniques, technology and standards.

As our process for analyzing hazards evolved we found that many standards used similar although not identical methods of analysis, we found that there was duplication of effort as hazards were re-evaluated using a differing method. Because of the need for regulatory compliance using a single hazard analysis method may not be possible. We are required by OSHA to use a national consensus standard such as ANSI/RIA R15.06.

IEC 61508 has two related standards that cover specific industry requirements, IEC 61511, used by the process industry, and IEC 62061, which covers machine safety. We found that some vendors were more familiar with sector specific standards. All provide similar protection, so each could be used where they are most applicable; however, the project needs to understand all three.

Personnel safety is specifically addressed within the standard. However, the standard lacks recommendations to derive a required SIL rating based on strictly monetary damages. The project also needed to address the issue of what was considered an unacceptable risk with regard to equipment damage, which would result in monetary losses and/or lost or degraded observing time.

Following the Final Design Review, the procurement of the hardware necessary to construct the different LICs and the GIC will take place. The majority of the construction will occur during the following year. Following each LIC construction is scheduled its testing at the factory of the specific subsystem. The LIC will then be taken to the telescope's site for installation with the subsystem and then verification once again at subsystem site assembly and then integration into the entire GIS with a verification and validation process.

Taking place over a three year time span, this process is somewhat stretched out due to the dependency of the subsystems' on-site installation, integration, and test. The goal being that the project has a fully integrated safety system in place and functional prior to commissioning.

ACKNOWLEDGEMENTS

The National Science Foundation (NSF) through the National Solar Observatory (NSO) funds the ATST Project. The NSO is operated under a cooperative agreement between the Association of Universities for Research in Astronomy, Inc. (AURA) and the NSF.

The ATST represents a collaboration of 20 plus institutions, reflecting a broad segment of the solar physics community. The NSO is the Principal Investigator (PI) institution, and the co-PI institutions are the High Altitude Observatory, New Jersey Institute of Technology's Center for Solar Research, University of Hawai'i's Institute for Astronomy, and the University of Chicago Department of Astronomy and Astrophysics.

REFERENCES

[1] Robert P. Hubbard, "System safety and hazard analysis for the Advanced Technology Solar Telescope", Proc. SPIE 7738, (2010).
[2] Robotics Industries Association, [ANSI/RIA R15.06 American National Standard for Industrial Robots and Robot Systems—Safety Requirements], Robotics Industries Association, Ann Arbor, (1999).
[3] International Electrotechnical Commission, [IEC 61508 Functional safety of electrical/electronic/programmable electronic safety-related systems], International Electrotechnical Commission, Geneva, (1998).

Facility level thermal systems for the Advanced Technology Solar Telescope

LeEllen Phelps*[a], Gaizka Murga[b], Mark Fraser[c], Tània Climent[b]
[a] National Solar Observatory, 950 N. Cherry Ave., Tucson, AZ 85719 USA;
[b] AEC IDOM 15 South 5th Street - Suite 400, Minneapolis, MN 55402
[c] M3 Engineering & Technology Corp., 2051 W. Sunset Rd., Tucson, AZ 85704 USA

ABSTRACT

The management and control of the local aero-thermal environment is critical for success of the Advanced Technology Solar Telescope (ATST). In addition to minimizing disturbances to local seeing, the facility thermal systems must meet stringent energy efficiency requirements to minimize impact on the surrounding environment and meet federal requirements along with operational budgetary constraints. This paper describes the major facility thermal equipment and systems to be implemented along with associated energy management features. The systems presented include the central plant, the climate control systems for the computer room and coudé laboratory, the carousel cooling system which actively controls the surface temperature of the rotating telescope enclosure, and the systems used for active and passive ventilation of the telescope chamber.

Keywords: ATST, solar telescope, thermal systems, energy management

1. INTRODUCTION

Maintaining an environment that closely matches atmospheric conditions along the optical path is a key factor to the success of the Advanced Technology Solar Telescope (ATST). Analyses have shown that even a white-painted enclosure requires an active exterior skin-cooling system to mitigate seeing which is driven by thermal nonuniformities that change the refractive index of the air.[1] Interior to the telescope enclosure, which has been designed to maximize natural ventilation,[2] systems are provided to actively ventilate and prevent thermal stratification within the enclosure when ambient wind conditions are insufficient for this purpose. In addition to the thermal management of the enclosure, systems are provided to produce and distribute chilled heat transfer fluid at three distinct temperatures, two of which track at an offset from ambient temperature while the third is provided in steady, standard chilled water temperature range. These systems provide means to remove the thermal loads, including both solar loads and waste heat from equipment, and actively control the surface temperatures of all elements of the ATST. This includes all of the optics, instruments, drives, and electronics, as well as the climate control systems for the building, computer room, and coudé laboratory.

Starting with the central plant equipment in the next section, the major facility thermal equipment and systems to be implemented in the ATST project will be presented along with the associated energy management features of each. In Section 3, the climate control systems for the computer room and coudé laboratory are presented with the carousel cooling system and enclosure ventilation systems following in Sections 4 and 5 respectively. Finally, along with a summary of the energy reduction strategies adopted by the project, the results of an overall energy analysis demonstrating that the ATST as designed should exceed the requirements of the ASHRAE 90.1 Energy Standard by 29.1% will be presented in Section 6. It should be noted that these results include the process loads associated with the use of the facility as a physics laboratory. When using the revised methods of calculation allowed in energy efficiency performance standard for new federal buildings which treat the process loads differently, the ATST facilities have demonstrated an energy savings of 69.3%.[3]

*lphelps@nso.edu; phone 1 520 318-8552; fax 1 520 318-8500; atst.nso.edu

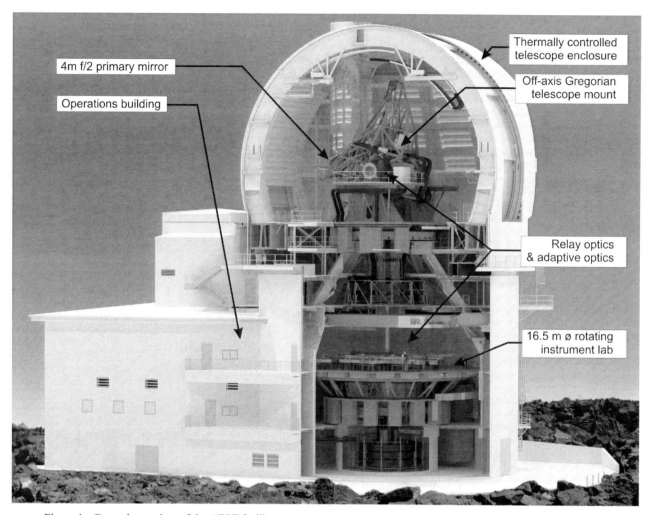

Figure 1 - General overview of the ATST facility

2. CENTRAL PLANT EQUIPMENT

A central plant in the remotely located Utility Building provides chilled heat transfer fluid at three different temperatures. The waste heat from the chillers is used to heat the operations building with any not needed used to maintain the surface temperature of the carousel cooling system near ambient rather than sub-cooled at night. Any unused chiller waste heat is exhausted at the fluid cooler. Ice storage tanks are incorporated into the coolant systems shifting some cooling load energy demands from day to night. Figure 2 shows the general arrangement of this equipment.

2.1 Chillers and Coolant Production

Implementing a modular design, there are six water-cooled chillers used to produce fluid at three different temperatures. One extra-low-temperature chiller system is used to produce coolant that tracks at an offset of 15°C (T_{amb}-15°C) lower than the ambient temperature. This coolant is generally used for applications where air is the direct cooling fluid in the ambient conditions of the telescope enclosure. The biggest consumer is the primary mirror. Two low temperature chiller systems are used to produce coolant that tracks at an offset of 4°C (T_{amb}-4°C) lower than ambient the ambient temperature. This coolant is generally used for applications in ambient conditions of the telescope chamber where it's used directly rather than being used in the production of chilled air. The biggest user is the carousel cooling system. The final three chillers are heat recovery chillers that produce a steady, standard 7°C coolant flow. In addition to space conditioning, coolant for the computer room, instruments, optics, coudé lab electronics, and for heat rejection of the other three low-temperature chillers.

The heat recovery chillers also produce 55°C hot water for space heating and reheat after dehumidification. At night, during ice production, any hot water produced that is not required for space heating needs is sent to the carousel cooling system and used to maintain the surface temperature of the enclosure near ambient rather than letting it sub-cool. Anything left after that is rejected via the fluid cooler.

Primary energy management issues considered in selecting the chillers:

- Chiller efficiency: water-cooled vs. air-cooled – The best available integrated part load value (IPLV) for conventional water-cooled chillers in the size range of interest is 0.46-0.47 kW/ton.[4] The best IPLV for conventional air-cooled chillers in the size range of interest is 0.80-0.83 kW/ton.[5] While the ATST low-temperature chillers are by no means conventional chillers that operate with IPLVs in these ranges, the relative efficiency between cooling technologies holds.

- Heat recovery for space heating – There are many more heating degree days than cooling degree days at the ATST Site. This provides a potential opportunity for heat recovery. There are two basic requirements for heat-recovery water heating: (1) hot water demand must be great enough to justify equipment and maintenance costs, and (2) the waste heat temperature must be high enough to serve as a useful heat source. The design layout of the mechanical equipment makes it possible to recover the waste heat stream and convert it into beneficial space heating.

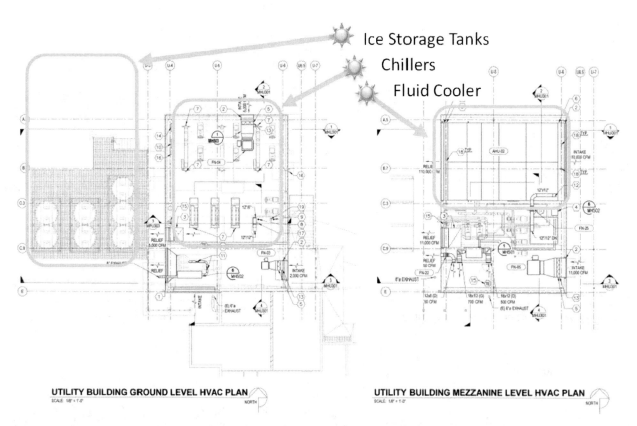

Figure 2 - General arrangement of the central plant equipment in the remotely located Utility Building

2.2 Fluid Cooler

Providing proper chiller condenser water temperature for the water-cooled chillers was a central challenge because of the site space limitations. A traditional evaporative cooling tower was not an option because of the exhaust water droplet drift associated with this type of equipment. So, a large air handler unit (AHU), acting as a fluid cooler, was implemented as a dry cooling tower. When excess water is available from the rain water collection system, it will be sprayed into the fluid cooler air stream prior to the cooling coil. The evaporative effects of this will increase the cooling capacity or lower the discharge air temperature. This allows higher cooling efficiency and, with the equipment arrangement, the carryover of water droplets into the exhaust air stream is avoided. The AHU was selected for an energy efficient configuration with an array of small, high efficiency variable frequency drive (VFD) motors.

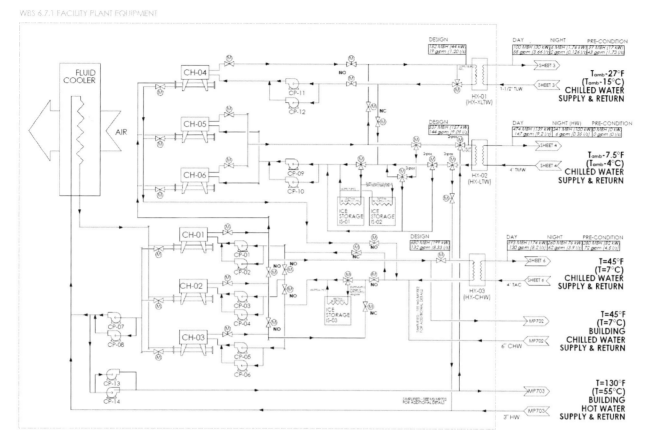

Figure 3 - Flow diagram of the central plant equipment

2.3 Ice Storage Tanks

In addition to shifting the power load to off-peak hours when it is less expensive, ice storage tanks allow the chillers to be optimized for and operate at full capacity. Ice storage is incorporated into two of the coolant systems. The 7°C is implemented as a partial-storage system while the T_{amb}-4°C system has full-storage capacity. The ice storage also provides future capacity without having to oversize the chillers. When conditions allow, in all but the coldest weather, the T_{amb}-4°C system is also used to produce the coolant for the T_{amb}-15°C system avoiding the need to run the extra-low-temperature chiller.

3. SPECIALTY CLIMATE CONTROL SYSTEMS

The computer room and the coudé lab are two areas that require special treatment with respect to their heating, ventilation, and air conditioning (HVAC) systems. The computer room has high internal thermal loads and requires humidity control. The coudé environmental system is intended to provide close control of design parameters within the coudé lab.[6] The coudé environmental system has been designed to provide highly filtered air with precise temperature control in a unidirectional flow pattern that is free of recirculation zones that are near any optics, optical path, or instrument.[7]

3.1 Computer Room

The configuration of the computer room includes full hot-aisle containment. The arrangement is shown in Figure 4. Each row of server racks will have a hydronic cooling unit installed and draw warm air from the contained hot aisle and exhaust it into the general computer room space at a load-neutral temperature. For the general occupancy, lighting, and other loads, a small fan coil unit is provided in the ceiling space that provides comfort conditioning as well as pressure and humidity control. Whether hot-aisle or cold-aisle containment is implemented, keeping the hot air generated by the computer equipment from mixing with the cool room air is an important energy efficiency strategy.

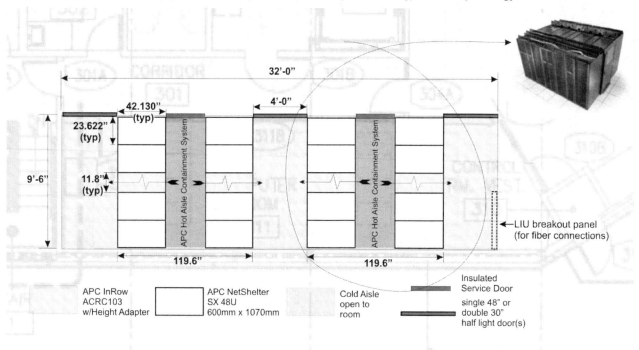

Figure 4 - Computer room arrangement plan

3.2 Coudé Instrument Lab

In addition to energy efficiency concerns, space limitations and vibration issues are the drivers for the HVAC equipment selection and installation for the coudé instrument lab. Two vertical 85,000 m^3/h AHUs utilizing an array of small high efficiency VFD motors, are central to this system. The units are mounted with spring isolators onto isolated foundations. They use an array of small, high-efficiency, direct variable frequency drive, airfoil plenum fans including coplanar silencers, a perforated inner wall backed by acoustically absorbent material. High velocity, low airflow resistance HEPA filters are used for the plenum ceiling in the space, providing full coverage in the room with an initial resistance 37 Pa rather than the 105 Pa projected from conventional HEPA filters.

4. CAROUSEL COOLING SYSTEM

The main objective of the Carousel cooling system is to reject solar radiation on the enclosure skin. In order to mitigate the enclosure thermal seeing effects, the skin temperature must be maintained at ambient to 4°C below ambient. To accomplish this, the Carousel cooling system will use water cooled plate coils placed on the surface of the enclosure cladding.

The layout of the plate coils has been optimized [1] to minimize surfaces receiving insolation and then optimized again[8] to prioritize the plate coils into zones and identify areas receiving such a minimal amount of insolation that they were not required. They are coated with a high performance ENERGY STAR rated roof coating[9] in order to minimize cooling system loads.

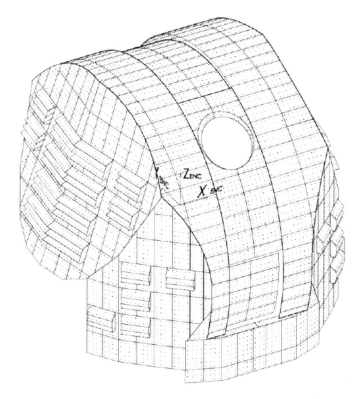

Figure 5 - Layout of carousel cooling system plate coils covering the enclosure cladding[10]

4.1 Plate Coil Thermal Performance Analysis

While the analyses performed to determine the design loads have been presented previously[11], once the cladding temperature was established, it was used to define zones of similar location and loads and to optimize the general layout. As the insolation incident angle changes throughout the observing day, the zones that require maximum coolant flow shift. A control valve will regulate the coolant flow to each zone allowing adjustment of plate coil temperature. Determining the most effective control approach will maximize both compliance with the performance requirements and energy efficiency for the system.

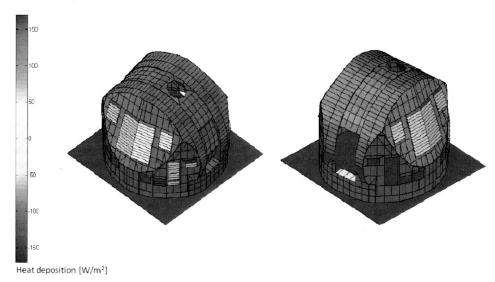

Figure 6 - Radiation model establishes heat deposition as it changes through an observing day

During nighttime hours, on average from 7 PM to 7 AM, the heat deposition data is negative. That means that as the carousel surface temperature is controlled near ambient temperature, instead of absorbing heat, the plate coil will transfer it to the ambient until the temperatures become balanced. In this case, an opportunity presents itself to use this to advantage and radiate excess heat from the chillers via the carousel cooling system as ice is produced for cooling during the next observing day.

Under certain weather conditions it is possible for frost, snow, or ice to form over the carousel surfaces. Circulating warm water through the plate coil system can prevent or accelerate the drying process. Using the chiller waste heat as a source for the warm water provides significant savings over any other available option.

4.2 Plate Coil Layout Optimization

After establishing 64 different zones for optimum surface temperature control, the possibility of removing plate coils placed in the lower part of the enclosure as shown in green in Figure 7 was investigated. In order to see if it is possible to trim back any plate coil coverage, the fluctuations of the cladding temperature in these zones were analyzed for the established design day (July, 1st 2003). This process identified a number of areas where reduced coverage was feasible.

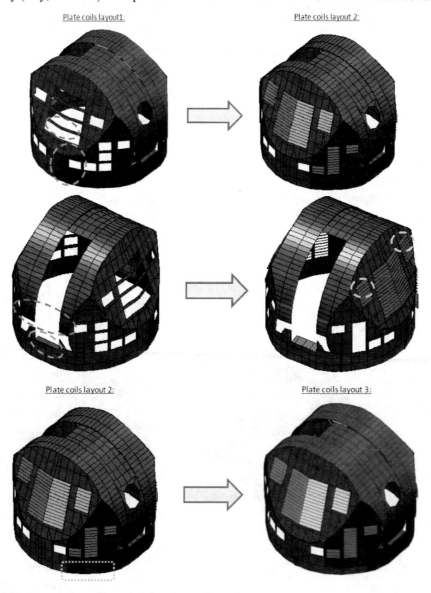

Figure 7 - Modifications performed in optimizing plate coil coverage

4.3 Thermal Coating Trade Study

The standard plate coil fabrication process is completed by painting the planar face that will be exposed to the sun with a high reflectivity/high emissivity coating. These properties will reduce the heat deposition by reflecting a large amount of the incoming solar radiation and by emitting as much thermal radiation as possible to the environment.

The coating initially considered, AZJ-4020 white epoxy thermal control coating manufactured by AZ Technology, [1] was compared with a more conventional coating Energy Seal Acu-Shield white acrylic elastomeric coating manufactured by Advanced Coating Systems, Inc. While the latter caused a larger heat deposition (+14%), its cost is far more economic (-27% of the cost of the plate coils). With the capital cost savings significantly higher than any future operating cost savings, the Acu-Shield, an ENERGY STAR rated roof coating, was selected.

5. ENCLOSURE VENTILATION SYSTEMS[12]

Minimum air velocity of 0.5 m/s is required throughout the enclosure to assure thermal performance of the telescope optical elements. Protection of the telescope mount from wind in excess of 5 m/s is required to avoid jitter. The passive and active telescope chamber ventilations systems are responsible for the thermal control so as to minimize the contribution to "dome seeing." To accomplish this, three systems are coordinated; the passive ventilation system, the active ventilation system, and the air scavenging system. The average wind velocity for the site is 5 m/s providing ample opportunity for natural air flushing via the passive ventilation system.

5.1 Passive Ventilation System

The passive ventilation system opens and closes the enclosure vent gates and the rear door to promote natural air flushing through the telescope chamber. A detailed CFD model of the enclosure was prepared in order to study the overall performance of the telescope chamber thermal control systems and obtain guidelines on control strategies.[12] Velocity maps and velocity values along a path coincident with the azimuth axis were analyzed in order to study the telescope chamber wind flushing.

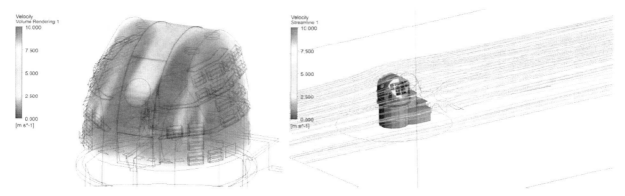

Figure 8 - Wind speed 5 m/s, direction 0°, all vent gates and rear door opened

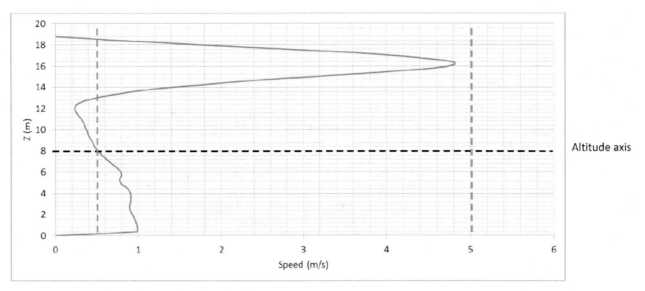

Figure 9 - Velocity profile (m/s) in azimuth axis

With an ambient wind speed of 5 m/s, air velocity close to 5 m/s is found in the azimuth axis at some positions but close to the elevation axis it drops to about 0.5 m/s. The number of air changes for this case is 26 volumes/hour. Numerous cases were analyzed to verify performance in all of the typical configurations expected. The case shown exceeds performance requirements, yet had the lowest number of air renovations of the cases explored. A strategy to make a more uniform air speed profile could be to close some of the vent gates and the rear door, and activate some lower enclosure fans to reduce the current speed.

5.2 Active Ventilation System

The active ventilation system consists of four airfoil propeller fans near the top of the enclosure. These provide a downward flow of air when the wind conditions are such that the passive ventilation system cannot met the minimum air velocity requirements. These fans are run in conjunction with an array of centrifugal fans at the base of the lower enclosure drawing air through openings in the telescope chamber floor. The fans were selected to minimize energy consumption with high efficiency VFD motors.

The CFD model of the enclosure was modified in order to study the overall performance of the active ventilation system with the top fans in varying orientations and varying numbers of the lower enclosure fans are activated.[12] Velocity maps and velocity values along a path coincident with the azimuth axis were analyzed in order to study the telescope chamber wind flushing.

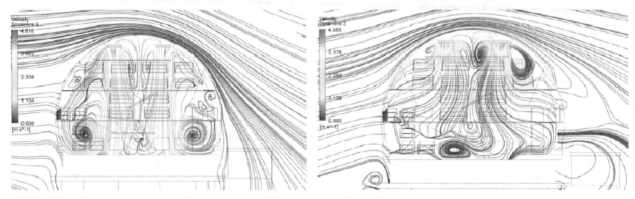

Figure 10 - Velocity streamlines activating all of the lower enclosure fans (left) and two lower enclosure fans (right)

These and additional results show that in order to have 25 volume renovations it should be necessary to activate 11 of lower enclosure fans. In order to optimize the speed profile in the proximity of the azimuth axis the propeller fans have been tilted 10° in its two horizontal axes.

5.3 Air Scavenging System

The air scavenging system is responsible for evacuating any warm air that has risen to the upper reaches of the telescope chamber, thereby avoiding temperature stratification of the air inside the carousel. Ducts at the top part of the enclosure are provided with four multileaf grilles or dampers provided to collect warm air from the upper part of the enclosure, which is conducted to the air handling units and exhausted to the lower enclosure. From there, the lower enclosure fans expel it out of the observatory.

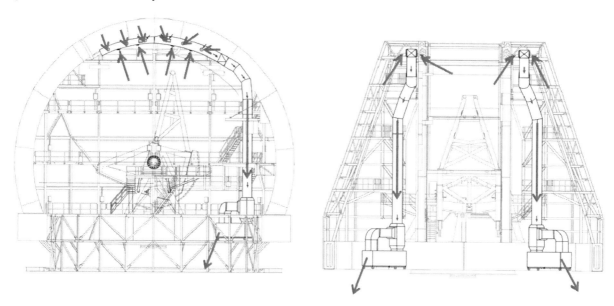

Figure 11 - Schematic view of the air scavenging system

6. ENERGY ANALYSIS[3]

The Advanced Technology Solar Telescope (ATST) project is located on Haleakalā, Hawaii at an elevation of 10,000 ft. and is divided into four sections: enclosure, lower enclosure, support and operations (S&O) and utility building. 10 CFR Part 433, Energy Efficiency Standards for the Design and Construction of New Federal Commercial and Multi-Family High-Rise Residential Buildings, requires new buildings to exceed the baseline building as outlined in Appendix G of ASHRAE Standard 90.1-2004, Energy Standard for Buildings Except Low-Rise Residential Buildings, by 30 percent. This analysis does not cover the enclosure, only the process loads from the carousel cooling system.

The ATST facility has typical office environments but also includes instrument prep labs, computer room, and mechanical equipment rooms. All have upper and lower temperature limits plus some spaces have humidity requirements that must be maintained. This study will look at the energy required to support these spaces, the Coudé Lab, and the telescope pier ventilation system.

A central plant in the Utility Building will provide chilled water at three different temperatures; waste heat from the chillers will be used to heat the building. Ice storage tanks will be incorporated in the chilled water system to shift some cooling load demands from day to night.

Most spaces have air handling units (AHU) or fan coils (FC) using chilled or heating water to regulate temperatures. Some of the mechanical spaces use ambient air for cooling instead of chilled water. Two low temperature chilled water systems will provide chilled water to heat exchangers that support the Enclosure and telescope instruments.

Using Carrier's HAP program version 4.5, the procedures in ASHRAE 90.1-2004 Appendix G, and La Paz, Bolivia TM2 weather data, these buildings have demonstrated an energy savings of 27.3 percent and a cost savings of 29.1 percent. Using the revised percent savings formula as called out in 10 CFR Part 433, these buildings have demonstrated an energy savings of 69.3 percent and a cost savings of 75.4 percent. Note that TM2 weather data is not available for Haleakalā, Hawaii, and the closest matching available city in elevation and temperature range was La Paz, Bolivia.

7. SUMMARY

Extensive efforts to find and implement all practical energy management strategies available have been productive. All of the following are now strategies and/or design features and/or operations strategies implemented throughout ATST:

- For traditional building components requirements exceed ASHRAE 90.1 by 30%
    - Building envelope, LED lighting, occupancy sensors, etc.
    - Economizer mode for HVAC systems using outside air when conditions allow
    - High performing ENERGY STAR rated roof coating
- High efficiency motors with VFDs on pumps and fans
- Fully contained hot aisles in Computer Room
- Regeneration on shutter drives
- Pre-cooler section on Fluid Cooler
- Ultrasonic humidifiers in building areas with humidity control
- Roller element bearings for TMA rather than hydrostatic
- Implement Energy Management Program based on Labs for the 21st Century
- Implement MECO Energy Efficiency Study recommendations
- Install concrete panels as thermal mass on lower enclosure rather than active ventilation[13]
- Concrete ground apron surrounding lower enclosure[1]
- Integrated wind speed as a control variable for carousel cooling economizer mode
- Implement economizer mode for Carousel Cooling System (v≥6m/s)
- Set back thermal systems in afternoon when site seeing begins to degrade
- Added nighttime radiative heat rejection capability to plate coil system

ACKNOWLEDGEMENT

This work utilizes data obtained by the Advanced Technology Solar Telescope (ATST) project, managed by the National Solar Observatory, which is operated by AURA, Inc. under a cooperative agreement with the National Science Foundation.

REFERENCES

[1] Phelps, L., Barr, J., Dalrymple, N., Fraser, M., Hubbard, J., Wagner, J., and Warner, M.. "The Advanced Technology Solar Telescope enclosure," Proc. SPIE 6267, 141 (2006).
[2] AEC Idom, "Facility Thermal Systems Design Services for the ATST Ventilation Analysis Report," 16124-TRE-048 (2012).
[3] Fraser, M., "ATST Energy Analysis Report," M3PN09077 (2012).
[4] http://www1.eere.energy.gov/femp/technologies/eep_wc_chillers.html..
[5] http://www1.eere.energy.gov/femp/technologies/eep_ac_chillers.html.
[6] Phelps, L., Rimmele, T., Hubbard, R., Elmore, D., "Advanced Technology Solar Telescope coudé lab thermal environment," Proc. SPIE 7733, 7733138 (2010).
[7] Fraser, M., "Coudé Environmental System Design Report," M3PN09077 (2012).
[8] AEC Idom, "Carousel Cooling System: Thermal, structural and control analysis report," 16124-TRE-046 (2012).
[9] Phelps. "Evaluation of thermal control coatings exposed to ambient weather conditions at Haleakala High Altitude Observatory." In *Ground-based and Airborne Telescopes II*. Edited by Stepp, Larry M.; Gilmozzi, Roberto. *Proc SPIE* **7012**, 701230 (2008).
[10] AEC Idom, "Facility Thermal System Design Report," 16124-TRE-045 (2012).
[11] Murga, G.; Marshall, H.; Phelps, L.; Hervás, A.; Larracoechea, I. "ATST enclosure mechanical and thermal models," in *Integrated Modeling of Complex Optomechanical Systems*. Edited by Andersen, T.; Enmark, A. *Proc SPIE* **8336**, 83360K (2011).
[12] AEC Idom, "Ventilation Analysis Report," 16124-TRE-048 (2012).
[13] Phelps, L., Warner, M., "Advanced Technology Solar Telescope lower enclosure thermal system," Proc. SPIE 7017, 701719 (2008).

Stray light and polarimetry considerations for the COSMO K-Coronagraph

Alfred G. de Wijn[*a], Joan T. Burkepile[a], Steven Tomczyk[a], Peter G. Nelson[b], Pei Huang[c], and Dennis Gallagher[a]

[a]High Altitude Observatory, National Center for Atmospheric Research, P.O. Box 3000, Boulder, CO 80303, USA; [b]Sierra Scientific Solutions, 1540 Patton Drive, Boulder, CO, USA; [c]Consultant to NCAR, 1484 N. Larkspur Ct., Lafayette, CO, USA

ABSTRACT

The COSMO K-Coronagraph is scheduled to replace the aging Mk4 K-Coronameter at the Mauna Loa Solar Observatory of the National Center for Atmospheric Research in 2013. We present briefly the science objectives and derived requirements, and the optical design. We single out two topics for more in-depth discussion: stray light, and performance of the camera and polarimeter.

Keywords: Sun, COSMO, Coronagraph, Polarimeter, Stray Light

1. INTRODUCTION

The solar corona is a million-degree plasma that is the source of the solar wind and the site of explosive activity such as Coronal Mass Ejections (CMEs) that drive space weather throughout the heliosphere. The corona is organized by the Sun's magnetic field into brighter magnetically 'closed' regions, where the plasma is contained by the field, and magnetically 'open' regions of very low density where the plasma and field are carried outward to form the solar wind. Coronal brightness and the distribution of closed and open regions vary over the 11-year solar cycle. Dynamo processes in the solar interior determine the quantity and distribution of magnetic flux into the corona and drive the sunspot activity cycle.

Much of what is known about CME properties and the density structure of the corona comes from white-light observations.[1,2,3,4,5,6,7] These observations are needed to understand CME formation, as well as the evolution of the global structure of magnetic field and density distribution of the corona. They also provide insight into the mechanisms responsible for coronal heating and solar wind acceleration. The K-Coronameter at the Mauna Loa Solar Observatory (MLSO) has been operating since 1980 and provides unique white-light observations of the low corona. The K-Coronameter utilizes a 1-D CCD detector to acquire scans of the low corona every ½ degree in azimuth and builds up an image of the corona over 360 degrees every 3 minutes. The new K-Coronagraph is scheduled to replace this aging instrument at MLSO in 2013. It will significantly improve our understanding of the formation and dynamics of CMEs and the global density structure of the corona through improved sensitivity, cadence, and field of view (FOV).

Even at the best sites and on the best days, the foreground sky brightness drowns out the corona. However, because the white-light K-corona is caused by electron scattering of photospheric light, it is linearly polarized tangential to the solar limb. Ground-based coronagraphs for white-light measurements exploit this by making a measurement of polarization brightness (pB), which can be related to coronal electron density.

The K-Coronagraph is one of the three instruments in the proposed COSMO suite. The two other instruments of COSMO are the Chromosphere and Prominence Magnetometer (ChroMag) and the Large Coronagraph for coronal magnetic field measurements. ChroMag is currently in a prototyping phase, while the Large Coronagraph is in the preliminary design phase.

2. K-CORONAGRAPH SCIENCE OBJECTIVES AND REQUIREMENTS

CMEs are explosive events driven by magnetic stresses in the solar atmosphere and are the primary driver of space weather at earth. CMEs form and accelerate low in the corona with speeds and accelerations that can vary over 3 orders

---

[*] dwijn@ucar.edu, phone +1 303 497 2171

Table 1. Instrument requirements for the COSMO K-Coronagraph.

Quantity	Units	Requirement	Goal	Mk4
FOV	R_{sun}	3	4	2.9
Lower Limit of the FOV	arcsec	50	25	120
Spatial Sampling	arcsec	6	3	5×9 to 5×23
Noise Level	pB/√Hz	3.9×10^{-9}	1.3×10^{-9}	5.4×10^{-8}
Map Time	s	15	8	180

of magnitude. The greatest acceleration occurs below 3 R_{sun} for most events. Observations of their onset and measurements of the rate-of-change of CME acceleration are needed to discriminate between the many models posited to explain their formation. This requires rapid image sequences of the very low corona with a FOV from 1.05 to 3.00 R_{sun} to track CMEs from their formation in the first coronal scale height (1.01 to 1.09 R_{sun}) to the height where they have acquired most of their acceleration (3 R_{sun}). CME structure sets the required spatial sampling at 6 arcsec or better.

Very high time cadence (13 to 40 s) images recorded by TRACE and YOHKOH SXT have allowed measurements of acceleration changes over very limited heights for a very small number of events due to their limited FOV.[8] A 15-s time cadence will provide the coverage that is needed to record accurate trajectories of the fastest CMEs.

Earth-directed "halo" CMEs are considerably fainter due to projection effects than those happening at the limb and ejected perpendicular to the line of sight. The noise level must be sufficiently low to detect coronal structures with intensities of 10^{-9} B_{sun}.

The instrument requirements from these objectives are summarized in Table 1.

3. OPTICAL DESIGN OVERVIEW

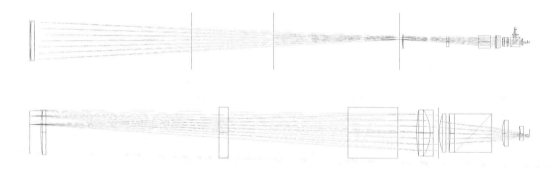

Figure 1. Overview of the COSMO K-Coronagraph optical design (top) and zoom-in on the back-end optics (bottom). The optical design shows the aperture stop and primary lens (O1) at the left and several baffles in the telescope tube. The zoom-in shows the prime focus (where the occulter is located) at the left, followed by the field lens, the bandpass filter, the modulator stack (here shown as a monolithic block), a doublet lens to correct O1 chromatic aberration, the Lyot stop, the first two elements of the camera lens, the polarimetric analyzer polarizing beam splitter, the last two elements of the camera lens, and finally the focal plane at the far right. The total optical track is 2734 mm.

Figure 1 shows an overview of the optical design as well as a more detailed view of the compact back-end. The instrument design meets or exceeds the requirements given in Table 1. The COSMO K-Coronagraph is a traditional Lyot coronagraph with a 20-cm fused silica primary (O1), a field lens just behind the occulter at prime focus, and a Lyot stop at the image of the aperture. The bandpass filter, the polarimetric modulator (in this model shown as a monolithic block), and a doublet lens that corrects the chromatic aberration of the primary are placed between the field lens and the Lyot stop. The first two elements of the camera lens system are placed immediately after the Lyot stop, followed by the polarizing beamsplitter analyzer for the polarimeter, and finally the last two elements of the camera lens. The O1 has focus capabilities in order to make a sharp image of the sun at the occulter. The camera focal planes move parallel to the optical axis to bring the occulter in focus.

The K-Coronagraph is a so-called dual-beam polarimeter, i.e., the modulated light is analyzed in two perpendicular directions and recorded simultaneously using two synchronized cameras. This technique allows for the elimination of nearly all crosstalk from intensity to polarization resulting from seeing, pointing jitter, etc. The camera lens system consists of four optical elements. The polarizing beamsplitter is placed in between the second and third optic, so that the first two elements are shared between the cameras.

4. STRAY LIGHT ANALYSIS

There are four main contributors to scattered light in any Lyot-type coronagraph: diffraction, ghosting, O1 surface roughness, and O1 surface contamination.

4.1 Diffraction

Light from the solar disk is diffracted by the telescope aperture, and some of the diffracted light spills over the occulter. The image at prime focus is truncated by both the occulter and the field stop. As a result, at the image of the aperture created by the field lens the diffracted light is concentrated in a ring at the edge of the aperture and at a spot in the center. The Lyot Stop blocks most of this light (as well as any light scattered off the edge of the aperture). The reduction in occulter edge diffraction by the Lyot Stop is given by:

$$\text{LS Attenuation} = \frac{P \times \Delta\beta^2}{2\pi(1-P)\frac{d}{0.001\lambda}} \times \left\{ \frac{1}{\left[\left(\frac{m \times s}{7200"} + \Delta\beta\right)\left(\frac{m \times s}{7200"}\right)\right]^2} + \frac{1}{\left[\left(\frac{(2+m)\times s}{7200"} + \Delta\beta\right)\left(\frac{(2+m)\times s}{7200"}\right)\right]^2} \right\}^2,$$

where P is the fractional projected area of Lyot Stop, and in the case of the K-Coronagraph, $m = 0.02$ (occulter over sizing in units of the solar disk), $d = 200$ mm (O1 diameter), $l = 0.735$ mm (center wavelength), $s = 1920$ arcsec (avg. solar diameter), and Db = angular distance from occulter edge (rad). In order to provide sufficient attenuation yet not block more light than necessary we chose $P = 0.85$.

4.2 Ghosting

The main risk of ghosting is associated with the double bounce inside the O1 lens since 1) it is uncoated; and 2) the solar disk light is very bright. A strong attenuation of this ghost by geometrical defocus is needed in order to not contribute significantly to the 2 ppm near-disk irradiance attenuation goal at the occulter. The solar disk ghost image at the occulter is about 137 cm in diameter compared to 1.78 cm for the main image. Due to the reflectance of uncoated fused silica (3.4% at 725 nm), we have an intensity attenuation of 1.2×10^{-3}, resulting in a total ghost level of $\sim1\times10^{-10}$ of that of the main disk at the detector, well below the required stray light levels.

4.3 Surface Roughness Analysis

The micro-roughness requirement is derived from the requirement that the integrated scattered light observed 48 arc seconds above the solar limb (0.05 radii) should be less than 2 ppm of the solar disk center. 2 ppm is approximately one third of the best scattered light level expected with a clean lens at this height, but ~4 times the brightness of the sky at MLSO under ideal sky conditions.

We select the common "ABC model" (also known as the K-correlation model) as a basis for our calculations:

$$S_2(f) = \frac{A_2}{\left[1 + (Bf)^2\right]^{(C+1)/2}}.$$

The subscript $_2$ refers to the fact that this is a model for a 2-D power spectrum, following the convention in Stover.[9] Note that the 1-D spectrum has the same form, except the power in the denominator is $C/2$; the slope of the 2-D spectrum is one power of frequency steeper than the 1-D spectrum of the same surface, at high spatial frequencies. In this expression, therefore, C is the slope of the 1-D power spectral density (PSD). A_2 gives an overall scaling and B describes at what frequency below which the power law becomes flat.

Among the best-characterized sets of fused silica optics of the appropriate dimension are those made for the Laser Interferometer Gravitational-wave Observatory. Their $\varnothing 250\times100$mm mirrors were polished to ~5 Å RMS over the frequency band of 4.3–7500 cm^{-1}. They measured the PSD of their micro-roughness over a very wide frequency range from 0.1 to 10,000 cm^{-1}. Those data show the 1-D PSD to have a slope C of ~1.3 with no evidence of a low-frequency

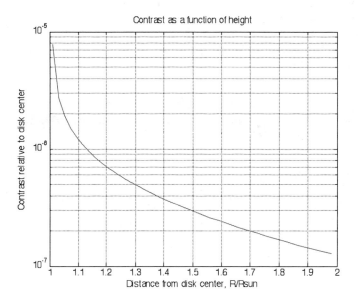

Figure 2. The predicted level of scattered light for a 5 Å RMS finish. The calculation includes solar limb darkening at 725 nm.

flattening (described by the B parameter). For our modeling we therefore set $C = 1.3$ and $B = 10^5$ µm. The scattered light level and the RMS micro-roughness (over a given band) then become only a function of the parameter A_2.

The value of A_2 is chosen such that the stray light requirement is met. The RMS surface roughness specification is then derived. To calculate the scattered light level we use the Modeled Integrated Scatter Tool (MIST) developed by the Optical Technology Division of the National Institute for Standards and Technology (NIST). We find that the spatial frequencies that contribute to the scattered light are from 3.23×10^{-4} to 1.25×10^{-2} µm^{-1}. In order to meet the 2 ppm requirement at 48 arcsec from the solar limb the RMS roughness over this bandwidth must be 5 Å or less.

Figure 3 shows the predicted level of scattered light for the specified surface finish quality. This calculation includes the

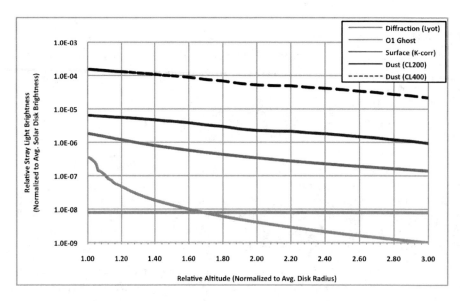

Figure 3. The predicted stray light contributions from diffraction, O1 ghost, surface roughness, and surface contamination.

effect of solar limb darkening at 725 nm. The 5 Å micro-roughness requirement is consistent with the requirements of other internally-occulted coronagraphs (such as the COR-1 instruments on the STEREO spacecrafts).

4.4 Surface Contamination Analysis

Stray light caused by scattering from lens surface contamination was modeled at two levels, CL200 and CL400. This is a Mie scattering calculation result, which can be calculated using MIST or approximated.[10]

4.5 Summary

In order to make comparisons between the results of the stray light models above and real-world coronal measurements, we must integrate all of the Point Source Transfer (PST) functions to form an overall stray light irradiance figure for each of the effects, and then scale the result to the solar disk brightness. The result is shown in Figure 3.

Diffraction would dominate the stray light at the low range of the FOV in the absence of the Lyot stop, but for the baseline stop of 85% of the aperture area it becomes small relative to other contributors. The O1 ghost is also insignificant. While the O1 surface roughness is critical, it is clear that O1 surface contamination is the most important factor in ensuring low levels of scattered light. A HEPA system was incorporated in the design to keep the O1 bathed in filtered air in order to reduce the build-up of contamination. Despite these efforts it is expected that the O1 lens will have to be cleaned on a regular basis.

5. INSTRUMENT MODEL

The K-Coronagraph performs a noise-dominated measurement. As such, a high photon flux is desired so that averaging can be employed to improve the signal-to-noise. The figure of merit of the camera thus is photo-electrons per second, i.e., the product of the well depth and the frame rate. Most scientific cameras have shallow wells and/or slow frame rates. Some custom cameras exist, but they tend to be expensive and difficult to implement. We selected the PhotonFocus MV-D1024E-160-CL-12 CMOS machine-vision camera for its deep wells (180 ke$^-$) and high frame rate (135 Hz). This camera provides a 12-bit readout. In order to verify that this industrial camera can still provide science-quality output, a sophisticated instrument model was developed.

5.1 Instrument Model

A known Stokes vector is sent into the model. It is passed through the telescope, modulated, analyzed, and converted into an electron count in the camera. The electron count is then sampled using a model of the Analog-to-Digital Converter (ADC). 506 readouts (approximately corresponding to a 15-s observation) are accumulated in each of the four modulation states, then summed, demodulated and calibrated. The experiment is repeated 500 times to gather statistical information on the noise of the measurement. The resulting measured Stokes vectors are compared with the input Stokes vector.

The parameter space is scanned in intensity from 2 to 18×10^{-6} B_{sun} with a 7-ms exposure (though the results can be scaled to different intensity/exposure ranges), and in degree of Stokes Q polarization from 0 to 10%. These conditions cover excellent to poor sky conditions. We assume that no U or V polarization enters the telescope.

5.2 Digitization

ADCs in the camera convert electrons captured in the wells into a digital readout. The process of digitization introduces a systematic error. A measurement at a specific intensity nearly always returns the same discrete data number (DN) if the combination of shot and read noise is much smaller than the least significant bit (LSB). The actual average readout then differs from the expected average readout. In order to reach 10^{-4} accuracy with a 12-bit camera, we must be able to determine an average to better than 0.1 DN. A simple calculation shows this translates into a requirement that the noise must be at least 11 e$^-$ for the MV-D1024E, easily satisfied by the 220 e$^-$ read noise. Note that if the MV-D1024E was an 8-bit camera, the noise would have to have been 350 e$^-$; this increase in noise required to remove the systematic effects of discretization disqualifies most 8-bit cameras for our application.

5.3 Bit Errors

In a perfect ADC, the value of each bit is exactly half that of the next more significant bit. However, in practice the bits exhibit small variations in size that we will call "bit errors". Bit errors manifest themselves as systematic offsets in the

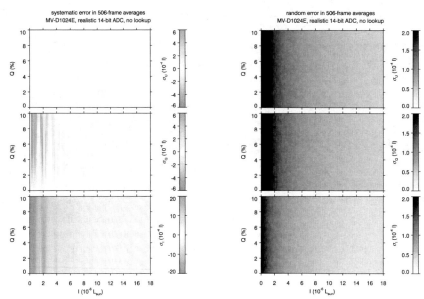

Figure 4. Systematic (left) and random (right) errors from a simulation that includes measured bit errors of an ADC in the MV-D1024E as a function of intensity and degree of polarization for a pure Stokes-Q signal. From top to bottom: error in Stokes U, Q, and I.

demodulated measurements. The ADCs in an MV-D1024E camera were characterized in the lab. The histograms show some values are favored over others. This can introduce a systematic error that we model here. Using the measured histograms we simulate the ADC in our instrument model, and derive the systematic and random errors resulting from the measurement. After discussion with PhotonFocus the camera was modified with a drop-in replacement 14-bit ADC (reading out only the 12 most significant bits), which reduced the systematic error by about a factor of two. However, the error remains at the 10^{-3} level (Figure 4).

Further improvement is made by applying a lookup table to correct individual readouts. The histogram used to encode the ADC characteristics is derived from different pixels than the one used for the correction. The latter also has 0.5%

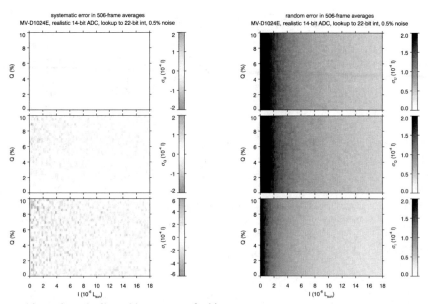

Figure 5. As Figure 4 but using a lookup table to correct for bit errors.

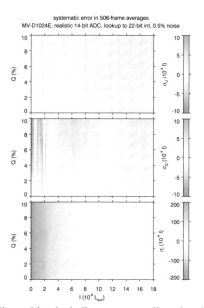
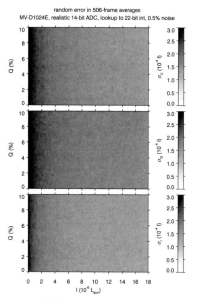

Figure 6. As Figure 5 but including sensor nonlinearity, dark current, a realistic modulation matrix, a telescope matrix derived from the ZEMAX model, and synthesized calibration.

normally distributed random noise added. The lookup table returns a 22-bit integer number, so that 1024 exposures can be added without risk of overflow in a 32-bit integer. Figure 5 shows that the errors are at the 10^{-4} level. Notice that systematic errors in I are similar in size to the random errors. Since the errors are determined on the basis of 500 model evaluations, the statistical noise on the determination of the systematic error is $1/\sqrt{500} \approx 0.044$ of the random error. We conclude that the noise in the lookup table dominates the systematic errors.

5.4 Calibration

The above tests all employed a perfect modulation and calibration. The functioning of the polarimeter and calibration system must also be verified. The model was extended to include dark current and sensor nonlinearity. The dark current is given in the MV-D1024E data sheet; for the nonlinearity we use the 3rd order polynomial fit given in the EMVA 1288 Standard test of the PhotonFocus MV-D1024E-160-CL-12 camera by AEON Verlag & Studio. It is corrected by applying a 4th order polynomial inverse in the lookup table. Furthermore, the modulation matrix is based on a realistic design (and thus non-ideal), and the telescope matrix is taken from the ZEMAX model. The analyzer is assumed to be a 99.9% polarizer in one beam, 99% in the other, but otherwise perfect.

First, 50000 sets of calibration observations are synthesized. In this process the calibration polarizer is modeled as a 99.9% polarizer and a diffuser intensity of 10^{-5} B_{sun} is used. These sets are then individually and independently processed to generate 50000 sets of calibration data consisting of the dark current, gain, and modulation matrix in each beam.

The calibration data are then used in a model run to calibrate the observations. Each model experiment uses a randomly chosen set of calibration data. The random noise is slightly increased compared to the perfect calibration (as can be expected). The systematic errors in Q are at the 10^{-3} level or below (Figure 6).

The systematic errors in I are large, at the 10^{-2} level. Since I is not a difference measurement, it is much more sensitive to errors in the determination of the dark current, gain, nonlinearity, etc. An absolute error of 1% is not unreasonable in real-world measurements. For the K-Coronagraph, I is not a measurement of interest, so this issue does not concern us.

ACKNOWLEDGMENTS

The National Center for Atmospheric Research is sponsored by the National Science Foundation.

REFERENCES

[1] Gosling, J.T., Hildner, E., MacQueen, R.M., Munro, R.H., Poland A.I., and Ross, C.L., "The speeds of coronal mass ejection events", Solar Phys., 48, 389-397 (1976)

[2] Howard, R.A., Sheeley, N.R., Michels, D.J., Koomen, M.J., "Coronal mass ejections – 1979-1981", J. Geophys. Res., 90, 8173-8191 (1985)

[3] Hundhausen, A.J., "Sizes and locations of coronal mass ejections: SMM observations from 1980 and 1984–1989", J. Geophys Res., 98, 13177-13200 (1993)

[4] Hundhausen, A.J., Burkepile, J.T., St.Cyr, O.C., "Speeds of coronal mass ejections: SMM observations from 1980 and 1984–1989", J. Geophys. Res., 99, 6543-6552 (1994)

[5] St.Cyr, O.C., Burkepile, J.T., Hundhausen, A.J., Lecinski, A.R., "A comparison of ground-based and spacecraft observations of coronal mass ejections from 1980–1989", J. Geophys. Res., 104, 12493-12506 (1999)

[6] Yashiro, S., Gopalswamy, N., Michalek, G., St.Cyr, O.C., Plunkett, S.P., Rich, N.B., Howard, R.A., "A catalog of white light coronal mass ejections observed by the SOHO spacecraft", J. Geophys. Res., 109, A7, A07105 (2004)

[7] Schmit, D.J., Gibson, S., deToma, G., Wiltberger, M., Hughes, W.J., Spence, H., Riley, P., Linker, J.A., Mikic, Z., "A novel metric for coronal MHD models", J. Geophys. Res., 114, A6, A06101 (2009)

[8] Alexander, D., Metcalf, T. R., Nitta, N. V., "Fast acceleration of a CME-related X-ray structure in the low solar corona", Geophys. Res. L., 29, 10, 41 (2002)

[9] Stover, J. C., "Optical Scattering – Measurement and Analysis", SPIE Press, Bellingham, WA, 2nd ed. (1995)

[10] Dittman, M. G., "Contamination scatter functions for stray-light analysis", Proc. SPIE, 4774, 99-110 (2002)

Quasi-static Wavefront Control for the Advanced Technology Solar Telescope

Luke C. Johnson[a], R. Upton[b], T. Rimmele[c], and S. Barden[c]

[a]National Solar Observatory,
950 Cherry Ave., Tucson AZ, USA;
[b]Sigma Space Corporation,
4600 Forbes Blvd., Lanham MD, USA;
[c]National Solar Observatory,
3010 Coronal Loop, Sunspot NM, USA

ABSTRACT

The Advanced Technology Solar Telescope (ATST) requires active control of quasi-static telescope aberrations in order to meet image quality standards set by its science requirements. Wavefront control is managed by the Telescope Control System, with many telescope subsystems playing key roles. We present the design of the ATST quasi-static wavefront and alignment control architecture and the algorithms used to control its four active mirrors. Two control algorithms are presented, one that minimizes force on M1 actuators and another that employs a neutral-pointing constraint on M2 to reduce pointing error. We also present simulations that generate typical daily active mirror trajectories which correct optical misalignments due to changing gravitational and thermal loads.

Keywords: active optics, quasi-static alignment, wavefront control, ATST

1. INTRODUCTION

ATST design is driven by use cases that highlight the most critical science requirements and drive performance of key telescope systems. To meet image quality standards set by its science requirements, ATST must be able to maintain tight alignment tolerances throughout the full range of telescope mount motion and daily temperature variations. As the telescope moves through an observing day, its elevation angle changes constantly, resulting in changing gravitational force vectors on M1 and M2 that will cause misalignments as the telescope flexes. Changing temperatures will also cause misalignment and wavefront error as supports thermally expand and mirror surfaces change shape.

Without active control, optical misalignments and distortions due to daily temperature changes and gravitational forces will cause wavefront error as large as 8 microns peak to valley in amplitude and degrade image quality below what is necessary to meet ATST's science requirements. The ATST Active Optics system is designed to control quasi-static wavefront error due to optical misalignments and thermal distortions to within the allocated error budget. It accomplishes this task through coordination of four active mirrors and two wavefront sensors, employing intelligent control methods to achieve its goals.

This paper describes the ATST active optics system design and the algorithms it uses to control wavefront error. The following section presents an overview of the Active Optics design, section 3 describes the algorithms used, and section 4 presents simulations that generate active mirror positions based on modeling of telescope flexure due to daily temperature and elevation angle fluctuations.

Corresponding author Luke Johnson: ljohnson@noao.edu

2. QUASI-STATIC WAVEFRONT CONTROL STRATEGY

ATST's active optics involves many telescope systems working together to control quasi-static wavefront error and alignment. Each telescope system can be considered as fitting into one or more of three roles, wavefront error measurement, wavefront error correction, and software control.

The wavefront error measurement systems are the low-order wavefront sensor (LOWFS) and the high-order adaptive optics (HOAO).[1] The LOWFS, designed to similar specifications as the LEST wavefront sensor[2] is used during seeing-limited observations and measures low frequency wavefront error. When active, it broadcasts wavefront measurements over the ATST Common Services Framework (CSF) event network at a rate of 30 Hz. The HOAO is used during diffraction-limited observations. It contains a high-order wavefront sensor (HOWFS), a deformable mirror (DM), and a control computer. During diffraction-limited operation, HOAO will offload low-order wavefront modes from the DM at a rate of 30 Hz. HOAO will also broadcast pupil motion measurements from the HOWFS during both diffraction-limited and seeing-limited observations.

The wavefront error correction systems are the M1 primary mirror, the hexapod-mounted M2 secondary, and the actively controlled flat mirrors M3 and M6. A summary of these mirrors is shown in Table 1. Duties are divided among the 4 active mirrors such that M1 and M2 control wavefront error and M3 and M6 control the "boresight alignment" of the telescope. Boresight alignment refers to maintaining alignment of the beam of light between mirrors M6 and M7. M6 steers light along the telescope's azimuthal axis down to the coudé room where M7 feeds light to the coudé level instrumentation. If the beam between M6 and M7 is not aligned with the coudé rotational axis then there will be pupil wobble on the coudé wavefront sensors and image movement as the coudé room rotates. Boresight alignment control still uses the methods discussed in Upton et al[3] so it will not be discussed further in this paper.

Table 1. ATST active mirrors

Mirror	Mirror surface	Optical plane	Degrees of freedom
M1	Off-axis parabola	Pupil	118 push-pull surface actuators
M2	Off-axis parabola	Pupil	6 - x, y, z, Rx, Ry, Rz (hexapod)
M3	Flat	Near image	2 - pupil positioning in x and y
M6	Flat	Near pupil	2 - image positioning in x and y

The Telescope Control System (TCS), Wavefront Correction Control System (WCCS),[4] active optics correction engine, M1 control system (M1CS), Top end optical assembly control system (TEOACS), and Feed Optics Control System (FOCS) comprise the software control entities involved in ATST quasi-static wavefront and alignment control. The ATST CSF employs a configuration-driven distributed control architecture. Within CSF, higher-level controllers send configurations to their subordinates as a list of attribute-value pairs that define desired configurations. The TCS is the top level controller and controls the configurations of the active mirror control systems and the WCCS. The WCCS contains the active optics correction engine, HOAO, and the LOWFS as its subcontrollers. During active correction, the WCCS configures the active optics corrections engine to process wavefront measurements and generate position errors of the active mirrors. WCCS then broadcasts position errors to the active mirror controllers using the CSF event network. WCCS also controls configurations of the HOAO and LOWFS controllers.

During observations, active optics operates in one of two correction modes, active or passive. In passive mode, wavefront sensors are inactive and active mirror controllers position their mirrors based on lookup tables that calculate the best mirror position based on telescope elevation angle, azimuth angle, temperature, and temperature gradient across M1. Lookup tables function as a database lookup for each parameter of interest, interpolating between the database entries that most closely bracket the lookup value of the desired parameter. Initially, lookup tables will be populated based on finite element modeling of the telescope mount assembly (TMA) and M1 mirror and cell assembly. They will then be refined and updated based on trending of active correction errors over long periods of ATST operations.

In active mode, one of the wavefront sensors will actively monitor the residual quasi-static error in the telescope by averaging wavefront measurements to filter out zero-mean atmospheric turbulence. The active

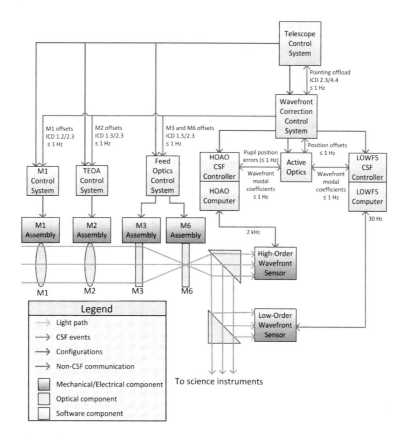

Figure 1. Signal flow diagram for systems involved in ATST active optics. (color online)

optics correction engine will further process the wavefront measurements to ensure it has a confident measure of quasi-static wavefront error and send position error events to the active mirror controllers that null out the measured quasi-static error. Position error events act as offsets to the active mirrors' internal lookup tables.

Figure 1 depicts the signal flow between all systems involved in active optics operations. The following section describes the algorithms used within the active optics correction engine to process wavefront sensor measurements into active mirror position errors.

3. ACTIVE OPTICS ALGORITHMS

The active optics correction engine performs calculations that filter wavefront error inputs and calculate active mirror position errors.

3.1 Wavefront error input filtering

When operating in active correction mode, the active optics correction engine takes wavefront error inputs from either LOWFS or HAO systems. Wavefront error measurements from these systems arrive as vectors of modal basis coefficients, either in the Zernike basis or M1 natural mode basis. Within either LOWFS or HOAO, wavefront measurements are averaged over a 1 second period in order to average out high temporal frequency atmospheric aberrations. However, wavefront error inputs from LOWFS or HOAO may still contain atmospheric "noise" so the active optics correction engine performs additional averaging to ensure that any corrections applied to active mirrors are true quasi-static error and not transient error due to atmospheric fluctuations.

To do this, the active optics engine tracks mean and signal to noise ratio of each coefficient in the incoming wavefront measurements. Each wavefront measurement arrives with a weight assigned to it based on the mean correlation peak of subaperture shift measurements used to calculate its modal coefficients. The weighted recursive mean estimator updates with each new wavefront measurement,

$$w_k = w_{k-1} + w_{in}$$
$$\bar{\phi}_k = \bar{\phi}_{k-1} + \frac{w_{in}}{w_k}\left(\phi_{in} - \bar{\phi}_{k-1}\right) \quad (1)$$

where ϕ_{in} is the latest wavefront measurement with weight w_{in}. Initial conditions are $\bar{\phi}_0 = \mathbf{0}$ and $w_0 = 0$

The recursive signal to noise estimator also updates with each wavefront measurement,

$$\mathbf{v}_k = \frac{\mathbf{v}_{k-1} + \left(\phi_{in} - \bar{\phi}_{k-1}\right)\left(\phi_{in} - \bar{\phi}_k\right)}{k-1}$$
$$\mathbf{SNR}_k = \frac{4k}{1+4k}\bar{\phi}_k\sqrt{\frac{k}{\mathbf{v}_k}} \quad (2)$$

where $\mathbf{v}_k$ is an array whose coefficients correspond to the sample variance of each element of the $\bar{\phi}_k$ array and $\mathbf{SNR}_k$ is an array whose elements correspond to signal to noise estimates of each wavefront modal coefficient. Initially, $\mathbf{v}_0 = 0$, $\mathbf{SNR}_0 = 0$, and $\mathbf{SNR}_1 = 0$.

As this process continues, the active optics engine will monitor $\bar{\phi}_k$ and $\mathbf{SNR}_k$, comparing each value to its own threshold every time they are updated. The two thresholds to compare are the significance threshold and the confidence threshold. Elements of $\bar{\phi}_k$ will be compared to the significance threshold. Any modal coefficients that exceed this threshold will be considered significant. Elements of $\mathbf{SNR}_k$ will be compared to the confidence threshold. Any elements of $\bar{\phi}_k$ corresponding to elements of $\mathbf{SNR}_k$ that exceed the confidence threshold will be considered confident. Initially, the significance threshold will be set at 50 nm rms and the confidence threshold to 100. These values will be fine-tuned during IT&C.

Whenever active optics determines that one or more modal coefficient measurements are both significant and confident, it will send those measurements to the selected compensation algorithm to calculate position errors for the active mirrors.

3.2 Wavefront compensation

In active correction mode, the active optics correction engine sends error offsets to M1 and M2 based on the wavefront measurements determined using the algorithm described in section 3.1. Two different compensation methods can be employed to calculate these postion errors, force-constrained least squares (FCLS) or neutral-pointing plus focus (NPF). From the point of view of an on-axis wavefront sensor, there is significant degeneracy between the rigid-body motions of the M2 hexapod and the surface bending modes of M1. These two methods use different ways to resolve this degeneracy and each has its own advantages and disadvantages.

3.2.1 Force-contrained least squares compensation

The FCLS method resolves degeneracy between M1 and M2 by penalizing the amount of force applied to M1 actuators. M2's hexapod is able to correct wavefront errors much larger in magnitude than the M1 axial actuators. For that reason, we prefer to correct high amplitude wavefront aberrations with M2 so that we do not saturate the M1 actuators.

This is done by adding a penalty matrix to the compensator that penalizes the amount of force applied to M1. If we assume that small optical misalignments can be modeled as a linear system then

$$\mathbf{g} = \mathbf{h}_0 + \mathbf{H}\mathbf{f} + \mathbf{n}, \quad (3)$$

Where $\mathbf{g}$ is the vector of wavefront coefficients measured by the wavefront sensor, $\mathbf{h}_0$ is the vector of wavefront coefficents when the system is in its unperturbed state, $\mathbf{H}$ is the force sensitivity matrix, $\mathbf{n}$ is a vector representing wavefront measurement noise, and $\mathbf{f}$ is a vector of forces applied to each active mirror corrective element. For example,

$$\mathbf{f} = \begin{bmatrix} \mathbf{f}_{M1} \\ \mathbf{f}_{M2} \end{bmatrix} \quad (4)$$

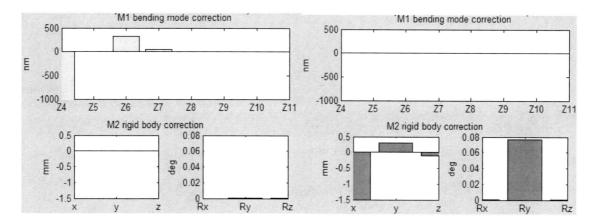

Figure 2. Simulated active optics corrections. Both solutions correct the same wavefront error to the same level of accuracy. Left: solution with $\mathbf{P} = \mathbf{I}$. Right: solution with $\mathbf{P}_{M1} = \rho\mathbf{I}$ and $\mathbf{P}_{M2} = \mathbf{0}$.

where $\mathbf{f}_{M1}$ is a 118-element array in which each element represents the force applied to its corresponding actuator and $\mathbf{f}_{M2}$ is a 6-element array representing the force applied to each of the M2 hexapod degrees of freedom.

We wish to solve equation 3 to determine, given a wavefront measurement $\mathbf{g}$, what is the $\mathbf{f}$ vector that minimizes both the wavefront error and the force applied to M1? To do this, we express the wavefront error and M1 total force in a sum of squares cost function,

$$J = ||\mathbf{Hf} - \mathbf{g} - \mathbf{h}_0||^2 + ||\mathbf{Pf}||^2, \qquad (5)$$

where

$$\mathbf{P} = \begin{bmatrix} \mathbf{P}_{M1} & \mathbf{0} \\ \mathbf{0} & \mathbf{P}_{M2} \end{bmatrix} \qquad (6)$$

and $\mathbf{P}_{M1}$ and $\mathbf{P}_{M2}$ are weighting matrices. The solution to our minimization problem is a canonical Tikhonov-regularized least squares solution,

$$\mathbf{f}_{FCLS} = \left[\mathbf{H}^T\mathbf{H} + \mathbf{P}^T\mathbf{P}\right]^{-1} \mathbf{H}^T \left(\mathbf{g} - \mathbf{h}_0\right). \qquad (7)$$

Choosing $\mathbf{P}$ properly allows us to penalize specific actuators or correction modes. For example, choosing $\mathbf{P}_{M1} = \rho\mathbf{I}$ and $\mathbf{P}_{M2} = \mathbf{0}$ adds an adjustable penalty factor, ρ, that equally penalizes all M1 actuators and does not constrain M2 movement. Or, if actuator number 23 is stuck, setting values in column 23 of $\mathbf{P}_{M1}$ much greater that all other elements of $\mathbf{P}_{M1}$ will find a solution that tries not to move actuator 23.

Figure 2 demonstrates how FCLS can be used to reduce forces applied to M1 actuators.

3.2.2 M2 Neutral pointing compensation

One drawback of using FCLS to prefer M2 rigid-body motion over M1 bending modes is that unconstrained motion of M2 changes the telescope pointing. This causes the higher-order active optics correction (M1 and M2) to work at cross-purposes to the boresight correction (M3 and M6). To reduce this effect, the NPF compensation method constrains M2 to only three degrees of freedom, rotation in x and y about the center of curvature of its parent,[5] and translation along the z-axis.

Tilting M2 about the center of curvature of its parent induces mostly astigmatism and translating M2 along the z-axis induces mostly defocus. Modeling optical misalignments due to telescope motion and temperature fluctuations predicts that defocus and astigmatism errors will dominate the quasi-static wavefront of the telescope. Based on these results, we designed the NPF compensation method to use M2 to correct defocus and astigmatism wavefront modes and M1 to correct aberrations of higher order.

Using the same solving process as the previous section, we can express the NPF solution as,

$$\mathbf{f}_{NPF} = \mathbf{S}_{NPF} \left[\mathbf{H}^T_{NPF}\mathbf{H}_{NPF} + \rho^2\mathbf{I}\right]^{-1} \mathbf{H}^T_{NPF} \left(\mathbf{g} - \mathbf{h}_0\right). \qquad (8)$$

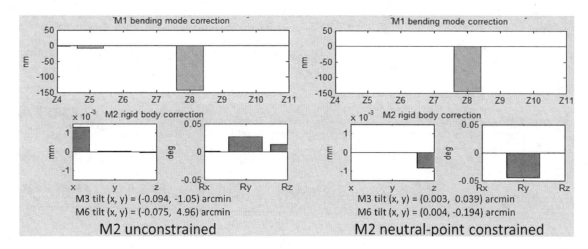

Figure 3. Simulated correction of 1.0 microns rms astigmatism at the M1 surface. Both solutions correct to the same level of accuracy. Left: FCLS method with unconstrained M2 motion. Right: NPF method. The NPF solution reduces M6 motion by a factor of 25.

where $\mathbf{S}_{NPF} = \begin{bmatrix} \mathbf{I} & \mathbf{0} \\ \mathbf{0} & \mathbf{S}_{M2} \end{bmatrix}$ and $\mathbf{S}_{M2}$ is a six by three matrix that maps neutral-pointing tilts to the typical six hexapod motions. $\mathbf{H}_{NPF}$ is the NPF sensitivity matrix. Instead of mapping the six hexapod degrees of freedom to wavefront sensor measurements, the NPF sensitivity matrix maps the three NPF degrees of freedom to wavefront sensor measurements. the NPF sensitivity matrix is related to the sensitivity matrix from the previous section, $\mathbf{H}_{NPF} = \mathbf{H}\mathbf{S}_{NPF}$. Additionally, in mapping $\mathbf{H}$ to $\mathbf{H}_{NPF}$, the first three rows of $\mathbf{H}$ are eliminated so that M1 defocus and astigmatism modes are in the null space of $\mathbf{H}_{NPF}$.

An example of how the NPF method reduces pointing error is shown in figure 3. While NPF reduces the amount of pointing error compared to FCLS, it does restrict the M2 degrees of freedom, possibly requiring more M1 actuator force than the FCLS solutions. From simulations of TMA flexure, it appears that the NPF method will be preferred because aberrations other than astigmatism and defocus are expected to be small in magnitude. However, final determination of the best method will be done during IT&C when actual quasi-static wavefront aberrations are known.

4. ACTIVE OPTICS SIMULATION RESULTS

This section presents simulation results that show M1, M2, M3, and M6 active motion over the course of a typical observing day. Initial analysis, performed by Ingersoll Machine Tools, Inc., used finite element modeling to calculate M1 and M2 displacements as a function of elevation angle and temperature. We took the finite element modeling results and input them into the QATST optical simulation tool to calculate both wavefront errors seen as a result of the the M1 and M2 deflections and corresponding M1 and M2 corrections that correct the wavefront error to the level necessary for ATST's seeing-limited error budget. In all cases, both the NPF and FCLS compensation methods reduced the residual wavefront error down to the level set by the simulated wavefront sensor noise, as shown in Figure 4.

QATST, originally developed by Robert Upton, uses a ZEMAX optical model of ATST in conjunction with MATLAB control software. The ZEMAX model generates wavefront measurements due to misalignment of telescope optics and the MATLAB software calculates the active optics control commands and feeds them back into the ZEMAX model. Sensor noise is simulated by adding 25 nm rms white noise to the wavefront sensor slope measurements before applying the compensation algorithm.

For each simulated time, the correction converges to wavefront error equal to the sensor noise after 3 or 4 iterations. These simulations demonstrate that the ATST quasi-static wavefront control solution should be able to correct quasi-static wavefront error to a level sufficient to meet science requirements under expected operating conditions. Our simulations also demonstrate the approach that will be used to generate initial lookup tables for the active mirrors during IT&C. As data is accumulated over many days of operations, the lookup tables will

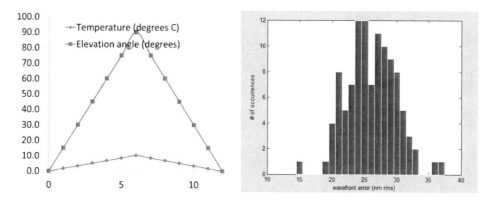

Figure 4. Left: Inputs to finite element analysis of telescope mount. Right: Histogram of results from 100 independent active optics simulations with 25 nm rms sensor noise.

be further refined based on trends identified in daily operations logs. These simulation tools will also be refined based on observation data.

Figure 5 shows mirror trajectories that correct quasi-static wavefront error due to the temperature and elevation angle curves shown in Figure 4. As mentioned earlier, both compensation methods correct wavefront error to the same level of accuracy using primarily M2 motion with M1 correcting a small amount of higher order residual error. The NPF method requires less M3 and M6 motion than FCLS but larger M2 motion. For both methods, M1 surface bending is very small and all M2 motion is well within the ±1 degree tilt range and the ±10 mm translation range of the M2 hexapod.

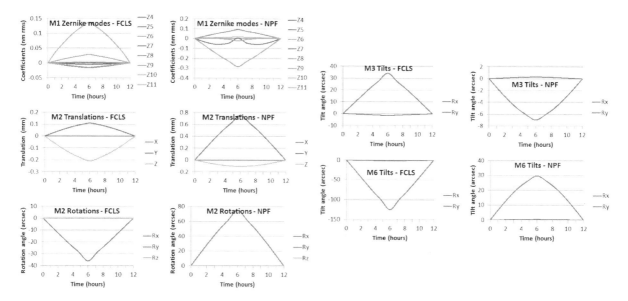

Figure 5. Simulated active mirror trajectories that correct quasi-static wavefront errors generated over a day of ATST observation using both the FCLS and NPF compensation methods (color online).

5. CONCLUSIONS

We have presented the active optics design of ATST and explained its operation in both passive and active correction modes.

We also presented simulation results which predict, in active correction mode, ATST will be able to control quasi-static wavefront errors to a level that is limited by wavefront sensor noise. Performance in passive correction mode will depend on the repeatablilty of telescope flexure in day-to-day operation and requires more detailed modeling to accurately predict. Current finite-element models are still in the preliminary stages and as the models are developed, our active optics modeling will improve as well. Our current state of modeling demonstrates that the design and control algorithms are sufficient to achieve error budget requirements but does not fully predict as-built ATST performance because it oversimplifies active mirror dynamics and day-to-day repeatability.

As more information about dynamic response and positioning accuracy of active mirrors becomes available, we plan to update our models to more accurately reflect the information known about ATST's mount and mirror systems. These updated models will provide more detailed analysis of expected operating parameters for risk management, error budget refinement and IT&C troubleshooting.

ACKNOWLEDGMENTS

ATST is a project of the National Solar Observatory, operated by the Association of Universities for Research in Astronomy, under a cooperative agreement with the National Science Foundation.

REFERENCES

1. K. Richards, "Adaptive optics real time processing design for the advanced technology solar telescope," in *Astronomical Telescopes: Adaptive Optics Systems III*, Proc. SPIE **8447**, 2012.
2. M. Owner-Petersen, T. Darvann, and O. Engvold, "Design of the LEST Slow Wavefront Sensor," in *Real Time and Post Facto Solar Image Correction*, R. R. Radick, ed., p. 63, Dec. 1993.
3. R. Upton and T. Rimmele, "Active reconstruction and alignment strategies for the Advanced Technology Solar Telescope," in *Society of Photo-Optical Instrumentation Engineers (SPIE) Conference Series, Society of Photo-Optical Instrumentation Engineers (SPIE) Conference Series* **7793**, Aug. 2010.
4. E. K. Kinney, K. Richards, L. C. Johnson, T. R. Rimmele, and S. C. Barden, "The wavefront correction control system for the advanced technology solar telescope," in *Astronomical Telescopes: Adaptive Optics Systems III*, Proc. SPIE **8447**, 2012.
5. A. M. Manuel and J. H. Burge, "Alignment aberrations of the New Solar Telescope," in *Society of Photo-Optical Instrumentation Engineers (SPIE) Conference Series, Society of Photo-Optical Instrumentation Engineers (SPIE) Conference Series* **7433**, Aug. 2009.

Optical Design of the COSMO Large Coronagraph

Dennis Gallagher[1], Steven Tomczyk[1], Haiying Zhang[2], Peter G. Nelson[3]

1. The National Center for Atmospheric Research High Altitude Observatory-Boulder, Colorado
2. Nanjing Institute of Astronomical Optics and Technology- Nanjing, China
3. Sierra Scientific Solutions- Boulder, Colorado

The Coronal Solar Magnetism Observatory (COSMO) is a facility dedicated to measuring magnetic fields in the corona and chromosphere of the Sun. It will be located on a mountaintop in the Hawaiian Islands and will replace the current Mauna Loa Solar Observatory (MLSO). COSMO will employ a suite of instruments to determine the magnetic field and plasma conditions in the solar atmosphere and will enhance the value of data collected by other observatories on the ground (SOLIS, ATST, FASR) and in space (SDO, Hinode, SOHO, GOES, STEREO, DSCOVR, Solar Probe+, Solar Orbiter). The dynamics and energy flow in the corona are dominated by magnetic fields. To understand the formation of Coronal Mass Ejections (CMEs), their relation to other forms of solar activity, and their progression out into the solar wind requires measurements of coronal magnetic fields. The COSMO suite includes the Large Coronagraph (LC), the Chromosphere and Prominence Magnetometer (ChroMag) and the K-Coronagraph. The Large Coronagraph will employ a 1.5 meter fuse silica singlet lens and birefringent filters to measure magnetic fields out to two solar radii. It will observe over a wide range of wavelengths from 500 to 1100 nm providing the capability of observing a number of coronal, chromospheric, and photospheric emission lines. Of particular importance to measuring coronal magnetic fields are the forbidden emission lines of Fe XIII at 1074.7 nm and 1079.8 nm. These lines are faint and require the very large aperture. NCAR and NSF have provided funding to bring the COSMO Large Coronagraph to a preliminary design review (PDR) state by the end of 2013.

In this paper we discuss the current state of the design of the Large Coronagraph. Several key technology drivers will be discussed, including design and fabrication of a 1.5 meter lens, methods for keeping the lens clean and the requirements for the tunable birefringent filter.

Large Coronagraph Overview

The LC will employ a Lyot style coronagraph design. The operating wavelength range for the LC will span 500-1100nm with a narrow band (~.1 to .15nm FWHM) tunable birefringent filter. The narrow band pass of the filter will allow observing many visible and near infrared coronal and chromospheric emission lines along with prominence emission lines of H-Alpha 656.3nm and He I 1083.0nm. The field of view of the LC will be 1 degree which will allow viewing of the entire solar corona out to 2 solar radii . The inner field will allow viewing down to 1.05 solar radii or ~50 arc-seconds above the solar limb. Stray light requirements are less than 5 millionths the brightness of the solar disk. The LC will operate at a single wavelength at a time and observe in various polarization states. The LC will use a Lyot style birefringent filter with a large etendue to isolate pass bands (~0.1 to 1.5 nm FWHM). The filter will be electro-optically tuned to perform line scans and pass band wave length changes.

A HgCdTe sensor, that would cover the entire pass band is currently under evaluation. A 4096x4096 sensor with 15μ pixels would meet LC's field of view requirements with a final operating effective focal length (EFL) of 3500mm (F/2.4)

Design trades for F/4.5-F6.0 operation at the focus of the main objective are being evaluated. A nominal design that is F/5.0 using several paraxial lenses is shown in figures 1a-1c.

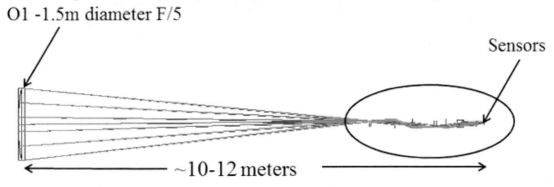

Figure 1(a) Current thin lens model. This model uses a F/5.0 objective A closer view of the optics is shown in the next figure 1(b)

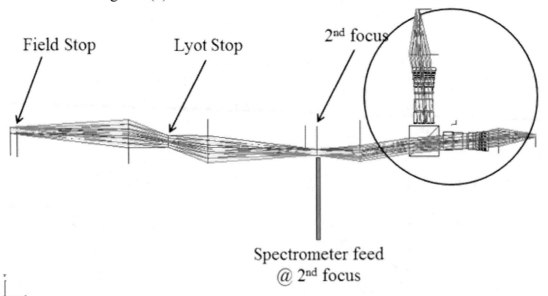

Figure 1(b): The occulter and field stop are located at the 1st focus. The next section images the pupil where the Lyot stop is placed. A 2nd intermediate focus is formed where a fiber will feed a spectrometer to be developed in collaboration with the University of Hawaii. The next section includes the polarizing beam splitter and birefringent filters. The exact design for this section is still under study. See figure 1(c)

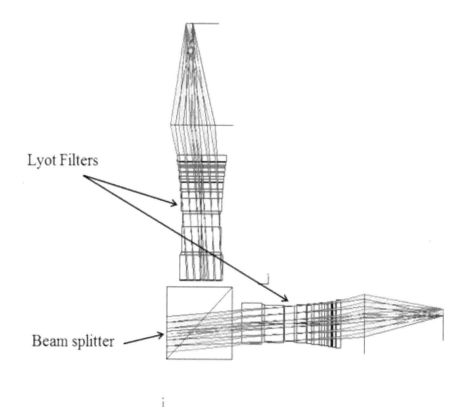

Figure 1(c): In this design, light is split into two polarization states and feeds two birefringent filters. A final imaging optic creates the final image with an EFL ~3500mm. The CoMP instrument (Tomczyk), currently observing at MLSO, uses a single birefringent filter where simultaneous observations at two wavelengths are done by using the polarizing beam as the analyzer for the thickest filter element. The exact design of the LC polarization analysis is still under study.

The basic concept of operations is to focus the objective at the occulter for a specific wavelength. The size of the occulter will need to be variable not only to adjust for the seasonal variations in apparent solar diameter, but also to adjust for the variations in effective focal length (EFL) due to chromatic aberration. The occulter for the LC must reject ~ 2.3 KW of energy which will require a special design for heat rejection. The Lyot stop, located at the image of the objective, may be variable in diameter to accommodate image size changes at various wave lengths due to chromatic aberration. The Lyot stop will be sized to 0.92 times the objective image size to block diffracted light coming from the edge of the objective. After the Lyot stop is a secondary focus. A fiber feed will be located at this focus to feed a spectro-polarimeter. A polarizing beam splitter will be used to split the light into two polarization states for simultaneous observations of each polarization state. The light beams will be image space telecentric as they pass through the beam splitter and the birefringent filters so the center of the pass band is uniform through field. Finally, the positions of the sensors will be adjusted to focus for different wave lengths. Other optics such as a ferroelectric liquid crystal polarization analyzer and the band pass filters are yet to be determined and are not shown in Figure 1.

Singlet lens

The main feature of the LC design is the large 1.5 meter aperture by 0.15 meter thick singlet lens. Currently this lens would be made from Corning C79-80 fused silica glass. Corning has made fused silica blanks as large as 1.88 meters in diameter and 0.23 meters thick. For a 1.5 meter blank corning fused silica has inclusions at levels an order of magnitude less than BK7 which was the glass used in many old coronagraph designs. Modeling has already determined that gravity and thermal loads on the lens during operation will allow for diffraction limited performance. Fused silica has a coefficient of thermal expansion eighteen times less than BK7 glass. When a lens bends, slight changes in refraction angles at each surface tend to cancel out and leave the wave front undistorted which is unlike reflection from a bending mirror where the deformed surface changes ray angle and is not cancelled. Stress in the lens has been calculated and polarization effects will be less than 1 part in 10000. For more information see Nelson et al. 2006.

The LC will occult the sun out to 1.05 solar radii which requires low scatter at angles greater than 50 arc-seconds. To achieve this, the objective lens will be super polished to <0.7 nm RMS for spatial periods 0.040-3.2 mm. The objective lens will not be coated to eliminate any light scattering from an optical coating.

The objective lens design needs to be good enough to image the limb of the sun so light does not get pass the occulter due to aberrations. Chromatic aberration for an F/5.0 singlet lens is ~200mm (500-1100nm) so focusing the lens will need to be done at each wave length. The other dominant aberration is spherical. Designing a lens with spherical surfaces will not meet the 5% oversize occulter where the occulter's edge is .4 mm beyond the image of the solar limb. The front surface of the lens will be aspheric by adding a conic constant departure of ~-0.8 to meet the imaging requirements. It will be easier to test a conic departure with a null test rather than testing for polynomial departures from a sphere. This figure is still challenging to polish where departure from spherical is ~0.5 mm. Figure 2 shows the spot diagrams spanning the LC wavelength coverage where the maximum spot size is ~83µm

Figure 2: Spot diagrams at an angle just above the edge of the occulter at 530.3, 656.3, and 1083.0nm. The scale line is 200μm long.

Cleanliness

A dominant source of stray light for a coronagraph is from particulates on the lens surfaces. Scattering is proportional to the particle size to the 4th power. The method of reducing dust on the operating MkIV K-coronagraph is to perform manual daily cleaning of the lens by blowing off small dust particles. This method of cleaning would be difficult on a 1.5 meter optic. One method is to turn the inside of the observatory into a clean room. Modeling has shown that a positive pressure HEPA system can eliminate particles from entering the observatory through the observation opening. 350 cubic feet per minute air flow from the observatory floor eliminates most particles. Figure 3 shows air flow simulations with trajectories for various sized dust particles. The small particles <10μm do not enter the observatory. Particles >20μm come into the dome but rapidly fall to the floor and do not reach the surface of the objective lens. The inside of the coronagraph will be sealed off and filled with helium so the inside surface of the objective is not exposed to outside air. There will have to be periodic cleaning of the objective lens over a yet to be determined time period, for example, the MkIV coronagraph objective lens at MLSO is removed and cleaned about 4 times per year. Other methods of air flow to keep the objective lens clean are currently being studied.

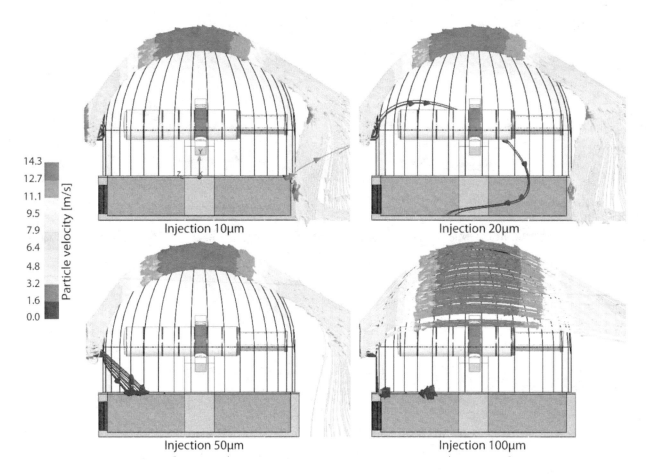

Figure 3: Simulations of air flow and particle trajectories with size of 10, 20, 50, and 100μm. This is a worse case simulation with a 10 m/sec wind speed directed at the observatory opening. The flow of particles does not reach the object lens with 350 cubic feet per minute air flow coming from the observatory floor.

Birefringent filter

One of the key elements of the LC is the narrow band birefringent filter. The requirement for the LC to have a 1 degree field of view places requirements on the filter etendue and the sensor pixel size. The area solid angle A for the LC system can be derived from the objective lens size and the 1 degree field of view. Throughout an optical system A is related to image plane size d and system focal ratio $\mathcal{F}$ by the following relation.

Equation (1)
$$A\Omega \approx \frac{\pi^2 d^2}{4\mathcal{F}^2}$$

Assuming an F/5 objective for the LC along with a Teledyne 4096x4096 sensor with 15μm pixels, a final F/ratio can be derived for the LC system from the field of view requirements. The LC final F/ratio becomes F/2.35 or EFL=3500mm corresponding to 0.88 arc-seconds per pixel. Since the LC design will be telecentric to preserve filter band pass center through field, the maximum ray angle at the focus can be derived from the beam F/ratio and is ~12 degrees. The band pass requirement for the LC birefringent filter is ~0.15nm FWHM for the near infra-red @1μm wavelength. This places requirements on the angular response of the filter which is driven by the thickest element of the birefringent filter. The thickness of the thickest filter element can be derived from the FWHM requirement and is given by

Equation (2) $$FWHM = \frac{\lambda^2}{2\varepsilon l}$$

Where ε is the crystal birefringence and l the crystal thickness. There are two crystal materials that have been considered for the LC birefringent filter: calcite, and lithium niobate. Calcite has a birefringence of ~.17 and lithium niobate has a birefringence of ~0.08 at 1μm. The element thickness d for a 0.15nm FWHM filter element at 1 μm wavelength is 20.7 mm thick for calcite and 44mm thick for lithium niobate. If the birefringent filter element is a wide field design then the fractional change in retardance with incidence angle i is given by Equation 3.

Equation (3) $$\frac{\delta R}{R} = 1 - \left(\frac{1}{4n_o^2}\right)\left(\frac{n_e - n_o}{n_e}\right)\sin^2(i)$$

Where n_o and n_e are the ordinary and extraordinary indices of the crystal. Figure 4 shows a plot of the transmission as a function of angle for calcite and lithium niobate wide field elements with a FWHM of 0.15nm at 1μ. Where the transmission falls to zero for the 1st time in each case is a shift in the center wavelength of one FWHM.

From Figure 4 it can be seen that lithium niobate offers a significant throughput advantage over calcite of a factor of ~5.9. Using calcite for the LC filter would require increasing the diameter of the filter elements to reduce incidence angle. The LC design has shown in figure 1 uses 100mm diameter lithium niobate crystals for its 5 stage filter design. Trying to find high quality calcite crystals with apertures greater than 100mm would be extremely difficult.

A 12 degrees maximum incident angle F/2.35 beam gives a wavelength shift ~ ½ FWHM in lithium niobate. This is pushing the limits of the throughput of a birefringent filter using 100 diameter lithium niobate crystals. The passband of the filter will widen by ~1/2 FWHM or the 0.15nm pass band will widen to ~0.225nm FWHM. There will also be an increase in pass band side lobes. A second 0.15nm FWHM element can be added to the filter stack to narrow the FWHM and reduce pass band side lobes.

At the short end of the LC operating wave length at 530nm the birefringent filter angular response will become larger where the pass band center will shift by slightly more than one FWHM for a 44mm thick lithium niobate crystal. The FWHM of a 44mm thick lithium niobate crystal at

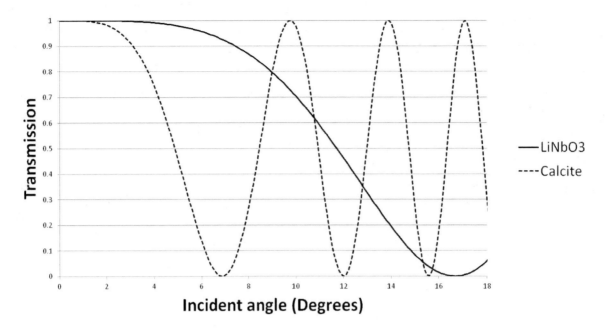

Figure 4: Transmission vs. incident angle for a 0.15nm FWHM wide field birefringent filter element at 1μm (20.7 mm total thickness calcite; 44 mm total thickness lithium niobate)

530nm is 0.036nm FWHM. For observing the coronal green line at 530.3nm a 0.036nm FWHM filter would be too narrow and reduce signal-to-noise since the green line is thermally broadened to about 0.1nm wide. To reduce the angular response, the polarizer on the thickest lithium niobate could be removed and the 5 stage birefringent filter operated as a 4 stage filter (.071nm FWHM).

Another advantage of using lithium niobate over calcite is lithium niobate can be electrically tuned if layers of transparent and conductive indium tin oxide are added to the crystal surfaces. This eliminates the need for tuning the filter with liquid crystals plates and reduces the number of optical elements in the filter. Tests are currently being performed on a 75mm diameter 32mm thick lithium niobate wide field element to better define the electrical tuning properties of lithium niobate.

Conclusion

The COSMO LC will provide unprecedented synoptic observations of the coronal magnetic field and plasma conditions. These observations will lead to significant advances in our understanding of the dynamic processes occurring in the solar atmosphere that ultimately impact the space environment at Earth and all interplanetary space and will complement current and future ground- and space-based solar missions.

The key element of the COSMO LC will be the large 1.5m fused silica singlet lens. Several key requirements for this lens have been defined in the paper. The next step will be to specify requirements for materials and manufacturing so the cost of the LC objective lens can be better defined with feedback from potential vendors. The purity of the lens material and the ability of achieving the required surface smoothness will be key to the COSMO LC performance.

Cleanliness of the LC objective lens has already been defined as a main driver of the design. The COSMO LC observatory will use airflow in the observatory and around the objective lens rather than use the more typical approach of routine lens cleaning. The cleanliness requirements of the main objective would require daily cleaning of the objective lens if air flow were not used. This would be impractical for a lens of this size.

The optical design of the LC is in a preliminary state, but key parts of the design have been defined from basic optics principles. The birefringent filter is also a key element of the COSMO LC system. It has been shown that lithium niobate will meet the filter requirements which flow down from the requirement to provide synoptic imaging of the sun out to 2 solar radii. Lithium niobate crystals are plentiful and are made in sizes that are required for the LC birefringent filter. Testing of lithium niobate crystal is currently being done for homogeneity, aperture size, electrical tuning capability, manufacturability, and cost.

Acknowledgments

The National Center for Atmospheric Research is sponsored by the National Science Foundation.

References

Nelson, Peter G.; Tomczyk, Steven; Elmore, David F.; Kolinski, Donald J. , The feasibility of large refracting telescopes for solar coronal research', 2008, in: Ground-based and Airborne Telescopes II. Edited by Stepp, Larry M.; Gilmozzi, Roberto. Proceedings of the SPIE, Volume 7012, pp. 701231-701231-12, DOI: 10.1117/12.789494

Tomczyk, Steven, Card, G.L., Darnell, T., Elmore, D.F., Lull, R., Nelson, P.G., Streander, K.V., Burkepile, J., Casini, R., Judge, P.G. An instrument to measure coronal emission line polarization, 2008, *Solar Physics*, **247**, 411.

Behavior of a horizontal air curtain subjected to a vertical pressure gradient

James Linden, LeEllen Phelps
National Solar Observatory, 950 N. Cherry Avenue, Tucson, AZ 85719, USA

ABSTRACT

We present the details on an experiment to investigate the behavior of an air curtain that is subjected to a transverse pressure gradient. The setup simulates the conditions that will be present in the Advanced Technology Solar Telescope (ATST), a 4-meter solar observatory that will be built on Haleakala, Hawaii. A test rig was built to replicate the region at which the optical path crosses a temperature and pressure boundary between the telescope mount region, which is at the ambient temperature and pressure, and a warmer, pressurized lab space directly below. Use of an air curtain in place of an optically-transmitting window at the interface would allow science observations at a wider range of scientific wavelengths. With the air curtain exhibiting transitional flow behavior across the boundary, and applied pressure gradients of up to 6.5 Pa, we found that the air curtain was able to hold a pressure gradient of 0.25 Pa. As the applied pressure was increased, transient turbulent regions formed at the interface, and predictable flow behavior only occurred in the region closest to the air curtain blower. Computer modeling is used to validate the test data, identify laminar regions of the air curtain where minimal image distortion would occur, and explore the relationship between the applied pressure, effective pressure difference, and air curtain profile.

Keywords: air curtain, fluid modeling, solar telescope, ATST, pressure differential

1. INTRODUCTION

The Advanced Technology Solar Telescope (ATST) is a four-meter off-axis Gregorian telescope that will be built on Maui, Hawaii. As of June 2012 the ATST has gone through final design reviews of the detailed construction documentation for several of the major work packages (e.g. Enclosure, Support & Operations Building, and Facility Thermal Systems) and is in the construction phase of the project.

In order to resolve magnetic solar fluctuations in the solar atmosphere at the fundamental length scales, the four-meter aperture must be able to resolve 0.1 arc-seconds in the near-infrared. Observations beyond 1.5 μm give significant advantage in determining magnetic field, temperature, and velocity properties of the upper layers of the solar atmosphere. Thus, the ATST Science Requirements Document[1] stipulates that the solar image passing from the enclosure into the Coudé Lab must be useable for science observation at wavelengths of 380 nm – 28 μm. Additionally it is required that multiple wavelengths be observed simultaneously by multiple instruments situated in the lab.

The ATST will use an air curtain to separate the telescope, an ambient and unsteady-state environment, from the Coudé Lab, which has tight tolerances on changes in temperature, humidity, and internal air velocity, as well as particle infiltration. The lab's clean-room environment is sustained by utilizing oversized air handler units to generate positive pressure within the room and strategic placement of air returns throughout the coudé floor for controlled exfiltration. These controlled louvers are designed to facilitate the movement of heat and particles away from the optical path and instrumentation. At the interface between the telescope mount and the lab environment, some optical telescopes use windows that are transparent in the visible spectrum. Though these windows prevent backflow of air through the optical path they are opaque in the infrared and therefore restrict the usable wavelengths. In the ATST the air curtain will be used in place of a window to prevent the updraft of air while allowing a greater range of observable wavelengths.

Successful implementation of an air curtain requires a minimal loss of image quality. In short, this requires that the air curtain remain laminar at the interface despite both a temperature and pressure gradient. Science requirements dictate that for days with excellent seeing conditions a total maximum wavefront distortion of less than 25 nm rms must be achieved. An investigation done by *Hubbard et al.*[2] found that the use of a double curtain can maintain a 8 °C temperature difference with a wavefront distortion of less than 30 nm rms. With adaptive optics, the group showed a wavefront distortion of 5 nm rms could be achieved. Rocklin and the ATST Optical Systems Engineering Group[3] extended the experiment by modifying Hubbard's setup to reflect design changes in the ATST. These changes included the use of a horizontal curtain rather than a vertical one, and the reduction of the optical image from 10 cm to 5.5 cm. Though Rocklin generated fit-curves that demonstrated the ability to remain below 5 nm rms for temperature gradients of up to 20 °C, the group concluded that the experimental results should only be viewed as qualitative due to the uniqueness of their experimental setup.

Investigations into the behavior of air curtains in the presence of pressure gradients often focus on industrial use of air curtains in doorway. In these studies, a small, dynamic pressure gradient occurs with an external gust of wind. *Rouaud and Havet*[4] found that dynamic spikes in the differential pressure cause a swirling effect that can allow the exchange of pollutants across the open boundary. Unfortunately, they also found that the optimal velocity for minimal mixing across the boundary corresponded to a "strongly unsteady flow," which is most likely not acceptable in the ATST given the need to maintain minimal optical distortion. Experiments by *Valkeapää et al.*[5] found that the angle of inclination with respect to the applied pressure is more important than increases in the air curtain momentum (velocity). As the air curtain in the ATST will be shielded from external gusts of wind, the region will likely experience a quasi-static pressure differential and while the former reference shows the benefits of a strong curtain, the latter indicates the need for novel geometry to hold the pressure gradient.

2. PRESSURE GRADIENT

In order to maintain a constant flow of heat and particles away from the optical path and instrumentation, the Coudé Lab is designed to be 15 Pa above the ambient pressure. This is achieved by supplying enough conditioned outside air to maintain the overpressure requirement. CFD analysis by Phelps[6] served to show the movement of air within the lab given several possible duct configurations. These analyses were done with the assumption that the air curtain would be able to hold the pressure differential.

In the current design, air from the telescope chamber flows along the optical path into the air knife assembly as shown in Figure 1. The air is pulled into the ducting where it is filtered, thermally adjusted, and blown into the torque tube and Coudé Lab below. The optical path passes directly downward through the open window and horizontal air curtain. Though the air flow below the air curtain is directed down and away from the curtain, reducing the apparent gradient, the air movement above the curtain will cause a negative pressure and the curtain will still operate in a pressure gradient. Initially, the following experiment was designed to investigate the pressure differential that the air curtain could hold. Because the air curtain did not hold a significant amount of air, the goal of the experiment shifted to a qualitative analysis of the air curtain behavior as the internal pressure increased.

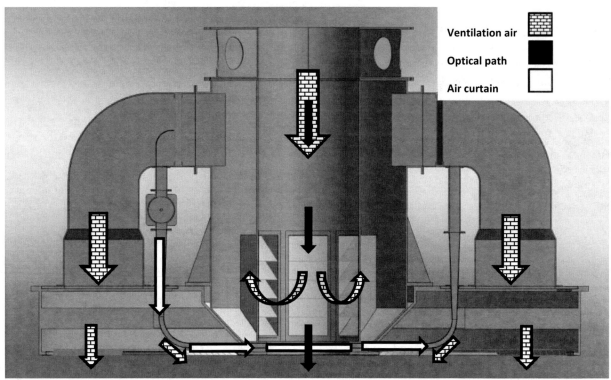

Figure 1: SolidWorks model of air flow and optical path in the air knife assembly

3. EXPERIMENT

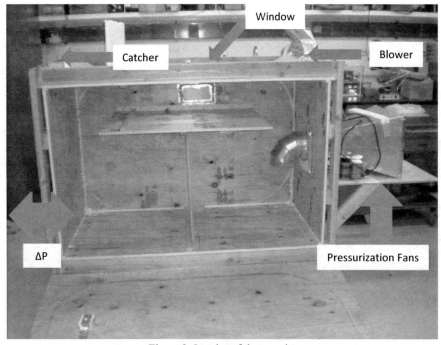

Figure 2: Interior of the test-rig

The thermally controlled test-rig used both in Hubbard's and Rocklin's experiments was modified and sealed. Figure 2 shows a side view of the box with the side wall hinged open. An array of four axial fans, each capable of pushing 1.6 m³/min, are used to accomplish the pressurization of the box. Power to the vans is controlled by a Staco variable autotransformer in order to maintain control of the air supplied into the box. Air flow from the pressurization fans is ducted away from the air curtain area to minimize turbulence at the window. Differential pressure measurements are taken across the box at the lower-left portion of the setup.

Figure 3: Blower and window with in-plane copper wires used for grid layout

Figure 4: View of air curtain and window from inside of the box

Figures 3 and 4 show the setup of the air curtain with respect to the window. A single curtain is situated on the high-pressure side of the window with a maximum volumetric flow rate of 9.5 m^3/min. A second blower of the same rating is used as a catcher by pulling air across the window. The hope in this design is that the blower-catcher combination improves the efficiency of a single-sided curtain. Additionally, the window is enlarged from a diameter of 15 cm, as used in the Hubbard and Rocklin experiments, to 33 cm, which reflects the size of the window in newer versions of the ATST designs. Air from the blower is ducted through a flow-straightening region for 38 cm and creates a curtain of cross-sectional area 35.56 cm x 1.91 cm. Flow straightening is achieved using 38 cm long straws in the initial test and removed to allow for a stronger curtain in the second test.

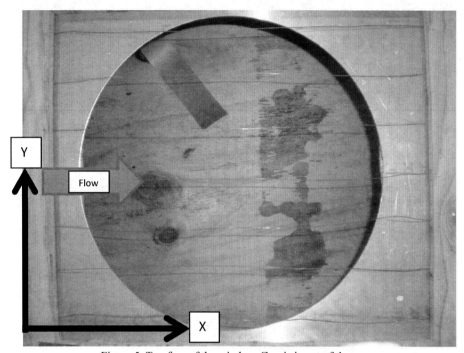

Figure 5: Top face of the window. Z-axis is out of the paper

A 7 x 8 grid layout on the face of the window allows evenly spaced vertical and horizontal velocity measurements to be used to generate velocity profiles in the plane of the air curtain (x-velocity) and perpendicular to the plane of the air curtain (z-velocity). A hand-held Alnor AVM410 velocity probe was used to record the x and y velocities at the center of each of the grid squares. Since the underside of the window was inaccessible, x-velocities could only be taken in the interior of the window. After initial data indicated that the air knife was underpowered for the larger window, the flow straighteners were removed and a second set of data was taken.

For each experiment the window was first covered by a wooden board to prevent air loss through the window. Power to the pressurization fans was adjusted until a desired steady-state pressure differential was achieved. The cover to the window was removed and the pressure was recorded with the air curtain on and with the air curtain off.

4. RESULTS

4.1 Experimental results

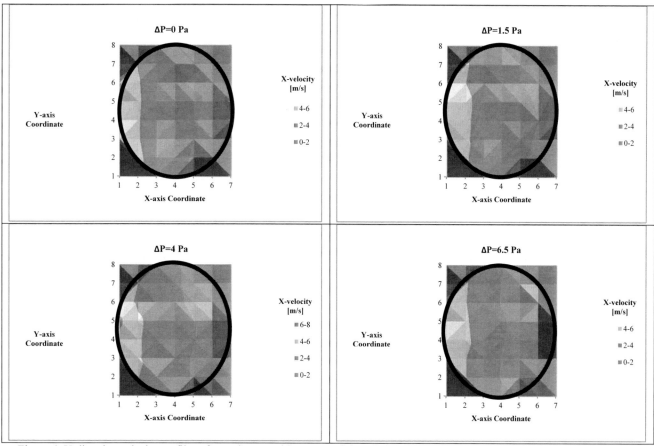

Figure 6: X-direction velocity profiles of experiments with straw straighteners. Pressure difference represents a total applied pressure rather than the static pressure held by the air curtain.

Figure 6 shows in-plane profiles of the air curtain with the straighteners (straws) present in the duct. When there is no pressure difference across the air curtain, air leaving the duct has a velocity of approximately 5.5 m/s and slows to less than 2 m/s by the last third of the window. The catcher appears to have a negligible effect on the curtain as a whole though a small increase in the velocity is seen just in front of the catcher. With 1.5 Pa of applied pressure, the curtain appears to increase in efficiency and nearly crosses the entire window with a velocity above 2 m/s. At 4 Pa the flow is similar to the 0 Pa case and at 6.5 Pa the air curtain only has a significant velocity for the first half of the window.

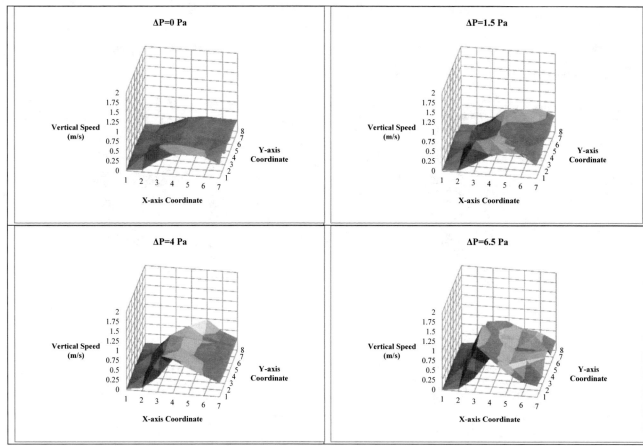

Figure 7: Z-direction velocity profiles of experiments with straw straighteners.

Intuitively it is expected that the applied pressure will break the planar flow of the curtain and push air out through the window. The z-direction plots are generated in order to show where this occurs as the pressure is increased. Figure 7 shows that with no pressure difference across the window a natural flow of air at 0.3 m/s – 0.5 m/s is seen in the window. As the pressure is increased the velocity profiles show sharp increases at the center of the window and fall off significantly thereafter. This seems to indicate that for the given geometry and power of the air curtain, it is only effective for a hole diameter of less than half of the current window size for the maximum tested pressure.

In order to increase the power and efficiency of the curtain, the straws were removed from the ducting and the experiment was repeated.

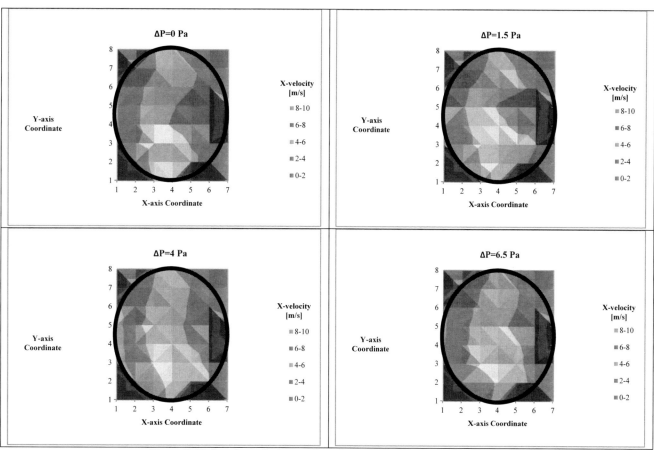

Figure 8: X-direction velocity profiles of experiments without straighteners.

With the straws removed air leaves the ducting at over 8 m/s and remains above 2 m/s across the entire window for all three test pressures, seen in figure 8. An additional point of interest is that the catcher still appears to only affect the air directly in front of its low pressure duct. Unlike the pitcher, the catcher is not unidirectional and therefore does not significantly improve the efficiency and usability of the air curtain.

Though the in-plane velocity profile looked positive, the z-direction profiles, seen in figure 9, indicate that the air curtain is turbulent for all applied pressures. Because the boundary is thermal as well as barometric, it is expected that this amount of turbulent mixing will have a significant impact on seeing and is unacceptable for the ATST. However, it should also be pointed out that only the turbulence within the optical beam path is problematic. The optical path is allowed only a minimal amount of turbulence, and at this location, this type of air flow at the window is unacceptable in the ATST.

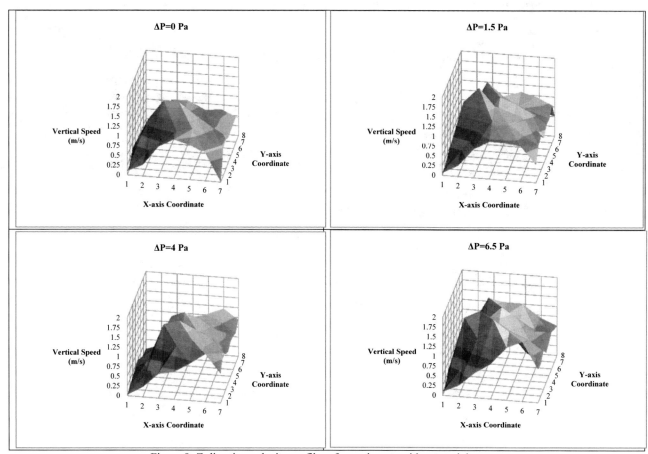

Figure 9: Z-direction velocity profiles of experiments without straighteners.

4.2 Simulation results (work in progress)

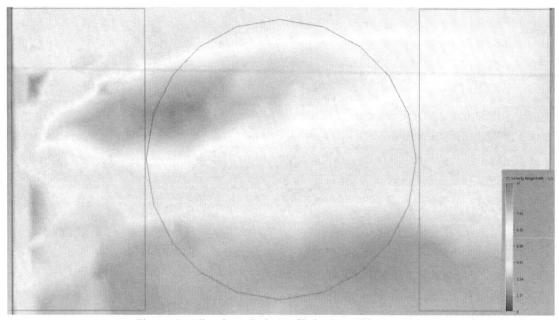

Figure 10: x-direction velocity profile in plane with air curtain

A simplified SolidWorks model of the test rig is used to do Computational Fluid Dynamics (CFD) analysis at the window. Figure 10 shows the velocity profile in the curtain just below the window for a case in which the rig has been pressurized to 1.5 Pascals. It is seen that the curtain profile tends to be pushed to one side of the window, which is unexpected since the rig is symmetric in this area. However, it is noticed that this result is also present in Figure 9, though less pronounced. Inspection of the velocity profile within the rig, shown in Figure 11, indicates that a natural recirculation may occur inside of the rig that causes the z-velocities to vary through the center yz-plane of the window. In the experimental setup, the wood platform internal to the rig would have dampened this effect and the inclusion of this extra geometry is considered important in the continual refinement of the CFD model.

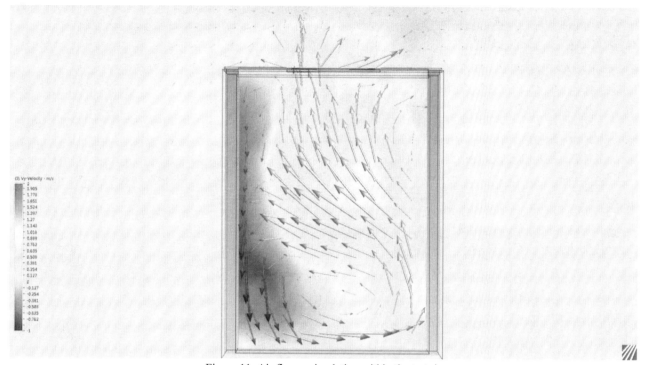

Figure 11: Air flow recirculation within the test rig

5. FUTURE WORK

A crucial requirement for the final design is that the air curtain be laminar. A fair amount of time was spent trying to characterize flow behavior using a smoke generator and a camera. Ultimately none of the photos taken could be used to definitively determine the laminar, transition, and turbulent flow regions across the window. Visualization of the flow using methods such as Schlieren photography, which uses a mirror and a polarizing filter, and ultimately a refined CFD model are vital in validating the assumptions made in this experiment regarding the flow behavior and the ability to use this experiment to refine the air curtain design. For laminar flow, it appears that flow straighteners are required, though higher supply pressure will be required than the fans used in the first part of this experiment could develop.

From the experiment it is clear that the air curtain's blower must be sized up by a large factor in order to hold the desired 15 Pa of pressure across the window. It may also be possible to have strategic placement of an air bleeds below the curtain in order to reduce the pressure just below the air curtain while keeping the lower part of the box at a higher pressure. Angling the air curtain toward the high pressure region should also be investigated given the results of *Valkeapää*'s paper. In the final design of the air curtain, turbulence at the window may also be reduced due to the presence of ducts just above the window. These may act to pull air up and away from the center of the window, reducing the image degradation in the optical path.

Once the configuration of the air knife is optimized, wavefront distortion analysis will be done with both a temperature and pressure gradient present.

ACKNOWLEGEMENTS

The National Science Foundation (NSF) through the National Solar Observatory (NSO) funds the ATST Project. The NSO is operated under a cooperative agreement between the Association of Universities for Research in Astronomy, Inc. (AURA) and the National Science Foundation (NSF).

The ATST represents a collaboration of 20 plus institutions, reflecting a broad segment of the solar physics community. The NSO is the Principal Investigator (PI) institution, and the co-PI institutions are the High Altitude Observatory, New Jersey Institute of Technology's Center for Solar Research, University of Hawai'i's Institute for Astronomy, and the University of Chicago Department of Astronomy and Astrophysics.

REFERENCES

[1] Rimmele, T., ATST Science Team, "ATST Science Requirements Document," Project Documentation Specification 0001 Revision B, (2005).
[2] Hubbard, R., et al., "Controlling wavefront distortion across a thermal boundary," Proc. SPIE 6267, 54, (2006).
[3] Rocklin, M., ATST Optical Systems Engineering Group, "Analysis of Wavefront Distortions across an Open Coudé Window," Project Documentation Technical Note 0092 Revision A, (2007).
[4] Rouaud, O., Havet, M., "Behavior of an Air Curtain Subjected to Transversal Pressure Variation," Journal of Environmental Engineering ASCE, Volume 132, Issue 2, pp. 263-270, (2006).
[5] Valkeapää, A., et al., "Air Leakage through Horizontal Curtains – an Experimental Study," aiha.org, Accessed 6/20/2011.
[6] Phelps, L., ATST Mechanical Group, "Reference Design Studies and Analysis (RDSA) Document for the ATST Coudé Environmental System," Project Documentation Technical Note 0109 Draft B, (2010).

ATST Telescope Mount – Telescope of Machine Tool

Paul Jeffers*[a], Günter Stolz[b], Giovanni Bonomi[b], Oliver Dreyer[c], Hans Kärcher[c]

[a]National Solar Observatory / ATST, 950 N Cherry Avenue, Tucson, AZ, USA 85719, [b]Ingersoll Machine Tools Inc, 707 Fulton Avenue, Rockford IL, USA 61103, [c]MT Mechatronics, Weberstr. 21, 55130 Mainz, Germany.

ABSTRACT

The Advanced Technology Solar Telescope (ATST) will be the largest solar telescope in the world, and will be able to provide the sharpest views ever taken of the solar surface. The telescope has a 4m aperture primary mirror, however due to the off axis nature of the optical layout, the telescope mount has proportions similar to an 8 meter class telescope.

The technology normally used in this class of telescope is well understood in the telescope community and has been successfully implemented in numerous projects.

The world of large machine tools has developed in a separate realm with similar levels of performance requirement but different boundary conditions. In addition the competitive nature of private industry has encouraged development and usage of more cost effective solutions both in initial capital cost and thru-life operating cost.

Telescope mounts move relatively slowly with requirements for high stability under external environmental influences such as wind buffeting. Large machine tools operate under high speed requirements coupled with high application of force through the machine but with little or no external environmental influences.

The benefits of these parallel development paths and the ATST system requirements are being combined in the ATST Telescope Mount Assembly (TMA). The process of balancing the system requirements with new technologies is based on the experience of the ATST project team, Ingersoll Machine Tools who are the main contractor for the TMA and MT Mechatronics who are their design subcontractors.

This paper highlights a number of these proven technologies from the commercially driven machine tool world that are being introduced to the TMA design. Also the challenges of integrating and ensuring that the differences in application requirements are accounted for in the design are discussed.

Keywords: ATST, Telescope, new technologies, bearings, drives.

1. INTRODUCTION

The technology utilised in telescope mounts is usually sourced from industry and then adapted to the unique and challenging requirements present in modern telescopes. These technologies are present in many different industries and applications. There is as telescopes get larger and more complex a drive to implement cost effective and reliable solutions. To identify these potential technologies the telescope industry has quite often looked to the defense sector.

There is always a risk in adopting a brand new technology especially when there will only be one production model. This combined with the long gestation period that is normal in telescope projects limits the opportunity to try out new solutions and perfect them. This is another reason to look to industry where the production rate and therefore the development process happens faster and involves more units of production.

The Advanced Technology Solar Telescope (ATST) concept was based on 'traditional' technologies that have been implemented very successfully in other telescopes of a similar scale. The Request for Proposal was however based around a performance specification, so not limiting the potential contractors to the technology solutions chosen in the concept design. A view of the ATST telescope reference design is shown in Figure 1.

*pjeffers@nso.edu; phone 1 520 318-8572;atst.so.edu

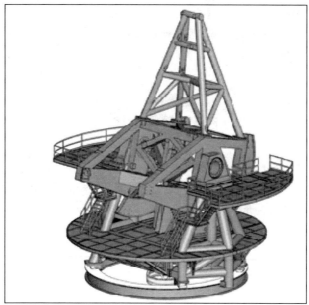

Figure 1. ATST Mount Reference Design

Ingersoll Machine Tools Inc. (IMT) was one of the proposers and was subsequently awarded the contract for the ATST Telescope Mount.

IMT had no previous experience in the telescope industry but is a world leader in the design and manufacture of large machine tools for heavy industry and aircraft builders. A typical machine tool developed and delivered by IMT is shown in Figure 2.

Figure 2. IMT Typical Machine Tool.

They operate in a very commercially competitive market and brought this experience to the ATST Project along with their design subcontractor MT Mechatronics GmbH (MTM) who has a long history in the telescope industry. The typical lead time for delivery to site of a new machine tool, including design, from placement of contract is approximately 18 months. At any one time within the Ingersoll facility there can be up to ten large machines in varying stages of manufacture, assembly and test.

In the following sections the similarities and contrasts between telescopes (using ATST as the example) and machine tools are considered as well as how these technologies / processes are being integrated into the ATST project.

2. TELESCOPE VERSUS MACHINE TOOL REQUIREMENTS

The driving requirements for telescopes are often defined in terms of:

- Tracking accuracy at low or zero speed;
- Pointing;
- Slew rate;
- Range of motion;
- Stability (or jitter); and
- Availability.

The repeatable errors in the telescope system are reduced using a pointing model that characterises the repeatable behavior of the telescope structure in terms of angular position in both azimuth and altitude axes. In some telescopes there are compensations also for thermal effects on the structure e.g. the lengthening of the seurrier truss with changes in ambient temperature.

The requirements on a large machine tool have similarities but also have some quite significant differences:

- Material removal rate;
- Cutting rate;
- Work volume and work piece weight capacity;
- Reliability/Availability;
- Accuracy of finished part and surface quality;
- High Accelerations and travel speed; and
- Number of controlled axes.

In machine tools the removal of repeatable errors is known as Volumetric Compensation. In essence it is the same idea as pointing compensation where a number of machine positions are matched to the commanded position and the position error is used to produce compensation look up tables. This is done in a number of ways and also by subsystem. For example linear encoders are quite often calibrated over their length using a laser interferometer to create specific look up tables for the encoder. This goes part way to reducing the position error but still leaves a number of structural deflections as residuals. IMT has developed a volumetric calibration procedure for large machine tools utilizing the Siemens "Space Error Compensation" software and partnering with API. This method uses the API laser tracker as the method of measuring the position error rather than in a telescope where you measure the offset to a known star.

A significant difference between the two sets of requirements is the disturbance or external influence on the system. In the telescope environment the wind excitation plays a significant part in the ability of the system to meet its stability requirements. There are some controls that can be introduced to reduce the impact of wind excitation or to limit the excitation in the first place e.g. wind screens. Moreover the telescope can often be located in an enclosure to limit external influences. However in the machine tool world external influences can be controlled in a much more effective manner. For example if thermal issues in the structure are a significant problem then in the extreme the machine tool can be situated in a thermally controlled environment that reduces stratification and temperature drift. The main excitation force that has to be dealt with in machine tools is internally generated and is directly related to the force and frequency created by the cutting tool. The balance of excitation i.e. more force = more material removal has to be balanced with efficiency and rate of material removal. These parameters are variable but are both testable and predictable in the shop environment.

Another significant difference in the requirements is the use of Interfaces. In machine tools the interfaces are generally limited to the:

- Material being worked including the volumetric requirements;
- Material handling system that delivers and removes the finished work;
- Machining data;
- Cutting tool; and
- Foundation that supports the machine.

This means that in the factory acceptance the testing of the machine tool is quite often a machining of a predefined test piece which is then measured for accuracy to the CAD file definition. It is a more straight forward pass or fail at acceptance versus the telescope situation where a number of different sub-system performance metrics are measured e.g. deflection and encoder position error. These subsystem performances are then assessed through the system engineering process for impact on the overall telescope system performance.

A major advantage of a machine tool supplier's involvement in a telescope project is that there is full agreement on the need to do a full assembly and test at the factory as this is normal practice. This is planned to be carried out for the ATST Telescope Mount and is also getting more and more common in telescope projects. Usually, the additional effort for the work shop assembly and testing is more than compensated for by less work on site.

Another major difference between the telescope and machine tool business is the documentation requirements and the client involvement in design. While telescopes (such as ATST) are unique structures, their design and realization have many similarities to R&D projects. Extensive documentation, including analyses reports, simulations results, drawings plus the usual project documentations, e.g. schedules or risk and hazard analyses, are discussed between client and contractor at numerous intermediate milestones and continuous progress reporting is normal before the final delivery.

In the commercially competitive market of machine tools with short times to market (compared to telescope projects), clients more often focus on the final machine and operational/maintenance manuals and are usually less involved in other details.

3. PERFORMANCE COMPARISON

Part of the process of approaching the ATST project involved assessing an existing machine tool of current design with the ATST Mount requirements. Table 1 converts the ATST angular requirements into equivalent linear requirements. A machine tool on the assembly floor at IMT was tested and shown to meet these linear requirements. This provided the confidence for both the Project and IMT that the technologies being proposed would indeed meet the ATST requirements.

Table 1. ATST Mount Linear Equivalent Performance Requirements

Origin	Requirement	Value (rotational)	Value (linear) (1arcsec≈21μm 1°≈74mm)
Operational	Slewing rate Mount AZ	0 to +/-2.0°/sec	0 to +/-148mm/sec
Operational	Slewing rate Coudé AZ	0 to +/-6.0°/sec	0 to +/-445mm/sec
Science	Tracking rate	0 to +/- 0.1°/sec	0 to +/-7mm/sec
Engineering	Tracking acceleration	0 to +/- 0.1°/sec²	0 to +/-7mm/sec²
Engineering	Acceleration limits	≤ 0.5°/sec²	≤ 37mm/sec²
Operational	Settling time	≤ 10sec	≤ 10sec
Engineering	Encoder resolution	≤ 0.02 arcsec	≤ 0.4μm
Science	Blind pointing	≤ 5arcsec rms (full range)	≤ 105μm
Science	Offset pointing accuracy	≤ 0.5arcsec rms (1°)	≤ 11μm
Science	Tracking image drift	≤ 0.05arcsec rms (1 minute) ≤ 0.1arcsec rms (30 minutes)	≤ 1μm (1 minute) ≤ 2μm (30 minutes)
Science	Jitter	≤ 0.075arcsec rms	≤ 1.6μm

4. ANATOMY OF A TELESCOPE / MACHINE TOOL

4.1 Bearing Systems

Traditionally smaller telescopes have used rolling element bearings where the larger 4m and 8m class telescopes have utilized hydrostatic bearing systems. The basic hydrostatic bearing consists of pads supplied with high pressure oil running on a track or journal. The negligible friction and capability to support high mass systems has made the hydrostatic bearing the default for large telescope concepts.

In the large machine tool industry the use of hydrostatic bearing systems was extensive through to end of 1990's. The advent of re-circulating bearing trucks on precision rails however provided an opportunity to reduce the capital cost and running cost of bearing systems. The penalty however was higher friction relative to the hydrostatic systems and potential stick slip effects at extremely low speeds. With the machine tools where high forces due to material removal are normal, this increased friction is not a significant problem for the motors / drives used. IMT have been using this linear rail bearing type of system for approximately 8 years in their large machine tools which are of similar mass magnitude to the ATST telescope mount. A typical linear rail and bearing truck arrangement is shown in Figure 3.

Figure 3. Machine tool linear rail and bearing truck arrangement.

A more recent development was the availability of bearing trucks and rails on a circular rail rather than a linear rail. Figure 4 shows a basic circular rail single truck arrangement.

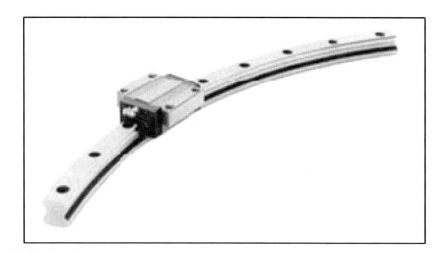

Figure 4. Circular Bearing Rail and Truck.

A more in-depth discussion and review of the application of this bearing system is given in paper Kärcher, H.J. et al.,"The azimuth axes mechanisms for the ATST telescope mount assembly"[1].

4.2 Motors & Drive

The motor and drive systems utilized in the telescope and machine tool industries are very similar. A common and recurring system is the rack and pinion. In machine tools the rack is normally straight and in telescopes circumferential. Figure 5 shows a typical machine tool application with the torque motor shown in Figure 6 and the rack and pinion drive in Figure 7.

Figure 5. Machine tool typical drive layout.

Figure 6. Machine tool torque motor

Due to the higher production rate of units, machine tool suppliers tend to standardize on a motor type and limit the number of variants they use. This allows for usage of a proven and known component and also more straightforward support for spares. As both the telescope and the machine tool are designed for stiffness especially in the drive train this approach is well adapted to the telescope design requirements.

Figure 7. Linear rack and pinion

Another advantage of this arrangement is that the lifetime of a machine tool is of the same order as a telescope and so due to the significantly higher number of machine tool units in the market using similar components; spares availability in the future is improved.

4.3 Structure

The structure of a machine tool is similarly designed for stiffness, but without the requirements to keep an open structure for flushing to avoid seeing or to reduce profiles for wind excitation. However the same types of structure both monocoque and lattice are used and the fabrication techniques are the same. IMT as a supplier of machine tools also maintain a selection of their machine tools on site that are used in the manufacturing process. A number of examples of machine tool designs are shown in Figures 8 and 9.

Figure 8, IMT machine tool structure

Figure 9. IMT machine tool during assembly.

4.4 Encoding system

The other main aspect of the machine tool that is common with the telescope is the Encoding system. The encoders are matched to the precision of the machine. Therefore different encoder types are used as well as from different manufacturers. This evaluation of the necessary precision prior to selecting a type of encoder is important in ensuring that the machine tool is cost effective. Also the environment plays a part in this selection as certain parts of the machine tool are exposed to high levels of contamination from waste material chips and or machining fluids. This environment makes use of non-optical based encoders very attractive when they meet the precision requirements.

In some very specialist areas the optical encoders do not meet the precision required over the spatial volume of the machine and in those applications, laser interferometer encoders are utilized.

4.5 Control systems

Machine tools use numerical control systems which are in general highly standardized products from such suppliers as Siemens, Heidenhain and Fanuc. This is mainly due to the large volume of machine tools manufactured each year by the industry and a desire for both standardization and for turnkey control systems. Telescopes however use very project-specific Control Systems often developed over time by specific suppliers on which they have invested significant development time to meet the specific needs of antenna and telescope control. In the case of the ATST telescope mount the motors and drives are as per a typical machine tool but the control system is provided by MTM and based on their extensive experience of antennae and telescope control.

5. CONCLUSIONS

The similarities between the telescope industry and large machine tools are not just in the technology but also the industry approach. As a potential supplier to telescope projects these companies are used to dealing with large high precision structures but are also used to operating in a commercial environment that ensures cost effective solutions are always being developed.

The ATST project is benefiting from this approach. The continued support of the technologies has a high confidence level due to the number of other 'machines' that are in the market place and use of the same style, size and even make of motors etc.

Another major benefit seen by the project is the assembly and test facility that is a normal part of the machine tool industries supply process. Having an existing facility large enough to fully assemble and test the ATST mount is a significant advantage.

6. REFERENCES

[1] Kärcher, H.J. et al., "The azimuth axes mechanisms for the ATST telescope mount assembly", Proc. SPIE 8444-08 (2012)

7. ACKNOWLEDGMENTS

The National Science Foundation (NSF) through the National Solar Observatory (NSO) funds the ATST Project. The NSO is operated under a cooperative agreement between the Association of Universities for Research in Astronomy, Inc. (AURA) and NSF.

The ATST represents a collaboration of 20 plus institutions, reflecting a broad segment of the solar physics community. The NSO is the Principal Investigator (PI) institution, and the co-PI institutions are the High Altitude Observatory, New Jersey Institute of Technology's Center for Solar Research, University of Hawai'i's Institute for Astronomy, and the University of Chicago Department of Astronomy and Astrophysics.

Performance introduction of a 2.5m telescope mount

Guomin Wang*, Bozhong Gu, Shihai Yang, Xiang Jiang, Zhiyong Zhang, Yu Ye, Jin Xu

[a]National Astronomical Observatories / Nanjing Institute of Astronomical Optics & Technology, Chinese Academy of Sciences, Nanjing 210042, China; [b]Key Laboratory of Astronomical Optics & Technology, Nanjing Institute of Astronomical Optics & Technology, Chinese Academy of Sciences, Nanjing 210042, China

ABSTRACT

In recent years, Nanjing Institute of Astronomical Optics and Technology (NIAOT) has made several telescopes for observatories all around the world. In 2011 NIAOT just finished the development of a 2.5m optical/infrared telescope mount. First part of this paper is to introduce the mount structure and their adjustment work. Second part is to give an introduction of the mount performance test methods and test results finished on NIAOT workshop.

Keywords: telescope, mount, performance

1. INTRODUCTION

The mount introduced in this paper is made for a 2.5m optical/infrared telescope which is a cooperation project between Nanjing Institute of Astronomical Optics and Technology, CAS (NIAOT, China) and Sagem Défense Sécurité (France). NIAOT is in charge of the telescope mount design and manufacture, including the corresponding control system. Sagem is in charge of the optical element design and manufacture. The telescope is a Ritchey-Chrétien Cassegrain with f/8 focus with one Cassegrain focus, two Nasmyth foci and two student Nasmyth foci. In order to get high image quality, M2 is supported by a kind of 5-dof active positioning system to compensate the misalignment between M1 and M2 due to thermal and gravity. M3 is supported by a turntable, which will transfer light to the different Nasmyth foci. Meanwhile, in order to achieve the change between Cassegrain focus and Nasmyth focus, the tertial mirror need to escape from one position to another position to let the light go to different focus. First part of this pape is to give an introduction of the mount structure. Second part is to give an introduction of the mount performance tested in NIAOT workshop.

2. BRIEF STRUCTURE INTRODUCTION

2.1 Azimuth axis structure

Here azimuth axis consists of basement, rotating table, supporting bearings, drive motors, feedback encoder and other subsidiaries. Fig.1 shows the structure arrangement of azimuth axis. Basement is the basic part which fix the telescope to the ground (pier) through the central bearing and support the weight of the whole telescope by the means of vertical bearing. In our structure, we use two angle contact ball bearings to center the telescope and oil bearing as vertical bearing. The envelop dimension is Φ3540mm×460mm. The bottom of the basement will connected to the concrete pier by means of total 12 anchor bolts embedded in the pier concrete in advance. Between the basement and concrete pier, there are corresponding 12 adjusting iron pads whose load capacity is 6 metric tons per unit. With the help of these iron pads, we can adjust the levellness of the top surface of oil bearing pads embedded in the basement. After finishing the level adjustment of basement, in fact it is the level of a plane formed by top surface of 12 oil bearing pads, we test the levellness using a optical split-image levelling instrument and calculate the plane inclination with the help of least square method. The inclination is 1.09 arcsecond.

---

*Guomin Wang, gmwang@niaot.ac.cn; phone 0086-025-85482303; fax 0086-025-85482303

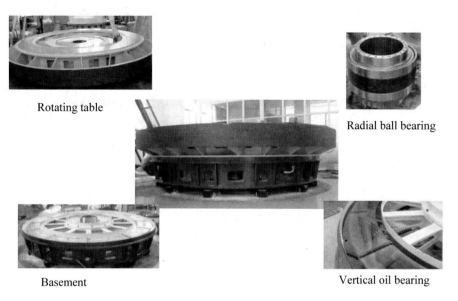

Fig.1 Component of azimuth axis structure

Rotating table is a large weldment sub-assembly with Φ4000mm diameter and 491mm height. It is centered to basement by the central bearing (radial ball bearing). A couple of angle-contact ball bearing is used as the center positioning bearing. The axial preload is applied to enhance the bearing stiffness and to increase its running accuracy. Externally pressurized hydrostatic oil bearing is used as axial bearing to support the telescope weight which is about 27 metric tons. Oil thickness of oil bearing is set to 0.1mm and can be controled from the system control GUI in the control room.

2.2 Elevation axis structure

The elevation shaft is supported by two set of angle contact ball bearings which was put on the top of forks. Here are two requirements need to check and adjust: one is the perpendicularity between azimuth axis and elevation axis and another one is the space cross of azimuth axis and elevation axis. With the help of cross hair set at the center of hollow elevation shafts and autocollimator, we finished the work to check the above requirements. After check, the error can be corrected through the spacer between the fork and elevation bearing houses. The final perpendicularity between azimuth axis and elevation axis is 1.1″ and the space cross error is 0.025mm. Fig.2 shows the adjustment work status finished in the workshop. It is a iterative work and depends on the maching accuracy of the components, such as the height error between two forks and two elevation hollow shaft to their corresponding reference surface

Fig.2 Adjustment figure

2.3 OSS structure

OSS (Optical Support Structure) structure is used to support the primary mirror and secondary mirror and to keep the stability of their relative position during the operation from horizon to zenith under the gravity load and outside disturbance load. The structure of OSS is shown as Fig.3. As the figure showing, the primary mirror cell is attached to center section directly. Secondary mirror cell is connected to center section using traditional way: through vanes, top-ring and eight supporting trusses. The mechanics performance of these parts are calculated and optimized using FEM software to ensure the system has the enough stiffness. On the other hand, after we mount the all the parts to OSS, we will test the relative offset between primary mirror and secondary mirror for M2 positioning system to compensate.

Fig.3 OSS structure

2.4 M2 positioning structure

As Fig.3 shows that the primary mirror cell is attached to center section directly and the secondary mirror cell is supported by center section through truss. They are not the Serrurier structure, so the relative deflection between primary mirror and secondary mirror is large due to the gravity. A 5-DoF active positioning unit is used to compensate such deflection. The detail structure of such 5-DoF unit is shown as Fig.4. The 5 motions are arranged at three layers: top layer for X translation, secondary layer for Y translation and third layer for Z translation. Third layer is connected to mirror cell by three units which can translate along main optical path. So, the Z translation (focus) is finished when three units moving at the same direction and same speed. Tilt around X axis (parallel to elevation axis) and around Y axis can be achieved by the combinational motions of these three units.

Fig.4 M2 positioning unit

2.5 M3 positioning structure

M3 positioning system is used to transfer the light to different foci according to the observing arrangement and to ensure the alignment between optical elements. Besides the four Nasmyth foci, this telescope is equipped with a Cassegrain focus. So, M3 positioning system will also have the ability to escape when the telescope uses the Cassegrain focus. The FOV requirement for Cassegrain focus is 1°. The third auxillary motion is that the tertiary mirror can be adjusted a certain distance along elevation axis. In order to accomplish the above three motion function, we proposed a structure shown as Fig.5. The structure is composed of three main sub-parts: escape unit, rotation unit and translation unit. The room left for M3 escape is very small. So, the escape motion consists of two linear movements. M3 cell is supported by a U shape welded supporting structure through 3 bolts. The supporting structure is optimized to ensure high stiffness and to keep M3 cell free from outside stress. The supporting structure is positioned by 4 ball rails through 4 corresponding blocks of rails mounted on a stiffing pedestal. Escape motion is driven by the combining of worm gear and ball screw. Linear encoder is used to feedback the position information to control system. There are switches and hard ends to ensure the safety of the motion. Besides the escape function, the tilt adjustment of the M3 will also be fulfilled by this system.

The motion of translation along Nasmyth axis is driven by slide screw along the ball screw rails. The position is controlled by linear encoder through motion control system. A spring is used to release the backlash of the slide screw.

Rotation system is composed of supporting roller bearing, spur gear drive, driving motor and position feedback tape encoder. Two set of pinions are arranged 180°separate to release the backlash of the system and driven by two servo motors. The diameter of the gear is 900mm with 30 gear ratio.

Fig.5 M3 positioning unit

3. PERFORMANCE INTRODUCTION

3.1 Mount pointing and tracking performance

Mount pointing and tracking performance was tested in NIAOT workshop through a kind of blind tracking and pointing. The tracking performance test include the test of azimuth axis, elevation axis, C1 derotator, N1 derotator and N2 derotator. Detail procedure is as following:

- Select several tracking targets from star catalogue around the sky, including star crossing the meridian.
- Order the mount to go along the trajectory of the selected targets.
- Get the actual trajectory from the encoder mounted in azimuth axis, elevation axis, C1 derotator, N1 derotator and N2 derotator.
- Computer the offset between the theoretical trajectory and actual trajectory.
- Calculate the tracking accuracy based on the motion offset.

Following tables list the three typical results: one is general target, one is a meridian crossing target and third one is near and cross the blind spot. The blind spot of this telescope is 1° in diameter.

Table 1 General target tracking performance

Target: RA 19h50m46.999 Dec 8d52m5.959s	
Test date and time: 2011-08-01 19:56:45.571 ~ 20:07:26.858	
Azimuth range: 105.0004176° → 107.141504° Elevation range: 38.1501793° → 40.4649158° N2 derotator range: -54.9543715 → -55.7715456 °	
Azimuth settling time (s)	7
Azimuth tracking accuracy RMS (″)	0.0716
Azimuth tracking accuracy PTV (″)	0.615
Elevation settling time (s)	12
Elevation tracking accuracy RMS (″)	0.026
Elevation tracking accuracy PTV (″)	0.193
N2 settling time (s)	10
N2 tracking accuracy RMS (″)	0.377

Table 2 Meridian crossing tracking performance

Target: RA17h59m42.13s Dec -9d46m22.6s (meridian crossing time: 21:25:44.259)	
Test date and time: 2011-08-01 21:09:09.371 ~ 21:26:45.360	
Azimuth range: 179.6999998° → 173.9567583° → 107.141504° Elevation range: 35.2093944° → 48.1489383° N2 derotator range: 5.758982° → 0.2280375°	
Azimuth settling time (s)	6
Azimuth tracking accuracy RMS (″)	0.0862
Azimuth tracking accuracy PTV (″)	0.689
Elevation settling time (s)	11
Elevation tracking accuracy RMS (″)	0.0338
Elevation tracking accuracy PTV (″)	0.371
N2 settling time (s)	10
N2 tracking accuracy RMS (″)	0.428

Table 3 Near blind spot target tracking performance

Target: RA 20h24m21.84s Dec 32d13m51.1s	
Test date and time: 2011-08-01 23:29:55.384 ~ 23:47:43.658	
Azimuth range: 100.6951859° → 65.0577265° Elevation range: 60.1324473 ° → 89.528207° (more than 89.5°) N2 derotator range: 62.3851103° → 73.028488°	
Azimuth settling time (s)	6
Azimuth tracking accuracy RMS (″)	0.089 (remove the part of blind spot)
Azimuth tracking accuracy PTV (″)	0.45 (remove the part of blind spot)
Elevation settling time (s)	11
Elevation tracking accuracy RMS (″)	0.0534
Elevation tracking accuracy PTV (″)	0.719
N2 settling time (s)	13
N2 tracking accuracy RMS (″)	0.734

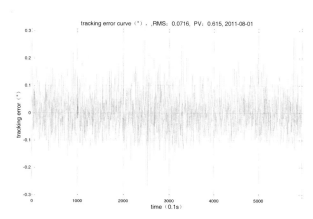

Fig.6 Azimuth tracking error curve of Test 1

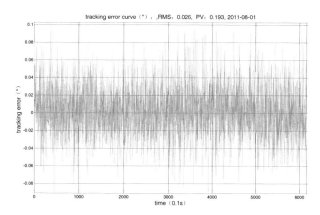

Fig.7 Elevation tracking error curve of Test 1

3.2 M2 positioning performance test

As introduced above M2 will be controlled in 5 degree of freedom: translation along X direction (parallel to elevation axis) and Y direction (perpendicular to X direction) and Z direction (along main optical path), tilt around X axis and Y axis. All these functions are achieved by translation motions, so we use a kind of digital gauge to check the motion performance. Test includes position accuracy and repeatability accuracy. Following table shows the test results:

Table 4 M2 positioning performance

Item		Requirement	Test results
Decenter	range	±6 mm	±10 mm
	resolution	1 μm	0.1 μm
	repeatability	±10 μm	± 3 μm
	speed	0.25 mm/s	0.25 mm/s
Tilt	range	±1°	±1.8°
	resolution	2 μrad	0.2 μrad
	repeatability	±10 μrad	±7.5 μrad
	speed	2 mrad/s	2 mrad/s
Focus	range	±8 mm	±15 mm
	resolution	1 μm	0.1 μm
	repeatability	±4 μm	± 3 μm
	speed	0.1 mm/s	0.1 mm/s

3.3 M3 positioning performance test

According to the configuration of the whole system, three motions are needed for M3 support structure to perform: escape motion between Cassegrain focus and Nasmyth focus, rotation between four Nasmyth foci, and translation to align M3 with main optical path. The detail structure is decribed in Section 2.5. Here just give the test results as following table 5:

Table 5 M3 positioning performance

Item		Requirement	Test results
Rotation	Time between different Nasmyth foci	< 2min	1.5min
	Nasmyth foci angular positioning and locking accuracy	< 10″	3″
	Nasmyth foci repeatability	2″ RMS	1.6″ RMS
Escape	Accuracy of re-insertion in the Nasmyth configuration	< 10″	4.35″
	M3 escape time	< 2min	1.8min
Translation	Translation adjustment along Nasmyth axis	Range: 5mm Resolution: 10um	Range: 6mm Resolution: 5um

4. CONCLUSIONS

This paper introduced the 2.5m telescope mount structure and its performance test finished in NIAOT workshop. During the mount test in NIAOT workshop, all the optical elements, such as M1, M2, M3, were simulated by dummy mirrors with the right size, weight, gravity CoG, inertia moment. So, the performance of the mount will not degrade when the optical elements is integrated on site. Through the workshop test, all the performance of the telescope mount can meet the requirements assigned by the general error budget.

REFERENCES

[1] Bozhong Gu, Guomin Wang, Jiang Xiang. Introduction of a 2.5m telescope mount. 2008, Proc. of SPIE, Vol.7018: 701846-1~701846-9.
[2] Guomin Wang, Zhiyong Zhang, Bozhong Gu. Analysis and optimization of a 2.5m telescope mount. 2008, Proc. of SPIE. Vol.7018: 701847-1~701847-10.

Installation and Verification of High Precision Mechanics in Concrete Structures at the Example of ALMA Antenna Interfaces

Volker Heinz[1], Max Kraus[1], Eduardo Orellana[2]

1 European Southern Observatory, Karl-Schwarzschild-Strasse 2, 85748 Garching, Germany
2 Bautek, Las Esteras Norte 2540, Quilicura, Santiago, Chile

Abstract

For the ALMA interferometer at the array operation facility near San Pedro de Atacama at 5.000 meters asl 192 concrete antenna foundations had to be equipped with coupling points for 66 antennas. These antennas will be frequently moved between the foundations and placed on these interfaces without further adjustment.

To position the ALMA antennas with the required accuracy, high precision inserts need to be installed in previously casted concrete foundations. Very tight mechanical tolerances have to be applied to civil structures, with standard tolerances of not less than millimeters. This is extremely difficult considering the material (mortar and steel in a concrete slab) to be used and the environmental conditions on site. Special tools had to be designed and an installation and alignment procedure developed, tested and improved. Important was to have a robust process, which allows highest precision installation without major re-machining for approx 600 interface blocks. Installation material, which could cope with the conditions, was specially tested for these requirements.

The geometry of the interface and other parameters such as horizontal and vertical stiffness must be verified after the installation. Special metrology tools to measure reliable at micron level at high altitude had been selected.

The experience and knowledge acquired will be beneficial for the installation of any opto-mechanical device in civil engineering structures, such as telescope and dome track rails, but also in optical interferometer installations. Metrology requirements and environmental conditions in most of these cases are equally challenging.

Keywords: telescope foundation, high altitude, high precision installation, precision metrology equipment, opto-mechanical interfaces

Introduction

To install telescopes and their subsystems, civil structures mainly foundations need to be erected first. On and in these high precision interfaces are installed to mount the opto-mechanical systems. It can be distinguished between open and enclosed interfaces, although their installation appears mostly in the open air. For large telescopes the interface to the concrete peer often is the Azimuth track, which can either serve as base for a hydrostatic bearing, but can also be a track for a roller type bearing. For smaller, especially mobile telescopes, this interface often is designed as a kinematic mount with clamping facilities. A general description of these interfaces and their installation and verification shall be the topic of this manuscript. At the example of the installation of the foundation inserts for the ALMA antenna at the Array observation site the process will be explained.

The Atacama Large Millimeter/Sub millimeter Array (ALMA) is an advanced instrument with a 54 plus 12-antenna interferometer array. The ALMA Array Operation Site (AOS) will be sited in the Altiplano of northern Chile at an elevation of 5.000 meters (16,500 feet) above sea level and the ALMA Operations Support Facility (OSF) will be sited at an elevation of 2.900 meters (9,600 feet) above sea level in the vicinity of the AOS site near the village of San Pedro de

Atacama in the Second Region of the Republic of Chile in South America. The ALMA AOS site is the highest, permanent, astronomical observing site in the world.

As an engineering project, ALMA is a collection of 54 plus 12 precisely-tuned mechanical structures each weighing more than 100 tons, cryogenically cooled super conducting electronics, and optical transmission of data at terabit rates- all operating together and continuously. Antennas will be transported between foundations regularly. 192 foundations will be built for the 50 antennas of the main array and the 16 antennas of the compact array. Each foundation consists of a concrete pad with recesses for the precision antenna interfaces. The high precision inserts for the antenna interfaces will be mounted after the concrete pad has cured. Creep and settling of the concrete prevents high precision alignment and installation for 6 months after casting.

For assembly and technical commissioning of the antennas also in the base camp area 11 antenna stations are installed.

ALMA Antenna Interface Design

The interface between the ALMA antenna and its foundation consists of three steel blocks, embedded with high precision into recesses of a previously casted concrete slab. It works as kinematic mount in Y- shape, which allows mounting the antenna in different thermal state.

Based on the following requirements the detailed design of the anchoring system was developed:

Requirement	Value/Tolerance
Global position of interface center in the array	Circle of ⌀50 mm in the closest configuration of the array
Rotation of interface against Astronomical North	0.2 Degree
Horizontality of interface/verticality of the Az axis	Local horizontal plane with 0.4 mm tilt/± 10"
Local Tilt of interface block	100 μm/meter

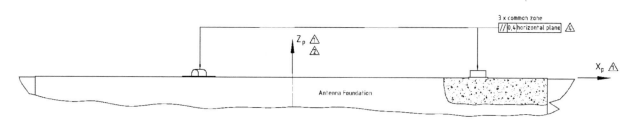

1 Interface Section View

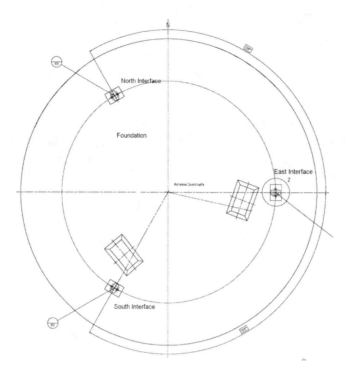

2 Top view of the foundation interface

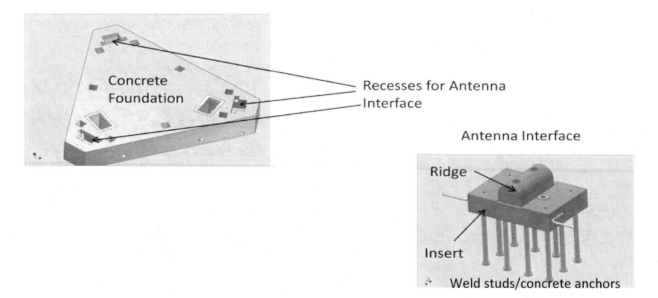

1 Foundation Recesses and Steel Inserts with Interfaces

For embedding the three inserts into the recesses of the concrete slab, they need to be positioned and aligned. Once having the steel blocks stable in place, the recess is being filled with mortar, a grout material of either cementitious or Epoxy base. To achieve the required position accuracy, stability and verticality of the Azimuth Axis of the antenna on the foundation, tight tolerances must be applied. For each individual insert block the local tilt with respect to the normal horizontal plane is 100 micron/meter. The plane defined by the three interface blocks must be parallel to the normal horizontal within 400 micron. As a fallback solution if the alignment goal could not be reached, the possibility for rectification was foreseen. The upper part of the interface, the ridge, can be removed from the insert and be machined on a mill or flat bed grinder.

For the installation of the antenna on the pad it is important to have also the embedded insert in the correct position and orientation

2 Antenna Installed on foundation with Interface detail

Tools and Materials

Alignment and Installation Tool

To meet the design requirements for positioning and alignment accuracy an interface alignment tool was designed. The structure consists of two parts, upper and lower half. While the lower part is fixed to the concrete with anchor bolts, the upper part, which carries the inserts and ridges during adjustment and holds them during grout curing, is used to fine adjust the interface to the correct position (center and North are marked on the foundation), and finally the height of the three inserts within 400 micron and the local tilt of 100 micron/meter of each embedded plate are adjusted. The advantage of this tool is that with its lateral stiffness the position of the three inserts relative to each other is well defined and fixed. No care needs to be taken to align the plates wrt each other. As the ridge, which defines the interface with the antenna has a rotational degree of freedom, the angular adjustment of the inserts is not critical and can be performed visually. Before moving to Chile with the equipment, a procedure was developed and tested during a trial installation in Germany.

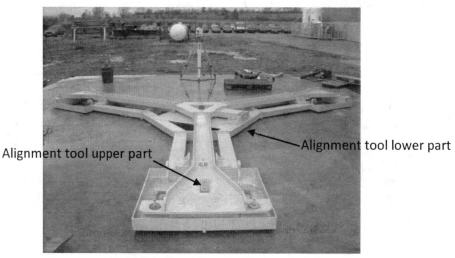

3 Interface Alignment Tool

As the installation sequence per pad was estimated with 4 days with only one tool the pure installation time would be 2 ½ years, not considering rest and bad weather periods. Therefore three tools were built.

Metrology tools

Not only for the installation of the interfaces were metrology tools required but also for the verification of the installation result and other parameters, such as the stiffness of the whole foundation. For the geometrical measurements and verification the following equipment was used.

1. Industrial Total Station

The main metrology tool is a Leica industrial total station, similar to a theodolite, but with an additional distance measurement and therefore a full 3D coordinate measuring system. It fulfills the necessary accuracy requirements for the tight installation tolerances, is robust enough for the harsh outdoor use under extreme conditions (low temperature, humidity, dust). In addition this instrument allows creating local coordinate systems, and measure positions within these coordinates, a function, which was used in part of the array where the absolute position tolerance for the interface center was a circle of 5o mm.

Angular measurement
Standard deviation
per ISO17123-3, 1 ƒ 0.5 0.15 m on
Units of measurement 360 sexa esimal, 400 on
360 decimal, 6400 mil
Data storage and interfaces
PCMCIA memory card
RS232 pro rammable interface
Motor and fine drives
Fine drives Coarse/fine, motorised, infinite,
slip couplin
Motor
Speed of rotation 45 /s 50 on/s
Positionin accuracy 0.8 0.2 m on
Temperature range
Wor in 20 C to 50 C 4 F to 122 F
Stora e 40 C to 70 C 40 F to 158 F
Point accuracy (total RMS H1 ƒ) $_{2)}$δ 0.3 mm 0.012
at 20 m 65 ft measurin volume
Distance measurement (integrated in the TDM5005 and TDA5005)
Standard deviation absolute 1 mm 2 ppm 0.04 2 ppm
per ISO17123-4, 1 ƒ over the entire measurement ran e
Typical distance accuracy
at 120 m 365 ft measurin volume₃
Reflective tape 0.5 mm 0.02
Corner cube reflector 0.2 mm 0.008
Units of measurement m, mm, feet, inch
Display 0 5 decimal places, dependent
 smallest selectable unit on the selected unit
Reflectors (selectable) Prisms, Corner Cube Reflectors CCR
 1.5 diameter , Leica reflective
tapes, 360 prisms
Measurement range with CCR
 dependent on atmospheric conditions 2 to 600 m 6 to 1 900 ft
Measurement range with reflective tapes
 dependent on tar et size 2 to 180 m 6 to 600 ft
ATR – Automatic Target Recognition Integrated in the TDA5005
Trac in speed lateral linear 3 m/s 10 ft/s

[1]

2. Precision Water Level

To measure the local tilt a high precision tooling water level, normally used for alignment of precision machine tool and turbine shafts, is being used:

With a resolution of 20 micron/meter sufficient accuracy of the instrument is assured to verify the tight tolerances of the ALMA installation. However this tool normally is used in a stable and temperature controlled environment. When used at the construction site for installation and verification utmost precaution had to be applied.

3. Hydrostatic Leveling System

For the stiffness verification a precision hydrostatic level system was applied (see chapter Quality Control and Verification. This tool consists of 4 small vessels with capacitive sensors on top, which are interconnected by hoses. With the sensors level differences in the micron range can be measured, the signals are converted in an electronics rack and via RS232 interface transmitted to a laptop with the measurement software.

- Global accuracy: 5 µm
- Repeatability: 2 micron
- Range: 5 mm

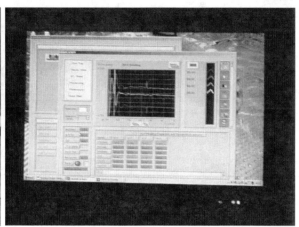

4 Hydrostatic Level Sensor and Control Screen

4. Laser Detector (Laser Trac™)

To measure small movements without mechanical contact a laser detector was used. This laser with a CCD allows to measure movements of several microns, however this instrument is extremely sensitive to sun and wind exposure and therefore best results for outdoor measurements can only be achieved during night time.

5 Laser Detector CCD installed on Interface block

Materials used

To embed the steel insert in the concrete foundation recess, a non shrinking mortar must be used. Two types of this material (grout) were available.
- Cement Based
- Epoxy Type Grout

The cement based grout has the following properties
+ Cheap
+ Easy to process
+ Higher Young Modulus
+ Can be installed at low temperatures, and material does not need to be prepared
− Longer curing time
− Higher tendency for cracking

While the Epoxy material has the following properties
+ Fast curing
+ No cracking, higher positive expansion
+ Higher probability of failsafe installation
− More expensive
− Material must be preconditioned before application
− Young modulus too low

After using cementitious grout for all foundations in the base camp area, a test installation at the high site has shown that Epoxy grout is necessary as the faster curing time has great advantages for the stability of the aligned setup and therefore direct proportional better installation results, which lead to the avoidance of rectification.

Although the nominal compression modulus (Young's modulus) of the product "Five Star DP Epoxy grout" is too low, an independent test [2] had shown that the material reaches our specifications to 90% for the compression modulus. This was approved by the ALMA project and the decision was taken to use mostly Epoxy grout Five Star DP. Only during the summer period with higher temperatures, less problematic to reach the specified tolerance, cement grout was used to reduce installation costs, in about 30% of the foundations.

On the cost side, the ratio between Cement based material and Epoxy is approximately a factor of 7. However considering the high installation manpower cost per day, it is definitely worth to safe the time due to faster curing, which justifies the higher material expenses. The additional benefit is the better accuracy achieved with Epoxy material, which saves the rectification costs and such very expensive machine hours in a precision workshop.

Special Conditions and Challenges of the ALMA Antenna Mechanical Interface Installation

To apply grout some basic thermal requirements must be fulfilled, a temperature level of minimum 13 Deg C must be maintained over at least 24 hours to allow a successful installation. High precision installations however require even more stable conditions and thermal control. The tolerances given for this installation can normally only be achieved under indoor and controlled conditions. Certainly temperature variations of up to 20 degrees during the alignment and curing process do not allow installations at micron level. Therefore it was necessary to assure the maximum stability. This could be achieved by designing a protected environment (thermally insulated shelters) and thermal control to the maximum extent possible. This was implemented at the Llano de Chajnantor at more than 5.000 meter asl and severe conditions like cold nights, strong radiation during daytime and heavy winds together with precipitation, mostly as snowfall. This thermal control is part of a specially developed and strictly followed installation procedure, which was first tested under normal conditions and by first experience adapted to the special requirements of a high altitude work site.

6 thermal shelter for the insert installation

As the following photo shows, building thermal shelters alone is not sufficient, also heating systems need to be installed to maintain a controlled environment. This includes power reliable power generation at 5.100 meters.

9 thermal shelter with alignment tool covered with snow

The diameter of the three insert center points from the center of the tool is approximately 6 meters. Considering a delta T of 20 deg C during the curing time of the grout for the inserts will cause a thermal expansion/shrinkage of 10^{-5} m/(m*deg)*20 deg*6 meter= 1.2 mm. This thermal deformation applied to a structure fixed to the concrete pad will cause buckling, which does not allow achieving the required installation tolerance. Goal was to reduce this effect as much as possible by maintaining the temperature inside the shelter during the 24 h curing period of the grout as stable as possible. As only limited active temperature control was possible through the thermometers of electrical heaters, a great part of the alignment success depended on the experience of the site staff.

Quality Control Verification of the Installation

Regular Inspection of Instruments

Quality Control Verification of the Installation

Regular Inspection of Instruments

Crucial for the high precision installation is the reliable measurement of the total station, which was used at the limits of its accuracy. Despite the high efforts a yearly re-certification was executed at the manufacturer's premises.

Not resulting in a formal certificate but giving a reliable re-calibration was the regular adjustment of the precision water level on a plane surface.

Due to limited usage and high complexity of re-calibrating for the further precision metrology equipment this service was not executed during the 2 years of its usage. However one sensor of the Hydrostatic leveling system had to be sent to the manufacturer after measuring results were in doubt.

Documentation archiving, CIDL

All project documentation was properly archived and logically structured. For redundancy purposes and general access the storage and release process was organized as follows:

During execution of work and before final acceptance of the installation, local copies and a backup on a shared drive were used to save the documents, once released for general access the package was uploaded to the ALMA documentation server. The complete documentation and its configuration control was handled through a Configuration Item Data List (CIDL)

Verification measurements and protocols

To hand over the installation to the ALMA project verification measurements were executed and registered. These protocols, part of the data package, contain the checklist for the correct and complete installation, the geometrical measurement values, and temperature logs for each individual foundation.

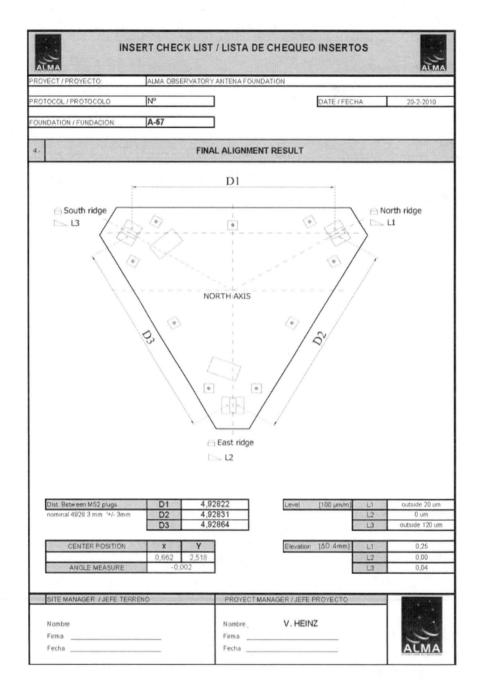

7 Extract from verification protocol

Stiffness tests of whole foundations

One important requirement for the finished antenna foundation is the stiffness, which has a direct impact on the eigenfrequency of the antenna system and thus its performance. To verify compliance a complete test for vertical and lateral stiffness was developed and executed. For this test it was necessary to load the foundation with a load as big as possible and to measure the deflection. This was achieved by loading the foundation with a dummy mass, which was also used to test the antenna transporter and which represents an exact antenna interface with the antenna weight. For the measurement two systems were used, the hydrostatic level and the laser detector. For vertical movements the high precision level has proven to be most suitable. With its accuracy it was sufficiently precise possible to measure the expected movement of 35 micrometer. One reference sensor was installed as far away as possible from the foundation and the other three were connected to the antenna. While lowering down and lifting the dummy the relative movement between reference and foundation sensors, was recorded. Due to the sensitivity of the system it is very susceptible to temperature drifts, which can easily be eliminated from the measurement result, because tests have shown that they are very similar for all sensors

8 Antenna dummy before lowering down on foundation with Sensors installed

The results of various tests had shown that the foundations did not reach the specified stiffness of 13 GN/m. As a consequence of this result the soil was analyzed and The actual soil stiffness was much lower than assumed in the FE analysis. With the new value for the soil stiffness the measured values could be confirmed with the CAD/FEM model.

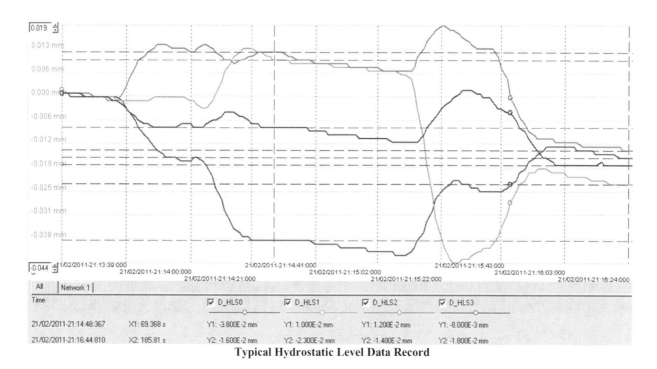

Typical Hydrostatic Level Data Record

Other precision interfaces for Telescopes and their Subsystems

Paranal Unit Telescope tracks

Other precision interfaces for Telescopes and their Subsystems

Paranal Unit Telescope tracks

9 Telescope Azimuth Track with Hydrostatic Bearing

AT plate mounts

10 Paranl Auxiliary Telescope Mounting Plate, Enclosure, Alignment tool

VLT Interferometer Delay Lines and Mirror Mounts

Summary

The outfitting of almost 200 antenna foundations for the ALMA project with antenna fixation and coupling devices had shown that with the correct preparation, procedures, strategies and equipment, very demanding alignment results under extremely severe environmental conditions can be achieved. The experience gained from this project is very valuable for further large astronomical installations. While the design of the antenna interface allows later rectification to reach the tolerance, other systems will not have the same possibility and the initial installation must be carried out with utmost precision to reach the tight tolerances required for opto-mechanical systems. Strict quality control and verification during installation are required to reach the alignment goal. The metrology equipment used for the ALMA foundations and the associated processes can easily be adapted to different interface installations with similar requirements and tolerances.

Lessons learned:

Do not underestimate extreme geographical and climatic conditions; the required time to achieve a good result is significantly longer than at lower altitude and decent temperature. Alignment work, having low and changing temperature and workers wearing thermal protection clothes, becomes much more difficult. All materials and equipment change properties and behavior under the Chajnantor conditions. Sufficient Instrument checks and calibration must be carried out to assure the required precision.

References

[1] Leica datasheet for the Industrial Total Station TDA 5005

[2] Dictuc Test results from the Five Star compression modulus qualification test

E-ELT Telescope Main Structure

Alfredo Orden Martínez, Angel Dilla Martínez, Noelia Ballesteros Pérez, Manuel Alcantud Abellán.
Empresarios Agrupados Internacional, S.A. Magallanes 3, 28015 Madrid, Spain

ABSTRACT

The European Extra Large Telescope is ESO's biggest astronomical telescope project.

The E-ELT is an active and adaptive telescope. It has an astigmatic optical solution (five mirrors, including two flat ones). The telescope structure is of alt-azimuth type able to support a primary mirror with an equivalent diameter of 40 m.

The telescope will be installed in a high-seismicity zone, in Cerro Armazones, Antofagasta Region, Chile, at an altitude of 3046 metres above sea level. This has significantly affected the boundary conditions and safety aspects considered during the project.

The scope of the paper describes the Telescope Main Structure configuration developed by Empresarios Agrupados (Spain) during the FEED Studies performed from June 2009 to July 2011 in the frame of ESO Contracts.

Most of the solutions implemented were extrapolated from existing installations in which Empresarios Agrupados has participated, adjusting for the extra large size of this new telescope.

1. INTRODUCTION

The E-ELT is an active and adaptive 40-m telescope that uses a Laser Guide Star as reference star. It has an astigmatic optical solution (five mirrors including two flat ones) equipped with the following foci:

- A through-focus on each Nasmyth platform.
- Two lateral foci on each Nasmyth platform to which the central section of the field with a diameter of 5 arcminutes is directed by a M6 flat mirror.
- A Coudé focus beneath the telescope to which the central section of the field, with a diameter of 20 arcseconds, is directed by a M6 flat mirror. Only the main structure on the Nasmyth A platform will be equipped with a Coudé path.

The Telescope Main Structure (MS) hosts the following major external units: M1 Unit, M2 Unit, M3 Unit, M4 Unit, M5 Unit, ADC, Calibration Units, PFS, Instrumentation, Coudé Feed and ELGS.

This combination of incredible technological power and gigantic size mean that the E-ELT will be able not only to detect the presence of planets around other stars but also to begin to make images of them, so:

- It will detect earth-size planets and image Jupiter-size ones…
- It will also probe the most distant objects…
- It will make detailed studies of the first galaxies and possibly measure the acceleration of the expansion of the Universe directly…

The E-ELT has the following advantages over other telescopes:

- E-ELT will gather 15 times more light than any other telescope today and take images 15 times sharper
- E-ELT will have several science instruments and it will be possible to switch from one instrument to another within minutes.
- Adaptive mirrors are incorporated into the optics of the telescope to compensate for the fuzziness in the stellar images introduced by atmospheric turbulence.

- The E-ELT will have a novel five-mirror design that no other telescope has
- The telescope and dome will also be able to change positions on the sky and start a new observation in a very short time

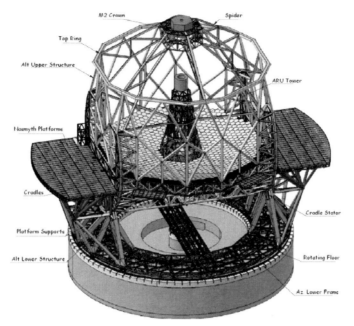

Figure 1: E-ELT Main Structure with Hosted Units

The Main Structure (MS) basically comprises the Azimuth Structure on which the Altitude Structure rests, the drives of both axes, the cable rotators, the M5 Switching Mechanism, the hydrostatic bearings system and the associated hydraulic system, cranes, brakes and limit stops for both axes rest. It also includes all equipment needed to access the telescope.

The MS (Figure 1) contains the telescope structure including all interfaces to the hosted units. It contains the exchange mechanisms for the M3, M5 and M1 segments. It also contains the Nasmyth platforms and the interface to the exchange mechanisms for the instrumentation, as well as all handling tools for exchanging mirrors to/from the main structure.

The site used to develop the FEED study is Cerro Armazones, Region of Antofagasta, Chile. It is located in the Atacama Desert at an altitude of 3046 m above sea level.

2. TELESCOPE MAIN STRUCTURE PERFORMANCES

The performances of the E-ELT Telescope Main Structure are summarised in Table 1:

Table1 - E-ELT MS Performances

Azimuth operational rotating range	+/- 270°
Altitude operational rotating range (operational)	20° to 89°
Altitude operational rotating range (maintenance)	0° to 90°
Altitude operational rotating range (safety)	-1° to 105°
Maximum acceleration (azimuth and altitude)	+/- 0.1 degree.s^{-2}

Maximum deceleration (azimuth and altitude)	+/- 10 degree.s^{-2}
Maximum velocity altitude axis	1 degree.s^{-1}
Maximum velocity azimuth axis	2 degree.s^{-1}
Eigenfrequencies	
Telescope Main Structure Life Time	20 years

3. TELESCOPE STRUCTURE

The telescope structure will be of the alt–azimuth type. It will comprise two main parts, the azimuth structure and the altitude structure, with the altitude axis located above the primary mirror.

The need to find a compromise between a strong, rigid structure that could sustain the optical units with their strict deformation and weight requirements made it necessary to undertake several studies to determine the most appropriate materials for the E-ELT.

After a detailed study of the advantages and disadvantages of different options, the study focussed on the possibility of design and manufacturing the upper part of the Altitude Structure in carbon fibre. Meetings with designers and manufacturers led those involved to conclude that the technical advantages offered by the strength and lightweight nature of this material were overshadowed by this high cost and the challenges involved in integrating it into the rest of the steel structure.

In addition and paradoxically, achieving balance in the Altitude Structure if carbon fibre were used in its upper part would have meant have to counterweight the structure to keep it balanced.

In the end, a structure made entirely of carbon steel (equivalent to S275 and S355) has been proposed. The total weight of the structure's rotating parts, including the hosted units, is 2800 t.

Besides the requirements imposed by the optical units, human accessibility to the machine and the assembly or disassembly of the optical units are other important aspects that have had to be taken into account in the structure design.

Aspects related to transport and final erection at site have also had to be considered in the final configuration adopted, due to the need to optimise the cost of the design and ensure that it could be installed and ready to operate within a tight schedule. The design developed is grounded in the premises that the necessary verifications, checks and tests will be done in shop where it will be easier and more economical to take any corrective action that could be required prior to transport to the site. This approach not only includes the manufacture and control of the assembly but also the preassembly, verification and testing of some of the most crucial subassemblies, including the ARU tower and the M1 Segments Handling Crane, the three Azimuth Rings, both Altitude Cradles and the M1 Cell.

3.1 Azimuth Structure

The Azimuth Structure (Figure 2) weighs about 1300 tonnes (including the hosted units) and supports the Altitude Structure Assembly and the Instrument Platforms. It transmits the reaction loads to the concrete pedestal through the Azimuth Tracks. The Azimuth Structure supports the Optical Instruments, adapters and Coudé train.

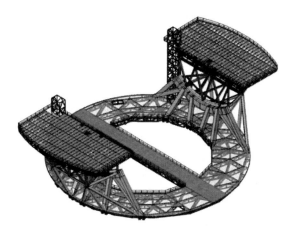

Figure 2. Azimuth Structure

The Azimuth Structure can be broken down into the following parts:

- Lower Frame
- Platform Supports
- Cradle Stator
- Rotating Floor
- Nasmyth Platforms and Auxiliary Platforms

The telescope structure is supported on three azimuth rings (Figure 3). The two biggest rings, the outer and the intermediate ones, of Ø51.5 m and Ø34 m of diameter, respectively, support the main vertical reactions during all operational modes. The horizontal (radial) loads are supported by the intermediate ring. The central or inner ring, Ø4.5 m, supports part of the rotating floor (central corridor).

The upper web of the azimuth rings is the running track for several axial hydrostatic azimuth bearings. In the intermediate azimuth ring the circumferential strip is the journal of the azimuth radial bearings that assures the rotation of the telescope centred in the azimuth axis. In the internal azimuth ring the circumferential strip is the journal of four radial azimuth bearings and assures that the telescope structure is centred with respect to the azimuth axis during erection.

The outer azimuth ring supports the rotor of the azimuth main drives and the azimuth brake disk. The intermediate azimuth ring supports the azimuth encoder tape. It also contains the holes for the azimuth locking pins. The three azimuth rings support the oil trays that collect the oil of the hydrostatic bearings.

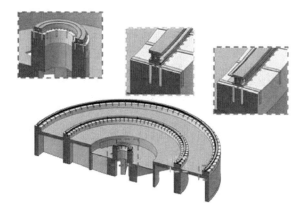

Figure 3 – Azimuth Rings

The design of the Azimuth Structure is conditioned mainly by the requirements linked to the position of the Altitude Structure and the influence of the Rotating Floor on the overall behaviour of the Azimuth assembly.

The Rotating Floor is a surface of more than 2,000 m2 whose central corridor has to support high loads. It transmits MS assembly loads to the foundations. The design of this floor has proven key and has conditions to a large extent the design of the Azimuth Structure, especially as regards the definition of the location of the drives and the Azimuth hydrostatic bearing system.

3.2 Altitude Structure

The total weight of the altitude structure (Figure 4), with the hosted units and mirror cell, is about 1500 tonnes. It is almost entirely assembled from steel tubular sections with a variety of diameters. The structure assembly contains:

- The structural elements of the altitude structure, from the interface with the azimuth structure to the interfaces with the main optomechanical units (M1, M2, M3, M4 and M5), the ADC (if any at intermediate focus), the adaptive optics calibration unit, and the launch telescopes (if required)
- All structural elements supporting the main optomechanical units (M1, M2, M3, M4 and M5), the ADC (if any at intermediate focus), the adaptive optic calibration unit, the LPS and their beam relay (if required). This includes the M5 switching mechanisms and local controls necessary for the change of focal station

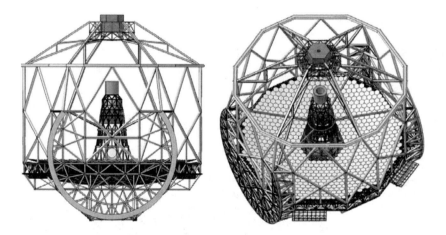

Figure 4 –View of the Complete Altitude Structure

The Altitude Structure breaks down into the following parts:

- M1 Cell
- ARU Tower
- Cradles
- Lower Structure
- Upper Structure
- Top Ring
- Spider
- M2 Crown
- Altitude Auxiliary Platforms

The design of the Adaptive Relay Tower (ARU) (Figure 5) depends largely on the interfaces with the optical mirrors. The interface for the central tower is the primary mirror M1.

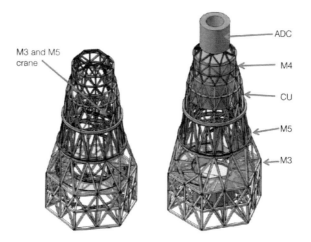

Figure 5 - Adaptive Relay Tower

The two large cradles are supported by radial and lateral hydrostatic bearings. Each cradle has ten locations of identical supports encompassing one radial and two lateral hydraulic pads. The pads are mounted on a common structure and in close proximity to the motor segments. The Altitude Cradle section facilitates integrating the radial and axial bearing in the Main Structure (Figure 6).

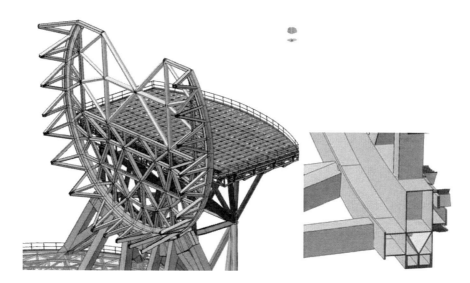

Figure 6 – Altitude Cradle detail

4. TELESCOPE STRUCTURE ANALYSIS

The optimisation of the MS mass, inertia and overall stiffness were key issues that were addressed during the design phase in the analyses. The telescope will be installed in a high-seismicity zone. This significantly affected the boundary conditions and safety aspects considered during the design.

The different loads for the analysis done are:

- Static Analysis – Operational Loads
    - Dead Weight Loads

- The static deflections and the out-of-plane deformation (flatness) of the Nasmyth platform interface flanges are very important because of the precise instruments hosted.
  Mean displacements and rotation are met for Nasmyth Platforms Interfaces.
- The movement of the secondary unit with respect to the primary one, both statically and dynamically, is also a potential major source of wavefront error.
- The minimisation of the deflections of the primary mirror cell is a key metric for the design of the altitude structure.
- The deformation requirements for the hosted unit rigid body and for the hosted unit interface points are met.

o Wind Loads

- A deployable windscreen protects telescope primary and other mirrors, other than the secondary one, from high wind speeds and from direct exposure to the wind, while many louvers increase the ventilation of the internal volume.
- The analyses against wind loads have been developed. The wind forces are applied to the structure based on the altitude angle of the telescope, on the area of the optical units, on the position and height of the wind screen and on the type of beam of the structure.
- Maximum displacement requirements of the optical units and precise instrumentation have been checked for each altitude angle of telescope.
- The analysed configuration behaves satisfactorily under wind loads. The rigid body displacement and rotation of the M2 Unit under static wind load operational conditions for all altitude angle positions between 90° and 20° are met

o Thermal Loads (-10°C to 15°C) – Gradient along axes

- The effects of differential temperatures on the structure were analysed. Two cases were checked, one with a gradient of around 3°C along the Y-axis and another one with a gradient of around 6°C along the Z-axis.
- The telescope structural stresses caused by thermal gradient loads are low stress values

o Local Analysis.

Some parts of the telescope require detailed, local analysis of FEMs, for example:

- Altitude and azimuth cradles, influence of the dead weight deformation of the structure in the cradle areas where the bearings are located.
  The structure behaves well with respect to cradle deformations due to the distribution of the weight of the altitude structure on the bearing pads
- Azimuth Ring. The performance of the telescope pier has been analysed using detailed finite element modeling to establish its impact on the eigenfrequencies of the telescope main structure and its ability to withstand the various load cases.
- Altitude and azimuth drive supports. In order to analyze the behaviour of the drive supports, it two detailed FEM models were developed with ANSYS (one for the altitude drive supports and the other for the azimuth ones).

These models were used to determine the values of the local stiffness in the structure that holds the stator of the altitude motors and the azimuth motors.

- o Dynamic Analysis

Modal analyses have been performed with blocked and free motors for any altitude angle position at 0°, 20°, 45°, 60° and 90°each.

The first eigenmode for locked rotor is found at 2.9 Hz with the telescope pointing to zenith, and drops to 2.5 Hz for horizon pointing. The dynamic frequencies of the Main Telescope Structure are higher than the required values.

Some frequencies are critical for the design:

- The dynamic behaviour of the platforms forms part of the interface to instrumentation
- Dynamic behaviour of the spider is also critical because M2 is mounted on it.
- Dynamic behaviour of the ARU tower due to the hosted units.

- Buckling analyses and fatigue analyses. High safety factors are obtained.
- Seismic Events - Survival Loads and Stresses

The maximum structure stresses are obtained in local areas given mainly by the earthquake, No-Collapse Earthquake or Damage Limitation Earthquake load events, acting simultaneously with the gravity load event.

5. MAIN DRIVES

The E-ELT main drives will use direct drive technology. The drive system of such a large structure must necessarily be based on distributed segmented permanent magnet linear motors. The motors, developed by Phase Motion Control (Italy), use FeNdB-type magnets suitably protected against shock and corrosion.

This enormous machine can only be controlled by placing the actuation points in the stiffest positions of the structure and as close as possible to structure nodes, where the maximum inertial loads will be encountered. In this sense, motor location is a consequence of structure design.

Placing motors close to hydrostatic bearings allows a fairly constant air-gap to be maintained despite the dimensions of the structure. This allows the motor to be designed with a relatively small air-gap and consequently higher efficiency for a defined motor dimension. Moving far from these locations would make control impossible and require a larger motor to keep the same torque and efficiency.

The current design considers the implementation of the Azimuth Motor over the Ø51.5 m outer azimuth ring. For the altitude motor there is only one possible location: at the edge of both altitude structure cradles. The altitude axis torque demand and the location of the motor close to the encoder tape support this decision.

For the azimuth axis, an axial single-gap configuration appears to be the most suitable option. Although a dual-gap configuration allows the axial force unbalance to be minimised, a single gap configuration allows significant cost saving in such a large structure and solves contact problems in case of an earthquake. The motor sectors must be located very close to the axial bearing pads in order to control the air-gap between motor sections and magnet segments. It is also essential to maintain a stiff connection between force generators (motor sections) and the main nodes of the structure.

A similar axial gap approach is followed for the altitude structure. For this application, a balanced, single radial air gap motor, with a rotor made of relatively thin cylindrical sections, and segmented, curved linear motors on either side, will be used. This topology is indifferent to axial movement and pivotal rotations around the centre of the motor, typical of a large, flexible structure, and is therefore ideally suited to altitude drives for ELT. Additionally, motors must be located close to bearing pads in this case also, for the same reasons as set out above.

From the power electronic point of view, the drive system of the E-ELT will be based on the latest technology with the drive and control units embedded into each drive segment (two drives per azimuth and one per altitude motor section). This dramatically simplifies the wiring as the same power and field bus wire connects all motor segments. The embedded power drives are based on Phase Motion Control production units with a peak power of 5 kW. The LCU controls all the units over an Ether Cat field-bus line.

The motor environmental protection was designed to be compliant with protection degree IP54 for the use of a telescope with dome.

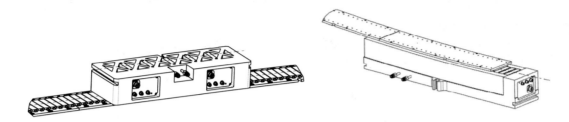

Figure 7 – Azimuth and Altitude Drive Segment and Rotor

Table 2 E-ELT Motors Performances

Parameter	Azimuth motor	Altitude motor
Peak Torque SIR 3% (max. 30s)	3,769,200 Nm	2,104,000 Nm
Number of single gap stator segments	24 (48 drives)	16
Single gap section mass	76 kg	169 kg
Air gap, mechanical	2.9 mm	2.9 mm
Average rotor diameter	50,760 mm	28,200 mm
Expected cogging torque	±5000 Nm	±2500 Nm
Expected torque ripple	37,692 Nm	21,040 Nm

6. ENCODER SYSTEM

6.1 Encoder configuration

The main axis encoder system will be based on high-accuracy incremental optical encoder technology. The current configuration of the encoder system is provided by Heidenhain based on ERA scanning heads (AK ERA 8480), AK ERA 8400C tapes and EIB 749 interpolation electronics.

Applicable requirements concerning the encoder system are related to the overall control system performances. The jitter measured in the encoder system shall be ≤0.02 arcsec without any disturbance and ≤0.3 arcsec considering wind disturbance.

The azimuth encoder configuration is based on a 360° circular ERA 8400C tape located in a slot machined in an external radial surface of the 34,500 mm diameter intermediate azimuth ring. It includes four AK ERA 8480 scanning heads and two EIB 749 electronic boxes, supported on the azimuth structure.

The dimensions of the azimuth ring make it necessary to install the azimuth tape along the external cylindrical surface of the ring. The tape shall be pulled tight by a dedicated system to ensure equal tensioning along the complete length of the tape.

Table 3 Azimuth encoder system configuration

Tape	Full circle
Slot diameter	34,507.94 – 1.725 mm
Tape mounting	Outside (tensioned)
Number of scanning heads	2 pairs separated by 60°
Nominal reference marks spacing	5000 Signal Periods
Number of signal periods	2,710,000
Signal period	0.4782 arcsec
Measuring step	0.1168 marcsec

The altitude encoder configuration is based on one 110° ERA 840xC tape segment per side, located in an external cylindrical surface of the altitude cradles at a radius of approximately 13,475 mm, two scanning heads per side and two EIB 749 electronic boxes.

The encoder slot is machined in the altitude cradles. The scanning heads are fixed to the azimuth structure.

Non-equidistantly-distributed-over-the-tape scanning heads are treated as one scanning head for error budget analysis because low frequency errors are not reduced by this configuration. Increasing the number of scanning heads could have a positive impact on the reduction of high-frequency errors.

Table 4. Altitude encoder system configuration

Tape	Segment
Number of tapes	One per side
Slot diameter	26,915.04 – 1.346 mm
Tape mounting	Outside
Number of scanning heads	Two per side
Nominal reference marks spacing	2000 Signal Periods
Number of signal periods over 360°	2,113,700
Signal period	0.613 arcsec
Measuring step	0.150 marcsec

6.2 Scanning Head Flexible Mountings

The azimuth encoder head will be mounted on a flexible device to account for the potential imperfections of the azimuth tape slot and the differential thermal expansion between the azimuth ring and the scanning head supporting structure. The flexible mounting ensures that the nominal gap of 0.75 mm between the scanning head and the tape is respected. As regards the azimuth axes the main concern is radial direction, so the flexible mounting will be adaptable in that direction.

The proposed azimuth scanning head flexible support will be basically composed of:

- An encoder base support where the encoder head is installed. The encoder base support has free axial movement given by two lateral recirculating ball-bearing guides. The two recirculating ball-bearing guides are attached to the rotating azimuth structure by one stiff structure
- A steel wheel installed in the front part of the encoder base support. This wheel is attached to the encoder base support through a central ball bearing
- A spring that applies a constant axial force of around F=20-30 N is installed in the back of the encoder base support,
- The wheel will contact tracks that will have to be machined at the same time as the tape slot and thus have the same potential imperfections
- A first error estimation of this device could be evaluated considering the preload of the spring, the radius where the tape is mounted, R=17.25 m, and the friction coefficient assumed (<0.005).

Detailed design should focus on providing a circumferential stiffness for the assembly better than K>1E+7 N/m. Based on these assumptions, the error estimate is 0.18 marc sec.

Comparing this value with the measuring step (0.11 marcsec), the error estimate for the conceptual concept described is well below one single signal period of the tape.

In a similar approach to the flexible mounting of the azimuth scanning head, the altitude one should account for radial variations of the distance between the scanning head and the tape and for axial variations as well.

Altitude encoder tapes will be installed on two slots machined in the cradles. Both cradles experiment certain out of plane deformations due, among other things, to the variation in the inertia of the altitude structure as it rotates from 0° to 90°.

This out of plane displacement is limited by the altitude lateral hydrostatic bearing to ±3 mm.

As a result, the conceptual design of the encoder altitude flexible mounting is similar to that proposed for the azimuth one but includes some features to enable the possibility of adjustment in the ±X direction. The design will be basically composed of a horizontal encoder base support where the encoder head is installed.

The horizontal encoder base support has an axial movement given by two lateral recirculating ball-bearing guides. The two recirculating ball-bearing guides are attached to one vertical axial support which is attached to the azimuth structure by two more recirculating ball-bearing guides, and by one stiff supporting assembly.

One horizontal and one vertical steel wheels are installed in the encoder base support. These wheels are responsible for maintaining the radial and lateral head-tape distance within tolerances. These wheels are attached to the encoder base support through a central ball bearing of around 20 mm in diameter.

A metallic spring that applies a constant axial force of around F=20 to 30 N is installed at the back of each of these two base supports

Similarly to what is assumed for azimuth, an error estimate has been calculated considering a circumferential stiffness K>0.5E+7 N/m and a friction coefficient of around 0.005 for the contact between the wheels and the tracks. As for the azimuth, these wheels roll over two tracks machined in the Cradles.

Considering a tape radius of 13.5 m, the error estimate is 0.9 marc sec.

The error budgets (tape joints, runout, interpolation, graduation tape, etc) for the azimuth and altitude configuration are shown in figure 8.

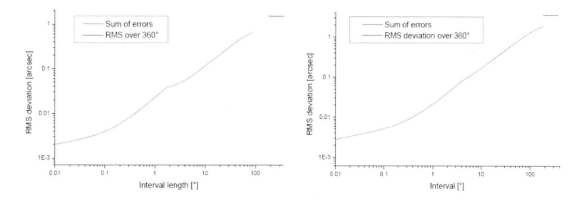

Figure 8 – Azimuth and Altitude error budget

7. HYDROSTATIC BEARINGS AND OIL SUPPLY SYSTEM

7.1 Hydrostatic Bearings

The purpose of the Hydrostatic Bearing System (Tomelleri s.r.l) is to support and guide the movement of the two main axes of the E-ELT, yielding the required performances (high dynamic stiffness, enough static stiffness, low friction, good damping, no stick slip and no hysteresis, low power consumption, high reliability availability and maintainability). The bearings must have enough tilting and floating capability in relation to the tolerances of the surface of the tracks and of the cradles, and be able to accept the loads in case of earthquake.

It is considered as a priority to spread the loads out uniformly over and through the various parts of the MS. It is vitally important to limit the deformations in the areas that receive these loads not only to assure that the hydrostatic bearings behave correctly but also to limit the effects these deformations may have on the operation of other systems such as the Main Drives or the Encoders.

In this sense it is necessary to discuss certain statements that strongly constrain the bearing system design and that have to be taken into account when defining the functionality of each type of bearing.

- The need to define accurately a centre of rotation of the MS during the erection at Site but that should serve as a permanent reference for performing any adjustment needed during the life of the Telescope like, for example, that necessary after a seismic event.

- Radial thermal expansion of the Main Structure that shall be absorbed by the hydrostatic bearing system and/or by stressing the Main Structure itself.

- Imperfections of the axial and radial tracks of both axes considering that manufacturing of these surfaces is quite demanding and that the erection at Site, no matter what the shop test results may have highlighted, will be complicated and could introduce additional sources of problems.

- Operational and extraordinary loads, like seismic ones, are quite different, and this makes it necessary to reach a compromise between the operational performances of every pad and their survival condition. Therefore, it is necessary to determine whether the bearings are being designed more to withstand an earthquake, thus penalizing their ordinary operational functionality.

The Telescope is supported over Axial hydrostatic bearings placed in 36 positions under the three azimuth tracks, and Radial hydrostatic bearings placed in 28 positions (Figure 9).

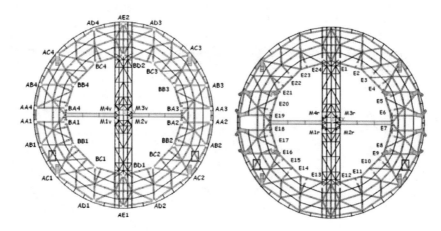

Figure 9- Azimuth bearings (Axial and Radial) layout

The axial pads are concentrated around the most loaded areas (below Nasmyth platforms — positions AA, AB, BA, AC and BB — with double pad supports). The other locations (positions BC, AD and AE — with single pad supports) have lower contributions.

The 34-metre diameter track is used for the radial constraint using 24 supports of equally distributed pads (typically single ones). Four radial bearings in the central 4-metre ring support are used to define the azimuth axis during the on-field machining phase for the inner and outer tracks. High stiffness is necessary not only for operation, but also because of manufacturing and installation issues.

The Altitude structure is supported over radial and lateral bearings placed in 10 positions, in each Cradle (Figure 10).

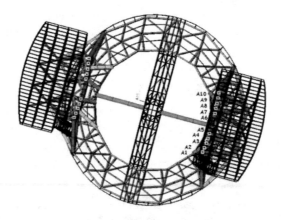

Figure 10 - Altitude bearings (Radial and Lateral) layout

7.2 Oil supply system (OSS)

The purpose of the hydraulic system is to provide an adequate flow of oil, so that the MS HSBs and the bearing track surfaces are always fully separated from each other by an oil film of approximately 70 μm. This oil film should be regulated as far as possible so that a constant film thickness is achieved at different operating temperatures. Vibrations and pulsations in the hydraulic system shall be minimized to the extent possible.

The OSS system will supply oil at enough viscosity, pressure and temperature to each of the telescope hydrostatic bearings. The system will allow the independent motion of each of these assemblies.

Three more systems are directly linked to the OSS: the oil cooling system, the oil recovery system and the OSS control system. The first one cools the oil so that its temperature at the inlet of each bearing is adequate. In addition, the oil is collected and re-channelled to the tank from where it will be pumped. The oil recovery system is used for this purpose. Finally, the OSS control system will supply and control all the OSS electrical components, such as motors, electro valves, electromagnets, etc.

The environmental operating conditions are considered to range from between -5°C (possibly -10°C in the most extreme case) to +15°C. The telescope structure temperature may differ from the ambient temperature by +/- 1.5°C and +/- 2°C (for night and day, respectively).

The thermal requirements are one of the main design constraints for this system, since the operation of the bearings is highly sensitive to the thermal conditions of the oil.

In order to ensure precise operation of the Telescope, oil will exit the bearings at +/-1°C with respect to the ambient temperature. Furthermore, the oil can undergo a 10 °C heatup due to friction as it crosses the bearing, which means that a cooling system must be designed to maintain the temperature of the oil at the bearing inlet around 10°C below the outlet temperature.

Taking into account the considered range of ambient temperature, the system will have to be capable of cooling the oil down to a temperature of around -15°C in the most extreme case.

Considering these premises, it is reasonable to envisage the following OSS breakdown structure:

- Oil tank

    - The oil tank has a capacity of 12,000 litres.

    - Recirculating compartment. One pump +heat exchanger to cool and filter the oil

    - Feed pump compartment. Three + one feeding pumps supply the oil to the hydrostatic bearings (500 l/min)

- Accumulators

    - Four accumulators, in two redundant groups, with 50 litres each one with a total volume of 200 litres, preloaded at about 80 bar)

- Electrical/control system

    - The OSS and all electrical boxes are connected to the OSS Control System based on a SIMATIC PLC for real-time control and monitoring of all HBS and supervising all functions.

Testing, characterization, and control of a multi-axis, high precision drive system for the Hobby-Eberly Telescope Wide Field Upgrade

Ian M. Soukup[b], Joseph H. Beno[b], Gary J. Hill[a], John M. Good[a], Charles E. Penny[b], Timothy A. Beets[b], Jorge D. Esguerra[b], Richard J. Hayes[b], James T. Heisler[b], Joseph J. Zierer[b], Gregory A. Wedeking[b], Michael S. Worthington[b], Douglas R. Wardell[b], John A. Booth[a], Mark E. Cornell[a], Marc D. Rafal[a]

[a]The University of Texas McDonald Observatory, 1 University Station C1402, Austin, Texas, USA 78712-0259
[b]The University of Texas Center for Electromechanics, 1 University Station R7000, Austin, Texas USA 78712

ABSTRACT

A multi-axis, high precision drive system has been designed and developed for the Wide Field Upgrade to the Hobby-Eberly Telescope at McDonald Observatory. Design, performance and controls details will be of interest to designers of large scale, high precision robotic motion devices. The drive system positions the 20-ton star tracker to a precision of less than 5 microns along each axis and is capable of 4 meters of X/Y travel, 0.3 meters of hexapod actuator travel, and 46 degrees of rho rotation. The positioning accuracy of the new drive system is achieved through the use of high-precision drive hardware in addition to a meticulously tuned high-precision controller. A comprehensive understanding of the drive structure, disturbances, and drive behavior was necessary to develop the high-precision controller. Thorough testing has characterized manufacture defects, structural deflections, sensor error, and other parametric uncertainty. Positioning control through predictive algorithms that analytically compensate for measured disturbances has been developed as a result of drive testing and characterization. The drive structure and drive dynamics are described as well as key results discovered from testing and modeling. Controller techniques and development of the predictive algorithms are discussed. Performance results are included, illustrating recent performance of several axes of the drive system. This paper describes testing that occurred at the Center for Electromechanics in Austin Texas.

Keywords: Center for Electromechanics, CEM, Hobby-Eberly Telescope, HET, HETDEX, University of Texas, McDonald Observatory, Telescope tracking systems, Large hexapod positioning systems, High payload hexapod positioning systems, precision positioning systems, hexapod design and controls

1. INTRODUCTION

The Hobby Eberly Telescope, designed to gather large amounts of light for spectroscopy, is undergoing a transformation to begin the search for "Dark Energy". Dark energy is a term used to describe unknown forces causing the acceleration of cosmic expansion[1,2]. To help define the evolution and significant properties of dark energy, the Hobby Eberly Telescope (HET) at the McDonald Observatory (MDO) was chosen to survey and map the locations of other galaxies in space. This Dark Energy Experiment (DEX), which is estimated to be completed over 100 nights of viewing, is now referred to as HETDEX[3]. The success of HETDEX is highly dependent on the dynamic positioning performance of the upgraded telescope tracking system. Disturbances, noise, drive structure uncertainties and random drive behavior can all contribute to performance degradation. The HETDEX tracker has the added complexity of positioning a large payload, 20 tons, and traversing a large range or travel: 4 meters of X/Y travel, 0.3 meters of hexapod actuator travel, and 46 degrees of rho rotation[4]. The controller development focused on system identification and error compensation that evolved from extensive testing. The testing followed a laddered approach where root level tuning and error compensation algorithms were implemented first through individual actuator or independent axis testing, followed by progressive grouping of subsystems or elements. This approach created layered error compensation algorithms that were executed serially during motion control. Two parallel test paths were implemented in order to reduce overall schedule length; one test path focused on the X/Y system, while the other focused on the hexapod, **Figure 1** and **Figure 2**. Examples of tuning, error compensation algorithms, system identification, and other significant test results will be discussed for each subsystem.

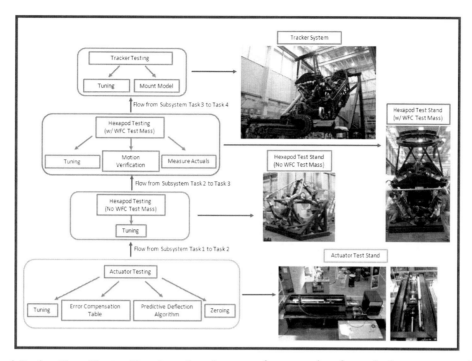

Figure 1. Hexapod Testing Flow Chart – The chart plots the upward progression from single actuator testing to the fully assembled tracker testing phase.

The images illustrate the flow of testing from the more basic subsystems to the fully assembled system. The lowest subsystems highlighted in the green boxes were the initial test cases for the hexapod and X/Y systems. The tests progressed for each system until the tracker was fully assembled; at which time the two paths merged to begin the "Tracker Testing" phase. Each subsystem test is identified as a sequentially number "Task" in Figure 1 and Figure 2. The "Tasks" contain the identifier for each test phase (e.g. "Actuator Testing") followed by the achievements, identified by the red arrows, that flowed from the completion of one task to the start of another.

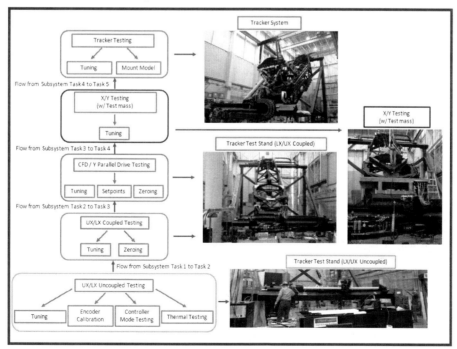

Figure 2. X/Y System Testing Flow Chart – These tests progressed in parallel with the hexapod testing to reduce schedule length

2. HEXAPOD TESTING

The hexapod actuators underwent extensive testing via a single actuator test stand. The single actuator test stand was configurable to apply four loading conditions to the actuators representing their fully assembled load states: 17-25kN compression, 4-7 kN compression, 3-5 kN tension and no load[5]. The single actuator test stand provided an ideal test space for characering each actuator in an isolated environment. Major accomplishments include 1) actuator performance tuning and 2) implementation of sensor error correction tables and algorithms.

2.1 Actuator Tuning and Performance – Single Actuator Test Stand

The hexapod actuator controller used a position to velocity cascaded control loop with a 1 kHz controller cycle time. The Danaher Motion AC servo drives were programmed to torque mode resulting in analog current commands being sent to each actuator motor. Feedback was obtained from an absolute Heidenhain EnDAT encoder with 32 bit resolution and 0.5 micron accuracy grade[6]. Controller gain tuning was performed on each actuator resulting in 0.75 to 1 microns RMS "following error" at 1 mm/s as measured by the dSPACE controller hardware, Figure 3. "Following error" is defined as the command position subtracted from the sensor position for each control cycle.

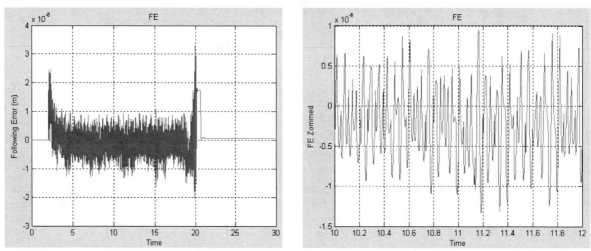

Figure 3. The plots illustrate actuator following error measurements made by the dSPACE controller. These plots were recorded from an actuator that was undergoing testing in the single actuator test stand.

2.2 Zeroing, Error Compensation, and Predictive Deflection Algorithms – Single Actuator Test Stand

Calibration of the actuator feedback device, Heidenhain encoder, began by defining the zero positions of each actuator. This was achieved by measuring the distance between the pin locations in our solid model for the actuators at their home position and comparing this to the pin distance measurements made by an API laser tracker for each actuator in the test stand. The difference between the model and the actual measurements generated the offsets that were applied to each actuator sensor. Limit switch positions were located and adjusted to design specifications as needed once the actuators' zero positions were defined.

Error compensation tables were developed for each actuator to adjust for sensor error as a function of actuator displacement. It was determined that in order to achieve the highest accuracy measurements from the laser tracker, precise alignment of the laser tracker to the actuator's axis of motion was required. This minimized the use of the encoders in the laser tracker and therefore the distance calculated by the laser tracker became largely a function of interferometry only. Sensor data and laser data was gathered by moving the actuator from one end of travel to the other (totaled 295 mm) while pausing every 1mm. During the pauses, measurements were automatically made and recorded by both the laser tracker and the dSPACE controller hardware. These measurements were repeated for each actuator at each load condition. Data from the measurements was used to calculate the error associated with the encoders as defined in equation 1.

$$error = laser - encoder \tag{1}$$

Error plots were generated as a function of actuator displacement for each load case, **Figure 4**. Repeatability was examined by taking several data sets for each load condition as shown in **Figure 4**.

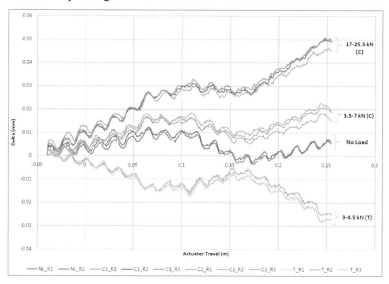

Figure 4. Actuator sensor error was recorded as illustrated in this figure for each load condition. Multiple runs are shown for repeatability verification.

The data shows very similar characteristics for each load case and confirms a high level of repeatability. By inspection each error curve demonstrated a low frequency roll with a superimposed higher frequency oscillation. The other imperative observation is that the slope of the data sets for each load condition is a function of actuator deflection. Deflection occurred due to the mechanical stiffness of the actuator reflecting the increase in load from one end of travel to the other.

Two methods of error compensation were evaluated to correct for the encoder error. 1) A 6th order polynomial curve fit with superimposed sine wave was generated, which resulted in a maximum 13 microns of error when measured against the averaged no load data sets. 2) A linear correction table function, 1-D lookup table generated in SimuLink, resulted in a maximum 6 microns of error when measured against the averaged no load data sets. Based on improved matching the 1-D lookup table was chosen over the polynomial curve fit.

Predictive deflection algorithms were developed in order to eliminate the error associated with the deflection due to mechanical stiffness. The stiffness of each actuator was calculated from the slope of the error curves in **Figure 5**; the delta was calculated using equation 2.

$$\delta = \frac{\Delta F}{k_{act}} \qquad (2)$$

The delta function was subtracted from the error curves, for each load case, resulting in a rotational shift of the curves so that the loaded conditions matched the no load condition, **Figure 5**.

The measured error, red line, is first shifted using the predictive stiffness algorithm which generated the blue line. The purple line is a no-load error measurement and shows a close match to the shifted curve; the purple line is for illustrative purposes and was used to generate the linear correction table. Next the linear correction table function takes the data from the blue line and generates the orange line, which is the final corrected error curve of the sensor. The error was reduced from over 40 microns to less than 5 microns.

2.3 Tuning, Measuring Actuals, and Motion Verification – Hexapod Test Stand

The hexapod was assembled onto the hexapod test stand, shown in **Figure 6**, with the addition of a Wide Field Corrector

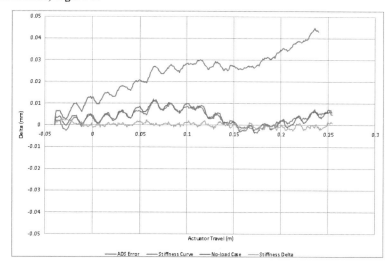

Figure 5. Shown is an example of a stiffness correction applied to the red line to generate the blue line The linear correction table then generates the orange line.

[WFC] test mass that had similar inertial and geometric properties to the new WFC[5]. The hexapod test stand not only enabled parallel testing of the X/Y system but provided a more manageable environment for testing and measuring.

Measurement features swept by the laser tracker were much more accessible in the test stand and therefore prevented fabrication of special tooling that would be required for making measurements with the hexapod installed on the tracker. Hexapod testing accomplished the following: performance tuning, motion verification, and measurement of as-built geometry (e.g. U-Joint Rotation Centers, Upper / Lower U-Joint Planes, and Upper / Lower U-joint Circle Centers).

"Following error" plots similar to **Figure 3** were generated for each actuator with command velocities ranging from 0.05mm/s to 3mm/s. The controller gains for the actuators were tweaked until the following error performance was less than 3 micron RMS for each actuator.

The laser tracker measured the rotation centers of the actuators, defined the two planes located at the upper and lower U-Joint centers, and positioned targets at the predicted WFC Control Point and Stationary-Image Rotation Point [SIRP]. These measurements generated new actuator offsets that replaced the original actuator offsets defining the actuator zero position. The locations of the U-Joint centers and U-Joint planes were entered into the hexapod motion controller so that the calculated actuator lengths based on the orientation of the upper-U joint plane would have higher fidelity.

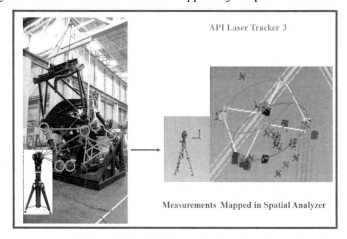

Figure 6. The images illustrate measurement points with the API Laser Tracker 3 and recording the points in Spatial Analyzer. With hexapod conveniently located on the hexapod test stand over two dozen features were measured to accurately define as built geometry and predict the WFC control point and SIRP.

Motion verification tests were performed to verify that motion control algorithms calculated valid commands and that the desired motion envelope was within the mechanical limitations of hexapod components. According to calculations there were six pre-determined conditions that resulted in the largest U-Joint articulation. These conditions were calculated through an iterative parametric analysis that utilized hexapod SolidWorks models and calculated component motion[7]. The six conditions were tested and visual inspection concluded that the U-Joint articulation abilities were more than adequate to achieve the desired motion envelope. To validate the motion control algorithms the laser tracker was used to measure the position of the hexapod given an X,Y,Z,Theta,Phi command. Measurement targets were strategically located to provide translation and rotation results of the WFC test mass SIRP. Initial tests indicated a delta of less than 20 arc seconds between the laser and the dSPACE command.

3. X/Y SYSTEM TESTING

X/Y independent testing was critical in discovering controller weaknesses, feedback sensor limitations, structural resonances, temperature dependences, torque requirements, and load sharing characteristics of the Y/CFD parallel drive operation. These tests drove design changes for the controller, feedback sensors, motor/gearhead combinations, grease lubrication, and defined the parallel drive set points.

3.1 Initial Testing and Tuning

Initial installation of the X/Y system required setting the linear encoders for each axis and calibrating the tape tension to achieve the desired X/Y position feedback accuracy from the encoder. This was accomplished by mounting the laser tracker so that it measured along the axis of motion; each axis was tensioned independently with different laser setups (i.e. 3 setups were required: lower X [LX], upper X [UX], and Y). The X axis trolleys and Y axis carriage were positioned from positive stop to negative stop; encoder measurements were recorded by the dSPACE controller and laser tracker measurements were recorded at each stop. The tape tension was adjusted for each axis until the measurements reported by the dSPACE controller were within 10 microns of the laser tracker readings.

At first the X axis controller used a position to velocity cascaded control loop with a 1 kHz controller cycle time similar to the controller for the hexapod actuators[6]. The Danaher Motion AC servo drives were programmed to torque mode resulting in analog current commands being sent to each actuator motor. Feedback was obtained from an incremental Heidenhain linear encoder with 0.3 micron resolution and +/- 5 micron accuracy grade. It was quickly determined that the following error for the Upper X axis [UX] and Lower X axis [LX] was substantially larger, measuring 15 to 20 um peak to peak, than the measurements for the hexapod, **Figure 7**.

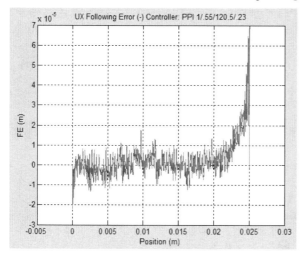 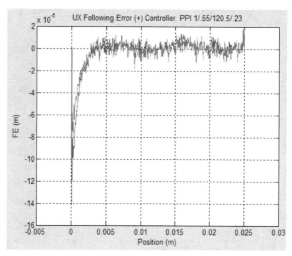

Figure 7. UX/LX Following error plots in torque mode show larger following error than what was achieved by the hexapod actuators.

Testing indicated that the larger following error was a result of decreased system bandwidth due to the following factors: reduced system stiffness, lower feedback resolution, and noisier feedback sensors. To address these issues the LX/UX and Y servo drives were configured to operate in velocity mode. A stiffer dynamic system was produced by allowing the drive to use feedback from a sensor at the back of the motor for velocity control while the linear encoder at the other end of the drive system closed only the position loop. To further enhance the controller performance motors with digital Heidenhain encoders were used to replace the existing motors that had resolver feedback. The new sensor improved sensor resolution and reduced sensor noise producing a better performing controller. The following error performance increased by a factor of two, reducing the error from 15-20 microns to less than 10 microns, **Figure 8**.

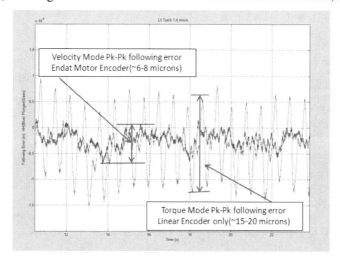

Figure 8. Image shows the increased performance with velocity control and Heidenhain digital motor encoders. The blue line illustrates the new performance with only 6-8 micron peak to peak as compared to the original following error traced by the green line.

3.2 Constant Force Drive/Y Drive Parallel Testing

Testing and tuning the parallel drive systems, Y drive and Constant Force Drive [CFD], was also performed before the tracker assembly was completed. The CFD operates as the active safety system for the tracker while the Y drive positions the SIRP along the Y axis of the primary mirror[8]. The CFD employs an analog controller to represent Proportional Integral Derivative [PID] control to the wire rope tension based on load cell feedback[8]. The Y drive is controlled with the equivalent control method as described for the X axes. It was determined that there was a delicate balance between the control shared by these two systems. Several scenarios were tested: 1) open loop CFD control that provided constant current to the CFD drive 2) high load percentage carried by the CFD with PID control and 3) low load percentage carried by the CFD with PID control. Testing showed that running the CFD with open loop constant current control generated very large spikes in following error. Although the motor received a constant current command the force felt by the carriage and Y-drive was erratic. It is believed that the force felt by the Y-drive varied due to the following two reasons:
- Stiction - the CFD is a large drive system including the largest Danaher AKM series servo motor, AKM74P, a hefty industrial 140:1 reduction bevel-helical gearbox (Renold Brand HC Series), as well as a large 10 inch diameter steel drum with overhead lift rated bearing units
- Inertia- the large system also possesses high rotational inertia which resulted in fluctuating inertial forces as the carriage accelerated to constant speed.

After discovering the inability to operate the CFD with open loop control, tests were conducted to determine the influences on performance while operating the CFD under PID control. Of particular interest was the relationship between CFD load set point and following error performance. Based on following error performance plots, the higher the load carried by the CFD the more it adversely influenced the Y-Drive performance. The higher load cases gave the CFD more force authority causing its control loop to overpower the Y-Drive controller and produced increased following error disturbances. At low set points the CFD still maintains all of its safety functionality, assuming a minimum value is maintained that provides tension to the cables, therefore a low load share of 10% or 5kN was chosen as the final set point. The effect of four different CFD load set points on Y drive following error performance is illustrated in **Figure 9**.

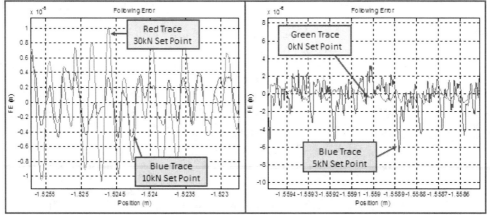

Figure 9. The Y-drive following error achieves smaller variation, increasing performance, with lower CFD set points.

3.3 System Identification

System identification models were compared against the empirical Danaher drive bode plots. The empirical bode plots were generated through response measurements during an automated frequency sweep. The goals of the models were to identify the components driving the lower frequency modes and to develop a design tool that could evaluate new drive system changes that were required to increase torque. Component FEA and catalogue stiffness values were used to accurately define the parameters in the rotational dynamic model. Given the different diameter changes along the screw the FEA models efficiently and accurately calculated the torsional stiffness of the screw with several different tracker positions, **Figure 10**.

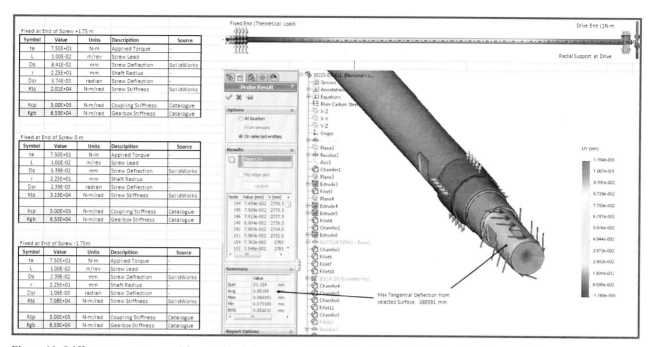

Figure 10. Stiffness parameters used for rotational dynamic model including FEA calculated torsional screw stiffness.

The transfer function defining the angular displacement of the motor to the input torque of the motor was computed using the parameters for the X and Y drive systems. The transfer function bode plots were generated using MatLab, **Figure 11**, and compared to the empirical results, **Figure 12**. The models and drives results correlated well providing validation of the calculated results. The validated models were used as a performance evaluation tool to weight changes to the drive systems to increase drive torque. Essentially, gearhead and motor changes were evaluated using the mathematical models before selections were made.

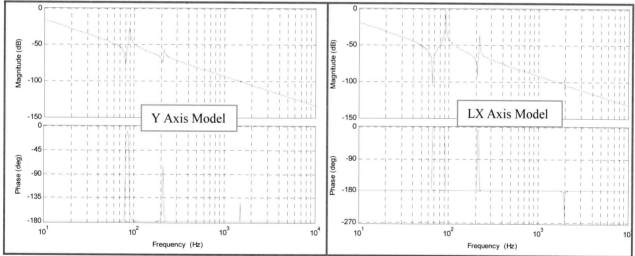

Figure 11. Y axis and LX axis bode plots generated from mathematical model of system.

Natural frequencies were recorded from the mathematical models and tabulated for comparison, **Table 1**. The Y axis evaluation cases were: case 1 – 20:1 Kollmorgen Planetary Gearhead, case 2 – 35:1 Kollmorgen Planetary Gearhead, case 3 – 100:1 Harmonic Drive Reducer. Case 1 is the validation case, but was determined to be undersized for the system torque requirements. Case 2 and Case 3 were two options for increasing the Y axes drive torque to acceptable levels. From the models it was determined that the harmonic drive system, Case 3, would yield the stiffest system, ω_1=96 Hz, however due to mechanical constraints the increase in stiffness was not substantial enough to warrant choosing the harmonic drive over the 35:1 Kollmorgen Planetary Gearhead, ω_1=81.2 Hz – Case 2.

Table 1. Evaluation list of higher torque options based on model derived natural frequencies

Y Axis Model							Model		
	Parameters Reflected to Motor								
	Ks (Nm/rad)	Kgb (Nm/rad)	Kc (Nm/rad)	Jgb (Kgm2)	GR	Mass (kg)	W1 (Hz)	W2 (Hz)	W3 (Hz)
Case 1	78.25	189.94	1250.00	2.60E-04	20	7711.39	89.50	214.80	1500.00
Case 2	25.55	53.32	408.16	2.50E-04	35	7711.39	81.20	203.70	1489.70
Case 3	3.13	42.00	50.00	2.10E-04	100	7711.39	96.00	337.40	1846.20
LX Axis Model							Model		
	Parameters Reflected to Motor								
	Ks (Nm/rad)	Kgb (Nm/rad)	Kc (Nm/rad)	Jgb (Kgm2)	GR	Mass (kg)	W1 (Hz)	W2 (Hz)	W3 (Hz)
Case 1	247.00	495.03	1100.00	9.00E-05	10	11000	92.31	218.04	1957.61
Case 2	24700.00	-	110000.00	-	-	11000	155.18	1215.94	-

The X axis evaluation cases were: case 1 – 10:1 Kollmorgen Planetary Gearhead, case 2 – no gearhead. Case 1 is the validation case and shows good correlation. From testing it was determined that Case 2 was undersized for the system torque requirements. The models indicated that no gearhead, Case 2, was the much stiffer option, ω_1=155 Hz compared to ω_1=92 Hz, but due to the need for increased torque the gearbox option, Case 1, was chosen.

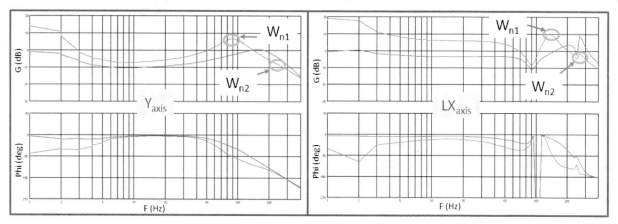

Figure 12. Empirical bode plots measured from Y and LX servo drives. The measured results correlated well to the mathematical models providing validation of the calculated model results.

The experimental bode plots show similar natural frequencies to the validation cases of from the mathematical models. The natural frequencies for the Y axis bode plot are ω_1=90 Hz and ω_2=250 Hz. The natural frequencies for the LX axis bode plot are ω_1=140 Hz and ω_2=260 Hz, **Figure 12**.

3.4 Temperature Tests

Significant viscosity changes at low temperatures were important discoveries made during testing; identifying the need for substantially higher drive torque. This was primarily a problem for the X axis since its running torque was much less than the other axes and therefore the changes at cold temperatures resulted in a significant percentage increase in the torque required by the X axis. It was determined that original screw grease viscosities reached the steep part of their exponential viscosity vs. temperature curve by 40-50°F. This was evident by the increase in drag while operating at higher speeds in colder weather. Using the original grease the slew drive reached the 3 amp maximum current at 50 mm/s for a 20°F test as compared to less than 1.5 amps at 70°F *(Figure not shown)*. After new grease was applied a 40°F test was completed that showed improvements in the viscous grease losses (i.e. losses were decreased by over 50% at 80mm/s). The slew drive could reach 80 mm/s with a 0.7 to 1.0 amp increase from the drive current required at 70°F, **Figure 13**.

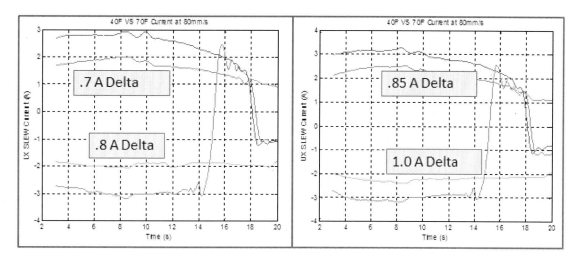

Figure 13. The increase in required drive current for LX and UX axis at 80mm/s for 70°F and 40°F ambient temperatures is shown.

4. TRACKER SYSTEM TESTING

Tracker system testing is defined as testing that occurs with the tracker in its fully assembled state. Performance was evaluated for each axis under the fully assembled configuration. Tracker system mount models were developed to compensate for structural deflection of the tracker mounting system. Future testing will include system performance testing (e.g. Open and Closed loop pointing accuracy, Tracking Accuracy, Guiding Accuracy, etc…)

4.1 Tracker System Performance Testing

The assembly of the tracker system increased payloads and resistances that resulted in adjusting control parameters to generate adequate following error performance. Sensor and control cables were installed at their full lengths, which in some cases was as long as 100 meters, to simulate the final configuration and test for increased noise and impedance problems that could develop. Large resistance forces were now in play such as the resistance from the energy chains which contained large cables and hoses that were forced to flex dynamically as the tracker moved. For the individual axes, goals of less than 10 micron RMS were targeted and testing to reach those goals is currently progressing at CEM in Austin TX.

4.2 System Mount Models

Tracker system mount models were developed to compensate for known deviations, inaccuracies, and imperfections in the mounting of the tracker system. Mount models generated for the tracker testing will be replaced once the tracker is installed at the McDonald Observatory and put into service; however for performance verification, test cell mount models were constructed. Mount models were generated for X/Y motion without hexapod movement, followed by mount models generated for hexapod motion without X/Y movement. The test setup did not include the WFC assembly, instead the WFC test mass was installed; throughout the following the SIRP location refers to a predefined point on the WFC test mass, not the actual SIRP.

Mount model generation involved superimposing the previously mapped feature locations that were measured during testing on the hexapod test stand, section 2.3, onto a few selected hexapod features locations that were measured once the hexapod was installed onto the tracker. These points generated the locations of critical geometry including the location of the SIRP relative to the Lower Hexapod Frame [LHF], which supports the hexapod.

X/Y mount models were constructed by using the laser tracker to measure the predicted location of the SIRP and the rotation of the LHF at discrete locations within travel limits. Data taken by the laser tracker was transformed into translations and rotations of the SIRP and the LHF. The measured data, compared to the commanded translations and rotations, generated errors that could be described by polynomial surface fit functions, **Figure 14**.

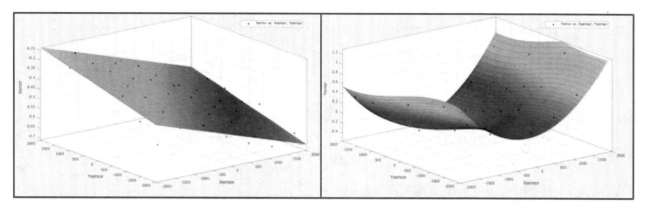

Figure 14. Error Plots with polynomial surface fits for the X and Y error of the tracker. The equations that describe the surfaces were used to generate mount model correction algorithms.

Algorithms from the polynomial surface fits were formulated and inserted into the tracker controller so that X,Y,Z, and Phi corrections were made given the command. For verification, the mount model algorithms and hexapod were activated and a reduced sampling of points was measured with the laser tracker. These points were again transformed and compared to the commands to determine the remaining error. The data comparing the error for X,Y,Z and Phi is listed in Table 2. The Mount Deflection data reflects the data sampled to create the mount models and therefore describes the deflected structure. The Mount Model Error is not zero and is a direct result of the fit error. Higher order surface polynomials would reduce this error but not substantially. The mount model does show remarkable improvements with error reduction over 90% in the Z and Phi degrees of freedom.

Table 2. The table lists the data of mount deflection and the error from the mount model. The mount model error shows considerable improvement compared to the mount deflection data.

	Mount Deflection				Mount Model Error			
	Xerror (mm)	Yerror (mm)	Zerror (mm)	Phi Error (ArcSec)	Xerror (mm)	Yerror (mm)	Zerror (mm)	Phi Error (ArcSec)
Maximum	-0.2615	1.2193	-4.706	160.37	0.0929	0.0924	-0.016	18.9208
Minimum	-0.6896	-0.4745	-7.847	-454	-0.1313	-0.1328	-0.2733	-30.97
Avg	-0.4578	0.2633	-6.2933	-142.8534	0.013	-0.0281	-0.1291	-1.5458
RMS	0.4695	0.5139	6.3467	239.5214	0.0496	0.0561	0.1406	11.4448
STDV	0.1043	0.4409	0.8222	192.2587	0.0478	0.0485	0.0554	11.34
Pk to Pk	0.4281	1.6938	3.141	614.37	0.2042	0.2252	0.2573	49.8908
Pk to Pk Impr.					0.2239	1.4686	2.8837	564.4792
% Improvement					52.30	86.70	91.81	91.88

Hexapod mount models were developed and implemented in a similar fashion. The tracker was locked at X=0, Y=0 and the hexapod was positioned within its travel limits. Measurements of the translation of the SIRP and rotation of the WFC test mass were measured with the laser tracker. Error plots were generated but algorithms have not yet been developed or implemented. Additional investigation is needed to verify that SIRP and rotation control points in the controller match the location being measured with the laser tracker. This testing is currently progressing at CEM in Austin TX.

5. SUMMARY

A multi-axis, high precision drive system has been designed and developed for the Wide Field Upgrade to the Hobby-Eberly Telescope at McDonald Observatory for initial use on the Dark Energy Experiment. A sequential laddered test approach was discussed with two parallel paths that involved hexapod testing and tracker subsystem testing.

Position control approaches were discussed while recent performance results were provided. Hexapod actuators demonstrated solid performance, less than 3 micron RMS, fully assembled. New controller methods, velocity drive control, and higher resolution feedback sensor were implemented to the X/Y axes to improve performance gains. After the changes the X/Y axes demonstrated acceptable performance at the subsystem testing level, less than 10 micron RMS, however performance roll-off started to occur during tracker testing.

Characterization measurements were illustrated and analytical methods for predicting drive dynamics and behavior were summarized. Temperature dependant test results were summarized and a reduction in viscous losses with new grease was shown, greater than 50% reduction at 80mm/s. Performance disturbances caused by excessive CFD load share were depicted while the optimum set-point was detailed, 5kN (10%).

Structural deflection compensation methods were demonstrated at both the hexapod actuator level and the tracker testing phase. Combinations of sensor error correction and deflection prediction decreased actuator positioning error from over 40 microns to less than 5 microns. Tracker mount model corrections for the X/Y plane resulted in error reduction over 90% in the Z and Phi degrees of freedom.

Tracker testing is ongoing at CEM with further improvements being accomplished. Future testing will include system performance testing such as open and closed loop pointing accuracy, tracking accuracy, guiding accuracy, etc. These tests will also take place at the CEM test facility in Austin, TX.

REFERENCES

[1] Hill, G. J., Gebhardt, K., Komatsu, E., Drory, N., MacQueen, P. J., Adams, J. J., Blanc, G. A., Koehler, R., Rafal, M., Roth, M. M., Kelz, A., Gronwall, C., Ciardullo, R. and Schneider, D. P., "The Hobby-Eberly Telescope Dark Energy Experiment (HETDEX): Description and Early Pilot Survey Results," ASP Conf. Series, 115-118 (2008).

[2] Hill, G. J., et al., "Current status of the Hobby-Eberly Telescope wide field upgrade", Proc. SPIE 8444-19 (2012).

[3] McDonald Observatory. (2008, January) HETDEX Illuminating the Darkness. [Online]. hetdex.org/hetdex/index.php

[4] Worthington, M. S., Beets, T. A., Good, J. M., Mock, J. R., Murphy, B. T., and South, B. J., "Design and development of a high precision, high payload telescope dual drive system", Proc. SPIE 7733, 201 (2010).

[5] Zierer, J. J., Beno, J. H., Weeks, D. A., Soukup, I. M., Good, J. M., "Design, testing, and installation of a high-precision hexapod for the Hobby-Eberly Telescope Dark Energy Experiment (HETDEX)", Proc. SPIE 8444, 176 (2012)

[6] Mock, J. R., Beno, J. H., Zierer, J. J., Rafferty, T. H., Cornell, M. E., "Tracker controls development and control architecture for the Hobby-Eberly Telescope dark energy experiment", Proc. SPIE 7733, 152 (2010)

[7] Wedeking, G. A., Zierer, J. J., Jackson, J. R., "Kinematic optimization of upgrade to the Hobby Eberly Telescope through novel use of commercially available three dimensional CAD package", Proc. SPIE 7733, 42 (2010).

[8] Mollison, N. T., Mock, J. R. Soukup, I. M., Beets, T. A., Good, J. M., Beno, J. H., Kriel, H. J., Hinze, S. E., and Wardell, D. R., "Design and development of a long-travel positioning actuator and tandem constant force actuator safety system for the Hobby-Eberly Telescope wide-field upgrade", Proc. SPIE 7733, 150 (2010).

Acknowledgement
HETDEX is run by the University of Texas at Austin McDonald Observatory and Department of Astronomy with participation from the Ludwig-Maximilians-Universität München, Max-Planck-Institut für Extraterrestriche-Physik (MPE), Leibniz-Institut für Astrophysik Potsdam (AIP), Texas A&M University, Pennsylvania State University, Institut für Astrophysik Göttingen, University of Oxford and Max-Planck-Institut für Astrophysik (MPA). In addition to Institutional support, HETDEX is funded by the National Science Foundation (grant AST-0926815), the State of Texas, the US Air Force (AFRL FA9451-04-2-0355), and generous support from private individuals and foundations.

Enclosure rotation on the Large Binocular Telescope

James Howard*[a], Robert Meeks[a], David Ashby[a], Warren Davison[b], James Wiese[a], Jeff Urban[a], Rick Hansen[a], Jared Schuh[a].

[a]Large Binocular Telescope, 933 N Cherry Ave, Tucson, AZ, USA 85721; [b]Steward Observatory, 933 N Cherry Ave, Tucson, AZ, USA 85721

ABSTRACT

After several years of operation the enclosure rotation system of the LBT is exhibiting wear and other performance issues that may impact operations. This paper reviews the system design and assumptions used, describes the current performance and observed symptoms, and discusses recent improvements made to improve performance and reliability.

The rotating enclosure of the LBT is a 2200 ton structure riding on four bogies with a total of 20 wheels. Identified deficiencies include wheel bearing capacities, bogie misalignment, and rail loading. These are partially due to excess enclosure weight, which was supposed to be 1500 tons, but also due to design errors.

The most serious problem was the failure of several wheel bearings. The bearings were not designed for field serviceability, so a crash program began to determine how to replace them. This got us back on sky quickly, but a review of the engineering calculations identified an error which led to the use of undersized bearings. A method of installing a larger bearing was found, and these have been installed.

One set of bogie wheels are misaligned so severely the structure makes loud popping and banging noises when the direction of building rotation changes. The bogie needs to be rotated about its vertical axis, but there was no provision in the design for this.

The circular rail the bogies roll on is wearing faster than expected. The rails are extremely difficult to replace, so the short term plan is to study the problem.

Keywords: Bogies, enclosure, rotation, LBT,

1. INTRODUCTION

The LBT enclosure was completed in 2000. It consists of a fixed portion and a moving portion. The fixed portion houses offices, storage, assembly areas, bedrooms, the kitchen, and the control room. The moving portion co-rotates with the azimuth axis of the telescope and provides the enclosure for the telescope as well as housing the telescope support machinery. It is a box 31 m by 29 m by 35 m tall, with two aperture shutter doors and four ventilation doors. It was designed to weigh 1500 tons, but ended up being 2200 tons.

Figure 1. LBT enclosure with the two aperture doors open.

*jhoward@lbto.org; phone 1 520 626-9847; fax 1 520 626-9333; lbto.org

Even without the excess weight, there were a few design problems which would have manifested themselves. The most serious problem is there is no easy way to service or upgrade the four bogies. The bogies were installed first and the whole enclosure was built on top of them. The four bogies ride on a total of 20 wheels, which give 80 tons per wheel by design and 110 in reality. Note this is far in excess of a typical railroad wheel load of 20 tons per wheel.

1.1 System Description

The rotating structure is supported by four pylons resting on top of four bogies. The whole rotating structure is attached to these pylons. The pylons are mostly covered today, but Figure 1 illustrates one in place during erection.

Figure 2. Photo taken during erection showing support pylon installed on top of a bogie.

Due to the need for an overhanging gantry crane for lifting telescope components through a hatch to the telescope, the structure is offset. Because of this offset, two of the bogies experience higher loads. These bogies have six wheels each, the other two have four.

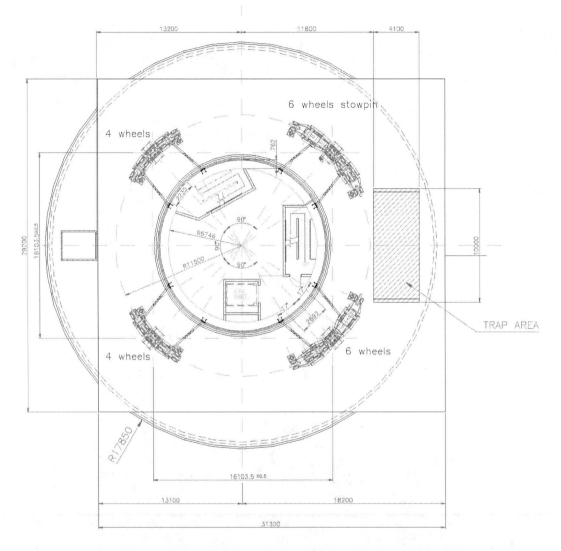

Figure 3. Plan view of the four bogies. Note offset to the side of the 6 wheel bogies to accommodate the trap door for lifting things up to the chamber from the ground floor.

The wheels are each 1 meter in diameter and ride on a track that is 23 meters diameter. The axles of each wheel are supported at each end by self-aligning spherical bearings. Originally the bearing on both the inside and outside was an SKF 22240.

In order for the wheels to roll in a circle, each is conical and the axis is tilted so it intersects the plane of the rail top surface in the exact center of the circle traced by the rail. There is a precision bore through each axle designed to fit an alignment telescope to facilitate this adjustment. The upper portion of the pier blocks the view of the telescope, so four sets of holes 90 degrees apart were added to allow the telescope to view a central target. Pointing adjustments are made by loosening the bogie to pylon interface and rotating the bogie about the vertical axis or moving it radially, both of which is much easier said than done. Also, the bearings supporting each end of the axles are mounted in eccentric bushings which allow limited adjustment of the steering of the wheel, as well as inclination with respect to the horizontal plane.

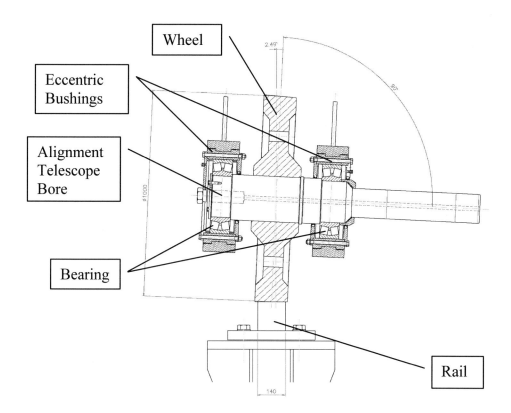

Figure 4. A wheel assembly

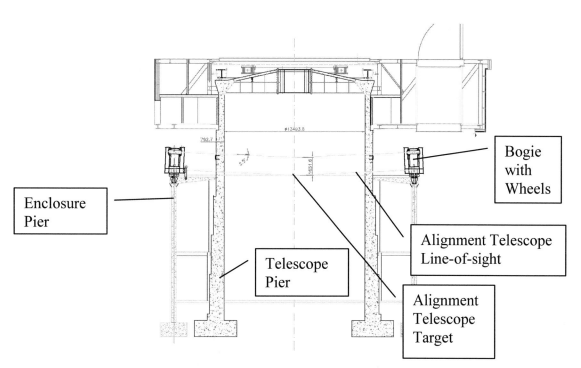

Figure 5. Cross section cut vertically through the pier illustrating the wheel alignment.

The bogies have pivots providing a wiffle tree arrangement which allow the load to be distributed evenly across all the wheels on a particular bogie. The wheel at either end of each bogie is driven, for a total of 8 driven wheels. The rest are idlers.

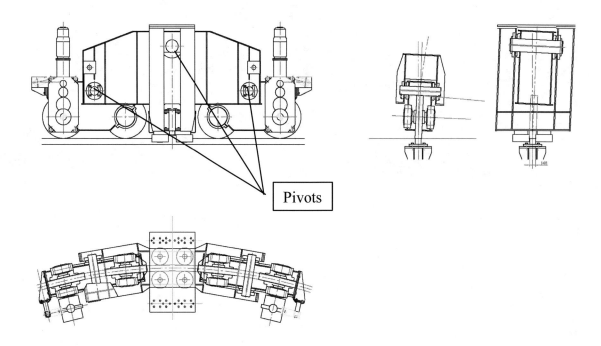

Figure 6. Four wheel bogie.

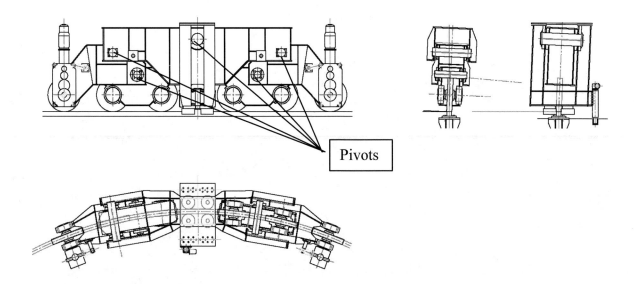

Figure 7. Six wheel bogie.

The rail is 140 mm wide, and unlike railroad track, has a rectangular cross-section. It is a high strength steel which was machined precisely and bolted to the reinforced pier top with a total of 556 M36 bolts. The track is in sections, with seven 40 degree segments, two 37.5 degree segments, and one 5 degree segment. The 5 degree segment was intended to be easily removed to facilitate the change of a wheel.

2. SYSTEM ISSUES

2.1 Alignment

There were a lot of problems aligning the bogies during enclosure construction. Unfortunately that phase predates the authors' involvement with the telescope and those who actually performed the work have moved on[1], so the process has had to be recreated. The forces involved are large and the space provided for actuators is small. It was not apparent until the telescope started operations there was a problem with the alignment of one of the four-wheeled bogies. It is rotated about a vertical axis by about 1 degree, which is enough to induce significant forces.

2.2 Rail wear

The rail is flaking off small, foil-like pieces of metal. The excessive weight is a major factor, probably exacerbated by the alignment issues described in the previous section. Also, the rectangular cross section adds high stress concentrations and it is very hard to get the wheels to roll flat on it.

2.3 Bearing loading

One of the outer bearings on a drive wheel failed in January 2011. Inspection of all the bearings showed that failure was imminent on all the outer bearings on the seven other drive wheels. The outer bearings on the idler wheels showed wear but were in much better shape. The inner bearings were all fine.

After some research, it was found an assumption made during the design phase was the thrust on the bearings was negligible. In fact, there is significant thrust due to the 2.5° axle inclination and the horizontal friction load between the wheel and rail arising from never perfect wheel alignment. The thrust is entirely taken by the outer bearing, and the original bearings have very little thrust load capability. Assuming a friction coefficient of 0.1, the thrust (F_{axial}) varies from 130 kN outward to 83 kN inward, depending on which way the friction force ($F_{friction}$) is acting. The radial load reaches a maximum of 693 kN for the inside bearing (F_{inner}) and 782 kN for the outside bearing (F_{outer}), where the outside bearing sees a greater force because it is closer to the wheel centerline in the vertical plane.

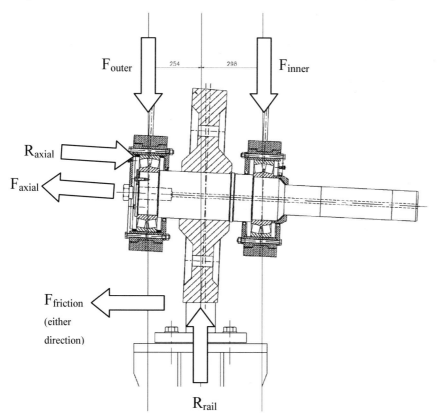

Figure 8. Forces acting on the wheel assembly.

2.4 Maintainability

The design of the bogies is such that a major job like replacing an idler wheel is extremely difficult. The drive wheels on either end of the bogies can be replaced by tilting the end of the bogie up, but getting an idler wheel out requires lifting the bogie a large fraction of a meter. Even lifting the bogies a millimeter for bearing changes has proved challenging.

3. SOLUTIONS

3.1 Bogie Alignment

The difficulty in managing such massive loads and forces with very little space has led to significant planning.

The first attempt was to install two double acting hydraulic cylinders to push and pull each end of the bogie. The cylinder capacity was limited by the building structure they were pushing against. The bolts clamping the joint between the pylon and bogie were all loosened, 20 tons of force were applied in tension on one end, and compression on the other. The building was then rotated in the hopes the wheels would slide into the new position. This attempt failed, and we determined we would need forces far higher than the structure could handle.

Figure 9. Hydraulic cylinders installed for the first attempt at moving the bogie.

The next idea was to try and separate the joint between the bogie and pylon with jacks and something with compliance in the horizontal plane. There was a provision in the design to allow four 100-ton jacks in the joint, and use a rotary bearing or a piece of soft elastomer sandwiched between the jacks and the pylon to allow for rotation[1]. Short 100 ton jacks are 85 mm tall, and the space they would fit is 95 mm tall, leaving only 10 mm for a bearing or elastomer. The building is overweight, with 110 tons per wheel or 440 tons at this joint, so the jack size would need to be increased or more jacks added. There is only room for four jacks, and the next jack size up, 150 tons, is 100 mm tall.

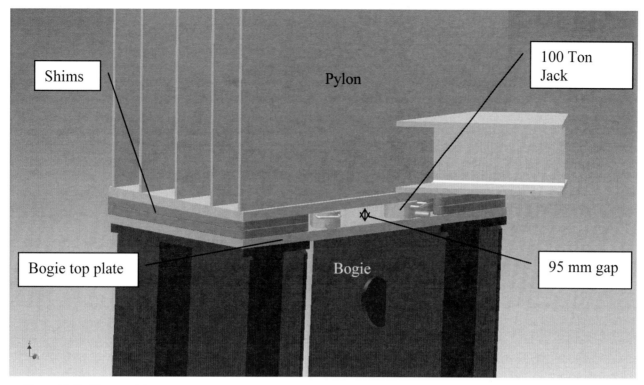

Figure 10. Jacking provision.

A possibility to overcome this is to weld structure to the bogie and the pylon which would allow for a proper jacking point with room for large jacks and bearings. Another possibility is to modify the 150 ton jacks so they are shorter, and also machine a cylindrical surface on the top and bottom which would let them roll. This would obviate the need for a bearing or elastomer, as the cylinders could be set radially from the center of the pylon to allow rotation.

A fallback position is to machine custom eccentrics for each axle to force each wheel to align properly.

If all this fails there is a final possibility. Use a 500 ton crane to lift the top of the pylon, then rotate the bogie into place.

3.2 Rail wear

There are no easy solutions to reduce the rail wear other than improving bogie alignment. Installing a reprofiled new rail or increasing the number of wheels to reduce the wheel load are extremely expensive and time consuming. The plan right now is to carefully monitor the rate or wear by periodically picking up the steel flakes that are shed off the rail and weighing them.

3.3 Bearing upgrade

In order to better handle the thrust and vertical loads, the outer bearing was upgraded to the largest one that could be made to fit without replacing significant parts of the bogie. The new bearing is an SKF 24144K, which is significantly wider than the original 22240. It required machining the eccentric bushings to accommodate the larger outside diameter, adding a stub to the axles to accommodate the increased width, and adding a sleeve for the increased inner diameter. The precise alignment telescope bore was used to register the stub into place. It took significant engineering resources to come up with this solution and 9 months to totally implement it on the 20 outer bearings. This provides a much better capability for handling thrust loads as well as the radial load, and should increase bearing life by a factor of five[2]. Interestingly, the cost of this bearing versus the original bearing is not significantly greater.

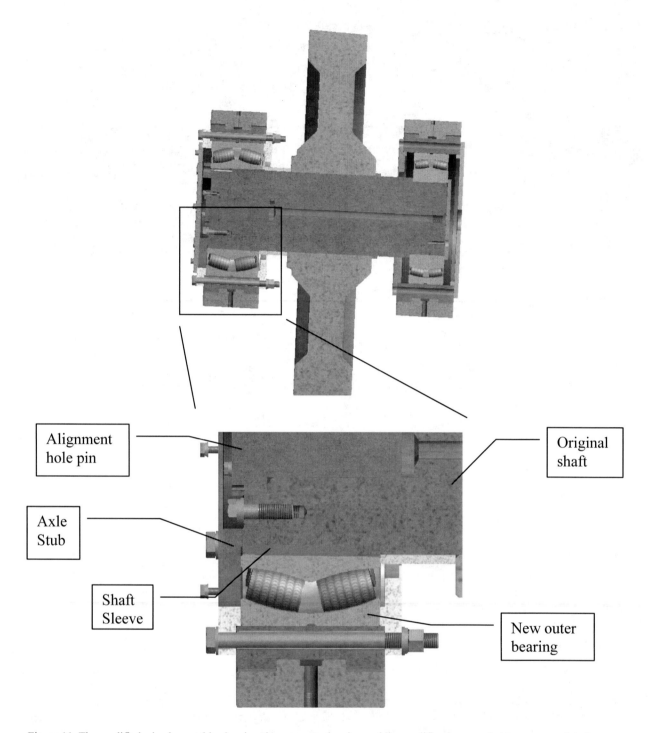

Figure 11. The modified wheel assembly showing the new outer bearing and the modifications needed to accommodate it.

3.4 Lubrication upgrade

The original bearing assembly did not include any provision for lubrication other than removing each individual cover and packing the bearing with grease. This was a factor in the shortened life of the original bearings, so the modified bearing assembly provides proper holes for supplying grease to the central part of the bearing. This will make maintenance an order of magnitude easier.

REFERENCES

[1] Williams, J. T., MMT Observatory, interview, Jan. 24, 2011.
[2] SKF Industries, Inc., [Engineering Data], SKF Industries, Philadelphia, PA, 5-52 (1980).

The 3,6 m Indo-Belgian Devasthal Optical Telescope: The hydrostatic azimuth bearing

Jonathan de Ville, Maxime Piérard, Christian Bastin

Advanced Mechanical and Optical Systems (AMOS s.a.), LIEGE Science Park, B-4031 ANGLEUR (Liège), BELGIUM

ABSTRACT

AMOS SA has been awarded of the contract for the design, manufacturing, assembly, tests and on site installation (Devasthal, Nainital in central Himalayan region) of the 3.6 m Indo-Belgian Devasthal Optical Telescope (IDOT).
The telescope has a Ritchey-Chrétien optical configuration with a Cassegrain focus equipped with one axial port and two side ports. The primary mirror is a meniscus active mirror. The mount is an Alt–Az type with for the azimuth axis a 5 m diameter hydrostatic track. This paper presents the solution adopted by AMOS to meet the specific requirements for the azimuth axis. The track is designed to be able to control the positioning of the telescope around the azimuth axis with an accuracy of 0.05 arc second for all tracking configurations. The challenge came from this tight accuracy with a mass in rotation weighting 125 tons. The azimuth track was mounted and tested in AMOS workshop; the tests and performances are also discussed.

Keywords: hydrostatic, bearing, IDOT

1. INTRODUCTION

The Telescope design[1] was presented 4 years ago at Marseille; since then it has been fully assembled[2] and tested at sub-system and system levels in AMOS assembly hall.

Figure 1 shows pictures of the telescope in AMOS assembly hall. The telescope size is: height 13 m, width 7 m and total weight 150 tons.

This papers provides an overview of the ARIES 3,6 m telescope hydrostatic azimuth bearing. After a short presentation of the system, track assembly, integration and testing will be developed. The final part will include this subsystem performances.

Figure 1: IDOT in AMOS integration hall

2. DESCRIPTION OF THE HYDROSTATIC AZIMUTH BEARING SUBSYSTEM

The 3.6m IDOT is equipped with two hydrostatic bearings for the Azimuth axis. The first is radial and is responsible of positioning stiffness of the rotor on the stator. The second is axial; it counteracts the telescope weight and allows a rotation with reduced friction. Rotating part around the Azimuth axis has a weight of 125 tons and an inertia varying around 900 000 kg m².

Figure 2: Azimuth hydrostatic bearing

2.1 Bearings requirements

Regarding the global tracking performances of the telescope, the azimuth axis is an essential part of the system. Indeed, it has a major influence on the wobbling of the structure which directly induces a depointing error and, the bearing friction generates errors in the axis control. Thus, the quality of both axial and radial bearings is critical in the project.

Hydrostatic bearings have to face the following conditions:
- Environment temperature: -5°C to +30 °C
- Relative humidity: 0 to 95 %
- Weight to support: 125 tons with non-uniform repartition (**Figure 6**)
- Speed rotation around Azimuth axis: up to 2°/s
- Tracking speed rotation around Azimuth axis: 0 to 300 arcsec/s
- Accelerations: max 0.5°/s²

2.2 Radial bearing

Radial bearing main characteristics are given in Table 1:

Number of pockets	6 (equally distributed)
Location Diameter	600 mm
Size of oil pocket	50° x 50 mm
Rotation	Rotating track (fixed pads)
Load	4 ton max
Control system	PM controller from HYPROSTATIK®: Adjusts the flow rate to keep the oil gap constant.
Pressure	Varying from 6 to 11 bars. Depends on figuring error of the cones (machining quality)

Table 1: Hydrostatic radial bearing characteristics

Figure 3: radial hydrostatic bearing stator (pockets)

It's composed of
1) Stator in which the pockets are machined (**Figure 3**).
2) Rotor
3) Oil flow controllers

The rotor and the stator have conical shapes (1/20). The oil is injected at the level of 6 pockets that are rectangular clearances in the main core of stator. The liquid flows then down in a collector and is sent back to the hydraulic aggregate by gravitation. For stiffness purpose, the gap has to be small (few tens of microns) and is mechanically determined in adjusting the relative height between the rotor and the stator. The sliding surfaces have consequently to be machined very accurately.

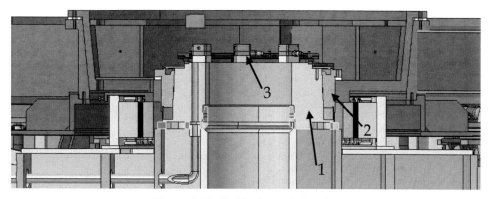

Figure 4: Radial hydrostatic bearing

Two major components of the Azimuth axis control are mounted close to this radial bearing: the Azimuth motor which is a direct drive torque motor used at maximum 15 000 Nm and the Azimuth encoder which is supplied by Heidenhain with an accuracy of 2 arcsec.

2.3 Axial bearing

Axial bearing main characteristics are given in **Table 2**

Number of pads	24 (equally distributed)
Size of pads	340 mm x 150 mm
Size of oil pocket	240 mm x 50 mm
Rotation	Rotating pads
Load	125 t
Location Diameter	4644 mm
Control system	PM controller from HYPROSTATIK®: Adjusts the flow rate to keep the oil gap as constant as possible
Track machining requirement	0.02 mm (flatness)
Rotor machining requirement (pads surface)	0.02 mm (flatness)

Table 2: Hydrostatic axial bearing characteristics

The axial bearing is composed of
1) Stator with the machined track
2) Rotor with the brass pads
3) Oil flow controllers

The stator is an annular flat track which has been machined very precisely. The rotor is composed of 24 pocket fixed to the structure and machined together. Similarly to radial bearing, oil is injected through the 24 pockets of the rotor, flows down to the collector and is sent back to hydraulic aggregate. In this configuration, the gap is only determined by the flow controllers.

The average pressure on each pocket depends of its position under the fork. Indeed, pockets placed under the lateral fork are more loaded which means that their average pressure is higher. It varies from 5 to 32 bars.

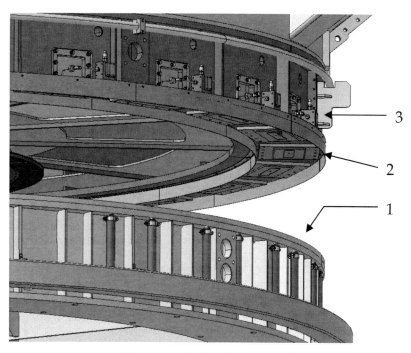

Figure 5: Axial bearing, exploded view

Figure 6 shows pressure distribution. As explained before, two areas below lateral forks are more loaded. Normally, dispersion should be smaller but imperfections in machining of the pockets on the rotor create deviation from theoretical behaviour. Moreover, in AMOS premises, the concrete floor is much more flexible than the support on ARIES site, thus behaviour slightly different is foreseen on site.

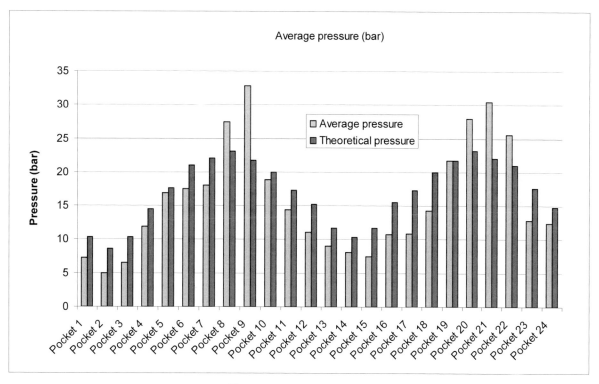

Figure 6: Average pressure

2.4 Hydrostatic circuitry

Oil supply on these systems is ensured by a hydraulic group with a tank of 1400 l. Inlet pressure on bearings flow controllers is 50 Bar and each pocket is equipped with a pressure sensor. During the oil cycle inside the system, the oil is naturally warmed. To avoid warming the telescope fork and get problems of seeing, the tank is cooled down in close loop and the oil temperature is kept at dome temperature.

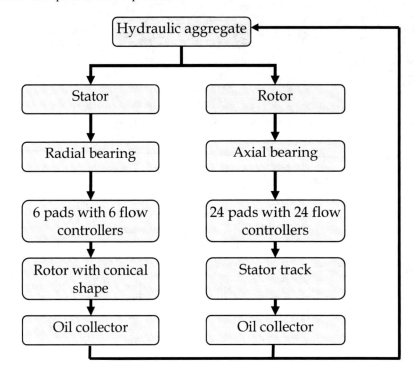

Figure 7: hydrostatic circuitry scheme

2.5 Flow controllers

Flow controllers are manufactured by HYPROSTATIK®. These equipments give a good stiffness to both bearings and create a gap, with a theoretical value of 53 µm, as constant as possible despite track and pocket machining imperfections. Moreover, they are practically unaffected by oil viscosity changes and allow to keep distance between bearing parts. Indeed, due to special design these components deliver an increasing oil flow through the pocket when the pressure increases which gives a higher stiffness to the system.

Mathematical model of the behaviour of the flow controller is derived from equation given by manufacturer [2] and theory of hydrostatic systems:

$$Q = Q_0 \left(1 + (k_r - 1)\frac{P}{P_p}\right)\frac{10^{-2}}{\mu}$$

With:
Q: flow rate through the pocket (m³/s)
Q0: flow rate for pocket pressure=0 (parameter given by manufacturer) (m³/s)
kr: parameter given by manufacturer
Pp: pump pressure
µ: oil dynamic viscosity (Pa.s)
P: pocket pressure

$$h = \sqrt[3]{\frac{\mu Q}{P}} * \alpha$$

h: height of the gap (m)
α: constant depending on pocket dimensions

3. ASSEMBLY AND INTEGRATION

3.1 Track adjustment

Performances of the global system are directly linked to azimuth axis performances and thus to hydrostatic bearing quality. To be able to reach all specification AMOS was constrained to realise several verifications and adjustment.

One of the first steps of ground interface structure (GIS) assembly was the track adjustment. Indeed, to avoid contacts between rotor and stator, the track had to be adjusted regarding several criterion:
- Track horizontality: it was measured tangentially and the highest result is 0.65 arcsec
- Track height variations: max. 30 μm.
- Track height variations for a rotation of 5° in azimuth: 5 μm

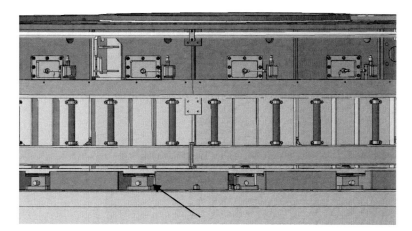

Figure 8: Levellers under the stator

The stator stands on 48 levellers that are adjusted to reach the required track flatness. After mounting of the rotor, the flatness is controlled in measuring the pressure during the rotation. Indeed, each pocket of the rotor is equipped with a pressure sensor which allows logging pressure during the movement.

As explained in §2.5, pressure is inversely proportional to the distance between fixed and mobile part of the hydrostatic pocket. Higher is the pressure, smaller is the distance which means that track is locally too high in this area. Such reasoning gives the global topology of the stator track and helps to determine the areas that have to be adjusted to reach performances. **Figure 9** shows a typical pressure curve measured before accurate adjustment at the level of 1 pocket while rotating. It corresponds to gap variations of ~10 μm PV.

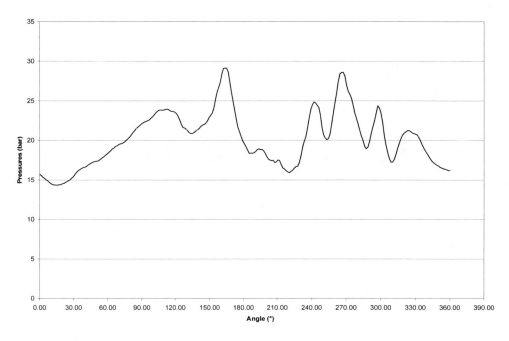

Figure 9: Pressure variation during rotation (pocket 12)

3.2 Performances

To reach performances on axis control, friction variations have to be as small as possible. When the gap between stator and rotor decreases, the friction increases. To check the track shape, motor torque can be analysed. It allows to put the highlight on areas were friction increases suddenly due to contact or local too small gap between track and rotor pads.

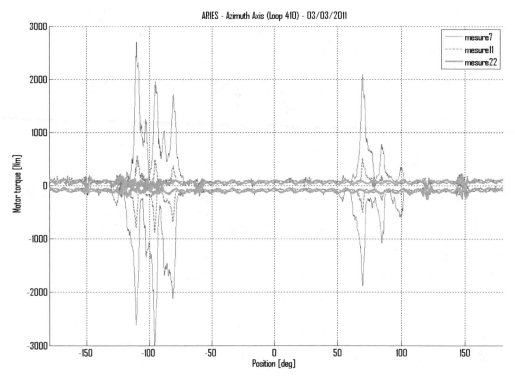

Figure 10: Azimuth motor torque

Figure 10, shows curve obtained initially on the system (blue curve– measure 7). Several peaks separated of 15 ° are observable on 2 zones separated of 180°. 15° is the angular distance between rotor pads so this curve shows that successive pads touch the same point on the track. These pads are situated under lateral fork which explains the 180° spaces between groups of peaks.

Figure 11 gives typical motor torque variations when the track is correctly adjusted. The variations that remain are mainly the motor cogging.

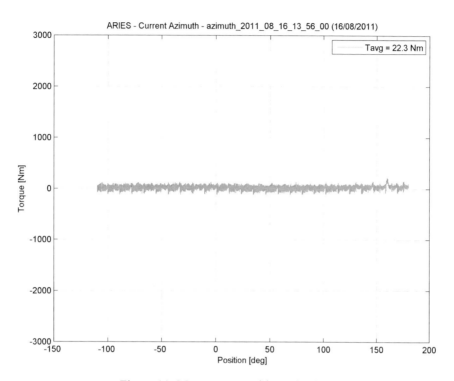

Figure 11: Motor torque with track adjusted

By matching this figure and **Figure 9**, it was possible to determine the areas to adjust on the track. Using levellers situated under the stator, track was adjusted to reduce motor torque to be under motor cogging. **Figure 10** shows the evolution regarding the track adjustment. Blue curve gives the initial behaviour, then red, then green and finally the results given in **Figure 11**.

The axis control loop is evaluated in measuring the following error i.e. the difference between the set-point and the actual position of the rotor.

With these adjustments, tracking performances have been measured on the telescope, while all other telescope systems were running. They can be summarised as follow:
- for speeds between 20 and 300 arcsec/s (i.e. elevation angle between 60 and 87.5deg), the azimuth following error (FE) is smaller than 0.05 arcsec rms
- for speeds between 0 and 20 arcsec/s, the FE is 0.02 arcsec rms
- for elevation angle between 15 and 30deg, the measurements shows large dispersion. The FE can be up to 0.04 arcsec rms.

3.3 Axis characterisation

Other parameters have been measured:
- Friction has a value between 61 and 170 Nm regarding axis speed (viscous friction). On other AMOS projects, classical bearing are used and friction has a value around 1200 Nm, the advantage of the hydrostatic system is obvious. Figure 12 shows the relation between axis speed and friction.

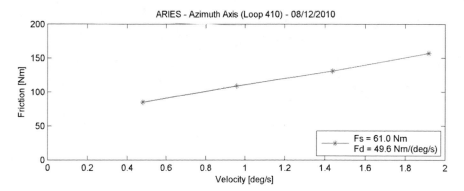

Figure 12: Friction Vs speed on hydrostatic bearing

- Azimuth axis wobbling is lower than 0.65 arcsec for any rotation of 70° and lower than 0.25 arcsec for any movement of 5°.
- Axial run-out is smaller than 7.5 μm

4. CONCLUSIONS

Hydrostatic bearings have been fully calculated, integrated and tested in AMOS premises. After several runs of adjustment, the choice of a hydrostatic system for the IDOT was confirmed as a good solution. Indeed, system is able to reach all performances that are required. With the tests performed in factory on the telescope completely mounted, AMOS has now the confirmation that risks are greatly reduced for the on-site installation. The small friction value was also decisive to help the control to reach its quality.

5. ACKNOWLEDGEMENT

This work has been performed under ARIES contract reference 1985-14-02. AMOS is very grateful towards ARIES team for having put their confidence in AMOS team for the design and manufacturing of the 3.6 m telescope.

6. REFERENCES

(1) Flebus C., Gabriel E., Lambotte S., Pierard M., Rausin F., Schumacher J.M. and Ninane N., "Opto-mechanical design of the 3,6 m Optical Telescope for ARIES", Proc. SPIE 7012-09 (2008).
(2) Ninane N., Flebus C. and Kumar B., "The 3.6 m Indo-Belgian Devasthal Optical Telescope: general description", Proc. SPIE 8444-67 (2012)
(3) Hyprostatic, "Technical information, PM controller and jet pump with examples of application"

Telescope positioning and drive system based on magnetic bearings, technical challenges and possible applications in optical stellar interferometry

Roland Lemke[a], Hans J. Kärcher[b], Lothar Noethe[c]

[a] Ruhr-Universität Bochum. Astronomisches Institut, Universitätsstr. 150, D-44801 Bochum
[b] MT-Mechatronics, Weberstraße 21, D-55130 Mainz
[c] European Southern Observatory, Karl-Schwarzschild-Str. 2, D-85748 Garching b. München

ABSTRACT

The paper describes the envisaged positioning and drive system for telescopes of the 2m class, gives some basic analysis for power consumption and positioning performance and proposes next steps for verification of the concept. A possible application could be in optical stellar interferometry where an additional delay could be spared if the positioning performance is better than 10nm. Fast re-positioning of the telescopes should allow for multiple baselines during one observing night giving the opportunity to obtain high spatial stellar images with high time resolution.

Keywords: Magnetic bearings, optical interferometry

1. INTRODUCTION

Over many years moveable telescopes have been discussed in interferometric instrumentation to spare delay lines and to allow for a very flexible coverage of the uv-plane. Positioning the telescopes within 50nm accuracy over 50 ms while tracking at about 4 mm/s seems to be possible with telescopes on linear magnetic bearings. With a star like (or elliptical) configuration of 6 2m class telescopes, establishing maximum baselines of a couple of 100 meters or even kilometers a high angular resolution as well as an instantaneous image quality in direct imaging is possible. In compacted observing configuration the interferometer would approach imaging properties like those of a monolithic telescope of approximately 5 m diameter. The telescopes will be equipped with an adaptive optics system to eliminate the effects of atmospheric turbulence. By re-positioning the telescopes over the full baseline range within minutes it will be possible to obtain a high angular resolution observation within one observing night, giving the opportunity to study objects in interferometric mode with a so far unrivaled time resolution.

Magnetic bearings are used in the machine tool industry for the support of fast rotating rotors and in the transportation industry for **mag**netic **lev**itated trains (MAGLEV principle). They avoid the risk of oil spill of hydrostatic respectively friction in roller bearings, but may have draw backs in regard of electro-magnetic pollution and energy consumption. For the application in telescope mounts only the MAGLEV type bearings shall be considered. Their advantages are the absence of any type of mechanical friction, the combination of the bearing with the drive functionality, and the fine controllability of movement and levitation. But obviously some effort for new technological development must be spent. The bearing principles exist, but the application in telescopes is new. The following paragraphs describe a conceptual design of a 2m class telescope mount, where the two main axes mechanics (bearings and drives) are combined with a linear bearing and drive for the movement of the telescope; the axes mechanisms and the linear bearings and drives follow all the same MAGLEV approach. Pros and cons are discussed in more detail and the next development steps are described.

2. SCIENCE CASE

Interferometric measurements are used to study stellar diameters, young stellar objects disks and jets and binary stars. Especially the search of exo-planets demands a high angular resolution in combination with special observing techniques like nulling and coronagraphy in order to overcome the contrast problem between the central star and the planet. High angular resolution could be achieved by large monolithic mirrors or by multiple small telescopes combined into an interferometer. Multiple small telescopes can be operated in the compact configuration similar to a monolithic telescope of equivalent collecting area. In wide mode it would give the highest resolution possible. As angular resolution increases with the baselines atmospheric turbulence starts to be a major problem. However it has been demonstrated that with adaptive optics it is possible to eliminate this problem. Moveable telescope would make delay lines which are a very complex part of an interferometer obsolete. The I2T (Interféromètre à 2 Téléscopes) used a beam combiner on an optical table which could be translated between the two telescope to compensate the optical path length. It was followed by the GI2T (Grand interféromètre à deux téléscopes). Proposed first by Labeyrie in 2005 the Optical Very Large Array Project started to study a concept for kilometric ground-based hypertelescopes. The telescopes should by mobile on a flat ground and forming an ellipse with the beam combiner in one of the centers for optimized uv-coverage. The original idea was based on telescopes on rails. The precision and stability problems related to this solution would, in our opinion, make it difficult to be reach the performance requirements for interferometric observations. Instead magnetic bearings usedd as a linear bearing could deliver a much higher accuracy. As they also do not require cables, they could be positioned at any location on their track. In addition fast movements between new baseline positions are possible, even without losing the fringes. Therefore, a new configuration could be arrranged within minutes. With these observing modes it will be feasible to obtain of a sufficient uv-coverage of a single source in a single observing night. Therefore a high angular resolution will be combined with a very high time resolution, enabling the study of exo-planet transits which last only hours or days.

3. TECHNICAL CONCEPTS

3.1 Magnetic bearings (principles)

Figure 1 explains the MAGLEV principle. It combines levitation magnets with spur magnets and a torque motor. It works for linear movements as well as for rotatory movements following the same principles. For the rotatory telescope axes all elements (the levitation and the spur magnets and the torque motor) are circular; for the linear telescope movement all three components are linear and arranged on both sides of the telescope base.

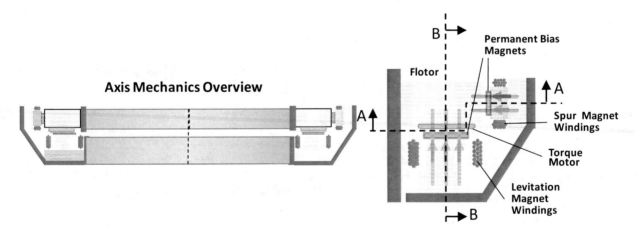

Figure 1: Explanation of the MAGLEV principle

The levitation magnets use other magnetic forces than the torque motors. The levitation function is based on the attractive force of the magnetic field itself, whereas the linear (respectively torque) motor is based on the Lorentz force of an electric current in a magnetic field. The physical behavior and the related controllability of both effects are rather different. Whereas the magnetic force behaves highly non-linearly and is difficult to control, the Lorentz force is, in first order, linear with the current and comparatively easy to control.

Levitation system

The levitation forces can only be controlled by the current in a field coil. However, generating the levitation forces only by field coils would cause high permanent energy consumption. Therefore so-called biased levitation magnets are used. The main levitation force is produced by permanent (rare earth) magnets, which waste no energy, and only the controlling part of the levitation force is produced by field coils. The permanent magnets have to be dimensioned such that energy saving and force controllability are balanced.

The permanent magnets and the field coils have to be arranged such that individual "flux cells" are generated (Figure 2). The flux in the cells is short-circuited by adequate shaping of the permeable iron on which they are arranged. The levitation as well as the spur magnets, follow the same principle, but the spur magnets have to generate much smaller forces. The levitation magnets have to counteract the forces introduced by the weight of the telescope above, whereas the spur magnets have to be designed only against some pretension force between the left and right side to be chosen according the maximal expected spur disturbing influences.

Obviously for the control of the levitation and spur forces gap sensors are required, which hold the levitation as well as the spur in the requested position (see also comments in the paragraph on the telescope concept).

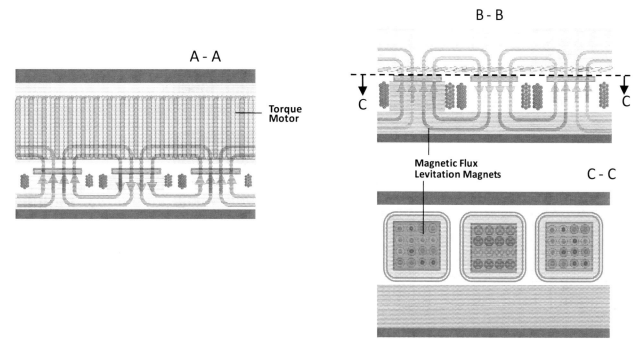

Figure 2: Magnetic flux in the levitation magnets and torque motors

Spur system

The position control requires linear and torque motors. As mentioned above, the driving forces of these motors iare produced by the Lorentz force due to a current in a magnetic field. The trick of the MAGLEV principle is to use the strong magnetic field produced be the levitation magnets at the same time as field magnets for the linear and the torque motors. Therefore electric coils have to be arranged in directions orthogonal to the directions of the movements. The choice whether the coils should be installed on the fixed or the moving part of the motor is free and can be based on practical considerations like the need for cable guiding, etc (see telescope concept below). A drawback of this concept is a cross coupling between the control of the levitation force via the biased levitation magnet and the control of the driving force via the current in the motor coils. But this must be tolerated considering the overall advantages of the integrated design particularly in regard of energy consumption.

3.2 Layout of a Interferometer with 6 Telescopes

Numerous publications about both radio and optical interferometers discussed the optimization of the layout of the telescope locations. These optimizations are mainly driven by the maximum achievable angular resolution and are heavily constrained by technical aspects like the mechanical stability and the atmospheric turbulence. Telescopes on magnetic bearings are requiring a locally linear track. Curves are possible, but as the light is collected in the telescope and sent by mirrors to a central point, where beam combining will take place, only linear tracks are considered. A star like configuration seems to be a good starting point for a prototype.

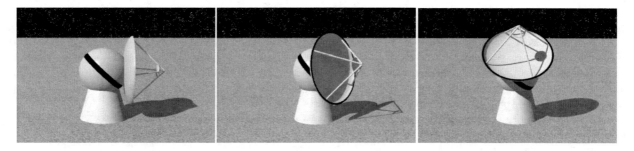

Figure 3: Artistic view of a single telescope on a slant axes mount

Figure 4: Artistic view of a star like configuration with the beam combining lab in the center. The shelters for the telescopes at the ends of the tracks are not shown.

Figure 3 shows an artistic view of a single telescope on a slant axes mount. 6 of these telescopes should be combined into a star like configuration as shown in Figure 4. For simplicity the shelters for the telescopes at the ends of each track are not shown.

3.3 Design concepts for the telescopes

The telescope mount should be as simple as possible. Figure 5 shows a so called "slant axes" design, with the two main axes slanted by 45 degree (one of them is the azimuth axis AZ; the other one is the slant axis SL, which replaces the altitude axis). The slant axes mount has two major advantages compared with the classical alt/azimuth mount: (1) the optical path to the interferometric line is much simpler (only two mirrors are necessary); (2) the mechanics (bearings and drive motors) for both axes can be identical (discussed below).

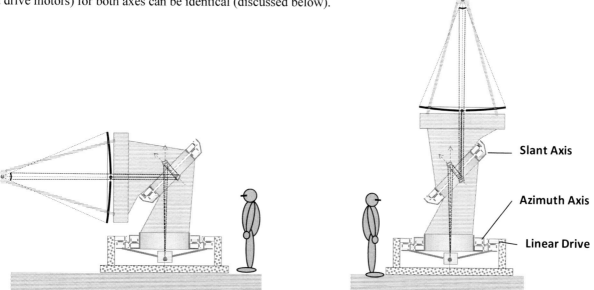

Figure 5: Design concept for a 2m class telescope on a slant-axes mount

Each telescope is carried by a "telescope carrier" (Figure 6), which combines the linear bearings and the drive system for the interferometric movement with the bearings and drives for the azimuth axis. The linear tracks for the carriers are located on a concrete tray, which comprises also the optical path to the central instrument of the optical array. The sketches don't show any kind of weather protection, which may be necessary but should be minimized by arranging telescope shelters at the ends of the linear tracks, where the telescopes could be stored during bad weather conditions.

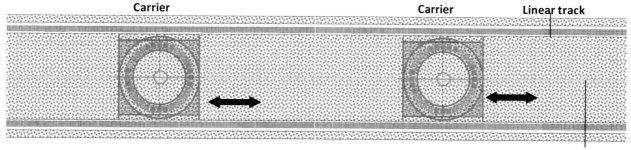

Figure 6: Telescope carriers and linear track

3.4 Magnetic bearings (details)

Figure 7 shows some details of the motor and bearing arrangements in the telescope carrier. The weight of the telescope is assumed as 10^5 N (equivalent to 10 to). The required dimensions for the levitation and spur magnets are gotten from formulas given in [7]. The stators on the one hand and the flotors and rotos on the other hand are arranged such that the permanent components of the biased magnets always act as tension elements. The linear bearing and drive mechanism and the azimuth bearing and drive mechanism are integrated into the compact carrier structure in the densest way. The slant axis bearing and drive system is identical with that for the azimuth axis. The coils of the linear motors are integrated into the track in the same way as in the MAGLEV trains. Power supply and cabling, as well as the arrangement of the position sensors have to be further elaborated.

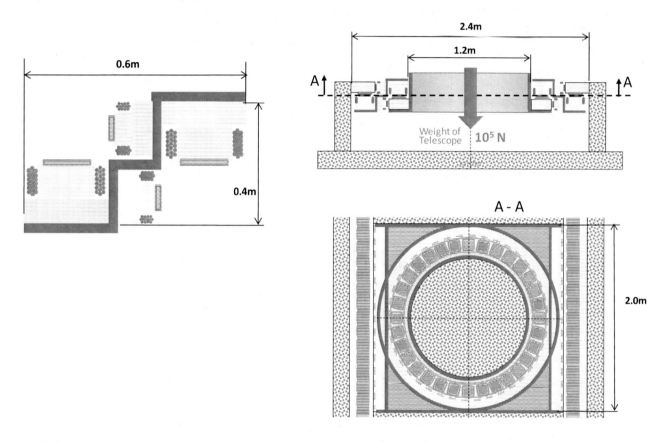

Figure 7: Details of the bearing arrangements in the telescope carriers

3.5 Position control

The control of all degrees of freedom of the bearings and the drives gets rather complex (Figure 8). Each bearing has, additionally to its rotatory respectively linear position, influence on the final optical position and phase by its magnetic axial, radial and tilt "run-out", which is measured by the gap sensors, which are needed anyway for the control of the magnetic bearing forces. These "run-out" degrees of freedom are also existent in mechanical bearings, but the run-out of mechanical bearings is but inherent and of statistical or highly non-linear behavior (backlash, friction) and not controllable as for the magnetic bearings. For the mechanical bearings the run-outs are corrected by pointing models, friction and backlash compensators in the control system and for the phase by the delay-line mechanisms. In the magnetic bearing case all these effects can be directly controlled by the bearing and motor controllers, but the complex nature of the system needs an adequate control strategy.

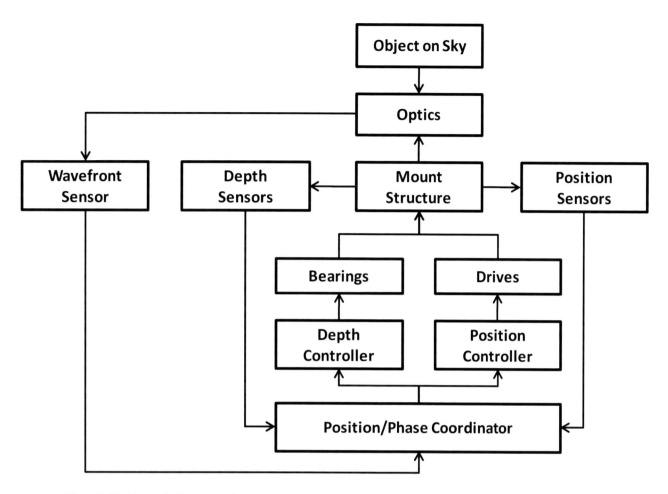

Figure 8: Position and phase control scheme

The control architecture should have a "cascaded" structure (Figure 8), with individual bearing respectively motor subsystem controllers, and as master the overall position/phase control system, which has to coordinate the actions of the bearing and motor controllers (called "Position/Phase Coordinator" in the diagram of Figure 8). The sensors of the subsystems may have the usual mechanically achievable accuracies (of the order of microns and arc seconds). The master control system uses the information from the wavefront sensor about the optical quality and distributes this information to the bearing and motor subsystems. The strategy for this distribution will probably be a "master/slave" concept and has to be further analyzed. Essential is the "controllability" of the position and phase, both affecting the optical quality, by the bearing magnets and the drive motors. The controllability may be limited by nonlinear effects as cogging and ripple, which are well known for the motors and are caused by deficiencies in the motor windings etc. This has to be further investigated, by analysis as well as bread-boarding and is proposed as the next step in the program.

4. CONCLUSION

We have presented a new approach to overcome mechanical limitations in telescope positioning by using a drive system based on magnetic bearings. We have to further investigate critical parameters as the power consumption and as well the problems arising with strong magnetic fields in the vicinity of scientific instrumentation. However, if the stability and position accuracy can be reached, new interferometric observing modes are possible allowing to obtain an image with high spatial resolution in a single observing night.

REFERENCES

[1] O. Lardière, et al., "The Optical Very Large Array project (OVLA): a concept for kilometric ground-based hypertelescopes", Bulletin de la Société Royale des Sciences de Liège, Vol. 74, 5-6, (2005)
[2] J. Dejonghe, et al., "The Optical Very Large Array (OVLA) prototype telescope: Status report - Perspective for large mosaic mirrors", Proc. SPIE Advanced Technology Optical/IR Telescopes VI (Kona), Vol. 3352, (1998)
[3] V. Borokowski, et al., "Hypertelescopes without delay lines", Bulletin de la Société Royale des Sciences de Liège, Vol. 74, 5-6, (2005)
[4] A. Labeyrie, "Kilometric arrays of 27 telescopes: studies and prototyping for elements of 0.2m, 1.5m and 12-25m size", Proc. SPIE 1998-Conference on Astonomical Interferometry (Kona), Vol. 3350, (1998)
[5] A. Labeyrie, et al., "An Introduction to Optical Stellar Interferometry"
[6] A. Glindemann, "Principles of Stellar Intereferometry"
[7] A. Traxler, E. Maslen, "Magnetic Bearings", Springer (2009)

Enclosure Design for the ARIES 3.6m Optical Telescope

A.K. Pandey[1], Vishal Shukla[1]*, Tarun Bangia[1], R. D. Raskar[2], R. R. Kulkarni[2], A.S. Ghanti[2]

[1]Aryabhatta research institute of Observational Sciences (ARIES), India

[2]M/s. Precision Precast Solutions Pvt. Ltd. Pune, India

Abstract

A 3.6-m, f/9 optical telescope is planned to be installed at Devasthal, India (Latitude:29° 21' 40" N, Longitude: 79° 41' 04" E, Altitude: 2450 m above msl). The telescope has Cassegrain focus and alt-azimuth mount. The design of the telescope enclosure and the auxiliary building includes a fixed base enclosure, a telescope pier, a rotating dome structure, an auxiliary building, ventilation and component handling systems. The design is optimized for thermal, mechanical, structural, as well as for telescope installation and maintenance requirements. The design aims to provide seeing limited images within the telescope enclosure. This paper presents design of the 3.6m telescope enclosure.

Keywords: Telescope enclosure, Auxiliary building, Ventilation system

1. Introduction

Aryabhatta Research Institute of Observational Sciences (acronym ARIES), an autonomous research institute under the Department of Science and Technology (DST), Government of India, has taken an initiative to establish a 3.6-m new technology Devasthal optical telescope (DOT) at Devasthal, Nainital. Belgium is also participating in the project under bi-lateral program of cooperation in science and technology with India.

The telescope is being installed at a good observing site Devasthal (Latitude:29° 21' 40" N, Longitude: 79° 41' 04" E, Altitude: 2450 m above msl) in India. Devasthal is about 65 km from Naintal and it has an added advantage of crucial geographical location for a number of time-critical observations of cosmic events. An extensive site characterization was conducted during 1980 – 2001[1]. Seeing measurements carried out over a period of two years close to ground level yielded a median ground level seeing estimate of about 1.1 arcsec; the 10 percentile values lie between 0.7 to 0.8 arcsec (mean = 0.75 arcsec) while for 35 percent of the time the seeing was better than 1 arcsec. The seeing measurements as well as number of yearly spectroscopic nights (~210), darkness of the per square arcsec sky (V ~ 21.8 mag) and other atmospheric parameters for Devasthal make this site comparable to international standards [2].

The f/9 configuration of the telescope has an alt-azimuth mount. It has Cassegrain focus fitted with a 30 arcmin wide field three-lens corrector, auto guiding unit and a derotator instrument interface. The telescope has two side ports and one main Cassegrain port[3]. The present contribution describes the design of the enclosure and other related building for the DOT.

*vishal@aries.res.in; phone +91-05942-233727; fax +91-05942-233439; www.aries.res.in

1.2 Functional requirement of the enclosure

The design of enclosure and associated building for optical telescope is an important and challenging task as it involves the balancing of optical, thermal, mechanical and structural requirements. The `seeing' within the enclosure may be affected by thermal disturbances induced by the structure. The other challenge due to space constraint is designing of the enclosure of the DOT to provide facility for telescope installation and maintenance within the building. The main purposes of the enclosure and associated building are;

- Protection of the telescope from dust and inclement weather.
- Keep the air temperature in the optical path of the telescope as close to ambient as possible during observations.
- To provide isolated telescope pier for the telescope mount.
- Keeping instrument room at a stable temperature.
- Provision of facilities for operational and maintenance personnel.

1.3 Environmental condition at Devasthal

The enclosure and associated building are designed to meet all the required performances throughout the following range of environmental condition.

Normal conditions:

Temperature:	0 to 30°C
Variation in temperature during the night:	< 2 deg C
Relative humidity:	0 to 90%
Wind:	< 3 m/s (during 75% nights)
Peak gusts:	<13 m/s
Pressure	810mb (±10mb)

Survival conditions (Dome closed):

Wind gusts:		55m/s
Air temperature	max:	40°C
	min:	-20° C
Diurnal temperature	:	20°C
Seismic Ground Acceleration:		Zone V
Design Precipitation event:		35mm rainfall/hour with 10m/s wind.
Max Relative humidity:		100%
Snow		~ 4 to 5 feet on the ground level

2. Building design

The site for 3.6-m DOT is located at about 3 km from the State Highway and connected by a about 5 meter wide road. The enclosure and associated building plan is done by M/s PPS Pvt. Ltd Pune in close interaction with the ARIES scientists and engineers. The geotechnical investigation of the site was done by M/s Nagadi consultants Private Ltd. (India) and Civil department of Pantnagar University (India). The recommendations of these agencies were taken into considerations while preparing the design. Since the wind direction at Devasthal is from SE and SW during observing period, the auxiliary building housing is planned at about 15 degree north – east direction. The site plan is shown in Figure 1.

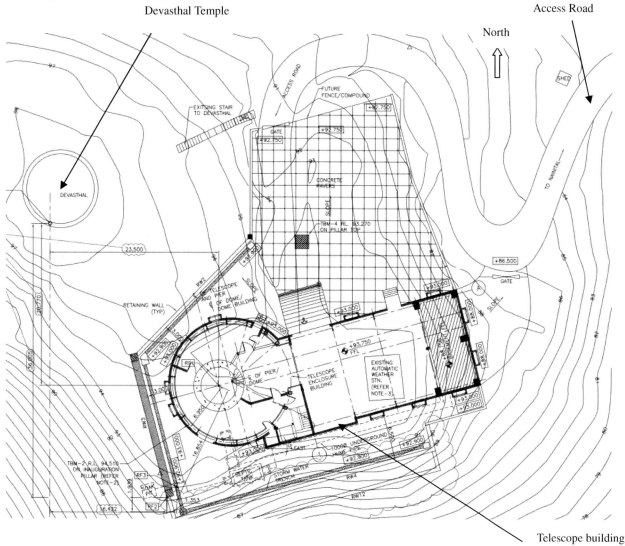

Figure 1: Devasthal Optical Telescope site plan showing telescope building

2.1 The layout of the enclosure and auxiliary building

The DOT facility consists primarily of interconnected but thermally isolated parts; telescope instrument room, control room, telescope pier, rotating dome and auxiliary building shown in figure 2. The ground level is divided into high resolution spectrograph, control and telescope accessories rooms. The auxiliary building contains space for mirror washing and coating. A rail system is included between the two buildings to facilitate a simple process for removing, cleaning, re-coating, and replacing the primary mirror.

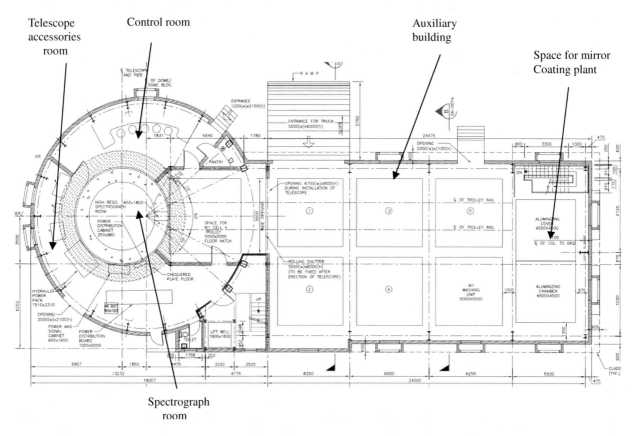

Figure 2: Floor plan at 0.0m level showing the auxiliary building, control room, telescope accessories room, spectrograph room, telescope pier

2.2 Telescope pier

A hollow concrete cylindrical pier will provide stable and rigid support for the 150 MT telescope mass at the top. The outer diameter of the pier is 7000mm and inner diameter is 5000mm. The top slab of the pier is 1000 mm thick. The base (foundation) of the pier is 9500 mm diameter and 1000 mm thick. The elevation of the pier is shown in Figure 3.The concrete grade for the construction of the pier is M25. The pier is completely isolated from other structures to avoid vibration transfer from other sources. Reinforced retaining wall all around the pier foundation with a clear gap of 150mm is provided to isolate the pier. The natural frequency of the pier and the telescope is 25.44Hz and 7.4Hz respectively. The combined frequency of the telescope and pier is 14.259Hz, which is significantly higher than that of the telescope to avoid the resonance.

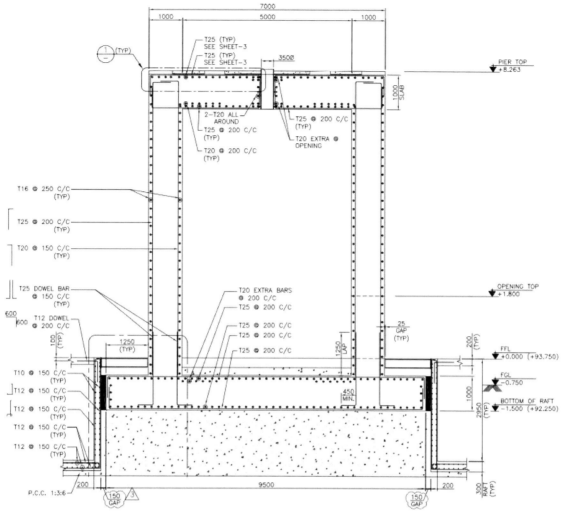

Figure 3: The pier elevation

2.3 The rotating dome and auxiliary building

The telescope enclosure building is separated in three parts (Figure 4) the dome, the non-rotating lower structure and an auxiliary building. The dome is the rotating insulated steel framed upper cylindrical structure with pitched roof. Considering the space limitation at the site an off-centered telescope pier has been planned to minimize the building size. Considering the gyration radius of the telescope (5749mm) and the space required for the assembly and future maintenance of the telescope, the diameter and height of the dome is kept as 16.5m and 13.0m respectively. A shutter opening is provided on dome surface to enable observations. Main structural components of dome are portal frames, the roof framing, slit shutter, dome ring beam. The rail and the wheels of the drive system are supported on bottom steel ring beam. The telescope floor level is at 11m height.

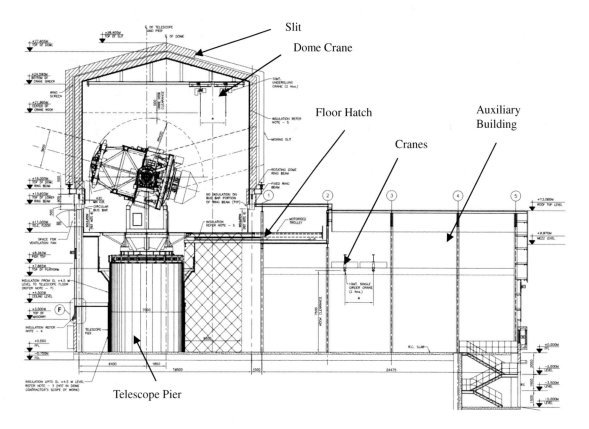

Figure 4: Telescope Building Section

The bottom ring beam is supported on the columns of a non-rotating lower steel structure which forms the second part of the building. In the present design the dome center-line diameter is 16.5m and the telescope pier is eccentric by around 1.85m with the dome center to get a space for a hatch of 5.5m x 5.5m to enable lifting of telescope parts from the ground floor. At ground floor there are various rooms viz. for equipment, control room, UPS systems etc (see figure 2). This building also houses a staircase and lift to have access on telescope floor.

The auxiliary building mainly accommodates aluminizing Plant, ventilation System and UPS room and component assembly & maintenance area.

3. Salient features of the enclosure

3.1 Dome drive

The proposed dome drive is of fixed type i.e. all the drive wheel assemblies and idler wheel assemblies are stationary and mounted on fixed ring beam. The rail is mounted on the bottom side of the rotating ring beam. The 170 MT dome will rest on 18 wheel assemblies. The dome will be rotated by 6 numbers of drive system. Other 12 wheel assemblies are idler wheel assemblies. The drive consist of variable- voltage, variable- frequency (VVVF) drive and a gear box to ensures smooth starting and stopping of the dome. Time taken to complete one dome rotation is 7.4 minutes. Interlocks are provided for dome drive system in such a way that dome drive motors shall not start when either dome is mechanically locked or overhead crane is in service. The salient features of the dome and drive system are given in Table 1.

Table 1: The salient features of the dome and drive system.

A	Mean dia of dome	16.5 m
B	Height of dome with slit	13.9 m
C	Total weight of dome	170 MT
D	Weight of Dome	140 MT
E	Weight of Slit	30 MT
F	Linear speed of dome	7 m/min.
G	Motor	1.5kW, 1395rpm
H	Gear box ratio	230
I	Wheel dia	400mm
J	Rail	CR 100
K	Spring Wire dia	48mm

3.2 Slit Cover Drive

The 30 MT slit cover is comprised of two parts moving in opposite direction. The slit is of inverted U shape as shown in Figure 5. Each part has two geared motors, thus there are total four drive motor limit switches provided on track of each slit to limit motion in either direction. Following interlocks are provided for slit cover operation; a) slit drive motors shall not operate when dome is rotating, b) if any of the four motors fails, the other three motors shall not start (we have to decouple the failed geared motor). Main design specifications of slit drive are given in Table 2.

Table 2: Main design specifications of slit drive.

A	Weight of the slit cover	30MT
B	Linear speed of the slit cover	6 m/min.
C	Travel length of rail	5.2 m on each side
D	No. of drive wheel assemblies	4 Nos.
E	No. of idler wheel assemblies	4 Nos.
F	Wheel diameter	200 mm
G	Motor	0.55 kW, 1360 rpm
H	Gear Ratio	138.4

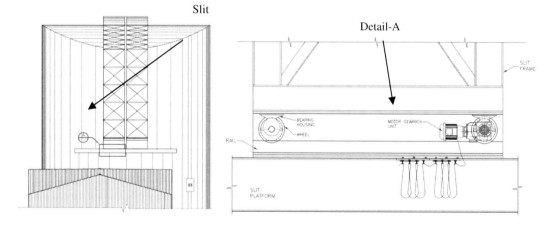

Figure 5: Arrangement of slit drive system

3.3 Hatch Cover

A hatch of 5.5 m x 5.5 m is provided on the telescope floor at 11.0 m level. It has a moving hatch cover. During erection period and thereafter during primary mirror aluminizing operation, this cover will be moved outside the circular structure on rails provided. The hatch cover shall move at 7m/min and stop at each end at desired location. The operation will be manual only with the help of control panel located on the telescope floor level. At all the four limits of the rail, limit switches are provided to control the desired motion of the hatch cover during opening or closing operation. When the overhead crane in the dome is in service the hatch cover drive shall not operate. The hatch cover drive consists of geared motors, wheel assemblies, electrical control panel, electrical and mechanical stoppers and other accessories as required for smooth functioning of the hatch cover motion. The main design specifications are given Table 3.

Table 3: Main design specifications of the hatch cover.

a	Speed of Hatch Cover	7 m/min.
b	Capacity of the Hatch Cover	15 MT
c	Self weight of Hatch Cover	6.5 MT
d	Travel Length	6374mm
e	Span	5900mm

3.4 Insulation

The entire dome structure, roof and slit will be insulated from inside with Phenotherm insulation panels having specifications as given in Table 4. The Phenotherm panels shall be CFC free and eco-friendly. The material is not easily ignitable and classified as per BS 476 – Part 5.

Table 4: The specifications of insulation panels

a	Material of Insulation	Phenotherm
b	Density	40 Kg/Cum
c	Thickness	75 mm and 50 mm
d	Compressive Strength	250 kN/Sqm.
e	Thermal Conductivity (K' Value)	0.018 W/M deg. K
f	Temperature Range	- 196 deg.C. To 130 deg.C.
g	Water Absorption	2% by volume
h	Panel size	about 1.2 m x 1.2 m

3.5 Wind Screen

A wind screen is provided to protect the telescope observation when wind velocity is more than 5meter per sec. It is located on the top of rotating ring beam of dome. The wind screen is similar to a rolling shutter but the shutter box installed at bottom and shutter moves in upward direction. The shutter movement is motorized with the facility to synchronize with telescope movement.

The wind screen shutter is made up of synthetic fabric with embedded strengthening bars or tubes. Two edges of the top end of the wind screen are connected to a dead weight. The motor and gear box will hold the wind screen from unwinding. The motor can rotate in both direction and release the screen. It will go up with the help of a counter weight, and to take the screen down the motor will pull the counter weight (see figure 6). The motor shaft is provided with extension of shaft on non-drive end so that in future encoder arrangement to sense the location / opening of the wind screen can be provided. The major parameters of the wind screen are given in following Table 5.

Table 5: The specifications of wind screen.

a	Opening Size	4200 mm wide x 9000 mm height
b	Material of Wind Screen	Synthetic Fabric
c	Wind screen to withstand at	15 m/sec wind speed
d	Operating Speed	3-6 m/min.
e	Manual Operation	By Crank or Ratchet
f	General Motor Rating	415 V, 3 ph, 0.55 kW

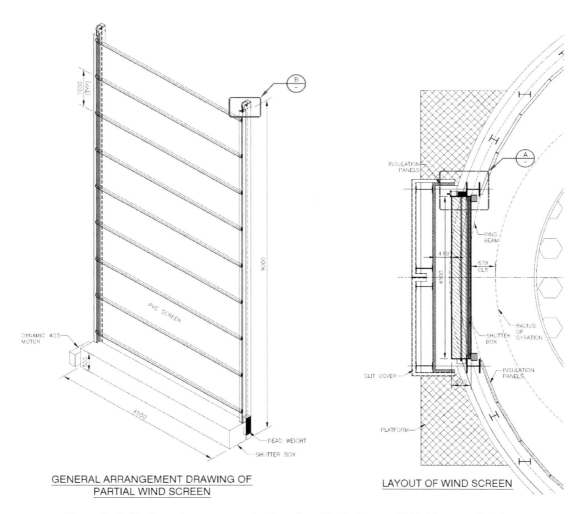

Figure 6: (Left) General arrangement drawing of partial wind screen. (Right) Layout of wind screen.

3.6 Telescope Floor and Building Temperature Control

To achieve best seeing inside the dome, the telescope enclosure must be designed to minimize the effects of convection, turbulence etc inside the dome. In order to achieve this goal a well ventilated telescope chamber, low thermal inertia construction, active ventilation using fans, and utilization of surface coating to control radiation losses etc are planned to be used[4].

The present enclosure and auxiliary building is designed using steel frame and metal clad panels so that the building can thermalise rapidly. The interior ventilation is designed to exhaust waste heat in prevailing downwind direction. Within the enclosure building the telescope floor and spectrograph room are properly insulated. The exterior of the building will be white in color to minimize solar heating. In spite of proper outside coating the inside area of the enclosure may get heated. Five numbers of ventilation systems are provided to remove hot air and maintain temperature equal to that of outside temperature. The ventilation plan is shown in Figure 7 and Figure 8.

3.6.1 Ventilation System for Telescope Accessories Room

There are number of equipment in the telescope accessories room such as Hydraulic Power Pack, Compressor, Air Dryer, Chiller, Electrical Control Panel etc. which generate heat during the normal operation. In order to exhaust the hot air from the room and release it at sufficient long distance from the telescope, centrifugal exhaust fans are located in the room at minus level in the auxiliary structure. Total heat generated in the telescope accessories room is estimated 35Kw, hence an air flow of about 15800 Cum/hr is required. There are 3 number of centrifugal exhaust fans (2 working + 1 stand-bye). From the telescope accessories room to ventilation room an underground hume pipe is laid. Two ventilation fans will suck the hot air from telescope accessories room and release it in the ventilation room. The capacity and static head of the exhaust fans are 7900 Cum/hr and 35 mm of WG (Water Gauge).

Three fans having capacity and static pressure of 13000 Cum/hr and 5 mm of WG are located near the hydraulic unit and chiller. These fans will assist in heat removal during any time / maintenance period. Eight exhaust fans having capacity and static head of 6000 Cum/hr and 5 mm of WG, respectively are provided in auxiliary building.

All the exhaust fans are provided with louvers which automatically close when fans are not working. The arrangement and construction of louvers is such that during rainy season water will not enter inside the room.

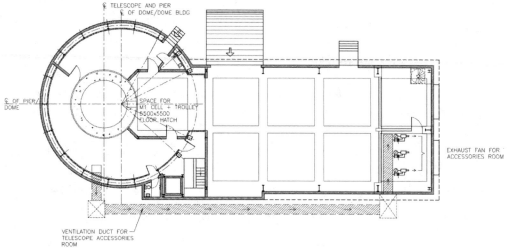

Figure 7: General arrangement of Ventilation system.

3.6.2 Ventilation System for Telescope Floor

During the day time the structure exposed to the Sun may get heated. The absorbed heat will be dissipated during night. Although dome is insulated from inside, some temperature difference is expected between atmospheric temperature and inside dome temperature. To remove the hot air from the inside of the dome and maintain the temperature equal to that of outside, 12 exhaust air fans (see Table 6) are provided on circumferential wall of the dome as shown in the Figure 8.

Table 6: The details for the exhaust fan

a	Type	Axial Flow
b	Capacity	75000 Cum/hr.
c	Static Pressure	3 mm of WG
d	Total Pressure	18 mm of WG

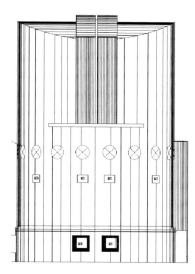

Figure 8: Ventilation system at telescope floor.

3.6.3 Centrifugal Blower for Pier

The high resolution spectrograph is planned inside the pier. Some electrical equipment may also be placed in the space inside the pier which may generate small amount of heat. The generated heat is proposed to be exhausted into telescope accessories room by exhaust fan. An opening of 300 mm is provided in the pier to exhaust the air from inside the pier. A centrifugal blower mounted on independent frame exhaust the air. Care is taken that blower vibrations are not transferred to the pier. The capacity and static head of the blower are 1300 Cum/hr and 10mm of WG, respectively.

3.7 Material Handling System

As per installation and maintenance requirements two 10 ton capacities under slung cranes are provided in the dome. The crane girder is suspended and supported from portal columns. Two 10 ton capacity single girder cranes operating in tandem in auxiliary structure are provided for loading/ unloading of telescope components, assembly operation and for loading component on the transfer trolley. After installation erection of the telescope these cranes will be utilized for maintenance purpose.

3.8 Schematic view of the enclosure

A schematic view of the enclosure described above is shown in Figure 9.

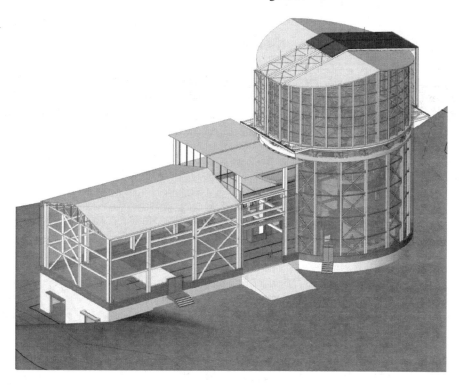

Figure 9: View of the enclosure building

4. Acknowledgements

The authors are thankful to Prof. Ram Sagar, Director ARIES for encouragement. Authors are also thankful to Prof. S.N Tandon, Prof T.P. Prabhu, Er. S.C. Tapade and all the scientists and engineers at ARIES for stimulating discussions and useful comments to improve the design of the above mentioned enclosure.

5. References

[1] Sagar R., Stalin C., Pandey A.K. et al (2000), A &AS, 144,349
[2] Stalin, C. S., Sagar, R., Pant, P., Mohan, V., Kumar, B., Joshi, Y. C., Yadav, R. K. S., Joshi, S., Chandra, R., Durgapal, A. K., and Uddin, W., "Seeing and micro thermal measurements near Devasthal top," Bull. Astron. Soc. India 29, 39–52 (2001).
[3] Ninane, N., Flebus, C., and Kumar, B., "The 3.6 m Indo-Belgian Devasthal Optical Telescope: general description," in [Ground-based and Airborne Telescopes IV], Proc. SPIE 8444 (2012).
[4] Blanco D & Johns M, "Thermal Design of the WIYN 3.5m Telescope Enclosure," Proc. SPIE Volume 2199, 743-753(1994.2199/743

An innovative alt-alt telescope for small observatories and amateur astronomers

M.Riva[a], S.Basso[a], R.Canestrari[a], P.Conconi[a], D.Fugazza[a], M.Ghigo[a], M.Landoni[a], G.Pareschi[a], P.Spanó[a], R.Tomelleri[b], F.M.Zerbi[a]

[a]I.N.A.F. - Osservatorio Astronomico di Brera
[b]Tomelleri s.r.l.

ABSTRACT

This paper want to show an innovative amateur oriented telescope with an unconventional alt-alt configuration. The goal is to make a telescope with good optical quality reducing production costs by adopting a gimbal based mounting to develop an alt-alt configuration suitable for a telescope. Reduce costs while preserving the optical quality is a necessary condition to allow small groups of amateur astronomers, schools and cultural clubs, with reduced economic resources, to acquire an astronomical instrument that encourages learning and advancing astrophysical knowledge. This unconventional mechanism for the realization of a telescope alt-alt provides significant advantages. The traditional rotary motors coupled with expensive precision bearings are replaced with two simple linear actuators coupled to a properly preloaded gimbal joint and the cell becomes the primary structure of the telescope. A second advantage would be secured by mechanical simplicity evident in the easy portability of the instrument. The frame alt-alt has some limitations on the horizon pointing but does not show the zenith blind spot of the alt-az mount. A dedicated alt-alt pointing and tracking model is under development to be compatible with commercial telescope softwares and with the proposed new mounting.

1. INTRODUCTION

Medium size alt alt telescopes have almost entirely replaced equatorial mounts in recent years due to their greater versatility, the presence of Nasmith gravity "invariant" foci and the possibility of being placed in various locations with a simple GPS GEO-location and reconfiguration of pointing model.

On the other hand this configuration exploit some disadvantage, like increased mass and costs due to heavy mechanics and electronics; moreover the required field derotation impose a mechanical blind spot at the zenith due to the high rotation speed (ideally infinite right at the zenith). This is often considered a problem for astronomical observations, since they are precluded in an area where astronomical events and atmospheric transmission are characterized by superior quality. An alt-alt configuration overcomes the problem transferring the criticality of the zenith point to the horizon where the astronomical observations are relatively less interesting relatively lose interest.[1]

INAF Brera has recently completed a feasibility study for a fast pointing 3-meter telescope[2,3] the alt alt configuration was really considered as a technically viable solution. Despite the promising results derived from the studies, the wide adoption of these configuration in astronomy is currently limited by the lack of an engineering development and prototyping. Thus the increased costs do not justify for the moment the quite better performances.

The amateur and professional telescopes markets have historically been separated by a large gap in costs and technology. Over the last ten years scientific cases like follow-up transient, long-term monitoring of variables, etc.., difficult to implement in large professional telescopes but adequately covered by an intermediate class of telescopes (60 cm 1.2 m) have been established. They are managed either by amateur groups or professional entities. Among these we can advise REM at La Silla, STELLA in the Canary Islands and SONG the multiple spectroscopic telescopes project funded by the European Union.

Further author information: (Send correspondence to M.Riva)
M.Riva: E-mail: marco.riva@brera.inaf.it, Telephone: +39 039 5971 105

Although the unit price still very high (from 300 to 800k according to size and performance) many telescopes of this size have been sold and installed over the last ten years by a few producers in the market. In addition several public small observatories in the Italian territory are now under procurement (Astronomy of Madonie Park, Antola Regional Observatory, etc..).

An expansion of this class of telescopes is desired at the international level (see "The Alt-Az Initiative" U.S.[4]) to increase their observational skills in astronomy in general and in the scientific cases mentioned above in detail. However the main obstacle to huge diffusion is still the cost.

This paper wants do describe an INAF funded project that proposes to cover a possible way to reduce costs through a radical simplification of the telescope mounting by developing the alt alt configuration.

2. OVERALL DESCRIPTION

The project baseline is an alt-alt amateur telescope with a 600mm diameter main mirror, the mounting key point is a gimbal joint that enable the required motion keeping an high pointing precision with reduced costs. The optical design currently almost frozen is a parabolic mirror coupled to a spherical one. This require a double lens corrector but allow the procurement of a relatively cheap main mirror considering the massive scale production of this elements. The development of a software pointing that takes account the alt alt motion is under development starting from a previously well tested alt-az based one.

Summary of Main technical characteristic:

- ALt alt mounting
- Parabolid Primary Mirror
- Pointing and tracking ($60min$) precision 1 arcsec
- Angular speed $5°/sec$
- Minimum declination angle $15°$

The proposed mounting is sketched in Figure 1 and was patented by the company in Italy TOMELLERI*.

Figure 1. Conceptual view of the alt alt telescope.

---

*Patent No. VR2010A000170 deposited in Verona on 03/09/2010

2.1 Support and drive.

The support of the cell of the primary mirror is made by means of three parts consisting in a gimbal joint and two linear actuators. The drivers are fixed nearly to the vertexes of an equilateral triangle disposed at the center of the lower surface of the mirror holder cell. The gimbal joint constraints four degrees of freedom of the cell: the three translations and the rotation along the vertical axis, while the two remaining degrees of freedom are constrained by two linear actuators. The motion of the two linear actuators allows to point the cell in all the orientations within the solid angle of a cone with the symmetry axis coincident with the telescope axis and an amplitude of approximatively $75°$.

2.2 Mirror's orientation measurement

The gimbal joint is made by of three parts, the base of the main shaft which is fixed to a column welded to the base plate of the frame, the orthogonal shaft which is fixed to the cell, and the intermediate body that is interposed between the other two. The intermediate body of the cardan shaft is free to rotate according to a declination axis whose angle of rotation can be measured directly with a precision rotary encoder (Figure 2); the second shaft rotation can be measured in the same way with a similar encoder (Heidenhain). The set of this two measurements provides uniquely the orientation of the cell, related directly to the direction of the axis of the mirror, and is not affected by any structural deformation of telescope support.

Figure 2. Precision angular encoder.

2.3 Mirror support

The mirror's cell support will be made adopting the concept of a whiffletree.[5] This solution keeps the basic principle of the kinematic three-point support system, but spread the load supported by each of the three points over a larger number of points on the mirror. To preserve the kinematic nature of the system, support points on the back of the mirror are grouped by twos or threes and mounted on pivots.

The three main constraints points are located near the cardan shaft and the two linear actuators in order to directly download the weight from the whiffle tree to the basement. The lateral support of the mirror is formed by three rods fixed tangentially to the lateral surface of the mirror in this way the mirror is constrained so isostatic by three axial constraints and by three lateral restraints.

A preliminary tree optimization has been done with a dedicated software called Plop. The best compromise obtained in terms of connection point Vs mirror weight for a $60cm$ consider a mirror thickness of $1/10 = 60mm$. The tree includes 18 connection point (Figure 3): 6 are located in an inner ring with $134mm$ radius and 9 in an outer with $252mm$ radius. The connection point are grouped in six triangle, one vertex is a inner connection point while the other are two outer connection points. Each pair of triangles is linked by a connecting bar. The baricenters of the three connecting bar are the three main connection points of the tree to the rest of the cell and are located an a radius of $180mm$.

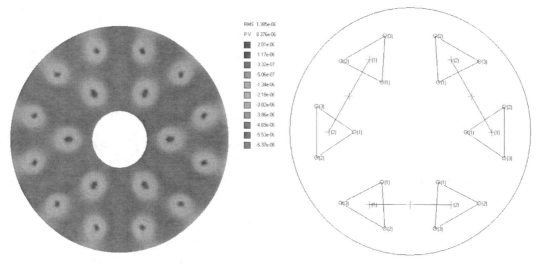

Figure 3. M1 gravity deformation (left) with 18 connection point whiffle three (right).

3. OPTICAL LAYOUT

The cost of each optical element strongly depends on the required surface quality and the complexity of its form as a result the demonstrator will be a compromise between quality optical image and the total cost of the optics. The optical layout has been then tailored to minimize cost through the adoption as much as possible of off-the-shelf components. Where this is not possible, the acquisition of raw optics with surfaces as close as possible to the nominal ones is foreseen; by means of the ion Beam Figuring which is a technology already developed at INAF-Brera the breadboard optics will be modified to get the desired shape.

In particular M1 will be a commercial[6] parabolic $600mm$ mirror with a focal length of $2400mm$ and a thickness of $6cm$. M2 is designed to be a spheric convex Mirror with a diameter of $250mm$ and a curvature radius of $5082mm$, the manufacturing of this mirror can take advantage of off-the-shelf spheric concave reference blank.[7] The field correction can be made through two S-BSL7 Ohara spheric lenses with a clear aperture of $125mm$. Finally M3 can be a simple flat elliptic mirror easy to found on the market.[6]

Summary of optical Constraints and assumptions:

- $600mm$ diameter parabolic primary mirror to Orion.uk
- Secondary mirror diameter 125 spherical radius (master to Edmund)
- field Correction: two elements in S-BSL7 Ohara useful diameter 125mm + mirror plane
- Distance between the corrector and M1 $< 450mm$
- Focal extraction of $350mm$ and $135mm$ from the top of the primary
- FOV 0.5^o over $400 - 700nm$

4. PRELIMINARY FEM

In order to dimension and optimize the telescope structure a simplified Finite Element Model has been set up. The whole model is done through 2D elements for the plate and 1D element for the truss rods. The optical element has been modeled through semi-rigid element with proper mass and inertia characteristics. The gravity displacement under X, Y and Z direction has been evaluated and introduced via a Matlab routine into the optical model to obtain the image displacement and quality degradation. Considering that within a certain amplitude the flexure can be directly compensated by the pointing model the optimization has been mainly driven by the optical quality. The model is made by 23268 elements and 23318 nodes.

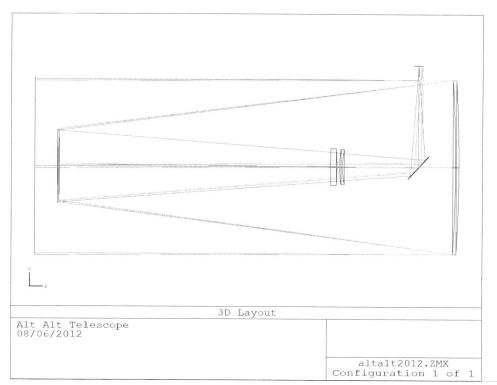

Figure 4. Telescope Optical Layout.

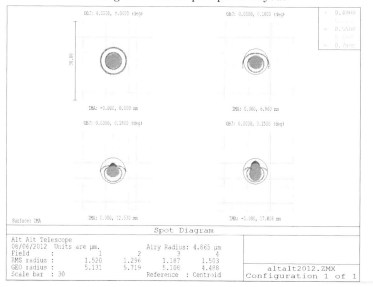

Figure 5. Telescope Optical quality.

5. POINTING AND TRACKING SOWFTWARE

The variation of parallactic angle combined with the physical behavior of the mountings, , induces a field rotation during observation that must be compensated to keep the star in the desired position on the focal plane of the instrument. Of course this rotation depends on a number of parameters: the latitude of the observer, the coordinates of the object and the focal position where the instrument is mounted. Different corrections are required if the telescope has an alt azimuth motion or an alt alt one as can be seen from Figure 7

We note that the meridian passage for an object at the zenith corresponds to a velocity of ∞. A comparison of the alt cases zenith at low speed (field) and even when observing the celestial equator this velocity vanishes

Figure 6. Telescope Finite Element model overall views.

because having an axis parallel to the axis of the earth acts as a matter of fact creating a pseudo-equatorial configuration. The following table shows the different behavior for an alt az telescope (above) and alt (below). On the left is shown the parallattic angle Vs the corner zone to the right while the speed of rotation of the field vs. the angle of azimuth.

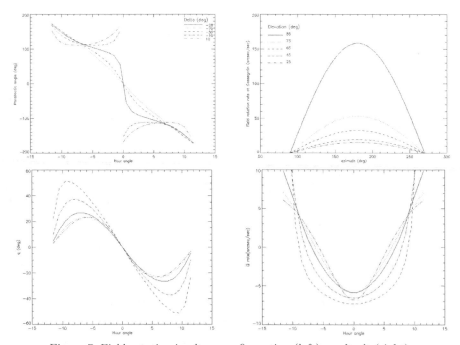

Figure 7. Field rotation in alt az configuration (left) or alt alt (right).

6. CONCLUSIONS

The weight of the gimbal mount structures is very low and the actuation systems are particularly simple since the movement is driven by the linear actuators worm. This particular notation keeps the overall complexity and cost quite reduced. Furthermore, considering the needs of amateur and small groups to conduct observations from different "viewpoints", portability is a major strength of the proposed system. The bred board under development is also conceived to demonstrate the scalability of these technologies to larger telescopes offering further developments and applications (even scientific type).

REFERENCES

[1] A. Meinel, "Astronomical telescopes," Applied Optics and Engineering, volume = 5, 1969.
[2] F. Vitali, G. Chincarini, M.Zannoni, and all., "A path to the stars: the evolution of the species in the hunting to the grbs," *Ground-based and Airborne Telescopes III* **7730**, pp. 77330W–1/ 77330W–14, SPIE, 2010.
[3] M. Riva, F. Bortoletto, and C. B. et al., "Mechanical design considerations for a 3m class fast pointing telescope," *Ground-based and Airborne Telescopes III* **7730**, pp. 77330X–1/ 77330X–14, SPIE, 2010.
[4] "The alt-az telescope initiative." http://www.altazinitiative.org/index.htm. Accessed: 30/09/2011.
[5] P. Bely, *The Design and Construction of Large Optical Telescopes*, Springer, 2003.
[6] "Orion optics uk." http://www.orionoptics.co.uk/OPTICS/opticsprofession.html. Accessed: 30/05/2012.
[7] "Edmund optics." http://www.edmundoptics.com/products/. Accessed: 30/05/2012.

Prototype Enclosure Design For
The Korea Microlensing Telescope Network (KMTNet)

N. Kappler, L. Kappler, TBR Construction & Engineering (United States); W.M. Poteet, H.K. Cauthen, CP Systems, Inc (United States); Byeong-Gon Park, Chung-Uk Lee, Seung-Lee Kim, Sang-Mok Cha, Korea Astronomy and Space Science Institute (Korea, Republic of)

ABSTRACT

TBR Construction and Engineering (TBR) has under development for the Korea Astronomy and Space Science Institute (KASI), a project to provide three 1.6 meter optical telescopes observatories in three southern countries: Chile, South Africa, and Australia. The contracting team has chosen to develop a full scale prototype of the observatory. This will become a functional assembly and testing facility for all three project telescopes in Tucson, Arizona. This prototyping concept is meant to allow the optics team to make changes to the observatory as needed for the scientific mission while minimizing the expense of making changes in remote countries.

Keywords: KMTNet, Microlensing, Observatory Construction, Dome, Enclosure Building

1. INTRODUCTION

TBR Construction & Engineering (TBR) is responsible to design and deliver three semi-identical observatories in three southern hemisphere locations. This project is known as the Korea Microlensing Telescope Network (KMTNet). TBR is located in Tucson, Arizona, USA. The KMTNet observatories will be in Chile, South Africa, and Australia. Many of the key suppliers, subcontractors, and consultants are located in Arizona, USA, including CP Systems Inc, M3 Engineering, and Vroom Engineering and Machine.

TBR proposed to build a local implementation of the KMTNet observatory in Tucson, in order to significantly reduce the risk of costly corrections in the field. In addition, the local implementation would serve as an accurate test facility for the interface between the Telescope System and the enclosure building.

Following the conclusion of the project Critical Design Review and the production of the 100% Construction Documents by M3 Engineering, TBR began construction on the local Tucson prototype building in October, 2011. Now that the Tucson prototype construction is complete, the many advantages that it offers can be discussed in this paper.

Figure 1 Photo of Tucson's Full Size KMTNet Observatory Prototype

2. ENCLOSURE BUILDING

The enclosure building has a 9.2 meter dome, an attached control room, computer room, equipment room, and a small break area. The equatorial telescope will mount to an isolated concrete pier foundation with three embedded anchors. This building's design criterion is to withstand 135 MPH wind load. See the following floor plan.

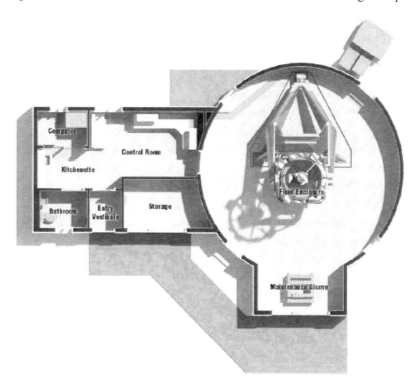

Figure 2 Floor Plan Rendering

3. ADVANTAGES OF FULL SCALE MOCK UP

3.1 Confirming Construction Details

One major benefit to constructing a prototype in Tucson is to be able to evaluate all the construction details in a field condition nearby. Since our project plan required TBR to hire building contractors who are local to the southern hemisphere, a major concern was that some construction details would be too specialized and difficult to communicate. This would inevitably result in an increase cost to the entire project. During the construction, we had our own team of builders and carpenters follow the construction documents very carefully. This experience allowed us to make notes where some details could be improved for ease of construction. Upon the completion of each construction discipline, our building architect and engineers walked through the work in progress and made field comments, and ultimately approved the recommended changes.

3.2 User Group Walk Through

Now that the construction in Tucson is complete we have a fully functional observatory building. We invited our customers, the project members at KASI, to make a site investigation and offer their comments and suggestions. This second look by the User Group and overall project team at the building design resulted in several positive changes.

3.3 Improving Design Choices

In our Critical Design we had a floor trench formed in concrete as a path for all cable and hoses to connect equipment and computers in the telescope chamber. Once all team members saw this trench in the Tucson prototype, a discussion was held on the pros and cons of such a trench. The result was that most team members agreed that the trench would be too cumbersome to access, as well as too prone to collecting dirt, water, and pests. The team therefore concluded to change the trench for a cable tray mounted overhead.

3.4 Adding Scope

The next improvement that we made was concerning the equipment enclosure. It was originally planned as an exterior concrete pad without walls and a roof. However, the Camera Vendor, Ohio State University, project team raised the requirement for their equipment to be housed in a climate controlled and weather proof enclosure. During the onsite review of the prototype we decided to incorporate the equipment pad, and make it an equipment room under the roof. Making improvements like these during the early phase of the project rather than later in a host country, thousands of miles away, helps to ensure the success of this project.

3.5 Mechanical Fit Check

We needed to perform a mechanical fit check of the actual telescope mechanism and the planned 9.2 meter Ash Dome. Vroom Engineering, in Tucson Arizona, fabricated the telescope structural components, the Base, Right Ascension ring, and the Optical Truss Assembly. Since Vroom Engineering is located within 2 miles of the prototype site, the frequent collaboration was more convenient. Vroom Engineering delivered the nearly complete steel structures to the prototype and our team, led by CP Systems, Inc, carefully craned the different pieces of steel into the chamber and assembled the telescope mechanism. To see video footage of this process please use the following QR code links:

Links to videos of setting telescope into Tucson Prototype (parts 1-4)

3.6 Mock up of the Camera Assembly

We wanted to demonstrate that the dome clearance was adequate to support safe access to the primary focus camera at the head ring. Unfortunately, the camera was not yet fabricated to support this early mechanical fit check. Therefore, we decided to build a mock-up of the camera and head ring out of plywood. This mock up would suffice to test the clearances between the telescope and the dome walls.

Figure 3 Mock Up of Camera, Corrector Lens, Head Ring

3.7 Result of the Mechanical Fit Test

Again, our project team rallied to the site of the prototype in Tucson to witness the mechanical fit. The KASI team, and the OSU team joined with TBR to demonstrate how the telescope fit comfortably inside the dome. The end result of the mechanical fit test was a complete validation of the dome size. We successfully determined that the telescope instrument, with the camera and head ring in place, would fit adequately inside the dome.

4. FLOOR PLAN OF THE KMTNET ENCLOSURE

4.1 Computer Room

The Computer Room is designed to house the computer racks for the KMTNet program. There are three computer systems to be located there: Telescope Control System, Camera System, and the Observer System. It is separated from the Control Room by insulated walls and a sliding glass door. This room has a dedicated Fan Coil to keep it under refrigeration at all times. It also has dedicated circuits for unique computer system.

During the course of construction of the Tucson Prototype, it was discovered that the Computer Room may be too small for all the equipment that would be located there. Making the Computer Room larger for the prototype was not a very big problem. We enlarged the room by a few inches on each side. When the impacted parties came on the site to inspect the building, the size of the Computer Room was verified and it was found to be acceptable. This was one benefit of building a full size mock up building. Changes like this are easy and low cost while the construction is in Tucson. They are much higher cost when the work is performed out of the county and by separate contractors.

4.2 Equipment Room

The equipment room will provide enclosed space that is protected from the weather, heated and cooled, for the cyrotigers and chillers for the Telescope and Camera systems. At first this equipment room was going to be an exterior concrete pad only. But as the design of the Camera System caught up with our building design, it became clear that this needed to be both protected from the weather and minimally heated and cooled. Therefore, during the prototype phase, we switched the design of this space to be incorporated under the Control Room roof. The Equipment Room has open air louvers on both sides, allowing fresh air to flow into the room both passively and forced by exhaust fans. The air from the Equipment Room is isolated from the rest of the building. We wanted to isolate this room from the telescope enclosure room to ensure that heat exhaust or other fumes were unable to contaminate the conditions of the telescope environment. A custom raceway for cabling and hoses was made to connect the Equipment Room to the telescope. The following photograph shows the external chase between the two areas.

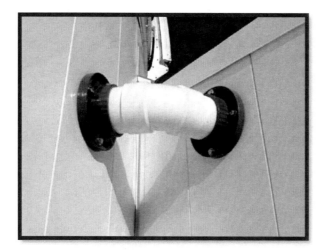

Figure 4 Isolated connection between equipment Room and Telescope Enclosure

4.3 Control Room

The Control Room is located between the Telescope Enclosure and the Computer Room. It has built in counter tops which will be used as desk space, chairs, and room for small furniture. It also has a break area, and a toilet room. This room has access to a cable tray that runs between the Computer Room and Enclosure. .

4.4 Enclosure Room and Ash Dome

The Enclosure Room, or chamber, is a radius room that houses the telescope pier foundation. It has a diameter of 30' and supports an Ash Dome. This is the largest astronomy dome that is available through Ash Manufacturing. A Key consideration for the Tucson Prototype was the installation of this dome. For this purpose, TBR set aside a crew of 4 skilled carpenters to join with the Ash Dome installation team. The dome was installed in 2 weeks in Tucson. Our crew received a great education in installing these domes. Our plan for the KMTNet project is to send this trained team overseas to install the domes on the final installation for the three KMTNet observatories. The following picture shows a in progress photo of the dome installation.

Figure 5 Photo of Ash Dome Work in Process

5. BUILDING SYSTEMS

5.1 Orientation of the Telescope Pier Foundation

As the KMTNet telescope is a primary focus instrument, we must orient the concrete pier and anchor bolts precisely to the polar access. One difference between our Tucson Prototype and the future actual KMTNet implementations is that we are building in the Northern Hemisphere. TBR decided to layout the telescope pier aligned to true North, or aligned to Polaris. In the final locations of the KMTNet observatories, the concrete pier will be oriented to true South.

At first, our construction technique was to use conventional construction survey GPS to obtain a North South reference line on site. We were suspicious that this would not be accurate enough for our telescope system. So the team members at CP Systems, Inc inspected our construction stakes to verify. Using a theodolite, they set up directly over one survey monument. Then they sighted Polaris. Transferring Polaris's position down to a new construction stake, we noticed that our conventional north south line was about 6 inches off in 100 feet. With the new Polaris mark we now had established our north south reference line which will be used all throughout construction of this observatory in Tucson.

5.2 Telescope Pier

This photograph shows the telescope concrete pier, aligned with a north-south reference line. Also note that the anchor bolts are cast in place with a custom template to ensure precise alignment.

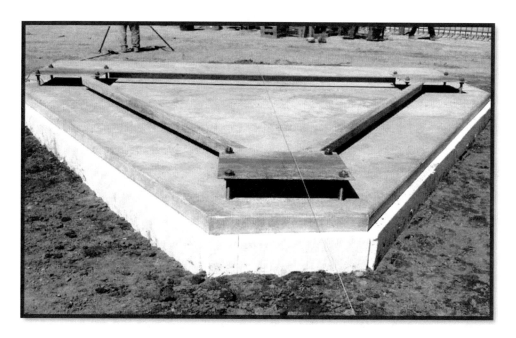

Figure 6 Concrete Pier for Telescope

5.3 Building Foundation

Typical of any mountain top observatory, we expect the three sites of the KMTNet observatories to be rocky soil or solid hard rock. Our design requires the telescope pier to be poured against existing intact rock. The building is then held down with a series of drilled rock anchors and a relatively shallow building footing. There are 24 drilled rock anchors holding down the building to counteract the expected wind load of 135 MPH.

However, in our Tucson prototype location, we have no existing rock to deal with. We have sandy and clay soil, which is easily excavated. For this condition, the construction of the foundation in Tucson is slightly different than the eventual foundations in the actual destination observatories. We omitted the rock anchors.

Figure 7 Building Foundation

5.4 Building Frame

The KMTNet building shell is comprised of light gauge structural steel studs, with steel stud track on the bottom and top of each stud. To create the radius walls in the chamber, we used a flexible steel stud track system. These flexible framing members are able to be formed to our custom radius. We created a plywood template cut to the radius that we needed. Then we used this plywood as a jig to form all the pieces of the track. These materials are easily accumulated in Tucson and shipped ready to install. The following photos show the flexible track system used to create the radius walls.

Figure 8 Flexible Track System   **Figure 9 Radius Walls at Chamber**

5.5 Exterior Walls.

The exterior stud walls all receive shear support from either exterior plywood panels or 20 gauge sheet metal panels.

Figure 10 Exterior Wall are sheared with plywood or sheet metal

The final finish on the exterior walls and roof are architectural metal panels. The white color was selected for its reflective properties. There are no exposed fasteners. Our structural engineers have calculated that this system can withstand the 135 MPH wind pressures. Another benefit to using these wall panels is they do not require painting in the field. This means there is less work to be done during the installation and less ongoing maintenance of the building. These panels have a 20 year guarantee on their paint finish. The metal panels are fabricated in the USA and accumulated in Tucson. They are shipped ready to install.

Figure 11 Photo of Architectural Wall Panels

5.6 Doors and Windows

To facilitate the passive cooling of the telescope chamber during observation, there is included 5 operable windows. These windows will be opened to allow passive air flow into the chamber at night. The windows are aluminum casement style and they swing on a hinge. Unlike most windows, these windows allow no light to come into the chamber when they are closed. They have laminated aluminum panels covering their glass openings. There are no windows other than in the enclosure chamber. The reason for this is to reduce the stray light that is allowed in the building.

The exterior doors are made from US style hollow metal frames and doors, supported by heavy duty commercial hardware. The largest door opening is in the alcove of the enclosure room. It is a pair of doors 8 feet wide and 8 feet tall. This opening is the intended opening for delivery of the primary mirror. All other doors are also heavy duty metal frames, doors, and hardware and intended to perform well with the expected windy conditions.

All the doors, frames, and hardware were accumulated in Tucson and can be shipped ready to install.

5.7 Interior Finishes

The interior walls will get sanded, filled, and factory painted white plywood panels, fastened to the stud walls with exposed fasteners. The reason this was chosen as the wall finish was to avoid the use of drywall and paint in the field. The panels provide the benefit of being removable, in case of additional circuits or utilities of any sort that would like to be in the wall cavity. Also the plywood is easy to surface mount conduit or other equipment to. All the plywood panels can be prefinished in Tucson and then shipped ready to install in the southern observatory sites.

6. DEPLOYING THE ENCLOSURE BUILDING TO THE FIELD LOCATIONS

Now that the full size mock up, or Tucson Prototype, of the KMTNet is complete, work has begun on the export plan for all the building materials to the remote southern hemisphere observatory locations. For this project, TBR decided to purchase three high cube shipping containers, rather than rent the container from the shipping company. The reason for this was so that the containers could be used for secured storage while at our site and when it arrives at the host observatory. Two containers will be sufficient for the building materials and one container for the Telescope System.

6.1 First Deployment to Chile

The first installation of the KMTNet Observatory will be at Cerro Tololo International Observatory, Chile. The forwarding agent, DHL Global Forwarding, arranged to pick up two containers in Tucson on May 24, 2012, and has provided an estimated time of arrival in Val Parasio, Chile as June 19, 2012 or approximately 4 weeks ocean shipping time. The containers with the building materials will arrive in Chile just as the Chilean building contractor has completed work on the foundation and is ready to start erecting the building.

7. CONCLUSION

Building a full size mock up observatory as a prototype in Tucson has proven to be an invaluable tool for the delivery of the KMTNet project. TBR Construction & Engineering, as the USA prime contractor located in Tucson, Arizona, responsible to design and deliver three observatory buildings and three 1.6-Meter telescopes for the KMTNet project in the Southern Hemisphere, understood the risk and expense of building for the first time overseas was considerable. To mitigate this risk, we decided to build a full size construction prototype, adhering to actual construction details as much as practicable, so that we would have a local working model. This model, or test facility, allows for us to realize every construction detail, every building technique required, before deploying it overseas. The initial advantages of the prototype effort are:

- Ability to refine the building construction details, to ensure workability for overseas contractors.
- Ability to perform advanced mechanical fit checks with the telescope system, prior to overseas deployment.
- Ability to discover constraints and problems with the enclosure building early on in the project lifecycle, allowing for team to make changes locally before deploying overseas.
- Ability to assemble and test fully operational telescope system, including optics, locally prior to overseas deployment.

With this very powerful tool in Tucson, we now look forward with renewed confidence to the deployment of the KMTNet project to Cerro Tololo International Observatory, Chile in 2012, South Africa Astronomical Observatory, Sutherland in 2013, South Africa, and Siding Springs Observatory, Australia in 2013.

8. ACKNOWLEDGMENTS

The authors acknowledge the design and construction team members who have contributed their many talents to the success of this project. They include M3 Engineering, CP Systems, Inc, Vroom Engineering, WRS Associates, Palo Verde Electric, Sun Mechanical, Earl Moyer, and Lance Lesny.

9. AUTHOR CONTACT INFORMATION

Nathan Kappler, nkappler@tbrconstruction.net, Larry Kappler, lkappler@tbrconstruction.net, Wade Poteet, wpoteet@earthlink.net, Harold Cauthen, fbcauthen@earthlink.net, Byeong-Gon Park, bgpark@kasi.re.kr, Chung-Uk Lee, leecu@kasi.re.kr, Seung-Lee Kim, slkim@kasi.re.kr, Sang-Mok Cha, chasm@kasi.re.kr

10. WEB LINKS

www.tbrconstruction.net
Web Camera at site of KMTNet construction progress at CTIO: http://kmtaxis.ctio.noao.edu
Web Camera at Tucson site of Prototype: http://50.79.221.209

Figure 12 Team Members Gathered in Tucson 2011

Initial Alignment and Commissioning Plan for the LSST

William J. Gressler[a], Jacques Sebag[a], Chuck Claver[b]

[a]National Optical Astronomy Observatory, 950 N. Cherry, Tucson, AZ, USA 85719
[b]LSST Corporation, 933 N. Cherry, Tucson, AZ, USA 85721

ABSTRACT

The planned construction and completion of the Large Synoptic Survey Telescope (LSST) Project consists of phased activities. The initial telescope construction period will transition to a multi-year commissioning phase, which will conclude with final hand off to science operations. The initial telescope alignment will utilize laser tracker fiducials and nodal aberration theory (NAT) to demonstrate Engineering First Light with a three-mirror optical system and test camera, prior to the integration of the science camera. This plan exploits the diffraction limited on-axis image quality of the three-mirror design. Commissioning consists of final integration of the three LSST subsystems (Telescope, Camera, and Data Management), followed by on-sky science verification to show compliance with the survey performance specifications.

Keywords: LSST, alignment, commissioning

1 INTRODUCTION

The planned construction and completion of the Large Synoptic Survey Telescope (LSST) Project [1] consists of a set of phased activities. The initial telescope construction period will transition to a multi-year commission phase, which will conclude with final hand off to science operations. The telescope integration plan includes optical tests of the primary/tertiary mirror assembly to verify the pre-shipping mirror support matrices. This test configuration will be used to build upon the zenith and initial analytic look-up tables for gravity deflection.

Integration of the secondary mirror assembly and a surrogate commissioning camera attached to the science camera rotator/hexapod will enable three-mirror testing. The initial three-mirror telescope alignment will utilize laser tracker targets and nodal aberration theory to exploit the diffraction limited on-axis imaging performance. It is highly advantageous to align the three-mirror optical system prior to the integration of the science camera on the telescope, which corrects the field of view (FOV) via three large lenses and includes the operational wavefront sensors. Engineering First Light will demonstrate on-axis image quality with the surrogate test camera and support the transition to the commissioning phase.

Commissioning is defined roughly by two phases: a technical period of systems integration and test, followed by a period focused on science verification. Systems integration and test activities are designed to complete the technical integration of the three LSST Project subsystems (Telescope, Data Management (DM), and the Camera). Final science verification tasks will be based solely on the measured on-sky performance of the LSST system and are designed to show compliance with the survey performance specifications.

2 LSST OPTICAL LAYOUT

The LSST optical layout consists of the primary (M1), secondary (M2), and tertiary (M3) mirrors. After reflecting off the three mirrors, light converges as it enters into the LSST science camera, which is mounted and suspended on a rotator/hexapod within the M2 center hole. The telescope's 3.5-degree field of view (FOV) is corrected over its operational wavelengths via the science camera which contains three refractive elements, a filter, and a 64-cm diameter focal plane array [2]. The main optical parameters of the three mirrors are summarized in Table 1.

Table 1: LSST three-mirror optical design parameters

	Radius of Curvature (mm)	Conic Constant	Distance (mm)	Clear Aperture Diameter (mm)	Aspheric Coefficient r^6	Aspheric Coefficient r^8
M1	-19835.0	-1.215	-6156.2	8360	$+1.381 * 10^{-24}$	0.0
M2	-6788.0	-0.222	+6390.0	3420	$-1.274 * 10^{-20}$	$-9.680 * 10^{-28}$
M3	-8344.5	+0.155	-4643.2	5016	$-4.500 * 10^{-22}$	$-8.150 * 10^{-30}$

The telescope optical design incorporates a unique M1M3, where the two aspheric mirrors are fabricated onto a single monolithic substrate [3]. This approach reduces the operational alignment degrees of freedom and supports the compact telescope design necessary to achieve the demanding slew and settle requirements to meet the LSST survey. The two optical surfaces will be polished and tested separately via center of curvature interferometry to achieve desired structure function performance specifications and relative alignment tolerances. Final optical testing will be performed viewing both surfaces simultaneously to verify bending mode compliance of both mirror surfaces given the single set of support forces.

In addition to optical surface testing, laser tracker targets attached around the mirror outer diameter will be measured to capture the optical axis geometry of M1. Twelve mounting flexures have been bonded to bosses which are part of the outer wall of the mirror substrate as shown in the zoomed section of Figure 1. The flexures support a shaft which holds the laser tracker retro-reflector target. The mounts are designed to capture the optical axis geometry of the M1 surface relative to the outer diameter. The flexure mounts are kinematic to enable removal for mirror coating and replacement such that the overall measured coordinate geometry is preserved.

Figure 1: M1M3 mirror showing laser tracker fiducial mount

The M2 optical axis geometry will likewise be defined by a set of laser targets attached to the M2 tangent link support pads which will be measured during final optical testing. Finally, the optical axis of the camera system will be measured via targets attached to the front lens. During initial telescope integration, all these laser tracker targets will all be measured via a laser tracker embedded within the M1M3 center hole. Measurement of these targets will enable rapid initial alignment of the three optical systems (M1M3, M2, and the camera) well within the capture range of the operational wavefront sensors located around the periphery of the science camera focal plane elements.

3 TELESCOPE SYSTEM INTEGRATION

The development of the Telescope and Site subsystem is planned to maximize factory testing of components and subsystem testing prior to summit delivery. This approach minimizes the amount of hardware mated for the first time on the summit and the level of subsystem commissioning required on the summit where these tasks are often less efficient. Limited by overall schedule logistics, manufacturing and practical realities, costs and overall efficiency assessments, the approach is to have all hardware pretested at factories and test facilities and mated with associated hardware and software prior to shipping to the site.

The overall integration flow diagram for summit activities is given in Figure 2. The diagram provides color reference to associate tasks with subsystems. The plan begins with site and facility construction and shows how each of the major elements and critical systems feed into the activities through final system testing and commissioning. A key objective of the plan is to support scheduling so that elements and services are available to support the integration flow.

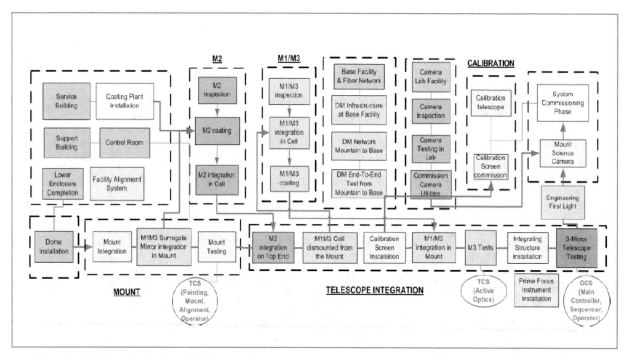

Figure 2: LSST integration flow diagram

The integration starts with the excavations and preparations of the site and construction of the summit facilities. Facility development is driven by dome construction and the coating facility installation. The dome provides the weather protection for the interior finishes in the facility's lower enclosure and coating facility is used early in the integration plan to coat the mirrors as they are delivered to the summit.

Once the dome is sufficiently complete, the telescope mount integration will start with all components inserted through the enclosure doors. A laser tracker will be used during the integration to monitor and position the elements. Mount pointing and alignment testing will be done on site before installing the mirrors and the camera, using surrogate masses and a small alignment telescope mounted on the mount structure. The M1M3 mirror cell will be used during the mount testing with a surrogate mirror in place of the M1M3 glass mirror. Telescope Control System (TCS) testing will be repeated with a more complete on-sky motion control than was previously tested at the mount vendor factory. At this phase of the integration, the dome and mount will be able to slew, to point, and track with the proper cadence and performance.

In parallel with the mount and dome completion, the mirrors will be received, assembled, and coated on the summit. The M2 will be coated first in the lower vessel of the coating chamber and then integrated into its mirror cell. This activity will occur while the M1M3 mirror cell is integrated and tested in the telescope with the surrogate M1M3 mirror. During this mount testing period, the glass M1M3 mirror will be stored inside its shipping container. Once mount testing is complete, the M1M3 surrogate mirror will then be removed from the telescope and mirror cell and replaced with the M1M3 mirror. This is a space intensive activity requiring the full maintenance facility. The M1M3 mirror will be coated in its cell and then transported back to the dome to be integrated in the telescope.

During the coating of the M1M3 mirror, the M2 system will be mounted in the telescope. In addition, the interferometer/null lens system used to test the M3 mirror optical surface from its center of curvature will be mounted on the telescope top end assembly as shown below in Figure 3. There is sufficient space to accommodate the interferometer hardware and the null lens beam projection through the center hole of M2.

The goal of this optical testing is to repeat the M3 optical surface test performed at zenith at the Steward Observatory Mirror Laboratory during final acceptance to verify the support matrices of both M3 and M1. This test will then be repeated with varying telescope elevation angles to build upon the active optics look-up tables previously populated with analytic values for gravity distortion. This testing (which does not required on-sky measurements) will also support refinement of the M2 active optics look-up tables during three-mirror testing described in the following section.

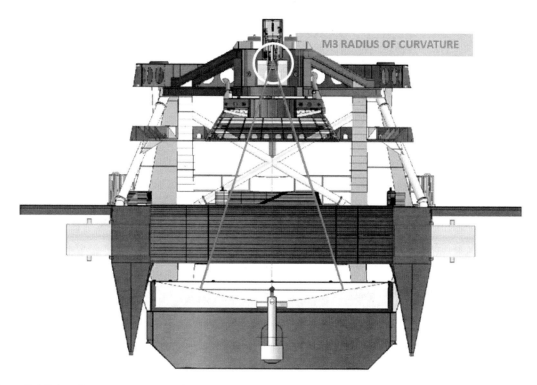

Figure 3: M3 interferometer set-up on telescope

4 THREE-MIRROR TELESCOPE TESTING

To expedite the integration and commissioning of the telescope system, the optical alignment process will be separated into two stages. During the first phase, the three-mirror system alone (without the science camera) will be aligned using a reduced portion of the total FOV. The aligned three-mirror telescope without the science camera has been designed to be diffraction-limited on-axis with negligible field curvature, but it suffers substantial aberrations with increasing field angle. Off-axis, the system produces 78 waves of field-linear third-order coma at 0.5-degree field. The LSST three-mirror system, with ray tracing to show the optical beam paths is shown in Figure 4.

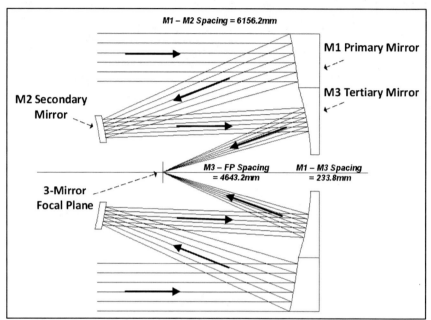

Figure 4: LSST three-mirror optical system layout

This large coma aberration has to be taken into account when planning the initial telescope alignment procedure. It becomes an issue for off-axis wavefront measurements, as the wavefront sensor (WFS) must have sufficient capture range to accommodate the large aberration. This aberration condition can also be used as a beneficial property, however, as will be described below where the approach to a nodal-based alignment process is described.

5 NODAL ABERRATION THEORY ALIGNMENT

The LSST three-mirror telescope alignment procedure prior to the installation of the large science camera is based on using nodal aberration theory (NAT) to determine the optimum position of the secondary mirror (M2) [4]. NAT reveals that aberration fields at the focal plane have an intrinsic characteristic nodal behavior. Developing an alignment approach based upon locating the nodal locations enables a linear systematic approach that is responsive to sparse data sampling. The types of aberrations remain unchanged in the presence of tilted and/or decentered mirrors. However, the field dependence of each aberration type develops characteristic nodal properties. For example, astigmatism becomes binodal; it develops two points in the field when there is no astigmatism.

The M2 position is determined taking into account the fixed fabrication position error of the tertiary mirror (M3) in the monolithic M1M3 and in the presence of M1 figure errors. NAT provides the analytical tools to determine the nodal behavior of third-order coma (annular Zernike polynomials Z7/Z8), third-order astigmatism (annular Zernike polynomials Z5/Z6), oblique spherical aberration (annular Zernike polynomials Z12/Z13), and fifth-order coma

(annular Zernike polynomials Z12/Z13), and fifth-order coma (annular Zernike polynomials Z14/Z15) aberrations in the LSST optical design. The alignment approach is built on the description of the nodal behavior of these aberrations to compute the best M2 displacements to make the nodes and midpoints coincident in the focal plane.

To perform the alignment, the optical axis of the primary mirror is defined to be the reference axis; the goal is that all mirror centers of curvature and aspheric vertices will lie on this line. The M2 hexapod will provide centering, piston, and tip/tilt positioning relative to the optical axis. The M3 rigid-body positions are fixed relative to the primary mirror and will not be adjustable, as the M1 and M3 are fabricated on the same substrate. Decenters and tips/tilts were applied to the M2 and M3 to analyze the effects caused by M2 misalignments and/or as-built fabrication errors of the M3. The nodal responses to these misalignments for the aberration types used to determine the state of alignment are summarized in Table2 and Table 3.

Table 2: Behavior of nodes and midpoints

	3rd order Coma Node	5th order Coma Node	Astigmatism Midpoint	Oblique Spherical Midpoint
M2 Decenter along Y axis (arcmin/mm)	1.76	-2.44	-1.1	-0.67
M2 Tilt about X axis (arcmin/arcmin)	-3.21	0.58	2.19	-0.99
M3 Decenter along Y axis (arcmin/mm)	2.53	1.49	0.68	1.01
M3 Tilt about X axis (arcmin/arcmin)	-5.03	-2.48	-0.39	-1.21

Table 3: Nodal displacements for binodal aberrations

	Astigmatism Node 1	Astigmatism Node 2	Oblique Spherical Node 1	Oblique Spherical Node 2
M2 Decenter along Y axis (arcmin/mm)	0.08	-2.25	-1.54	0.20
M2 Tilt about X axis (arcmin/arcmin)	-0.09	4.47	0.54 (x)	-0.54 (x)
M3 Decenter along Y axis (arcmin/mm)	0.19 (x)	-0.19 (x)	0.10	1.93
M3 Tilt about X axis (arcmin/arcmin)	1.55 (x)	-1.55 (x)	-1.68	-0.75

The process starts with on-axis WFS measurements to reduce third-order coma, followed by multiple WFS measurements in the FOV, determination of the nodes and midpoints locations, and computation of the optimal M2 displacements. Wavefront measurements at a sparse set of locations in the FOV will retrieve the node locations. One on-axis WFS will be located near the center of the FOV, with the other off-axis WFS deployed away from the center of the FOV, up to a radius of 0.5-degree. Multiple fields containing a number of usable bright stars will be available with this size FOV. Observations near the meridian will decrease the telescope elevation change while tracking.

The WFS system is a Shack-Hartman design and includes a guiding camera to image the telescope focus as shown in Figure 5. The system includes a camera mass simulator that will be attached to the camera rotator/hexapod on the telescope. Both of these mechanisms will be used initially to position the on-axis WFS. The guide camera images provide an initial alignment near the optical axis by visually reducing the third-order coma. This on-axis WFS position is set as reference and the positions of all other off-axis WFS measurements are measured relative to this reference. Laser tracker targets located on the system will monitor the different wavefront positions.

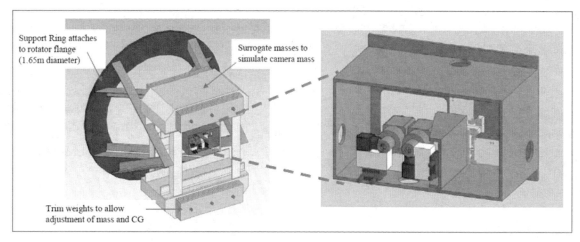

Figure 5: WFS system used for NAT alignment

Simulation code was developed to test the alignment approach. The code was designed to: (1) determine the positions of the nodes and midpoints for a given set of misalignment; and (2) determine the misalignments for a given set of nodes and midpoint positions. It was used to evaluate the sensitivity of the methods with respect to accuracy errors in the measurements necessary to perform the alignment. Possible error sources include: errors in the mirror radii of curvature and vertex-to-vertex spacing between the mirrors used in the NAT equations, M2 hexapod error to position the secondary mirror, and WFS location error and measurement error that affect the determination of the node position.

Wavefront measurement error is the largest source of error in the determination of the node positions. At each field point, the wavefront is expected to be computed as the mean value of 20 consecutive measurements. For each of the 20 measurements, an error was generated randomly using a Gaussian distribution with a 0.050 and 0.025 wave standard deviation assumed as the wavefront accuracy, and the error was added to the Zernike polynomials. The resulting position errors are listed in Table 4.

Table 4: Nodal position sensitivity

Position Errors for 0.050 Wave WFS Error

Node Position Error (mm)	Number of Field Points		
	8	4	3
3rd Coma (Z7/Z8)	0.011	0.029	0.074
3rd Astigmatism (Z5/Z6)	0.120	0.554	0.776
5th Oblique Spherical (Z12/Z13)	5.571	125.682	NA
5th Coma (Z14/Z15)	0.375	0.833	2.385

Position Errors for 0.025 Wave WFS Error

Node Position Error (mm)	Number of Field Points		
	8	4	3
3rd Astigmatism (Z5/Z6)	0.079	0.147	0.496
5th Oblique Spherical (Z12/Z13)	1.343	84.328	NA
5th Coma (Z14/Z15)	0.196	0.483	0.922

The sensitivity analysis shows the baseline NAT alignment method will meet the criterion identified to achieve the alignment in the presence of measurement errors. A reasonable wavefront measurement error was found to be acceptable for third-order coma, third-order astigmatism, fifth-order coma node, and midpoint position determination. Although it can have a non-negligible impact on the determination of the midpoint node position for fifth-order oblique spherical aberration, that error may be counterbalanced by using the fifth-order oblique spherical mirror figure changes to identify that midpoint position independently.

6 ENGINEERING FIRST LIGHT

Engineering First Light (EFL) will be achieved with the three-mirror telescope optical system and the surrogate camera mass WFS system as shown previously in Figure 5. Four months of on-sky time is dedicated prior to the start of the Commissioning phase. EFL will demonstrate a number of attributes to include: verifying on-axis active optic control of all the mirror surfaces, verifying the performance of the M1M3 thermal control system, verifying the TCS is integrated with the Observatory Control System (OCS), and provide for early testing of the survey operational scheduler.

7 COMMISSIONING: SYSTEMS INTEGRATION AND TEST

The commissioning phase of the LSST project is defined by roughly two phases: a period dominated by the technical activities of Systems Integration and Test, followed by a period that is focused on Science Verification. These two phases represent a continuum and the plan is meant to be flexible to take advantage of opportunities as they occur. The three LSST subsystem teams (Telescope and Site, Data Management (DM), and the Camera) will integrate into one group for commissioning that maintains the technical expertise from each subsystem. The core commissioning team will be supported by an external oversight committee to help balance the community pressure to make data available as early as possible and to review the quality of commissioning data and recommend whether and when the data should be released.

Systems Integration and Test assumes pre-conditions of all three subsystems. The telescope is essentially fully functional at the start of the commissioning phase. Using a surrogate camera, the telescope will be delivering images meeting system requirements over a limited FOV and can point and track to its open loop requirements. The dome mechanisms are assumed to be working including azimuth tracking, elevation tracking of the entrance opening, ventilation louvers, and HVAC system for daytime thermal conditioning. The "Engineering First Light" milestone coincides with the start of Commissioning.

The DM Base Facility computing and data archiving infrastructure is expected to be installed and working. The DM processing pipelines for Instrumental Signature Removal, Calibration Data Products, Transient Alert Processing, and nightly photometry and astrometry processing will be installed and functional. These pipelines will have been tested and verified with two data types: (1) simulated data generated for the internal Data Challenges used in the project development phase, and (2) legacy data from previous telescope surveys (e.g. CFHT Legacy Survey, SLOAN Digital Sky Survey, and other surveys currently in the early stages of operation). In addition, the 100Gb/sec link between the Summit and Base facilities will be installed and functional.

The Observatory Control System (OCS) development cycle enables annual releases with increasing levels of functionality at each stage. The methodology makes portions of the OCS available to other subsystem developers that need to interface to this package. At the start of commissioning, the OCS software will have had its fourth annual release. The following software functionality will be fully implemented: communications middleware with messaging, engineering and facility database, scheduler, operator GUIs, monitor software with status displays, maintenance support software, and application control software with sequencer to synchronize the subsystems.

The LSST science camera will be integrated and tested at the SLAC National Accelerator Laboratory prior to shipment to the summit. The camera will be unpackaged in the camera service space within the observatory and the set of verification tests will be re-run to verify no functional or performance damage occurred during shipment. The instrument will then be connected to the Summit Facility network. With the camera on the network, the data path between the Camera and DM processing pipelines will be established and verified with live pixel data from the science focal plane array. During this initial testing the command and control of all Camera functions with the OCS will be verified. Successful completion of these tests will signify the Camera is ready to enter the Commissioning phase and start Systems Integration and Test.

The LSST science camera is integrated with the telescope top end integrating structure (TEIS) and brought to the dome floor via the camera cart. The assembly is transferred from the cart to the dome crane fixture for installation into the telescope (which is horizon pointing). The camera and TEIS are inserted into the telescope top end assembly using guide rods and the dome crane as shown in Figure 6.

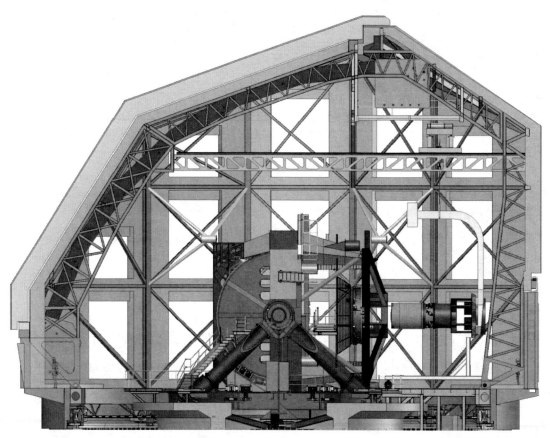

Figure 6: LSST camera installation on the telescope

Initial on-sky camera/telescope activities will focus on active optics control. Initial alignment of the systems will use the laser tracker and analytic look-up tables. The laser tracker will be used to map the camera rotator/hexapod deflection versus elevation angle. Through-focus sequence images will be acquired to determine best fit plane with respect to the focal plane array, with hexapod adjustments made to minimize residuals as a function of telescope elevation.

The active optics look-up tables for the full FOV will be updated using the entire science camera focal plane array as a wavefront sensor using intra and extra focal images. The camera wavefront sensors will be calibrated to the science focal plane array to determine the focus offset and the active optics solution offsets from the full focal plane array and the four system WFS alone. Finally, the optical reconstruction pipeline to convert wavefront error to misalignment and mirror surface errors will be verified for each degree of freedom in the active optics system. All image data will be processed through the DM pipelines.

8 COMMISSIONING: SCIENCE VERIFICATION

The science verification period is structured around demonstrating that the survey functional and performance specifications given in the Science Requirements Document (SRD) [5] and the LSST system requirements are being met. For planning purposes, the science verification period is structured into four periods: (1) a four-month period where the emphasis is in verifying compliance with single visit requirements; (2) verifying the functional performance of the LSST scheduler and autonomous operation; (3) verification of the full survey performance requirements for image stacks and area coverage; and (4) final science verification and acceptance tests for operation readiness.

The science verification framework includes time for engineering related activities throughout, but is more heavily proportioned at the beginning, transitioning to something near early operational levels by the conclusion. The science verification matrix shown in Figure 7 shows the methods to be used to verify each of the SRD performance requirements.

Requirement Title	SRD Variable	Inspection / Specpcification	Demonstration / Integral Testing	Lab Test / Mesurments	On-Sky Direct Measurement	On-Sky Statistical Measurement	Modeling / Simulaiton	On-Sky Measurement coupled with Extended Analysis	On-Sky Mesurement coupled with model	Extrapolation from On-Sky Measurements
Single Image Performance										
Filter Complement (ugrizy)		X								
Number of Filters	Nfilters	X								
Filter Change Time	Tfmax		X							
Filter out of band	Fleak		X	X						
	FleakTot		X	X						
Filter Temperal Stability							X			
r-Band Reference Depth	D1				X	X				X
	DF1		X				X			
	Z1		X				X			
Filter deapths (ugrizy)	DB1_filt				X	X				X
variation of Dpeth over FOV	DF2				X	X				
	Z2				X	X				
Minimum Exposure Time	Etmin	X	X							
Delivered Image Quality	S1							X	X	
	SF1					X				
	SX					X				
Image budget at Airmass=2	SXE							X	X	
Pixel Size	pixSize	X							X	
Image Spatial Profile	SRn							X	X	
Image Ellipticity Distribution	SE1					X			X	
	EF1					X				
	SE2					X				
Photometric Repeatability	PA1					X			X	
	PF1					X			X	
	PA2					X			X	
Phtometric Sptatial Uniformity	PA3					X		X		
	PF2					X				
	PA4					X				
Phtometric Band-to-band calibration	PA5(u)						X	X	X	
Absolute Phtometry	PA6				X				X	
Realtive Astrometry	AM1,2,3					X		X		
	AF1,2,3					X				
	AD1,2,3					X				
Crossband Relative Astrometry	AB1					X		X	X	
	ABF1					X				
	AB2					X				
Absolute Astrometry	AA1					X			X	
Time Recording Accuracy	TACREL	X	X							
	TACABS	X	X							
Full Survey Performance										
Sky Area	Asky	X					X			
Total Number of Visits (10-year)	Nv1						X			X
Visit Distribution	Nv1_filt						X			X
Idealize Stacked depth										
Distribution of Visits in Time	RVA1						X			X
Astrometric Parallax	SIGpara				X				X	
y-band Parallax	SIGparaRed				X				X	
ProperMotion	SIGpm				X				X	
Ellipticity Residual Correlations	TE1							X	X	X
	TE2							X	X	X
	TE3							X	X	X
	TE4							X	X	X
Data Release Cadence	DRT1	X	X							
Transient Alert Latency	OTT1	X	X							
Number of transients	transN		X							

Figure 7: Science verification matrix

9 CONCLUSION

The initial alignment and commissioning plan for the LSST consists of a phased set of activities. The development of the Telescope and Site subsystem is planned to maximize factory testing of components prior to summit delivery. Initial telescope integration includes optical tests of the M1M3 assembly to verify pre-shipping mirror support matrices. The initial three-mirror telescope alignment will utilize laser tracker fiducials and nodal aberration theory to exploit the diffraction limited on-axis image quality. It is highly advantageous to align the three-mirror optical system prior to the integration of the science camera on the telescope, which corrects the FOV and includes the operational wavefront sensors. Engineering First Light will demonstrate on-axis image quality with a test camera and support the transition to final system integration and test and science verification.

10 ACKNOWLEDGEMENTS

LSST project activities are supported in part by the National Science Foundation through Governing Cooperative Agreement 0809409 managed by the Association of Universities for Research in Astronomy (AURA), and the Department of Energy under contract DE-AC02-76-SFO0515 with the SLAC National Accelerator Laboratory. Additional LSST funding comes from private donations, grants to universities, and in-kind support from LSSTC Institutional Members.

11 REFERENCES

1. Krabbendam, V., Sweeney, D. "The Large Synoptic Survey Telescope Final Design Status", SPIE 8444-09, 1 Jul 2012.
2. Kahn, S., et al., "Overview of the 3.2 Gigapixel Camera for the Large Synoptic Survey Telescope", SPIE 8446-239, 2 Jul 2012.
3. Tuell, M., et al., "Fabrication of the LSST Monolithic Primary-Tertiary Mirror", SPIE 8450-174, 5 Jul 2012.
4. Sebag, J., et al., "LSST Alignment Plan Based on Nodal Alignment Theory", Publications of the Astronomical Society of the Pacific, 124:380-390, Apr 2012.
5. Z. Ivezic, et al., "LSST: From Science Drivers to Reference Design to Anticipated Data Products", Astrophysics (astro-ph): arXiv:0805.2366v2, 7 Jun 2011.

Dark Energy Camera installation at CTIO - overview.

Timothy M. C. Abbott[*], Freddy Muñoz, Alistair R. Walker, R. Chris Smith, Andrés Montane, Brooke Gregory, Roberto Tighe, Patricio Schurter, Nicole van der Bliek, Germán Schumacher

Cerro Tololo Inter-American Observatory, Casilla 603, La Serena, Chile.

ABSTRACT

The Dark Energy Camera (DECam) has been installed on the V. M. Blanco telescope at Cerro Tololo Inter-American Observatory in Chile. This major upgrade to the facility has required numerous modifications to the telescope and improvements in observatory infrastructure. The telescope prime focus assembly has been entirely replaced, and the f/8 secondary change procedure radically changed. The heavier instrument means that telescope balance has been significantly modified. The telescope control system has been upgraded. NOAO has established a data transport system to efficiently move DECam's output to the NCSA for processing. The observatory has integrated the DECam high-pressure, two-phase cryogenic cooling system into its operations and converted the Coudé room into an environmentally-controlled instrument handling facility incorporating a high quality cleanroom. New procedures to ensure the safety of personnel and equipment have been introduced.

Keywords: telescope, instrument, imager, upgrade, infrastructure, installation

1. INTRODUCTION

Beginning this year, the V. M. Blanco 4-m telescope at Cerro Tololo Inter-American Observatory will be host to and platform for the Dark Energy Survey[1] (DES). This is a large area optical survey, 525 nights spanning 5 years to map 5000 square degrees, intended to measure the dark energy equation of state parameter, w, to a precision of ~5%. The DES Collaboration consists of more than 120 senior scientists in 16 institutes in six countries (USA, UK, Spain, Brazil, Germany, Switzerland). The DES Collaboration has built the Dark Energy Camera[2,3] (DECam) in order to carry out the DES. As a facility instrument at CTIO, DECam will also be available by the NOAO user community and Chilean astronomers, and a DES-provided reduction pipeline will be operated by NOAO for this mode. The lead institution for camera construction is Fermilab, using mostly DOE funding.

DECam incorporates ~520 million pixels, it entirely replaces the prime focus cage on the Blanco telescope and is accompanied by significant additions to the telescope infrastructure. These include an upgraded telescope control system[4] (TCS), a high-pressure, re-circulating liquid nitrogen cryogenic system[5,6,7], an upgraded glycol cooling system, a new system for flat-field calibrations[7], and a mechanism for handling the f/8 secondary mirror.

The construction of DECam has culminated in the delivery of the various instrument components to Cerro Tololo in Chile where they are being assembled for installation on the telescope. For several months, many staff at CTIO have changed from their normal roles supporting Blanco operations to being part of a construction crew augmented by visiting personnel, largely from Fermilab, KPNO and UCL, but also from a number of other institutes.

The project has been regularly reviewed in order to satisfy the requirements of the primary governing agencies, the National Science Foundation and the Department of Energy, synthesizing their differing approaches into a mutually acceptable process. In additional, CTIO has carried out its own reviews of the planning for installation, the Integration and Installation Review of April 2011 and the Installation and Safety Review of October 2011. In response to an incident in which the f/8 secondary mirror was damaged, CTIO held a review of safety procedures in March 2012 which also covered many aspects of the DECam installation. Each of these reviews produced a catalog of recommendations which were then integrated into the installation planning.

---

[*] tabbott@ctio.noao.edu; phone: +56 51 205200

2. PREPARATIONS

Because fabrication and test of instrument components have taken place off-site, considerable effort has been put into bringing CTIO staff up to speed on use, trouble-shooting and maintenance of the various subsystems. This has included sending key staff to assist in the work at Fermilab and at University College, London and considerable involvement in bringing up the individual systems in Chile, especially in the areas of imager testing, optical assembly and alignment, and cryogenic system installation and test.

Before the telescope was taken offline for the installation of DECam itself, a number of work packages were completed. These were:

1. The 24 primary mirror radial supports were modified in order to resolve a long standing problem in which they would fail at the rate of 1-2 per year. The ultimate cause proved to be a minor design flaw in the mechanism attaching them to the mirror edge. As they failed, the primary mirror developed unpredictable lateral movements on its cell, adversely affecting delivered image quality. Since their replacement in April 2009 none have failed, the mirror demonstrates repeatable, flexure-dominated movements in the telescope and the issue is considered resolved[9]. These residual movements can be compensated with the hexapod which is part of DECam and on which ride the imager and optical corrector.

2. The aluminizing chamber was upgraded and used to apply a new coating to the primary mirror in March 2011. The resulting coating is probably the best yet achieved on the Blanco, showing near nominal pure aluminum reflectivity.

3. The telescope operating system (TCS) was upgraded to resolve issues of obsolescence and to achieve the performance required by the Dark Energy Survey – most notably that the time taken slew between image fields should be defined by the inter-exposure time, approximately 17 seconds. The new TCS was tested before the shutdown, but considerable work is anticipated in tuning it with the installation of DECam, which will leave the telescope approximately 10 tons heavier than before.

4. The Coudé room was converted into an instrument maintenance facility incorporating a high quality (approximately class 1000) cleanroom large enough to permit work on the DECam imager. This conversion also included the introduction of humidity control, as part of measures to control electrostatic discharge, and oxygen deficiency monitors.

5. A new air compressor was installed, and a dryer system to supply compressed air for a range of subsystems.

6. The ground floor garage space of the Blanco building was converted into an area suitable for assembly of the new prime focus cage, assembly and alignment of the corrector, and integration of the two with the hexapod. This included a new 15-ton crane and a clean-tent in which the optics could be worked on without the introduction of dust or other contaminants.

7. A new 40T glycol chiller has been installed and the glycol circulation system upgraded to support the needs of DECam.

8. An f/8 secondary mirror handling system was installed and tested on the dome floor. Previously, switching between the prime focus imager and the f/8 Cassegrain instruments was achieved through rotating the prime focus cage through 180° about an axis perpendicular to the optical axis of the telescope – one end of the cage fed the prime focus imager, the other carried the secondary mirror. With the new cryogenic system this is no longer possible because large, vacuum-jacketed liquid-nitrogen lines now cross the telescope top rings and a rotation like this would cut them off. Instead, we have elected to perform a partial rotation of the prime focus cage while near the north-west horizon and mount the secondary mirror in front of the DECam corrector when it is needed.

9. The Blanco computer room was modernized and expanded to house the systems required to support DECam. This includes raised floors and new active temperature control.

10. The Blanco control room was relocated and expanded to house four separate workstations, including the 8-headed DECam control station.

11. The DECam liquid nitrogen supply system was installed and tested by cooling the imager in the Coudé room. This system uses a storage vessel where nitrogen is liquefied by a cryo-cooler and pumped up through vacuum-jacketed hoses to the telescope prime focus. The nitrogen in these hoses is mixed phase at a pressure of 100 PSI. This installation involved constructing a new room to house the helium compressors which supply the cryo-cooler. The storage vessel is mounted on the reinforced roof of the old control room on the dome floor in order to minimize the head height with respect to the telescope prime focus.

12. A new all-sky mid-infrared cloud camera (RASICAM) was delivered to CTIO, installed and tested. This will be used to monitor observing conditions during the survey.

13. The shutter was delivered to CTIO and tested.

14. The filter-handling mechanism was delivered and tested.

15. Four filters have been received and mounted in their frames. Two more are en route.

16. The DECam imager was delivered and tested at CTIO at the end of 2011.

17. The DECam control software suite, SISPI, was tested with the imager on site, and the new "Data Transport System" (DTS) software was delivered from NOAO/Tucson and tested, transferring data from Cerro Tololo to Fermilab

18. The hexapod was delivered and tested in July/August 2011.

19. The new prime focus cage was assembled using the hexapod, a dummy corrector and a dummy imager in November 2011

20. The optical corrector was delivered, assembled and aligned[9] in January 2012.

21. A new flat-field calibration screen and illumination system is being installed in the dome.

22. Each prime focus spider leg was instrumented with a strain gauge to be used later for balancing the tensions on each.

23. The new prime focus cage was installed with the telescope locked at the zenith, therefore appropriate platforms were designed and fabricated to allow access to the relevant points. These were built with safety rails and kick plates and with the capacity to allow as many as 12 workers on the telescope at a time (see Figure 1).

24. During the installation of DECam, the primary mirror would be on the dome floor near, but not under, the telescope. A robust steel cover, capable of withstanding significant impact (a 10kg tool dropped from the top ring of the telescope), was built to protect the mirror while work proceeded on the telescope.

A variety of equipment was fabricated, purchased or rented for the purposes of the shutdown. These include: a boom lift or "cherry picker" which has proved invaluable for access to all parts of the telescope while it is immobile, a prime focus cage lifting tool and new slings, shackles etc. for heavy lifting, the new ground-floor crane mentioned above, fiber-glass temporary flooring for additional access on the telescope, and a remotely readable, heavy-duty scale capable of measuring all the components being installed, separately and combined. New personal safety and rescue equipment was purchased and a professional at-height rescue team hired for the shutdown itself. Additional safety training for our own staff was provided. Before use, all heavy equipment was professionally certified as in good condition.

Figure 1, the DECam prime focus cage in place on the top end of the telescope. The yellow bars on the side of the prime focus cage are intended to lend rigidity during the exercise and were later removed. The at-height work platforms are visible.

3. SAFETY AND ORGANIZATION

Dismantling a ~40 year old, working telescope for the purposes of a major upgrade combines some of the demands of a normal shutdown for realuminization of the primary mirror or minor upgrades with some of those of a full-scale telescope construction project. As such it creates the need for some new, formal rules in work procedures to compensate for the changes in accustomed, and often informal, daily practice.

The telescope was taken offline for installation of the DECam cage on 20 February 2012. Distressingly, we experienced an incident on that same day in which the f/8 secondary mirror was dropped some distance onto its apex and damaged, injuring two members of our personnel in the process. The usability of the mirror is currently being assessed; the injured personnel are fully recovered. While the direct causes of this incident were identified as inadequate communications and the improper use of mirror handling equipment, and considerable attention had already been applied to safety and procedure development for the installation shutdown, we were nevertheless prompted to perform a thorough review of all our operating and safety procedures, including the external review mentioned above. This section describes aspects of procedures and safety as now practiced at CTIO, some of which were introduced on the recommendation of the review panel, and some of which were already in place.

During the shutdown, overall on-site authority and responsibility is held by the "capataz" (Spanish, trans.: foreman) who oversees work as directed by the installation project manager who in turn communicates with the observatory director, the CTIO instrument scientist and the DES collaboration and world at large. Work packages are assigned to teams, each of which has its own leader. Communications on-site are largely via walkie-talkie radios with a dedicated channel, but telephone, public address, email, and video-conference facilities are also available. The capataz authorizes all entry to the Blanco building, access between work areas and use of heavy equipment. Any use of heavy equipment is only permitted after appropriate training. Team leaders keep the capataz informed of the state of their work package. The capataz anticipates and resolves conflicts between teams. The capataz is bilingual.

A shift is begun with a meeting of all the teams working that shift. That meeting is led by the capataz who ensures all teams are aware of the plans of all other teams, and who generates a widely distributed report after that meeting. Likewise, the shift ends with another meeting in which progress is documented, plans for the next shift collated and another report is circulated. If two shifts work back-to-back the first shift's end-of-shift meeting serves as the next shift's beginning-of-shift meeting. If two shifts are separated in time (e.g. overnight) the capataz for the following shift attends the previous shift's end-of-shift meeting.

All work is subject to a job hazard analysis (JHA). This is a formalized study and discussion of the risks, to personnel or to equipment, associated with that work, and their means of amelioration. If a specific task is not identified as requiring its own JHA, then a general JHA for work in the building applies. If a special-purpose JHA is used, all workers gather to review and discuss it before the task is performed, and subsequently sign off on it. Any heavy lifting has an associated critical lift plan (CLP) with a similar motivation and process as the JHA. Any complex procedure has an associated procedure description document. Visiting personnel involved in the installation undergo a thorough orientation and are provided with a complete package of information. All of these documents are archived and available on-line.

An open stop-work policy has been instituted, in which any worker may call a stop to work, for any reason, until their concerns are addressed. All observatory systems are regularly inspected for safety and recommendations distributed and followed-up. New lockout/tagout procedures have been implemented. Work which is identified to carry any significant risk to personnel is attended by other personnel who are qualified to make the appropriate emergency response if it is required.

4. THE INSTALLATION

This section provides an overview of the installation process, further detail may be found in a companion paper[10].

A zenith orientation of the telescope for the prime focus cage exchange was selected in spite of the height of the work area so that the majority of forces involved were symmetrically balanced around the vertical. Making the exchange near the horizon carried a number of additional risks resulting from the awkwardness of the access. The telescope was originally assembled pointed at the zenith, but the prime focus cage, spider and top rings were assembled at ground level then installed as a unit on 6 of the 8 Serrurier legs, with the final 2 legs installed afterwards. Instead, we elected to leave the top rings in place and disconnect the spider legs at the zenith. We could then remove the prime focus cage and lower it vertically, using the 50T dome crane, down through the telescope with the primary mirror removed as if for realuminization.

Before moving to zenith, the prime focus cage was stripped of as much mass as possible with the telescope near the northwest horizon, and the access platforms were installed. All material removed was weighed to provide input to the rebalancing calculations necessary before the telescope can be moved again once DECam has been installed. The f/8 secondary mirror was removed and replaced with a dummy. A corrosion-removing compound was applied to all the linkages between the prime focus cage and the top ring (these had not been touched for some decades). The telescope was moved to the zenith, clamped in position and the drives disabled.

Before removing the primary mirror we established its fiducial position with respect to the telescope using a tool to locate the mirror's vertex and by aligning a pair of crosshairs with its optical axis. These crosshairs were strung between the legs of the Serrurier truss with tunable and precisely relocatable mounts. The crosshairs were then used to align the new cage using a laser sight from below and an optical target on its front surface. The goal was to achieve centration better than 1 mm and tilt less than 100 arcsec.

The instruments, Cassegrain cage and primary mirror were removed. DECam is significantly more massive than the old prime focus cage. It was decided to compensate by adding counterweights on the bottom of the Cassegrain cage. In all, the weight of the telescope is increased by ~10%. This required that the Cassegrain cage be reinforced to carry sufficient lead and it was removed from the building for that purpose. The primary mirror was moved as far from the telescope as possible on the dome floor and protected with the steel cover. Additional protection against falling objects was provided by draping the south side of the telescope with a tarpaulin (in the event, only small tools were ever dropped, none approached anywhere near the mirror, and no damage occurred).

The old prime focus cage was exchanged for DECam over a period in which two shifts were worked per day. An earthquake-unsafe condition was identified in which the dome crane would be attached to a prime focus cage while it was either attached to the telescope itself, or being moved up or down inside the telescope. During these phases, a significant earthquake could have had disastrous consequences. Therefore such phases were required to be completed within one working day so that the dome crane was never left attached to the telescope nor was weight left hanging for any longer than absolutely necessary.

The DECam prime focus cage is designed to be identical to the old cage in terms of its attachment to the telescope. The spider has 12 legs – 4 upper and lower diagonals each and 4 center legs. Each upper and lower diagonal is attached to the cage at one end and to the telescope top ring at the other. The center legs are attached to the cage at one point and to the rings at two points, one above and one below. The resulting assembly is overconstrained, and careful attention must be paid to alignment, interactions between attachment points, and the tensions on each leg must be balanced. The tensions ultimately applied to the spider legs were tuned by use of the strain gauges and an influence matrix with the boundary condition that none should go into compression at any telescope orientation. Unlike like the old cage, the new cage is electrically isolated from the spider by means of G10 cuffs around the pins which connect them, additional care was necessary to avoid breaking these cuffs during installation.

Removing the old cage proved straight-forward. Worries that bolt threads might be corroded, bound or damaged proved unfounded. Similarly, although some hammering was necessary, the retaining pins were removed without significant difficulty.

It was decided to install the DECam cage "empty" without the corrector and its delicate and finely aligned optics, nor with the hexapod, until we had established and practiced the procedure more thoroughly. This configuration would also be significantly lighter. This too went very smoothly. We practiced some of the maneuvers required to align it, then removed it and integrated the optics. The turnbuckles used to attach the spider legs to the telescope top ring were cleaned thoroughly and one thread reworked. The final cage installation proceeded without incident and the final position was displaced from the previously established optical axis less than 600 microns and tilted less than 7arcsec. The entire procedure took 31 calendar days (with no work at the weekends) from starting to remove the old cage to final alignment of the DECam cage.

Reinstalling the reinforced Cassegrain cage and loading it with lead counterweights proved more time consuming than anticipated. Moving 11 tons of 7-30kg lead plates around by hand is an arduous process! It was also found necessary to rework some of the bolt connections between cage and telescope to be sure of their robustness. With careful accounting for all the changes in loading of the telescope, counterweights calculated to leave the telescope ~500kg bottom heavy were installed and the telescope released from its declination restraints. Starting with the declination axis, we drove the telescope small distances from zenith and measured the imbalance using motor drive current. We corrected the north/south balance at zenith and then drove larger distances in order to measure the imbalance on the telescope's long axis. It proved necessary to remove ~400kg from the bottom of the Cass cage to achieve a balance of better than 8kg out of 110,000 kg. Declination friction was found to have increased by ~20%. Similar maneuvers were then carried out in the hour angle axis. At the time of writing, the telescope is balanced and can be moved to any required position.

The next phase is to dress the telescope with the liquid nitrogen, glycol, compressed air and vacuum lines and other utilities. This is a significant task as the flexible, vacuum-jacketed nitrogen lines alone are ~100mm in diameter and when empty weigh 7.4 kg per meter. IGUS cable wraps have been purchased to allow free movement around the hour angle, declination and top ring axes. The wraps on all three axes must be supported on the sides of their normal planes of operation to allow for use at all telescope angles, although the top ring wrap will only be operated when the telescope is at the northwest horizon, when the f8 secondary is removed or installed. Thus a complex assembly of mounting brackets is being installed on the telescope and suitably low friction panels are being constructed against which the wraps will slide. The empty hoses and brackets will add some 600 kg to the total weight of the telescope.

The installation ends with fine tuning the TCS and installing the imager and then we can commission the instrument. We anticipate the future need to remove the imager from the telescope from time to time for maintenance purposes thus suitable handling gear has been built. This will be used to install the imager on the prime focus cage with the telescope pointed near the northwest horizon. This station is where all work on previous prime focus instruments has been carried out and the platform used is being upgraded to handle the DECam imager.

First light for DECam is anticipated around the end of September 2012.

REFERENCES

[1] Flaugher, B. "The Dark Energy Survey", Int. J. Modern. Phys. A20, 3121 (2005)
[2] DePoy, D., et al., "The Dark Energy Camera (DECam)," Proc. SPIE 7014 (2008).
[3] Flaugher, B. et al., "Status of the Dark Energy Survey Camera (DECam) Project", Proc. SPIE, Volume 7735 (2010)
[4] Warner, M., Cantarutti, R., Schumacher, G., Mondaca, E., Estay, O., Martinez, M., Aguirre, V., Alvarez, R., Leiva, R., Abbott, T. M. C., van der Bliek, N. S., Proc. SPIE, these proceedings (2012)
[5] Schmitt, R.L. et al., "Cooling the Dark Energy Camera Instrument," Proc. SPIE, Volume 7014 (2008).
[6] Cease, H. et al., "Cooling the Dark Energy Camera CCD array using a closed loop two phase liquid nitrogen system", Proc. SPIE, Volume 7735 (2010)
[7] Rheault, J.-P., DePoy, D. L., Behm, T.W., Kylberg, E. W., Cabral, K., Allen, R., Marshall, J. L., "Spectrophotometric calibration system for DECam", Proc. SPIE, these proceedings (2012).
[8] Abbott, T. M. C., et al., "Improving the Blanco Telescope's delivered image quality", Proc SPIE, Volume 7733, 131 (2010)
[9] Doel, et al., "Assembly, alignment, and testing of the DECam wide field corrector optics", Proc. SPIE, these proceedings (2012).
[10] Munoz, F, "Dark Energy Camera installation at CTIO - Technical challenges", Proc. SPIE, these proceedings (2012)

Dark Energy Camera installation at CTIO: technical challenges

Freddy Muñoz A.*, Andres Montane, Roberto Tighe, Michael Warner, Timothy M.C. Abbott

Cerro Tololo Inter-American Observatory, Colina El Pino S/N, Casilla 603 - La Serena, Chile.

ABSTRACT

Dark Energy Camera (DECam) is a new prime focus, wide-field imager for the V. M. Blanco 4-m telescope at Cerro Tololo Inter-American Observatory (CTIO) and is the science instrument for the Dark Energy Survey Collaboration (DES). The innstrument includes a large, five-lens optical corrector mounted on a hexapod mechanism for fine alignment, filters, and a 520 Megapixel focal plane; all integrated in a new cage structure identical to the original telescope prime focus assembly. Installation of this 8,600 kg instrument requires the removal of the primary mirror from the telescope, the removal of the old prime focus cage, the new integration of new cage, and fine adjustment of large, over-constrained mechanisms followed by telescope reassembly. A large facility shutdown was scheduled for this upgrade and several tools, fixtures, monitoring systems and procedures were developed in order to identify and then recover the optical alignment of the telescope, to control the distribution of stresses during the tuning, and to maintain the balance of the telescope with significant added mass. The final goal has been to keep the high performance of the telescope for both the existing f/8 Ritchey-Chretien mounted instruments and the new DECam now in commissioning. Blanco allowed a flip of prime focus structure so that the f/8 secondary mirror, mounted on the back of the cage, points towards the primary mirror for Ritchey-Chretien observations. DECam will maintain this capability by attaching the existing f/8 mirror cell to the front of the new cage when not observing with prime focus. The challenges presented in handling large elements, real-time monitoring, alignment, verification and feedback are described.

Keywords: DECam installation, dark energy, Blanco telescope, CTIO,

1. INTRODUCTION

The DES collaboration will probe the origin of the accelerated expansion of universe, with a science program designed to make a significant improvement in the constraints on the dark energy equation of state. The DECam instrument, in development since 2008 for this purpose, will replace the entire prime focus cage (PFC) of the Blanco telescope at CTIO. The original PFC held in position both the f/8 mirror assembly (M2) for Cassegrain observing; and the MOSAIC-II prime focus imager plus optics, with eight 2048×2048 SITe CCDs, 36×36 arcmin field of view. These units faced in opposite directions on the prime focus cage (PFC) and each could be pointed towards the primary mirror (M1) as the top-ring structure allowed a 180o flip of the cage, giving a wide range of observing capabilities for Blanco. On replacing the old PFC and installing the new DECam, MOSAIC-II has been decommissioned but F/8 and Cassegrain instruments will continue operation.

The DECam installation represents a major upgrade of the telescope, as the Blanco PFC has not been removed since it was commissioned, in the early 1970s. Significant considerations addressed by this upgrade included the movement and handling of large and delicate equipment through the facilities, the availability of cranes and headroom in the building, development of clean areas for integration, and most relevant, the dismantling and reintegration of a large part of the telescope.

Before removing the PFC from the telescope, the Cassegrain cage was removed and relocated outside the building where it was reinforced to carry the additional counterweight required by the heavier DECam PFC. The primary mirror was also removed and safely stowed. Once the PFC had been removed from the telescope, all optical references on the telescope would be lost without careful preparation to ensure that they could be reacquired.

---

* fmunoz@ctio.noao.edu, Telephone +56 51 205200

2. DECAM INSTRUMENT PROJECT

DECam is a collaborative project led by Fermi National Accelerator Laboratory (FNAL). The imager incorporates 74 CCDs (62 for science and 12 for guiding and alignment), provided by Lawrence Berkeley National Laboratory (LBNL). This focal plane array totals 519 million pixels, illuminated by an optical corrector containing five large lenses polished by SESO Co., France, and installed and aligned in their cells by the University College of London (UCL). The corrector is a steel barrel divided into a cone section holding the largest front lens and its cell C1 (930 mm diameter) and the body, holding the four smaller cells, C2 through C5 (540 mm diameter). The CCD vessel built at FNAL is attached to the top of the corrector and includes the focal plane array, cryogenics lines, vacuum utilities, and a low noise CCD readout system housed in four cooled electronic crates. The DES 600 mm diameter filters (g,r,i,z,Y) were manufactured by Asahi Spectra Co., Japan, and are housed in the filter changer mechanism assembly (FCMA).[1] The FCMA and shutter fits into the body section of the corrector, between the C3 and C4 cells. The FCMA was designed by University of Michigan and is finally integrated with the imager shutter, developed by Argelanger Institute for Astronomy and Bonn University. The imager-filters-corrector assembly works as a single unit, supported by a hexapod developed by ADS Intl., Italy. The hexapod gives six degrees of freedom for the instrument, and can provide tip-tilt, focus, and lateral adjustments. See fig.1 for an overview of instrument components.

The instrument is installed in a cylindrical steel structure developed by FNAL: the DECam prime focus cage. This cage has similar geometry to the old Blanco PFC, to ensure a good fit in the telescope. The f/8 mirror assembly, previously mounted to the back of the PFC, is now to be mounted on front of the new cage. A handling platform that can install and remove the f/8 secondary mirror and cell at this new position has been built by Argonne National Laboratory, including a cylindrical counterweight (f/8-CW) to replace the mirror when is not in use and maintain balance of the prime focus cage and the telescope. The 180° flip is no longer necessary in this new configuration.

The final integration of these elements took place in the Blanco telescope building. The imager, electronics and FCMA were assembled in a clean room, and are currently being tested. They will be commissioned once the DECam cage is installed, and the rebuilt telescope is moving again. The DECam cage now installed on the telescope incorporates the f/8-CW unit, hexapod, corrector and optics, but a dummy imager. Auxiliary components such as the top counterweight for balance, and large rigid pieces attached to each side of the cage handling (strongbacks, since removed) were also included.

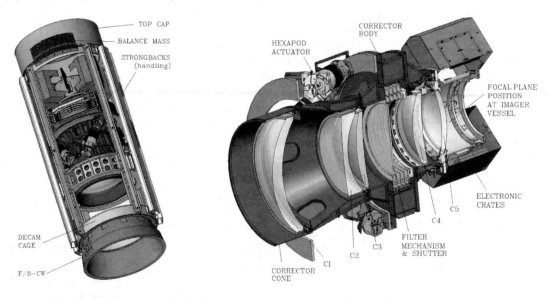

Figure 1. Left side: New DECam cage, including f/8-CW, hexapod, and handling equipment. Right side: Instrument section view shows the internal of the instrument, including optics. When operation, DECam will weight 8,660 kg and contain 74 CCD, yielding a 2.2º field of view. The survey of the southern skies will cover 5000 square degrees, over 525 nights in the next 5 years.

3. BLANCO TELESCOPE

The V. M. Blanco is an equatorial mount telescope, with a 4 m primary mirror, in the Chilean Andes. It is a twin of the 4m-Mayall telescope at Kitt Peak National Observatory (KPNO) in Arizona; both were built in the early 70s. Referring to fig.2, the major components of the telescope are the support framing (orange), the right ascension bearing structure or "horseshoe" (blue), and the main superstructure of the telescope (white) reaching to 19 m above the main floor below. The superstructure carries all the major systems for observing: the lower Cassegrain cage and instrumentation, the primary mirror cell, and the top-end assembly with PFC, with a total weight of 110 tons. The Blanco is housed within a 32 m diameter dome, with a double 50/5 ton radial bridge crane mounted between the two arch girders above.

The original design of the telescope provided a headroom clearance of just 1 m between the 50 ton hook and the PFC top flange. The radial bridge crane, depending on the hook in use, has just 8.3m to 9.5m effective radius of movement around the telescope. The dome carries an additional 3 ton jib crane, for light tasks in areas outside the bridge crane hook ranges. Understanding all equipment and space constraints before drafting a DECam installation procedure was a key consideration. The amount of old equipment removed, new equipment installed and auxiliary gear such carts, man-lifts, crates and lifting fixtures being moved around pushed the Blanco facility to the limits of its capacity.

A complete finite-element analysis of the building structural strength was performed in order to verify that the framing can withstand all equipment movements associated with the DECam installation. The results of this provided several counter-intuitive constraints. Finally a fully choreographed procedure involving all these parameters proved to be an invaluable tool to avoid unsafe interferences or conditions. The analyis was based on accurate weight and geometry CAD files of the building with all relevant elements, and detailed descriptions and procedures from the different DECam collaboration teams involved.

3.1 Secondary assembly and telescope simulator

The position for the new DECam cage is defined by the geometry of the telescope top-end structure. Description of these components is relevant as many issues were identified by the FNAL team during cage installation testing. The understanding of its behavior became one of the most critical tasks for the CTIO crew, not only for installation but for alignment as well. The original design of the 4m telescope top-end consists of two concentric steel rings, the outer rigidly connected to the center section of the telescope by the 7-m, eight-legged Serrurier truss, and an inner ring connected to the outer ring by two shafts, defining the flip axis. The inner ring holds the PFC by means of an array of 12 supporting spider legs (figure 2). A replica of this top-ring structure was built at Fermilab - the telescope simulator - and was used to test the behavior of the new cage and imager. In 2010 the telescope simulator was populated with the DECam cage and the hexapod, plus a copy of the corrector unit with dummy optics weights. Once the cage was installed in the rings, the simulator was oriented for an imager installation rehearsal. Liquid nitrogen supply lines, f/8 counterweight handling and mock observing were also tested.[2]

3.2 PFC mechanical connection

The upper and lower spider legs are identical, and the cage is symmetrical with respect to the top-ring horizontal mid-plane. The center spider legs have a triangular profile and provide connections to both upper and lower sides of the top ring. Each upper and lower spider leg has a 2 inch, 16 threads-inch turnbuckle and a clevis-end, connected to the inner ring block bracket with a rigid pin. Each pinned joint is shared with one of the center leg yoke-ends, with short 1.5 inch, 16 thread/inch turnbuckles. The section view in lower left of figure 2 shows how these elements are connected. The center leg triangle functions as a parallel mechanism; it can only be operated in the horizontal plane, and defines a position of the cage on that plane. The full array of spider legs is an over-constrained system which in principal does not allow free adjustment beyond that given by the mechanical distortion of the elements and construction tolerances in threads, pins and bearings. The position of each leg is identified by a new coordinate system as shown in the lower right of figure 2, and position legends are stamped in the hardware. This design was first developed for the Mayall telescope commissioning. Both telescopes were designed to use the center spider legs to help support the PFC. However when the Mayall was built, the secondary assembly could not consistently maintain both f8 - prime focus alignment during a major

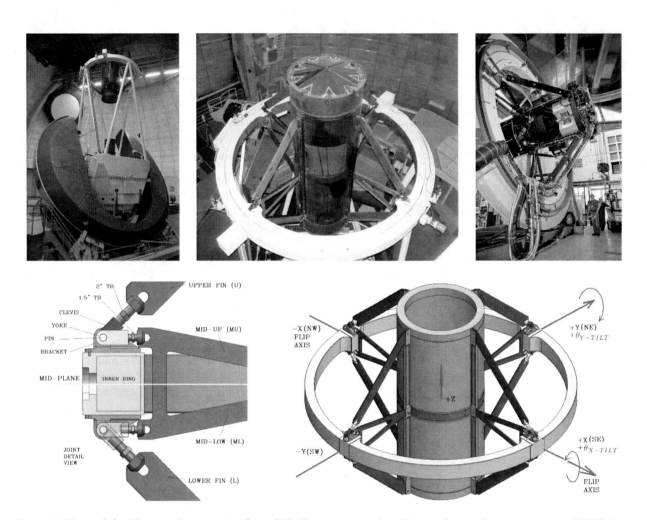

Figure 2. Upper left: Blanco telescope view from SW. Upper center: detail view of secondary structure and PFC, image from south facing north. Upper right: telescope simulator at SiDet laboratory facilities (FNAL, Illinois). Lower left: PFC spider legs ring joint section view. Lower right: coordinate system reference.

flip. The center legs over-constrain the assembly, while the upper and lower legs hold the weight. The center legs were disconnected from the Mayall PFC, and optical alignment in both positions could then be maintained. Those legs were replaced by aluminum members used as cable guides and not pinned at the PFC. The Blanco, built two years later, changed the way that these components were integrated, by assembling and aligning both PFC and top rings as a single unit at ground level, then lifting the completed structure up to the top of the trusses. The original design goals could be attained with good optical performance.

Experience gained with the Fermilab telescope simulator exposed critical issues when installing the DECam cage. The procedure followed was to lift the cage up to the simulator rings with the structure pointed at the zenith, the center spider legs attached to the cage, but not the top rings. The upper and lower legs, first attached to their ring brackets by the pin joint, were connected to the cage rails. After several attempts to install the center legs it was determined that any small angular misalignment would prevent connection to the yokes. The original NOAO drawings show a tolerance of just few thousands of an inch for mating these large elements. Schedule pressure drove FNAL to move to a KPNO-type center leg connection, keeping the triangle steel legs but removing the rigid yoke joint. However the center leg stiffness is necessary to keep the large mass of DECam in its position when the telescope is observing. A CTIO crew travelled to FNAL to solve this step in the simulator, now a major concern for the installation plan. The strategy to resolve the mechanical installation is addressed in sections 4.2 and 4.5 respectively, where simulator experience and CTIO alignment procedure are described.

4. PFC REMOVAL AND DECAM INSTALLATION

The DECam cage was installed with the telescope at the zenith. The installation rehearsal using the telescope simulator provided valuable experience for the installation on the telescope. However, the low headroom between the dome bridge crane and the telescope structure would not permit lifting the cage over the rings, then lowering to the ring mid-plane level as in the simulator. A large boom-crane to access the top of the telescope from outside of the building, through the dome shutter was not available. The cages were therefore lifted up and down through the telescope with the primary mirror removed.

4.1 Mechanical planning and alignment prior installation

The normal Blanco schedule includes regular shutdowns for recoating the primary mirror. The telescope is locked at zenith, then both Cassegrain cage and the primary mirror on its cell are removed. The mirror is placed in a vacuum chamber, located in the dome and close to the base structure, for re-aluminization. Before the mirror is removed from its cell, the relative position of the two is precisely measured so that the same configuration can be reacquired afterwards. The Cassegrain cage is stowed on one side of the telescope, waiting for reassembly.

No realuminization was planned during the DECam installation shutdown, so the primary mirror and cell were not separated. This reduced the number of steps required when reassembling the telescope and maintained the original mirror-cell alignment. The coating chamber is not designed for both the primary mirror and its cell, so a new steel cover was designed to protect against possible damage caused by other tasks in the dome. The DECam prime focus cage is approximately 3 tons more massive than the old cage. This must be counterweighted by 5.5 tons of lead on the bottom of the Cassegrain cage. This cage was not designed to carry such a load and was therefore reinforced. To avoid polluting the dome, the necessary welding was done outside the building, which meant it had to be removed from the dome. This involved moving the aluminizing chamber out of the way of the fully opened hatches, uncovering the shaft through the building through which such large components may be removed. All this was done for the first time since the telescope was built and much effort was expended in planning the choreography.

A set of optical tools to locate the vertex of the mirror and its optical axis were developed and tested before the PFC removal. A vertex tool with crosshairs, a sighting telescope and a target on the cover of the prime focus corrector were used to locate the optical axis. Kevlar-wire crosshairs on precisely relocatable mounts were then aligned on this reference axis in two positions on the telescope trusses: just below the top rings (upper) and close to the mirror (lower). These crosshairs were later used to refine the DECam cage alignment without the primary mirror, after lifting and mechanical connection to the telescope. A good repeatability for the installation of the vertex tool and the Kevlar wires is key to making this procedure effective, and the crosshairs were measured to be 300 μm decentered and 14 arcsec tilted for the upper and 300 μm decentered, 23 arcsec tilted for the upper and lower crosshairs respectively. Both are within the DECam installation specification requiring $\pm$ 1mm decenter and less than 100 arcsec tilt.

Figure 3. Left: the prime focus center reference (small cross in the center of the picture). Center: looking up at tripod vertex tool used to find the vertex point of primary mirror parabola, the PFC structure is above. Right: the sighting telescope, mounted below the mirror cell. These elements are critical as they locate and measure the optical axis to the telescope, the baseline for DECam installation.

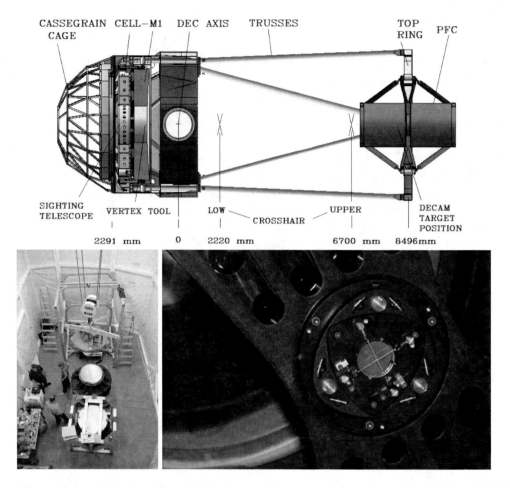

Figure 4. Top: Blanco declination axis outline, with major components and dimensions. The DECam was lifted through the empty structure, from floor level up to the top-rings. The Cassegrain cage and primary mirror must be removed, after the tools in figure 3 were proved to be able to recover the optical axis. Lower left: Corrector body and cone integration in a temporary clean area. Lower right: DECam target installed in front of C1, the optical reference for installation.

The DECam corrector arrived at CTIO with body and cone separated. By January 2012, a UCL team assembled an alignment frame for integration and alignment in a temporary clean area located in the Blanco building at ground-level (figure 4. Once the corrector lenses were inspected after shipping, and the body and cone were mated and aligned, a target was installed on its axis.[3] The target included a beamsplitter-crosshair pair mounted on an adjustable XY/tip-tilt stage, and became the instrument optical reference for installation. During installation, before the primary mirror was reinstalled, the target was aligned with the Kevlar-wire crosshairs in the telescope using a laser projected from below. After the primary mirror was installed, the laser was replaced with the sighting telescope precisely mounted under the primary mirror and the alignment adjusted accordingly, thus tracing the overall alignment back to the normal to the primary at its vertex.

The DECam prime focus cage has no tilt adjustment by design. The hexapod limited to ranges ±100 arcsec in tilt, ±11mm in the horizontal (XY plane) and ± 23mm in focus (Z axis). The tilt became a hard constraint for mechanical alignment as the residual combined tilt of the cage due to manufacturing errors plus the cage installation had to be less than 100 arcsec for the hexapod to operate within its nominal range. The X, Y and Z adjustments of the cage had looser tolerances because the actuator ranges are larger in these dimensions. However, the design of the hexapod means that the further away it is from X, Y, Z, origin the less range it has for tip/tilt correction. Therefore the installation sought a decenter of less than 1.8 mm radial distance, and Z within ±9mm of nominal focal plane.

4.2 Telescope simulator experience and auxiliary tools

During the cage installation rehearsal on the telescope simulator, the array of spider legs over-constrained the system to the point that it did not allow proper alignment. The turnbuckle threads were damaged when adjusted, and the cage moved to unexpected positions. This was the result of driving the spider legs into compression. If all legs are kept under tension, the cage will self-align to the line connecting anchor points, but under compression buckling pushes legs against each other, inducing residual torque on the system and rotating the cage to unexpected positions. The over-constraints made the adjustment a non-intuitive process. To help control the stresses in the spider legs during installation, we instrumented each with a strain gauge pair with real-time readout. The exact length of the legs in their original configuration was recorded as a function of the turnbuckle adjustment, and gauge blocks were made based on those values. These blocks therefore recorded the old PFC cage position and provided a starting point for the telescope simulator. The center of the inner ring circle was located and transfered to the ground with a plumb laser, as was a starting reference for cage lifting. The horizontal mid-plane of the rings was also used as reference and must align with the center of the installed cage.

Installing the cage begins with lifting the cage until it is centered in the top rings and connecting the four lower spider legs, allowing them to take the load. This is quick and permitted us to release the crane rigging, which is essential to avoid excess earthquake risks during long-term tasks in Chile. Adding a pair of orthogonal center legs helped avoid unexpected tilt movements. The handling and connection of legs on the telescope simulator was perfomred with guide ropes. In order to avoid center and lower legs falling down when their joints were released, special tools were developed to keep their original position relative to the cage even when ring pin-joints were not present. These leg-holders proved to be useful not only when removing but also when reinstalling, as they allowed fine control for alignment. The first set was developed for use on the simulator, and an improved design was used during DECam installation at CTIO.

The load to be transferred between the crane and the spider legs being connected was unknown at the simulator. It is necessary to control the load being borne by the lifting hook, as increasing this value reduces the stresses on connected legs, makes turnbuckle operation easier, and reduces the risk of damaging threads. The headroom available with the simulator allowed us to use a standard crane scale to solve this situation, but the headroom at CTIO is too short for most standard models in market. Ultimately, we selected a custom, remotely readable, 25 ton capacity, 10-inch headroom load cell plus appropriate shackles. Once the cage was centered and aligned in the mid-plane of the top ring, the lower spider legs were positioned using leg-holders, connected, and adjusted to a nominal PFC value. Center and upper legs could be then placed in position using leg-holders. As the cage was properly plumb and centered, these last installed spider legs and their joints hardware were within the construction tolerance ranges of ring brackets bores before they were connected. Pin joints were connected requiring only minimal alignment on leg-holders and turnbuckles, resulting on a smooth and successful installation of DECam empty cage in the simulator rings, without compression stresses.

4.3 Old PFC removal from telescope

The telescope simulator was located just 3 m above the ground level at Fermilab facilities. But at the Blanco telescope we required appropriate access facilities to reach the top rings. Two stories of custom-fit, light aluminum platforms were designed and built by CTIO, and installed in the Serrurier truss, around the PFC, before the telescope was moved to the zenith (see figure 8). To do this, it was necessary to remove a measured amount of mass from the PFC in order to kept the total telescope imbalance below 3,000 foot-pounfd torque, the safe limit of the motor drives. The MosaicII imager, PFC light baffles and electronics crates compensated for the extra mass of the platforms. These platforms allowed the installation crews approach the PFC and DECam at the top rings, 17m above the ground, without excessive risk.

An incomplete set of the strain gauges mentioned above were installed on some of spider legs before the removal. Using these, it was possible to make a direct estimate the old PFC mass, based on different loads measured at zenith and near the horizon positions. This method gave an expected weight of 5800 ± 200 kg, less 550kg of material removed from the old cage. The gauge-blocks, mid-plane location and crane position were rechecked before the installation was started. The 50 ton capacity bridge crane hook was attached to the cage using slings and 50% of the calculated weight transferred to it to reduce shear-stress on the spider leg pins. The upper legs were removed first and dismantled, leaving the center leg upper joints free, but not the lower. The over-restraint of the cage was thus reduced, but still safe against earthquake if the crane was disconnected. Once the center legs were removed, the cage was no longer earthquake safe, thus to minimize the risk removal of the cage from this point to placement on the ground below was required to take no longer than an extended working day. The load on the hook was increased to 90% of the cage weight and the lower legs retaining pins removed. They remained connected to the rings and leg-holders could be used to safely rotate them away from the ring bracket, allowing access to the center-lower leg to ring assembly. Once these were also disconnected, the cage without spider legs was moved straight down through the telescope, and the old PFC left the Blanco top rings. At the floor level, the cage was rotated to horizontal with a custom spreader-bar, placed on a cart, and moved away from the telescope work area (figure 5). The full complement strain gauges were installed on the spider legs and recalibrated. The removed cage was found to weigh 5150 ± 20 kg. Brackets, turnbuckles and spider leg threads were cleaned for re-use with the new cage (a spare set was available in the form of the same hardware used at Fermilab on the simulator, but ultimately not used).

4.4 DECam Installation: dry-run

The DECam cage was first installed in the telescope "empty", that is without the corrector or hexapod, in order to reduce the risk to expensive components. The "strongbacks" which give the unsupported cage additional rigidity were installed and the f/8-CW made sure the assembly was bottom-heavy. The DECam cage was moved beneath the telescope and lifted in the same way the old PFC was removed. Spider leg holders were attached before lifting. Once at the top rings, the inner ring mid-plane of the rings was located using a laser level, and the empty DECam was adjusted to this position using crane. The lower spider legs were connected to the cage rail and leg-holders first, then to the top ring brackets.

Temporary loose-fit pins were used for quick connection between the lower spider legs and their brackets. The center legs were connected to the cage and to the upper top ring bracket. Still not aligned to any reference, this allowed the cage rigging to be removed and work to stop for the night. Resuming next day, the rigging was connected to the hook again, and the 50/90% iteration on load transfering to the crane was repeated to continue with upper spider installation and hardware connections with original pins. Any misalignment would mean that the upper bracket bores and their pins could not be connected. All the joints were successfully connected and the new cage was installed, over-constrained, in the top rings of the Blanco, in the same position as the old cage within $\pm 1/32$ in turns of the turnbuckles. The next challenge was to refine the alignment and study the behavior of the spider under stress.

Figure 5. Right and Center: PFC being removed from top rings. Left: DECam cage dry-run installation. The near crew are operating leg-holders (red) to fine align center legs prior to installing temporary pin joints.

4.5 FEA modeling and monitoring as a baseline alignment

The empty DECam cage spider was not carefully tuned, and we established an alignment precision of 2 mm decenter (south), 100 arcsec tilt (south west) between the DECam cage geometry and the reference axis defined by the Kevlar crosshairs.

The procedure for correcting the offsets was developed using the telescope simulator. The upper half of the spider must be disconnected (both upper and center-upper legs) and the tensions in the lower and center-lower legs must be equalized to an average value given by the free-weight of the instrument. Otherwise, further adjustment will change the distribution of tensions in the system but the cage will not move as is highly over-constrained. Deflections were made with a maximum of $\pm 1/16$ of a turn on the turnbuckles. Any more and we risked damaging the threads.

A finite element analysis (FEA) model of the DECam cage was made to investigate the behavior of the cage with the upper half of the spider released. Although the model was for an empty cage, the weight assumed in the model was that of the full instrument mass without strongbacks, calculated as 7660 ± 20kg according scale readouts. All spider legs were included as flat beams with geometry, section and position descriptions. The DECam cage lower section, where the hexapod is mounted, was defined as a rigid unit, the top of the cage was allowed to flex. Spider leg joints with the inner ring were given different degrees of freedom depending on X or Y position. The resulting FEA gave an expected value of 2.70 tons of tension in all lower legs when balanced, and around 300 kg for the center-lower leg connections (figure 6).

The prescription for aligning the cage is as follows, the kevlar crosshairs setup are installed, and the guiding laser tool at floor level pointed and aligned to these reference points, repeating the previous optical axis of the telescope. The DECam target at the cage partially reflects the beam, and the spot can be located around the laser setup to measure starting decenter and tilt (resulting 2mm south in decenter, 100 arcsec southwest tilt for the standard gauge-blocks installation). At the telescope platform level, the alignment crew coordinates with the optical engineer and used live strain gauge readouts to perform iterations to reduce the misalignment while monitoring stresses. After the alignment is checked and confirmed to be within installation parameters, a 1200 kg tension preload should then be applied to the system, in order to force the spider legs always to be in tension at any telescope orientation.

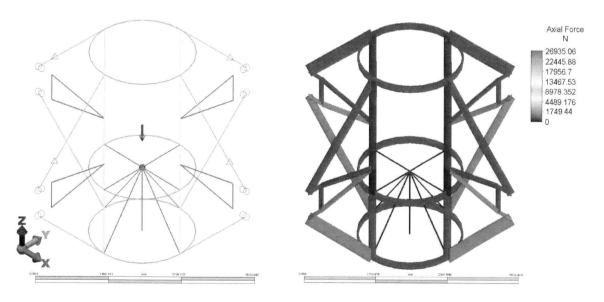

Figure 6. Left: DECam cage FEA model, with geometry parameters. Right: FEA results, axial forces expected on legs. This model has the upper spider legs and center-upper legs joints released, as the alignment required was guided using the four center-lower legs turnbuckles adjustment.

The alignment was guided by the four lower-middle leg turnbuckles (+/-XML, +/-YML), while the four lower legs supported the cage weight. Iterating the model allowed us to determine the appropriate directions for the corrections.

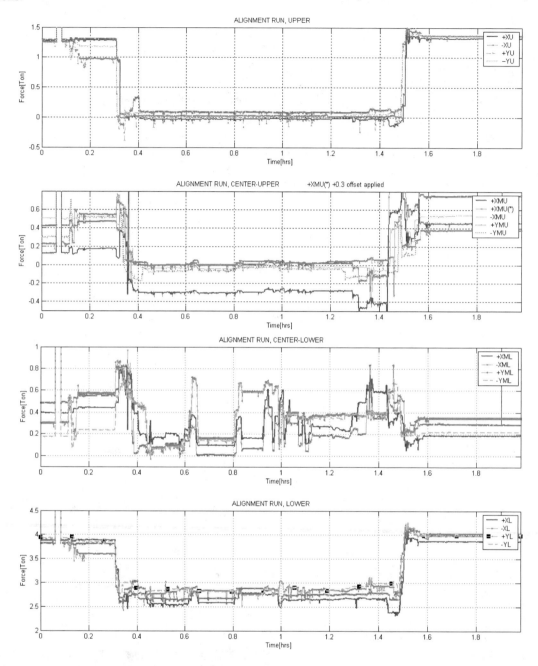

Figure 7. Load cell readouts during alignment (starting at 0.3hrs). The upper spider legs and center-upper leg joints are not connected during alignment, keeping a zero stress condition. The +XMU sensor bonding partially failed during run, giving a -300 kg force readout offset even when unloaded. The +XMU(*) represents the correction applied. Center-lower legs are used to lead the cage to desired positions: correcting a 2mm south decenter by tensioning joints +YML and -XML joints to tension, forcing the cage towards the north. Tension applied is followed immediately by an equal reaction at opposite -YML and +XML joints, then zeroed in order to release cage move it. +YML requires larger adjustment than -XML, since tension in this axis also corrects the 100 arcsec, south-west tilt measured (a -YML component). Preloading starts at 1.45 hrs.

4.6 Conclusion: DECam full instrument installed

The empty DECam cage was removed from the telescope following the same procedure as for the old PFC. The cage was moved to the ground level of the Blanco building the integration of hexapod and corrector, plus dummy imager and counterweights for balance. A final check of the hexapod coordinate system was made under the telescope. The instrument was lifted into position in the top ring and installed following the same procedure as for the empty cage using the same set of gauge blocks.

Table 1. Final alignment and stress results

North(+)/South(-) decenter	-0.53 ± 0.31 mm
East(+)/West(-) decenter	-0.08 ± 0.31 mm
Focus (+up)	-3.8 ± 1.0 mm
North/South axis tilt, (+: west)	-5.1 ± 14 arcsec
East/West tilt, (+: north)	-7.4 ± 14 arcsec
Upper legs axial force, preloaded	1.350 ± 0.100 ton
Center-upper legs axial force, preloaded	0.400 ± 0.050 ton
Center-upper legs axial force, preloaded	0.300 ± 0.050 ton
Center-upper legs axial force, preloaded	3.800 ± 0.100 ton

The same misalignment between DECam cage and optical axis was found, with south decenter and southwest tilt. The alignment prescription above was followed. After partial confirmation with the guiding laser tool, the optical alignment fixtures were removed and the primary mirror and cell reinstalled in the telescope. The reinforced Cassegrain cage and lead counterweights were then installed. The final alignment measures were obtained with the accurate sighting telescope from below the primary mirror cell where Cassegrain instruments are mounted (Table 1). These results are all within specification, allowing the full range of motion in the hexapod. At the time of writing, the telescope has been balanced and is able to move all over the sky. We are currently installing the LN2 feeds and other utilities. First light for the instrument is expected around the end of September 2012.

ACKNOWLEDGMENTS

Many NOAO colleagues and others contributed to this successful, high-complex upgrade of the Blanco. We thank the KPNO team which supported CTIO during critical long days: Will Goble, Mike Hawes, Fred Wortman and Jose Montes; and Fermilab's Ken Schultz experience both at telescope simulator and Blanco.

REFERENCES

[1] B. Flaugher et al., "Status of the Dark Energy Survey Camera (DECam) Project", Proc. SPIE, 7735-12 (2010)

[2] H. Thomas Diehl, et al., "Testing the Dark Energy Camera on a Telescope Simulator", Proc. SPIE, 7735-125 (2010)

[3] Timothy M.C. Abbott, et al., "Dark Energy Camera installation at CTIO - Overview", Proc. SPIE, 8444-156 (2012)

Figure 8. Blanco telescope with new Dark Energy Camera installed, balanced and moving again, after 4 months shutdown.

Korea Microlensing Telescope Network: Science cases

Byeong-Gon Park[*a], Seung-Lee Kim[a], Jae Woo Lee[a], Byeong-Cheol Lee[a], Chung-Uk Lee[a], Cheongho Han[b], Minjin Kim[a,f], Dae-Sik Moon[c], Hong-Kyu Moon[a], Soo-Chang Rey[d], Eon-Chang Sung[a], Hwankyung Sung[e]

[a]Korea Astronomy and Space Science Institute, Daejeon, 305-348, Korea;
[b]Chungbuk National University, Cheongju, 371-763, Korea;
[c]University of Toronto, Toronto, ON M5S 3H4, Canada;
[d]Chungnam National University, Daejeon 305-764, Korea;
[e]Sejong University, Seoul 143-747, Korea;
[f]The Observatories of the Carnegie Institution for Science, Pasadena, CA 91101, U.S.A.

ABSTRACT

We present the science cases with the Korea Microlensing Telescope Network (KMTNet) which consists of three wide-field 1.6 m telescopes distributed in Chile, South Africa, and Australia, respectively, providing unique continuous sky coverage with the three telescopes. The primary scientific goal of the KMTNet project is to explore the structure and diversity of planetary systems and variable objects. Since the system is mainly optimized to conduct gravitational microlensing surveys, it will enable detections of very low-mass exoplanets, potentially down to the mass of Mars that are inaccessible by other ground-based techniques. In addition to the primary science, it is possible to conduct a variety of other observational programs with the KMTNet system, including photometric studies of nearby galaxies and galaxy clusters, discovery of supernovae and their follow-up observations, and observations of near-Earth objects. We expect synergies between the KMTNet project with other similar or complementary projects in the southern sky, such as SkyMapper.

Keywords: Optical Telescope, Wide-field Photometric Survey, Extrasolar Planets, Gravitational Microlensing

1. INTRODUCTION

Korea Microlensing Telescope Network (KMTNet) project is a wide-field imaging survey being conducted in Korea Astronomy and Space Science Institute (KASI). The specifications of the KMTNet systems and details of the sites are described in Kim et al. (2010)[1]. As the name of the project implies, the project primarily aims to discover extrasolar planets (exoplanets) based on the analyses of gravitational microlensing phenomena. Focusing at a selected 4 degree by 4 degree field towards the Galactic bulge, we will observe the field once every ten minutes with exposure time of 2 minutes in I band. This high-cadence, 24 hours uninterrupted observation strategy will enable us to derive an unbiased census of exoplanets with a mass range down to Mars mass. Especially we aim to discover the first Earth-mass planets in the habitable zone from ground-based observations. The microlensing community predicted that the KMTNet-like system could detect hundreds of extra-solar planets per year (see Exoplanet Community Report 2009[2] and Gaudi et al. 2010[3]). We will also discover exoplanets by transit method, and detect a variety of variable objects towards the Galactic Bulge.

In addition to the exoplanets and variable objects survey, we anticipate a variety of science cases which can take advantage of the wide field of view (FOV), 24 hours continuous observation system. Although the observation time for the Galactic bulge is already allocated for primary sciences, we released a Call For Proposals (CFP) for sciences other

---

[*] bgpark@kasi.re.kr; phone 82-42-865-3207; fax 82-42-865-3272

than exoplanet search in the season when the Galactic bulge is not observable. This paper introduces some of the proposed science cases together with the primary science cases though the final list of science proposals is not yet fixed.

2. SCIENCE POLICY OF THE KMTNET PROJECT

2.1 Observation time allocation

Observation time for the primary sciences, i.e., the Galactic bulge monitoring, has been already allocated until 2018. The non-bulge season will be open to the public astronomical community, though there are some regulations to be followed (If interested, contact the author for more information) and final selection is subject to a review process. The following graph shows the bulge observing time by solid line between February 20th and October 22nd as well as the non-bulge one by dashed line at CTIO in Chile. This pattern is almost identical to the other two sites, South African Astrophysical Observatory and Siding Spring Observatory because of their similar latitudes.

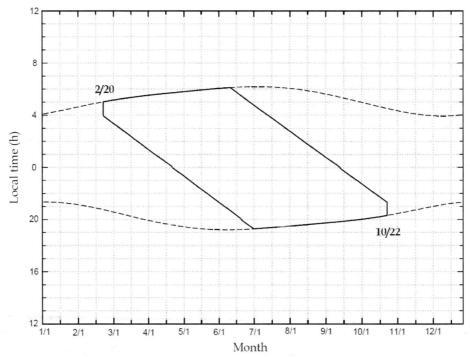

Figure 1. Observation time chart of KMTNet. Solid line denotes bulge observing season and dashed line denotes non-bulge season.

2.2 KMTNet rules for external members

As most survey projects, KMTNet project also has internal rules and regulations for the members who use the system for their specific sciences. Detailed rules for proprietary period of the data and regulations for the publication can be found in the CFP document and are not included in this short paper. Most important regulation in KMTNet project is related with the detection of exoplanets, regardless of accidental or planned. As the primary goal of KMTNet project is the search for exoplanets, we strongly urge that all exoplanet candidates discovered based on the KMTNet data should be informed to and strictly kept within the KMTNet primary science team before being announced to outside members/community privately or publicly. The primary science members will examine the observational data of the planet candidates and will decide further actions.

3. SCIENCE CASES OF THE KMTNET

3.1 Monitoring of the Galactic bulge

These are primary sciences of the KMTNet project which utilize all the data obtained during the period when the Galactic bulge is observable (aka, Bulge season). During the Bulge season, a 4 by 4 degree field centered at the Galactic coordinate $(l, b) = (+2.0, -2.0)$ shown in Figure 2 (Gaudi 2006; unpublished) will be continuously monitored once in ten minutes with I filter during 24 hours a day. We allocate all the observation time where the Galactic bulge is seen for the following primary science cases of the KMTNet project.

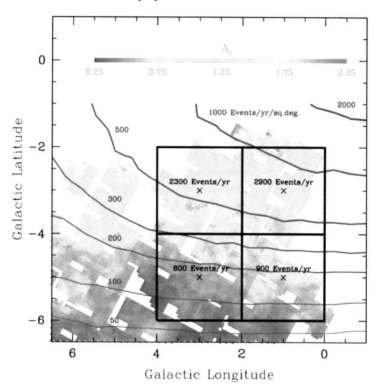

Figure 2. KMTNet field of view in Galactic bulge along with expected microlensing event rates.

3.1.1 Search for the earth-mass planets using the microlensing method

As of May 30th 2012, there are 770 exoplanets listed in the Extrasolar Planets Encyclopedia (http://exoplanet.eu; Schneider et al. 2011[4]). Among them the number of planets detected by microlensing method is only 15. Considering that microlensing is an effective method to detect both low- and high- mass planets, small and large semi-major axis, even unbound planets, and planetary systems with distance as far as that of extragalaxies, the apparent sparseness is not due to inherent insensitiveness of the method to exoplanets but due to ineffectiveness of the observing systems and strategy. In KMTNet, we will conduct 24 hours continuous monitoring of a single wide field towards Galactic bulge at a frequency of 6 exposures per hour. From this observing strategy, we can obtain complete light curves for all microlensing events at frequencies of typical intensive follow-up mode observation. So the efficiency of the planet detection will be much higher than the cases of conventional survey / follow-up mode observations where most light curves are not completely covered and only those selected high-amplification events are monitored intensively. From a numerical simulation using the characteristics of the KMTNet system and that of the sites, we expect to detect tens of earth-mass planets per year, with the lower limit of the planetary mass down to that of Mars as shown in Figure 3 (Gaudi et al. 2010[3]).

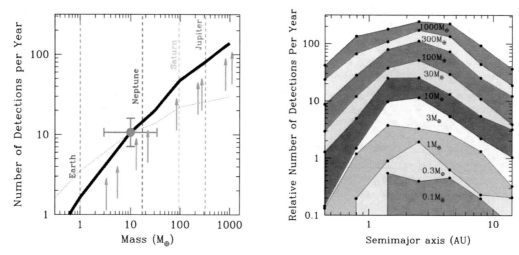

Figure 3. Expected number and lower mass limit of planets detected with KMTNet project.

3.1.2   Search for the transiting exoplanets and variable objects

In addition to the microlensing exoplanets mentioned above, the 24 hours monitoring data of the Galactic bulge will enable us to discover a huge number of variable objects by analyzing time-varying phenomena produced by planets and stars themselves with various methods. Such objects are shown in Figure 4, which are transiting exoplanets, eclipsing binaries of W UMa and Algol types, and pulsating stars including low-amplitude, short-period variables such as δ Scuti stars. Because the times of minimum/maximum light in variable stars act as an accurate clock, we expect to detect the circumbinary/circumstellar exoplanets for stellar systems with timing accuracies of about 10 seconds. Also, it will be possible to detect additional planets or exomoons by using the transit timing variation with known transiting planets. From the observational results and theoretical modeling, we plan to achieve a better understanding of the properties of star-star and star-planet systems.

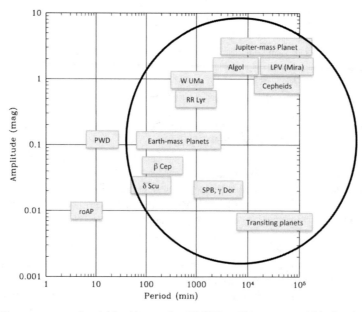

Figure 4. Discovery map of variable objects using KMTNet. Object types within the circle can be detected.

3.2 Wide-field survey for the non-bulge time

In this section we introduce summaries of some science proposals received by the time of this writing. The opportunities to submit a proposal is still open, and the final list of proposals is subject to a deliberation process. Some of the important milestones of the KMTNet project are listed at the end of this paper.

3.2.1 Photometric survey of stars in Magellanic Clouds (PI: Hwankyung Sung)

The Large and Small Magellanic Clouds (MCs; LMC and SMC) are the most important astrophysical laboratories to conduct test on the stellar evolution theories and the study of galaxy evolution. As these two galaxies are so close and have large angular sizes, the photometric studies of the MCs have been very limited in space or depth. With the wide FOV of the KMTNet and good observing condition at the CTIO, it will be possible to observe whole regions of the MCs and to provide a homogeneous photometric catalogue of astronomical objects in MCs. The specific aims of this study are:

a. Provision of a homogeneous catalogue for the astronomical objects in MCs down to $V = 23$ mag.
b. Three-dimensional structure of MCs using several distance indicators.
c. Spatial distribution of Herbig Ae/Be stars from $H\alpha$ photometry and current star formation activities of MCs.
d. Star formation histories of MCs from the analyses of color-magnitude diagrams.
e. Spatial distribution and the luminosity function of planetary nebulae.
f. Interactions between LMC and SMC, between LMC and MWG, and between SMC and MWG from the spatial distribution of faint stars belonging to MCs.

3.2.2 Search for exoplanets in the Large/Small Magellanic clouds (PI: Jae Woo Lee)

So far more than 770 exoplanets are discovered but not in extragalaxies. We try to detect the first extragalactic planet in the MCs using transit method from high-precision, high-cadence photometric survey. Since most of the known transiting planets show transit durations of 100 ~ 300 minutes, the time resolution of about 10 minutes is needed to detect a transit-like feature. With the KMTNet system, we can attain ~ 1% photometric precision for stars down to $I = 21$ mag with such high time resolution. In this study, we will conduct the round-the-clock monitoring of the variability of millions of stars which are mainly F- and G- type main sequence and blue horizontal branch stars in the range of $I = 19 \sim 21$ mag in MCs. Most observation will be done using I filter with exposure time of 10 minutes, and complementing observations with V band at start, middle and end of the night for data calibration using color information. From these data, we expect to detect tens of exoplanets in the MCs. From the data set accumulated for several years, we can also adopt timing method with which we can discover exoplanets with longer orbital periods than those discovered by other techniques, by precisely measuring the times of maxima and minima of the light curves of variable stars like RR Lyrae type pulsating stars and eclipsing binaries.

3.2.3 Search for ultra-faint dwarf galaxies using both KMTNet and SkyMapper (PI: Eon-Chang Sung)

The Sloan Digital Sky Survey (SDSS) has been immensely successful in detecting new Milky Way satellite galaxies over the past seven years. It was instrumental in finding examples of the least luminous galaxies we know in the Universe, uncovering apparent inconsistencies between cold dark matter theory and dwarf galaxy properties, providing first evidence for a possible lower mass limit of dark matter halos in visible galaxies, and reopening the discussion about the building block scenario for the Milky Way halo. Nonetheless, these results are still drawn only from a relatively small number of galaxies distributed over an area covering about 29% of the sky, which leaves us currently with more questions than answers. The study of these extreme stellar systems is a multi-parameter problem: ages, metallicities, star formation histories, dark matter contents, population fractions and spatial distributions must be determined. Progress in the field is discussed and drawn attention to some of the limitations that currently hamper our ability to fully understand the phenomenon "ultra-faint dwarf galaxy". In this context, the joint work of KMTNet and SkyMapper represents a new initiative to systematically search and scrutinize optically elusive Milky Way satellite galaxies in the southern hemisphere. In doing so, the program aims at investigating some of the challenging questions in stellar evolution, galaxy formation and near-field cosmology.

3.2.4 Wide-field and deep survey of nearby southern clusters of galaxies (PI: Soo-Chang Rey)

Thanks to KMTNet's wide field of view, it is time to implement imaging survey of extensive area of clusters of galaxies in the southern sky with modern instrument. As part of potential long-term survey of nearby (D < 50 Mpc) well-known clusters of galaxies, we propose a wide-field and deep survey of Fornax cluster as a first step of the project. By imaging the 400 square degree region (100 fields) enclosed within the five times virial radius of the Fornax cluster, in three SDSS filters (g, r, i), we can provide an unprecedented view of structure of Fornax cluster using sample from giant to dwarf galaxies. We will secure galaxies with brightness comparable to the limiting magnitude (r = 23.1 AB mag) of SDSS. Furthermore, we also request extremely deep (limiting surface brightness of ~ 28 mag arcsec^{-2} for r' band) survey for the central region (16 square degree, i.e., four fields) of Fornax cluster. This will allow us to detect the diffuse intracluster light (ICL) that permeates clusters as a valuable tool for studying the hierarchical nature of cluster assembly. In order to complete whole survey, about 285 hours observing time (without overhead) is required. By combining data available at other wavelengths, it will offer unique constraints on the formation of large-scale structure and also provide important clues for theories of galaxy formation and evolution. Our proposed survey will be implemented in the close collaboration with researchers in various countries (Germany, Australia, UK, USA) and ongoing project (e.g., SkyMapper).

3.2.5 Deep wide-field imaging of nearby galaxies (PI: Minjin Kim)

We propose to obtain deep wide-field images of the nearby bright galaxies. The structural information of galaxies derived from the photometric data provides a crucial clue to understand galaxy formation. Conventionally, it is known that bulges (or elliptical galaxies) and disks are well described by simple Sérsic profiles with indices of 4 and 1, respectively. However, deep images obtained with modern telescopes and detectors yield that the shapes of galaxies are more complicated than we expected (e.g., pseudobulge, outer disk, truncated disk; Freeman 1970[5]; Kormendy & Kennicutt 2004[6]; Thilker 2005[7]). It suggests that different galaxies underwent different evolutionary history. Therefore, their evolution might have been affected by environment via minor/major merging, secular evolution, and cooling flow. However, origins of such features are not well understood yet. In order to explore those processes in detail, it is required to obtain very deep and wide imaging data for nearby galaxies (e.g., Martínez-Delgado et al. 2012[8]). We plan to image a comprehensive sample of galaxies in B, V, R, and I filters, for 1 ~ 2 hours exposure time in each filter. Using this dataset, we will look for diffuse, low surface brightness structures including outer disks, truncated disks, tidal features and faint companions. The multicolor data will enable us to estimate star formation history of those features.

3.2.6 Supernovae research using the KMTNet (PI: Dae-Sik Moon)

One of the interesting and important sciences ideally and competitively utilizing the unique resource of the KMTNet is the discovery of many supernovae (SNe), especially rare young and/or rapidly declining ones of prime scientific interests, and high-cadence follow-up photometric monitoring observations of the discovered SNe. The network of the three wide-field telescopes providing the continuous sky coverage is exceptionally well suited for such observations. Based on the previous results of other SNe search programs, we expect the KMTNet to discover in the range of 500 ~ 1,000 SNe per six months with 2 hours observing time at each telescope. This includes about 10 young SNe of early detection within approximately one day of explosions. Consequently, the KMTNet SNe program will enable many interesting research projects of SNe and related objects, such as: early shock breakout of SN explosions; systematic study of the nature of massive core-collapse star explosions; statistical analyses of SN progenitor systems and the properties of host galaxies; and (potentially) something unexpected and unknown.

3.2.7 KMTNet DEEP-SOUTH: KMTNet deep ecliptic patrol of the southern sky (PI: Hong-Kyu Moon)

As of June 2012, more than 9,000 Near-Earth Objects (NEOs) have been catalogued, yet less than 5% of them are registered for their physical and mineralogical properties. We propose an effective and prompt solution to the pressing demand for characterizing the population. Based on 24-hour round-the-clock photometric census utilizing the KMTNet, we will investigate their physical properties including orbits, sizes, shapes, and rotational status at the same time. We will also discriminate surface mineralogy of the objects employing SDSS and the Cousins colors. This unique and homogeneous dataset is expected to provide a multi-dimensional view of the whole population with a significantly improved resolution. The DEEP-SOUTH dataset will complement to space-borne database, e.g., *Near-Earth Object Wide-field Infrared Survey Explorer* (*NEOWISE*; Mainzer et al. 2011[9]) and the *Asteroid catalogue using AKARI* (*AcuA*; Usui. et al. 2012[10]). Our photometric census is focused on opposition in either side of the ecliptic, while the proposed targeted survey is designed to discover the Atens and Inner Earth Objects (IEOs) in the morning and the evening "sweet spots". Since there is no dedicated telescope on the line on other side of the equator, it has been an anticipation from the science community that a network of wide-field telescopes such as the KMTNet to actively participate as a core facility for NEO search and characterization (UK NEO Task Force, 2000[11]; Chapman et al., 2001[12]). A conservative estimate for the opposition census indicates that this network of telescopes to characterize 70 ~ 90% of km-sized NEOs; discover several thousand main belt asteroids (MBAs) and dozens of comets as byproducts, after five years of operation with 2 ~ 3 hours of observing time at each site.

4. SUMMARY

We introduce here two primary science cases and summaries of seven science proposals for the KMTNet project. At the time of writing this paper, a CFP is opened and we are still collecting proposals to use KMTNet. We expect the first KMTNet telescope to be commissioned in early 2014, while all three telescopes will be operational from 2015. The project is planned to close at the end of 2018.

Here is a list of important milestones:

- May 16, 2012: Release of the Call For Proposals
- July 15, 2012: Deadline to submit Proposals
- August 31, 2012: Review of the submitted Proposals
- September 30, 2012: Selection of the secondary KMTNet sciences
- Late-October, 2012: 1st Workshop of the PI's

ACKNOWLEDGEMENTS

This work is supported by Korea Astronomy and Space Science Institute.

REFERENCES

[1] Kim, S.-L., Park, B.-G., Lee, C.-U., et al., "Technical specifications of the KMTNet observation system", Proc. of SPIE 7733, 77333F (2010)
[2] JPL Publication 09-3, [Exoplanet Community Report], ed. Lawson, P. R., Traub, W. A., and Unwin, S. C. , NASA/JPL, Pasadena (2009)
[3] Gaudi, B. S., Beaulieu, J. P., Bennett, D. P. et al., "The Demographics of Extrasolar Planets Beyond Snow Line with Ground-based Microlensing Surveys", White Papers for the Astro2010 PSF Science Frontier Panel (2010)
[4] Schneider, J., Dedieu, C., Le Sidaner, P., et al., "Defining and cataloging exoplanets: the exoplanet.eu database", A&Ap 532, A79 (2011)
[5] Freeman, K. C., "On the disks of spiral and S0 galaxies" Astrophys. J. 160, 811-830 (1970)
[6] Kormendy, J. and Kennicutt, R. C., "Secular evolution and the formation of pseudobulges in disk galaxies" Annu. Rev. Astron. Astrophys. 42, 603-683 (2004)
[7] Thilker, D. A., Bianchi, L., Boissier, S., et al., "Recent star formation in the extreme outer disk of M83" Astrophys. J. 619, L79–L82 (2005)
[8] Martínez-Delgado, D., Romanowsky, A. J., Gabany, R. J. et al., "Dwarfs gobbling dwarfs: a stellar tidal stream around NGC 4449 and hierarchical galaxy formation on small scales" Astrophys. J. 748, L24-L27 (2012)
[9] Mainzer, A., Bauer, J., Grav, T., et al. "Preliminary Results from NEOWISE: An Enhancement to the Wide-Field Infrared Survey Explorer for Solar System Science," Astrophys. J. 731, 53-65 (2011)
[10] Usui, F., Kuroda, D., Müller, T. G., et al., "Asteroid Catalog Using Akari: AKARI/IRC Mid- Infrared Asteroid Survey", Publications of the Astron. Soc. of Japan 63, 1117-1138 (2011)
[11] UK NEO Task Force, [Report of the Task Force on Potentially Hazardous Near Earth Objects], Near Earth Object Information Centre, London, 37 (2000)
[12] Chapman, C. R., Marsden, B. G., Ostro, S. J., et al., "Asteroid Comet Impact Hazard: NEO Survey Workshop Report," NASA/ARC, Moffett Field, 60 (2001)

Design and Development of a Wide Field Telescope

Il Moon[1*], Sangon Lee[1,2], Juhee Lim[1,3], Ho Soon Yang[1], Hyug-Gyo Rhee[1], Jae Bong Song[1], Yun Woo Lee[1], Jong Ung Lee[4], Ho Jin[3]

[1] Korea Research Institute of Standards and Science, Daejeon, South Korea
[2] Department of Applied Optics and Electromagnetics, Hannam University, Daejeon, South Korea
[3] School of Space Research, Kyung Hee University, Yongin, South Korea
[4] Department of Laser and Optical Information Engineering, Cheongju University, Cheongju, South Korea

ABSTRACT

A prototype of large wide field telescope is a Cassegrain telescope which covers 2° field of view with two hyperbolic mirrors, a 0.5 m primary mirror and a 0.2 m secondary mirror with multiple correction lenses. To fulfill the optical and mechanical performance requirements in design and development phase extensive finite element analyses using NX NASTRAN and optical analyses with CODE V and PCFRINGE have been conducted for the structure of optical system. Analyses include static deformation (gravity and thermal), frequency, dynamic response analysis, and optical performance evaluations for minimum optical deformation. Image motion is also calculated based on line of sight sensitivity equations integrated in finite element models. A parametric process was performed for the design optimization to produce highest fundamental frequency for a given weight, as well as to deal with the normal concerns about global performance.

Keywords: Wide field telescope, Opto-mechanics, FEA, Line of sight (LOS), Dynamic response

1. INTRODUCTION

The wide field telescope is a Cassegrain type telescope designed 2° field of view on the 81.1 mm ×81.1 mm size of image plane. The telescope is composed of two hyperbolic mirrors, 5 lenses, a filter and a window. The primary mirror (M1) is the 500 mm diameter clear aperture mirror with a 140 mm diameter central hole. The 200 mm convex secondary mirror (M2) is located 309.58 mm from M1 along the z-axis. A novel and unique optical design was conducted to obtain the long working distance from the last lens to focal plane for the space of the CCD camera module This optical design was composed of five lenses as the aberration corrector and the focus compensator. The filter and the window were mounted on the same structure on the CCD camera module.

The environments, such as gravity, temperature, wind, humidity, vibration and earthquake, cause a certain amount of thermal expansion or displacement by static and dynamic loads on the structures. Under severe environmental conditions, the dynamic response of a wide field telescope is an important design concern for estimating the image motion. Depending on the purpose of the wide field telescope optical systems, the telescope structure should be optimized with the image motion of less than 9 μm, 1 pixel size of CCD, at the image plane [1]. Because the image motion is very sensitive to the relative motion of the optical elements, in this research the telescope structure was optimized with their mirrors, lenses and mounting system under the dynamic loads. Through several tradeoff studies, the proper shapes and materials for the optical system were selected and optimized to meet design specification. The global coordinate system used in this study is defined as the x-axis being the elevation axis, the z-axis being the optical axis and the y-axis being defined by a right-hand coordinate system.

The image motion by the acoustic vibration due to wind needs to analyze in the dynamic response mode. A special modeling technique may be required to calculate the image motion analyses. In order to obtain the nodal dynamic response (image motion), the physical point (nodal point) of the optical elements should be composed to the finite element model. The LOS, which connected with each optical element and with the physical image point integrated in the finite element analysis model, is a useful method for estimating the image motions for the optical systems. [2, 3] The result of the dynamic response analysis (time history) is expressed with a time series. Detailed optical design and opto-mechanical performance analyses including the image motion due to wind loads are presented herein.

2. OPTICAL DESIGN OF WIDE FIELD TELESCOPE

In the optical design, the most important requirement for the wide field telescope is 2 ° large field of view (FOV). Other conditions are aperture size of the primary mirror, 500 mm, and image size which fit to 4096 × 4096 array CCD with 9 μm size pixel. The range of wavelength is 365 through 900 nm. To achieve these requirements, we chose the Cassegrain telescope optical configuration like as shown in Figure 1 (a) and designed by using the optical design program Code V. This system is composed of two hyperbolic mirrors, five lenses, the filter, and the window. The primary mirror (M1) is a hyperbolic mirror, close to a parabolic mirror, made of Zerodur glass with a 500 mm diameter clear aperture and a 140 mm diameter central hole. The secondary mirror (M2) is a 200 mm hyperbolic convex mirror made of Zerodur glass and is located 309.58 mm from M1 along the z-axis. Five lenses used for the aberration corrector and the focus compensator to get better image in all over the FOV. We shaped all lenses into the spherical surface to fabricate with high quality and to test easily than the aspherical lens [3]. The distance between the fifth lens and the image plane is fixed to 72.74 mm for the CCD camera module (the field derotator, the chopper, band pass filters, etc.). The 5 band pass filters, using for wavelength 365 nm, 440 nm, 550 nm, 700 nm, and 900 nm, are mounted on the filter wheel in the CCD camera module to choose depend on purpose. The window is located in front of the CCD to protect the image sensor. The entrance pupil diameter (EPD) is 500 mm on the M1 and the EFL is 1493.46 mm at wavelength 550 nm, that is to say f-number of 2.99. The largest RMS spot size of the designed telescope is 7.9886 μm, which is smaller than a pixel size of CCD, as shown on Figure 1 (b).

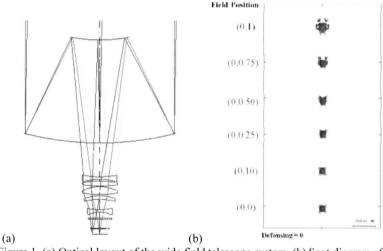

Figure 1. (a) Optical layout of the wide field telescope system, (b) Spot diagram of design.

The optical performance of the wide field telescope with MTF, finite ray aberration and depth of focus are shown in Figure 2. As shown in Figure 2 (a), we obtain the MTF including all fields, 0.0, 0.2, 0.5, 0.7, and 1.0 degree, has more than 0.6 at frequency of 60 cycles/mm. The depth of a focus is about 0.03 mm with MTF larger than 0.5 at Nyquist frequency of 56 cycles/mm. Finite ray aberrations of this optical system are illustrated in Figure (2). It is indicated that the distortion is 22 percent at full field.

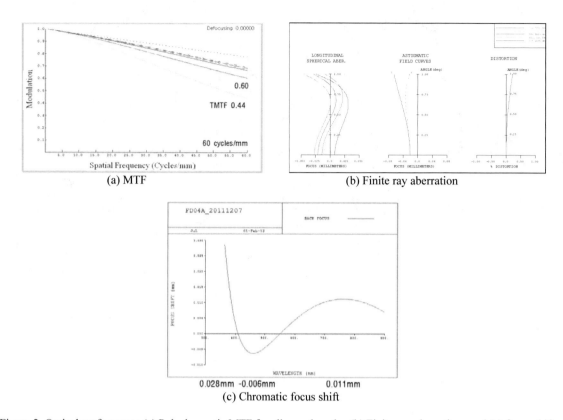

Figure 2. Optical performance (a) Polychromatic MTF for all wavelengths, (b) Finite ray aberrations and (c) focus shift.

3. OPTO-MECHANICAL DESIGN OF WIDE FIELD TELESCOPE

Opto-mechanical structures of the wide field telescope are composed of the primary mirror (M1) assembly, the secondary mirror (M2) assembly, the yoke interface, the assembly of lenses, and the CCD camera part in order. The components of the M1 assembly are the primary mirror, its M1 support system installed on M1 cell and the inner barrel. M1 support system is composed of six point axial supports connected on two point hindle system and three lateral supports located equally spaced at the periphery of the mirror. Through an iterative design procedure their designs were optimized. The M2 assembly includes the secondary mirror, its mounting system, alignment system consisted of linear stepper motor for focusing mechanism with three adjustment mechanism for tilt, spiders for M2 assembly and the top ring with adjustable bolting location for decenter. The three points tilting system are only used for an alignment and install, and defocusing system are used for defocusing with change of environment including operation temperature. The components of the yoke interface are the outer wall and two side walls on the x-axis, each of side wall has a hole to combine the yoke on the same height as the center of the telescope of gravity along the z-axis. The assembly of five lenses, L1, L2, L3, L4, and L5, are mounted on one lens barrel optimized under temperature and the lens barrel is placed under the inner barrel connected with M1 cell. The components of the CCD camera module are the CCD cover, a group of filters, a window, a detecting system (chopper, shutter, and field de-rotator), and a CCD. The material for telescope structure was chosen with design considerations of stiffness, lightweight, thermal stability, ease of fabrication, dynamic stability, and cost. Under these considerations, aluminum and invar were mainly used to fulfill the opto-mechanical performance criteria. The opto-mechanical structure layout of the wide field telescope is shown in Figure 3.

Figure 3. Opto-mechanical layout of the wide field telescope.

To analyze the telescope optical performance and image motion, the high-fidelity finite element model was established by using the program NX NASTRAN for each of static and dynamic load cases. For the dynamic analysis a group of the M1 assembly (G1) composes with the primary mirror, and its M1 support system, a group of the M2 assembly (G2) includes the secondary mirror, its support system, alignment system, and support spiders, and the CCD camera part (G3) are modeled as lumped mass elements at the center of gravity of the each group and assumed to be rigidly connected to the inner barrel, the upper barrel and CCD cover, respectively. The yoke and its interface are rigidly connected, and five lenses, L1, L2, L3, L4, and L5, also modeled as lumped mass elements at the center of gravity of the each element and assumed to be rigidly connected to the lens barrel. The finite element model for the dynamic analysis is shown in Figure 4.

Figure 4. Finite element model of wide field telescope for the dynamic analyses.

As a sample design of primary mirror and its support system, the surface error due to gravity and temperature are shown in Figure 5. The values are well controlled for the global performance criteria.

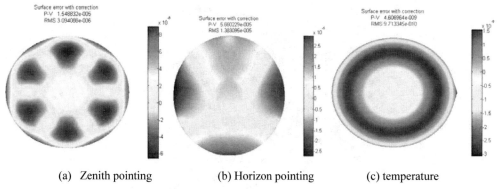

(a) Zenith pointing (b) Horizon pointing (c) temperature

Figure 5. Optical performance due to gravity and temperature on the primary mirror.

A frequency analysis was also conducted, and the first three mode shapes are shown in Fig. 6. The deformed shape at 144 Hz natural frequency rotates around the elevation x-axis, as shown in Figure 6 (a), the second mode of 197 Hz as translation along elevation axis, is shown in Figure 6 (b), and the third modes of 244 Hz as translation along optical axis is shown at Figure 6 (c), respectively. These frequencies were used to evaluate the dynamic response due to acoustic excitation.

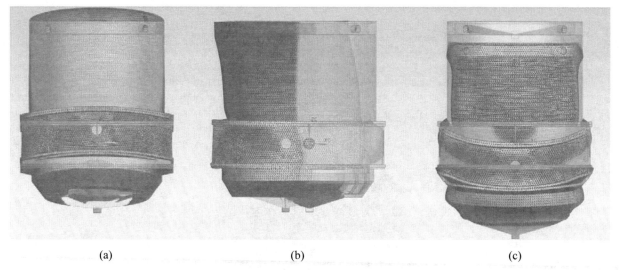

(a) (b) (c)

Figure 6. Fundamental frequencies of the wide field telescope.
(a) 144 Hz natural frequency, (b) 197 Hz second mode and (c) 244 Hz third mode.

4. DYNAMIC RESPONSE AND IMAGE MOTION ANALYSES

Many of latest large telescope projects have performed and reported some level of wind buffeting analysis [4-8]. In these references, they used a wind load obtained from a measurement of the wind speed at the observatory or scaled from it. Most were related to the wind load on the secondary mirror. For this study, we adapted more practical measurement data and assumptions.

For the practical approach to the image motion study, an optical sensitivity analyzed by the optical design program Code V was taken for the LOS. The sensitivity for this paper is associated with only the image motion on the image plane related with rigid body motions of each optical element, not to the optical performance evaluation. The sensitivity of the wide field telescope on the image plane was calculated, and the results are expressed in Equation (1).

$$\begin{aligned}
T^x_{image} &= 2.908 T^x_{G1} + 3.011 R^\beta_{G1} - 2.115 T^x_{G2} - 1.057 R^\beta_{G2} \\
&\quad - 0.789 T^x_{L1} + 0.007 R^\beta_{L1} + 0.469 T^x_{L2} - 0.008 R^\beta_{L2} + 0.886 T^x_{L3} \\
&\quad - 0.015 R^\beta_{L3} + 0.226 T^x_{L4} - 0.004 R^\beta_{L4} - 0.570 T^x_{L5} - 0.006 R^\beta_{L5} \\
&\quad - T^x_{G3} - 0.002 R^\beta_{G3}
\end{aligned}$$

$$\begin{aligned}
T^y_{image} &= 2.908 T^y_{G1} - 3.006 R^\alpha_{G1} - 2.115 T^y_{G2} + 1.057 R^\alpha_{G2} \\
&\quad - 0.789 T^y_{L1} - 0.007 R^\alpha_{L1} + 0.469 T^y_{L2} + 0.008 R^\alpha_{L2} + 0.886 T^y_{L3} \\
&\quad + 0.015 R^\alpha_{L3} + 0.226 T^y_{L4} + 0.004 R^\alpha_{L4} - 0.570 T^y_{L5} + 0.006 R^\alpha_{L5} \\
&\quad - T^y_{G3} + 0.002 R^\alpha_{G3}
\end{aligned}$$
(1)

In Eq. (1), the capital R and T represent the tilt and decenter, which are applicable to the G1, G2, five lenses, and G3. The superscript x and y express the coordinates. A set of the prescribed displacements referred to as constraints, for this case the image motion, provides a convenient boundary condition to solve the problem. The constraint equations are related to the nodal degrees of freedom to the displacement of one or more other freedoms. A typical form is similar to Eq. (1), and the left-hand sides of the constraints, the displacement of the image on the image plane, may be referred to as the slave or dependent degrees of freedom because the displacement is completely defined by the right-hand sides of the equations expressed by the tilt and decenter of G1, G2, five lenses, and G3. The multi-point constraints (MPCs) were constructed for the image translations of the x-axis and the y-axis by using the finite element analysis program NX NASTRAN with Eq. (1) to calculate the image motion. The other constraint for the current analyses is the rigidly fixed constraint on the yoke interface surface between the barrel and the yoke.

For the excitation forcing function, data on the wind velocity were recorded for 1 year at 8 AM in the morning, at 4 PM in the afternoon and at midnight, as shown Figs. 7 (a), (b) and (c), respectively. The measured wind velocity is sometimes too fast to sustain a stable image motion. For the safety of the telescope, we had to clarify whether we could observe the object under a certain level of wind speed. As a sample design study, the excitation functions at 8 hours, 16 hours and 24 hours, were calculated and were determined by using the NX NASTRAN. A parametric study was conducted and yielded an optimal performance under dynamic performances of the secondary mirror, because the secondary assembly was the weakest part of the wide field telescope structure. To predict the upper limit for the worst case, we applied the dynamic load to the vertex of the secondary mirror.

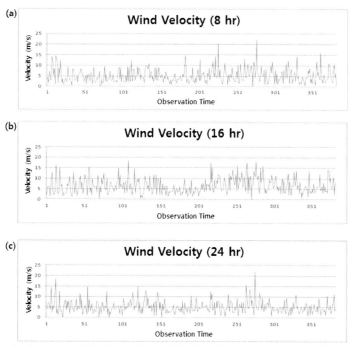

Figure 7. Profile of wind velocity recorded for 1 year. (a) profile of wind velocity recorded at 8 hour in the morning,(b) profile of wind velocity recorded at 16 hour in the afternoon, and (c) profile of wind velocity recorded at 24 hour at midnight.

From the excitation force at 8 hours, 16 hours, and 24 hours, the LOS along the x-axis and y-axis of the image motion responses were calculated. As one of the sample case of the x-axis image motion response, the excitation force at 8 hours is calculated and shown as Fig. 8. The responses of the excitation force along the x-axis, y-axis, and z-axis are shown in Figures 8 (a), (b), and (c), respectively. The magnitudes of the image motions from the excitation force along the x-axis, y-axis, and z-axis are 0.232 µm, 0.00007 µm, and 0.0003 µm with RMS (root mean square) values of 32 nm, 0.0092 nm and 0.0398 nm, respectively.

Fig. 9 shows the sample case of the y-axis image motion response to the excitation force at 8 hour. The responses of the excitation force along the x-axis, y-axis, and z-axis are shown in Figures 9 (a), (b), and (c), respectively. The maximum magnitudes of the motions from the excitation force along the x-axis, y-axis, and z-axis are 0.0000693 µm, 0.607 µm, and 0.00161 µm with RMS values of 0.0106 nm, 105nm and 0.434 nm, respectively.

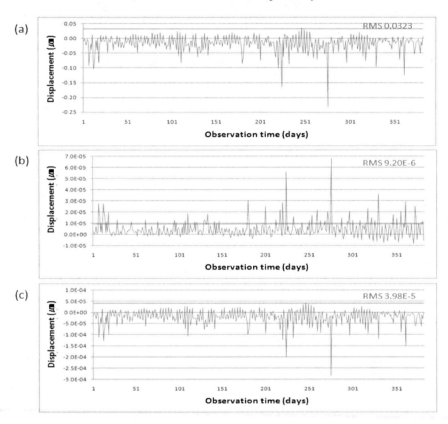

Figure 8. X-axis image motion response to wind at 8 hour. (a) response of image motion to excitation along X, (b) response of image motion to excitation along Y, and (c) response of image motion to excitation along Z.

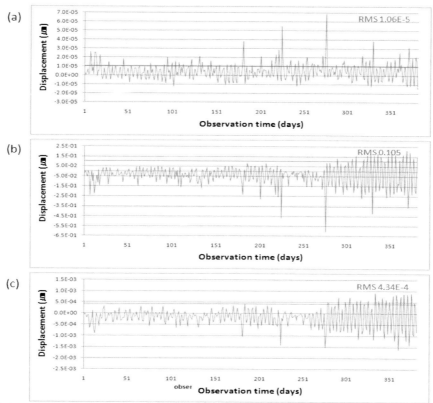

Figure 9. Y-axis image motion response to wind at 8 hour. (a) response of image motion to excitation along X, (b) response of image motion to excitation along Y, and (c) response of image motion to excitation along Z.

Other responses of the image motion on the x-axis and y-axis are summarized in Table 1. As shown in table 1, all the magnitudes of the image motion are less than 9 μm, the limit of the image motion. These results indicate that the design of the wide field telescope structure is relatively stiff enough to accommodate the operational wind condition for the image motion.

Table 1. Results of the image motion responses

Image motion	Excitation force	Hour	Min	Max	Magnitude	Average	RMS
X	x	8	-0.232	0.035	0.232	-0.017	0.032
		16	-0.153	0.043	0.153	-0.025	0.042
		24	-0.241	0.030	0.241	-0.016	0.030
	y	8	-8.13E-06	6.83E-05	6.83E-05	5.03E-06	9.20E-06
		16	-7.83E-06	4.68E-05	4.68E-05	7.35E-06	1.17E-06
		24	-4.91E-06	6.34E-05	6.34E-05	4.68E-06	8.32E-06
	z	8	-2.83E-04	4.06E-05	2.83E-04	-2.09E-05	3.98E-05
		16	-1.83E-04	4.97E-05	1.83E-04	-3.05E-05	5.15E-05
		24	-2.94E-04	3.80E-05	2.94E-04	-1.94E-05	3.87E-05
Y	x	8	-1.42E-05	6.93E-05	6.93E-05	5.19E-06	1.06E-05
		16	-2.55E-05	5.68E-05	5.68E-05	7.60E-06	1.56E-05
		24	-1.16E-05	6.22E-05	6.22E-05	4.84E-06	9.99E-06
	y	8	-0.607	0.216	0.607	-0.040	0.105
		16	-0.483	0.242	0.483	-0.059	0.139
		24	-0.521	0.121	0.521	-0.038	0.089
	z	8	-1.61E-03	9.32E-04	1.61E-03	-1.57E-04	4.34E-04
		16	-2.23E-03	1.39E-03	2.23E-03	-2.31E-04	7.11E-04
		24	-1.69E-03	8.63E-04	1.69E-03	-1.49E-04	4.86E-04

* Unit : μm

5. CONCLUSION

Optical performance and the image motion analyses of the wide field telescope were computed from static and the dynamic responses by using the LOS image sensitivity integrated in the finite element analysis method. The results of the dynamic response based on the excitation load from the recorded wind velocity indicate that the wide field telescope system design requirement was satisfactorily fulfilled by maintaining the image motion of less than 9 μm during the environment wind excitation. The overall performance of a wide field telescope well meets the design specification of the image size less than 2 pixels of CCD camera as well.

REFERENCES

[1] ProLine PL16803 LDR CCD manual.
[2] E. Huang, Gemini Report, TN-O-G0017 (1992).
[3] J. H. Burge, "An easy way to relate optical element motion to system pointing stability," Proc. SPIE, 6288, 62880I-1 (2006).
[4] M. K. Cho, NIO Report, RPT-GSMT-004 (2001).
[5] M. Johns, Optical Sensitivity Equations for use in the Dynamical Response Analysis for the Giant Magellan Telescope, Document 1114, GMT (2005).
[6] G. Z. Angeli, M. K. Cho, M. Sheehan, and L. M. Stepp, "Characterization of wind loading f telescope," in Integrated Modeling of Telescopes, T. Anderson, ed., Proc. SPIE 4757, 72-83 (2002).
[7] M. K. Cho, L. Stepp, and S. Kim, "Wind buffeting effects on the Gemini 8-m primary mirrors," in Optomechanical Design and Engineering 2001, A. E. Hatheway, ed., Proc. SPIE 4444, 302-314 (2001).
[8] M. K. Cho, L. M. Stepp, G. Z. Angeli, and D. R. Smith, "Wind loading of large telescope," in Large Ground-Based Telescopes, J.M. Oschman and L. M. Stepp, eds., Proc. SPIE 4837, 352-367 (2002).

Achieving high precision photometry for transiting exoplanets with a low cost robotic DSLR-based imaging system

Olivier Guyon*[a,b], Frantz Martinache[a]

[a]Subaru Telescope, National Astronomical Observatory of Japan, 650 N. A'ohoku Place, Hilo, HI 96720, USA;
[b]Steward Observatory, University of Arizona, 933 N. Cherry Ave, Tucson, AZ 85721, USA

ABSTRACT

We describe a low cost high precision photometric imaging system, which has been in robotic operation for one and half year on the Mauna Loa observatory (Hawaii). The system, which can be easily duplicated, is composed of commercially available components, offers a 150 sq deg field with two 70mm entrance apertures, and 6-band simultaneous photometry at a 0.01 Hz sampling. The detectors are low-cost commercial 3-color CMOS array, which we show is an attractive cost-effective choice for high precision transit photometry. We describe the design of the system and show early results.

A new data processing technique was developed to overcome pixelization and color errors. We show that this technique, which can also be applied on non-color imaging systems, essentially removes pixelization errors in the photometric signal, and we demonstrate on-sky photometric precision approaching fundamental error sources (photon noise and atmospheric scintillation). We conclude that our approach is ideally suited for exoplanet transit survey with multiple units. We show that in this scenario, the success metric is purely cost per etendue, which is at less than $10000s per square meter square degree for our system.

Keywords: Exoplanet transit, photometry

1. INTRODUCTION

We describe a small experimental low-cost robotic wide field imaging system for astronomy and atmospheric science based on low-cost digital single lens reflex (DSLR) camera with CMOS detectors[1-6]. The project is a collaboration with the VYSOS project[7], and is located at the Mauna Loa Observatory, an excellent site for nighttime astronomy.

This project is aimed at exploring a low-cost approach to perform a scientifically useful all-sky 3-color imaging survey. The system was installed on Mauna Loa observatory in December 2010, and started fully robotic operation in early 2011 with a single camera. It was upgraded to a 2-camera system in summer 2011. The first year of observations (until summer 2012) was used to evaluate scientific performance and system reliability, as well as develop data reduction algorithms which will process the large volume of data acquired. This project focuses especially on:

- Low surface brightness observations of the night sky, large size features (atmospheric and astronomical)
- High precision photometry in 3 colors simultaneously (variable stars, exoplanet transits)

All hardware and software for the system are open source, and is meant to help other amateur and professional astronomers duplicate and improve this approach. The low cost of the system, its ability to produce well calibrated 3-color photometry (under study) and ease of making color images make this system suitable for both astronomical research and outreach.

*guyon@naoj.org; phone 1 818 293-8826

2. SYSTEM OVERVIEW

2.1 Introduction

Figure 1. Left: Image of the robotic camera system on Mauna Loa observatory. Mauna Kea, to the North, can be seen in the background on the left. The cameras are mounted on an equatorial motorized mount on top of a metallic frame. The domes on the right host the VYSOS project robotic telescopes. Right: View pointing South. The Mauna Loa solar observatory is in the background. Openings for the two objectives can be seen on the front of the cameras enclosure. Note that the system does not have a dome, and simply points down during daytime or bad weather. The small shiny box on the right of the camera system is hosting humidity, temperature and cloud sensors.

The imaging system consists of two commercial digital cameras: models Canon 500D and 550D. The Canon 500D native IR-blocking filter in front of the sensor was replaced with a filter offering higher transmission in the red. The 550D is unmodified and includes the native IR-blocking filter. The two cameras combination therefore offers six different colors. Each camera is equipped with a 85mm focal length lens at F1.2 (Canon EF 85mm f/1.2L II USM). The cameras are pointed at the same direction, and mounted on a 2-axis equatorial mount (Orion Atlas EQ-G). The system (camera + mount) is computer controlled with a laptop. The data is stored on the laptop hard drive, and copied to an external hard drive. The main system characteristics are listed in table 1.

Table 1. Imaging system characteristics.

	Camera 1	**Camera 2**
Camera body	Canon 500D	Canon 550D
field of view	150 sq deg (10x15 deg)	
Aperture	85mm F1.2 lens (70mm diam)	
Pixel size	10"	8"
Detector	CMOS color, 15 Mpix	CMOS color, 18 Mpix
Sensitivity	photon noise limited on background, mV~15.5 point source sensitivity in 4mn	

The data volume is approximately 5 GB per night, stored locally. The data is physically retrieved every ~2 month by copying it to an external drive (2 month of data = 300 GB). The average total power consumption is about 25W (including approximately 10W for laptop) at night, and 15W during the day.

2.2 Detector and sensitivity

In this section, we report the results of calibration and performance measurement for the Canon 500D camera body (camera #1 of the system), with a 15Mpix CMOS sensor. Detector characteristics were measured and used to derive the system sensitivity and optimal photon-noise limited exposure times and ISO setting.

The **detector noise** was measured by differencing two short exposure dark frames at 800 ISO. The RMS deviation between the two images is divided by sqrt(2) to compute the readout noise. The **Gain** was measured by differencing two images taken with a background level of ~1200 ADU (after subtracting bias), using green pixels only. A quadratic subtraction of readout noise is performed to isolate and measure photon noise. Scaling to other ISO values is done assuming linearly of gain with ISO setting.

Table 2. Information on video and audio files that can accompany a manuscript submission.

	ISO 100	**ISO 200**	**ISO 400**	**ISO 800**	**ISO 1600**
Readout noise [ADU]	10.9	11.6	13.94	19.87	32.27
Gain [e-/ADU]	1.36	0.68	0.34	0.17	0.09
Readout noise [e-]	15.8	7.91	4.74	3.38	2.74
(RON=photon noise) level [ADU]	161.5	92.08	66.11	67.16	88.49
Minimum exposure time (no Moon)	190.0 s	54.2 s	19.4 s	9.9 s	6.5 s
Number of exposures per hr	18.95	66.4	185.6	363.6	553.8
Saturation level [e-/frame]	22282.2	11141.1	5570.6	2785.3	1392.6
Saturation level [e-/hr]	4.22E+005	7.40E+005	1.03E+006	1.01E+006	7.71E+005
Dynamical range for 1 hr	6542	11471	16026	15702	11951

Measurement results are shown in table 2. The 4th entry of the table shows the count level for which readout noise is equal to photon noise. Exposures should be sufficiently long to ensure that the background counts are above this level to ensure photon-noise limited performance.

The sky background level was measured under dark sky (no moon), with the 85mm lens at F1.2 and ISO 800 setting. Values are given below for each of the 3 color channels.

- RED pixels: 8.9 cnt / sec / pix @ 800 ISO = 1.513 e- / sec / pix

- GREEN pixels: 11.1 cnt / sec / pix @ 800 ISO = 1.887 e- / sec / pix

- BLUE pixels: 6.8 cnt / sec / pix @ 800 ISO = 1.156 e- / sec / pix

The blue pixels have the smallest counts. The minimum exposure times to ensure photon-noise limited performance are given by combining the sky background count levels with the previously derived minimum count level to ensure photon-noise limited sensitivity. Table 2 shows that under dark conditions, this exposure time ranges from 6.5 sec at ISO 1600 to 190 sec at ISO 100.

The dynamical range is the ratio between noise and saturation for a single pixel. All values given for a 1hr observation, assuming that the exposure time is chosen such that readout noise = sky background photon noise under dark conditions, and that readout time is much less than exposure time.

The best dynamical range is achieved by co-adding exposures taken at ISO 400 or ISO 800, with individual exposures of approximately 20 sec (@ ISO 400) and 10 sec (@ ISO 800). In practice, this optimal exposure time cannot be sustained for a long survey: at 20 sec per exposure, 10 hr observation per night, the shutter lifetime (rated at 90000 exposures) corresponds to 50 nights of observation. **Single frame exposure time is therefore a compromise between shutter lifetime and dynamical range** . The imaging system currently operates at ISO 100 with exposure times longer than 200 sec in dark time to optimize shutter lifetime, and operation without shutter is being explored. At ISO100, 120sec, saturation is reached on the best frames for mV~10.5 (approximate).

2.3 Optics and image quality

The system is acquiring images at F1.2 (lens fully open) to maximize sensitivity. Image quality is good through the field, with PSF FWHM of 1.5 to 2 pixel. Figure 2 shows a sample image: the PSF degradation at the edge of the field is noticeable but moderate (FWHM ~ 2pix).

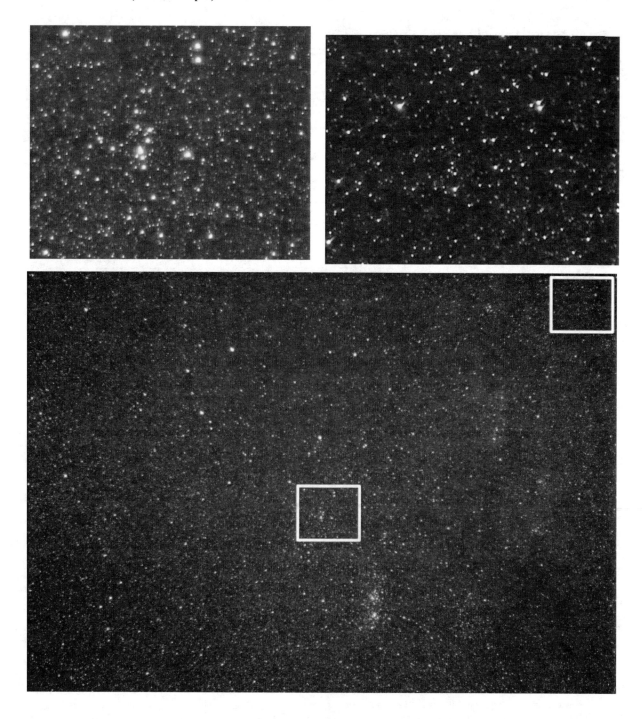

Figure 2. Sample image from camera #2 (287 sec exposure at ISO 100). The full field image (10x15 deg) is shown in the bottom. Details of the central and corner regions are shown at the top.

3. HARDWARE IMPLEMENTATION

3.1 Weatherproofing

The durability of the system is a strong requirement. In order to keep the system simple and low cost, it does not have a dome: the camera points down when it rains/snows or when there are clouds. The system includes several sensors to determine if the weather is suitable for observing:

- Four webcams acquire images every minute, and are used to automatically confirm nighttime (which is primarily derived from the Sun altitude below the horizon, according to the computer clock). Visual inspection of the webcam images can also identify snow/ice on the mount or camera.

- Temperature probes facing the sky and the ground are used to detect clear sky at night: if the sky is clear, the upward looking temperature probe is colder than the downward looking probe (thermal radiation to the sky). If this temperature difference is larger than a preset limit, then the sky is deemed clear and observing can start.

- A humidity sensor

In addition to these sensors, the weather information provided by the VYSOS observatory and the the Mauna Loa observatory are downloaded every minute from the network. Decision to observe is made from all sensor values.

The system is designed to minimize the long-term impact of weather :

- no exposed plastic (to avoid UV degradation of plastics)
- mount has been sealed against water with silicone
- camera is sealed (cover) except at the front (lens) which points down when weather is bad
- use of weather-resistant materials when possible: Aluminum instead of steel when possible, Stainless instead of standard steel for bolts/nuts, use of Kapton tape when tape must be exposed.

3.2 Mount and Tracking

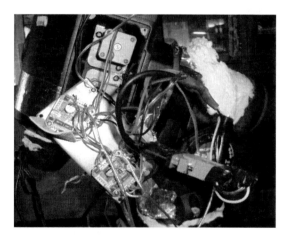

Figure 3. Mount stepper motor driver/controller modification.

The prototype 1 implementation (until summer 2011) performance was satisfactory except for pointing, due in part to the equatorial mount electronics and to the fact that our prototype 1 was mounted on the side wall of a wooden building. While the mount worked reliably, the electronics driving the stepper motors and the communication protocol to the electronics did not easily allow high performance tracking. This problem, combined with the fact that the mounting on a wooden wall did not provide a very stable reference, let to large drifts in pointing (approximately 1" to 5" per mn). This issue is addressed in our second prototype (in use since summer 2011) by:

- Mounting the unit on a sturdy metal frame, directly bolted to a ground concrete pad

- Replacing the native mount electronics with stepper controller+driver circuits offering more fine control of pointing and tracking (allowing for example small updates in tracking speed without introducing unwanted jumps/interruptions in the tracking)
- Implementing closed loop guiding: the images acquired are continuously used to refine pointing and tracking

The original mount electronics were replaced in July 2011 by two EZHR17EN EZstepper circuits (vendor: Allmotion) to improve tracking: the original circuit had unexplained behaviors and somewhat unreliable communication issues with the host computer. The EZstepper also provides more flexibility, and homing can be done without requiring the computer polling limit switch status.

4. DATA ACQUISITION SOFTWARE

4.1 Low level system architecture

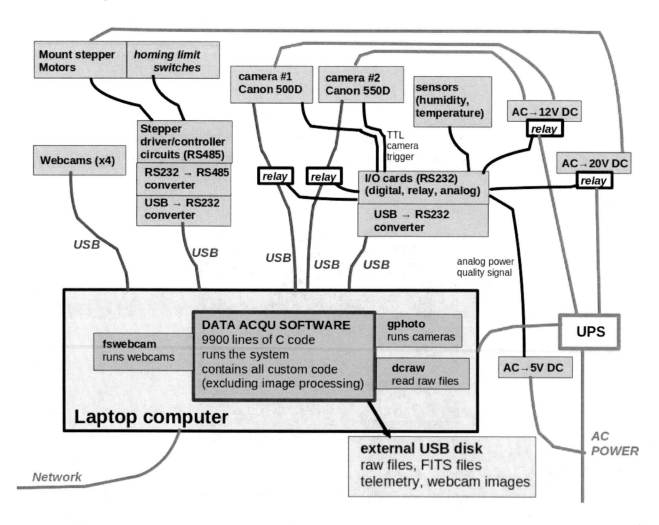

Figure 4. System architecture. A single laptop computer controls all hardware.

Figure 4 shows the overall system architecture. A single laptop computer controls all hardware. The cameras are linked to the computer by USB cables. When driven in USB mode, the gphoto software is used to acquire images and download them. The USB channel can also be de-activated (with a relay) to allow off-line camera control from a TTL trigger signal. This second mode of operation allows for higher duty cycle, but the images are then only available at the end of the observation (morning), when they are downloaded from the camera memory card through the USB cable. The system makes extensive use of open source software for low-level control of the cameras and webcams: gphoto and dcraw are used to control the cameras and read RAW files, and fswebcam drives the webcams.

4.2 Scheduling

The data acquisition software monitors external sensors (humidity, temperature, light level coming from webcams) to decide if the system should be observing. An ASCII file contains priority fields (ranked) to be observed. If the file is empty or none of the fields is accessible (too low, or too close to the moon), a random pointing is chosen, and will be adopted for the next few exposures – after which a new random pointing will be chosen.

4.3 Astrometry

Each frame is referenced to an astrometric grid by comparison with the Hipparcos/Tycho catalog. The data acquisition software already provides a good estimate of the astrometic pointing, easing comparison with the catalog. The astrometric solution for each frame consists of 30 parameters: pointing (2 parameters), field rotation, plate scale and distortion coefficients to 4th order (26 parameters). Over the 10x15 deg field, agreement between the photocenters and catalog is within about a quarter of a pixel RMS, and likely dominated by pixelization effects (the photocenter is not an accurate estimate of the star location in the image).

5. PHOTOMETRY

5.1 Description photometric data reduction approach

Our photometric measurement is differential: other stars in the field are used to construct a reference against which the target star is compared. Choosing the optimal of PSF(s) used for comparison with the target star is essential to compensate for error terms correlated with other sources (variable extinction due to clouds and airmass, color effects, detector non-linearity). The choice of the comparison PSFs is therefore critical to achieving photometric precision, and is complicated by the detector's undersampling of the PSF, discussed in the next paragraph.

The main challenge to precision photometry with a low-cost DSLR-based system is to overcome errors due to PSF sampling, which are particularly serious in our system, as the pixel size is comparable to the PSF size, and the pixels are colored (25% of pixels are red-sensitive, 50% are green-sensitive and 25% are blue-sensitive). This issue could be mitigated by defocusing the image, thus spreading light of each star on many pixels. For example, defocusing star images to a 35 pixel diameter disk, a previous study[2] reports achieving 1% photometric accuracy in each of the 3 detector color channels over 90 sec exposures with a 203 mm telescope, and measuring transit depth to 0.1% (1 millimagnitude) for a 1hr duration transit. While this scheme is appropriate for photometry of a small number of bright stars, it is not suitable for a transit survey aimed at monitoring a large number of stars, as the combined loss of angular resolution (crowding limit) and faint-end sensitivity (mixing starlight with background) would have a large impact on the survey performance. In addition to the PSF sampling issue, a large number of variables can affect the measured apparent flux from stars (for example airmass, color extinction effects, PSF variations). Comparison PSF(s) used for differential photometry must be chosen to include these effects, either by choosing stars which are subjected to the same errors, or by understanding, modeling and compensating for these effects.

Our approach to solving these challenges relies on an image-based identification of suitable reference PSFs. The wide field image delivered by the system offer a large number of potential reference targets. We perform image cross-correlations between our target and other stars in the field to select PSFs which have the intensity distribution amount pixels as the target for each image of the sequence. **This step implicitly selects PSFs that have the same color, experience the same optical aberrations, fall on the same fractional pixel position and experience the same local detector defects as the target,** without having to compute explicitly these parameters. An optimal linear combination of selected candidate PSFs is then perform to construct a template against which the target images are compared to produce the photometric light curve. While similar schemes have previously been applied to lightcurves[8,9], the strong pixelization issues in our system require image-level processing before a lightcurve is created.

5.2 Example data

The scheme described in the previous section was implemented and used to produce a lightcurve for HD54743, observed on 2011-04-15 (UT). Thirty consecutive 65-sec exposures were acquired at ISO100. The images were acquired during bright time, with a relatively strong background (level = 2500 ADU, 2600 ADU and 1900 ADU per pixel per exposure in R, G and B respectively). This dataset is especially challenging for photometry, as the PSF is strongly undersampled (in part due to the fact that the camera's anti-aliasing filter has been removed), the detector is a color CMOS array, and the tracking is relatively poor, with a drift of approximately 5" per minute (0.5 pixel between consecutive frames). We note that the poor tracking was later solved by replacing the electronics in the mount, as detailed in section 3.2.

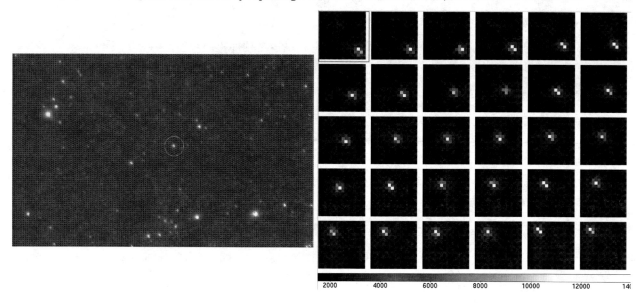

Figure 5. Left: Single frame (detail) showing star HD54743, circled. Right: Sequence of consecutive images of HD54743, showing the part of the detector where the star image is located.

Figure 5 shows that tracking errors move the star's image across different pixels, and that the combination of these tracking errors, the undersampling of the PSF by the detector and the fact that the detector is a color array, create a highly variable star image from frame to frame. Not surprisingly, the aperture photometry extracted from these frames is therefore a poor estimate of the actual star flux. Figure 6 (left) shows that aperture photometry leads to ~10% RMS variation in the measured flux from frame to frame. Our proposed algorithm reduces the error to about 2% per color channel.

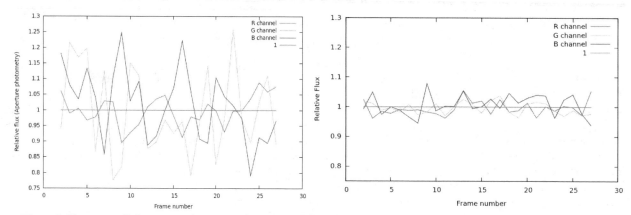

Figure 6. Photometry light curve with aperture photometry (left) and with our proposed data reduction scheme (right).

6. CONCLUSIONS

We have demonstrated that a relatively inexpensive robotic imaging system can be built and operated using mostly commercially available components. The total cost of our system is approximately $10k for an etendue of about 1 square degree square meter – significantly more affordable than hardware typically used for professional astronomy. The system has demonstrated reliability and fully robotic operation for more than a year, and can easily be duplicated.

A major challenge of using a color CMOS array for photometry is the pixelization of the undersampled PSF on the color matrix of pixels. Conventional aperture photometry is not suitable for these detectors, but an algorithm that carefully selects comparison PSFs using image-based correlations with the target star was successful at removing most of the pixelization errors. In fact, the 2% photometric accuracy over 1min reported in this paper approaches the photon noise limit and scintillation limit. When reaching these fundamental limits, the scientific merit of the system is purely measured by cost vs. etendue, and according to this metric, our proposed system is highly competitive. We therefore conclude that a particularly cost-effective strategy for exoplanet transit surveys would be to deploy a large number of units similar to the one described in this paper.

ACKNOWLEDGMENTS

This project was made possible thanks to Bo Reipurth and Josh Walawender of the Variable Young Stellar Object Survey (VYSOS) project, and the Mauna Loa observatory. Paul Stewart (University of Sydney) provided valuable help to install the system. Josh Walawender allowed the system to use VYSOS's weather monitoring and provided assistance in setting up the hardware at the site.

REFERENCES

[1] Fiacconi, D. and Tinelli, L., "Light curve analysis of XX Cygni from data taken using DSLR," Open European Journal on Variable Stars 114, 1–+ (2009).

[2] Littlefield, C., "Observing Exoplanet Transits with Digital SLR Cameras," Journal of the American Association of Variable Star Observers (JAAVSO) 38, 212–+ (2010).

[3] Hoot, J. E., "Photometry With DSLR Cameras," Society for Astronomical Sciences Annual Symposium 26, 67–+ (2007).

[4] Buil, C., "Spectroscopy, CCD and Astronomy," http://www.astrosurf.org/buil/ (2011).

[5] Clark, R. N., "Digital Camera Sensor Performance Summary," http://www.clarkvision.com (2011).

[6] Pokhvala, S. M., Zhilyaev, B. E., Reshetnyk, V. M., "High-speed multicolor photometry with CMOS cameras," Advances in Astronomy and Space Physics (2012).

[7] Walawender, J., Reipurth, B., and Paegert, M., "Computer Infrastructure for the Variable Young Stellar Objects Survey," Proceedings, Telescopes from Afar, (2011).

[8] Kov´cs, G., Bakos, G., and Noyes, R. W., "A trend filtering algorithm for wide-field variability surveys," MNRAS 356, 557–567 (2005).

[9] Tamuz, O., Mazeh, T., and Zucker, S., "Correcting systematic effects in a large set of photometric light curves," MNRAS 356, 1466–1470 (2005).

An active surface upgrade for the Delingha 13.7-m radio telescope

Dehua Yang[*,1,2], Yong Zhang[1,2], Guohua Zhou[1,2], Aihua Li[1,2,3], Kunxin Chen[1,2], Zhenchao Zhang[1,2], Guoping Li[1,2], Yingxi Zuo[3], Ye Xu[3]

1. National Astronomical Observatories / Nanjing Institute of Astronomical Optics & Technology, Chinese Academy of Sciences, Nanjing 210042, China;
2. Key Laboratory of Astronomical Optics & Technology, Nanjing Institute of Astronomical Optics & Technology, Chinese Academy of Sciences, Nanjing 210042, China;
3. Graduate University of Chinese Academy of Sciences, Beijing 100049, China;
4. Qinghai Station of Purple Mountain Observatory, Chinese Academy of Sciences, Delingha 817000, China.

ABSTRACT

An upgrade program is proposed for the Delingha 13.7-m radio telescope to implement active surface for multi-beam observation at 3 mm wavelength. The upgrade involves three critical development aspects. One is the displacement actuator, which must fit the existing position, space and connections of the panels and backup structure, meanwhile, must be as compact and lightweighted as possible. The second is that a new sub-reflector is necessitated by the multi-beam observation, where a new hyperbolic surface figure is optimized. The third, more crucial and difficult, is to realize active control of the actuators and real-time closed-loop of the full active surface. This paper is to present the progress of the development work, test and experiments associated with the three areas. With one of the spare panels of the telescope, an experiment system is carried out with six sets of actuator and control electronics. Another experiment system of a novel laser-based closed-loop measurement concept is also conducted with four smaller dummy panels. Both experiment setups have output expected results and further experiments are going on with them. In this paper, based on the two experiments, we will describe the special design and test of the actuator, including the design of its special mounting and connecting mechanisms. The design and manufacture and measurement of the new hyperbolic sub-reflector will be detailed as well as the principle, simulation and realization of the laser-based measurement system. Besides, the control strategy of the large scale use of the hundreds of actuators and EMI suppression are also covered.

Keywords: Radio telescope, active surface, actuator, control, closed-loop, measurement

1. INTRODUCTION

As shown in Fig. 1.1, the Delingha 13.7-m telescope, residing in a random of 20.7 m in diameter, is the only largest radio telescope operating at 3-millimeter waveband in service in China[1]. The dish is located at the Qinghai station west to the city of Delingha, Qinghai Province on the Tibetan Plateau at an altitude of approximately 3200 m (latitude 32°44'.4N and longitude 97°33'.6E). Founded in the early 1980s, Qinghai station thanks to its high altitude, dry atmosphere, and clear weather conditions has been an excellent observatory for millimeter-wave astronomy. The 13.7-m radio telescope, as the primary facility at Qinghai station, is equipped with an indigenously developed SIS super-conductor receiver working at the band of 85-115 GHz. It adopts a classic Cassegrain system with a normal focal length of 5080 mm and hyperbolic sub-reflector of 1080 mm in diameter. Fig. 1.2 shows the panel pattern of the telescope, there are 24 panels in the inner ring and 48 in the outer to form a full aperture of 13.7 m. All the 72 panels are supported by 480 screw adjustors.

In 2008, a detailed report proposed the surface measurement and analyzed accuracy limitation of the 13.7-m dish[2]. Thereafter, the surface of the primary reflector as well as that of the sub-reflector was measured by photogrammetry. According the measured surface data, the gravitational deflection of the dish is up to 0.1 mm RMS, and thereby the panels in the primary reflector were adjusted with predicted values so as to have achieved a surface accuracy better than 0.07 mm RMS as the telescope points to any elevation. However, thermal deformation under a differential temperature

---

[*] Correspondence: Email: dhyang@NIAOT.ac.cn; Telephone: +86-25-85482-255; Fax: +86-25-85405562; http://www.niaot.ac.cn; Nanjing Institute of Astronomical Optics & Technology (NIAOT), 188 Bancang Str., 210042 Nanjing, P. R. China

of up to 16°C over the reflector is unpredictable.

Thereafter, decision has been made to upgrade the telescope to a better performance by implementing with multi-beam observation and an active surface. Last summer, the telescope was already equipped with a multi-beam receiver. Also, an active surface upgrade program has been undertaking ever since[1, 2]. The active surface upgrade program is to replace the adjustors with linear displacement actuators to compensate for gravitational and/or thermal deflections, hopefully by close-loop control, to achieve/maintain higher surface accuracy. Besides, a new hyperbolic sub-reflector is implemented to increase the field of view to enable multi-beam devices.

This paper is to present the progress of the active surface upgrade program of the 13.7-m telescope. In Section 2, we review the development of a compact precision linear actuator to replace the screw adjustors according to the given technical requirements and constraints imposed by the existing telescope structure[4]. Emphasis is focused on the exploration and experiment of a close-loop concept based on laser angle metrology for an active surface, which is elaborated in Section 3[5]. In Section 4, fabrication of a new CFRP-based sub-reflector is reviewed. The tested close-loop concept is promising to be applicable to the active surface with assertive experiment results. Before coming to a conclusion of the active surface upgrade program, we cover the preliminary design of the control strategy of the large number of linear actuators and consideration on suppression of EMI throughout the upgrade of the Delingha 13.7-m radio telescope.

Fig. 1.1 Back view of the 13.7-m telescope

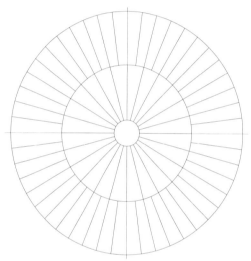

Fig. 1.2 Panel pattern

2. LINEAR ACTUATOR DESIGN AND TEST

The 13.7-m telescope is composed of 72 precise panels in its primary reflector. As illustrated in Fig.2.1, each inner panel weighing 14 kg has eight adjustors, while each outer panel 20 kg six adjustors. As seen in the close view in Fig. 2.2, the adjustor's installation space varies from 120 to 160 mm in height. Unlike building a brand new telescope, apart from the limited various space for the installation of actuators, more constraints are imposed on the actuator design for the upgrade, such as the flexibility of the lightweighted panel with six and/or eight supports instead of normal four, the various interfaces of the actuator where a flexible design of the actuator mounting is require. All the parameters and conditions are constrains for the development of a new motorized linear displacement actuator substitution of the screw actuators of the active surface upgrade.

2.1 Design of the linear actuator

We reviewed and compared a couple of frequently used design principles of linear displacement actuators according to the special application in the Delingha 13.7-m telescope[3]. Finally, we chose the concept of "double nuts + screw" to build the displacement actuator, thanks to its positive advantages such as easy manufacturing, self-locking, compact size, and especially eliminable and adjustable backlash, though it is sliding-friction based with relatively smaller transmission ratio and relatively lower transmission efficiency. As the panel either weighs no greater than 14kg and 20 kg and is supported by at least three or four actuators (an actuator replaces two adjustors), on average, an actuator is deemed to output a mild norm force of 120 N. Fig. 2.3 shows the 2D section drawing (in the left) and the actuator (in the right) of the double-nuts backlash elimination concept. The backlash in the bolt-nut pair is to be adjusted and eliminated by turning a rear "backlash elimination nut" and afterwards the nut is to be fixed by small screws. The backlash between the drive nut and the housing thus can be eliminated by turning and fixing the bearing stop nut. By these measures, all backlashes in the actuator mechanism are possible to be controlled within a reasonably allowed threshold. The advantages are clear that no extremely strict precision machining is required, the backlash is to be adjusted and eliminated even after being worn so as to further prolong life cycle of the actuators.

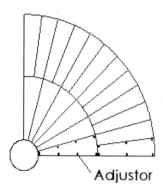

Fig. 2.1 Distribution illustration of adjustors

Fig. 2.2 Close view of the panel adjustors

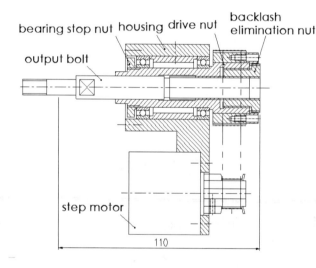

Fig. 2.3 Left: Section view of the folded double-nut backlash-elimination actuator, Right: Photo of the actuator

In order to minimize the actuators dimension to fit the existing installation space, referred in Fig. 2.3, a folded structure is introduced with a synchronous belt mechanism. The final dimension of the actuator is enclosed within a box of 95x44x110 mm3.

According to the active surface upgrade application of the 3-mm observed waveband of the 13.7-m telescope and the practical design of the actuator, the detail reasonable technical specifications are listed in Table 2.1.

2.2 Test of a single linear actuator

Comprehensive tests have been carried out for the actuator, including its required displacement resolution, backlash, and loads tests of axial 12 kg.f and/or lateral 6 kg.f. Fig. 2.4 shows the result data curve from a resolution test at step of 3 μm under an axial load of 12 kg.f. Backlast eliminating performance test is shown in Fig.2.5, where the backlash is around 3 μm. Fig. 2.6 shows the results of linearity test, where, the maximum error is less than 5 um after linear fitting. These fundamental parameters well meet the requirement by the 13.7 radio telescope.

Table 3.1 Technical specification of the actuator

Theoretical displacement resolution	0.75 μm
Stroke	±5 mm
Accuracy	10 μm
Backlash	<20 μm
Axial load capacity	120 N
Lateral load capacity	60 N
Dimensions	95x44x110 mm^3
Weight	~880 g

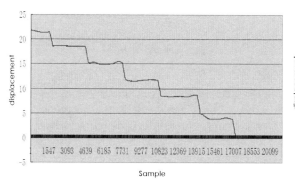

Fig. 2.4 Test of resolution of 3 μm   Fig. 2.5 Test of backlash at displacement step of 12 μm

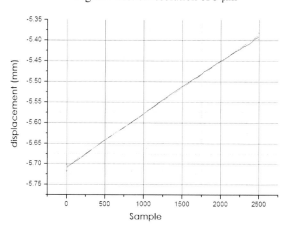

Fig. 2.6 Linearity test and fitting   Fig. 2.7 An outer panel tested with six actuators

2.3 Integrated test with a spare panel

Further, we carried out test of the actuators integrated with one of the two spare panels of the Delingha 13.7-m telescope. Two testing mounts were constructed to simulate panels in the two special horizontal and vertical orientations in a dish. As seen Fig. 2.7, it is the spare outer ring panel which is supported by six actuators under test to simulate a horizontal panel.

Each actuator is installed in the normal direction of the panel, due to the curvature of the panel, the axes of the actuators are not parallel with each other, as results in induced lateral over-constraints between the actuators. As the actuators are working/thrusting, their output rods will experience great lateral force which may exceed permitted threshold not only for the actuators themselves but also for the panel. To release such over-constraint effect, a longer thinner output shaft is adapted for the actuators as seen in Fig. 2.8 Left. In order to fit the varying installation interfaces of the original panel adjustors, a flexible blade-based mounting mechanism is developed for the actuators. As seen in Fig. 2.8, the fixture is actually a pair of cross blades. Peanut shaped holes are designed for easy adjustment and installation.

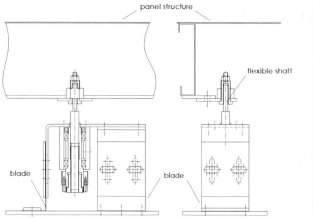

Fig. 2.8 Blade-based flexure mounting for the actuators

Three typical elevations of the panel have been tested: the panel faces to the horizon, 45°, and 90°. All the actuators are able to equidirectionally output displacement of 6 mm at the same time. A single actuator is able to output 3 mm displacement. Meanwhile, as the panel turns from the horizon to the vertical direction, it shifts 0.16 mm, which conforms to negligible degradation of image Strahl ratio of the telescope and is quite acceptable for a perfect reflecting surface of the Delingha 13.7-m radio telescope[5]. The test confirms the adapted longer thinner output shaft of the actuators functions well with appropriate lateral stiffness and flexibility.

The actuator-panel assembly integrated test confirms that: 1) The redesign of the output shaft of the actuators functions well with good lateral stiffness and flexibility; 2) The specially designed "blade-based mounting" of the actuators is well practical and applicable to the varying installation interfaces of the original panel adjustors; and 3) the actuator works to actively deform the panel surface well with a real panel on a real mounting.

3. CLOSE LOOP CONTROL AND EXPERIMENT

An active surface is one of the key technologies for constructing large precision telescopes, especially for the millimeter/sub-millimeter radio telescopes. There emerged a couple of detection method for the active surface control, the most successful three methods are: the rangefinder based detecting technique like the laser metrology system used by the Green Bank Telescope; the photogrammetric system; and the microwave holography. However, by far, all of the measurements fail to build a real-time close loop for the active surface correction, either due to unbearable time-consumption, or the incompetent precision. In this sense, there in no any real-time close-loop active surface accomplished in the world.

Seeing the reality of the existing active surfaces, we turned to active optics and sought any inspiration to have figured out a novel laser-angle-based metrology system for an active surface[6, 7]. By a series of experiment and improvement, it achieves an accuracy of microns and has an applicable time-response, as its similar precedent used in a closed-loop active optics would take an order of magnitude of minutes. The laser-angle metrology technique is well suitable for the upgrade of the active surface of the 13.7-m, on the other hand, the technique is wished to be implemented in a real radio telescope as the first realization of closed-loop control of an active surface[4].

3.1 Prototype of the laser-angle-based metrology

In principle, the laser-angle-based metrology technique can be deemed as a simplified Shack-Hartmann-type wave-front sensor, which is extensively used by active optics. It projects the laser spots emitted by the laser transmitter modules installed and well aligned on each panel on a CCD camera or any other photonic imaging detector, say, Position

Sensitive Device-PSD. By calculating the positions of the spots on detector, the normal angle of deflection and axial displacement of the panel can be deduced[7]. The laser angle metrology system is hence an integration of both angle and range measurements. And the precision of lateral displacement of the reflector can be directly related to the precision of the laser spot detection. Fig. 3.1 shows the experiment system of the laser-angle-based metrology prototype system, which is build with four panels, five linear actuators, and a CCD camera system as detector.

Clearly, the laser angel metrology system is not suitable for the initial calibration of the surface shape[7]. However, the laser-angel metrology system is well valid for maintaining the active surface shape. The whole system features in easy implementation and automation, efficiency, and low cost.

Fig. 3.1 Prototype experiment of the laser angle metrology system

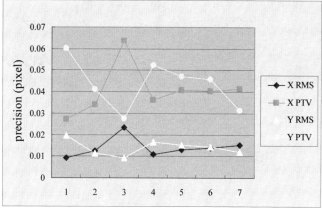
Fig. 3.2 Precision of the laser spot detection

3.2 Test results

Basically, to ensure a radio telescope to work with diffraction-limited performance, the RMS accuracy of the antenna's surface should be less than 1/20 wavelength, considering the actual condition of installation. For example, for a sub-millimeter telescope working at 0.2 mm, the accuracy of the surface shape is required to be down to 10 μm RMS; for the to be upgraded 13.7-m telescope, the active surface would be maintained at a surface RMS of 30 μm.

We carried out careful measurements of the laser spot detection precision with different detectors and distances between the panel and the detector; afterwards, the accuracy of the surface shape maintenance, and the full aperture accuracy were evaluated.

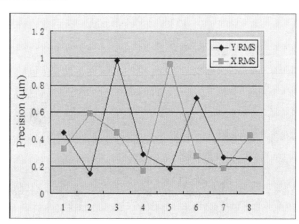
Fig. 3.3 Precision detected by PSD

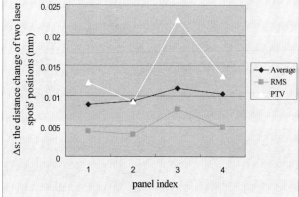
Fig. 3.4 Precision of Δs for four panels with PSD

As illustrated in Fig. 3.2, the precision of the spot measurement can be up to 0.02 pixel RMS and 0.06 pixel PTV, which is conforming to the precision of angle measurement of 0.11 arcsecs RMS, where, the distance between the panel and the screen is 2315 mm. By substitution of a PSD for the CCD camera, as plotted in Fig. 3.1, the precision of the laser spot detection by a PSD is improved to 0.8 μm RMS, and the precision of angle measurement is improved to 0.07 arcsec.

The precision of range measurement is represented by the precision of the distance change of two laser spots' positions, Δs. We measured the distance change of the two laser spots from a same panel, respectively. The experiment with improved devices, which is hardly influenced by gravity, shows that the accuracy can be up to 5 μm RMS as seen in Fig. 3.4.

3.3 Correction efficiency of the laser-angle-based metrology concept

We corrected a randomly distorted initial configuration (PV=4.469 mm, RMS=1.007 mm) of the four panels. Fig.3.5 plots the correction iteration process. It is seen that the precision of surface shape decreases from 1 mm RMS down to 4 μm RMS only by three iteration steps within an amazingly short time-cost (three steps less than 15 minutes) and the surface is maintained very well thereafter.

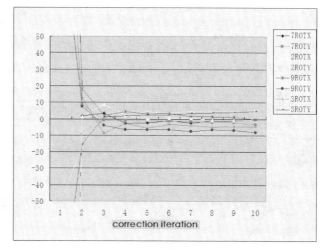

Fig. 3.5 Correction iteration from a random distortion of the four panels (spot #7: panel #1, #2: panel, #9: panel #3, #3: panel #4)

3.4 Close-loop detection simulation of the 13.7-m telescope

First, we generate a normal angle error distribution over a full aperture of 50 m in diameter, as shown in Fig. 3.6 Left, to simulate the full aperture detection and correction capability of the laser-angle-based metrology system. The full aperture detection accuracy, without considering the errors from the mechanical process and initial calibration, is about 4 μm RMS, as shown in Fig. 3.6 Right.

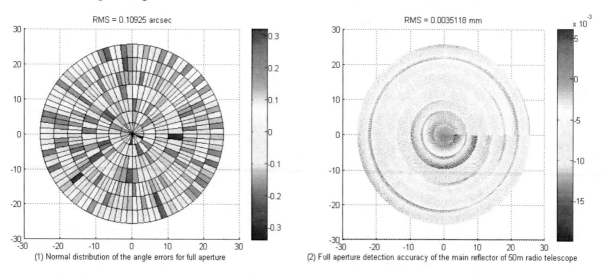

Fig. 3.6 Left: Normal distribution of the angle errors over a 50-meter radio telescope; Right: Full aperture correction simulation.

For the 13.7-m telescope active surface upgrade, the high accuracy laser angle real-time measurement is also applicable to the real-time close-loop maintenance of the active surface during the two periods of active surface calibration. For the both kinds of panel in the inner ring and the outer ring of the 13.7-m telescope, two stiffness matrice derived from the laser angle measurements are accurately constructed, respectively. Correction simulation based on the real-time close-loop metrology, gravitational deformation and thermal distortion by far have been successfully carried out with a residual error only of microns. An experiment carried out with the spare panel and with the latest improved laser module together with corresponding laser wavelength filters is underway at lab, promising results of even higher test precision are expected thereby.

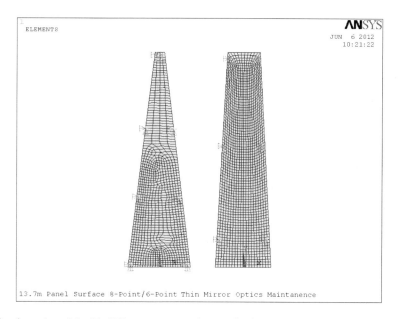

Fig. 3.7 Two kinds of panel model with different support schemes displayed at different scales for better illustration by ANSYS

4. A NEW SUB-REFLECTOR

In accordance with the active surface upgrade, a new sub-reflector is required to realize a wider Field of View to support multi-beam observation. A new CFRP-based sub-reflector is under fabrication, whose governing function is given in Fig. 4.1. The convex working surface is required to be less than 38 mm. A dedicated mould was build for the replication of the new CFRP sub-reflector. Fig. 4.2 shows the error contour over the fitted surface of the mould measured by a 3D laser tracker, where, RMS surface error in X axis is 25 μm, and that along the surface normal direction is 22.4 μm.

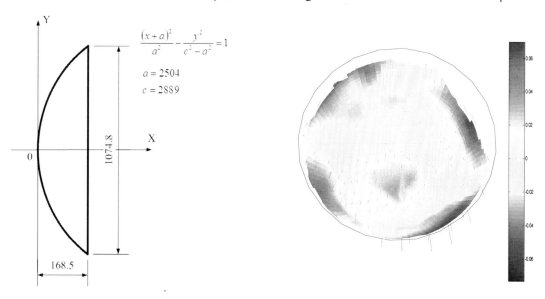

Fig. 4.1 Theoretical governing equation of the new sub-reflector   Fig. 4.2 Measured contour of the mould

Fig. 4.3 shows the finished new CFRP sub-reflector with a lightweighted aluminum alloy frame for interfacing with its motorized adjustors on back. In addition, an optical crosshair is integrated at center and four auxiliary marks are attached on brim. As seen in Fig. 4.4, it was measured by a 3D laser tracker. The normal error of the convex working surface is 0.713 mm RMS, which are far behind the required surface accuracy and hence further finishing process is necessary to achieve the required 0.038 mm RMS. Finally, the working convex surface will be aluminized with the CFRP sub-reflector as substrate. The aluminizing process has been successfully tested with a smaller sample convex CFRP panel of 400 mm in diameter.

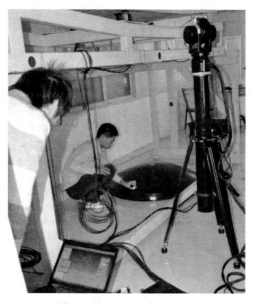

Fig. 4.3 The new CFRP sub-reflector

Fig. 4.4 Measuring by a laser tracker

5. CONTROL DESIGN

A preliminary control strategy for the active surface of the 13.7-m telescope has been considered based on the successful experience of the active optics control of the LAMOST[8], in which the first combination of both deformable active optics and segmented mirror active optics has been realized in the world.

As sketched in Fig. 5.1, the control architecture of the active surface of the 13.7-m Delingha radio telescope is dived into eight sub-systems, as detailed in Fig. 5.2, each of which covers 30 actuators out of the total 240. The control strategy is basically Ethernet-and-CAN-based. The master control computer receives and analyzes commands from an upper astronomical observation control computer, and commands the actuator controllers to drive the actuators to required displacements. Afterwards, the actuators feeds back their status as they finish execution of commanded displacements to the master control computer.

Clearly, EMI is of great importance for a millimeter radio telescope with an active surface, the following considerations are taken during the control design to suppress EMI to the greatest extent.

Above all, the motors and drivers must be chosen with conformation to EMI standards, and more careful measures will be adopted to reduce electrical noise, such as:

- Power and control cables separated to improve interference immunity. The brake control cable must be separately shielded.

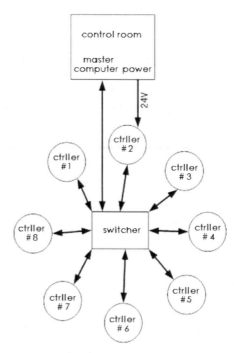

Fig. 5.1 Actuator control architecture

- Ground all shielding with large areas (low impedance), with metallic connector housings or shield connection clamps.

- Install all feedback cables with an adequate cross-section, and use the requested cable material to reach maximum cable length.

- Use differential inputs for analog signals. Use twisted-pair shielded signal lines, connecting shields on both ends.

- Wires between drives and filters and external brake resistors must also be shielded.

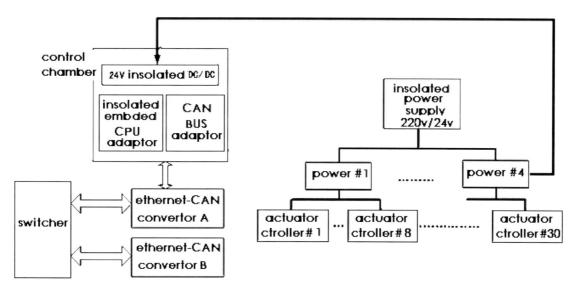

Fig. 5.2 Control logic within each sub-system out of the total eight

6. CONCLUSION

An active surface upgrade program was proposed for the 13.7-m Delingha radio telescope, which has been in operation for 20 years, to carry out multi-beam observation at 3 mm wavelength. We have developed critical technologies associated with the active surface upgrade through the past three years. In this paper we reported the progress by far of the technique developments with design, test and experiments about the special linear actuator, the new CFRP-based sub-reflector, the novel laser-angle-based closed-loop measurement system, and the preliminary control strategy of the large scale use of the hundreds of actuators. According to the finished work not elaborated in full detail in this paper, the folded precision linear actuator with capable accuracy and load capacity is well capable for the future batch use in the active surface of the 13.7-m radio telescope; The novel laser-angle-based real-time close-loop detection concept is promising for practical implementation in the active surface upgrade, as the experiments and simulations worked out very convincing results. In the coming weeks we will go on with detail design of engineering realization, including the refining of the sub-reflector, the large-scale control strategy of the actuators, and further, the full close-loop real-time control of the active surface.

7. ACKNOWLEDGEMENT

The research is sponsored by the "National Natural Science Foundation of China" under Grant No. 10833004, Grand No. 11073035, Grand No. 11078020, and Grand No. 10973025.

REFERENCES

[1] http://english.dlh.pmo.cas.cn/
[2] Ji Yang, Road map towards an accuracy upgrade of the surface of the Delingha 13.7-m radio telescope, Purple mountain observatory, (2008)
[3] Guohua Zhou, Aihua Li, Dehua Yang, et al.. Development of a compact precision linear actuator for the active surface upgrade of the Delingha 13.7-m radio telescope, submitted to the same conference.

[4] Yong Zhang, Jie Zhang, Dehua Yang, et al.. Progress of the active reflector antenna using laser angle metrology system, submitted to the same conference.
[5] Dehua Yang, Daniel Okoh, and Guohua Zhou, et al. positioning error and support mechanism for a 30-m THz radio telescope, Research in Astronomy and Astrophysics, vol. 11, n 6, 725-736, (2011)
[6] Zhang, Y., Yang, D. & Li, Y. et al., "The Measurement of Segmented Radio Telescope Antenna Based on Parallelogram Panels," China Patent: CN200910212710.X., http://www.soopat.com/Patent/200910212710, (2010).
[7] Zhang, Y., Yang, D. & Li, Y. et al., "A new efficient laser angle metrology system for maintaining the large radio telescope active reflecting antenna," Proc. SPIE 7733, 77333M-1-8 (2010).
[8] http://www.lamost.org

Author Index

Numbers in the index correspond to the last two digits of the six-digit citation identifier (CID) article numbering system used in Proceedings of SPIE. The first four digits reflect the volume number. Base 36 numbering is employed for the last two digits and indicates the order of articles within the volume. Numbers start with 00, 01, 02, 03, 04, 05, 06, 07, 08, 09, 0A, 0B...0Z, followed by 10-1Z, 20-2Z, etc.

Abbaki, S., 6C
Abbott, Timothy M. C., 45, 46
Abe, L., 5I
Abfalter, Christian, 1B
Abrams, Don Carlos, 6J
Acernese, F., 3C
Acton, D. Scott, 23
Ade, Peter A. R., 15
Agabi, A., 5I
Ageorges, Nancy, 1B, 2V
Agócs, Tibor, 6J
Aguiar, M., 2Y
Ahmadi, Aida, 5C
Ahmed, Sara, 1Q
Aigrain, S., 5I
Aja, B., 2Y
Al Marri, S., 1U
Al Muqbel, Y., 1U
Al Naimiy, H., 1U
Alagoz, Enver, 4P
Alapini, A., 5I
Alcantud Abellán, Manuel, 3W
Alcock, Charles A., 0D
Almaguer, Jaime, 4L
Anderson, David, 0H
Andersson, B-G., 10
Andrew, John, 0J
Angers, Mathieu, 0W
Angilè, Francesco E., 15
Aniol, Peter, 1B, 2V
Aoki, Tsutomu, 6B
Arai, Akira, 6G
Arcidiacono, Carmelo, 1C, 4Z, 56, 5R
Arendt, J., 09
Argomedo, Javier, 1C, 4Z, 56, 5R
Ariño, Javier, 08
Aristidi, E., 5I
Arndt, Kirk, 4P
Artal, E., 2Y
Asada, Keiichi, 1J
Asaki, Yoshiharu, 3K
Asano, Kentaro, 6B
Asayama, Shin'ichiro, 2K, 3F
Ashby, David S., 1A, 3Y, 4G
Ashley, Michael C. B., 1Q, 1R, 5H
Assus, P., 6C
Augarten, Yael, 1R

Ávila, Gerardo, 4F
Bader, S., 09
Bagish, Alan, 4H
Bagnasco, John, 24
Bähr, Jürgen, 17
Bahrami, Mehdi, 6H
Ballesteros Pérez, Noelia, 3W
Banavar, R. N., 5O
Bangia, Tarun, 41
Barden, Samuel C., 07, 3O
Barbieri, M., 5I
Barentine, J., 09
Baril, Marc, 1S
Barkats, Denis, 2K
Barone, F., 3C
Barreiro, R. B., 2Y
Barriga, P., 1Y
Barwig, Heinz, 2V
Basso, S., 42
Bastaits, Renaud, 25
Bastin, Christian, 2U, 3Z
Battye, R., 2Y
Baum, S., 0X
Bauman, Steven, 0W, 1S
Bautista, Ruben, 4L
Bazhanov, Y. V., 6E
Becker, M., 58
Becklin, Eric E., 10, 14
Beets, Timothy A., 0K, 3X
Behera, Bagmeet, 17
Bender, Ralf, 1B, 2V, 58
Benedict, Tom, 1S
Beno, Joseph H., 0K, 3X, 4O, 5J
Benton, Steven J., 15
Bettonvil, Felix C. M., 06, 0I
Beyerlein, Matthias, 58
Bezawada, Naidu, 0F
Bhatia, Ravinder, 2K
Bida, Thomas A., 14, 19, 1E, 5I
Biddick, Chris, 4G
Bilbao, Armando, 0P
Billet, J., 1W
Blackhurst, E., 2Y
Blair, Samantha, 2K
Blake, Simon, 3A
Blanc, P.-E., 5I
Blanco Alonso, Oscar, 4L

Bloom, Joshua S., 5L
Blundell, Raymond, 1N
Bogner, Stephanie, 58
Boisson, Catherine, 3A
Bolli, P., 5W
Bondoux, E., 5I
Bonner, Colin S., 1R
Bonomi, Giovanni, 0A, 3T
Booth, John A., 0K, 3X, 4O
Borgnino, Julien, 31, 6C
Born, Andrew J., 0F
Bouchez, Antonin H., 1H, 23
Boutsia, Konstantina, 1D
Breckenridge, Craig, 0W
Bredthauer, Greg, 4H
Brewster, Rick, 14, 2S
Brink, Janus D., 4I
Brooks, David, 0C
Brown, M., 2Y
Brunetto, Enzo, 1F
Brynnel, Joar G., 1A, 1D
Buckley, David A. H., 2W, 6A
Buffa, Franco, 2E
Bulau, Scott, 07, 3L
Bunton, John D., 2A
Burgdorf, Martin J., 2S
Burgett, William, 0H
Burkepile, Joan T., 3N
Burlegih, Matthew, 0E
Burton, Michael G., 1R
Busatta, A., 0Q, 3B
Busonera, G., 5W
Bustos, E., 69
Butler, Nathaniel R., 5L
Bycroft, Luke, 1R
Byrnes, P. W. G., 3D
Cabak, Gerald, 4R
Cabral, Alexandre, 4F
Cagigas, J., 2Y
Calisse, P. G., 3F
Cameron, Robert A., 18
Canestrari, R., 39, 42
Cano Infantes, Diego, 6J
Cano, J. L., 2Y
Canzian, Blaise, 09
Cao, Wenda, 03
Capaccioli, Massimo, 1C
Caputa, Kris, 1P
Carel, JL, 1W
Carlberg, Raymond G., 5C
Casas, F. J., 2Y
Casey, Sean C., 14
Cassali, Mark, 1F
Catala, Laure, 6A
Cauthen, H. K., 43, 5S
Cavadore, C., 1W
Cayrel, M., 1X
Centrone, Mauro, 2Q

Ceseña, Urania, 5L
Cha, Sang-Mok, 43, 5S
Chadid, Merieme, 64
Chadwick, Paula, 3A
Chanan, Gary, 22
Chandrasekharan, Srinivasan, 4P, 4Q
Chatila, Amjad, 4G
Chauvin, Eric, 2M
Chauvineau, B., 6C
Chávez, Arturo, 4L
Chazelas, Bruno, 0E
Chen, Chien-Ping, 1J
Chen, Kunxin, 4B
Chen, Ming-Tang, 1J
Chen, Yi, 5B
Cheng, Jingquan, 5B
Cho, Myung K., 20, 24
Chuprakov, Sergey A., 6I
Chylek, Tomas, 19, 5I
Cisse, E. M., 6C
Clark, David, 5L
Clark, Dusty L., 33, 5P, 5Q
Clavel, Bruno, 2M
Claver, Charles F., 44, 4P, 4Q
Climent, Tània, 3M
Cochran, Leighton, 0U
Coelho, João, 4F
Coetzee, Chris, 2W, 4I
Collins, Peter L., 14, 19
Colorado, Enrique, 5L
Conconi, P., 42
Cook, Kem H., 0D
Corbard, T., 6C
Córdova, Antolín, 5L
Cornell, Mark E., 0K, 3X, 4S, 5J
Cornwell, Tim J., 2A
Corral, Luis, 4L
Corredor, Andrew, 20, 24
Cortes, German, 2M
Cortés, P. C., 3F
Côté, Patrick, 2Z
Cowley, David J., 4R
Craig, Simon, 07
Cramer, Claire, 1L
Crampton, David, 2Z
Crause, Lisa A., 4I
Crawford, Steven M., 2W, 6A
Croner, Ernest, 1Q
Crouzet, N., 5I
Cruise, W., 5K
Cui, Xiangqun, 1O, 61
Cushing, Norman J., 1A, 1D
Daban, Jean-Baptiste, 5I, 6J
Dall'Ora, Massimo, 1C, 4Z, 56, 5R
D'Almeida, E., 6C
Danyo, G., 09
da Silva Bento, Joao, 0E
Davies, R. D., 2Y

Davis, R. J., 2Y
Davison, Warren, 3Y
de Alba, Durruty, 4L
de Bernardis, Paolo, xxix
Dee, Kevin M., 6J
de Frondat, Fatima, 3A
DeGroff, William T., 19
Deiana, Gian Luigi, 2E
Delabre, Bernard-Alexis, 4F
Delacroix, Alex, 1Q
de Gurtubai, A. G., 62
de la Fuente, Eduardo, 4L
de la Fuente, L., 2Y
della Valle, A., 62
De Lorenzi, S., 0Q, 3B
de Mooij, Ernst, 0I
Deng, Yuanyong, 05
Dent, William, 2K
De Rosa, R., 3C
de Ville, Jonathan, 2U, 3Z
Devlin, Mark J., 15
Devost, Daniel, 2Z
Devries, Joe, 0J
de Wijn, Alfred G., 3N
Di Cianno, A., 4E
Di Marcantonio, Paolo, 4F
Di Rico, G., 4E
Dickinson, C., 2Y
Diego, J. M., 2Y
Dierickx, Philippe, 1F
Dilla Martínez, Angel, 3W
Dimmler, M., 1Y, 1Z
Ditsler, Jennifer, 07
DiVarano, Igor, 35
Dober, Brad, 15
Doel, Peter, 0C
Doi, Mamoru, 6B
D'Orsi, Sergio, 1C, 4Z, 56, 5R
Dournaux, Jean-Laurent, 3A
Dreyer, Oliver, 0A, 0L, 3T
Dribusch, Christoph, 20, 24
du Plessis, Charl A., 4I
Du, Fujia, 1O, 5M
Dugue, M., 5I
Dumas, Delphine, 3A
Dunham, Edward W., 10, 11, 14, 19, 51
DuVall, Eugene, 2K
Ebihara, Yusuke, 1Q
Egner, Sebastian, 1B
Eigmüller, Philipp, 0E
Els, S. G., 1U, 68
Engfer, Christian, 12
Epchtein, Nicolas, 5E
Erikson, Anders, 0E, 5I
Esguerra, Jorge D., 0K, 3X, 5J
Espoz, Juan, 4H
Etxeita, Borja, 4N
Fantei-Caujolle, Y., 5I

Fappani, Denis, 0C
Farah, Alejandro, 5L
Farahani, Arash, 0U, 1H
Farinato, Jacopo, 1C, 4Z, 56, 5R
Fasola, Gilles, 3A
Feain, Ilana, 2A
Fernandez, Esther, 4N
Fernández-Cobos, R., 2Y
Fiaschi, M., 4E
Filgueira, Jose, 1H
Fischer, David, 0M
Fissel, Laura M., 15
Flaugher, Brenna, 0C
Flebus, Carlo, 1V, 2U, 4V
Fleck, Burkhard, 58
Fletcher, Murray, 1P
Flores, Jorge L., 4L
Fodil, M., 6C
Fourez, Julien, 0C
Fox, Ori D., 5L
Fox-Machado, Lester, 5L
Fraser, Mark, 3M
Fressin, F., 5I
Fruth, T., 5I
Fucik, Jason, 1Q
Fugazza, D., 42
Fujishiro, Naofumi, 6G
Fukui, Yasuo, 15
Fulton, B. J., 5V
Furesz, Gabor, 0D
Gabriel, Eric, 2U
Gai, Mario, 4K, 5A
Gajadhar, Sarah, 1S
Gale, David M., 53
Gallagher, Dennis, 3N, 3P
Gandilo, Natalie N., 15
García, Benjamín, 4L, 5L
García-Torales, Guillermo, 4L
Gaudiomonte, F., 5W
Geary, John C., 0D
Gebhardt, Karl, 0K
Genolet, Ludovic, 0E
Génova-Santos, R., 2Y
Georgiev, Leonid N., 5L
Geyl, R., 1W
Ghanti, A. S., 41
Ghigo, M., 42
Ghiretti, Paolo, 38
Gibson, J. Duane, 32, 33, 5Q
Giordano, Christophe, 64
Giordano, G., 3C
Glaese, Roger M., 0V
Goad, Mike, 0E
Gomes, Ricardo, 4F
Gomez, Alberto, 4N
Gómez, Celia, 0P
Gómez-Reñasco, F., 2Y
Goncharov, Alexander V., 6H

Goncalves, I., 5I
Gong, Xuefei, 1O, 5F
González, J. Jesús, 5L
Good, John M., 0K, 3X, 4O
Goode, Philip R., 03, 0Y
Goodrich, Bret, 07
Gössl, Claus, 1B, 2V
Gouvret, Carole, 5I, 6J
Graham, James R., 5C
Grainge, K., 2Y
Green, Richard F., 1A, 1D
Greenshaw, Tim, 39, 3A
Gregory, Brooke, 45
Gressler, William J., 0J, 44
Grimes, Paul, 1N
Grünewald, Sandra, 17
Grupp, Frank U., 1B, 2V, 58
Gu, Bozhong, 3U, 4X, 50
Gu, XueDong, 26, 2F
Guillot, T., 5I
Guisa, Gerardo, 5L
Gundersen, Joshua O., 15
Gunnels, Steve, 0N
Gupta, Alok, 28
Gusick, Michael, 4G
Güsten, Rolf, 10
Gutiérrez, Leonel, 5L
Guyon, Olivier, 4A
Häuser, Marco, 0K, 4J
Hall, Helen, 13
Halman, Mark, 5C
Hammerschlag, Robert H., 06
Han, Cheongho, 47
Hansen, Eric R., 07
Hansen, Rick, 3Y
Hardy, Tim, 1P
Hargrave, Peter C., 15
Harper, S., 2Y
Hart, Michael, 4S
Hartl, Mike, 1B
Hartmann, Peter, 37
Hastings, Peter, 0F
Hauchecorne, A., 6C
Hay, Stuart G., 2A
Hayes, Richard J., 0K, 3X, 5J
Hebbeker, Thomas, 30
Hegwer, Steve L., 07
Heinz, Volker, 38, 3V
Heisler, James T., 0K, 3X
Heller, C., 09
Herranz, D., 2Y
Herrera, Joel, 5L
Herreros, J. M., 2Y
Herter, Terry L., 10
Hervet, Olivier, 3A
Hileman, Edward, 0J
Hill, Gary J., 0K, 3X, 4J, 4O, 4S
Hill, John M., 1A, 1D

Hinton, James, 3A
Hirart, David, 4L
Hiriart, David, 0D
Ho, Kevin K. Y., 1S, 5K
Ho, Paul T. P., 0D, 1J
Hobson, M., 2Y
Hopp, Ulrich, 1B, 2V, 58
Horner, Scott D., 14
Horville, David, 3A
Howard, James, 3Y
Hoyland, R. J., 2Y
Hu, Shao-Ming, 58
Hu, Yi, 1O
Huang, Pei, 3N
Huang, Shanjie, 04
Huang, Yau-De, 1J
Hubbard, Robert P., 07
Hudek, John S., 21
Huet, Jean-Michel, 3A
Hughes, Gareth, 17
Hull, Charles, 0N, 0U, 24
Ibarra, Salomon Eduardo, 4L
Ichizawa, Shunsuke, 6G
Iguchi, Satoru, 2O, 3H, 3I
Ikeda, Yuji, 6G
Ikhlef, R., 6C
Ilyin, Ilya, 35
Inatani, J., 3H
Inoue, Makoto, 1J
Irbah, A., 6C
Isogai, Mizuki, 6G
Iwamuro, Fumihide, 5Z
Izumiura, Hideyuki, 5Z
Jacoby, George, 1H
Jager, R., 3F
Jägers, Aswin P. L., 06
Jakob, Holger, 14, 2R
Jedamzik, Ralf, 37
Jeffers, Paul F., 0A, 0L, 3T
Jégouzo, Isabelle, 3A
Jensen, Stephen C., 14
Ji, Haisheng, 05
Jiang, Xiang, 3U, 5T
Jin, Ho, 48
Jin, Zhenyu, 05
Johns, Matt, 0N, 0U, 1H, 24
Johnson, Luke C., 3O
Jones, Patricio, 4H
Jordán, Andrés, 0E
Jouve, JC, 1W
Junquist, R., 0X
K., Sanjith K., 28
Kaiser, Nicholas, 0H
Kamizuka, Takafumi, 6B
Kampf, Dirk, 1B
Kappler, L. G., 43, 5S
Kappler, N., 43, 5S
Kärcher, Hans J., 0A, 0L, 0M, 2R, 3T, 40

Kato, Natsuko M., 6B
Kawakita, Hideyo, 6G
Kawara, Kimiaki, 6B
Keas, Paul J., 1I
Keil, Stephen L., 0Z
Keilig, Thomas, 2R
Kemp, Simon, 4L
Kenworthy, Matthew, 0I
Kern, Jonathan, 0N, 24
Killebrew, Jana L., 14
Kim, Minjin, 47
Kim, Seung-Lee, 43, 47, 5S
Kim, Young-Soo, 20, 24
Kino, Masaru, 5Z
Kissil, Andy, 2M
Kissler-Patig, Markus, 1F, 1I
Kiuchi, Hi., 2O
Klein, Christopher R., 5L
Klein, Jeffrey, 15
Knee, L. B. G., 3F
Koch, Dietmar, 2G
Kohno, Kotaro, 6B
Kolesnikov, Alexy, 0U
Konishi, Masahiro, 6B
Korotkov, Andrei L., 15
Koshida, Shintaro, 6B
Kou, Songfeng, 4U, 5T, 6D
Krabbe, Alfred, 1O, 2R
Krabbendam, Victor, 4Q, 69
Krämer, Ewald, 12
Kraus, Max, 3V
Kriel, Herman J., 0K
Kuhn, Jeff R., 0Y
Kuijken, Konrad, 56
Kulesa, Craig A., 5H
Kulkarni, R. R., 41
Kumar, Brijesh, 1T, 1V
Kumar, T. S., 5O
Kurita, Mikio, 5Z
Kutyrev, Alexander S., 5L
Lachenmann, Michael, 2S
Lagos, Jose, 38
Lagrange, T., 1W
Lalcare, F., 6C
Lampater, Ulrich, 11, 14, 2R
Lamprecht, Jürgen, 58
Landoni, M., 42
Lang-Bardl, Florian, 1B, 2V, 58
Langlois, Maud, 5E
Laporte, Philippe, 3A
Lasenby, A., 2Y
Lattanzi, Mario G., 5A
Lauscher, Markus, 30
Laux, Uwe, 35
Law, Nicholas M., 5C
Lawrence, Jon S., 1R, 5H
Lazo, Fransisco, 4L, 5L
le Poole, Rudolf, 0I

Le Van Suu, A., 5I
Leck, Ron, 0K, 5J
Leckie, Brian M., 1P
Lee, Byeong-Cheol, 47
Lee, Chung-Uk, 43, 47, 5S
Lee, David, 0F
Lee, Hanshin, 0K, 4J, 4S
Lee, Jae Woo, 47
Lee, Jong Ung, 48
Lee, Sangon, 48
Lee, William H., 0D, 5L
Lee, Yun Woo, 48
Lehner, Matthew J., 0D
Leibold, Torsten, 59
Lemke, Roland, 40
Leplan, H., 1W
Lesueur, P., 6C
Leveque, S., 1Y
Levine, Stephen E., 19
Levinson, Rebecca Sobel, 55
Li, Aihua, 4B, 4D, 4Y, 61
Li, Cheng, 4U
Li, Guoping, 4B, 4D, 4Y, 5B
Li, Hongmei, 61
Li, Hui, 5X, 5Y
Li, Xiaoyan, 1O, 4W, 5M
Li, Yeping, 61
Li, Yinzhu, 04
Li, Zhengyang, 1O
Liang, Ming, 4P, 4Q
Lim, Juhee, 48
Lima, Jorge, 4F
Lin, M., 6C
Linden, James S., 3S
Liske, J., 1I
Little, J. K., 1A
Liu, Genrong, 61, 6D
Liu, Guangqian, 04
Liu, Liyong, 1K, 64
Liu, Zhong, 05
Loewen, Nathan, 0W
Loop, David, 0W
López, B., 3F
López, C., 3F
López, Eduardo, 5L
López-Caniego, M., 2Y
López-Caraballo, C., 2Y
López-Fernández, José Antonio, 2D
López-Pérez, J. A., 5W
Lorentz, Tom, 08
Lorenzetti, Dario, 2Q
Lotz, Paul J., 19
Lou, John, 2M
Lousberg, Grégory, 0G
Love, Jonathan, 4I
Lovis, Christophe, 4F
Luna, Esteban A., 5L
Luong-Van, Daniel M., 1R

Lutz, Thorsten, 12
Lykke, Keith, 1L
MacQueen, Phillip J., 0K
Maffei, B., 2Y
Magrin, Demetrio, 1C, 4Z, 56, 5R
Maihara, Toshinori, 5Z
Mandushev, Georgi I., 14
Manuel, Eric, 0L
Marchiori, Gianpietro, 0Q, 2J, 3B
Marcum, Pamela M., 10, 14
Maree, J., 1U
Marongiu, Pasqualino, 2E
Marrero, J., 1Y
Marshall, Heather, 08
Marti-Canales, Javier, 2K
Martin, D. Christopher, 1Q
Martin, F., 6C
Martin, Sara F., 06
Martinache, Frantz, 4A
Martin-Cocher, Pierre L., 1J
Martinez Fiorenzano, A. F., 62
Martínez, Benjamín, 4L, 5L
Martinez, J., 5V
Martínez-Gonzalez, E., 2Y
Marty, Laurent, 1C, 4Z, 56, 5R
Masi, Silvia, xxix
Matsushita, Satoki, 1J, 2K, 3K
Matthews, Tristan G., 15
Mattioli, Massimiliano, 2Q
McBride, Dennis, 21
McBride, William R., 07
McCarthy, Patrick, 1H
McConnachie, Alan, 0W, 2Z
McCormac, James, 0E
McCulloch, M., 2Y
McDaid, Scott, 1R
McGraw, John T., 1L, 69
McLaren, Campbell, 1R
McLean, Ian S., 14
McMullin, Joseph P., 07, 3F
McPherson, Alistair, 1F
Mediavilla, A., 2Y
Meeks, Robert L., 3Y, 4G
Meftah, M., 6C
Mégevand, Denis, 4F
Mekarnia, D., 5I
Melhuish, S., 2Y
Melkumyan, David, 17
Mello, Melinda J., 5N
Melsheimer, Frank, 0D
Meurer, Christine, 30
Meyer, Allen W., 14
Miccolis, Maurizio, 2K
Michel, Fabrice R., 2U
Middendorf, Lukas, 30
Miller, Douglas L., 1D, 59
Miller, Walter E., 14
Minezaki, Takeo, 6B

Mitsch, Wolfgang, 1B, 2V
Miyata, Takashi, 6B
Mohanachandran, K., 28
Moitinho, André, 4F
Mokrani, Bilal, 25
Molinari, Emilio, 62
Mollison, Nicholas T., 0K
Moncelsi, Lorenzo, 15
Montane, Andrés, 45, 46
Moon, Dae-Sik, 47
Moon, Hong-Kyu, 47
Moon, Il-Kweon, 20, 48
Morand, F., 6C
Moore, Anna M., 1Q
Moreau, Vincent, 0G, 0H
Moretto, Gil, 0Y, 5E
Morgan, Jeffrey S., 0H
Moriera, Walter, 0K
Morita, Koh-Ichiro, 2K, 3K
Moritani, Yuki, 5Z
Morokuma, Tomoki, 6B
Morsiani, Marco, 2E, 5W
Motohara, Kentaro, 6B
Mroczkowski, Tony K., 15
Müller, M., 1Y, 1Z
Muñoz A., Freddy, 45, 46
Murga, Gaizka, 08, 0P, 3M, 4N
Murillo, Francisco, 5L
Murillo, José Manuel, 5L
Murowinski, Richard, 5C
Nakamura, Tomohiko, 6B
Nakanishi, K., 3H
Nakos, T., 3F
Napier, Peter, 2K
Navarro, Ramon, 0I
Navarro, Silvana G., 4L
Neill, Douglas R., 0J, 0R, 0T
Nelson, Jerry, 4R
Nelson, Peter G., 3N, 3P
Netterfield, C. Barth, 15
Neveu, Marion, 0E
Ngan, Wayne, 5C
Ni, Jijun, 61
Niggemann, Tim, 30
Nigoche-Netro, Alberto, 4L
Nikolic, Bojan, 3K
Ninane, Nathalie, 1V, 2U, 4V
Niu, Dongsheng, 4U, 50, 5T
Niu, Yong, 4W
Noethe, Lothar, 40, 4Z, 56
Nordsieck, Kenneth H., 2W
Norton, Timothy, 0D
Novak, Giles, 15
Núñez, Juan Manuel, 4L, 5L
Nutter, David, 15
Nye, Ralph A., 51
Oceguera-Becerra, Tomas, 4L
Ochoa, José Luís, 4L, 5L

O'Connor, James E., 4I
O'Donoghue, Darragh E., 2W, 4I
Oliver, Richard C., 5I
Olmi, Luca, 15
O'Mahony, Neil, 6J
Omar, Amitesh, 1T, 2U
Omarov, Ch. T., 5U
Orden Martínez, Alfredo, 3W
Orellana, Eduardo, 3V
Ortiz, D., 2Y
Ortolani, S., 62
Otten, Gilles, 0I
Ottogalli, Sebastien, 6J
Oya, Igor, 17
Padin, Steve, 2M
Padovani, P., 1I
Paine, Scott N., 1J
Paluszek, Heiko, 2G
Pan, Gao-feng, 5Y
Pandey, A. K., 1T, 4I
Panteleev, Sergey, 21
Pareschi, G., 39, 42
Park, Byeong-Gon, 43, 47, 5S
Park, Kwijong, 20
Park, Won-Hyun, 20, 24
Parr-Burman, Philip, 0F
Pascale, Enzo, 15
Pazder, Jonh, 0W
Pedichini, Fernando, 2Q
Pei, Chong, 4X
Pelaez Santos, A. E., 2Y
Penney, Charles E., 3X, 5J
Perez, Frank, 4H
Perna, C., 3B
Pernechele, Claudio, 2E
Pernet, B., 1W
Peron, S., 5I
Perry, Dave M., 0K, 4S
Petit, P.-Y., 5I
Pfüller, Enrico, 11, 12, 13, 14
Pfund, Johannes, 58
Phelps, LeEllen, 3M, 3S
Phillips, John Peter, 4L
Phillips, N., 3F
Piccirillo, L., 2Y
Pickering, Timothy E., 6A
Pickles, A. J., 5V
Piérard, Maxime, 2U, 3Z, 4V
Pillai, Rajeev R., 28
Pinto, Patricio, 4H
Pirnay, Olivier, 0G
Pisano, G., 2Y
Pisanu, Tonino, 2E, 5W
Pistor, Christoph, 4R
Plainchamp, P., 1W
Planesas, P., 3F
Poidevin, Frédérick, 15
Poiet, G., 6C

Pollacco, Don, 0E
Pollak, Daniela, 38
Pont, F., 5I
Poppi, Sergio, 2E, 5W
Porter, D., 5Q
Poteet, W. M., 43, 5S
Poutriquet, F., 1W
Preumont, André, 25
Prochaska, J. Xavier, 4R, 5L
Qian, Xuan, 1K, 64
Queloz, Didier, 0E
Quirós, Fernando, 5L
Radiszcz, M., 3F
Rafal, Marc D., 0K, 3X, 4O, 4S, 5J
Rafferty, Tom H., 0K, 4S
Ragazzoni, Roberto, 1C, 4Z, 56, 5R
Rakich, Andrew, 1D, 54, 59
Ramiller, Chuck, 0K, 4S
Ramos-Larios, Gerardo, 4L
Rampini, Francesco, 2J, 3B
Ramsay, Suzanne, 1F
Ramsey, Lawrence W., 0K
Rapchun, David A., 5L
Raskar, R. D., 4I
Rasmussen, Paul, 2M
Rauer, Heike, 0E, 5I
Raybould, Keith, 1H
Reach, William T., 10
Rebolo, R., 2Y
Redding, David, 2M
Reinacher, Andreas, 14, 2R
Ren, Changzhi, 4T, 4W, 5T
Renaud, C., 6C
Rey, Soo-Chang, 47
Reyes-Ruiz, Mauricio, 0D
Rhee, Hyug-Gyo, 48
Rho, Jeonghee, 14
Ricci, Masimo, 2Q
Richer, Michael G., 0D, 5L
Riddle, R., 68
Rimmele, Thomas R., 07, 3O
Riva, Alberto, 4K, 5A
Riva, M., 42
Rivet, J.-P., 5I
Roberts, Lewis C., 60
Robini, A., 5I
Roda, J., 5W
Rodrigues, Gonçalo, 25
Roellig, Thomas L., 14
Röser, Hans-Peter, 11, 12, 13, 2T
Romano, R., 3C
Román-Zúñiga, Carlos, 5L
Rosing, W., 5V
Rouzé, M., 6C
Rubiño-Martin, J. A., 2Y
Ruch, E., 1W
Ruder, Micheal, 1B, 2V
Sagar, Ram, 1T

Saito, H., 3H
Saito, Masao, 2O, 3H, 3I
Sakhamuri, Nagarjun, 28
Sako, Shigeyuki, 6B
Salbi, Pegah, 5C
Salmon, Derrick, 0W, 1S
San Juan, J., 62
San Vicente, Ander, 4N
Sanchez de la Rosa, V., 2Y
Sand, D., 5V
Sanquirce, Rubén, 4N
Santos, Nuno C., 4F
Sarazin, Marc, 31
Sass, Craig, 4I
Saunders, R., 2Y
Saunders, Will, 5H
Savage, Maureen L., 14
Savage, Richard D., 0K, 4S
Savini, Giorgio, 15
Schaller, Skip, 5P
Schechter, Paul L., 55
Schinckel, Antony E. T., 2A
Schipani, Pietro, 1C, 4Z, 56, 5R
Schlenstedt, Stefan, 17
Schmider, F.-X., 5I
Schmoll, Jürgen, 3A
Schöck, M., 68
Schoening, Bill, 0J
Schuh, Jared, 3Y
Schumacher, German, 0J, 45
Schumacher, Johannes, 30
Schurter, Patricio, 45
Schwanke, Ullrich, 17
Scott, Douglas, 15
Scott, P., 2Y
Sebag, Jacques, 0J, 0R, 44, 69
Sebring, P., 0X
Sebring, T., 0X
Sedghi, B., 1Y, 1Z
Serra, Giampaolo, 2E, 5W
Shang, Zhaohui, 1O
Shariff, Jamil A., 15
Shectman, Stephen, 0N, 1H, 24
Sheehan, Michael, 0N, 0V, 1H, 24
Shelton, Chris, 60
Shimko, Steve, 07
Shimono, Atsushi, 5Z
Shipsey, Ian P., 4P
Shukla, Vishal, 28, 41
Siegel, Benjamín, 4N
Sierra, Gerardo, 4L
Simon, R., 3F
Simons, Doug, 2Z
Sims, Geoff, 1R, 5H
Sivanandam, Suresh, 5C
Skidmore, W., 68
Slagle, J. H., 1A
Sliepen, Guus, 06

Smith, Carey, 0N
Smith, Chris R., 45
Smith, David R., 2N, 3G
Smith, Erin C., 14
Smith, Michael P., 4J
Snellen, Ignas A. A. G., 0I
Sol, Hélène, 3A
Soler, Juan Diego, 15
Song, Jae Bong, 48
Song, Xiaoli, 4W
Souccar, Kamal, 2N, 3G
Soukup, Ian M., 0K, 3X, 4O, 5J
Soyano, Takao, 6B
Spano, Paolo, 0W, 42
Spyromilio, Jason, 1F
Sramek, Richard, 2K
Steinbring, Eric, 1J, 1P, 5C
Stephan, Maurice, 30
Stepp, Larry, 1G
Stiemer, Siegfried, 0W
Stobie, Brian, 0F
Stolz, Günter, 3T
Storey, John W. V., 1R, 5H
Strassmeier, Klaus G., 35
Strümpfer, Francois, 4I
Strydom, Ockert J., 4I
Stuik, Remko, 0I
Stutzki, C., 0X
Sugimoto, Masahiro, 2K
Sun, Jing-hai, 5Y
Sung, Eon-Chang, 47
Sung, Hwankyung, 47
Süß, Martin, 0M, 2G
Sust, Eberhard, 2D
Sütterlin, Peter, 06
Szentgyorgyi, Andrew, 0D
Szeto, Kei, 0W, 2Z
Szulagyi, J., 5I
Takahashi, Hidenori, 6B
Takato, Naruhisa, 6B
Tamura, Yoichi, 6B
Tanabe, Toshihiko, 6B
Tanaka, Masuo, 6B
Tarusawa, Ken'ichi, 6B
Taylor, Charles A. III, 0K
Taylor, Trey, 4S
Temi, Pasquale, 14
ter Horst, Rik, 0I
Terán, J. V., 2Y
Teufel, Stefan, 11, 14
Thanjavur, Karun, 1S
Thiele, Hans, 1B, 2V
Thomas, J., 5K
Thomas, Nicholas E., 15
Thuillier, G., 6C
Tighe, Roberto, 45, 46
Tokovinin, A., 69
Toledo, I., 3F

Tomczyk, Steven, 3N, 3P
Tomelleri, R., 42
Travouillon, T., 68
Trebisky, Thomas, 32, 5Q
Truch, Matthew D. P., 15
Tucker, Carole E., 15
Tucker, Gregory S., 15
Turner, J., 69
Uchiyama, Mizuho, 6B
Umbriaco, Gabriele, 1C, 4Z, 56, 5R
Upton, R., 3O
Urban, Jeff, 3Y
Valentini, A., 4E
Valentini, G., 4E
Valyavin, Gennady, 5L
van der Bliek, Nicole S., 45
Vargiu, Giampaolo, 2E
Vattiat, Brian L., 0K, 4J, 4S
Vauglin, Isabelle, 5E
Vega, Oscar, 4N
Vega-Moreno, A., 2Y
Velur, Viswa, 1Q
Veillet, Christian, 0W
Venetiou, Alexander J., 19, 1E
Vermeulen, R. C., 2B
Vernin, Jean, 64
Vettolani, G., 3B
Vielva, P., 2Y
Viera-Curbelo, T., 2Y
Villa, E., 2Y
Vitali, Fabrizio, 2Q
Vizcargüenaga, Alberto, 4N
Vlahko, V. B., 6E
Wagner, Jörg, 2R
Wagner, R. M., 1A
Walker, Alistair R., 45
Walker, Simon, 0E
Wang, Daxing, 1O, 5M
Wang, Guomin, 3U, 4U, 50, 5T, 6D
Wang, Hai, 5B
Wang, Hongshuai, 1K, 64
Wang, Lifan, 1O
Wang, QiMing, 26, 2F
Wang, Shiang-Yu, 0D
Wang, Yiping, 1K, 64
Wang, Yuefei, 61
Wardell, Douglas R., 0K, 3X
Ward-Thompson, Derek, 15
Warner, Mark, 07
Warner, Michael, 0J, 0R, 46, 69
Wassila, Dali-Ali, 31
Watson, Alan M., 5L
Watson, R. A., 2Y
Weatherwax, Allan, 1Q
Wedeking, Gregory A., 0K, 3X
Weeks, Damon A., 4O
Weis, Ulrich, 0A
Wen, Haikun, 1O, 5F

Werz, Alexander, 37
West, Richard, 0E
Westerhoff, Thomas, 37
Wheatley, Peter J., 0E
White, Richard, 39, 3A
Whiteis, Peter G., 5N
Whyborn, Nicholas, 2K, 3F
Wiecha, Oliver, 0J, 69
Wiedemann, Manuel, 11, 12, 14, 2T
Wiese, James, 3Y
Wiid, Eben, 4I
Wilkinson, Martin, 4I
Williams, G. Grant, 32
Williams, Theodore B., 2W
Williams, Timothy R., 3L
Wilson, Richard, 6A
Woche, Manfred, 35
Woeger, Friedrich, 07
Wolf, Jürgen, 11, 12, 13, 14, 2R, 2S, 2T
Woodward, John T. IV, 1L
Woody, David, 2M
Worthington, Michael S., 0K, 3X
Wu, MingChang, 26, 2F
Wu, Zhixu, 61
Xin, Bo, 4P, 4Q
Xu, Fangyu, 04
Xu, Jin, 3U, 4T, 4W
Xu, Lingzhe, 1O, 5M
Xu, Ye, 4B
Xue, JianXing, 26
Yagoubov, Pavel, 2K
Yang, Dehua, 4B, 4D, 4Y, 5B
Yang, Ho-Soon, 48
Yang, Qing-ge, 5Y
Yang, Shihai, 3U, 5G
Yang, Shi-mo, 5X
Yao, Yongqiang, 1K, 64
Yatagai, H., 3F
Ye, Yu, 3U, 4T, 4U, 4X
Yen, Wei-Ling, 0D
Yin, Jia, 1K, 64
Yoneda, Mizuki, 6B
Yonehara, Atsunori, 6G
Yoshida, Michitoshi, 5Z
Yoshii, Yuzuru, 6B
Yoshikawa, Tomohiro, 6G
Young, Erick, 10
Yousif, A., 1U
Yuan, Xiangyan, 1O
Zacchiroli, G., 5W
Zazueta, Salvador, 4L
Zech, Andreas, 3A
Zeng, Yizhong, 61
Zerbi, Filippo M., 42, 4F
Zhang, Haiying, 3P
Zhang, Jie, 4Y
Zhang, Ru, 5F
Zhang, Xin-yu, 5X, 5Y

Zhang, Yong, 4B, 4Y, 5B, 61
Zhang, Yue, 1O, 5M
Zhang, Zhenchao, 4B, 4D
Zhang, Zhi-Wei, 0D
Zhang, Zhiyong, 3U, 4X, 5T
Zhantayev, Zh. Sh., 5U
Zhao, BaoQing, 2F
Zhao, Jianlin, 5M
Zhao, Qing, 26
Zhou, Guohua, 4B, 4D, 4Y, 5B
Zhu, Ming, 26
Zhu, Wen-bai, 5Y
Ziad, Aziz, 31, 6A, 6C
Zierer, Joseph J., 0K, 3X, 4O
Zimmer, Peter, 1L, 69
Zinnecker, Hans, 1O
Zirzow, Daniel C., 1L
Zitelli, V., 62
Zolwoker, Jeff, 2M
Zoonemat Kermani, Saeid, 19
Zuo, Yingxi, 4B